Springer Proceedings in Mathematics & Statistics

Volume 410

This book series features volumes composed of selected contributions from workshops and conferences in all areas of current research in mathematics and statistics, including data science, operations research and optimization. In addition to an overall evaluation of the interest, scientific quality, and timeliness of each proposal at the hands of the publisher, individual contributions are all refereed to the high quality standards of leading journals in the field. Thus, this series provides the research community with well-edited, authoritative reports on developments in the most exciting areas of mathematical and statistical research today.

Rajesh Kumar Sharma · Lorenzo Pareschi ·
Abdon Atangana · Bikash Sahoo ·
Vijay Kumar Kukreja
Editors

Frontiers in Industrial and Applied Mathematics

FIAM-2021, Punjab, India, December 21–22

Springer

Editors
Rajesh Kumar Sharma
Department of Mathematics
Institute of Technology
Nirma University
Ahmedabad, India

Abdon Atangana
Institute for Groundwater Studies
University of the Free State
Bloemfontein, South Africa

Vijay Kumar Kukreja
Department of Mathematics
Sant Longowal Institute of Engineering
and Technology
Longowal, India

Lorenzo Pareschi
Dipartimento Di Matematica
University of Ferrara
Ferrara, Italy

Bikash Sahoo
Department of Mathematics
National Institute of Technology Rourkela
Rourkela, Odisha, India

ISSN 2194-1009 ISSN 2194-1017 (electronic)
Springer Proceedings in Mathematics & Statistics
ISBN 978-981-19-7271-3 ISBN 978-981-19-7272-0 (eBook)
https://doi.org/10.1007/978-981-19-7272-0

Mathematics Subject Classification: 00B25, 11Yxx, 37-XX, 37Mxx, 46-XX, 60-XX, 60-06, 65-06, 74-06, 76-06, 92-06, 93-06

This Springer imprint is published by the registered company Springer Nature Singapore Pte Ltd.
The registered company address is: 152 Beach Road, #21-01/04 Gateway East, Singapore 189721, Singapore

Contents

Computational and Experimental Investigation of Flow and Convective Heat Transfer Along Rough Surfaces

C. Özman, T. Saner, F. Gül, M. Diederich, A. C. Benim, and U. Janoske

Abstract Flow and heat transfers along rough surfaces are investigated. A test facility is established, where rough surfaces generated by additive manufacturing can be tested. The computational work follows two goals. On the one hand, a computational tool is developed that can analyze the characteristics of a rough surface and generate rough surfaces with prescribed characteristics. On the other hand, Computational Fluid Dynamics (CFD) is applied for the analysis of flow and heat transfer along rough surfaces. The present focus is on the validation of turbulence models. Within this context, two alternative treatments, namely the wall functions (WF)-based approach and roughness resolving (RR) approach are assessed. Turbulence is modeled within a RANS (Reynolds Averaged Numerical Simulation) framework. All of the considered four turbulent viscosity models, using WF, showed a similar agreement with the measurements. Quantitatively, the realizable k-ε model is observed to deliver a better accuracy, in general, which is, then, also applied in RR calculations.

C. Özman · T. Saner · F. Gül · M. Diederich · A. C. Benim (✉)
Center of Flow Simulation, Düsseldorf University of Applied Sciences, Düsseldorf, Germany
e-mail: alicemal@prof-benim.com

C. Özman
e-mail: cansu.oezman@hs-duesseldorf.de

T. Saner
e-mail: taner.saner@hs-duesseldorf.de

F. Gül
e-mail: fethi.guel@hs-duesseldorf.de

M. Diederich
e-mail: michael.diederich@hs-duesseldorf.de

U. Janoske
Chair of Fluid Mechanics, Department of Mechanical Engineering, University of Wuppertal, Wuppertal, Germany
e-mail: janoske@uni-wuppertal.de

R. K. Sharma et al. (eds.), *Frontiers in Industrial and Applied Mathematics*,
Springer Proceedings in Mathematics & Statistics 410,
https://doi.org/10.1007/978-981-19-7272-0_1

The RR approach showed a fair qualitative performance, which was, however, quantitatively not as good as the WF approach. This is attributed to the idealized geometry on the one hand and possible limitations on the RANS turbulence modeling approach on the other hand. The analysis will be deepened in the future work.

Keywords Fluid dynamics · Heat transfer · Surface roughness

1 Introduction

The additive manufacturing (AM) technology allows compact and lightweight heat exchangers to be produced, which are of importance in different areas, e.g. in aviation [1]. Surfaces generated by AM are, however, not smooth compared to conventional procedures, but exhibit roughness patterns, depending on the particular technique [2]. The main body of the existing knowledge in convective heat transfer [3] refers to smooth surfaces, while rough surfaces received comparably less attention. The purpose of the present research is to achieve a better understanding of forced convection along rough surfaces. Two roughness categories can be defined as (1) regular roughness, also called technical roughness or periodic roughness, where roughness elements consist of regular arrays of well-defined shapes such as pins and fins, and (2) irregular roughness, also called arbitrary or random roughness, where such regularities cannot be presumed.

In boundary layers, under certain conditions, the flows may be described by ordinary differential equations, employing boundary layer assumptions and similarity parameters [4, 5]. In most engineering applications, like the present one, this is prohibited by the prevailing flow conditions, geometry, and boundary conditions that necessitate the solution of the full Navier–Stokes equations [4]. Turbulent flows are characterized by extremely small time and length scales. Their direct numerical solution, the so-called Direct Numerical Simulation (DNS), is, thus, very expensive [6]. The common engineering approach is the solution of the time-averaged equations the so-called Reynolds Averaged Numerical Simulation (RANS) [6, 7]. The terms resulting from averaging are approximated using the so-called turbulence models [6]. Combinations of scale resolving and modeling approaches are the Unsteady RANS (URANS), Detached Eddy Simulation (DES), and Large Eddy Simulation (DES) [6, 8]. These turbulence modeling strategies were alternately applied in the previous studies on rough surfaces that are outlined below. The near-wall turbulence, with re-laminarization and its consequences on the mathematical modeling and discretization, requires special attention. An engineering approach is to approximate this flow by the so-called wall functions (WF) [6], derived under simplifying assumptions. For rough surfaces, the challenges in modeling the near-wall flow are increased, additionally, by the complex wall topology.

In modeling flow near rough walls, a straightforward approach is the full geometric resolution of the surface by the computational grid. In the case of regular roughness, the surface topology is easier to grasp. Turbulent flows over regular arrays of obstacles with well-defined shapes were presented by many researchers [9–12]. For irregular roughness, additional challenges exist due to the capturing and discretization of the surface topology. Numerical analysis of forced convection over irregularly rough surfaces was also presented by several authors, previously [13–16]. Obviously, the direct resolution of roughness is nearly impracticable for many engineering applications due to grid resolution requirements near walls. A remedy is provided by the WF modeling of near-wall turbulent flow [6], which allows an incorporation of the roughness effects through model constants. This is the main approach in engineering applications [21, 22]. A fundamental difficulty in using this approach is that the model is designed for sand grain roughness (SGR). For other roughness types, the model should be employed using an equivalent SGR. For this conversion, different procedures were proposed, including elaborate modifications of the WF [17–22]. However, given the large variety of roughness types, the proposed modifications are found not to be generally applicable with sufficient accuracy [23].

2 Experimental

A test system is constructed for the experimental part (Fig. 1). Arbitrary rough surfaces are produced by the SLS printing technique. The roughness is mapped by using a laser scanner. A high-power vacuum blower is used to manipulate the airflow. To obtain fully developed flow, a bell mouth and flow conditioner are used. Measurements are taken for different Reynolds numbers. The mass flow rate is measured by an orifice flow meter, cross-checking the results with a high-precision hot wire anemometer. The test section is uniformly heated with cartridge heaters. The pressure drop is measured between the inlet and outlet by a differential pressure transducer. RTD-type thermometers are placed in the channel as well as on the heated surface to obtain temperature measurements. The velocity profile in the boundary layer is measured with a specially designed hot wire anemometer (Dantec Dynamics, 55P15).

3 Surface Analysis and Reconstruction

A surface analysis program is created in the MATLAB environment [24]. For surface generation, a power spectrum density-based or autocorrelation-based concept can alternatively be used. The generated surface can then be written out as an STL file. The two concepts can also be used for the analysis of a measured surface.

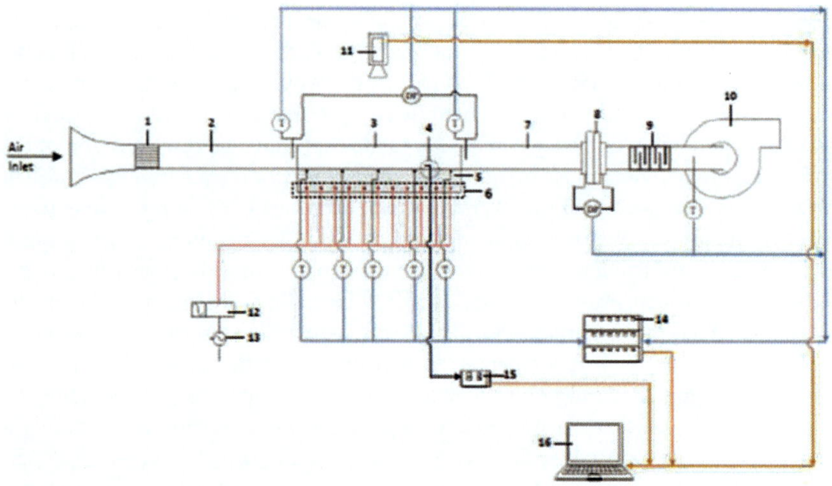

Fig. 1 Test rig: T: thermocouple, DB: differential pressure transducer, 1: flow straightener, 2: entrance length, 3: test section, 4: hot wire anemometer, 5: aluminum block, 6: cartridge heaters, 7: calming length, 8: orifice plate and flanges, 9: mixing chamber, 10: high-pressure blower with frequency converter, 11: thermographic camera, 12: proportional control solid state relay, 13: power grid, 14: main data logger, 15: data logger of hot wire anemometer, and 16: computer

4 Mathematical and Numerical Flow Modeling

Incompressible flow with constant material properties described by Navier–Stokes equations is computationally modeled by the general-purpose CFD software ANSYS Fluent 18.0 [25], based on the Finite Volume Method. Lattice Boltzmann Method (LBM)-based procedures that may be more convenient in DNS analysis of the present problem are to be considered in the future work [26]. A coupled procedure is adopted for the solution of Navier–Stokes equations. For the discretization of the convective terms, the second-order accurate upwind scheme [25] is used for all variables. Within the scope of the current work, turbulence is modeled within the RANS framework, using different turbulent viscosity models [6, 25] including the Standard k-ε model (STD-KE) [6, 25], Renormalization Group Theory k-ε model (RNG-KE) [6, 25], Realizable k-ε model (REL-KE) [6, 25], and the Shear Stress Transport k-ω model (SST-KO) [6, 25] (k: turbulence kinetic energy; ε: dissipation rate of k; ω: specific dissipation rate). DNS, LES, DES, and URANS studies are planned for the future. Flows in straight pipes and channels are considered. For the treatment of the near-wall flow, two approaches are applied: The wall function (WF) approach and the roughness resolving (RR) approach. The WF-based calculation is performed in 2D. The RR calculations are performed in 3D. In the RR calculations, the REL-KE is used as the turbulence model. In doing so, the so-called Enhanced Wall Treatment based on Two-Layer Model [25, 27, 28] is employed to account for the re-laminarization effects. Due to space limitations, the governing equations of the models are not provided here, as they can be obtained through the cited sources. For convenience, a very basic discussion on WF and roughness modeling is presented below. Please note that, at the current stage, the WF model implemented in the used software [25] is used with its default settings.

4.1 *Roughness Modeling via Wall Functions: A Brief Overview*

For pressure gradient-free, unidirectional boundary layer flow along a straight wall, the time-averaged velocity (u) as a function of distance from the wall (y) can be described by a logarithmic function in the turbulent, near-wall region [3, 4]. A similar relationship can be derived for the time-averaged temperature (T) utilizing the Reynolds analogy [3, 4]. This is the basis for the so-called "wall functions" (WF). They are normally expressed in terms of dimensionless quantities indicated by a "+" in the superscript, where the non-dimensionalization is done by the wall shear stress and material properties.

Experiments indicate that the effect of wall roughness can be expressed by a shift (ΔB) in the logarithmic law of the wall [4, 6, 25]

$$u^+ = \frac{1}{\kappa} \ln\left(E\,y^+\right) - \Delta B \tag{1}$$

where κ denotes the Von Karman constant [4] ($\kappa = 0.41$), and E = 9.0. For tightly packed, uniform SGR, with a roughness height of *k*, analysis of experimental data reveals that the roughness effect can be classified into three categories in terms of dimensionless *k*: (1) hydrodynamically smooth, for $k^+ \leq 2.25$; (2) transitional, for $2.25 < k^+ < 90$; and (3) fully rough $k^+ > 90$ [4]. In the first regime, the effect of roughness is neglected ($\Delta B = 0$). For the remaining regimes, expressions of the form [29]

$$\Delta B = f\left(k^+\right) \tag{2}$$

are assumed. Presently, the empirical expressions from Cebeci and Bradshaw [29] are employed to relate ΔB to k^+, which are based on Nikuradse's SGR data [4].

5 Results

5.1 Validation of Turbulence Models for Wall Functions-Based Modeling

Turbulent fully developed pipe flow (diameter: D) with SGR is calculated for different Reynolds numbers (Re = $5 \cdot 10^3$, $1 \cdot 10^4$, $2 \cdot 10^4$, $3 \cdot 10^4$, $4 \cdot 10^4$, $5 \cdot 10^4$, $7.5 \cdot 10^4$, $1 \cdot 10^5$, $2 \cdot 10^5$, $5 \cdot 10^5$, $1 \cdot 10^6$) and for different values of the relative roughness height (k/D = 0.001, 0.002, 0.004, 0.008, 0.016, 0.033) using the above-mentioned turbulence models. The predictions are compared with the empirical data in terms of the friction factor (λ) in Fig. 2.

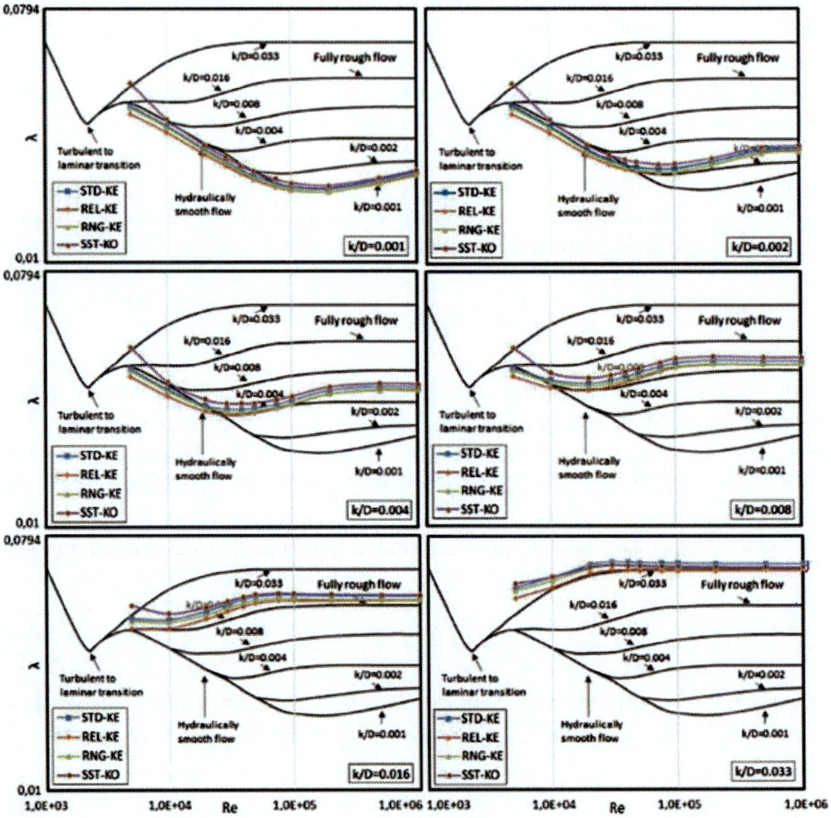

Fig. 2 Predicted friction factors (λ) as a function of Reynolds number (Re) for different values of relative roughness (k/D) for fully developed pipe flow, compared with empirical data (black lines, reproduced from Ref. [30])

In Fig. 2, one can see that all models show a fair qualitative agreement with the data, whereas quantitative differences exist. The differences between the models among one another are larger for low Re and get smaller with increasing Re. For low roughness (k/D $\leq$ 0.004), the models underpredict for low Re, and overpredict for high Re, except for SST-KO, which constantly overpredicts. For high roughness (k/D $\geq$ 0.008), all models overpredict for the whole Re range. Comparing the models with each other, a very clear distinction cannot be made. As a general trend, one can see that STD-KE and RNG-KE perform rather similar to each other, and SST-KO and REL-KE tend to have slightly higher and lower values, respectively. Overall, a better quantitative agreement of REL-KE with the measurements can be attested.

5.2 *Validation of Roughness Resolving Approach Based on SGR*

A reasonable first step toward the validation of a roughness resolving modeling approach can be the investigation of SGR, for which much data exists. This is attempted in the present study, as a basis. Most of the existing data on SGR is, however, for pipe flow. From a practical point of view, a planar channel flow is more amenable for a roughness resolving treatment An engineering approach to utilize the pipe data for different channel shapes is provided by the concept of hydraulic diameter [3, 4]. Since this is an engineering approximation, its accuracy in converting the pipe data to a planar channel is analyzed first, for the presently considered case. For this purpose, the planar channel flow is calculated, and the results are compared with the pipe data, using the concept of hydraulic diameter. These calculations are performed for the relative roughness of k/D = 0.033, using the turbulence models REL-KE and SST-KO. The results obtained for Re = $5{\cdot}10^3$, $1{\cdot}10^4$, $2{\cdot}10^4$, $3{\cdot}10^4$, $4{\cdot}10^4$, and $5{\cdot}10^4$ are compared with pipe data in Fig. 3.

One can see in Fig. 3 that the deviation of the channel analogy to the pipe is larger for low Re and gets reduced for increasing Re. One can also observe that REL-KE-CHANNEL agrees better with the experiments, with quite good agreement for large Re.

In an attempt of obtaining a surface resolving solution for SGR, one shall first recognize that SGR represents, principally, an irregular roughness. In conceptual considerations, SGR is normally assumed to be represented by a tightly packed, regular array of perfect spheres with a uniform diameter (k). It should be stated that this is a rather strong idealization of the reality. Not only because of the assumed perfect spherical shape, but also for the connection of the spheres to the wall. The spheres touch the wall at a point, leaving much free space in the next-to-wall region, which may not well correspond to reality. This is especially problematic for the thermal problem, since the roughness elements are thermally disconnected from the wall, as the ideal point contact allows no finite heat transfer area between the spheres

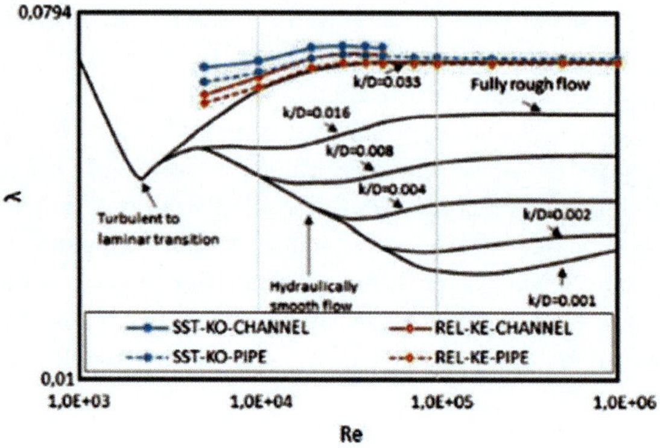

Fig. 3 Predicted friction factors (λ) as a function of Reynolds number (Re) for k/D = 0.033 obtained for fully developed pipe flow and fully developed channel with equivalent hydraulic diameter, compared with empirical data for pipe (black lines, reproduced from Ref. [30])

and the wall. Still, this idealization is applied for the sake of a systematic approach as a first step of the intended more detailed study.

In applying this idealization for SGR, inline and staggered arrangements of the spheres are considered to find out how much role the difference plays. The calculations are performed for four Reynolds numbers, i.e. for $\mathrm{Re} = 1 \,.\, 10^4, 2 \,.\, 10^4, 4 \,.\, 10^4$, and $5 \,.\, 10^4$. Since REL-KE shows a more consistent behavior between pipe and channel, and a better agreement with the experiments, the REL-KE model is used in these calculations. Applying local grid adaptions, it is ensured that y^+ values of the next-to-wall cells are everywhere well below unity. Detailed views of surface grids on the roughness elements are displayed in Fig. 4.

Detailed views of the predicted vector fields in a longitudinal plane near the wall, through the roughness elements, are displayed in Fig. 5, for inline and staggered arrangements of the roughness elements.

In Fig. 5, the recirculation zones with comparably low velocities within the spaces between the roughness elements can be observed, which also exhibit different patterns

Fig. 4 Surface grids on SGR elements. Left: inline arrangement, right: staggered arrangement

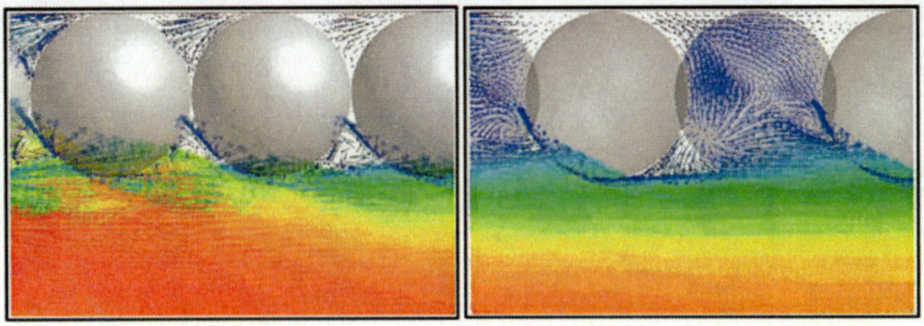

Fig. 5 Detail velocity vectors near the wall in a longitudinal plane through roughness elements. Left: inline arrangement, right: staggered arrangement (k/D = 0.033, Re = 50,000, REL-KE)

for the different arrangements of the roughness elements (a color scale is not provided, since it is about a qualitative discussion. Besides the vector length being proportional to the velocity magnitude, red and blue colors mark the maximum and minimum values of the velocity magnitude). One can see that the velocity starts to increase just above the roughness elements, which can, in a way, be interpreted as a kind of "lifting" of the boundary layer by the roughness elements. This can be seen to be in accordance with the empirically considered shift of the assumed boundary layer velocity profile as expressed in Eq. (1).

The thermal problem is posed as the heating of the fluid (air, Prandtl number = 0.7) by a hot wall. Isothermal boundary conditions are applied (Air inlet temperature: 25 °C, channel wall temperature: 50 °C). It is assumed that all solid surfaces including the planar wall and spheres have the same, constant wall temperature.

Detailed views of predicted temperature fields in a longitudinal plane near the wall through the roughness elements are displayed in Fig. 6, for both arrangements of the roughness elements. In the figure, the distribution of wall heat flux on the roughness elements is also displayed (a color scale is not provided, since it is about a qualitative discussion. Red and blue colors mark the maximum and minimum values of the corresponding variable). In the vicinity of the wall, up to a level of approx. mid-height of the roughness elements, the temperature distribution is quite uniform, which is the result of the mixing and homogenization caused by the recirculation zones. This, in return, seems to lead to comparably low local heat flux values, due to the prevailing low temperature difference. This can be observed on the wall-side surfaces of the spheres. On the flow side of the spheres, the roughness elements are subject to a unidirectional flow with high velocity, higher temperature gradients occur in the layers next to the roughness elements, and maximum heat flux values on the surfaces of roughness elements are observed here, on their upstream sides, due to the impingement effect.

Friction factors predicted by the surface resolving calculations are compared with the experiments in Fig. 7, where the wall functions-based solution is also indicated.

Fig. 6 Detail temperature fields near the wall in a longitudinal plane through roughness elements, as well as wall heat flux distributions on the solid surfaces. Left: inline arrangement, right: staggered arrangement (k/D = 0.033, Re = 50,000, REL-KE)

One can see that the RR calculations show a fair agreement with the measurements. This is, however, not as good as that of the WF approach, which is, however, empirically tuned to achieve the best accuracy exactly for SGR.

Predicted Nusselt numbers (Nu) are compared with experimental values in Fig. 8. One can see that the results by the WF approach show a fairly good agreement with measurements. The Nu values obtained by the RR approach are, however, strongly overpredicting the experimental values.

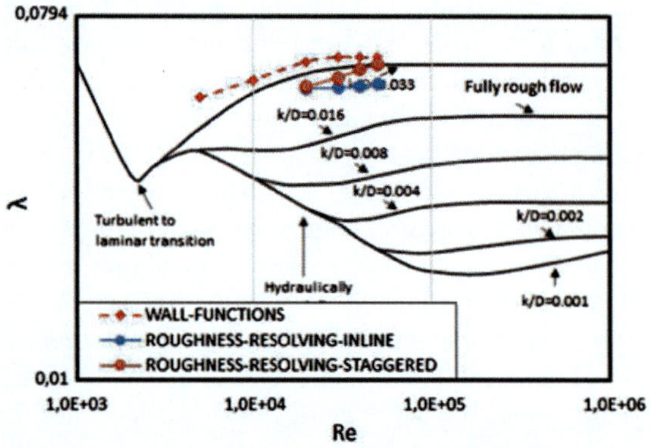

Fig. 7 Predicted friction factors (λ) as a function of Reynolds number (Re) for SGR, k/D = 0.033 obtained for fully developed pipe flow and fully developed channel flow with equivalent hydraulic diameter, compared with empirical data for pipe (black lines, reproduced from Ref. [30])

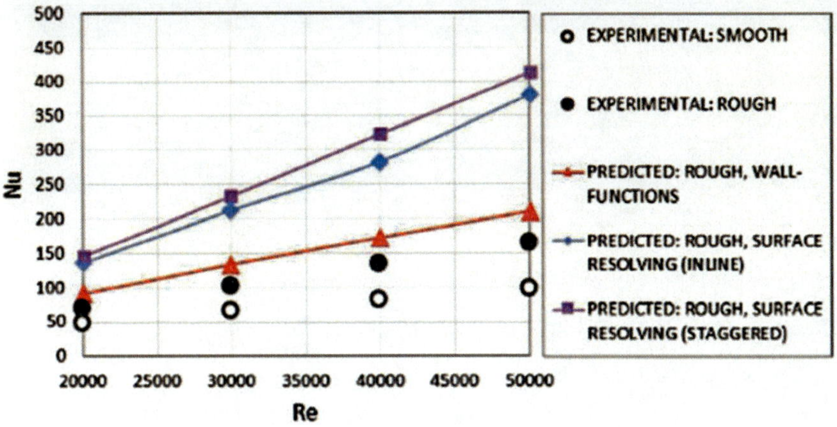

Fig. 8 Predicted and measured Nu as a function of Re for channel flow with SGR, k/D = 0.033

5.3 Roughness Resolving Approach for Surface with Irregular Roughness

As the WF approach is tuned for SGR, the RR approach gains more value for non-SGR, irregular roughness patterns. In applying the RR approach, the discretization of a measured irregularly rough surface often represents a great challenge, due to very complex shapes on the rough surface.

An amenable approach is to re-construct the surface by keeping the main characteristics of the rough surface, but smoothing it, at the same time, at a certain level, by applying some kind of filtering to remove the too-spiky structures, to allow sufficient grid smoothness and stability.

Since this means some loss of topology information, it is a trade-off between accuracy and practicability, the optimal point of which is to be explored in the future studies.

Flow in a rectangular channel with an irregular surface roughness is considered for Re = 50,000. The generated computational surface grid for the measured, and subsequent to smoothing, re-constructed rough surface is displayed in Fig. 9.

The predicted distribution of wall shear stress in magnitude is presented in Fig. 10, where alternating changes between high and low values at the peaks and valleys can be observed (a color scale is not provided, since it is about a qualitative discussion. Red and blue colors mark the maximum and minimum values of the corresponding variable).

The calculated overall friction coefficient is observed to overpredict the measured value by about 20%, which may, at least partially, be caused by the smoothing applied to the surface.

Fig. 9 Surface grid for irregularly rough surface of channel flow

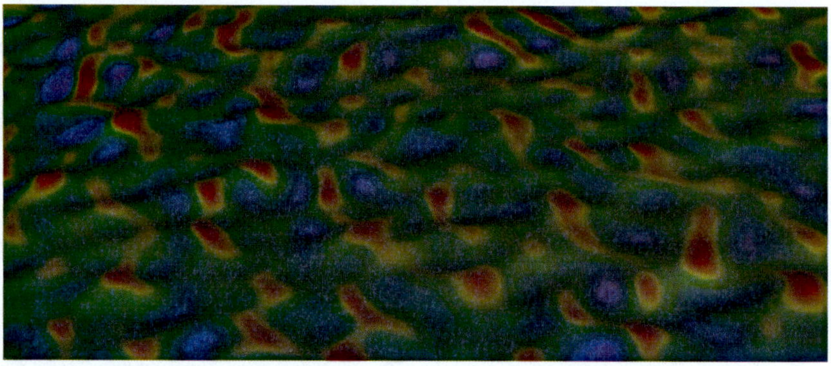

Fig. 10 Distribution of wall shear stress magnitude on the rough surface

6 Conclusions

The WF-based roughness modeling is observed to provide a fair accuracy in the transitional region of pipe flow with SGR, where the realizable k-ε turbulence model showed a slightly better quantitative accuracy compared to alternative two-equation models. Based on the measured SGR data in pipes and measurements performed in rectangular channels, the RR approach is observed to be less accurate in comparison. Improvements in the geometry representation and turbulence modeling are expected to lead to a better accuracy. This is to be explored in the future work.

Acknowledgements Funding by the Ministry for Economic Affairs, Innovation, Digitalization and Energy of the Government of the Federal State of North Rhine Westphalia (MWIDE NRW), Germany, is gratefully acknowledged [Project 005-2010-0082_061]. Colleagues at ZIES (Centre Innovation Energy System, Prof. Adam) and ISAVE (Inst. Sound & Vibration Eng., Prof. Kameier) of the Düsseldorf University of Applied Science are gratefully acknowledged for their support in test rig installation and instrumentation.

References

1. Gisario, A., Kazarian, M., Martina, F., Mehrpouya, M.: Metal additive manufacturing in the commercial aviation industry: a review. J. Manuf. Syst. **53**, 124–149 (2019)
2. Towsend, A., Senin, N., Blunt, L., Leach, R.K., Taylor, J.S.: Surface texture metrology for additive manufacturing: a review. Precis. Eng. **46**, 34–47 (2016)
3. Kays, W.M.: Convective Heat and Mass Transfer. McGraw-Hill, New York (2005)
4. Schlichting, H.: Boundary Layer Theory, 7th edn. McGraw-Hill, New York (1979)
5. Ali, R., Farooq, A., Shahzad, A., Benim. A. C., Iqbal. A., Razzaq, M.: Computational approach on three-dimensional flow of couple-stress fluid with convective boundary conditions. Physica A **553**, 124056 (2020)
6. Durbin, P.A., Petterson Reif, B.A.: Statistical Theory and Modeling for Turbulent Flows, 2nd edn. Wiley, Chichester (2011)
7. Benim, A.C., Cagan, M., Nahavandi, A., Pasqualotto, E.: RANS predictions of turbulent flow past a circular cylinder over the critical regime. In: Proceedings of the 5th IASME/WSEAS International Conference on Fluid Mechanics and Aerodynamics, pp. 232–237, Athens, Greece, August 25–27 (2007)
8. Benim, A.C., Nahavandi, A., Stopford, P.J., Syed, K.: URANS, LES and DES analysis of turbulent swirling flows in gas turbine combustors. WSEAS Trans. Fluid Mech. **1**(5), 465–472 (2006)
9. Xie, Z., Castro, I.P.: LES and RANS for turbulent flow over arrays of wall mounted obstacles. Flow Turbul. Combust. **91**, 291–312 (2006)
10. Ashrafian, A., Andersson, H.I.: Roughness effects in turbulent channel flow. Progr. Comput. Fluid Dyn. Int. J. **6**(1/2/3), 1–20 (2006)
11. MacDonald, M., Hutchins, N., Chung, D.: Roughness effects in turbulent forced convection. J. Fluid Mech. **861**, 138–162 (2019)
12. Benim, A. C., Diederich, M.: Prediction of roughness effects on wind turbine aerodynamics. In: E3S Web of Conferences, vol. 128, p. 09004 (2019)
13. Wang, Z. J., Chi, X., Shih, T., Bons, J.: Direct simulation of surface roughness effects with a RANS and DES approach on viscous adaptive Cartesian grids. In: AIAA 2004-2420 (2004)
14. Yoon, S., Na, S., Wang, Z. J., Bons, J., Shih, T.: Flow and heat transfer over rough surfaces: usefulness of 2D roughness-resolved simulations. In: AIAA 2006-0025 (2006)
15. Busse, A., Lützner, M., Sandham, N.D.: Direct numerical simulation of turbulent flow over a rough surface based on a surface scan. Comput. Fluids **116**, 129–147 (2015)
16. Forooghi, P., Strip, M., Frohnapfel, B.: A systematic study of turbulent heat transfer over rough walls. Int. J. Heat Mass Transf. **127**, 1157–1168 (2018)
17. Coleman, H.W., Hodge, B.K., Taylor, R.P.: A re-evaluation of Schlichting's surface roughness experiment. J. Fluids Eng. **106**, 60–65 (1984)
18. Sigal, A., Danberg, J.E.: New correlation of roughness density effect on the turbulent boundary layer. AIAA J. **28**(3), 554–556 (1990)
19. Craft, T. J., Gerasimov, A. V., Iacovides, H., Launder, B. E.: Progress in the generalization of wall-function treatments. Int. J. Heat Fluid Flow **23**, 148–160 (2002)

20. Suga, K., Craft, T.J., Iacovides, H.: An analytical wall-function for turbulent flows and heat transfer over rough walls. Int. J. Heat Fluid Flow **27**, 852–866 (2006)
21. Ashrafian, A., Johansen, S. T.: Wall boundary conditions for rough walls. Progr. Comput. Fluid Dyn. Int. J. **7**(2/3/4), 230–236 (2007)
22. Chedevergne, F.: Analytical wall function including roughness corrections. Int. J. Heat Fluid Flow **73**, 258–269 (2018)
23. Diederich, M.: Modelling of turbulent flow over a surface with technical roughness. M.Sc. thesis, Department of Mechanical Process Engineering, Düsseldorf University of Applied Sciences (2018)
24. Saner, T.: Statistical analysis of 3D arbitrary surface roughness. M.Sc. thesis, Department of Mechanical and Process Engineering, Düsseldorf University of Applied Sciences (2021)
25. ANSYS Fluent 18.0, Theory Guide. www.ansys.com
26. Aslan, E., Taymaz, I., Benim, A. C.: Investigation of LBM curved boundary treatments for unsteady flows. Eur. J. Mech. B/Fluids **51**, 68–74 (2015)
27. Chen, H. C., Patel, V. C.: Near-wall turbulence models for complex flows including separation. AIAA J. **26**(6), 641–648 (1988)
28. Wolfshtein. M.: The velocity and temperature distribution of one-dimensional flow with turbulence augmentation and pressure gradient. Int. J. Heat Mass Trans. **12**, 301–318 (1969)
29. Cebeci, T., Bradshaw, P.: Momentum Transfer in Boundary Layers. Hemisphere, NY (1977)
30. Miller, D. S.: Internal Flow Systems, 2nd edn. British Hydromechanics Research Association (1990)

Distribution of Noise in Linear Recurrent Fractal Interpolation Functions for Data Sets with α-Stable Noise

Mohit Kumar, Neelesh S. Upadhye, and A. K. B. Chand

Abstract In this study, we construct a linear recurrent fractal interpolation function (RFIF) with variable scaling parameters for data set with α-stable noise (a generalization of Gaussian noise) on its ordinate, which captures the uncertainty at any missing or unknown intermediate point. The propagation of uncertainty in this linear RFIF is investigated, and a method for estimating parameters of the uncertainty at any interpolated value is provided. Moreover, a simulation study to visualize uncertainty for interpolated values is presented.

Keywords Fractals · Random fractal interpolation function · Recurrent fractal interpolation · Stable distribution · Stable noise

1 Introduction

In 1986, Barnsley [1] introduced the notion of fractal interpolation function (FIF) based on the theory of iterated function system (IFS), which can produce nowhere differentiable self-similar continuous functions. In 1989, Barnsley et al. [3] generalized this FIF technique to recurrent FIF (RFIF) by using recurrent IFS (RIFS), which can generate even more complex locally self-similar functions. Thereafter, RFIF is widely used for obtaining missing or unknown values at any intermediate points of a prescribed deterministic data set. However, if the provided data set contains noise on its ordinate, then capturing uncertainty at these interpolated values is essential, but incapable of doing so. This motivates us to study the fractal interpolation for noisy data sets.

Over the last three decades, many researchers have constructed fractal functions for deterministic data sets in various ways (for instance, see [2, 4, 7, 12, 13]) and discussed their analytical properties. At present, fractal interpolation is an advanced

M. Kumar (✉) · N. S. Upadhye · A. K. B. Chand
Department of Mathematics, Indian Institute of Technology Madras,
Chennai 600036, Tamil Nadu, India
e-mail: mohittripathi.5678@gmail.com

R. K. Sharma et al. (eds.), *Frontiers in Industrial and Applied Mathematics*,
Springer Proceedings in Mathematics & Statistics 410,
https://doi.org/10.1007/978-981-19-7272-0_2

approach to approximate and analyze a wide range of scientific data that include irregularities or self-similarities. However, fractal interpolation for data with uncertainty has received little attention from researchers (see, [5, 6]). In this study, we use data sets with α-stable noise (a generalization of Gaussian noise) on its ordinate and extend this RFIF technique to capture the uncertainty at any missing or unknown intermediate values.

The paper is organized as follows. Section 2 recalls definitions and some basic results related to RFIF and α-stable distribution. In Sect. 3, the construction of a RFIF with variable scaling for α-stable noisy data is discussed and the parameter estimation of the uncertainty at any intermediate point of this RFIF is given. Section 4 discusses numerical experiments to validate and visualize analytical results. Section 5 concludes with a brief overview of our theoretical developments.

2 Preliminaries

In this section, we briefly describe the basic notions of RIFS, RFIF, and α-stable distribution. The details are given in [2, 11, 13].

2.1 *Basics of RIFS*

Definition 1 Let (K, d) be a complete metric space and $W_i : K \to K$ $(i = 1, 2, \ldots, N)$ be contraction maps. Also, let $P = \left(p_{ij}\right)_{N\times N}$ be an $N \times N$ irreducible row-stochastic matrix. Then $\{K; P; W_i : i = 1, 2, \ldots, N\}$ is called a recurrent iterated function system.

Further, the recurrent structure of the RIFS is given by a connection matrix $C = \left(c_{ij}\right)_{N\times N}$ which is defined by

$$c_{ij} = \begin{cases} 1, & p_{ji} > 0, \\ 0, & p_{ji} = 0. \end{cases} \tag{1}$$

This C is also an irreducible matrix. Let $\mathcal{H}(K)$ be the set of all nonempty compact subsets of K, and h be the Hausdorff distance in $\mathcal{H}(K)$ defined by

$$h(A, B) = \max\{\max_{a\in A} \min_{b\in B} d(a, b), \max_{b\in B} \min_{a\in A} d(a, b)\}, \ A, B \in \mathcal{H}(K).$$

Then $(\mathcal{H}(K), h)$ is a complete metric space. Let us denote the product space

$$\tilde{\mathcal{H}}(K) := \underbrace{\mathcal{H}(K) \times \cdots \times \mathcal{H}(K)}_{N \text{ times}} = \mathcal{H}(K)^N,$$

and define a metric $\tilde{h}$ on $\tilde{\mathcal{H}}(K)$ by

$$\tilde{h}\left((A_1, A_2, \dots, A_N), (B_1, B_2, \dots, B_N)\right) := \max\{h(A_i, B_i) : i = 1, 2, \dots, N\},$$

for all $(A_1, A_2, \dots, A_N), (B_1, B_2, \dots, B_N) \in \tilde{\mathcal{H}}(K)$. Then $\left(\tilde{\mathcal{H}}(K), \tilde{h}\right)$ is also a complete metric space. Now, we define a transformation $W : \tilde{\mathcal{H}}(K) \to \tilde{\mathcal{H}}(K)$ by

$$W(\mathbf{B}) := \begin{pmatrix} \bigcup_{j=1}^{N} c_{1j} W_1(B_j) \\ \bigcup_{j=1}^{N} c_{2j} W_2(B_j) \\ \vdots \\ \bigcup_{j=1}^{N} c_{Nj} W_N(B_j) \end{pmatrix} = \begin{pmatrix} \bigcup_{j\in\Lambda(1)} W_1(B_j) \\ \bigcup_{j\in\Lambda(2)} W_2(B_j) \\ \vdots \\ \bigcup_{j\in\Lambda(N)} W_N(B_j) \end{pmatrix},$$

for all $\mathbf{B} = (B_1, B_2, \dots, B_N) \in \tilde{\mathcal{H}}(K)$. Here we considered

$$c_{ij} W_i(B_j) = \begin{cases} W_i(B_j) & \text{if } c_{ij} = 1, \\ \emptyset & \text{if } c_{ij} = 0, \end{cases}$$

for all $i, j = 1, 2, \dots, N$ and $\Lambda(i) = \{j : c_{ij} = 1\}$ for all $i = 1, 2, \dots, N$. Alternatively, W can be represented in a matrix as $W = \left(c_{ij} W_i\right)_{N\times N}$, i.e.

$$W = \begin{pmatrix} c_{11}W_1 & c_{12}W_1 & \dots & c_{1N}W_1 \\ c_{21}W_2 & c_{22}W_2 & \dots & c_{2N}W_2 \\ \vdots & \vdots & \vdots & \vdots \\ c_{N1}W_N & c_{N2}W_N & \dots & c_{NN}W_N \end{pmatrix}.$$

This transformation W is a contraction map on $\tilde{\mathcal{H}}(K)$ and hence there exists a unique fixed point $\mathbf{A} = (A_1, A_2, \dots, A_N) \in \tilde{\mathcal{H}}(K)$ such that $W(\mathbf{A}) = \mathbf{A}$, which is called an invariant set or an attractor or a recurrent fractal of the RIFS. Moreover, $A_i = \bigcup_{j\in\Lambda(i)} W_i(A_j)$ for all $i = 1, 2, \dots, N$. Usually, making a slight abuse of notation, we often call $A = \bigcup_{i=1}^{N} A_i$ as the attractor of the RIFS.

We first utilize this RIFS theory to construct a fractal function associated with a deterministic data set and then consider a noisy data set for generating a random fractal function with variable scaling based on the notion of RIFS.

2.2 RFIF with Variable Scaling for Deterministic Data Set

Let us take an initial data set $\mathcal{D} = \{(t_i, y_i) : i = 0, 1, \dots, N\}$ in $\mathbb{R}^2$, where $t_0 < t_1 < \dots < t_N$. We denote intervals $I := [t_0, t_N]$, and $I_i := [t_{i-1}, t_i]$ for all $i =$

$1, 2, \ldots, N$. Also, let us consider intervals $J_j := [t_{l(j)}, t_{r(j)}]$, where $l(j), r(j) \in \{0, 1, \ldots, N\}$ with $l(j) < r(j)$ for all $j = 1, 2, \ldots, N$. Now, we define homeomorphisms $L_k : J_k \to I_k$ by $L_k(t) = a_k t + b_k$ for $k = 1, 2, \ldots, N$, which map end points of J_k to end points of I_k such that $L_k(t_{l(k)}) = t_{k-1}$ and $L_k(t_{r(k)}) = t_k$. Therefore, we have

$$a_k = \frac{t_k - t_{k-1}}{t_{r(k)} - t_{l(k)}} \text{ and } b_k = \frac{t_{r(k)}t_{k-1} - t_{l(k)}t_k}{t_{r(k)} - t_{l(k)}}.$$

Also, for all $t, t^* \in J_k$, we have $|L_k(t) - L_k(t^*)| \leq |a_k||t - t^*|$. If we consider the length of J_k to be greater than the length of I_k, that is $|t_k - t_{k-1}| < |t_{r(k)} - t_{l(k)}|$, then $|a_k| < 1$ and L_k becomes a contraction.

Define continuous maps $F_k : J_k \times \mathbb{R} \to \mathbb{R}$ by $F_k(t, y) = c_k t + d_k(t)y + e_k$, where d_k are real-valued continuous functions defined on I and satisfying

$$\|d_k\|_\infty := \sup\{|d_k(t)| : t \in I\} < 1. \tag{2}$$

In addition, each F_k satisfying join-up conditions $F_k\left(t_{l(k)}, y_{l(k)}\right) = y_{k-1}$ and $F_k\left(t_{r(k)}, y_{r(k)}\right) = y_k$. Therefore, we get

$$c_k = \frac{y_k - y_{k-1}}{t_{r(k)} - t_{l(k)}} - \frac{d_k(t_{r(k)})y_{r(k)} - d_k(t_{l(k)})y_{l(k)}}{t_{r(k)} - t_{l(k)}},$$

$$e_k = \frac{t_{r(k)}y_{k-1} - t_{l(k)}y_k}{t_{r(k)} - t_{l(k)}} - \frac{t_{r(k)}d_k(t_{l(k)})y_{l(k)} - t_{l(k)}d_k(t_{r(k)})y_{r(k)}}{t_{r(k)} - t_{l(k)}}.$$

Moreover, $|F_k(t, y) - F_k(t, y^*)| \leq |d_k(t)||y - y^*|$, $t \in J_k$ and $y, y^* \in \mathbb{R}$. Hence, F_k is a contraction with respect to y-variable.

Next, we consider $W_k : J_k \times \mathbb{R} \to I_k \times \mathbb{R}$ by $W_k(t, y) = (L_k(t), F_k(t, y))$ for all $k = 1, 2, \ldots, N$. We can easily check that $W_k(t_{l(k)}, y_{l(k)}) = (t_{k-1}, y_{k-1})$ and $W_k(t_{r(k)}, y_{r(k)}) = (t_k, y_k)$. Moreover, all W_k are contractions with respect to some metric, equivalent to the Euclidean metric in $\mathbb{R}^2$. Let us define a row-stochastic matrix $P = \left(p_{ij}\right)_{N \times N}$ by

$$p_{ij} = \begin{cases} \frac{1}{N_i}, & I_i \subset J_j, \\ 0, & \text{otherwise}, \end{cases}$$

where N_i denotes the number of j such that $I_i \subset J_j$ for $i = 1, 2, \ldots, N$. We can make P an irreducible matrix by selecting J_k's appropriately. Therefore, we can construct a RIFS $\{I \times \mathbb{R}; P; W_k : k = 1, 2, \ldots, N\}$ associated with $\mathcal{D}$.

Remark 1 In this RIFS, we employed function contractivity factors (or variable scaling parameters) d_k, which describe fractal objects better than constant contractivity factors and provide more flexibility to fractal functions. For detailed information, see [13].

Using (1), we obtain the connection matrix $C = \left(c_{ij}\right)_{N\times N}$, where

$$c_{ij} = \begin{cases} 1, & I_j \subset J_i, \\ 0, & \text{otherwise.} \end{cases} \tag{3}$$

Let $\mathcal{C}(I)$ be the collection of real-valued continuous functions defined on I. Define a metric d_∞ on $\mathcal{C}(I)$ by $d_\infty(f, g) :=\| f - g \|_\infty = \sup\{|f(t) - g(t)| : t \in I\}$. Then $(\mathcal{C}(I), d_\infty)$ is a complete metric space. Further, let us define

$$\mathcal{C}^*(I) := \{f \in \mathcal{C}(I) : f(t_i) = y_i,\ i = 0, 1, \ldots, N\}.$$

Then $(\mathcal{C}^*(I), d_\infty)$ is also a complete metric space. Now, we define an operator $T : \mathcal{C}^*(I) \to \mathcal{C}^*(I)$ by

$$Tg(t) := F_k\left(L_k^{-1}(t), g\left(L_k^{-1}(t)\right)\right),\ t \in I_k \text{ and } k = 1, 2, \ldots, N.$$

Here T is known as the Read-Bajraktarević operator, which is a contraction on $(\mathcal{C}^*(I), d_\infty)$. Therefore, T has a unique fixed point $f_\mathcal{D} \in \mathcal{C}^*(I)$ such that

$$f_\mathcal{D}(t) = Tf_\mathcal{D}(t) = F_k\left(L_k^{-1}(t), f_\mathcal{D}\left(L_k^{-1}(t)\right)\right),\ t \in I_k \text{ and } k = 1, 2, \ldots, N. \tag{4}$$

This $f_\mathcal{D}$ is called a linear RFIF with variable scaling parameters associated with $\mathcal{D}$. Let $A := \{(t, f_\mathcal{D}(t)) : t \in I\}$, and $A_i := \{(t, f_\mathcal{D}(t)) : t \in I_i\}$ for all $i = 1, 2, \ldots, N$. Then $A = \bigcup_{i=1}^N A_i$. Moreover,

$$\begin{aligned} A_i &= \{(t, f_\mathcal{D}(t)) : t \in I_i\} = \{\left(t, F_i\left(L_i^{-1}(t), f_\mathcal{D}\left(L_i^{-1}(t)\right)\right)\right) : t \in I_i\} \\ &= \{(L_i(t), F_i(t, f_\mathcal{D}(t))) : t \in J_i\} = \{W_i(t, f_\mathcal{D}(t)) : t \in J_i\} \\ &= \bigcup_{j\in\Lambda(i)} W_i\left(A_j\right). \end{aligned}$$

Thus, $\mathbf{A} = (A_1, A_2, \ldots, A_N)$ is an attractor of the RIFS $\{I \times \mathbb{R}; P; W_i : i = 1, 2, \ldots, N\}$ associated with $\mathcal{D}$.

In the subsequent section, we define α-stable distributions and some of its properties required for further study.

2.3 α-Stable Distribution

An α-stable distribution, also known as stable distribution, belongs to the family of heavy-tailed distributions and is a generalization of Gaussian distribution. A complete description of a stable distribution requires the following four parameters: an index of stability or tail index $\alpha \in (0, 2]$, a skewness parameter $\beta \in [-1, 1]$, a scale parameter

$\sigma > 0$, and a location parameter $\mu \in \mathbb{R}$. Generally, a stable distribution does not have closed form formulae for its probability density function (PDF) or cumulative distribution function (CDF) [11]. However, it can be described by its characteristic function.

Definition 2 A random variable X follows a stable distribution, denoted by $X \sim S_\alpha(\beta, \sigma, \mu)$, if its characteristic function has the form

$$\begin{aligned}\phi_X(t) &= \mathbb{E}\left[e^{itX}\right] \\ &= \begin{cases} \exp\left(it\mu - |\sigma t|^\alpha \left\{1 + i\beta \text{sign}(t) \tan\left(\frac{\pi\alpha}{2}\right)\left[|\sigma t|^{1-\alpha} - 1\right]\right\}\right), & \alpha \neq 1, \\ \exp\left(it\mu - |\sigma t| \left\{1 + i\beta \text{sign}(t) \frac{2}{\pi} \ln |\sigma t|\right\}\right), & \alpha = 1, \end{cases}\end{aligned}$$

for $t \in \mathbb{R}$, where $\text{sign}(t) = \begin{cases} \frac{t}{|t|}, & t \neq 0, \\ 0, & t = 0. \end{cases}$

Remark 2 Several parameterizations for α-stable distributions are available in the literature, but Nolan's [8] parameterization is used here for numerical reasons.

For $\alpha = 2$, the Gaussian distribution is obtained, i.e. $X \sim \mathcal{N}(\mu, 2\sigma^2)$. The nth moment of a non-Gaussian ($\alpha \neq 2$) stable random variable X is finite iff $n < \alpha$. When $\beta = 0$, the distribution is symmetric about its location parameter μ.

Property 1 If $X \sim S_\alpha(\beta, \sigma, \mu)$ and $0 \neq a, b \in \mathbb{R}$, then

$$aX + b \sim S_\alpha(\text{sign}(a)\beta, |a|\sigma, a\mu + b).$$

Property 2 For all $i = 0, 1, 2, \ldots, N$, if $X_i \sim S_\alpha(\beta_i, \sigma_i, \mu_i)$ are independent and $\omega_i \in \mathbb{R}$, then $\sum_{i=0}^N \omega_i X_i \sim S_\alpha(\beta, \sigma, \mu)$, where

$$\sigma^\alpha = \sum_{i=0}^N |\omega_i \sigma_i|^\alpha, \qquad \beta\sigma^\alpha = \sum_{i=0}^N \text{sign}(\omega_i)\beta_i |\omega_i \sigma_i|^\alpha,$$

$$\mu = \begin{cases} \sum_{i=0}^N \omega_i \mu_i + \tan\left(\frac{\pi\alpha}{2}\right)\left(\beta\sigma - \sum_{i=0}^N \omega_i \beta_i \sigma_i\right) & \alpha \neq 1, \\ \sum_{i=0}^N \omega_i \mu_i + \frac{\pi}{2}\left(\beta\sigma \ln \sigma - \sum_{i=0}^N \omega_i \beta_i \sigma_i \ln |\omega_i \sigma_i|\right) & \alpha = 1. \end{cases}$$

For more detailed information, the reader can see [9–11].

In the following section, we construct a linear RFIF with variable scaling for any given α-stable noisy data set and determine the probability distribution of any interpolated value of this RFIF.

3 RFIF for Noisy Data Set

Consider a data set $\Delta = \{(t_i, y_i, \epsilon_i) : i = 0, 1, \dots, N\}$, where $t_0 < t_1 < \cdots < t_N$ and $\epsilon_i \sim S_\alpha(\beta_i, \sigma_i, 0)$ is the α-stable noise in the value of y_i. We assume that these ϵ_i's are independent. First, we construct RIFS for this noisy data set. Let $Y_i := y_i + \epsilon_i$, using Property 1, we have $Y_i \sim S_\alpha(\beta_i, \sigma_i, y_i)$ for all $i = 0, 1, \dots, N$. These Y_i's are also independent. Let Y be a real-valued continuous random variable. Define $\mathcal{F}_k : J_k \times \mathbb{R} \to \mathbb{R}$ (a random analog of F_k) by $\mathcal{F}_k(t, Y) = C_k t + d_k(t)Y + E_k$ satisfying $\mathcal{F}_k\left(t_{l(k)}, Y_{l(k)}\right) = Y_{k-1}$ and $\mathcal{F}_k\left(t_{r(k)}, Y_{r(k)}\right) = Y_k$ for all $k = 1, 2, \dots, N$. Therefore,

$$\begin{aligned} C_k =& \frac{Y_k - Y_{k-1}}{t_{r(k)} - t_{l(k)}} - \frac{d_k(t_{r(k)})Y_{r(k)} - d_k(t_{l(k)})Y_{l(k)}}{t_{r(k)} - t_{l(k)}}, \\ E_k =& \frac{t_{r(k)}Y_{k-1} - t_{l(k)}Y_k}{t_{r(k)} - t_{l(k)}} - \frac{t_{r(k)}d_k(t_{l(k)})Y_{l(k)} - t_{l(k)}d_k(t_{r(k)})Y_{r(k)}}{t_{r(k)} - t_{l(k)}}. \end{aligned} \tag{5}$$

Define $\mathcal{W}_k : J_k \times \mathbb{R} \to I_k \times \mathbb{R}$ by $\mathcal{W}_k(t, Y) = (L_k(t), \mathcal{F}_k(t, Y))$ for all $k = 1, 2, \dots, N$, and construct RIFS $\{I \times \mathbb{R}; P; \mathcal{W}_k : k = 1, 2, \dots, N\}$ associated with Δ, which is a random analog to the RIFS $\{I \times \mathbb{R}; P; W_k : k \in \mathbb{N}_N\}$ associated with $\mathcal{D}$. There exists a unique [up to distribution] RFIF $f_\Delta : I \to \mathbb{R}$ such that

$$\begin{aligned} f_\Delta(t) &= \mathcal{F}_k\left(L_k^{-1}(t), f_\Delta\left(L_k^{-1}(t)\right)\right) \\ &= C_k L_k^{-1}(t) + d_k\left(L_k^{-1}(t)\right) f_\Delta\left(L_k^{-1}(t)\right) + E_k, \;\; t \in I_k, k = 1, \dots, N. \end{aligned} \tag{6}$$

Apparently, this f_Δ is a random analog of $f_\mathcal{D}$. Next, we write f_Δ in explicit form to find its distribution. We can see that $I = \bigcup_{k=1}^N I_k$ and $I_k = L_k(J_k) = \bigcup_{j \in \Lambda(k)} L_k(I_j)$. Therefore, I is the attractor of RIFS $\{I; P; L_k : k = 1, 2, \dots, N\}$. Hence, for any given point $t \in I$, there exists a sequence $\{k_n\}_{n\in\mathbb{N}}$, where each $k_n \in \{1, 2, \dots, N\}$, such that

$$\lim_{n\to\infty} L_{k_1} \circ L_{k_2} \circ \cdots \circ L_{k_n}(s) = t, \;\; \text{for } s \in I. \tag{7}$$

By recursively applying (6), we can easily obtain the following expression:

$$f_\Delta(T_0(s)) = D_n(s) f_\Delta(s) + \sum_{j=1}^{n} D_{j-1}(s)\left(C_{k_j} T_j(s) + E_{k_j}\right), \tag{8}$$

where

$$T_j(s) = \begin{cases} L_{k_{j+1}} \circ \cdots \circ L_{k_n}(s) & \text{for } j = 0, 1, \dots, n-1, \\ s & \text{for } j = n, \end{cases}$$

and

$$D_j(s) = \begin{cases} 1 & \text{for } j = 0, \\ \prod_{i=1}^{j} d_{k_i}(T_i(s)) & \text{for } j = 1, 2, \ldots, n. \end{cases}$$

We can rewrite (7) as $\lim_{n\to\infty} T_0(s) = t$. Also, we get $\lim_{n\to\infty} D_n(s) = 0$ by using (2). Since f_Δ is a continuous function, as n approaches ∞ in (8), we obtain

$$f_\Delta(t) = \sum_{j=1}^{\infty} D_{j-1}(s)\left(C_{k_j} T_j(s) + E_{k_j}\right), \; s \in I. \tag{9}$$

Using (5), we can rewrite (9) as

$$\begin{aligned} f_\Delta(t) = \sum_{j=1}^{\infty} D_{j-1}(s) \Bigg[& \left(\frac{t_{r(k_j)} - T_j(s)}{t_{r(k_j)} - t_{l(k_j)}}\right) Y_{k_j-1} + \left(\frac{T_j(s) - t_{l(k_j)}}{t_{r(k_j)} - t_{l(k_j)}}\right) Y_{k_j} - \\ & \left(\frac{t_{r(k_j)} - T_j(s)}{t_{r(k_j)} - t_{l(k_j)}}\right) d_{k_j}(t_{l(k_j)}) Y_{l(k_j)} - \left(\frac{T_j(s) - t_{l(k_j)}}{t_{r(k_j)} - t_{l(k_j)}}\right) d_{k_j}(t_{r(k_j)}) Y_{r(k_j)} \Bigg]. \end{aligned} \tag{10}$$

For each $k_j \in \{1, 2, \ldots, N\}$, we have $Y_{k_j-1}, Y_{k_j}, Y_{l(k_j)}, Y_{r(k_j)} \in \{Y_0, Y_1, \ldots, Y_N\}$. Therefore, by equating coefficients of each Y_i in (10), we get

$$f_\Delta(t) = \sum_{i=0}^{N} \omega_i Y_i, \; t \in I, \tag{11}$$

where ω_i depends on the sequence $\{k_j\}$ of t. We can easily see that the linear RFIF $f_\Delta(t)$ is a random variable for each $t \in I$. Now, we determine the probability distribution of $f_\Delta(t)$. By using Property 2 in (11), we get

$$f_\Delta(t) \sim S_\alpha(\beta, \sigma, \mu),$$

where

$$\sigma = \left(\sum_{i=0}^{N} |\omega_i \sigma_i|^\alpha\right)^{1/\alpha}, \quad \beta = \frac{\sum_{i=0}^{N} \text{sign}(\omega_i)\beta_i |\omega_i \sigma_i|^\alpha}{\sigma^\alpha},$$

$$\mu = \begin{cases} \sum_{i=0}^{N} \omega_i y_i + \tan\left(\frac{\pi\alpha}{2}\right)\left(\beta\sigma - \sum_{i=0}^{N} \omega_i \beta_i \sigma_i\right) & \alpha \neq 1, \\ \sum_{i=0}^{N} \omega_i y_i + \frac{\pi}{2}\left(\beta\sigma \ln\sigma - \sum_{i=0}^{N} \omega_i \beta_i \sigma_i \ln|\omega_i \sigma_i|\right) & \alpha = 1. \end{cases}$$

Moreover, initial data set $\mathcal{D}$ is a realization of the noisy data set Δ. Therefore, by using (11), we get

$$f_{\mathcal{D}}(t) = \sum_{i=0}^{N} \omega_i y_i.$$

Hence, the location parameter μ of $f_\Delta(t)$ becomes

$$\mu = \begin{cases} f_D(t) + \tan\left(\frac{\pi\alpha}{2}\right)\left(\beta\sigma - \sum_{i=0}^{N} \omega_i \beta_i \sigma_i\right) & \alpha \neq 1, \\ f_D(t) + \frac{\pi}{2}\left(\beta\sigma \ln \sigma - \sum_{i=0}^{N} \omega_i \beta_i \sigma_i \ln |\omega_i \sigma_i|\right) & \alpha = 1. \end{cases}$$

Thus, $f_\Delta(t)$ is an α-stable random variable for each $t \in I$.

Remark 3 If α-stable noise in the data set Δ is symmetric, i.e. $\epsilon_i \sim S_\alpha(0, \sigma_i, 0)$ for all $i = 0, 1, \ldots, N$, then $f_\Delta(t)$ is also a symmetric α-stable variate and its location parameter is $f_D(t)$ that is $f_\Delta(t) \sim S_\alpha(0, \sigma, f_D(t))$ for all $t \in I$, where $\sigma = \left(\sum_{i=0}^{N} |\omega_i \sigma_i|^\alpha\right)^{1/\alpha}$. Moreover, if $\alpha = 2$, then $\epsilon_i \sim \mathcal{N}(0, 2\sigma_i^2)$ for $i = 0, 1, \ldots, N$ and $f_\Delta(t) \sim \mathcal{N}(f_D(t), \sigma^2)$, where $\sigma^2 = \frac{1}{2}\sum_{i=0}^{N} \omega_i^2 \sigma_i^2$.

4 Simulation

In this section, we present a simulation study through a numerical example to illustrate the propagation of uncertainty in a linear RFIF with variable scaling parameters for a given α-stable noisy data set.

Let $\Delta = \{(t_0, y_0, \epsilon_0), (t_1, y_1, \epsilon_1), (t_2, y_2, \epsilon_2), (t_3, y_3, \epsilon_3), (t_4, y_4, \epsilon_4)\}$ be a given data set, where

$$t_0 = 0,\ t_1 = 0.3,\ t_2 = 0.5,\ t_3 = 0.7,\ t_4 = 1;$$
$$y_0 = 2.3,\ y_1 = 1.6,\ y_2 = 3.8,\ y_3 = 2.9,\ y_4 = 1.2;$$

and

$$\epsilon_0 \sim S_{1.8}(0.3, 0.4, 0),\ \epsilon_1 \sim S_{1.8}(-0.3, 0.5, 0),\ \epsilon_2 \sim S_{1.8}(0.5, 0.7, 0),$$
$$\epsilon_3 \sim S_{1.8}(0.7, 0.6, 0),\ \epsilon_4 \sim S_{1.8}(-0.2, 0.3, 0).$$

For this data set, we have $N = 4$; $I = [0, 1]$; and

$$I_1 = [0, 0.3],\ I_2 = [0.3, 0.5],\ I_3 = [0.5, 0.7],\ I_4 = [0.7, 1].$$

Now, let us take $J_1 = [0.3, 0.7],\ J_2 = [0.5, 1.0],\ J_3 = [0, 0.5],\ J_4 = [0, 0.5]$. Then, by using (3), we can form the connection matrix

$$C = \begin{pmatrix} 0 & 1 & 1 & 0 \\ 0 & 0 & 1 & 1 \\ 1 & 1 & 0 & 0 \\ 1 & 1 & 0 & 0 \end{pmatrix}.$$

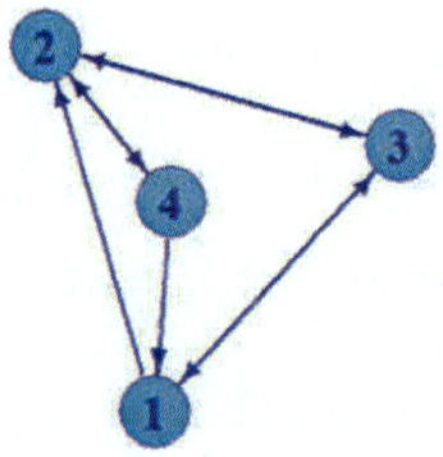

Fig. 1 The directed graph of C

From Fig. 1, we can observe that the directed graph of C is strongly connected, implying that C and therefore P is irreducible.

By using the given data set Δ, we can form the deterministic data set

$$\mathcal{D} = \{(0, 2.3), (0.3, 1.6), (0.5, 3.8), (0.7, 2.9), (1, 1.2)\},$$

and for this data set, we can construct the RIFS $\{I \times \mathbb{R}; P; W_1, W_2, W_3, W_4\}$. If we consider the variable scaling factors:

$$d_1(t) = \frac{1}{3}e^{-5t} + 0.5, \ d_2(t) = \frac{1}{2}\sin(3t) + 0.4,$$
$$d_3(t) = \frac{1}{8}e^{2t}\cos(3t) + 0.6, \ d_4(t) = \frac{1}{2}e^{-5t} + 0.3.$$

Then, we can calculate other parameters of the above RIFS:

$$a_1 = 0.75, \ a_2 = 0.4, \ a_3 = 0.4, \ a_4 = 0.6;$$
$$b_1 = -0.225, \ b_2 = 0.1, \ b_3 = 0.5, \ b_4 = 0.7;$$
$$c_1 = -3.1505, \ c_2 = 10.1011, \ c_3 = -3.2077, \ c_4 = -2.3119;$$
$$e_1 = 2.3261, \ e_2 = -6.8658, \ e_3 = 2.1325, \ e_4 = 1.06.$$

Further, by using (4), we can calculate the values of RFIF $f_{\mathcal{D}}$, whose graph is shown in Fig. 2. In this figure, the red colored dots represent the data points of $\mathcal{D}$, and the RFIF $f_{\mathcal{D}}$ passing through these points is shown in the blue curve. Moreover, we also represent the 95% lower and upper quantile bands of the linear RFIF f_Δ in Fig. 2, which imply that any realization of the RFIF f_Δ will lie between these bands with a probability of 0.95.

Now, we consider an arbitrarily point $t = 0.58$ in I. If we select $s = 0.3$, then we can obtain a sequence $\{k_n\}$ of t such that

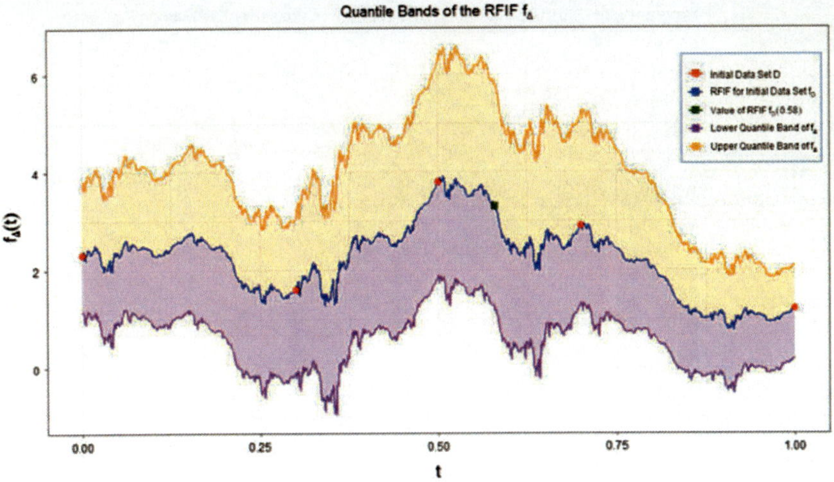

Fig. 2 95% Quantile band of the RFIF f_Δ and graph of the RFIF $f_\mathcal{D}$ along with the points of the data set $\mathcal{D}$

$$\{ 3, 1, 3, 1, 3, 1, 2, 3, 2, 4, 2, 3, 1, 3, 2, 4, 1, 3, 2, 4, 2, 4, 2, 4, 1, 2, 4, 2, 4, 2, 3, 2, 4,$$
$$1, 3, 2, 4, 1, 2, 3, 2, 3, 1, 3, 1, 3, 1, 3, 2, 4, 1, 2, 4, 2, 4, 2, 4, 1, 2, 4, 2, 4, 2, 4, 2, 4,$$
$$1, 2, 4, 1, 2, 3, 2, 3, 1, 2, 3, 2, 4, 2, 4, 2, 3, 2, 3, 1, 3, 2, 4, 1, 2, 4, 1, 2, 3, 1, 3, 2, 4,$$
$$2, 4, 2, 3, 1, 3, 1, 3, 2, 4, 2, 4, 1, 3, 2, 4, 2, 3, 1, 3, 1, 2, 4, 2, 4, 1, 2, 3, 1, 2, 3, 2, 3,$$
$$2, 3, 2, 4, 2, 4, 1, 3, 1, 3, 1, 3, 1, 3, 2, 4, 2, 3, 2, 3 \},$$

with a maximum tolerance error of 0.001 in (7). This sequence is called a fractal code of t. By utilizing (10) and (11), we can compute the coefficients of Y_i as follows:

$$\omega_0 = -0.259217,\ \omega_1 = 0.472842,\ \omega_2 = 0.631665,$$
$$\omega_3 = 0.257078,\ \omega_4 = -0.010617.$$

Hence, the distribution of RFIF f_Δ at point $t = 0.58$ is given as follows:

$$f_\Delta(0.58) \sim S_{1.8}(0.31393, 0.56363, 3.3099). \tag{12}$$

Next, we consider 8000 random samples of the data set Δ. For each realization, we form a RFIF with variable scaling (as we have constructed for the data set $\mathcal{D}$). Therefore, we have 8000 realizations of the RFIF f_Δ and thus we have 8000 realizations of $f_\Delta(0.58)$.

In Fig. 3(i), we represent the histogram of these 8000 random samples of $f_\Delta(0.58)$. In the same figure, we have fitted an empirical PDF to these observed values and also plotted the PDF of the analytically estimated distribution of $f_\Delta(0.58)$, which is given in (12). Here, we can see that the analytically estimated PDF of $f_\Delta(t)$ is very

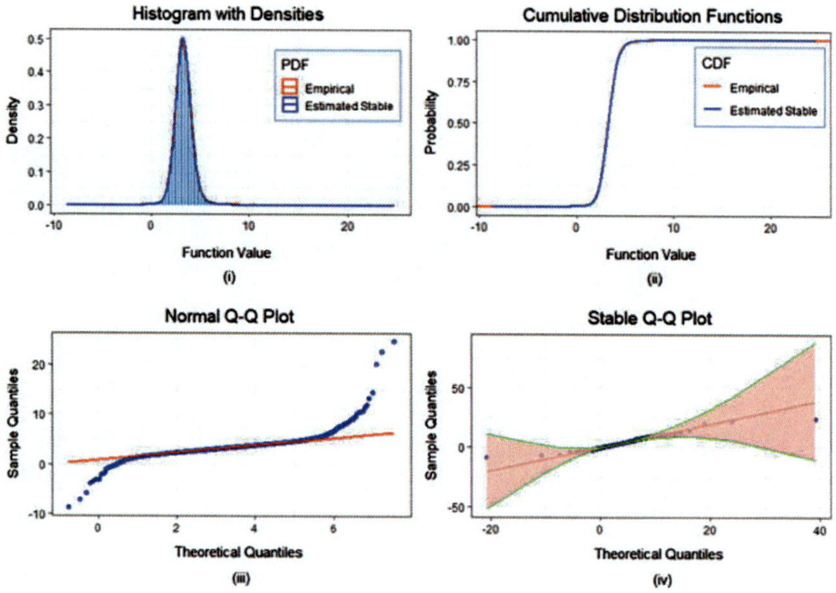

Fig. 3 (i) Histogram with Empirical & Estimated PDFs, (ii) Empirical & Estimated CDFs, (iii) Normal Q-Q Plot, and (iv) Stable Q-Q Plot with 95% Confidence Bands

close to its empirically fitted PDF. A similar conclusion can be drawn from the CDFs plot displayed in Fig. 3(ii).

Moreover, a normal quantile-quantile plot for observed samples of $f_\Delta(0.58)$ is shown in Fig. 3(iii). We can observe here that both tails deviate from the red color reference line, indicating that the distribution of $f_\Delta(0.58)$ has heavier tails than the normal distribution.

Further, a stable quantile-quantile plot is exhibited in Fig. 3(iv). In the same figure, we have displayed 95% confidence band for the simulated values of $f_\Delta(0.58)$, which represents the variation in the estimate of $f_\Delta(0.58)$ from its location based on the noisy data Δ. Here, we can see that nearly all the observed samples of $f_\Delta(0.58)$ fall along the reference line, implying that $f_\Delta(0.58)$ follows the same distribution as we specified in (12). Therefore, our analytically estimated distribution for $f_\Delta(t)$ in (12) is valid. Moreover, $t = 0.58$ is an arbitrarily chosen point in I; therefore, for any $t \in I$, we can similarly estimate and validate the distribution of $f_\Delta(t)$.

5 Concluding Remarks

A commonly used tool for analyzing uncertainty at any point is the estimation of the probability distribution at that point. If the data is collected from a process that has fractal properties and contains α-stable noise, in that case, the recurrent fractal interpolation technique efficiently determines uncertainty at any intermediate point in this noisy data set. Moreover, for any given data set with α-stable noise on its ordinate, the probability distribution of a recurrent fractal interpolation function at any interpolated value is also an α-stable. And the remaining parameters of this distribution can be estimated analytically.

Acknowledgements The authors would like to acknowledge IC&SR, IIT Madras, for the funding support from the IoE research project [Project Number = SB20210848 MAMHRD008558].

References

1. Barnsley, M.F.: Fractal functions and interpolation. Constr. Approx. **2**(1), 303–329 (1986)
2. Barnsley, M.F.: Fractals Everywhere, 2nd ed. Morgan Kaufmann (2000)
3. Barnsley, M.F., Elton, J.H., Hardin, D.P.: Recurrent iterated function systems. Constr. Approx. **5**(1), 3–31 (1989)
4. Chand, A.K.B., Kapoor, G.P.: Generalized cubic spline fractal interpolation functions. SIAM J. Numer. Anal. **44**, 655–676 (2006)
5. Kumar, M., Upadhye, N.S., Chand, A.K.B.: Distribution of linear fractal interpolation function for random dataset with stable noise. Fractals **29**(4), 1–12 (2021)
6. Luor, D.C.: Fractal interpolation functions for random data sets. Chaos Solitons Fractals **114**, 256–263 (2018)
7. Navascués, M.A., Chand, A.K.B., Viswanathan, P., Sebastián, M.V.: Fractal interpolation functions: a short survey. Appl. Math. **5**, 1834–1841 (2014)
8. Nolan, J.P.: Numerical calculation of stable densities and distribution functions. Commun. Stat. Stoch. Models **13**(4), 759–774 (1997)
9. Nolan, J.P.: Parametrizations and modes of stable distributions. Stat. Prob. Lett. **38**, 187–195 (1998)
10. Nolan, J.P.: Univariate Stable Distributions: Models for Heavy Tailed Data. Springer Nature (2020)
11. Samorodnitsky, G., Taqqu, M.: Stable non-Gaussian random processes: stochastic models with infinite variance. CRC Press (1994)
12. Vijender, N., Chand, A.K.B., Navascués, M.A., Sebastián, M.V.: Quantum Bernstein fractal functions. Comp. Math. Methods **3**, 1–13 (2021)
13. Wang, H.Y., Yu, J.S.: Fractal interpolation functions with variable parameters and their analytical properties. J. Approx. Theory **175**, 1–18 (2013)

Oblivious Transfer Using Non-abelian Groups

Maggie E. Habeeb

Abstract The field of non-commutative group-based cryptography has flourished in recent years, resulting in numerous non-commutative key exchange protocols, digital signature schemes, and secret sharing schemes. In this paper, we propose two 1-out-of-n oblivious transfer protocols using conjugation of group elements. The protocols are obtained by modifying the Ko-Lee key exchange protocol and the Anshel-Anshel-Goldfeld key exchange protocol.

Keywords Oblivious transfer · Cryptography · Group theory

1 Introduction

Rabin first introduced the notion of oblivious transfer in [14] in 1981. It is a method in which Alice sends a message to Bob (of his choice) in a manner such that Alice does not know which message Bob received and Bob only receives his desired message. More formally, we may define a 1-out-of-2 oblivious transfer as follows. Alice has two messages m_0 and m_1 and Bob has a choice bit $c \in \{0, 1\}$. The goal of the oblivious transfer protocol is to send the message m_c to Bob where

- Alice does not learn which message Bob received;
- Bob learns nothing of the message m_{1-c}.

A 1-out-of-2 oblivious transfer can be generalized to a 1-out-of-n oblivious transfer in which the sender has n messages and the receiver is to obtain exactly one of them.

Oblivious transfer protocols are indispensable in the sense that they can be used as cryptographic primitives. Cryptographic protocols based on oblivious transfer result in several advantages over number theoretic-based protocols (see [8] for more details). Oblivious transfer has been utilized in [8, 9] for secure multiparty computation and commital protocols. Secure Multiparty Computation gives several mutually

M. E. Habeeb (✉)
California University of Pennsylvania, California, PA 15419, USA
e-mail: Habeeb@calu.edu

R. K. Sharma et al. (eds.), *Frontiers in Industrial and Applied Mathematics*,
Springer Proceedings in Mathematics & Statistics 410,
https://doi.org/10.1007/978-981-19-7272-0_3

distrustful parties the opportunity to perform a computation together without compromising the privacy of their inputs or the validity of their outputs (see [8]). Commital protocols are essential in the theory of zero-knowledge proofs, as they allow one participant to commit to x without the other participant learning x until the protocol is complete (see [9]). These examples exhibit the importance of oblivious transfer protocols.

Chou and Orlandi [3] introduced a simple and efficient 1-out-of-n oblivious transfer protocol based on the Diffie-Hellman key exchange protocol. The security of this protocol is based on the difficulty of the discrete log problem. However, with the polynomial time quantum algorithms put forth by Shor in [16], the security of this protocol is at risk with the advent of the quantum computer. In the recent years, the field of post-quantum cryptography has flourished resulting in many new non-commutative cryptographic protocols (see [4, 13]). In this paper, we provide non-commutative analogs of this protocol based on conjugation by a group element.

2 Preliminaries

In this section, we give a brief background of the relevant non-abelian protocols and a description of the oblivious transfer protocol introduced in [3]. Throughout Sects. 2.1 and 2.2, given two elements x, y in a group G, we denote $x^y = y^{-1}xy$.

2.1 Ko-Lee Key Exchange Protocol

The Ko-Lee key exchange protocol introduced in [10] is a non-abelian analog of the Diffie-Hellman key exchange, where conjugation is used instead of exponentiation. The protocol is based on the search conjugacy problem which can be stated as follows.

Definition 1 (*Search Conjugacy Problem*) In a group G, given $g \in G$ and $h = a^{-1}ga$, find such an a.

Let G be a non-abelian group, u an arbitrary element in G, and S and T two commuting subgroups of G. Suppose that two people, Alice and Bob, want to agree on a shared key.

The group G, the element $u \in G$, and the subgroups S and T are published. Then,

1. Alice chooses a secret element $s \in S$ and sends u^s to Bob.
2. Bob chooses a secret element $t \in T$ and sends u^t to Alice.
3. Alice computes $(u^t)^s = u^{ts}$, and Bob computes $(u^s)^t = u^{st}$. The shared key is $u^{st} = u^{ts}$.

The shared key is straightforward to determine for Alice and Bob since each knows secret elements s and t, respectively. If the secret elements are not known, and only the public data is available, then the search conjugacy problem can be used to determine s and t from u^s and u^t, which is assumed infeasible.

2.2 *Anshel-Anshel-Goldfeld Protocol*

The Anshel-Anshel-Goldfeld (AAG) protocol [1], which is also known as the commutator key exchange, utilizes conjugation of group elements but does not require commuting subgroups as in the Ko-Lee protocol.

Let G be a finitely presented non-abelian group and S and T be two finitely generated subgroups of G with generators $\{s_1, \cdots, s_k\}$ and $\{t_1, \cdots, t_l\}$, respectively. Suppose Alice and Bob want to agree on a shared key.

The group G, subgroups S and T, and their generators are published. Then,

1. Alice chooses a private element $a = s_{i_1}s_{i_2}...s_{i_m} \in S$ and sends $t_1^a, t_2^a..., t_l^a$ to Bob.
2. Bob chooses a private element $b = t_{j_1}t_{j_2}...t_{j_n} \in G$ and sends $s_1^b, s_2^b..., s_k^b$ to Alice.

From this, Alice can compute

$$s_{i_1}^b s_{i_2}^b...s_{i_m}^b = (b^{-1}s_{i_1}b)(b^{-1}s_{i_2}b)...(b^{-1}s_{i_m}b) = b^{-1}s_{i_1}s_{i_2}...s_{i_m}b = a^b$$

while Bob computes

$$t_{j_1}^a t_{j_2}^a...t_{j_n}^a = (a^{-1}t_{j_1}a)(a^{-1}t_{j_2}a)...(a^{-1}j_{i_n}a) = a^{-1}t_{j_1}t_{j_2}...t_{j_n}a = b^a.$$

Alice and Bob can both compute the commutator, $[a, b] = a^{-1}b^{-1}ab$, from the a^b and b^a as their shared secret since each knows the secret elements a and b, respectively. For an adversary to obtain the shared key, it is not enough to solve the search conjugacy problem since there are several instances that must be satisfied at once. Rather, one would need to solve the simultaneous conjugacy search problem which can be stated as follows.

Definition 2 *(Simultaneous Search Conjugacy Problem)* In a group G, given $g_1, g_2, \ldots, g_n \in G$ and $h_1 = a^{-1}g_1a$, $h_2 = a^{-1}g_2a$, ..., $h_n = a^{-1}g_na$, find such an a.

2.3 *Chou and Orlandi Oblivious Transfer*

The OT protocol introduced by Chou and Orlandi in [3] is based on the well-known Diffie-Hellman key exchange. Given a group G and a generator g, Alice and Bob can agree on a shared key by

1. Alice chooses private element $a \in Z_p^*$ with p a prime and publishes $A = g^a$.
2. Bob chooses private element $b \in Z_p^*$ with p a prime and publishes $B = g^b$.
3. Both parties can compute $g^{ab} = A^b = B^a$.

By observing that Alice can compute a different key from the value $(BA^{-1})^a = g^{ab-a^2}$ and that Bob cannot compute this without knowing a, the Diffie-Hellman key exchange can become an OT protocol. We only introduce the 1-out-of-2 OT protocol as the motivation for the non-abelian analog. Suppose now that Alice has two messages m_0 and m_1 and Bob wishes to receive m_c where $c \in \{0, 1\}$. The group G and a generator $g \in G$ are public.

1. Alice chooses private element $a \in G$ and publishes $A = g^a$.
2. Bob chooses private element $b \in G$. If $c = 0$, Bob computes $B = g^b$. If $c = 1$, Bob computes $B = Ag^b$.
3. Bob publishes B.
4. Alice creates two keys $k_0 = B^a$ and $k_1 = (BA^{-1})^a$.
5. Alice sends encrypted messages m_0 and m_1 based on the keys k_0 and k_1, respectively.

Bob can recover the key k_c corresponding to his choice bit c, but cannot compute k_{1-c}. To see this if $c = 0$ then $k_0 = B^a = g^{ab}$, which Bob can compute. On the other hand, $k_1 = (BA^{-1})^a = g^{ab-a^2}$ which Bob cannot compute. Similarly, if $c = 1$ Bob can compute $k_1 = (BA^{-1})^a = (Ag^bA^{-1})^a = g^{ab}$ but he cannot compute $k_0 = B^a = g^{ab+a^2}$.

3 Non-abelian Oblivious Transfer

Motivated by the OT protocol in [3], we establish two 1-out-of-n oblivious transfer protocols; one motivated by the Ko-Lee key exchange and the other motivated by the Anshel-Anshel-Goldfeld protocol. Throughout subsections 3.1 and 3.2, given two elements x, y in a group G, we denote $x^y = y^{-1}xy$.

3.1 Oblivous Transfer Based on Ko-Lee Protocol

We begin by establishing a 1-out-of-2 oblivious transfer protocol based on the Ko-Lee key exchange. Let G be a non-abelian group with commuting subgroups S and T. The group G, $g \in G$ and subgroups S and T are public. Suppose Alice has two messages m_0 and m_1 and Bob wishes to receive m_c based on his choice bit $c \in \{0, 1\}$.

1. Alice chooses $a \in S$. Bob chooses $b \in T$ and his choice bit c.
2. Alice makes public $A = g^a = a^{-1}ga$.
3. Bob computes $B = g^b = b^{-1}gb$ if $c = 0$ and $B = Ag^b$ if $c = 1$.
4. Alice computes two keys: $k_0 = B^a$ and $k_1 = (A^{-1}B)^a$.
5. Alice sends encrypted messages m_0 and m_1 dependent on the keys.

Clearly, Bob can recover k_0 when $c = 0$ since a and b commute and

$$k_0 = B^a = (g^b)^a = a^{-1}b^{-1}gba = b^{-1}a^{-1}gab = A^b,$$

but not k_1 since Bob does not know a and

$$k_1 = (A^{-1}B)^a = (a^{-1}g^{-1}ab^{-1}gb)^a = a^{-2}g^{-1}ab^{-1}gba.$$

Similarly, Bob can recover k_1 when $c = 1$, but not k_0.

We can extend this 1-out-of-2 OT protocol to a 1-out-of-n OT protocol by making a simple modification as follows. Let G be a non-abelian group with commuting subgroups S and T. The group G, $g \in G$ and subgroups S and T are public. Alice has messages $m_0, m_1, \ldots, m_{n-1}$ and Bob wishes to receive m_c.

1. Alice chooses $a \in S$. Bob chooses $b \in T$ and his choice $c \in \{0, 1, \ldots, n-1\}$.
2. Alice makes public $A = g^a = a^{-1}ga$.
3. Bob computes $B = A^c g^b = a^{-1}g^c ab^{-1}gb$.
4. Alice computes n keys: $k_i = (A^{-i}B)^a$ for $i \in \{0, 1, \ldots, n-1\}$.
5. Alice sends encrypted messages $m_0, m_1, \ldots, m_{n-1}$ dependent on the keys.

For Bob's choice c he can compute k_c since a and b commute and

$$k_c = (A^{-c}B)^a = (A^{-c}A^c g^b)^a = g^{ba} = A^b,$$

but he cannot compute k_i for $i \neq c$ since

$$k_i = (A^{-i}B)^a = (A^{-i}A^c g^b)^a = (a^{-1}g^{-i+c}ab^{-1}gb)^a = a^{-2}g^{-i+c}ab^{-1}gba$$

and Bob does not know a.

3.2 *Oblivious Transfer Based on Anshel-Anshel-Goldfeld Protocol*

To avoid using commuting subgroups, we introduce a 1-out-of-n OT protocol based on the Anshel-Anshel-Goldfeld key exchange protocol. Suppose that Alice has messages $m_1, m_2, \ldots, m_n$ and Bob wishes to receive m_c. A group G and $a_1, \ldots, a_k, b_1, \ldots, b_m$ are made public with $m \geq n$.

1. Alice picks a secret word $a \in G$ as a word in $a_1, \ldots, a_k$ and sends $b_1^a, \ldots, b_m^a$ to Bob.
2. Bob picks a secret word $b \in G$ as a word in $b_1, \ldots, b_m$ and sends $b_c^a a_1^b, \ldots, b_c^a a_k^b$ to Alice.
3. Alice computes n keys: $(b_i^{-1})^a b_c^a a_1^b, \ldots, (b_i^{-1})^a b_c^a a_k^b$ for each $1 \leq i \leq n$.
4. Alice then follows AAG protocol for each $1 \leq i \leq n$ to obtain the commutators based on the n keys in step 3.
5. Alice sends encrypted messages $m_1, \ldots, m_n$ based on the keys.

Alice will compute a^b only in the case when $i = c$, resulting in the shared key, k_c, being the commutator $a^{-1}b^{-1}ab$. Since Alice published

$$b_1^a, \ldots, b_m^a,$$

and Bob has b written in the generators $b_1, \ldots, b_m$, he can determine b^a. Hence, Bob can recover k_c by computing $(b^a)^{-1}$ and multiplying on the right by b. Bob can then decrypt his choice message, m_c.

When $i \neq c$ Alice will compute in step 3.

$$(b_i^{-1})^a b_c^a a_1^b, \ldots, (b_i^{-1})^a b_c^a a_k^b$$

and will now compute a word other than a^b, resulting in a different key that Bob is unable to compute. To see this, suppose that $a = a_{i_1} a_{i_2} ... a_{i_l}$. To compute k_i, Alice proceeds by first computing

$$(b_i^{-1})^a b_c^a a_{i_1}^b (b_i^{-1})^a b_c^a a_{i_2}^b \ldots (b_i^{-1})^a b_c^a a_{i_l}^b$$

and then multiplying on the left by a^{-1}, resulting in

$$k_i = a^{-1} (b_i^{-1} b_c)^a a_{i_1}^b (b_i^{-1} b_c)^a a_{i_2}^b \ldots (b_i^{-1} b_c)^a a_{i_l}^b.$$

In this case, k_i is no longer the commutator, $a^{-1}b^{-1}ab = (b^a)^{-1}b$, and it not computable by Bob. Thus, Bob would only be able to decrypt his choice message, m_c.

4 Security Considerations

To ensure that these protocols are secure, they must be implemented in an appropriate platform group, which has proved to be a difficult open research problem. The platform group must have the following properties to ensure the security of the protocols.

1. The group should have a normal form of group elements to effectively hide information, ensuring the protocol remains "oblivious."
2. For the OT protocol based on the Ko-Lee key exchange, the conjugacy search problem should be intractable. If the conjugacy search problem is solvable in polynomial time, then an adversary could easily obtain the various keys and decrypt each message.
3. For the OT protocol based on the AAG key exchange, the simultaneous conjugacy search problem should be intractable. If the simultaneous conjugacy search problem is solvable in polynomial time, then an adversary could easily obtain the various keys and decrypt each message.

We would like to note that although these properties are necessary for the security of the protocol, they are not necessarily sufficient.

Initially, braid groups were proposed as platform groups for the Ko-Lee key exchange protocol and the Anshel-Anshel-Goldfeld protocol, but the protocols were later found to be susceptible to length-based attacks (see [7]) as well as linear algebra attacks. Due to Cheon and Jun (see [2]) and B. Tsaban (see [18]), most groups that admit faithful linear representations of small dimensions are susceptible to linear algebra attacks. Hence, these groups are no longer in consideration as platforms.

Since then Thompson's groups and polycyclic groups were proposed (see [5, 17]). Unfortunately, these groups were still susceptible to heuristic attacks (see [11, 15]). Recently, in [6], the computational complexity of the conjugacy search problem in certain metabelian groups was analyzed. In [6], a length-based conjugacy search was performed on generalized metabelian Baumslag-Solitar groups, which is a heuristic attack based on the original length-based attacks. The experiments indicated that these groups are resistant to these types of search algorithms and a conjugating element cannot be found in sufficient time. It is not known if these groups are susceptible to other attacks.

5 Conclusions

We have presented two 1-out-of-n oblivious transfer protocols based on the Ko-Lee key exchange protocol and the Anshel-Anshel-Anshel-Goldfeld protocol. These protocols are based on the conjugacy search problem and the simultaneous conjugacy search problem. Currently, there is one other non-commutative oblivious transfer protocol (see [12]) based on the group factorization problem. In addition, we outlined some requirements necessary for the platform group to ensure the security of the protocols. Determining the appropriate platform group has been a long-running difficult research problem.

References

1. Anshel, I., Anshel, M., Goldfeld, D.: An algebraic method for public key cryptography. Math. Res. Lett. **6**, 287–291 (1999)
2. Cheon, J., Jun, B.: A polynomial time algorithm for the braid diffie-hellman conjugacy problem. CRYPTO 2003. Lecture Notes in Computer Science, vol. 2729, pp. 212–224 (2003)
3. Chou, T., Orlandi, C.: The simplest protocol for oblivious transfer. In: Proceedings of the 4th International Conference on Progress in Cryptology, LATIN-CRYPT 2015, pp. 40–58 (2015)
4. Fine, B., Habeeb, M., Kahrobaei, D., Rosenberger, G.: Aspects of nonabelian group based cryptography: a survey and open problems. JP J. Algebr. Number Theory Appl. **21**(1), 1–40 (2011)
5. Garber, D., Kahrobaei, D.D., Lam, H.T.: Length based attack for polycyclic groups. J. Math. Cryptol. De Gruyter 33–44 (2015)
6. Gryak, J., Kahrobaei, D., Martinez-Perez, C.: On the conjugacy problem in certain metabelian groups. Glasgow Math. J. **61**(2), 251–269 (2019)
7. Hughes, J., Tannenbaum, A.: Length based attacks for certain group based encryption rewriting systems, workshop seci02 sécurité de la communication sur internet (2002)

8. Ishai, Y., Prabhakaran, M., Sahai, A.: Founding cryptography on oblivious transfer-efficiently. Advances in Cryptology-CRYPTO: CRYPTO 2008. Lecture Notes in Computer Science, vol. 5157, pp. 572–591 (2008)
9. Kilian, J.: Founding cryptography on oblivious transfer. In: Proceedings of the 20th Annual ACM Symposium on Theory of Computing, 2–4 May 1988, Chicago, Illinois, USA, pp. 20–31 (1988)
10. Ko, K.H., Lee, S.J., Cheon, J.H., Han, J.W., Kang, J., Park, C.: New public key cryptosystem using braid groups. Advances in Cryptology, CRYPTO 2000. LNCS, vol. 1880, pp. 166–183. Santa Barbara, CA (2000)
11. Kotov, M., Ushakov, A.: Analysis of a certain polycyclic group based cryptosystem. J. Math. Cryptol. **9**(3), 161–167 (2015)
12. Li, J., Li, X., Wang, L., He, D., Niu, X.: Oblivious transfer protocols based on group factoring problem. Advances in Broad-Band Wireless Computing, Communication and Applications. BWCCA 2016. Lecture Notes on Data Engineering and Communications Technologies, vol. 2 (2017)
13. Myasnikov, A., Shpilrain, V., Ushakov, A.: Group Based Cryptography. Advanced Courses in Mathematics, CRM Barcelona. Birkhauser Verlag, Basel (2008)
14. Rabin, M.O.: How to exchange secrets with oblivious transfer. Technical Report TR-81. Aiken Computation Laboratory, Harvard University (1981)
15. Ruinskiy, D., Shamir, A., Tsaban, B.: Length based cryptanalysis: the case of Thompson's group. J. Math. Cryptol. **1**, 359–372 (2007)
16. Shor, P.W.: Polynomial time algorithms for prime factorization and discrete logarithms on a quantum computer. J. Sci. Statist. Comput. **26**, 1484 (1997)
17. Shpilrain, V., Ushakov, A.: Thompson's group and public key cryptography. Lecture Notes Computer Science, vol. 3531, pp. 151–164 (2005)
18. Tsaban, B.: Polynomial-time solutions of computational problems in noncommutative-algebraic cryptography. J. Cryptol. **28**, 601–622 (2015)

Solution of Population Balance Equation Using Homotopy Analysis Method

Prakrati Kushwah and Jitraj Saha

Abstract In this paper, homotopy analysis method (HAM) is used to obtain the analytic solution for fragmentation population balance equation. Different sample problems are solved using HAM and their series solution is obtained. A detailed analysis of the series solution and the region of convergence of the solution is also studied. It is observed that the convergence region of the series solution can be adjusted with the help of certain parameters involved in HAM.

Keywords Homotopy analysis method · Population balance equation · Analytic approximations · Fragmentation · Convergence

1 Introduction

The events where two or more particles collide among each other and undergo certain changes in their physical properties are known as particulate processes. These changes can be in their mass, volume, size, entropy or some other properties of particles. Population balance equations (or PBEs) are basically integro-partial differential equations which represent the change in the particle properties present in a system due to particulate process. Various examples of these events can be seen in different fields of science and engineering like the formation of stars, growth of gas bubbles in solids, merging of drops in atmospheric clouds, and so on [1].

PK thanks Ministry of Education (MoE), Govt. of India for their funding support during her PhD program. JS thanks NITT for their support through seed Grant (file no.: NITT/R & C/SEED GRANT/19–20/P-13/MATHS/JS/E1) during this work.

P. Kushwah (✉) · J. Saha
Department of Mathematics, National Institute of Technology Tiruchirapalli,
Tiruchirapalli 620 015, Tamil Nadu, India
e-mail: kprakrati1256@gmail.com

J. Saha
e-mail: jitraj@nitt.edu

R. K. Sharma et al. (eds.), *Frontiers in Industrial and Applied Mathematics*,
Springer Proceedings in Mathematics & Statistics 410,
https://doi.org/10.1007/978-981-19-7272-0_4

In this paper, the PBE considered represents particle fragmentation. Fragmentation is a process where a particle break into two or more smaller (or daughter) fragments. The total number of particles present in the closed system increases due to fragmentation. Let $c(t, x)$ denote the number density of particles of volume $x \geq 0$ at time $t \geq 0$ in a system undergoing particulate process (fragmentation). In this regard, for $(t, x) \in [0, T] \times [0, \infty)$ where $T < \infty$, the general fragmentation population balance equation is written as [2]

$$\frac{\partial c(t,x)}{\partial t} = 2\int_{x}^{\infty} F(x, y-x)c(t,y)\mathrm{d}y - c(t,x)\int_{0}^{x} F(x-y, y)\mathrm{d}y, \tag{1}$$

with the initial data

$$c(0, x) = c_0(x) \geq 0, \quad \text{for all} \quad x \geq 0. \tag{2}$$

The left hand side (lhs) of equation (1) gives the time evolution of particle number density $c(t, x)$. In the right hand side (rhs), the function $F(x, y)$ represents the rate at which the particles of size $x + y$ breaks into particles of size x and size y. The first term indicates the inclusion (or birth) of $x-$size particles in the system, and the second term indicates the removal (death) of $x-$size particles from the system.

Several numerical, semi-analytical, and analytical methods have been devised over the years to solve PBE. For fragmentation problems, one can refer to the articles [2, 3] for exact solutions and the articles [4, 5] for numerical solutions. To our knowledge, the homotopy analysis method (HAM) has not been used to solve the fragmentation problem to date, and this is the first attempt to solve above mentioned PBE with HAM.

HAM was first introduced by Liao [6] in 1992 to solve different linear and nonlinear differential equations appearing in the physical systems. Over the years, HAM has received sincere attention from the researchers due to its ability to solve different complicated real-life problems. This is an analytic method which uses the concept of homotopy from topology to generate a convergent series solution to the considered problem. In HAM, we construct deformation equations with the help of initial guess of the solution, auxiliary linear operator, auxiliary parameter, and auxiliary function, and we have great freedom to choose all of these. Because of this freedom, this method is very flexible and convenient to use as compared to other methods like Adomian decommission method, artificial parameter method, perturbation method, etc.

The outline of this paper is as follows. In Sect. 2, the preliminary idea of homotopy analysis method is discussed. In Sect. 3, HAM is applied to solve the particulate problem (1)–(2) mentioned in Sect. 1 and its convergence is discussed. In Sect. 4, some sample examples are solved using the software Mathematica 12.2, and the efficiency of the method is discussed. Finally, the conclusion of the present work is discussed in Sect. 5.

2 Preliminaries: Homotopy Analysis Method

The key methodology of HAM is to approximate the solution $c(t, x)$ in terms of a series of functions. In this regard, we consider $c_0(t, x)$ as an initial guess of the solution. Let us now define an unknown function $v(t, x; q)$, where t and x are independent variables and $q \in [0, 1]$ is the embedding parameter. The underlying idea of HAM is that a continuous mapping is described to relate the solution $c(t, x)$ and the unknown function v, with the aid of the embedding-parameter q. Thus, the initial guess $c_0(t, x)$ of the solution $c(t, x)$ is so chosen that $v(t, x; q)$ varies from $c_0(t, x)$ to $c(t, x)$ as q varies from 0 to 1. Representing this mathematically, we can write $v(t, x; 0) = c_0(t, x)$ and $v(t, x; 1) = c(t, x)$. To ensure the above relation, a linear operator $\mathscr{L}[v(t, x; q)]$, an auxiliary parameter $\hbar$ ($\neq 0$) and an auxiliary function $H(t, x)$ are needed to be defined wisely. Under all these considerations, let us discuss the homotopy analysis method.

Let the initial assumption to the solution is independent of t and coincides with the initial data (2), that is

$$c_0(t, x) = c(0, x). \tag{3}$$

Choose the linear operator with

$$\mathscr{L}[v(t, x; q)] = \frac{\partial v(t, x; q)}{\partial t} \tag{4}$$

such that

$$\mathscr{L}[f(x, y)] = 0 \iff f(x, y) = 0. \tag{5}$$

Let us consider the generalized problem in the following operator form

$$\mathscr{N}[c(t, x)] = 0. \tag{6}$$

Using embedding parameter, we can construct a homotopy

$$\begin{aligned} \mathscr{H}[v(t, x; q); q, \hbar, H] :=& (1 - q)\mathscr{L}[v(t, x; q) - c_0(t, x)] \\ & - q\hbar H(t, x)\mathscr{N}[v(t, x; q)] = 0. \end{aligned} \tag{7}$$

For $q = 0$, Eq. (7) along with (5) gives

$$\mathscr{L}[v(t, x; 0) - c_0(t, x)] = 0 \quad \text{imples} \quad v(t, x; 0) = c_0(t, x). \tag{8}$$

Again for $q = 1$, since $\hbar \neq 0$ and $H(t, x) \neq 0$, relation (7) becomes

$$\mathscr{N}[v(t, x; 1)] = 0, \tag{9}$$

which replicates the original problem (6), provided

$$v(t,x;1) = c(t,x). \tag{10}$$

According to Eqs. (8) and (10), $v(t,x;q)$ varies from the initial guess $c_0(t,x)$ to the exact solution $c(t,x)$ as the embedding parameter q varies from 0 to 1. The equation (7) is called *zero-order deformation equation.*

The freedom to choose $\mathscr{L}$, $c_0(t,x)$, $\hbar$, $H(t,x)$, enables us to adjust all the parameters properly such that the solution of deformation equation exists for $q \in [0,1]$. The $m-$th order derivative of $c_0(t,x)$ with respect to embedding parameter q is defined as

$$c^{[m]}(t,x) := \frac{c_0^{[m]}(t,x)}{m!} = \frac{1}{m!}\frac{\partial^m v(t,x;q)}{\partial q^m}\bigg|_{q=0}. \tag{11}$$

By Taylor's theorem, $v(t,x;q)$ can be expanded in a power series of q as

$$v(t,x;q) = v(t,x;0) + \sum_{m=1}^{\infty}\frac{c_0^{[m]}(t,x)}{m!}q^m = f(x) + \sum_{m=1}^{\infty}c^{[m]}(t,x)q^m. \tag{12}$$

In general, the above series will converge for $q = 1$, and hence using relation (9), we have

$$c(t,x) = f(x) + \sum_{m=1}^{\infty}c^{[m]}(t,x). \tag{13}$$

We define the vector $\vec{c}_n := \{c_0(t,x), c_1(t,x), c_2(t,x), \ldots, c_n(t,x)\}$. Differentiating zero-order deformation (7) $m-$times with respect to q, then dividing it by $m!$, and finally setting $q = 0$, we get the following $m-$th order deformation equation:

$$\mathscr{L}[c_m(t,x) - \chi_m c_{m-1}(t,x)] = \hbar H(t,x) R_m(\vec{c}_{m-1}, t, x) \tag{14}$$

with initial condition $c_m(0,x) = 0$, where $\chi_m := \begin{cases} 0, & \text{when } m \le 1, \\ 1, & \text{when } m > 1. \end{cases}$ and

$$\begin{aligned} R_m(\vec{c}_{m-1}, t, x) &= \frac{1}{(m-1)!}\frac{\partial^{m-1}\mathscr{N}[v(t,x;q)]}{\partial q^{m-1}}\bigg|_{q=0} \\ &= \frac{1}{(m-1)!}\frac{\partial^{m-1}}{\partial q^{m-1}}\mathscr{N}\left[\sum_{n=0}^{\infty}c_n(t,x)q^n\right]_{q=0}. \end{aligned} \tag{15}$$

Thus, the solution to problem is reduced to

$$c_m(t,x) = \chi_m c_{m-1}(t,x)] + \int_0^t \hbar H(t,s) R_m(\vec{c}_{m-1}, s, x). \tag{16}$$

where R_m is given by (15).

3 HAM Based Solutions and Convergence Theorem

For (1)–(2) the operator $\mathcal{N}$ is given by

$$\begin{aligned}\mathcal{N}[v(t,x;q)] = {} & \frac{\partial v(t,x;q)}{\partial t} - 2\int_x^\infty F(x, y-x) v(t,y;q)\mathrm{d}y \\ & - v(t,x;q)\int_0^x F(x-y, y)\mathrm{d}y.\end{aligned}$$

Using (15), we can calculate R_m for $m = 1, 2, 3, \ldots$ as shown below

$$\begin{aligned}R_1(\vec{c}_0, t, x) &= \mathcal{N}[c_0(t,x)]|_{q=0} \\ &= -2\int_x^\infty F(x, y-x) c_0(t,y)\mathrm{d}y \\ &\quad - c_0(t,x)\int_0^x F(x-y,y)\mathrm{d}y.\end{aligned} \tag{17}$$

For $m = 2$,

$$\begin{aligned}R_2(\vec{c}_1, t, x) &= \left[\frac{\partial}{\partial q}\mathcal{N}[c_0(t,x) + c_1(t,x)q]\right]_{q=0} \\ &= \frac{\partial c_1(t,x)}{\partial t} - 2\int_x^\infty F(x, y-x) c_1(t,y)\mathrm{d}y \\ &\quad - c_1(t,x)\int_0^x F(x-y,y)\mathrm{d}y.\end{aligned} \tag{18}$$

For $m = 3$,

$$\begin{aligned}R_3(\vec{c}_2, t, x) &= \frac{1}{2!}\left[\frac{\partial^2}{\partial q^2}\mathcal{N}\left[c_0(t,x) + c_1(t,x)q + c_2(t,x)q^2\right]\right]_{q=0} \\ &= \frac{\partial c_2(t,x)}{\partial t} - 2\int_x^\infty F(x, y-x) c_2(t,y)\mathrm{d}y \\ &\quad - c_2(t,x)\int_0^x F(x-y,y)\mathrm{d}y.\end{aligned} \tag{19}$$

Likewise the mth order representation is

$$R_m\left(\vec{c}_{m-1}, t, x\right) = \frac{\partial c_{m-1}(t, x)}{\partial t} - 2\int_x^{\infty} F(x, y-x)c_{m-1}(t, y)\mathrm{d}y - c_{m-1}(t, x)\int_0^x F(x-y, y)\mathrm{d}y. \tag{20}$$

Thus, the solution to fragmentation equation (1)–(2) for $H(t, x) = 1$ is written as

$$c_m(t, x) = \chi_m c_{m-1}(t, x)] + \int_0^t hR_m\left(\vec{c}_{m-1}, s, x\right), \tag{21}$$

where R_m is given by (20).

Theorem 1 *As long as the series* (13) *converges, where $c_m(t, x)$ is governed by the high order deformation equation* (14) *under the conditions* (15) *and* (16)*, it must be the exact solution of* (1)–(2).

Proof If the series $\sum_{m=0}^{\infty} c_m(t, x)$ converges, then we can write

$$v(t, x) = \sum_{m=0}^{\infty} c_m(t, x), \quad \text{that is} \quad \lim_{m\to\infty} c_m(t, x) = 0. \tag{22}$$

In this context

$$\sum_{m=0}^{n}\left[c_m(t, x) - \chi_m c_{m-1}(t, x)\right] = c_1 + (c_2 - c_1) + \cdots + (c_n - c_{n-1}) = c_n(t, x),$$

and hence in accordance with relation (22), we get

$$\sum_{m=0}^{\infty}\left[c_m(t, x) - \chi_m c_{m-1}(t, x)\right] = \lim_{n\to\infty} c_n(t, x) = 0.$$

Due to the linearity property of $\mathscr{L}$, we have

$$\sum_{m=0}^{\infty}\mathscr{L}\left[c_m(t, x) - \chi_m c_{m-1}(t, x)\right] = \mathscr{L}\sum_{m=0}^{\infty}\left[c_m(t, x) - \chi_m c_{m-1}(t, x)\right] = 0.$$

Thus recalling (14), we obtain

$$\sum_{m=0}^{\infty} \mathscr{L}\left[c_m(t,x) - \chi_m c_{m-1}(t,x)\right] = \hbar H(t,x) \sum_{m=1}^{\infty} R_m\left[\vec{c}_{m-1}\right] = 0.$$

Since $\hbar \neq 0$ and $H(t,x) \neq 0$, therefore

$$\sum_{m=1}^{\infty} R_m\left[\vec{c}_{m-1}\right] = 0. \tag{23}$$

Recalling (22), we get

$$\begin{aligned}
\sum_{m=1}^{\infty} R_m\left[\vec{c}_{m-1}\right] &= \sum_{m=1}^{\infty}\left[\frac{\partial c_{m-1}(t,x)}{\partial t} - 2\int_x^{\infty} F(x, y-x)c_{m-1}(t,y)\mathrm{d}y - c_{m-1}(t,x)\right. \\
&\quad \left.\int_0^x F(x-y, y)\mathrm{d}y\right] \\
&= \sum_{m=1}^{\infty} \frac{\partial c_{m-1}(t,x)}{\partial t} - 2\int_x^{\infty} F(x, y-x) \sum_{m=1}^{\infty} c_{m-1}(t,y)\mathrm{d}y \\
&\quad - \sum_{m=1}^{\infty} c_{m-1}(t,x) \int_0^x F(x-y, y)\mathrm{d}y \\
&= \frac{\partial}{\partial t}\sum_{m=1}^{\infty} c_{m-1}(t,x) - 2\int_x^{\infty} F(x, y-x) \sum_{m=1}^{\infty} c_{m-1}(t,y)\mathrm{d}y \\
&\quad - \sum_{m=1}^{\infty} c_{m-1}(t,x) \int_0^x F(x-y, y)\mathrm{d}y \\
&= \frac{\partial}{\partial t}\sum_{m=0}^{\infty} c_m(t,x) - 2\int_x^{\infty} F(x, y-x) \sum_{m=0}^{\infty} c_m(t,y)\mathrm{d}y \\
&\quad - \sum_{m=0}^{\infty} c_m(t,x) \int_0^x F(x-y, y)\mathrm{d}y.
\end{aligned}$$

Combining (22) and (23), we have

$$\frac{\partial v(t,x)}{\partial t} = 2\int_x^{\infty} F(x, y-x)v(t,y)\mathrm{d}y - v(t,x)\int_0^x F(x-y, y)\mathrm{d}y. \tag{24}$$

Now, from the initial conditions of $c_m(t,x)$, it holds that

$$v(0,x) = \sum_{m=0}^{\infty} c_m(0,x) = c_0(0,x) + \sum_{m=1}^{\infty} c_m(0,x) = c_0(0,x) = c(0,x).$$

Hence, from last two expressions, one can observe that $v(t, x)$ must be the exact solution of (1)–(2).

4 Numerical Examples and Discussions

In this section, we will consider some examples discussed in [3] and will discuss using graphs how the approximate solution is approaching the exact solution.

Example 1 Let us consider (1)–(2) with $K(x, y) = 1$ and the initial condition $c(0, x) = \exp(-x)$. The exact solution to this problem is $c(t, x) = (1 + t)^2 \exp(-x(1 + t))$.

Using the recursive scheme (21) along with (20), we obtain $c_m(t, x)$ for $m \geq 1$ as

$$
\begin{aligned}
c_1(t, x) &= ht(e^{-x}x - 2e^{-x}),\\
c_2(t, x) &= h(\frac{1}{2}t^2(hx(e^{-x}x - 2e^{-x}) - 2he^{-x}(x - 1)) + ht(e^{-x}x - 2e^{-x})),\\
c_3(t, x) &= h(\frac{1}{6}h^2t^3e^{-x}x^3 - h^2t^3e^{-x}x^2 + h^2t^2e^{-x}x^2 + h^2t^3e^{-x}x + 2h^2t^2e^{-x}\\
&\quad - 4h^2t^2e^{-x}x - 2h^2te^{-x} + h^2te^{-x}x).
\end{aligned}
$$

We have great freedom to choose $\hbar$, so we will now look for a set of values of $\hbar$ for which the solution obtained by HAM converges to the exact solution. From Fig. 1, we can observe that when $\hbar$ is near -0.5, the graph is flat. So, method will converge to the exact solution when $\hbar$ is near -0.5. Now, we will analyze graphs for different values of $\hbar$ for which the solution obtained by HAM converges to exact solution.

During the computation of graphs for several values of $\hbar$ near -0.4 and -0.2, it was observed from Fig. 2 that for $\hbar = -0.4$ the graph of 4–th, 6–th, 8–th and 10–th

Fig. 1 $\hbar$ against $c(x, t)$ for $t = 1$ and $x = 1$ for different order of approximation for Eqs. (1)–(2)

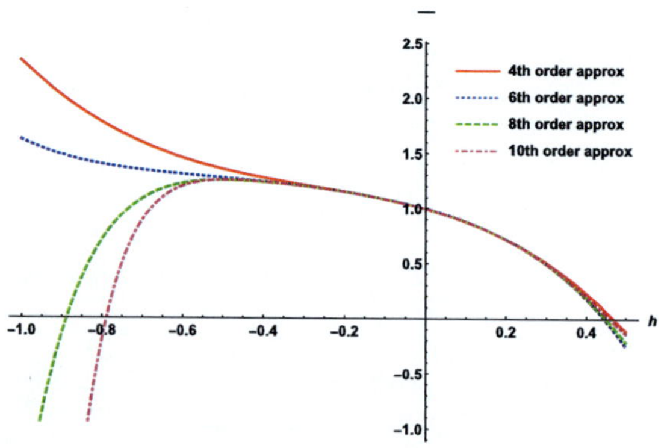

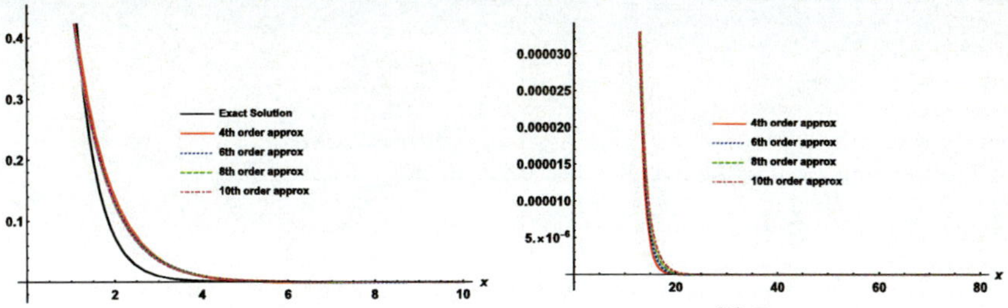

(a) Number density (b) Error

Fig. 2 Graph with $\hbar = -0.4$ for different order of approximation for Eqs. (1)–(2)

Table 1 Error for Example 1 when $\hbar = -0.4$

m–th order of approximation	4	6	8	10
Error	7.7606×10^{-5}	7.6782×10^{-5}	7.6403×10^{-5}	7.0951×10^{-5}

order approximation nearly coincides with the exact solution. Here we compute the numerical number density with the particle size distribution in Fig. 2a. On the other hand, Fig. 2b represents the numerical error curve obtained for different order approximations. For a detailed quantitative error analysis, we present the numerical errors in Table 1. To calculate the error, we recall the formula given in [5]. The following results supports that HAM predicts the solution with high accuracy even for a small number of approximate terms.

Example 2 Next consider (1)–(2) with $K(x, y) = x + y$ and the initial condition $c(0, x) = \exp(-x)$. The exact solution to this problem is $c(t, x) = \exp(-tx^2 - x)(1 + 2t(1 + x))$.

Again using the recursive scheme (21) along with (20), we obtain $c_m(t, x)$ for $m \geq 1$ as

$$c_1(t,x) = ht(e^{-x}x^2 - 2e^{-x}(x+1)),$$

$$c_2(t,x) = h(\frac{1}{2}t^2(hx^2(e^{-x}x^2 - 2e^{-x}(x+1)) - 2he^{-x}x^2(x+1)) + ht(e^{-x}x^2 - 2e^{-x}(x+1))),$$

$$c_3(t,x) = h(\frac{1}{6}h^2t^3e^{-x}x^6 - h^2t^3e^{-x}x^5 - h^2t^3e^{-x}x^4 + h^2t^2e^{-x}x^4 - 4h^2t^2e^{-x}x^3 - 4h^2t^2e^{-x}x^2 + h^2te^{-x}x^2 - 2h^2te^{-x}x - 2h^2te^{-x}).$$

Similar to the previous example, we will make use of graphs to investigate the value of $\hbar$ for which the series solution obtained by HAM method converges to exact solution. From Fig. 3, it is observed that graph is flat in between -0.4 and 0.

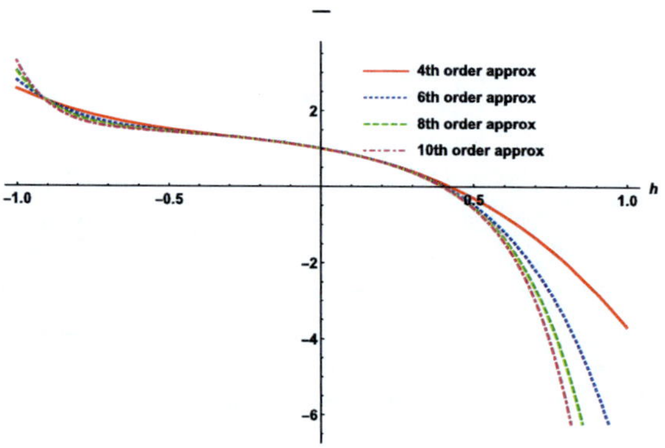

Fig. 3 $\hbar$ against $c(x,t)$ for $t=1$ and $x=1$ for different order of approximation for Eqs. (1)–(2)

Table 2 Error for Example 2 when $\hbar = -0.125$

m−th order of approximation	4	6	8	10
Error	2.9075×10^{-5}	2.8762×10^{-5}	2.8501×10^{-5}	2.8417×10^{-5}

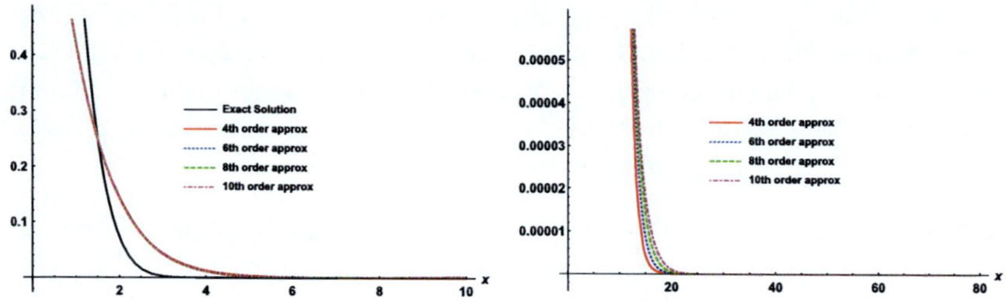

(a) Number density (b) Error

Fig. 4 Graph with $\hbar = -0.125$ for different order of approximation for Eqs. (1)–(2)

Therefore, the approximate solution will converge to exact solution somewhere in between these two points.

For a detailed investigation, we will plot the solution with respect to x for different values of $\hbar$ and it is observed that for $h = -0.125$ graph coincides with the exact solution. For a qualitative analysis of the method, we will plot the numerical solution as well as the error graph. For quantitative analysis, we present the error Table 2. Like before, it is observed that the HAM produces very accurate results for a very small number of terms and $\hbar = -0.125$ (Fig. 3).

5 Conclusion

The homotopy analysis method is applied to fragmentation PBE. A recursive scheme in the form of series solution is obtained to estimate the solution of PBEs. The convergence analysis shows that approximate solution will converge to exact solution. Error estimate for the sample problems is minimal which guarantees the accuracy of the method.

Acknowledgements The authors thank Dr. Kamalika Roy, Department of Mathematics, National Institute of Technology Tiruchirappalli for her valuable suggestions and inputs while writing the codes.

References

1. Saha, J., Kumar, J.: The singular coagulation equation with multiple fragmentation. Z. fur Angew. Math. Phys. **66**, 919–941 (2015)
2. Ziff, R.: New solutions to the fragmentation equation. J. Phys. A. Math. Gen. **24**, 2821 (1991)
3. Ziff, R., McGrady, E.: The kinetics of cluster fragmentation and depolymerisation. J. Phys. A. Math. Gen. **18**, 3027 (1985). http://www.ncbi.nlm.nih.gov. (National Center for Biotechnology Information)
4. Saha, J., Bück, A.: Improved accuracy and convergence analysis of finite volume methods for particle fragmentation models. Math. Methods Appl. Sci. **44**, 1913–1930 (2021)
5. Singh, R., Saha, J., Kumar, J.: Adomian decomposition method for solving fragmentation and aggregation population balance equations. J. Appl. Math. Comput. **48**, 265–292 (2015)
6. Liao, S.: Beyond perturbation Introduction to homotopy analysis mathod. CRC Press (2004)

Certain Properties and Their Volterra Integral Equation Associated with the Second Kind Chebyshev Matrix Polynomials in Two Variables

Virender Singh, Waseem A. Khan, and Archna Sharma

Abstract This paper presents several properties associated with the two-variable extension of the Chebyshev matrix polynomials of the second kind. In particular, we establish a three-term recurrence relation for these two-variable matrix polynomials and show that these two-variable matrix polynomials satisfy some second-order matrix differential equations. We derive their hypergeometric matrix representation and an expansion formula which links these generalized Chebyshev matrix polynomials with the Hermite matrix polynomials and the Laguerre matrix polynomials. We also drive their Volterra integral equation.

Keywords Hermite · Laguerre and Chebyshev matrix polynomials · Hypergeometric matrix functions · Matrix recurrence relations · Differential equations · Volterra integral equation

1 Introduction and Preliminaries

Development in the ideology of generalized and multivariate forms of special function serves as a foundation for massive problems in physical mathematics that have been explained exactly and also finds expansive practical applications. For example, the generalized Hermite polynomials are employed to solve quantum mechanics and optical beam transport problems [6]. An expansion to the matrix structure of the

V. Singh (✉)
Department of Applied Mathematics, Galgotias College of Engineering and Technology, Greater Noida, Uttar-Pradesh 201306, India
e-mail: virender.singh@galgotiacollege.edu; virenderamu2015@gmail.com

W. A. Khan
Department of Mathematics and Natural Sciences, Prince Mohammad Bin Fahd University, P.O Box 1664, Al Khobar 31952, Saudi Arabia

A. Sharma
School of Vocational Studies and Applied Sciences, Gautam Buddha University, Greater Noida, Uttar-Pradesh 201312, India

R. K. Sharma et al. (eds.), *Frontiers in Industrial and Applied Mathematics*, Springer Proceedings in Mathematics & Statistics 410,
https://doi.org/10.1007/978-981-19-7272-0_5

classical families of Hermite, Jacobi, Laguerre, Chebyshev, and Gagenbauer matrix polynomials have been mentioned with sincere aforethought in many papers [1, 3, 5, 10, 11, 14, 17, 19–25] for matrix in $\mathbb{C}^{N\times N}$. A close affinity between orthogonal matrix polynomials and second-order matrix equations appears in [7, 10, 11, 19]. The hypergeometric matrix function and hypergeometric matrix differential equation along with its general solution is well discussed by Jódar and Cortès in [12, 13]. The integral representation method is used to define second kind Chebyshev matrix polynomials in [4]. Also, the development of matrix function and a few families of bilinear and bilateral generating matrix functions for second form Chebyshev matrix polynomials are derived in [2].

Throughout the paper, D_0 refers to the complex plane cut along the negative real axis and $\sigma(A)$ (in particular spectrum of A), refers to the set of all the eigenvalues of A, where A is a positive stable matrix in the complex plane $\mathbb{C}^{N\times N}$ of all square matrices of common order N. If C is a matrix in $\mathbb{C}^{N\times N}$, its 2-norm $\parallel C \parallel$ is defined by

$$\parallel C \parallel = \sup_{x\neq 0} \frac{\parallel Ax \parallel_2}{\parallel x \parallel_2}$$

where for a vector y in $\mathbb{C}^N$, $\parallel y \parallel_2 = (y^T y)^{1/2}$ gives the Euclidean norm of y.

From [9, p. 558], we get

$$f(A)g(A) = g(A)f(A).$$

Here, $f(t)$ and $g(t)$ are holomorphic functions of the complex variables t, defined in an open set Ω of the complex plane.

For $\sigma(A) \subset D_0$ and $\log(t)$ marks the principal logarithm of t, then $A^{1/2} = \sqrt{A} = \exp(\frac{1}{2}\log(A))$ marks the image by $t^{1/2}$ of the matrix functional calculus acting on the matrix A then $Re(t) > 0$, for all $t \in \sigma(A)$.

The hypergeometric matrix function ${}_2F_1(U, V; W; z)$ has been defined as follows (see [18]):

$${}_2F_1(U, V; W; z) = \sum_{n=0}^{\infty} \frac{(U)_n(V)_n}{(W)_n} \frac{z^n}{n!}, \tag{1}$$

for matrices U, V and W in $\mathbb{C}^{N\times N}$ such that $C + nI$ is invertible for all integers $n \geq 0$ and for $\mid z \mid < 1$. Here,

$$(C)_n = C(C+I)...(C+(n-1)I) = \Gamma(C+nI)\,[\Gamma(C)]^{-1}, \quad n \geq 1;\ (C)_0 = I, \tag{2}$$

is the matrix version of Pochhammer symbol.

From (2), it can be seen that

$$(C)_{n-r} = (-1)^r (C)_n [(I - C - nI)_r]^{-1}; \quad 0 \leq r \leq n, \tag{3}$$

where $C + nI$ is invertible for all $n > 0$.

From the relation [21, p. 30], we find

$$\frac{(-1)^r}{(n-r)!}I = \frac{(-n)_r}{n!}I = \frac{(-nI)_r}{n!}; \quad 0 \le r \le n, \tag{4}$$

and from [21, p. 36], we get

$$(-nI)_{2k} = 2^{2k}\left(-\frac{1}{2}nI\right)_k \left(-\frac{1}{2}(n-1)I\right)_k. \tag{5}$$

Defez and Jódar [8] found that, if $G(n, m)$ and $H(n, m)$ are matrices in $\mathbb{C}^{N\times N}$ for $n \ge 0$ and $m \ge 0$, then the following relations hold see [18, pp. 57]:

$$\sum_{n=0}^{\infty}\sum_{m=0}^{n} G(n, m) = \sum_{n=0}^{\infty}\sum_{m=0}^{\infty} G(n, m+n), \tag{6}$$

$$\sum_{n=0}^{\infty}\sum_{m=0}^{[\frac{n}{2}]} H(n, m) = \sum_{n=0}^{\infty}\sum_{m=0}^{\infty} H(n, m+2n), \tag{7}$$

$$\sum_{n=0}^{\infty}\sum_{m=0}^{\infty} H(n, m) = \sum_{n=0}^{\infty}\sum_{m=0}^{[\frac{n}{2}]} H(n, m-2n). \tag{8}$$

With a matrix A (positive stable), the two-variable Chebyshev matrix polynomials (of second kind) are presented in [4]

$$F(\xi, \eta, z, A) = \left(1 - \xi z\sqrt{2A} + \eta z^2 I\right)^{-1} = \sum_{m=0}^{\infty} U_m(\xi, \eta, A) z^m, \ \|\xi z\sqrt{2A} - \eta z^2 I\| \le 1. \tag{9}$$

And the explicit representation for the generalized Chebyshev matrix polynomials of two variables [4] is as follows:

$$U_m(\xi, \eta, A) = \sum_{k=0}^{[\frac{m}{2}]} \frac{(-1)^k (m-k)!(\xi\sqrt{2A})^{m-2k}\eta^k}{k!(m-2k)!}. \tag{10}$$

Now in view of Eq. (10), it may remark that $U_m(\xi, \eta, A)$ is an even function of ξ for even m and odd function of ξ for odd m, we have

$$U_m(-\xi, \eta, A) = (-1)^m U_m(\xi, \eta, A) \tag{11}$$

$$U_{2m}(0, \eta, A) = (-1)^m \eta^m \tag{12}$$

$$U_{2m+1}(0, \eta, A) = 0 \tag{13}$$

$$U_{2m}(0, 0, A) = 0 \tag{14}$$

$$U_{2m+1}(0, 0, A) = 0 \tag{15}$$

$$\frac{\partial}{\partial \xi} U_{2m}(0, \eta, A) = 0 \tag{16}$$

$$\frac{\partial}{\partial \xi} U_{2m+1}(0, \eta, A) = \frac{(-1)^m}{m} \sqrt{2A} \eta^m \tag{17}$$

The integral representation satisfied by these family of matrix polynomials is

$$U_m(\xi, \eta, A) = \frac{1}{m!} \int_0^\infty e^{-z} z^m H_m \left(\xi, \frac{\eta}{z}, A \right) dz, \tag{18}$$

where $H_m(\xi, \eta, A)$ are called the two-variable Hermite matrix polynomials, see [4].

$$H_m(\xi, \eta, A) = m! \sum_{k=0}^{[\frac{m}{2}]} \frac{(-1)^k (\sqrt{2A})^{m-2k}}{k!(m-k)!} \xi^{m-2k} \eta^k; \; m \geq 0, \tag{19}$$

And the mth Laguerre matrix polynomials $L_m^{(A)}(\xi, \eta)$ for two variables [11] are defined by

$$L_m^{(A)}(\xi, \eta) = \sum_{k=0}^{m} \frac{(-1)^k \xi^k \eta^{m-k}}{k!(m-k)!} (A+I)_m [(A+I)_k]^{-1}, \tag{20}$$

Here, A is a matrix in $\mathbb{C}^{N \times N}$ with $-k$ not being an eigenvalue of A.

The following relations [4, 10, 21] also holds application in proving our results in Sect. 4:

$$\left(\xi\sqrt{2A}\right)^m = m! \sum_{k=0}^{[\frac{m}{2}]} \frac{\eta^k}{k!(m-2k)!} H_{m-2k}(\xi, \eta, A), \tag{21}$$

and

$$\xi^m I = m! \sum_{k=0}^{m} \frac{(-1)^k}{k!(m-k)!} (A+I)_m [(A+I)_k]^{-1} \eta^{m-k} L_k^{(A)}(\xi, \eta). \tag{22}$$

The article is an effort to further emphasize the significance of using generating functions to extract those properties associated with two-variable extensions of second form Chebyshev matrix polynomials. In the next section, we ascertain that these families of matrix polynomials follow some recurrence relations and appear as finite series solutions of matrix differential equations. Section 3 displays the hypergeometric matrix representation for the second kind Chebychev matrix polynomials with two variables. Section 4 is characterized by the link between generalized Chebyshev matrix polynomials and Hermite matrix polynomials and also with Laguerre matrix polynomials. Section 5 is to find Volterra integral equation associated with the second kind of Chebyshev matrix polynomials in two variables.

2 Matrix Differential Recurrence Relations

This section of the paper suggests establishing some pure and differential matrix recurrence relations satisfied by generalized two-variable Chebyshev matrix polynomials of the second kind $U_m(\xi, \eta, A)$. Now, taking term wise partial differentiation of Eq. (9)

$$(-1)(-z\sqrt{2A})\left(1-\xi z\sqrt{2A}+\eta z^2 I\right)^{-1} = \sum_{m=0}^{\infty}\frac{\partial}{\partial\xi}U_m(\xi, \eta, A)z^m.$$

Hence, it follows that

$$\frac{\partial}{\partial\xi}U_m(\xi, \eta, A) = \sqrt{2A}\left(1-\xi z\sqrt{2A}+\eta z^2 I\right)^{-1} U_{(m-1)}(\xi, \eta, A).$$

Continuing term wise differentiation for $0 \le r \le m$, we get

$$\frac{\partial^r}{\partial\xi^r}U_m(\xi, \eta, A) = (\sqrt{2A})^r\left(1-\xi z\sqrt{2A}+\eta z^2 I\right)^{-r} U_{(m-r)}(\xi, \eta, A).$$

Theorem 1 *The following relation holds true*

$$\xi\frac{\partial}{\partial\xi}U_m(\xi, \eta, A) + 2\eta\frac{\partial}{\partial\eta}U_m(\xi, \eta, A) - mU_m(\xi, \eta, A) = \mathbf{0}. \tag{23}$$

Proof To prove the above Eq. (23), we have partially differential equation (9) with respect to ξ, η and z respectively, we get

$$\frac{\partial}{\partial\xi}F(\xi, \eta, z; A) = z\sqrt{2A}F^2(\xi, \eta, z; A), \tag{24}$$

$$\frac{\partial}{\partial\eta}F(\xi, \eta, t; A) = -z^2 I F^2(\xi, \eta, z; A), \tag{25}$$

and

$$\frac{\partial}{\partial z}F(\xi, \eta, z; A) = (\xi\sqrt{2A} - 2\eta z I)F^2(\xi, \eta, z; A). \tag{26}$$

From the above three Eqs. (24), (25), and (26), we can easily frame that

$$\xi\frac{\partial F}{\partial \xi} + 2\eta\frac{\partial F}{\partial \eta} - z\frac{\partial F}{\partial z} = \mathbf{0}, \tag{27}$$

which in view of (9) and equating the coefficient of z^n gives

$$\xi\frac{\partial}{\partial \xi}U_m(\xi, \eta, A) + 2\eta\frac{\partial}{\partial \eta}U_m(\xi, \eta, A) - mU_m(\xi, \eta, A) = \mathbf{0}. \tag{28}$$

Theorem 2 *The following partial matrix differential equations hold true*

$$\xi\sqrt{2A}\frac{\partial}{\partial \xi}U_m(\xi, \eta, A) - m\sqrt{2A}U_m(\xi, \eta, A) = 2\eta\frac{\partial}{\partial \xi}U_{m-1}(\xi, \eta, A),$$

and

$$\xi\sqrt{2A}\frac{\partial}{\partial \eta}U_m(\xi, \eta, A) - 2\eta\frac{\partial}{\partial \eta}U_m(\xi, \eta, A) = (m+1)U_{m-1}(\xi, \eta, A). \tag{29}$$

Proof Commencing with (24) and (26), we have

$$\left(\xi\sqrt{2A} - 2\eta z I\right)\frac{\partial F}{\partial \xi} - z\sqrt{2A}\frac{\partial F}{\partial z} = \mathbf{0},$$

which on using (9) becomes

$$\sum_{m=0}^{\infty}\xi\sqrt{2A}\frac{\partial}{\partial \xi}U_m(\xi, \eta, A)z^m - \sum_{m=0}^{\infty}m\sqrt{2A}U_m(\xi, \eta, A)z^m = \sum_{m=1}^{\infty}2\eta\frac{\partial}{\partial \xi}U_{m-1}(\xi, \eta, A)z^m.$$

Keeping in view that $\frac{\partial}{\partial \xi}U_0(\xi, \eta, A) = \mathbf{0}$, we arrive at a differential matrix relation of the form

$$\xi\sqrt{2A}\frac{\partial}{\partial \xi}U_m(\xi, \eta, A) - m\sqrt{2A}U_m(\xi, \eta, A) = 2\eta\frac{\partial}{\partial \xi}U_{m-1}(\xi, \eta, A). \tag{30}$$

Similarly, the above procedure when applied with (25) and (26), gives

$$\xi\sqrt{2A}\frac{\partial}{\partial \eta}U_m(\xi, \eta, A) - 2\eta\frac{\partial}{\partial \eta}U_{m-1}(\xi, \eta, A) = (m-1)U_{m-1}(\xi, \eta, A). \tag{31}$$

This ends the proof.

Theorem 3 *The following three-term matrix recurrence relations hold true*

$$U_m(\xi, \eta, A) = \xi\sqrt{2A}U_{m-1}(\xi, \eta, A) - \eta U_{m-2}(\xi, \eta, A),$$

and

$$mU_m(\xi,\eta,A)+mu\sqrt{2A}U_{m-1}(\xi,\eta,A)-(3m-4)\eta U_{m-2}(\xi,\eta,A)=\mathbf{0}. \quad (32)$$

Proof From (24) and (26), on applying (9), one gets

$$\left(I-\xi z\sqrt{2A}+\eta z^2 I\right)^{-2}=\sum_{m=1}^{\infty}\left(\sqrt{2A}\right)^{-1}\frac{\partial}{\partial\xi}U_m(\xi,\eta,A)z^{m-1}, \quad (33)$$

$$\left(\xi\sqrt{2A}-2\eta zI\right)\left(I-\xi z\sqrt{2A}+\eta z^2 I\right)^{-2}=\sum_{m=1}^{\infty}mU_m(\xi,\eta,A)z^{m-1}. \quad (34)$$

It is apparent to write $I-\eta z^2 I-z(\xi\sqrt{2A}-2\eta zI)=I-\xi z\sqrt{2A}+\eta z^2 I$. Thus, by multiplying (33) by $I-\eta z^2 I$ and (34) by z followed by their difference yields

$$(m+1)\sqrt{2A}U_m(\xi,\eta,A)=\frac{\partial}{\partial\xi}U_{m+1}(\xi,\eta,A)-\eta\frac{\partial}{\partial\xi}U_{m-1}(\xi,\eta,A). \quad (35)$$

Substituting value of $\eta\frac{\partial}{\partial\xi}U_{m-1}(\xi,\eta;A)$ from (35) into (30), we get

$$\xi\sqrt{2A}\frac{\partial}{\partial\xi}U_m(\xi,\eta,A)=2\frac{\partial}{\partial\xi}U_{m+1}(\xi,\eta,A)-(m+2)\sqrt{2A}U_m(\xi,\eta,A). \quad (36)$$

Replacing m by $(m-1)$ in (36) and putting the resulting expression for $\frac{\partial}{\partial\xi}U_{m-1}$ (ξ,η,A) into (30), gives

$$\left(\left(\xi\sqrt{2A}\right)^2-4\eta I\right)\frac{\partial}{\partial\xi}U_m(\xi,\eta,A)=m\xi\left(\sqrt{2A}\right)^2U_m(\xi,\eta,A)-2\eta(m+1)\sqrt{2A}U_{m-1}(\xi,\eta,A). \quad (37)$$

Now by multiplying (30) by $\left(\left(\xi\sqrt{2A}\right)^2-4\eta I\right)$ and substituting for

$$\left(\left(\xi\sqrt{2A}\right)^2-4\eta I\right)\frac{\partial}{\partial\xi}U_m(\xi,\eta,A)$$

and

$$\left(\left(\xi\sqrt{2A}\right)^2-4\eta I\right)\frac{\partial}{\partial\xi}U_{m-1}(\xi,\eta,A).$$

From (37) to obtain

$$U_m(\xi,\eta,A)=\xi\sqrt{2A}U_{m-1}(\xi,\eta,A)-\eta U_{m-2}(\xi,\eta,A), \quad (38)$$

which is the three-term matrix recurrence relationship satisfied by second kind Chebyshev matrix polynomials in two variables.

After this, we have to apply a similar approach to (25) and (26), and we obtain

$$mU_m(\xi,\eta,A)+m\xi\sqrt{2A}U_{m-1}(\xi,\eta,A)-(3m-4)\eta U_{m-2}(\xi,\eta,A)=\mathbf{0}. \tag{39}$$

Now, we introduce the two-variable Chebyshev matrix differential equation as follows:

Theorem 4 *The following matrix differential equation holds true*

$$\left((\xi\sqrt{2A})^2-4\eta I\right)D_\xi^2U_m(\xi,\eta,A)+\xi(\sqrt{2A})^2D_\xi U_m(\xi,\eta,A)-m(m+1)(\sqrt{2A})^2U_m(\xi,\eta,A)=\mathbf{0}, \tag{40}$$

where $D_\xi^2=\frac{\partial^2}{\partial\xi^2}$ *and* $D_\xi=\frac{\partial}{\partial\xi}$.

Proof To obtain the above result, we have to replace m by $(m-1)$ in (36) and then differentiate with respect to ξ, we achieve

$$\xi\sqrt{2A}D_\xi^2U_{m-1}(\xi,\eta,A)=2D_\xi^2U_m(\xi,\eta,A)-(m+1)\sqrt{2A}D_\xi U_{m-1}(\xi,\eta,A). \tag{41}$$

Again differentiating (30) with respect to ξ, we have

$$\xi\sqrt{2A}D_\xi^2U_m(\xi,\eta,A)-m\sqrt{2A}D_\xi U_m(\xi,\eta,A)=2\eta D_\xi^2U_{m-1}(\xi,\eta,A). \tag{42}$$

By putting value of $D_\xi U_{m-1}(\xi,\eta,A)$ from (30) and $D_\xi^2U_{m-1}(\xi,\eta,A)$ from (42) into (41) and by a little rearrangement of terms, we get the Chebyshev matrix differential equation for two variable as follows:

$$\left(\left((\xi\sqrt{2A})^2-4\eta I\right)D_\xi^2+\xi(\sqrt{2A})^2D_\xi-m(m+1)(\sqrt{2A})^2\right)U_m(\xi,\eta,A)=0. \tag{43}$$

Theorem 5 *The following relation holds true*

$$\frac{\partial^r}{\partial\xi^r}U_{m-r}(\xi,\eta,A)-(-1)^r\left(\sqrt{2A}\right)^r\frac{\partial^r}{\partial\eta^r}U_m(\xi,\eta,A)=0. \tag{44}$$

Proof Differentiating (9) with respect to ξ and η, we get

$$\left(z\sqrt{2A}\right)\left(I-\xi z\sqrt{2A}+\eta z^2I\right)^{-2}=\sum_{m=0}^{\infty}\frac{\partial}{\partial\xi}U_n(\xi,\eta,A)z^m, \tag{45}$$

and

$$-z^2I\left(I-\xi z\sqrt{2A}+\eta z^2I\right)^{-2}=\sum_{m=0}^{\infty}\frac{\partial}{\partial\eta}U_m(\xi,\eta,A)z^m. \tag{46}$$

Iteration (45) and (46), for $0\le r\le m$, leads to the relation (44).

3 Hypergeometric Matrix Representation

In this section, we represent two-variable Chebyshev matrix polynomials of second kind $U_m(\xi, \eta, A)$ in terms of hypergeometric matrix function ${}_2F_1$ defined in (1).

From relation (4) for $r = 2j$, we obtain

$$\frac{1}{(m-2j)!}I = \frac{(-m)_{2j}}{m!}I = \frac{(-mI)_{2j}}{m!}; \qquad 0 \leq 2j \leq m, \tag{47}$$

and from relation (3), we can write

$$(m-j)! = (I)_{m-j} = (-1)^j (I)_m [(-mI)_j]^{-1}; \qquad 0 \leq j \leq m. \tag{48}$$

Now with the aid of (47), (48), and relation (5), the explicit representation for $U_m(\xi, \eta, A)$ transforms as follows:

$$U_m(\xi, \eta, A) = \sum_{j=0}^{[\frac{m}{2}]} \frac{(-1)^j (m-j)! \left(\xi\sqrt{2A}\right)^{m-2j} \eta^j}{j!(m-2j)!}$$

$$= \sum_{j=0}^{[\frac{m}{2}]} \frac{(-1)^j (I)_{m-j} \left(\xi\sqrt{2A}\right)^{m-2j} \eta^j (-mI)_{2j}}{j!m!}$$

$$= \sum_{j=0}^{[\frac{m}{2}]} \frac{(-mI)_{2j}\left[(-mI)_j\right]^{-1} \left(\xi\sqrt{2A}\right)^{m-2j} \eta^j}{j!}$$

$$= \sum_{j=0}^{[\frac{m}{2}]} \frac{2^{2j}\left(-\frac{1}{2}mI\right)_j \left(-\frac{1}{2}(m-1)I\right)_j \left[(-mI)_j\right]^{-1} \left(\xi\sqrt{2A}\right)^{m-2j} \eta^j}{j!}$$

$$U_m(\xi, \eta, A) = \left(\xi\sqrt{2A}\right)^m \sum_{j=0}^{[\frac{m}{2}]} \frac{\left(-\frac{1}{2}mI\right)_j \left(-\frac{1}{2}(m-1)I\right)_j \left[(-mI)_j\right]^{-1} \eta^j \xi^{-2j} (2A^{-1})^j}{j!}$$

$$U_m(\xi, \eta, A) = \left(\xi\sqrt{2A}\right)^m {}_2F_1\left[-\frac{1}{2}mI, -\frac{1}{2}(m-1)I; -mI; \frac{2\eta A^{-1}}{\xi^2}\right], \quad \left\|\frac{2\eta A^{-1}}{\xi^2}\right\| < 1, \tag{3.3}$$

which gives another representation in the form of hypergeometric matrix function ${}_2F_1$.

4 Expansion of Two-Variable Chebyshev Matrix Polynomials of Second Kind in Series of Hermite Matrix Polynomials and Laguerre Matrix Polynomials of Two-Variable

Working with (9) and (7) and using the identity (21), we have the series

$$\sum_{j=0}^{\infty} U_j(\xi, \eta, A) z^j = \sum_{j=0}^{\infty} \sum_{p=0}^{[\frac{j}{2}]} \frac{(-1)^p (j-p)! \eta^r \left(\xi\sqrt{2A}\right)^{j-2p}}{p!(j-2p)!} z^j$$

$$= \sum_{j=0}^{\infty} \sum_{p=0}^{\infty} \frac{(-1)^p (j+p)! \eta^p \left(\xi\sqrt{2A}\right)^{j}}{p! j!} z^{j+2p}$$

$$= \sum_{j=0}^{\infty} \sum_{p=0}^{\infty} \sum_{q=0}^{[\frac{j}{2}]} \frac{(-1)^p (j+p)! \eta^{r+s} H_{j-2q}(\xi, \eta, A)}{p! q! (j-2q)!} z^{j+2p}$$

$$= \sum_{j=0}^{\infty} \sum_{p=0}^{\infty} \sum_{q=0}^{[\frac{j}{2}]} \frac{(-1)^p (1)_{j+p} \eta^{p+q}}{p! q! (j-2q)!} H_{j-2q}(\xi, \eta, A) z^{j+2p}$$

$$= \sum_{j=0}^{\infty} \sum_{p=0}^{\infty} \sum_{q=0}^{\infty} \frac{(-1)^p (1)_{j+p+2q} \eta^{p+q}}{p! q! j!} H_j(\xi, \eta, A) z^{j+2p+2q}$$

$$= \sum_{j=0}^{\infty} \sum_{p=0}^{\infty} \sum_{q=0}^{p} \frac{(-1)^{p-q} (1)_{j+p+q} \eta^{p}}{(p-q)! q! j!} H_j(\xi, \eta, A) z^{j+2p}. \tag{49}$$

Now since $(1)_{j+p+q} = (j+p+1)_q (1)_{j+p}$, we have

$$\sum_{j=0}^{\infty} U_j(\xi,\eta,A)z^j = \sum_{j=0}^{\infty}\sum_{p=0}^{\infty}\sum_{q=0}^{p} \frac{(-1)^p(-1)^{-q}}{(p-q)!}\frac{(j+p+1)_q(1)_{j+p}\eta^p}{q!j!}H_j(\xi,\eta,A)z^{j+2p}$$

$$= \sum_{j=0}^{\infty}\sum_{p=0}^{\infty}\sum_{q=0}^{p} \frac{(-1)^p(-p)_q(j+p+1)_q(1)_{j+p}\eta^p}{p!q!j!}H_j(\xi,\eta,A)z^{j+2p}$$

$$\sum_{j=0}^{\infty} U_j(\xi,\eta,A)z^j = \sum_{j=0}^{\infty}\sum_{p=0}^{\infty} \frac{(-1)^p}{p!j!}{}_2F_0\left(-p, j+p+1; -; 1\right)(1)_{j+p}\eta^p H_j(\xi,\eta,A)z^{j+2p}.$$

Finally, by using (1) and (8), it reduces to

$$= \sum_{j=0}^{\infty}\sum_{p=0}^{[\frac{j}{2}]} \frac{(-1)^p}{p!(j-2p)!}{}_2F_0\left(-p, j-p+1; -; 1\right)(1)_{j-p}\eta^p H_{j-2p}(\xi,\eta,A)z^j. \tag{50}$$

Conclusively, an identification of coefficient of z^j implies an expansion of two-variable Chebyshev matrix polynomials as a series of Hermite matrix polynomials as

$$U_j(\xi,\eta,A) = \sum_{p=0}^{[\frac{j}{2}]} \frac{(-1)^p}{p!(j-2p)!}{}_2F_0\left(-p, j-p+1; -; 1\right)(1)_{j-p}\eta^p H_{j-2p}(\xi,\eta,A). \tag{51}$$

Once again, working with (9) and (7) and using (20), we have the series

$$\sum_{j=0}^{\infty} U_j(\xi,\eta,A)z^j = \sum_{j=0}^{\infty}\sum_{q=0}^{[\frac{j}{2}]} \frac{(-1)^q(j-q)!\eta^q\left(\xi\sqrt{2A}\right)^{j-2q}}{q!(j-2q)!}z^j$$

$$= \sum_{j=0}^{\infty}\sum_{q=0}^{\infty} \frac{(-1)^q(j+q)!\eta^q\left(\sqrt{2A}\right)^j\xi^j}{q!j!}z^{j+2q}$$

$$= \sum_{j=0}^{\infty}\sum_{q=0}^{\infty}\sum_{p=0}^{j} \frac{(-1)^{p+q}(j+q)!\eta^q\left(\sqrt{2A}\right)^j}{p!q!(j-p)!}\eta^{j-p}(A+I)_j[(A+I)_p]^{-1}L_p^{(A)}(\xi,\eta)z^{j+2q}, \tag{52}$$

which on using (6) becomes

$$\sum_{j=0}^{\infty} U_j(\xi,\eta,A)z^j = \sum_{j=0}^{\infty}\sum_{q=0}^{\infty}\sum_{p=0}^{\infty} \frac{(-1)^{p+q}(j+q+p)!\left(\sqrt{2A}\right)^{j+p}}{p!q!j!}\eta^{j+q}$$

$$\times (A+I)_{j+p}\left[(A+I)_p\right]^{-1} L_p^{(A)}(\xi,\eta)z^{j+p+2q}. \tag{53}$$

From (8), we have

$$\sum_{j=0}^{\infty} U_j(\xi,\eta,A)z^j = \sum_{j=0}^{\infty}\sum_{p=0}^{\infty}\sum_{q=0}^{\left[\frac{j}{2}\right]} \frac{(-1)^{p+q}(j-q+p)!\left(\sqrt{2A}\right)^{j+p-2q}}{p!q!(j-2q)!}\eta^{j-q}$$

$$\times (A+I)_{j+p-2q}[(A+I)_p]^{-1} L_p^{(A)}(\xi,\eta)z^{j+p}. \tag{54}$$

Now, we know that

$$(j+p-q)! = (1)_{j+p-q} = (-1)^q(1)_{j+p}\left[((-j-p)I)_q\right]^{-1}$$

and

$$(A+I)_{j+p-2q} = 2^{-2q}(A+I)_{j+p}\left[\left(\frac{1}{2}((1-j-p)I-A)\right)_q\right]^{-1}\left[\left(-\frac{1}{2}((j+p)I+A)\right)_q\right]^{-1}.$$

Therefore,

$$\sum_{j=0}^{\infty} U_j(\xi,\eta,A)z^j = \sum_{j=0}^{\infty}\sum_{p=0}^{\infty}\sum_{q=0}^{\left[\frac{j}{2}\right]} \frac{(-1)^{p+q}\left(\sqrt{2A}\right)^{j+p-2q}}{p!q!(j-2q)!}(-1)^q(1)_{j+p}\eta^{j-q}$$

$$\times\left[((-j-p)I)_q\right]^{-1} 2^{-2q}(A+I)_{j+p}\left[\left(\frac{1}{2}((1-j-p)I-A)\right)_q\right]^{-1}$$

$$\times\left[\left(-\frac{1}{2}((j+p)I+A)\right)_q\right]^{-1}[(A+I)_p]^{-1}L_p^{(A)}(\xi,\eta)z^{j+p}.$$

Next, using identities (4) and (5), we arrive at

$$\sum_{j=0}^{\infty} U_j(\xi, \eta, A)z^j = \sum_{j=0}^{\infty}\sum_{p=0}^{\infty}\sum_{q=0}^{[\frac{j}{2}]} \frac{1}{q!}\left(-\frac{1}{2}jI\right)_q \left(-\frac{1}{2}(j-1)I\right)_q \left[((-j-p)I)_q\right]^{-1}$$

$$\times \left[\left(\frac{1}{2}((1-j-p)I-A)\right)_q\right]^{-1} \left[\left(-\frac{1}{2}((j+p)I+A)\right)_q\right]^{-1}$$

$$\times \frac{(-1)^p \left(\sqrt{2A}\right)^{j+p-2q} \quad (1)_{j+p}}{j!p!} (A+I)_{j+p}[(A+I)_p]^{-1}\eta^{j-q} L_p^{(A)}(\xi, \eta) z^{j+p}$$

$$\sum_{j=0}^{\infty} U_j(\xi, \eta, A)z^j = \sum_{j=0}^{\infty}\sum_{p=0}^{j} {}_2F_3\left[-\frac{1}{2}(j-p)I, -\frac{1}{2}(j-p-1)I; -nI,\right.$$

$$\left.\times \frac{1}{2}((1-j)I-A), -\frac{1}{2}(jI+A); \frac{\left(\sqrt{2A}\right)^{-2}}{\eta}\right] \frac{j!(-1)^p \left(\sqrt{2A}\right)^j}{p!(j-p)!}$$

$$\times (A+I)_j[(A+I)_p]^{-1}\eta^{j-p} L_p^{(A)}(\xi, \eta)z^j, \tag{55}$$

where $\left\| \frac{(\sqrt{2A})^{-2}}{\eta} \right\| < 1$. Finally, on comparing the coefficient of z^j on both sides, we easily get the expansion of Chebyshev matrix polynomials in terms of Laguerre matrix polynomials as

$$U_j(\xi, \eta, A) = j!\left(\sqrt{2A}\right)^j \sum_{p=0}^{j} {}_2F_3\left[-\frac{1}{2}(j-p)I, -\frac{1}{2}(j-p-1)I; -jI,\right.$$

$$\left.\times \frac{1}{2}((1-j)I-A), -\frac{1}{2}(jI+A); \frac{\left(\sqrt{2A}\right)^{-2}}{\eta}\right] \frac{(-1)^p}{p!(j-p)!}$$

$$\times (A+I)_j \left[(A+I)_p\right]^{-1} \eta^{j-p} L_p^{(A)}(\xi, \eta). \tag{56}$$

5 Volterra Integral Equation of Chebyshev Matrix Polynomials of the Second Kind $U_j(\xi, \eta, A)$

We have matrix differential Eq. (43), whose solution is Chebyshev matrix polynomials of the second kind of two variables

$$\left(\left((\xi\sqrt{2A})^2 - 4\eta I\right)D_\xi^2 + \xi(\sqrt{2A})^2 D_\xi - m(m+1)(\sqrt{2A})^2\right)U_m(\xi, \eta, A) = 0. \quad m \geq 0 \tag{57}$$

in view of Eqs. (12) and (13), we have

$$\left(U_m(\xi, \eta, A)\right)_{(\xi=0)} = \begin{cases} 0 & m = 2r+1 \\ (-1)^m(\eta)^m & m = 2r \end{cases} \tag{58}$$

Also, from Eqs. (16) and (17), we have

$$\left(\frac{\partial}{\partial\xi}U_m(\xi, \eta, A)\right)_{(\xi=0)} = \begin{cases} 0 & m = 2r \\ \frac{(-1)^m}{m}\sqrt{2A}(\eta)^m & m = 2r+1 \end{cases} \tag{59}$$

Now, we deal with a problem of obtaining the Integral equation from the above matrix differential equation along with the initial conditions given by (58) and (59). Now we consider the case for even m, then Eq. (57) becomes

$$\left(\left((\xi\sqrt{2A})^2 - 4\eta I\right)D_\xi^2 + \xi(\sqrt{2A})^2 D_\xi - 2r(2r+1)(\sqrt{2A})^2\right)U_{2r}(\xi, \eta, A) = \mathbf{0}. \quad r \geq 0 \tag{60}$$

Now, we have

$$\frac{\partial^2}{\partial\xi^2}U_{2r}(\xi, \eta, A) = (\sqrt{2A})^2\left(1 - \xi z\sqrt{2A} + \eta z^2 I\right)^{-2} U_{(2r-2)}(\xi, \eta, A). \tag{61}$$

Integrating equation (61) and using the initial conditions, given by Eqs. (58) and (59) at $\xi = 0$

$$\frac{\partial}{\partial\xi}U_{2r}(\xi, \eta, A) = (\sqrt{2A})^2\left(1 + \eta z^2 I\right)^{-2}\int_{0.}^{\xi} U_{(2r-2)}(x, \eta, A)dx. \tag{62}$$

and

$$U_{2r}(\xi, \eta, A) = (\sqrt{2A})^2\left(1 + \eta z^2 I\right)^{-2}\int_0^{\xi}(\xi - x)U_{(2r-2)}(x, \eta, A)dx + (-1)^r\eta^{2r} \tag{63}$$

Using Eqs. (62) and (63) in Eq. (57).

$$(\sqrt{2A})^2\left(1 + \eta z^2 I\right)^{-2}\left((\xi\sqrt{2A})^2 - 4\eta I\right)U_{(2r-2)}(\xi, \eta, A) + (\sqrt{2A})^4\left(1 + \eta z^2 I\right)^{-2}$$

$$\times \int_0^{\xi} \Big(\xi - (\xi - x)(2r(2r+1))\Big) U_{(2r-2)}(x, \eta, A)dx$$
$$+ (-1)^r (2r(2r+1))(\sqrt{2A})^2 \eta^{2r} = 0 \tag{64}$$

Replacing r by $r+1$ in Eq. (64)

$$(\sqrt{2A})^2 \left(1 + \eta z^2 I\right)^{-2} \Big((\xi\sqrt{2A})^2 - 4\eta I\Big) U_{2r}(\xi, \eta, A) + (\sqrt{2A})^4 \left(1 + \eta z^2 I\right)^{-2}$$
$$\times \int_0^{\xi} \Big(\xi - (\xi - x)((2r+2)(2r+3)\Big) U_{2r}(x, \eta, A)dx$$
$$+ (-1)^{r+1} ((2r+2)(2r+3))(\sqrt{2A})^2 \eta^{2r+2} = 0 \tag{65}$$

Next we consider the matrix differential equation (57) for $m = 2r+1$ that is

$$\Big(\Big((\xi\sqrt{2A})^2 - 4\eta I\Big) D_\xi^2 + \xi(\sqrt{2A})^2 D_\xi - (2r+1)(2r+2)(\sqrt{2A})^2\Big) U_{2r+1}(\xi, \eta, A) = 0. \tag{66}$$

Following the same arguments, the matrix differential equation (66) reduces to

$$(\sqrt{2A})^2 \left(1 + \eta z^2 I\right)^{-2} \Big((\xi\sqrt{2A})^2 - 4\eta I\Big) U_{(2r-1)}(\xi, \eta, A) + (\sqrt{2A})^4 \left(1 + \eta z^2 I\right)^{-2}$$
$$\times \int_0^{\xi} \Big(\xi - (\xi - x)((2r+1)(2r+2))\Big) U_{(2r-1)}(x, \eta, A)dx$$
$$+ (-1)^{r+1} (2r+1)(2r+2)(\sqrt{2A})^2 \eta^{2r+1} = 0 \tag{67}$$

Replacing r by $r+1$ in Eq. (67)

$$(\sqrt{2A})^2 \left(1 + \eta z^2 I\right)^{-2} \Big((\xi\sqrt{2A})^2 - 4\eta I\Big) U_{(2r+1)}(\xi, \eta, A) + (\sqrt{2A})^4 \left(1 + \eta z^2 I\right)^{-2}$$
$$\times \int_0^{\xi} \Big(\xi - (\xi - x)((2r+3)(2r+4))\Big) U_{(2r+1)}(x, \eta, A)dx$$
$$+ (-1)^{r} (2r+3)(2r+4)(\sqrt{2A})^2 \eta^{2r+3} = 0 \tag{68}$$

Finally, combining both equations for even and odd, we get Volterra integral equation of Chebyshev matrix polynomials of the second kind of two variables in the form

$$(\sqrt{2A})^2 \left(1 + \eta z^2 I\right)^{-2} \Big((\xi\sqrt{2A})^2 - 4\eta I\Big) U_m(\xi, \eta, A) + (\sqrt{2A})^4 \left(1 + \eta z^2 I\right)^{-2}$$
$$\times \int_0^{\xi} \Big(\xi - (\xi - x)((m+2)(m+3))\Big) U_m(x, \eta, A)dx$$
$$+ (-1)^{[m/2]+1} (m+2)(m+3)(\sqrt{2A})^2 \eta^{m+2} = 0 \tag{69}$$

6 Conclusion

This section is the epitome of the results obtained in the foregoing sections. In this manuscript, we have studied and introduced the second kind Chebyshev matrix polynomials in two variables and obtained its significant properties. We obtained the solution of Volterra integral equation for the second kind Chebyshev matrix polynomials in two variables with the help of the matrix differential equation. We concluded the present research work by giving some comments on the results of Sects. 2–5.

Therefore, the so obtained results in this article seem to be useful in the generalized Chebyshev matrix polynomials of the second kind $U_j^{(m)}(\xi, \eta, A)$ and the first kind Chebyshev matrix polynomials for two variables $T_j(\xi, \eta, A)$ can be exploited to develop new properties (matrix recurrence relations and matrix differential equations) in a similar manner as shown in the present paper.

References

1. Aktaş, R., Çekim, B., Çevik, A.: Extended Jacobi matrix polynomials. Util. Math. **92**, 47–64 (2013)
2. Altin, A., Çekim, B.: Generating matrix functions for Chebyshev matrix polynomials of the second kind. Hacettepe J. Math. Stat. **41**(1), 25–32 (2012)
3. Altin, A., Çekim, B.: Some miscellaneous properties for Gegenbaur matrix polynomials. Util. Math. **92**, 377–387 (2013)
4. Batahan, R.S.: A new extension of Hermite matrix polynomials and its applications. J. Linear Algebra Appl. **419**(1), 82–92 (2006)
5. Çekim, B., Altin, A., Aktaş, R.: Some new results for Jacobi matrix polynomials. Filomat **27**(4), 713–719 (2013)
6. Dattoli, G., Lorenzutta, S., Maino, G., Torre, A., Cesarano, C.: Generalized Hermite polynomials and super-Gaussian forms. J. Math. Anal. Appl. **203**(3), 597–609 (1996)
7. Defez, E., Jódar, L.: Chebyshev matrix polynomials and second-order matrix differential equations. Util. Math. **61**, 107–123 (2002)
8. Defez, E., Jódar, L.: Some applications of the Hermite matrix polynomials series expansions. J. Comp. Appl. Math. **99**(1–2), 105–117 (1998)
9. Dunford, N., Schwartz, J.T.: Linear Operators, Part I. General Theory. Interscience, New York (1963)
10. Jódar, L., Company, R.: Hermite matrix polynomials and second order matrix differential equations. J. Approx. Theory Appl. **12**(2), 20–30 (1996)
11. Jódar, L., Company, R., Navarro, E.: Laguerre matrix polynomials and systems of second order differential equations. Appl. Numer. Math. **15**(1), 53–63 (1994)
12. Jódar, L., Cortès, J.C.: Closed form general solution of the hypergeometric matrix differential equation. Math. Comput. Model. **32**(9), 1017–1028 (2000)
13. Jódar, L., Cortès, J.C.: On the hypergeometric matrix function. J. Comp. Appl. Math. **99**(1–2), 205–217 (1998)
14. Jódar, L., Defer, E.: On Hermite matrix polynomials and Hermite matrix functions. J. Approx. Theory Appl. **14**(1), 36–48 (1998)
15. Kargin, L., Kurt, V.: Chebyshev-type matrix polynomials and integral transforms. Hacettepe J. Math. Stat. **44**(2), 341–350 (2015)

16. Khan, S., Al-Gonah, A.A.: Multi-variable Hermite matrix polynomials: properties and applications. J. Math. Anal. Appl. **412**, 222–235 (2014)
17. Metwally, M.S., Mohamed, M.T., Shehata, A.: On Chebyshev matrix polynomials, matrix differential equations and their properties. Afr. Mat. **26**(5–6), 1037–1047 (2015)
18. Ranville, E.D.: Special Functions. The Macmillan Company, New York (1960)
19. Sayyed, K.A.M., Metwally, M.S., Batahan, R.S.: Gegenbauer matrix polynomials and second order matrix differential equations. Div. Math. **12**(2), 101–115 (2004)
20. Sayyed, K.A.M., Metwally, M.S., Mohamed, M.T.: Certain Hypergeometric matrix function. Sci. Math. Japonicae 177–183 (2009)
21. Shehata, A.: A new extension of Gegenbauer matrix polynomials and their applications. Bull. Int. Math. Virtual Inst. **2**, 29–42 (2012)
22. Singh, V., Khan, M.A., Khan, A.H.: Study of Gegenbauer matrix polynomials via matrix functions and their properties. Asian J. Math. Comput. Res.. **16**(4), 197–207 (2017)
23. Singh, V., Khan, M.A., Khan, A.H.: On a multivariable extension of Laguerre matrix ploynomials. Electron. J. Math. Anal. Appl. **6**(1), 223–240 (2018)
24. Singh, V., Khan, M.A., Khan, A.H.: Generalization of multi-variable modified Hermite matrix polynomials and its applications. Honam Math. J. **42**(2), 269–291 (2020)
25. Singh, V., Sharma, A., Khan, A.H.: Two variable modified Hermite matrix polynomials properties and its applications. In: AIP(American Institute of Physics) Conference Proceeding, vol. 2253, pp. 020008 (2020). https://doi.org/10.1063/5.0018997
26. Srivastava, H.M., Manocha, H.L.: A treatise on generating functions. Halsted Press, Wiley, New York-Chichester-Brisbane and Toronto (1984)

Blow-up Analysis and Global Existence of Solutions for a Fractional Reaction-Diffusion Equation

R. Saranya and N. Annapoorani

Abstract This paper is concerned with the blow-up phenomena and global existence of a fractional nonlinear reaction-diffusion equation with a non-local source term. Under sufficient conditions on the weight function $a(x)$ and when the initial data is small enough, the global existence of solutions is proved using the comparison principle. We establish a finite time blow-up of the solution with large initial data by converting the fractional PDE into a simple ordinary differential inequality using the differential inequality technique. Moreover, by solving the obtained ordinary differential inequality, an upper bound of the blow-up time is also deduced.

Keywords Blow-up · Global existence · Fractional partial differential equation

1 Introduction

In this paper, we consider the following fractional nonlinear reaction-diffusion equation

$$
\begin{aligned}
\frac{{}^C\partial^\alpha u(x,t)}{\partial t^\alpha} &= D \triangle u(x,t) + a(x) f(u) + u^p(x,t), && x \in \Omega,\ t \in (0,T),\\
u(x,0) &= u_0(x), && x \in \Omega, \\
u(x,t) &= 0, && x \in \partial\Omega, t \in (0,T),
\end{aligned} \tag{1}
$$

where Ω is a bounded convex region in $\mathcal{R}^n (n \geq 1)$ with smooth boundary $\partial\Omega$ and $D > 0$ is the diffusion coefficient. $\frac{{}^C\partial^\alpha u}{\partial t^\alpha}$ is the Caputo fractional derivative of order $0 < \alpha < 1$ which is defined with respect to the time variable as

R. Saranya (✉) · N. Annapoorani
Department of Mathematics, Bharathiar University, Coimbatore 641046, India
e-mail: saranyabumaths@gmail.com

R. K. Sharma et al. (eds.), *Frontiers in Industrial and Applied Mathematics*, Springer Proceedings in Mathematics & Statistics 410,
https://doi.org/10.1007/978-981-19-7272-0_6

$$\frac{{}^C\partial^\alpha u}{\partial t^\alpha} = \frac{1}{\Gamma(1-\alpha)} \int_0^t (t-s)^{-\alpha} \frac{\partial u(x,s)}{\partial s} \, \mathrm{d}s. \tag{2}$$

Suppose that the nonlinearity $f(u) = u^l(x,t)\left(\int_\Omega u^{l+1}(x,t)\mathrm{d}x \right)^m$ is a non-negative continuous function. The nonlinear terms $a(x)f(u) + u^p$ represents the reaction-kinetics. The exponents $l \geq 0$, $m > 0$ and $p > 1$, and the weight function $a(x) \in C(\overline{\Omega})$ satisfies

(A1) $a(x) > 0, x \in \Omega.$
(A2) $0 \leq C_1 < a(x) < C_2 < \infty, \forall\, x \in \overline{\Omega}.$

Fractional derivative is an arbitrary order derivative which incorporates memory phenomena such that it concatenates both integral and differential operators. Fractional Calculus is once thought of esoteric in nature. But in recent few decades, it has been accustomed to model biological, physical and engineering processes. Besides, most visual phenomena of quantum mechanics, fluid dynamics, ecological systems and numerous models are controlled by fractional differential equations within their domain of existence.

Reaction-diffusion equation of type (1) emerges naturally in various mathematical models from dynamics of bio-reactors and bio-sensors, population dynamics, combustion theory, compressible reactant gas model and so on [2–4, 6, 11, 12]. In chemical systems, those equations illustrate the production of the material, by chemical reaction, which competes with the diffusion of that material. Systems of such equations generally comprise numerous interacting components as chemical reactions and are widely used to trace out the formation of patterns in a variety of processes in the applied sciences.

The nonlinear processes lead to the study of new problems in the areas of partial differential equations and analysis. The blow-up of the solution in the nonlinear evolution problem is one of the most remarkable properties that differs from the linear ones. The singularities that occur in linear problems are often known as fixed singularities whereas in nonlinear problems, they are known to be movable singularities as it depends on the initial data and other properties of the problem.

In recent decades, there are many works established which concerns the global existence and blow-up phenomena of local and non-local reaction-diffusion equation [9, 10, 17]. For $p = l = 0$, the blow-up phenomena of time-fractional diffusion equation (1) with variable exponents is discussed by Manimaran and Shangerganesh [15], where the non-local source term determines human-controlled distribution function. The global existence and lower and upper bounds of the blow-up time of the solution are obtained for (1) when $\alpha = 1$ by Ma and Fang [13]. Cao et al. [7] established a finite time blow-up and long-time behavior of the solution of a time-fractional diffusion equation with local source term. Pinasco [16] discussed the blow-up solution for the parabolic and hyperbolic problems with a non-local source term using Kaplan's eigenvalue method and established a local existence theory for the respective problems with a fixed-point technique. Ma et al. [14] investigated the blow-up phenomena of a reaction-diffusion equation with weighted exponential nonlinearity when $\alpha = 1$. Tao et al. [18] established a global existence by constructing suitable sub- and super-solutions and obtained lower bounds of the blow-up time by Kaplan's eigenvalue method.

Motivated by the above works, we analyze the blow-up phenomena of the problem (1) according to the conditions on the exponents l, m and p. Moreover, we prove the global existence of solutions to the problem (1). The outline of this paper is as follows: In Sect. 2, finite time blow-up of the solution is established for $0 \le l < 1$ and $l > 1$. In Sect. 3, we establish a comparison principle for (1) by defining sub- and super-solutions. Also, global existence is proved using comparison principle for $l, m \ge 1$ and $p > 1$.

2 Blow-Up of Solutions in Finite Time

In this section, we derive the energy functionals of the problem (1) which is important to derive the blow-up phenomena of the solution. The positive initial energy obtained in Lemmas 2 and 3 leads to the blow-up of the solution obtained in Theorems 1 and 2, respectively. Using maximum principle and monotonicity condition as in Cao et al. [7], the Caputo fractional derivative of order $0 < \alpha < 1$ can be written as

$$\frac{{}^C\partial^\alpha u}{\partial t^\alpha} \le \frac{t^{1-\alpha}}{\Gamma(2-\alpha)} \frac{\partial u}{\partial t}. \tag{3}$$

Lemma 1 (Jensen's Inequality [5]) *Suppose that Φ is a real valued function on Ω and let χ and φ be non-negative Riemann-integrable functions on Ω. Then,*

$$\Phi\Big(\int_\Omega \chi(x)\varphi(x)\mathrm{d}x\Big) \le \Big(\int_\Omega \Phi(\chi(x))\varphi(x)\mathrm{d}x\Big),$$

where $\int_\Omega \varphi(x)\mathrm{d}x = 1.$

Lemma 2 *Let the assumptions (A1)–(A2) hold true and the exponents $l, m \ge 1$ and $p > 1$. Define an energy function*

$$\mathcal{E}(t) := -D\int_\Omega |\nabla u|^2\mathrm{d}x + \frac{C_2}{m+1}\Big(\int_\Omega u^{l+1}\mathrm{d}x\Big)^m + \int_\Omega u^{p+1}\mathrm{d}x. \tag{4}$$

Then for $u_0(x) \ge 0$, $\mathcal{E}'(t) > 0$ which implies $\mathcal{E}(t) > \mathcal{E}(0)$.

Proof Multiplying (1) by u_t and integrating over Ω, we use $(A1) - (A2)$ to get

$$\begin{aligned}
\int_{\Omega} \frac{{}^{C}\partial^{\alpha} u}{\partial t^{\alpha}} u_t \mathrm{d}x &= D \int_{\Omega} \Delta u u_t \mathrm{d}x + \int_{\Omega} a(x) f(u) u_t \mathrm{d}x + \int_{\Omega} u^p u_t \mathrm{d}x \\
&\leq -D \int_{\Omega} \nabla u . \nabla u_t \mathrm{d}x + C_2 \Big(\int_{\Omega} u^l u_t \mathrm{d}x \Big) \\
&\quad \Big(\int_{\Omega} u^{l+1} \mathrm{d}x \Big)^m + \int_{\Omega} u^p u_t \mathrm{d}x \\
&= -\frac{D}{2} \frac{d}{dt} \int_{\Omega} |\nabla u|^2 \mathrm{d}x + \frac{C_2}{l+1} \frac{d}{dt} \Big(\int_{\Omega} u^{l+1} \mathrm{d}x \Big) \\
&\quad \Big(\int_{\Omega} u^{l+1} \mathrm{d}x \Big)^m + \frac{1}{p+1} \frac{d}{dt} \int_{\Omega} u^{p+1} \mathrm{d}x \\
&= -\frac{D}{2} \frac{d}{dt} \int_{\Omega} |\nabla u|^2 \mathrm{d}x + \frac{C_2}{(l+1)(m+1)} \\
&\quad \frac{d}{dt} \Big(\int_{\Omega} u^{l+1} \mathrm{d}x \Big)^{m+1} + \frac{1}{p+1} \frac{d}{dt} \int_{\Omega} u^{p+1} \mathrm{d}x.
\end{aligned}$$

Thus, we have

$$\begin{aligned}
&\frac{\Gamma(2-\alpha)}{t^{1-\alpha}} \int_{\Omega} \left| \frac{{}^{C}\partial^{\alpha} u}{\partial t^{\alpha}} \right|^2 \mathrm{d}x \\
&\quad \leq \frac{1}{2} \frac{d}{dt} \Big(-D \int_{\Omega} |\nabla u|^2 \mathrm{d}x + \frac{C_2}{m+1} \Big(\int_{\Omega} u^{l+1} \mathrm{d}x \Big)^{m+1} + \int_{\Omega} u^{p+1} \mathrm{d}x \Big).
\end{aligned}$$

This implies from (4) that

$$\mathcal{E}'(t) \geq \frac{2\Gamma(2-\alpha)}{t^{1-\alpha}} \int_{\Omega} \left| \frac{{}^{C}\partial^{\alpha} u}{\partial t^{\alpha}} \right|^2 \mathrm{d}x > 0.$$

Thus for $u_0(x) \geq 0$, $\mathcal{E}(t) > \mathcal{E}(0) > 0$.

Theorem 1 *Let the assumptions* $(A1)-(A2)$ *hold true for the weight function* $a(x)$ *and the exponents* $l, m \geq 1$ *and* $p > 1$*. If* $u(x,t)$ *is a non-negative solution of the problem (1), then for sufficiently large initial data* $u_0(x) \geq 0$*, there exists a finite time* t_* *such that*

$$\lim_{t \to t_*} u(x,t) = \infty.$$

Define an auxiliary function

$$\Psi(t) = \int_{\Omega} u^2(x,t) \mathrm{d}x. \tag{5}$$

Then the upper bound for the blow-up time can be deduced as

$$t_* \leq \left(\frac{2\alpha \Psi^{\frac{2-(m+1)(l+1)}{2}}(0)}{K((m+1)(l+1)-2)} \right)^{\frac{1}{\alpha}}. \tag{6}$$

Proof Differentiating (5) with respect to t, we have

$$\Psi'(t) = 2\int_{\Omega} uu_t \mathrm{d}x.$$

Using the fact $\frac{1}{\Gamma(2-\alpha)} \leq 2$ and in view of (3),

$$\begin{aligned}
\Psi'(t) &\geq \frac{2\Gamma(2-\alpha)}{t^{1-\alpha}} \int_{\Omega} u \frac{{}^C\partial^{\alpha} u}{\partial t^{\alpha}} \mathrm{d}x \\
&\geq \frac{1}{t^{1-\alpha}} \int_{\Omega} u \left(D \triangle u + a(x) u^l(x,t) \left(\int_{\Omega} u^{l+1}(x,t) \mathrm{d}x \right)^m + u^p(x,t) \right) \mathrm{d}x \\
&\geq \frac{1}{t^{1-\alpha}} \left(-D \int_{\Omega} |\nabla u|^2 \mathrm{d}x + C_1 \int_{\Omega} u^{l+1} \left(\int_{\Omega} u^{l+1} \mathrm{d}x \right)^m \mathrm{d}x + \int_{\Omega} u^{p+1} \mathrm{d}x \right) \\
&= \frac{1}{t^{1-\alpha}} \left(-D \int_{\Omega} |\nabla u|^2 \mathrm{d}x + C_1 \left(\int_{\Omega} u^{l+1} \mathrm{d}x \right)^{m+1} + \int_{\Omega} u^{p+1} \mathrm{d}x \right) \\
&= \frac{1}{t^{1-\alpha}} \left(-D \int_{\Omega} |\nabla u|^2 \mathrm{d}x + \frac{C_2(m+1)}{m+1} \left(\int_{\Omega} u^{l+1} \mathrm{d}x \right)^{m+1} + \int_{\Omega} u^{p+1} \mathrm{d}x \right) \\
&\quad + \frac{C_1 - C_2}{t^{1-\alpha}} \left(\int_{\Omega} u^{l+1} \mathrm{d}x \right)^{m+1} \\
&= \frac{1}{t^{1-\alpha}} \left(-D \int_{\Omega} |\nabla u|^2 \mathrm{d}x + \frac{C_2}{m+1} \left(\int_{\Omega} u^{l+1} \mathrm{d}x \right)^{m+1} + \int_{\Omega} u^{p+1} \mathrm{d}x \right) \\
&\quad + \frac{mC_2}{(m+1)t^{1-\alpha}} \left(\int_{\Omega} u^{l+1} \mathrm{d}x \right)^{m+1} + \frac{C_1 - C_2}{t^{1-\alpha}} \left(\int_{\Omega} u^{l+1} \mathrm{d}x \right)^{m+1}.
\end{aligned}$$

Suppose that $\left(\frac{mC_2}{m+1} + (C_1 - C_2) \right) = K > 0$. Then from Lemma 2 and using Jensen's inequality, we have

$$\begin{aligned}
\Psi'(t) &\geq \frac{1}{t^{1-\alpha}} \left(\mathcal{E}(t) + K \left(\int_{\Omega} u^{l+1} \mathrm{d}x \right)^{m+1} \right) \\
&\geq \frac{K}{t^{1-\alpha}} \left(\int_{\Omega} u^2 \mathrm{d}x \right)^{\frac{(m+1)(l+1)}{2}}
\end{aligned} \tag{7}$$

$$\Psi'(t) \geq \frac{K}{t^{1-\alpha}} \Psi^{\frac{(m+1)(l+1)}{2}}(t). \tag{8}$$

From (7), we see that $\Psi'(t) > 0$ which implies that $\Psi(t) > \Psi(0)$. Now rearranging and integrating (8) with respect to the time variable from 0 to t, we get

$$\Psi^{1-\frac{(m+1)(l+1)}{2}}(0) - \Psi^{1-\frac{(m+1)(l+1)}{2}}(t) \geq \frac{K(m+1)(l+1)-2}{2\alpha}t^{\alpha}.$$

If $\lim_{t\to t_*}\Psi(t) = \infty$, then we obtain the upper bound of the blow-up time t_* as in (6). Thus, we see that the solution of the problem (1) becomes unbounded in L^2 norm.

Lemma 3 *Suppose that* $0 \leq l < 1, m, p > 1$ *and the assumptions* $(A1) - (A2)$ *hold true. Define an energy function*

$$\mathcal{F}(t) := -D\int_{\Omega}|\nabla u|^2 \mathrm{d}x + \frac{2C_2}{(l+1)(m+1)}\left(\int_{\Omega}u^{l+1}\mathrm{d}x\right)^m + \int_{\Omega}u^{p+1}\mathrm{d}x. \quad (9)$$

Then for $u_0(x) \geq 0$, $\mathcal{F}'(t) > 0$ *which implies that* $\mathcal{F}(t) > \mathcal{F}(0)$.

Proof Following the same procedure as in the proof of Lemma 2, we have

$$\begin{aligned}
\int_{\Omega}\frac{{}^{C}\partial^{\alpha}u}{\partial t^{\alpha}}u_t \mathrm{d}x &\leq -\frac{D}{2}\frac{d}{dt}\int_{\Omega}|\nabla u|^2\mathrm{d}x + \frac{C_2}{(l+1)(m+1)}\frac{d}{dt}\left(\int_{\Omega}u^{l+1}\mathrm{d}x\right)^{m+1} \\
&\quad + \frac{1}{p+1}\frac{d}{dt}\int_{\Omega}u^{p+1}\mathrm{d}x \\
&= \frac{1}{2}\frac{d}{dt}\Bigg(-D\int_{\Omega}|\nabla u|^2\mathrm{d}x + \frac{2C_2}{(l+1)(m+1)}\left(\int_{\Omega}u^{l+1}\mathrm{d}x\right)^{m+1} \\
&\quad + \int_{\Omega}u^{p+1}\mathrm{d}x\Bigg).
\end{aligned}$$

Thus from (9), we obtain $\mathcal{F}'(t) \geq 0$. This suggests $\mathcal{F}(t) > \mathcal{F}(0)$.

Theorem 2 *Let* $0 \leq l < 1$, $m, p > 1$ *and the assumptions* $(A1) - (A2)$ *hold true for the weight function* $a(x)$. *Then the solution of the problem (1) blows up in finite time* t_* *such that the upper bound can be deduced as*

$$t_* \leq \left(\frac{2\alpha\Psi^{\frac{2-(m+1)(l+1)}{2}}(0)}{M((m+1)(l+1)-2)}\right)^{\frac{1}{\alpha}}. \quad (10)$$

Proof Using the auxiliary function defined as in (5) and following the same procedure as in Theorem 1, we have

$$\begin{aligned}\Psi'(t) &\geq \frac{1}{t^{1-\alpha}}\left(-D\int_{\Omega}|\nabla u|^2\mathrm{d}x + C_1\left(\int_{\Omega}u^{l+1}\mathrm{d}x\right)^{m+1} + \int_{\Omega}u^{p+1}\mathrm{d}x\right)\\ &= \frac{1}{t^{1-\alpha}}\left(-D\int_{\Omega}|\nabla u|^2\mathrm{d}x + \frac{2C_2}{(l+1)(m+1)}\left(\int_{\Omega}u^{l+1}\mathrm{d}x\right)^{m+1}\right.\\ &\quad\left.+\int_{\Omega}u^{p+1}\mathrm{d}x\right)\\ &\quad+\frac{1}{t^{1-\alpha}}\left((C_1-2C_2)\left(\int_{\Omega}u^{l+1}\mathrm{d}x\right)^{m+1} + \frac{2C_2K}{K+1}\left(\int_{\Omega}u^{l+1}\mathrm{d}x\right)^{m+1}\right),\end{aligned}$$

where $K = lm + l + m$. Assuming for suitable values of C_1, C_2, l, m and p, the constant

$$\frac{C_1(K+1) - 2C_2}{K+1} = M > 0.$$

Now by Lemma 3 and Jensen's inequality, we have

$$\begin{aligned}\Psi'(t) &\geq \frac{M}{t^{1-\alpha}}\left(\int_{\Omega}u^2\mathrm{d}x\right)^{\frac{(l+1)(m+1)}{2}}\\ &= \frac{M}{t^{1-\alpha}}\Psi^{\frac{(l+1)(m+1)}{2}}(t).\end{aligned} \tag{11}$$

Inequality (11) shows that $\Psi(t) > \Psi(0)$. Also by appropriate choice of constants l and m, we see that $\frac{(m+1)(l+1)}{2} > 1$. Now, integrating (11) from 0 to t, we get

$$\Psi^{1-\frac{(m+1)(l+1)}{2}}(0) - \Psi^{1-\frac{(m+1)(l+1)}{2}}(t) \geq \frac{M(m+1)(l+1)-2}{2\alpha}t^{\alpha}.$$

If $\lim_{t\to t_*}\Psi(t) = \infty$, then the blow-up time of the solution of the problem (1) is obtained as in (10).

3 Global Existence

This section discusses the global existence by constructing appropriate upper and lower solutions to the problem (1). We prove the comparison principle by defining the sub- and super-solutions of the solution to the problem (1) as follows.

Definition 1 A smooth function $w(x, t)$ is called the super-solution to the problem (1) on $(0, T)$ provided

$$\begin{aligned}
&\frac{{}^{C}\partial^{\alpha} w}{\partial t^{\alpha}} \geq D \Delta w(x,t) + a(x) w^{l}(x,t)\left(\int_{\Omega} w^{l+1}(x,t)\mathrm{d}x\right)^{m} + w^{p}(x,t), \\
&w(x,0) \geq w_0(x), && x \in \Omega, \qquad (12)\\
&w(x,t) \geq 0, && x \in \partial\Omega, t \in (0,T).
\end{aligned}$$

By reversing the inequalities (12), we can define the sub-solution $v(x,t)$ to the problem (1) similar to Definition 1. To prove the comparison principle, we are in need of the following lemmas from fractional calculus.

Lemma 4 ([1]) *Let X be a Hilbert Space and $u : [0,T] \to X$. Then for $0 < \alpha < 1$,*

$$2(u(t), {}_0^C D_t^{\alpha} u(t)) \geq {}_0^C D_t^{\alpha} |u(t)|^2.$$

Lemma 5 (Gronwall type lemma [8]) *Assume that $\alpha, T, \epsilon_1, \epsilon_2 \in \mathcal{R}_+$ and let $u : [0,T] \to \mathcal{R}$ is a continuous function satisfying the inequality*

$$|u(t)| \leq \epsilon_1 + \frac{\epsilon_2}{\Gamma(\alpha)} \int_0^T (t-s)^{\alpha-1} |u(s)| \mathrm{d}s$$

for all $t \in [0,T]$. Then,

$$|u(t)| \leq \epsilon_1 E_{\alpha}(\epsilon_2 t^{\alpha}), \qquad \forall t \in [0,T].$$

Here, E_α represents the Mittag-Leffler function of order α.

Next to prove the global existence, we present the comparison principle to the problem (1).

Theorem 3 *If $w(x,t)$ and $v(x,t)$ be the super- and sub-solutions to the problem (1), then for any $(x,t) \in \Omega \times (0,T)$, $w(x,t) \geq v(x,t)$.*

Proof Define $z(x,t) = v(x,t) - w(x,t)$. Suppose that for some $t_1 \in (0,T)$, $z(x,t_1) \geq 0$. We prove by contradiction that there does not exist such t_1 such that we prove $w \geq v$. Now using the definitions of $w(x,t)$ and $v(x,t)$, we can write

$$\frac{{}^{C}\partial^{\alpha} z}{\partial t^{\alpha}} \leq D \Delta z + a(x)\left(v^{l}\left(\int_{\Omega} v^{l+1}\mathrm{d}x\right)^{m} - w^{l}\left(\int_{\Omega} w^{l+1}\mathrm{d}x\right)^{m}\right) + (v^{p} - w^{p}). \tag{13}$$

Let $z^+ = \max(0, z(x,t))$. Multiply (13) by z^+ and integrate over Ω to get

$$\begin{aligned}\int_{\Omega} \frac{{}^{C}\partial^{\alpha} z^{+}}{\partial t^{\alpha}}.z^{+}\mathrm{d}x \leq\; & D\int_{\Omega} \Delta z^{+}.z^{+}\mathrm{d}x + \int_{\Omega} a(x)\Bigg(v^{l}\Bigg(\int_{\Omega} v^{l+1}\mathrm{d}x\Bigg)^{m} \\ & -w^{l}\Bigg(\int_{\Omega} w^{l+1}\mathrm{d}x\Bigg)^{m}\Bigg).z^{+}\mathrm{d}x + \int_{\Omega}(v^{p}-w^{p}).z^{+}\mathrm{d}x \\ \leq\; & -D\int_{\Omega} |\nabla z^{+}|^{2}\mathrm{d}x + C_{1}\int_{\Omega}\Bigg(v^{l}\Bigg(\int_{\Omega} v^{l+1}\mathrm{d}x\Bigg)^{m} \\ & -w^{l}\Bigg(\int_{\Omega} w^{l+1}\mathrm{d}x\Bigg)^{m}\Bigg).z^{+}\mathrm{d}x + \int_{\Omega}(v^{p}-w^{p}).z^{+}\mathrm{d}x. \quad (14)\end{aligned}$$

For $p > 1$, from [16], we write

$$v^{p} - w^{p} \leq p\psi^{p-1}(v-w), \tag{15}$$

where $\psi \in \mathcal{R}^{n}$ is bounded in $\Omega \times (0, T)$. Using (15) in the last term of the inequality (14), we have

$$\int_{\Omega}(v^{p}-w^{p}).z^{+}\mathrm{d}x \leq C_{3}\int_{\Omega}(z^{+})^{2}\mathrm{d}x. \tag{16}$$

For $a, b, c, d > 0$, if $(ac - bd) > 0$, we have $(a + b)(c - d) \geq (bc - ad)$. Using (15) we write

$$\begin{aligned}\int_{\Omega}\Bigg(v^{l}\Bigg(\int_{\Omega} v^{l+1}\mathrm{d}x\Bigg)^{m} - w^{l}\Bigg(\int_{\Omega} w^{l+1}\mathrm{d}x\Bigg)^{m}\Bigg).z^{+}\mathrm{d}x &\leq C_{4}\int_{\Omega}(v^{l}-w^{l}).z^{+}\mathrm{d}x \\ &\leq C_{5}\int_{\Omega}(z^{+})^{2}\mathrm{d}x, \quad (17)\end{aligned}$$

where $C_4 := \sup_{t\geq 0}\phi(t)$ and $\phi(t) := \Bigg(\Bigg(\int_{\Omega} v^{l+1}\mathrm{d}x\Bigg)^{m} + \Bigg(\int_{\Omega} w^{l+1}\mathrm{d}x\Bigg)^{m}\Bigg)$. Since the first term in the RHS of inequality (14) is strictly negative and inserting (16) and (17) in (14), we have

$$\frac{1}{2}\frac{{}^{C}\partial^{\alpha}}{\partial t^{\alpha}}\int_{\Omega}(z^{+})^{2}\mathrm{d}x \leq C_{6}\int_{\Omega}(z^{+})^{2}\mathrm{d}x.$$

Taking I^{α} on both sides and using Gronwall type of lemma, we get

$$\int_{\Omega}(z^{+})^{2}\mathrm{d}x \leq \Bigg(\int_{\Omega}(z^{+}(x,0))^{2}\mathrm{d}x\Bigg)E_{\alpha}(C_{6}t^{\alpha}).$$

The definitions of sub- and super-solution imply that $z^{+}(x, 0) \leq 0$. Hence, $v(x, t) \leq w(x, t)$.

Consider the eigenvalue problem

$$\begin{aligned} \triangle \chi + \lambda\chi &= 0, \qquad x \in \Omega \\ \chi &= 0, \qquad x \in \partial\Omega, \end{aligned} \tag{18}$$

where $\chi_1(x)$ is the first eigenfunction corresponding to the eigenvalue λ_1 and $\int_\Omega \chi_1(x)\mathrm{d}x = 1$. Next, we propose the main theorem of global existence of solutions to the problem (1).

Theorem 4 *Let the exponents $l, m \geq 1$ and the assumptions $(A1) - (A2)$ hold true. If the initial data $u_0(x) \leq \eta^{-\beta_1}\chi_1(x)$, where β is an arbitrary positive constant, $\chi_1(x) > 0$ and $\eta > 0$ is large enough, then the solution to the problem (1) exists for all $t > 0$.*

Proof We construct a function $w(x,t)$ as in [13]. Let $\eta_1 > 0$ and $\beta_1 > 0$ be the constants to be determined later such that

$$w(x,t) = (\eta_1 + t)^{-\beta_1}\chi_1(x). \tag{19}$$

We claim that $w(x,t)$ is the super-solution to the solution of the problem (1). Now, we compute

$$\begin{aligned} &\frac{{}^C\partial^\alpha w}{\partial t^\alpha} - D\triangle w - a(x)w^l\left(\int_\Omega w^{l+1}\mathrm{d}x\right)^m - w^p \\ &\geq \frac{t^{-\alpha}}{\Gamma(1-\alpha)}\left(\frac{1}{(\eta_1+t)^{\beta_1}} - \frac{1}{\eta_1^{\beta_1}}\right)\chi_1(x) + \frac{D\lambda_1\chi_1(x)}{(\eta_1+t)^{\beta_1}} \\ &\quad - \frac{C_2\chi_1^l}{(\eta_1+t)^{\beta_1 l(1+m)-\beta_1 m}}\int_\Omega\left(\chi_1^{l+1}\mathrm{d}x\right)^m - \frac{\chi_1^p}{(\eta_1+t)^{p\beta_1}} \\ &\geq \frac{\chi_1}{(\eta_1+t)^{\beta_1}}\left[\frac{t^{-\alpha}}{\Gamma(1-\alpha)}\left(1 - \frac{(\eta_1+t)^{\beta_1}}{\eta_1^{\beta_1}}\right) + D\lambda_1 - \frac{C_2}{(\eta_1+t)^{\beta_1(l+lm+m-1)}}\right. \\ &\quad \left. - \frac{1}{(\eta_1+t)^{\beta_1(1-p)}}\right]. \end{aligned}$$

Choosing η_1 sufficiently large, we have

$$\frac{{}^C\partial^\alpha w}{\partial t^\alpha} - D\triangle w - a(x)w^l\left(\int_\Omega w^{l+1}\mathrm{d}x\right)^m - w^p \geq 0. \tag{20}$$

Also by the hypotheses, the initial data satisfies

$$w(x,0) = \eta^{-\beta_1}\chi_1(x) \geq u_0(x). \tag{21}$$

Hence, the inequalities (20)–(21) show that $w(x,t)$ is a super-solution to the problem (1.1) and it exists globally. By Comparison Principle, $u(x,t)$ exists for all $t > 0$.

Remark If we set a function $v(x,t) = \eta_2(T-t)^{-\beta_2}\chi_1(x)$ with the initial data $u_0(x) \geq \eta_2 T^{-\beta_2}\chi_1(x)$, where $\beta_2 > 0$ and $\eta > 0$ sufficiently large,

$$\frac{{}^C\partial^\alpha v}{\partial t^\alpha} - D\,\Delta\, v - a(x)v^l\bigg(\int_\Omega v^{l+1}\mathrm{d}x\bigg)^m - v^p$$

$$\leq \frac{\eta_2\beta_2\chi_1(x)}{\Gamma(1-\alpha)(\alpha+\beta_2)t^{\alpha+\beta_2}} + \frac{D\eta_2\lambda_1\chi_1}{(T-t)^{\beta_2}} - C_2\frac{\eta_2^{l+lm+m}\chi_1^l}{(T-t)^{-\beta_2(l+lm+m)}} - \frac{\eta_2^p\chi_1^p}{(T-t)^{p\beta_2}}$$

$$\leq \eta_2\chi_1\bigg[\frac{\beta_2}{\Gamma(1-\alpha)(\alpha+\beta_2)t^{\alpha+\beta_2}} + \frac{D\lambda_1}{(T-t)^{\beta_2}} - C_2\frac{\eta_2^{K-1}\chi_1^{l-1}}{(T-t)^{-\beta_2 K}} - \frac{\eta_2^{p-1}\chi_1^{p-1}}{(T-t)^{p\beta_2}}\bigg].$$

For large values of η_2 and β_2, we see that

$$\frac{{}^C\partial^\alpha v}{\partial t^\alpha} - D\,\Delta\, v - a(x)v^l\bigg(\int_\Omega v^{l+1}\mathrm{d}x\bigg)^m - v^p \leq 0.$$

Hence with the choice of initial data taken, $v(x,t)$ is a sub-solution to the problem (1) and it blows up at finite time $t^* \leq T$.

References

1. Alikhanov, A.A.: A priori estimates for solutions of boundary value problems for fractional order equations. Differ. Equ. **46**, 660–666 (2010). https://doi.org/10.1134/S0012266110050058
2. Allegretto, W., Fragnelli, G., Nistri, P., Papini, D.: Coexistence and optimal control problems for a degenerate predator-prey model. J. Math. Anal. Appl. **378**, 528–540 (2011). https://doi.org/10.1016/j.jmaa.2010.12.036
3. Bebernes, J., Bressan, A.: Thermal behavior for a confined reactive gas. J. Differ. Equ. **44**, 118–133 (1982). https://doi.org/10.1016/0022-0396(82)90028-6
4. Bebernes, J., Eberly, D.: Mathematical Problems from Combustion Theory, Applied Mathematical Sciences. Springer, N.Y. (1989). https://doi.org/10.1007/978-1-4612-4546-9
5. Borwein, P., Erdlyi, T.: Polynomials and Polynomial Inequalities. Springer, New York (1995). https://doi.org/10.1007/978-1-4612-0793-1
6. Calsina, À., Perelló, C., Saldaña, J.: Non-local reaction-diffusion equations modelling predator-prey coevolution. Publ. Mat. **38**, 315–325 (1994). https://www.jstor.org/stable/43736486
7. Cao, J., Song, G., Wang, J., Shi, Q., Sun, S.: Blow-up and global solutions for a class of time fractional nonlinear reaction-diffusion equation with weakly spatial source. Appl. Math. Lett. **91**, 201–206 (2019). https://doi.org/10.1016/j.aml.2018.12.020
8. Diethelm, K.: The Analysis of Fractional Differential Equations: An Application-Oriented Exposition Using Differential Operators of Caputo Type. Springer. Berlin Heidelberg (2010). https://doi.org/10.1007/978-3-642-14573-5
9. Furter, J., Grinfeld, M.: Local vs. nonlocal interactions in population dynamics. J. Math. Biol. **27**, 65–80 (1989). https://doi.org/10.1007/BF00276081

10. Hu, B.: Blow-up Theories for Semilinear Parabolic Equations. Lecture Notes in Mathematics. Springer, Heidelberg (2018). https://doi.org/10.1007/978-3-642-18460-4
11. Ivanauskas, F., Laurinavičius, V., Sapagovas, M., Nečiporenko, A.: Reaction-diffusion equation with nonlocal boundary condition subject to PID-controlled bio-reactor. Nonlinear Anal. Model. Control. **22**, 261–272 (2017). https://doi.org/10.15388/NA.2017.2.8
12. Khramchenkov, M.G.: Dispersion and chemical reactions in porous media. Fluid Dyn. **36**, 166–168 (2001). https://doi.org/10.1023/A:1018896114067
13. Ma, L., Fang, Z.B.: Blow-up analysis for a nonlocal reaction-diffusion equation with robin boundary conditions. Taiwan. J. Math. **21**, 131–150 (2017). https://doi.org/10.11650/tjm.21.2017.7380
14. Ma, L.,Fang, Z.B.: Blow-up phenomena of solutions for a reaction-diffusion equation with weighted exponential nonlinearity. Comput. Math. Appl. **75**, 2735–2745 (2018). https://doi.org/10.1016/j.camwa.2018.01.005
15. Manimaran, J., Shangerganesh, L.: Blow-up solutions of a time-fractional diffusion equation with variable exponents. Tbil. Math. J. **12**, 149–157 (2019). https://doi.org/10.32513/tbilisi/1578020574
16. Pinasco, J.P.: Blow-up for parabolic and hyperbolic problems with variable exponents. Nonlinear Anal. **71**, 1094–1099 (2009). https://doi.org/10.1016/j.na.2008.11.030
17. Tang, G., Li, Y., Yang, X.: Lower bounds for the blow-up time of the nonlinear nonlocal reaction diffusion problems in $\mathcal{R}^N$ $(N \geq 3)$, Bound. Value Probl. **1**, 1–5 (2014). https://doi.org/10.1186/s13661-014-0265-5
18. Tao, X., Fang, Z.B.: Blow-up phenomena for a nonlinear reaction-diffusion system with time dependent coefficients. Comput. Math. Appl. **74**, 2520–2528 (2017). https://doi.org/10.1016/j.camwa.2017.07.037
19. Zhou, Y., Peng, L.: Weak solutions of the time-fractional Navier-Stokes equations and optimal control. Comput. Math. Appl. **6**, 1016–1027 (2017). https://doi.org/10.1016/j.camwa.2016.07.007

Ways of Constructing Multiplicative Magic Cubes

Narbda Rani and Vinod Mishra

Abstract In this paper, the methods for construction of multiplicative magic cubes of order *n* from the existing additive magic cube of order *n* has been introduced. Also, the modified Trenkler's formula for the multiplicative magic cubes of odd and doubly even order has been instigated. Moreover, the newly defined power method has been proposed for the construction of multiplicative magic cubes. In all the methods, the conditions for obtaining multiplicative magic cubes consisting of either odd or even or composite numbers as their elements have been elaborated.

Keywords Additive magic cube · Multiplicative magic cube · Magic cube formula · Trenkler's formula

1 Introduction

A magic cube (or additive magic cube (AMC)) of order n is a three-dimensional array of n^3 numbers in which the sum of n elements of each row, each column, each pillar, and each of the four space diagonals is the same [3]. The fixed sum is known as magic sum and is given by $\frac{n(n^3+1)}{2}$. For instance, the first, second, and third layer of a third-order magic cube are, respectively, given below

$$\begin{bmatrix} 10 & 26 & 6 \\ 24 & 1 & 17 \\ 8 & 15 & 19 \end{bmatrix} \quad \begin{bmatrix} 23 & 3 & 16 \\ 7 & 14 & 21 \\ 12 & 25 & 5 \end{bmatrix} \quad \begin{bmatrix} 9 & 13 & 20 \\ 11 & 27 & 4 \\ 22 & 2 & 18 \end{bmatrix}$$

N. Rani (✉) · V. Mishra
Department of Mathematics, Sant Longowal Institute of Engineering and Technology, Longowal, Sangrur 148106, Punjab, India
e-mail: narmadasharma1990@gmail.com

R. K. Sharma et al. (eds.), *Frontiers in Industrial and Applied Mathematics*, Springer Proceedings in Mathematics & Statistics 410,
https://doi.org/10.1007/978-981-19-7272-0_7

A multiplicative magic cube (MMC) of order n is a three-dimensional array of n^3 numbers with the property that the product of n elements along each row, column, pillar, and each space diagonal is the same [1]. The generation of multiplicative magic cubes opens a new direction to the application areas concerning image processing, cryptography, stenography, game theory, etc., which motivate the researchers for their study and utilization. Trenkler [4] has introduced the formula for the construction of an additive magic cube of odd, singly even, and doubly even order separately. Uko and Barron [2] has generalized the Trenkler's formula for the magic cubes and derive sufficient conditions to generate regular magic cubes. They illustrate three new formulas for the construction of odd order magic cubes that differ from each other and from the magic cubes generated with Trenkler's rule. Trenkler [5] has demonstrated several ways to construct additive and multiplicative magic cubes and provide the formula for the construction of multiplicative magic cubes from the existing additive magic cubes. He has also given an algorithm for constructing magic cubes [6]. In this paper, the various ways of constructing multiplicative magic cubes from the existing additive magic cubes of order n ($n \neq 2$) have been introduced. Also, the Trenkler's rules for constructing odd and doubly even order magic cubes have been modified.

2 Ways of Constructing Multiplicative Magic Cubes of Odd and Doubly Even Order

2.1 *Power Method*

Let $A_n = \left\{a_n(i, j, k) \,\middle|\, 1 \leq i, j, k \leq n\right\}$ be an additive magic cube (AMC) and M_n be a multiplicative magic cube (MMC) of order n. Then the first formula for constructing magic cubes of order n with magic constant $\sigma(M_n) = 2^{\frac{n(n^3+1)}{2}}$ is given as

$$M_n = \left\{m_n(i, j, k) = 2^{a_n(i,j,k)} \,\middle|\, 1 \leq i, j, k \leq n\right\}$$

By using this formula, one can obtain a multiplicative magic cube of smallest magic constant. But if there is no problem with the larger magic constant, then the above formula is generalized as below

$$M_n = \left\{m_n(i, j, k) = r^{a_n(i,j,k)} \,\middle|\, 1 \leq i, j, k \leq n\right\}$$

where r is any real (or complex) number. It is a multiplicative magic cube with magic constant depicted below

$$\sigma(M_n) = r^{\frac{n(n^3+1)}{2}} = r^{\sigma(A_n)}$$

The following table shows the effect of the elements of a multiplicative magic cube on changing the values of variable r:-

S.no.	Value of r	Effect on the elements of MMC
1	r is even number	All entries of the required MMC are even
2	r is odd number	All entries of the required MMC are odd
3	r is mixed type number	All the elements of the resulting MMC are mixed numbers

Example depicting the above formula is as below
Let A_n be the following AMC of order three with magic constant 42.

$$\begin{bmatrix} 8 & 15 & 19 \\ 24 & 1 & 17 \\ 10 & 26 & 6 \end{bmatrix} \quad \begin{bmatrix} 12 & 25 & 5 \\ 7 & 14 & 21 \\ 23 & 3 & 16 \end{bmatrix} \quad \begin{bmatrix} 22 & 2 & 18 \\ 11 & 27 & 4 \\ 9 & 13 & 20 \end{bmatrix}$$

Layer 1 Layer 2 Layer 3

Then the required MMC for $r = 3$ with magic sum 3^{42} is given by

$$\begin{bmatrix} 3^8 & 3^{15} & 3^{19} \\ 3^{24} & 3^1 & 3^{17} \\ 3^{10} & 3^{26} & 3^6 \end{bmatrix} \quad \begin{bmatrix} 3^{12} & 3^{25} & 3^5 \\ 3^7 & 3^{14} & 3^{21} \\ 3^{23} & 3^3 & 3^{16} \end{bmatrix} \quad \begin{bmatrix} 3^{22} & 3^2 & 3^{18} \\ 3^{11} & 3^{27} & 3^4 \\ 3^9 & 3^{13} & 3^{20} \end{bmatrix}$$

Layer 1 Layer 2 Layer 3

The above formula can be modified by adding or subtracting 1,2,3,...,$\sigma(A_n)$ from the powers of each entry of a MMC.

Thereby, $M_n = \left\{ m_n(i, j, k) = r^{a_n(i,j,k)-\sigma(A_n)} \middle| 1 \le i, j, k \le n \right\}$ where r and $\sigma(A_n)$ are any real number and magic sum of the additive magic cube A_n respectively. The addition of $1, 2, 3, \ldots, \sigma(A_n)$ in the powers makes the magic constant larger whereas the subtraction decreases the value of the multiplicative magic constant.

2.2 *Modified Trenkler's Formula for MMC*

Trenkler [5] has introduced formula for the construction of MMC of odd, singly even, and doubly even order by using his formulas for AMC. The formula for odd order MMC is represented as below

Let $M_n = \left\{ m_n(i, j, k) = \alpha n^2 + \beta n + \gamma + 1 \middle| 1 \le i, j, k \le n \right\}$ be an AMC of order n, for $\alpha = (i - j + k - 1)\ (mod\ n)$, $\beta = (i - j - k)\ (mod\ n)$ and $\gamma = (i + j + k - 2)\ (mod\ n)$. Then, $Q_n = \left\{ q_n(i, j, k) = 2^\alpha \cdot 3^\beta \cdot 5^\gamma \middle| 1 \le i, j, k \le n \right\}$ is the required MMC. If in the above formula 3 is replaced by $(2\beta + 1)$ for $\beta =$

$1, 2, ..., n-1$ and 5 by the numbers $(2n + 2\gamma - 1)$ for $\gamma = 1, 2, ..., n-1$ uniquely, then the MMC with smaller magic constant has been obtained.

Modification in the formula: By studying the above formula, we have analyzed that the expressions used for replacing 3 and 5 are not the fixed one and there is no restriction of such kind needed if we should not confine to the construction of normal magic cubes. So, it is possible to construct the multiplicative magic cube by using the formula

$$P_n = \left\{ p_n(i, j, k) = a^u \cdot b^v \cdot c^w \;\middle|\; 1 \le i, j, k \le n \right\} \quad (1)$$

where $u = (i - j + k - 1)\ (mod\ n)$, $v = (i - j - k)\ (mod\ n)$, $w = (i + j + k - 2)\ (mod\ n)$ and a, b, c are any three distinct real numbers. By using this formula, the construction of a MMC of any order n $(n \neq 2)$ with magic constant $\sigma(P_n) = a^\kappa b^\kappa c^\kappa = (abc)^\kappa = (abc)^{\frac{n(n-1)}{2}}$ has been instigated, where κ is the sum of numbers $0, 1, 2, 3, ..., n-1$.

Example of the 4th order MMC is represented by putting $a = 2, b = 4, c = 6, n = 4,$ *and* $k = 6$ *as under*:

$$\begin{bmatrix} 384 & 4608 & 3456 & 2 \\ 72 & 13824 & 128 & 96 \\ 3456 & 2 & 384 & 4608 \\ 128 & 96 & 72 & 13824 \end{bmatrix} \quad \begin{bmatrix} 1152 & 864 & 8 & 1536 \\ 55296 & 32 & 24 & 288 \\ 8 & 1536 & 1152 & 864 \\ 24 & 288 & 55296 & 32 \end{bmatrix}$$

Layer 1 Layer 2

$$\begin{bmatrix} 3456 & 2 & 384 & 4608 \\ 128 & 96 & 72 & 13824 \\ 384 & 4608 & 3456 & 2 \\ 72 & 13824 & 128 & 96 \end{bmatrix} \quad \begin{bmatrix} 8 & 1536 & 1152 & 864 \\ 24 & 288 & 55296 & 32 \\ 1152 & 864 & 8 & 1536 \\ 55296 & 32 & 24 & 288 \end{bmatrix}$$

Layer 3 Layer 4

which is a MMC with magic constant = $2^6 \cdot 4^6 \cdot 6^6 = 12, 230, 590, 464$.

Moreover, in (1), if u, v, w take the values as given below then again the MMC is obtained.

$$u = \left\{ (i - j + k - 1) - n\left[\frac{i - j + k - 1}{n} \right] \right\} (mod\ n)$$

$$v = \left\{ (i - j - k) - n\left[\frac{i - j - k}{n} \right] \right\} (mod\ n)$$

$$w = \left\{ (i + j + k - 2) - n\left[\frac{i + j + k - 2}{n} \right] \right\} (mod\ n)$$

Here, $[x]$ denotes the integer part of x and $(\text{mod } n)$ gives the remainder after division by n. The magic cubes obtained by using this formula are not normal magic cubes. The behavior of elements of the MMC constructed above depends entirely on the value of the variables a, b, and c. This method has also been tested by taking different values of a, b, and c like $a = 2, b = 3, c = 4$ and $a = 3, b = 4, c = 7$, etc., and for MMC of various orders. See the table below for checking the different behavior of a, b, c and the entries of a MMC:

S.no.	Value of a	Value of b	Value of c	Order of MMC	Effect on magic constant	Effect on the elements of MMC
1	Even	Even	Even	Odd	Even	All $p_n(i, j, k)$ are even except one entry when $p_n(i, j, k) = 1$
2	Even	Even	Even	Even	Even	All $p_n(i, j, k)$ are only even
3	Odd	Odd	Odd	Odd/ Even	Odd	All $p_n(i, j, k)$ are only odd
4	Even	Odd	Even	Odd	Even	All $p_n(i, j, k)$ are either odd or even
5	Even	Odd	Even	Even	Even	All $p_n(i, j, k)$ are only even
6	Odd	Even	Even	Odd	Even	All $p_n(i, j, k)$ are either odd or even
7	Odd	Even	Even	Even	Even	All $p_n(i, j, k)$ are only even
8	Even	Even	Odd	Odd	Even	All $p_n(i, j, k)$ are either odd or even
9	Even	Even	Odd	Even	Even	All $p_n(i, j, k)$ are only even

2.3 *Modified Trenkler's Formula for MMC of Doubly Even Order*

Trenkler [5] has introduced the following formula for the construction of MMC of doubly even order, i.e., for $n \equiv 0 \ (mod\ 4)$, from the AMC.

If $M_n = \left\{ m_n(i, j, k) \;\middle|\; 1 \le i, j, k \le n \right\}$ is the AMC, then its each entry $m_n(i, j, k)$ is given as below

$$m_n(i, j, k) = \begin{cases} (i-1)n^2 + (j-1)n + k & \text{if } \phi(i, j, k) = 1 \\ (\bar{i}-1)n^2 + (\bar{j}-1)n + \bar{k} & \text{if } \phi(i, j, k) = 0 \end{cases}$$

Then, its corresponding MMC, $Q_n = \left\{ q_n(i, j, k) \;\middle|\; 1 \le i, j, k \le n \right\}$ is defined as below

$$q_n(i, j, k) = \begin{cases} 2^{(i-1)} \cdot 3^{(j-1)} \cdot 5^{(k-1)} & \text{if } \phi(i, j, k) = 1 \\ 2^{(\bar{i}-1)} \cdot 3^{(\bar{j}-1)} \cdot 5^{(\bar{k}-1)} & \text{if } \phi(i, j, k) = 0 \end{cases}$$

where $\phi(i,j,k) = \left\{ i + \widetilde{i} + j + \widetilde{j} + k + \widetilde{k} \right\} (mod\ 2)$ and $\bar{x} = n + 1 - x$

$$\widetilde{x} = \begin{cases} 0 & \text{for } 1 \le x \le \frac{n}{2} \\ 1 & \text{for } \frac{n}{2} < x \le n \end{cases}$$

Modification in the formula: Let $A_n = \left\{ a_n(i,j,k) \;\middle|\; 1 \le i,j,k \le n \right\}$ be an AMC of order n, $n \equiv 0\ (mod\ 4)$, where each entry is defined as

$$a_n(i,j,k) = \begin{cases} (k-1)n^2 + (j-1)n + i & \text{if } f(i,j,k) = 1 \\ (n-k)n^2 + (n-j)n + (n-i) + 1 & \text{if } f(i,j,k) = 0 \end{cases}$$

$$\text{and } f(i,j,k) = \left\{ i + \left[\frac{2(i-1)}{n}\right] + j + \left[\frac{2(j-1)}{n}\right] + k + \left[\frac{2(k-1)}{n}\right] \right\} (mod\ 2) \quad (2)$$

Here, $[x]$ denotes the integer part of x. Then the resulting MMC, $P_n = \left\{ p_n(i,j,k) \;\middle|\; 1 \le i,j,k \le n \right\}$ of order $n \equiv 0\ (mod\ 4)$ is demonstrated by the following formula:

$$p_n(i,j,k) = \begin{cases} u^{(k-1)} \cdot v^{(j-1)} \cdot w^{i} & \text{if } f(i,j,k) = 1 \\ u^{(n-k)} \cdot v^{(n-j)} \cdot w^{(n-i)} & \text{if } f(i,j,k) = 0 \end{cases}$$

and the value of the function $f(i,j,k)$ is same as given in equation (3). Also, u, v, and w are any three distinct real numbers. If all the three variables u, v, and w are odd, then we get the MMC with only odd elements and having odd magic constant. Similarly, if u, v, and w are even, then all the entries of MMC are even including its magic constant. Moreover, if one or two of u, v and w is/ are taken to be even and the remaining as odd numbers, then the mixture of both odd as well as even numbers as the entries of MMC of order $n \equiv 0\ (mod\ 4)$.

Example of P_4 *on putting* $u = 3$, $v = 5$, *and* $w = 7$ *with magic constant* $(3 \cdot 5 \cdot 7)^6 = 65,664,686,390,625$ *is as below*

$$\begin{bmatrix} 7 & 231525 & 46305 & 875 \\ 165375 & 245 & 1225 & 1323 \\ 23625 & 1715 & 8575 & 189 \\ 2401 & 675 & 135 & 300125 \end{bmatrix} \quad \begin{bmatrix} 385875 & 105 & 525 & 3087 \\ 147 & 11025 & 2205 & 18375 \\ 1029 & 1575 & 315 & 128625 \\ 1125 & 36015 & 180075 & 9 \end{bmatrix}$$

Layer 1 Layer 2

$$\begin{bmatrix} 128625 & 315 & 1575 & 1029 \\ 441 & 3675 & 735 & 55125 \\ 3087 & 525 & 105 & 385875 \\ 375 & 108045 & 540225 & 3 \end{bmatrix} \quad \begin{bmatrix} 189 & 8575 & 1715 & 23625 \\ 6125 & 6615 & 33075 & 49 \\ 875 & 46305 & 231525 & 7 \\ 64827 & 25 & 5 & 8103375 \end{bmatrix}$$

Layer 3 Layer 4

2.4 Power Method for MMC of Order $n \equiv 0\ (mod\ 4)$

In this method, the construction of MMC of order $n \equiv 0\ (mod\ 4)$ from the existing AMC of order $n \equiv 0\ (mod\ 4)$, defined by the equation (2) and (3), has been represented using the following formula:

If $R_n = \left\{ r_n(i, j, k) \,\middle|\, 1 \le i, j, k \le n \right\}$ is the required MMC, then its all elements $r_n(i, j, k)$ are defined by

$$r_n(i, j, k) = \begin{cases} m^{(k-1)n^2+(j-1)n+i} & \text{if } g(i, j, k) = 1 \\ m^{(n-k)n^2+(n-j)n+(n-i)+1} & \text{if } g(i, j, k) = 0 \end{cases}$$

where $g(i, j, k)$ is equal to the function $f(i, j, k)$ as given by the equation (3). The magic constant of MMC is

$$\sigma(R_n) = m^{\frac{n(n^3+1)}{2}}$$

Example of R_n with magic constant 2^{130} is depicted as below by substituting $m = 2$ in the above formula in (4),

$$\begin{bmatrix} 2^1 & 2^{60} & 2^{56} & 2^{13} \\ 2^{63} & 2^6 & 2^{10} & 2^{51} \\ 2^{62} & 2^7 & 2^{11} & 2^{50} \\ 2^4 & 2^{57} & 2^{53} & 2^{16} \end{bmatrix} \quad \begin{bmatrix} 2^{48} & 2^{21} & 2^{25} & 2^{36} \\ 2^{18} & 2^{43} & 2^{39} & 2^{30} \\ 2^{19} & 2^{42} & 2^{38} & 2^{31} \\ 2^{45} & 2^{24} & 2^{28} & 2^{33} \end{bmatrix}$$

Layer 1 Layer 2

$$\begin{bmatrix} 2^{32} & 2^{37} & 2^{41} & 2^{20} \\ 2^{34} & 2^{27} & 2^{23} & 2^{46} \\ 2^{35} & 2^{26} & 2^{22} & 2^{47} \\ 2^{29} & 2^{40} & 2^{44} & 2^{17} \end{bmatrix} \quad \begin{bmatrix} 2^{49} & 2^{12} & 2^{8} & 2^{61} \\ 2^{15} & 2^{54} & 2^{58} & 2^{3} \\ 2^{14} & 2^{55} & 2^{59} & 2^{2} \\ 2^{52} & 2^{9} & 2^{5} & 2^{64} \end{bmatrix}$$

Layer 3 Layer 4

3 Conclusion

In this paper, it can be concluded that there are various ways of constructing multiplicative magic cubes of several orders from the existing additive magic cubes. It is also explained very well that the change in behavior of the elements of MMC including its magic constant on making some changes in the variables used in the construction formulas. Hence, it becomes possible to obtain a MMC consisting of either even or odd or composite numbers as its elements.

Conflict of interests The authors declare that they have no known competing financial interests or personal relationships that could have appeared to influence the work reported in this paper.

Acknowledgements One of the authors, Narbda Rani is thankful to Sant Longowal Institute of Engineering and Technology(SLIET), Longowal, Punjab, India, for providing financial support through fellowship.

References

1. Andrews, W.S.: Magic squares and cubes. Bull. Am. Math. Soc. **16**, 85–87 (1909). https://doi.org/10.1090/S0002-9904-1909-01866-X
2. Livinus, U.U., Terry, L.B.: A generalization of Trenkler's magic cubes formula. Recreat. Math. Mag. **8**, 39–45 (2018). https://doi.org/10.1515/rmm-2017-0019
3. Michel, R., Taubenfeld, G., Berman, A.: A connection between random variables and latin k-cubes. Discrete Math. **146**, 313–320 (1995). https://doi.org/10.1016/0012-365X(94)00073-7
4. Trenkler, M.: A construction of magic cubes. Math. Gazette **84**(499), 36–41 (2000). https://doi.org/10.2307/362147
5. Trenkler, M.: On Additive and Multiplicative Magic Cubes. Jan Dlugosz University of Czestochowa, Scientific Issues, Mathematics XIII (2008)
6. Trenkler, M.: An algorithm for making magic cubes. πME J. **12**(2), 105–106 (2005)

Novel q-Rung Orthopair Fuzzy Hamacher Dual Muirhead Mean Operator for Multi-attribute Decision-Making

Sukhwinder Singh Rawat and Komal

Abstract Real-life multi-attribute decision-making (MADM) has some major issues related to the space of the problem, inter-dependency among attributes, flexibility in the aggregation process, etc. So, our objective is to deal with these issues by adopting suitable tools and techniques like the q-rung orthopair fuzzy set (q-ROFS) for handling space-related difficulty. Dual Muirhead mean (DMM) is applied to address the inter-dependency among attributes, and for a flexible aggregation process, the Hamacher t-norm (TN) and t-conorm (TCN) are utilised. By fusing these approaches, this paper proposes two novel aggregation operators (AOs) named q-rung orthopair fuzzy Hamacher dual Muirhead mean (q-ROFHDMM) and q-rung orthopair fuzzy Hamacher weighted dual Muirhead mean (q-ROFHWDMM) operators. The essential properties of these AOs and special cases are explored as well. Finally, the q-ROFHWDMM operator has been used to construct a MADM method. The study also examines a practical example of selecting an enterprise resource planning (ERP) system, as well as sensitive and comparative analysis.

Keywords Dual Muirhead mean · Hamacher t-norm and t-conorm · Multi-attribute decision-making · q-Rung orthopair fuzzy set

1 Introduction

MADM is a prominent technique that is used to find the best option from a set of available options that depends on various attributes. Several MADM techniques exist in the literature to handle real-life MADM problems. Most of the real-life MADM problems have some common issues that need to be resolved for meaningful and realistic decision-making (DM). Among many, two major challenges faced by decision-makers are (i) expressing the assessment values of an alternative with respect to multiple attributes and (ii) considering the interactional behaviour of these attributes

S. S. Rawat (✉) · Komal
Doon University, Dehradun, India
e-mail: sukhwinderawat@gmail.com

R. K. Sharma et al. (eds.), *Frontiers in Industrial and Applied Mathematics*, Springer Proceedings in Mathematics & Statistics 410,
https://doi.org/10.1007/978-981-19-7272-0_8

in the DM process. To address the first problem, Zadeh introduced the notion of the fuzzy set [1], which assigns a membership degree to every element in order to express the impreciseness and vagueness of that element in the set. But the Zadeh fuzzy set does not address the sense of dissatisfaction. As a result, Atanassov [2] introduced intuitionistic fuzzy set (IFS) in 1986, which used both membership degree(μ) and non-membership degree (ν) with conditions $\mu, \nu \in [0, 1], 0 \leq \mu + \nu \leq 1$. In 2013, Yager discovered that the condition $0 \leq \mu + \nu \leq 1$ (IFS) on μ and ν is violated in many real-life DM problems. To overcome this drawback, Yager [3] extended the space of intuitionistic fuzzy numbers and proposed the Pythagorean fuzzy set (PFS) by making use of the conditions $0 \leq \mu, \nu \leq 1;\ 0 \leq \mu^2 + \nu^2 \leq 1$. Further, it is observed by many researchers that there are still many real-life DM problems in which assessment values (μ, ν) violate the PFS condition. For example, if (0.8, 0.7) is the assessment data provided by the decision-maker, then we get $0.8^2 + 0.7^2 \geq 1$. Therefore, more extended decision space is required. To further extend the decision space of fuzzy information (μ and ν), a generalised orthopair fuzzy set, i.e. q-ROFS, has been introduced by Yager in 2017 [4] . Its membership and non-membership degrees satisfy the conditions $\mu, \nu \in [0, 1];\ 0 \leq \mu^q + \nu^q \leq 1;\ q \geq 1$. As the AOs-based MADM approaches provide both comprehensive values and ranking orders of the alternatives, and also the DM process of these approaches is more intuitive than the classical ones such as TOPSIS, AHP, TODIM, PROMETHEE, etc [5]. Several AOs and their utilisation provide various MADM methods for q-rung orthopair fuzzy numbers (q-ROFNs). For instant, Liu and Wang [6] developed weighted geometric (WG) and weighted average (WA) operators; Liu and Liu [7] proposed Bonferroni mean (BM) and geometric BM (GBM) operators; Wei et al. [8] introduced generalised Heronian mean and geometric Heronian mean operators; Wei et al. [9] developed Maclaurin symmetric mean (MSM) and geometric MSM (GMSM) operators. Rawat and Komal recently used Muirhead mean (MM), Hamacher TN and TCN for q-ROFNs and introduced some AOs as well as a MADM approach based on them. The MM and DMM are aggregation functions which address the inter-dependency of multiple attributes through the correlation of their arguments for every permutation [11]. Various well-known means, like arithmetic mean (AM), geometric mean (GM), GBM and GMSM, are some special cases of DMM [12]. Hamacher TN and TCN are conjunctive and disjunctive aggregation functions [13]. Also, they are strictly decreasing and increasing with parameter γ, respectively, which helps to model conjunction and disjunction among arguments and provides flexibility in the aggregation process [14]. Consequently, many researchers utilised Hamacher TN and TCN-based arithmetic operations to develop some AOs for various fuzzy numbers like intuitionistic fuzzy numbers (IFNs), Pythagorean fuzzy numbers (PFNs), complex IFNs and q-ROFNs [15–18].

The focus of this article is to develop some novel Hamacher TN and TCN-based DMM operators for generalised orthopair fuzzy numbers. This fusion of Hamacher norms and DMM operator provides both interrelationship among multiple attributes and flexible aggregation process due to the additional parameter γ in Hamacher norms. The structure of the paper is as follows: In Sect. 2, definitions of q-ROFS, Hamacher TN and TCN, MM and DMM operators are discussed briefly. Section

3 introduces the q-ROFHDMM and q-ROFHWDMM operators with their essential properties and special cases. Further, in Sect. 4, the q-ROFHWDMM operator-based MADM approach has been developed, and a real-life DM problem has been examined through this. This section also provides sensitive and comparative analyses. Finally, some concluding remarks are given in Sect. 5.

2 Preliminaries

2.1 q-Rung Orthopair Fuzzy Set (q-ROFS)

Definition 1 ([4]) *The q-ROFS $\Im$ on a universal set U is defined as*

$$\Im = \{\langle x, (\mu_{\Im}(x), \nu_{\Im}(x))\rangle \mid x \in U\} \tag{1}$$

where $\mu_{\Im}(x) : U \to [0, 1]$ is membership and $\nu_{\Im}(x) : U \to [0, 1]$ is non-membership functions that holds, $0 \le (\mu_{\Im}(x))^q + (\nu_{\Im}(x))^q \le 1$ for all $q \ge 1$. The degree of hesitancy of x in $\Im$ is defined as $\pi_{\Im}(x) = (1 - (\mu_{\Im}(x))^q - (\nu_{\Im}(x))^q)^{1/q}$ and the q-rung orthopair fuzzy number (q-ROFN) can be written as $(\mu_{\Im}, \nu_{\Im})$.

Definition 2 ([6]) The basic arithmetic operations on any two q-ROFNs, $\aleph_1 = (\mu_1, \nu_1)$, and $\aleph_2 = (\mu_2, \nu_2)$, are as follows:

1. $\aleph_1 \oplus \aleph_2 = \left((\mu_1^q + \mu_2^q - \mu_1^q \mu_2^q)^{1/q}, \nu_1 \nu_2\right)$,
2. $\aleph_1 \otimes \aleph_2 = \left(\mu_1 \mu_2, (\nu_1^q + \nu_2^q - \nu_1^q \nu_2^q)^{1/q}\right)$,
3. $\lambda \aleph_1 = \left((1 - (1 - \mu_1^q)^\lambda)^{1/q}, \nu_1^\lambda\right)$,
4. $\aleph_1^\lambda = \left(\mu_1^\lambda, (1 - (1 - \nu_1^q)^\lambda)^{1/q}\right)$.

For comparing any two q-ROFNs, we have a score function (S) and an accuracy function (A) as follows:

Definition 3 ([6]) Let $\aleph = (\mu_\aleph, \nu_\aleph)$ be a q-ROFN, then the score value of $\aleph$ is obtained by $S(\aleph) \in [-1, 1]$ which is defined as

$$S(\aleph) = \mu_\aleph^q - \nu_\aleph^q \tag{2}$$

The accuracy value of $\aleph$ is obtained by $A(\aleph) \in [0, 1]$ which is defined as

$$A(\aleph) = \mu_\aleph^q + \nu_\aleph^q \tag{3}$$

Definition 4 For any two q-ROFNs say $\aleph = (\mu_\aleph, \nu_\aleph)$ and $\kappa = (\mu_\kappa, \nu_\kappa)$:

1. If $S(\aleph) > S(\kappa)$, then $\aleph \succ \kappa$
2. If $S(\aleph) = S(\kappa)$, then
   (a) If $A(\aleph) > A(\kappa)$, then $\aleph \succ \kappa$;
   (b) If $A(\aleph) = A(\kappa)$, then $\aleph = \kappa$.

2.2 Hamacher t-Norm (TN) and t-Conorm (TCN)

Hamacher TN (T) as product ($\otimes$) and Hamacher TCN (T^*) as sum ($\oplus$) are defined as follows [13]:

$$T(\imath, \jmath) = \imath \otimes \jmath = \frac{\imath\jmath}{\gamma + (1-\gamma)(\imath + \jmath - \imath\jmath)},$$
$$T^*(\imath, \jmath) = \imath \oplus \jmath = \frac{\imath + \jmath - \imath\jmath - (1-\gamma)\imath\jmath}{1 - (1-\gamma)(\imath\jmath)}; \ \gamma > 0.$$

For $\gamma = 1$, the Hamacher TN and TCN becomes algebraic TN and TCN:

$$T(\imath, \jmath) = \imath \otimes \jmath = \imath\jmath, \quad T^*(\imath, \jmath) = \imath \oplus \jmath = \imath + \jmath - \imath\jmath.$$

Similarly, for $\gamma = 2$, the Hamacher TN and TCN becomes Einstein TN and TCN:

$$T(\imath, \jmath) = \imath \otimes \jmath = \frac{\imath\jmath}{1 + (1-\imath)(1-\jmath)}, \quad T^*(\imath, \jmath) = \imath \oplus \jmath = \frac{\imath + \jmath}{1 + \imath\jmath}.$$

2.3 Hamacher Operations for q-ROFNs

If $\aleph_1 = (\mu_1, \nu_1)$ and $\aleph_2 = (\mu_2, \nu_2)$ are any two q-ROFNs and $\gamma > 0$, then the following arithmetic operations for q-ROFNs are defined using Hamacher TN and TCN [19]:

$$\aleph_1 \oplus \aleph_2 = \left(\left(\frac{(\mu_1)^q + (\mu_2)^q - (\mu_1)^q(\mu_2)^q - (1-\gamma)(\mu_1)^q(\mu_2)^q}{1 - (1-\gamma)(\mu_1)^q(\mu_2)^q} \right)^{1/q}, \right.$$
$$\left. \frac{\nu_1\nu_2}{\left(\gamma + (1-\gamma)\left((\nu_1)^q + (\nu_2)^q - (\nu_1)^q(\nu_2)^q\right)\right)^{1/q}} \right)$$

$$\aleph_1 \otimes \aleph_2 = \left(\frac{\mu_1\mu_2}{\left(\gamma + (1-\gamma)\left((\mu_1)^q + (\mu_2)^q - (\mu_1)^q(\mu_2)^q\right)\right)^{1/q}}, \right.$$
$$\left. \left(\frac{(\nu_1)^q + (\nu_2)^q - (\nu_1)^q(\nu_2)^q - (1-\gamma)(\nu_1)^q(\nu_2)^q}{1 - (1-\gamma)(\nu_1)^q(\nu_2)^q} \right)^{1/q} \right)$$

$$\lambda\aleph_1 = \left(\left(\frac{(1+(\gamma-1)\mu_1^q)^\lambda - (1-\mu_1^q)^\lambda}{(1+(\gamma-1)\mu_1^q)^\lambda + (\gamma-1)(1-\mu_1^q)^\lambda}\right)^{1/q}, \frac{(\gamma)^{1/q}\nu_1^\lambda}{\left((1+(\gamma-1)(1-\nu_1^q))^\lambda + (\gamma-1)(\nu_1^q)^\lambda\right)^{1/q}}\right)$$

$$\aleph_1^\lambda = \left(\frac{(\gamma)^{1/q}\mu_1^\lambda}{\left((1+(\gamma-1)(1-\mu_1^q))^\lambda + (\gamma-1)(\mu_1^q)^\lambda\right)^{1/q}}, \left(\frac{(1+(\gamma-1)\nu_1^q)^\lambda - (1-\nu_1^q)^\lambda}{(1+(\gamma-1)\nu_1^q)^\lambda + (\gamma-1)(1-\nu_1^q)^\lambda}\right)^{1/q}\right)$$

For $\gamma = 1$ Hamacher operations becomes algebraic operations and for $\gamma = 2$ they changes to Einstein operations.

2.4 *Muirhead Mean (MM)*

Definition 5 ([11]) The MM operator for n numbers say $\varsigma_1, \varsigma_2, ..., \varsigma_n$ and a parameter vector $P = (p_1, p_2, ..., p_n) \in \Re^n$ is defined as

$$\mathrm{MM}^P(\varsigma_1, \varsigma_2, ..., \varsigma_n) = \left(\frac{1}{n!}\sum_{\pi\in S_n}\prod_{j=1}^{n}\varsigma_{\pi(j)}^{p_j}\right)^{\frac{1}{\sum_{j=1}^n p_j}} \tag{4}$$

where S_n is the symmetric group of degree n.

2.5 *Dual Muirhead Mean (DMM)*

Definition 6 ([11]) The DMM operator for n numbers say $\varsigma_1, \varsigma_2, ..., \varsigma_n$ and a parameter vector $P = (p_1, p_2, ..., p_n) \in \Re^n$ is defined as

$$\mathrm{DMM}^P(\varsigma_1, \varsigma_2, ..., \varsigma_n) = \frac{1}{\sum_{j=1}^n p_j}\left(\prod_{\pi\in S_n}\sum_{j=1}^{n}p_j\varsigma_{\pi(j)}\right)^{\frac{1}{n!}} \tag{5}$$

where S_n is the symmetric group of degree n. Some special cases of the DMM operator for different values of P are as follows [12]:

1. If $P = (1, 0, 0..., 0)$, then the DMM operator becomes the GM operator

$$\mathrm{DMM}^{P}(\varsigma_1, \varsigma_2, ..., \varsigma_n) = \left(\prod_{i=1}^{n} \varsigma_i\right)^{\frac{1}{n}}.$$

2. If $P = (1, 1, ..., 1)$ or $(1/n, 1/n, ..., 1/n)$, then the DMM operator becomes the AM operator

$$\mathrm{DMM}^{(1,0,0,...,0)}(\varsigma_1, \varsigma_2, ..., \varsigma_n) = \frac{1}{n}\sum_{i=1}^{n} \varsigma_i.$$

3. If $P = (p_1, p_2, 0, 0, ..., 0)$, then the DMM operator becomes the GBM operator

$$\mathrm{DMM}^{(p_1,p_2,0,0,...,0)}(\varsigma_1, \varsigma_2, ..., \varsigma_n) = \frac{1}{p_1 + p_2} \prod_{i,j=1\ i\neq j}^{n} (p_1\varsigma_i + p_2\varsigma_j)^{\frac{1}{n(n-1)}}.$$

4. If $P = (\overbrace{1, 1, ..., 1}^{k}, \overbrace{0, 0, ..., 0}^{n-k})$, then the DMM operator becomes the GMSM operator

$$\mathrm{DMM}^{(\overbrace{1, 1, ..., 1}^{k}, \overbrace{0, 0, ..., 0}^{n-k})}(\varsigma_1, \varsigma_2, ..., \varsigma_n) = \frac{1}{k}\left(\prod_{1\leq i_1\leq ...\leq i_k\leq n} \sum_{j=1}^{k} \varsigma_{i_j}\right)^{\frac{1}{C_n^k}}.$$

3 q-Rung Orthopair Fuzzy Hamacher Dual Muirhead Mean Operators

3.1 The q-ROFHDMM Operator

Definition 7 Let $\varsigma_i = (\mu_i, \nu_i)$ be any q-ROFN and $P = (p_1, p_2, ..., p_n) \in \Re^n$ be a parameter vector such that $\sum_{j=1}^{n} p_j > 0$, then q-ROFHDMM operator on such n q-ROFNs is defined as

$$\text{q-ROFHDMM}^{P}(\varsigma_1, \varsigma_2, ..., \varsigma_n) = \frac{1}{\sum_{j=1}^{n} p_j}\left(\bigotimes_{\pi\in S_n} \bigoplus_{j=1}^{n} (p_j\varsigma_{\pi(j)})\right)^{\frac{1}{n!}} \tag{6}$$

where S_n is the symmetric group of degree n.

Theorem 1 *For any collection* $\{\varsigma_1, \varsigma_2, ..., \varsigma_n\}$ *of q-ROFNs, the aggregated value on applying the q-ROFHDMM operator is also a q-ROFN and it is defined as*

$$\left(\left(\frac{\left(\left(\prod_{\pi\in S_n}(\phi_2+(\gamma^2-1)\varphi_2)\right)^{\frac{1}{n!}}+(\gamma^2-1)\left(\prod_{\pi\in S_n}(\phi_2-\varphi_2)\right)^{\frac{1}{n!}}\right)^{\frac{1}{\sum_{j=1}^{n}p_j}}-\left(\left(\prod_{\pi\in S_n}(\phi_2+(\gamma^2-1)\varphi_2)\right)^{\frac{1}{n!}}-\left(\prod_{\pi\in S_n}(\phi_2-\varphi_2)\right)^{\frac{1}{n!}}\right)^{\frac{1}{\sum_{j=1}^{n}p_j}}}{\left(\left(\prod_{\pi\in S_n}(\phi_2+(\gamma^2-1)\varphi_2)\right)^{\frac{1}{n!}}+(\gamma^2-1)\left(\prod_{\pi\in S_n}(\phi_2-\varphi_2)\right)^{\frac{1}{n!}}\right)^{\frac{1}{\sum_{j=1}^{n}p_j}}+(\gamma-1)\left(\left(\prod_{\pi\in S_n}(\phi_2+(\gamma^2-1)\varphi_2)\right)^{\frac{1}{n!}}-\left(\prod_{\pi\in S_n}(\phi_2-\varphi_2)\right)^{\frac{1}{n!}}\right)^{\frac{1}{\sum_{j=1}^{n}p_j}}}\right)^{1/q},\right.$$
$$\left.\left(\frac{\gamma\left(\left(\prod_{\pi\in S_n}(\psi_2+(\gamma^2-1)\chi_2)\right)^{\frac{1}{n!}}-\left(\prod_{\pi\in S_n}(\psi_2-\chi_2)\right)^{\frac{1}{n!}}\right)^{\frac{1}{\sum_{j=1}^{n}p_j}}}{\left(\left(\prod_{\pi\in S_n}(\psi_2+(\gamma^2-1)\chi_2)\right)^{\frac{1}{n!}}+(\gamma^2-1)\left(\prod_{\pi\in S_n}(\psi_2-\chi_2)\right)^{\frac{1}{n!}}\right)^{\frac{1}{\sum_{j=1}^{n}p_j}}+(\gamma-1)\left(\left(\prod_{\pi\in S_n}(\psi_2+(\gamma^2-1)\chi_2)\right)^{\frac{1}{n!}}-\left(\prod_{\pi\in S_n}(\psi_2-\chi_2)\right)^{\frac{1}{n!}}\right)^{\frac{1}{\sum_{j=1}^{n}p_j}}}\right)^{1/q}\right) \quad (7)$$

where
$\phi_2 = \prod_{j=1}^{n}(1+(\gamma-1)\mu_{\pi(j)}^{q})^{p_j}$,
$\varphi_2 = \prod_{j=1}^{n}(1-\mu_{\pi(j)}^{q})^{p_j}$,
$\psi_2 = \prod_{j=1}^{n}\left(1+(\gamma-1)(1-\nu_{\pi(j)}^{q})\right)^{p_j}$,
$\chi_2 = \prod_{j=1}^{n}(\nu_{\pi(j)}^{q})^{p_j}$.

Proof The Eq. (7) is proved using mathematical induction and Hamacher operations of q-ROFNs, as discussed in Sect. 2.3:

$$p_j\varsigma_{\pi(j)} = \left(\left(\frac{(1+(\gamma-1)\mu_{\pi(j)}^{q})^{p_j}-(1-\mu_{\pi(j)}^{q})^{p_j}}{(1+(\gamma-1)\mu_{\pi(j)}^{q})^{p_j}+(\gamma-1)(1-\mu_{\pi(j)}^{q})^{p_j}}\right)^{1/q},\right.$$
$$\left.\frac{\gamma^{1/q}\nu_{\pi(j)}^{p_j}}{\left((1+(\gamma-1)(1-\nu_{\pi(j)}^{q}))^{p_j}+(\gamma-1)(\nu_{\pi(j)}^{q})^{p_j}\right)^{1/q}}\right)$$

Suppose we have two q-ROFNs $\varsigma_{\pi(1)} = (\mu_{\pi(1)}, \nu_{\pi(1)})$ and $\varsigma_{\pi(2)} = (\mu_{\pi(2)}, \nu_{\pi(2)})$, then

$$p_1\varsigma_{\pi(1)} \oplus p_2\varsigma_{\pi(2)}$$
$$= \left(\left(\frac{(1+(\gamma-1)\mu_{\pi(1)}^{q})^{p_1}-(1-\mu_{\pi(1)}^{q})^{p_1}}{(1+(\gamma-1)\mu_{\pi(1)}^{q})^{p_1}+(\gamma-1)(1-\mu_{\pi(1)}^{q})^{p_1}}\right)^{1/q},\right.$$
$$\left.\frac{\gamma^{1/q}\nu_{\pi(1)}^{p_1}}{\left((1+(\gamma-1)(1-\nu_{\pi(1)}^{q}))^{p_1}+(\gamma-1)(\nu_{\pi(1)}^{q})^{p_1}\right)^{1/q}}\right)$$
$$\oplus \left(\left(\frac{(1+(\gamma-1)\mu_{\pi(2)}^{q})^{p_2}-(1-\mu_{\pi(2)}^{q})^{p_2}}{(1+(\gamma-1)\mu_{\pi(2)}^{q})^{p_2}+(\gamma-1)(1-\mu_{\pi(2)}^{q})^{p_2}}\right)^{1/q},\right.$$

$$\left.\frac{\gamma^{1/q}\nu_{\pi(2)}^{p_2}}{\left((1+(\gamma-1)(1-\nu_{\pi(2)}^q))^{p_2}+(\gamma-1)(\nu_{\pi(2)}^q)^{p_2}\right)^{1/q}}\right)$$

$$=\left(\left(\frac{\prod_{j=1}^{2}(1+(\gamma-1)\mu_{\pi(j)}^q)^{p_j}-\prod_{j=1}^{2}(1-\mu_{\pi(j)}^q)^{p_j}}{\prod_{j=1}^{2}(1+(\gamma-1)\mu_{\pi(j)}^q)^{p_j}+(\gamma-1)\prod_{j=1}^{2}(1-\mu_{\pi(j)}^q)^{p_j}}\right)^{1/q},\right.$$

$$\left.\frac{\gamma^{1/q}\prod_{j=1}^{2}\nu_{\pi(j)}^{p_j}}{\left(\prod_{j=1}^{2}(1+(\gamma-1)(1-\nu_{\pi(j)}^q))^{p_j}+(\gamma-1)\prod_{j=1}^{2}(\nu_{\pi(j)}^q)^{p_j}\right)^{1/q}}\right)$$

Assuming that it is also true for $j = n-1$,

$$\sum_{j=1}^{n-1}p_j\varsigma_{\pi(j)}=\left(\left(\frac{\prod_{j=1}^{n-1}(1+(\gamma-1)\mu_{\pi(j)}^q)^{p_j}-\prod_{j=1}^{n-1}(1-\mu_{\pi(j)}^q)^{p_j}}{\prod_{j=1}^{n-1}(1+(\gamma-1)\mu_{\pi(j)}^q)^{p_j}+(\gamma-1)\prod_{j=1}^{n-1}(1-\mu_{\pi(j)}^q)^{p_j}}\right)^{1/q},\right.$$

$$\left.\frac{\gamma^{1/q}\prod_{j=1}^{n-1}\nu_{\pi(j)}^{p_j}}{\left(\prod_{j=1}^{n-1}(1+(\gamma-1)(1-\nu_{\pi(j)}^q))^{p_j}+(\gamma-1)\prod_{j=1}^{n-1}(\nu_{\pi(j)}^q)^{p_j}\right)^{1/q}}\right)$$

Now, the target is to show that this is also true for $j = n$.

$$\sum_{j=1}^{n-1}p_j\varsigma_{\pi(j)}\oplus p_n\varsigma_{\pi(n)}$$

$$=\left(\left(\frac{\prod_{j=1}^{n-1}(1+(\gamma-1)\mu_{\pi(j)}^q)^{p_j}-\prod_{j=1}^{n-1}(1-\mu_{\pi(j)}^q)^{p_j}}{\prod_{j=1}^{n-1}(1+(\gamma-1)\mu_{\pi(j)}^q)^{p_j}+(\gamma-1)\prod_{j=1}^{n-1}(1-\mu_{\pi(j)}^q)^{p_j}}\right)^{1/q},\right.$$

$$\left.\frac{\gamma^{1/q}\prod_{j=1}^{n-1}\nu_{\pi(j)}^{p_j}}{\left(\prod_{j=1}^{n-1}(1+(\gamma-1)(1-\nu_{\pi(j)}^q))^{p_j}+(\gamma-1)\prod_{j=1}^{n-1}(\nu_{\pi(j)}^q)^{p_j}\right)^{1/q}}\right)$$

$$\oplus\left(\left(\frac{(1+(\gamma-1)\mu_{\pi(n)}^q)^{p_n}-(1-\mu_{\pi(n)}^q)^{p_n}}{(1+(\gamma-1)\mu_{\pi(n)}^q)^{p_n}+(\gamma-1)(1-\mu_{\pi(n)}^q)^{p_n}}\right)^{1/q},\right.$$

$$\left.\frac{\gamma^{1/q}\nu_{\pi(n)}^{p_n}}{\left((1+(\gamma-1)(1-\nu_{\pi(n)}^q))^{p_n}+(\gamma-1)(\nu_{\pi(n)}^q)^{p_n}\right)^{1/q}}\right)$$

$$=\left(\left(\frac{\prod_{j=1}^{n}(1+(\gamma-1)\mu_{\pi(j)}^q)^{p_j}-\prod_{j=1}^{n}(1-\mu_{\pi(j)}^q)^{p_j}}{\prod_{j=1}^{n}(1+(\gamma-1)\mu_{\pi(j)}^q)^{p_j}+(\gamma-1)\prod_{j=1}^{n}(1-\mu_{\pi(j)}^q)^{p_j}}\right)^{1/q},\right.$$

$$\left.\frac{\gamma^{1/q}\prod_{j=1}^{n}\nu_{\pi(j)}^{p_j}}{\left(\prod_{j=1}^{n}(1+(\gamma-1)(1-\nu_{\pi(j)}^q))^{p_j}+(\gamma-1)\prod_{j=1}^{n}(\nu_{\pi(j)}^q)^{p_j}\right)^{1/q}}\right)$$

$$=\sum_{j=1}^{n}p_j\varsigma_{\pi(j)}$$

Then, taking the product of for all permutations, we get

$$\prod_{\pi\in S_n}\sum_{j=1}^{n} p_j \varsigma_{\pi(j)}$$

$$= \left(\left(\frac{\gamma \prod_{\pi\in S_n}(\phi_2 - \varphi_2)}{\prod_{\pi\in S_n}(\phi_2 + (\gamma^2-1)\varphi_2) + (\gamma-1)\prod_{\pi\in S_n}(\phi_2 - \varphi_2)} \right)^{1/q}, \left(\frac{\prod_{\pi\in S_n}(\psi_2 + (\gamma^2-1)\chi_2) - \prod_{\pi\in S_n}(\psi_2 - \chi_2)}{\prod_{\pi\in S_n}(\psi_2 + (\gamma^2-1)\chi_2) + (\gamma-1)\prod_{\pi\in S_n}(\psi_2 - \chi_2)} \right)^{1/q} \right)$$

and

$$\left(\prod_{\pi\in S_n}\sum_{j=1}^{n} \varsigma_{\pi(j)}^{p_j}\right)^{\frac{1}{n!}}$$

$$= \left(\left(\frac{\gamma \left(\prod_{\pi\in S_n}(\phi_2 - \varphi_2)\right)^{\frac{1}{n!}}}{\left(\prod_{\pi\in S_n}(\phi_2 + (\gamma^2-1)\varphi_2)\right)^{\frac{1}{n!}} + (\gamma-1)\left(\prod_{\pi\in S_n}(\phi_2 - \varphi_2)\right)^{\frac{1}{n!}}} \right)^{1/q}, \left(\frac{\left(\prod_{\pi\in S_n}(\psi_2 + (\gamma^2-1)\chi_2)\right)^{\frac{1}{n!}} - \left(\prod_{\pi\in S_n}(\psi_2 - \chi_2)\right)^{\frac{1}{n!}}}{\left(\prod_{\pi\in S_n}(\psi_2 + (\gamma^2-1)\chi_2)\right)^{\frac{1}{n!}} + (\gamma-1)\left(\prod_{\pi\in S_n}(\psi_2 - \chi_2)\right)^{\frac{1}{n!}}} \right)^{1/q} \right)$$

Finally,

$$\frac{1}{\sum_{j=1}^{n} p_j}\left(\prod_{\pi\in S_n}\sum_{j=1}^{n}(p_j \varsigma_{\pi(j)})\right)^{\frac{1}{n!}}$$

$$= \left(\left(\frac{\left(\left(\prod_{\pi \in S_n} (\phi_2 + (\gamma^2 - 1)\varphi_2) \right)^{\frac{1}{n!}} + (\gamma^2 - 1) \left(\prod_{\pi \in S_n} (\phi_2 - \varphi_2) \right)^{\frac{1}{n!}} \right)^{\frac{1}{\sum_{j=1}^{n} p_j}} - \left(\left(\prod_{\pi \in S_n} (\phi_2 + (\gamma^2 - 1)\varphi_2) \right)^{\frac{1}{n!}} - \left(\prod_{\pi \in S_n} (\phi_2 - \varphi_2) \right)^{\frac{1}{n!}} \right)^{\frac{1}{\sum_{j=1}^{n} p_j}}}{\left(\left(\prod_{\pi \in S_n} (\phi_2 + (\gamma^2 - 1)\varphi_2) \right)^{\frac{1}{n!}} + (\gamma^2 - 1) \left(\prod_{\pi \in S_n} (\phi_2 - \varphi_2) \right)^{\frac{1}{n!}} \right)^{\frac{1}{\sum_{j=1}^{n} p_j}} + (\gamma - 1) \left(\left(\prod_{\pi \in S_n} (\phi_2 + (\gamma^2 - 1)\varphi_2) \right)^{\frac{1}{n!}} - \left(\prod_{\pi \in S_n} (\phi_2 - \varphi_2) \right)^{\frac{1}{n!}} \right)^{\frac{1}{\sum_{j=1}^{n} p_j}}} \right)^{1/q}, \left(\frac{\gamma \left(\left(\prod_{\pi \in S_n} (\psi_2 + (\gamma^2 - 1)\chi_2) \right)^{\frac{1}{n!}} - \left(\prod_{\pi \in S_n} (\psi_2 - \chi_2) \right)^{\frac{1}{n!}} \right)^{\frac{1}{\sum_{j=1}^{n} p_j}}}{\left(\left(\prod_{\pi \in S_n} (\psi_2 + (\gamma^2 - 1)\chi_2) \right)^{\frac{1}{n!}} + (\gamma^2 - 1) \left(\prod_{\pi \in S_n} (\psi_2 - \chi_2) \right)^{\frac{1}{n!}} \right)^{\frac{1}{\sum_{j=1}^{n} p_j}} + (\gamma - 1) \left(\left(\prod_{\pi \in S_n} (\psi_2 + (\gamma^2 - 1)\chi_2) \right)^{\frac{1}{n!}} - \left(\prod_{\pi \in S_n} (\psi_2 - \chi_2) \right)^{\frac{1}{n!}} \right)^{\frac{1}{\sum_{j=1}^{n} p_j}}} \right)^{1/q} \right) \quad (8)$$

which illustrates that Eq. (7) holds.

Now, to show that Eq. (7) or (8) is a q-ROFN, we will prove the following:
(i) $0 \leq \mu' \leq 1$
(ii) $0 \leq \nu' \leq 1$
(iii) $0 \leq (\mu')^q + (\nu')^q \leq 1$
where μ' is the membership degree and ν' is the non-membership degree of Eq. (8).

Proof (i) and (ii). For any $\gamma > 0$, $q \geq 1$ and $P \in \Re^n$ s.t. $\sum_{j=1}^{n} p_j > 0$, we have $\phi_2, \varphi_2, \psi_2, \chi_2 \geq 0$ with $\phi_2 \geq \varphi_2, \psi_2 \geq \chi_2$ and the q-ROFN (μ', ν') can be written as $\left(\left(\frac{E^* - F^*}{E^* - F^* + \gamma F^*} \right)^{1/q}, \left(1 - \frac{G^* - H^*}{G^* - H^* + \gamma H^*} \right)^{1/q} \right)$. where

$E^* = \left(\left(\prod_{\pi \in S_n} (\phi_2 + (\gamma^2 - 1)\varphi_2) \right)^{\frac{1}{n!}} + (\gamma^2 - 1) \left(\prod_{\pi \in S_n} (\phi_2 - \varphi_2) \right)^{\frac{1}{n!}} \right)^{\frac{1}{\sum_{j=1}^{n} p_j}}$,

$F^* = \left(\left(\prod_{\pi \in S_n} (\phi_2 + (\gamma^2 - 1)\varphi_2) \right)^{\frac{1}{n!}} - \left(\prod_{\pi \in S_n} (\phi_2 - \varphi_2) \right)^{\frac{1}{n!}} \right)^{\frac{1}{\sum_{j=1}^{n} p_j}}$,

$G^* = \left(\left(\prod_{\pi \in S_n} (\psi_2 + (\gamma^2 - 1)\chi_2) \right)^{\frac{1}{n!}} + (\gamma^2 - 1) \left(\prod_{\pi \in S_n} (\psi_2 - \chi_2) \right)^{\frac{1}{n!}} \right)^{\frac{1}{\sum_{j=1}^{n} p_j}}$,

$H^* = \left(\left(\prod_{\pi \in S_n} (\psi_2 + (\gamma^2 - 1)\chi_2) \right)^{\frac{1}{n!}} - \left(\prod_{\pi \in S_n} (\psi_2 - \chi_2) \right)^{\frac{1}{n!}} \right)^{\frac{1}{\sum_{j=1}^{n} p_j}}$.

Since $E^*, F^*, G^*, H^* \geq 0$ s.t. $E^* \geq F^*$ and $G^* \geq H^*$. Therefore, it is easy to show that μ' and ν' satisfy the conditions (i) and (ii), respectively.

Proof (iii). Conditions (i) and (ii) $\Rightarrow 0 \leq (\mu')^q + (\nu')^q$. For $(\mu')^q + (\nu')^q \leq 1$, we know that $\mu_{\pi(j)}^q + \nu_{\pi(j)}^q \leq 1$ or $\mu_{\pi(j)}^q \leq 1 - \nu_{\pi(j)}^q$. Now by using $\mu_{\pi(j)}^q \leq 1 - \nu_{\pi(j)}^q$ and Eq. (8) for μ' and ν', we will get

$$(\mu')^q + (\nu')^q \leq 1. \qquad \text{Q.E.D}$$

Some important properties such as idempotency, monotonicity, boundedness and commutativity of the q-ROFHDMM operator are given below.

Property 1 (Idempotency) *If all the considered q-ROFNs are equal, that is,* $\varsigma_i = \varsigma = (\mu, \nu)$ *for all* $i = 1, 2, ..., n$, *then*

$$\text{q-ROFHDMM}^P(\varsigma_1, \varsigma_2, ..., \varsigma_n) = \varsigma = (\mu, \nu).$$

Property 2 (Monotonicity) *If $\varsigma_i = (\mu_i, \nu_i)$ and $\varsigma_i' = (\mu_i', \nu_i')$ for $i = 1, 2, ..., n$ are any two collection of q-ROFNs s.t. $\mu_i \leq \mu_i'$, $\nu_i \geq \nu_i'$ for all i, then*

$$q\text{-}ROFHDMM^P(\varsigma_1, \varsigma_2, ..., \varsigma_n) \leq q\text{-}ROFHDMM^P(\varsigma_1', \varsigma_2', ..., \varsigma_n').$$

Property 3 (Boundedness) *For any collection $\varsigma_i = (\mu_i, \nu_i)$ for $i = 1, 2, ..., n$ of q-ROFNs, if $\varsigma^- = \left(\min_{i=1}^{n}(\mu_i), \max_{i=1}^{n}(\nu_i)\right)$ and $\varsigma^+ = \left(\max_{i=1}^{n}(\mu_i), \min_{i=1}^{n}(\nu_i)\right)$, then*

$$\varsigma^- \leq q\text{-}ROFHDMM^P(\varsigma_1, \varsigma_2, ..., \varsigma_n) \leq \varsigma^+.$$

Property 4 (Commutativity) *For any permutation of $\varsigma_i (i = 1, 2, ..., n)$ say $\varsigma_i' (i = 1, 2, ..., n)$, the aggregated value remains unaffected. That is*

$$q\text{-}ROFHDMM^P(\varsigma_1', \varsigma_2', ..., \varsigma_n') = q\text{-}ROFHDMM^P(\varsigma_1, \varsigma_2, ..., \varsigma_n).$$

Now, some special cases of the q-ROFHDMM operator w.r.t γ and P are discussed hereafter.

1. For $\gamma = 1$, q-ROFHDMM operator becomes q-rung orthopair fuzzy dual Muirhead mean(q-ROFDMM) operator.
2. For $\gamma = 2$, q-ROFHDMM operator becomes q-rung orthopair fuzzy Einstein dual Muirhead mean(q-ROFEDMM) operator.
3. For $P = (1, 0, 0, ..., 0)$, q-ROFHDMM operator becomes q-rung orthopair fuzzy Hamacher geometric averaging(q-ROFHG) operator.
4. For $P = (1, 1, ..., 1)$ or $P = (1/n, 1/n, ..., 1/n)$, q-ROFHDMM operator becomes q-rung orthopair fuzzy Hamacher arithmetic averaging(q-ROFHA) operator.
5. For $P = (1, 1, 0, 0, ..., 0)$, q-ROFHDMM operator becomes q-rung orthopair fuzzy Hamacher geometric Bonferroni mean(q-ROFHGBM) operator.
6. For $P = (\overbrace{1, 1, ..., 1}^{k}, \overbrace{0, 0, ..., 0}^{n-k})$, q-ROFHDMM operator become q-rung orthopair fuzzy Hamacher geometric Maclaurin symmetric mean(q-ROFHGMSM) operator.

3.2 The q-ROFHWDMM Operator

Definition 8 Consider a set of q-ROFNs $\{\varsigma_1, \varsigma_2, ..., \varsigma_n\}$, a parameter vector $P = (p_1, p_2, ..., p_n) \in \Re^n$ such that $\sum_{j=1}^{n} p_j > 0$, and a weight vector $\omega = (\omega_1, \omega_2, ..., \omega_n)^T$, where $\omega_i \in [0, 1]$ corresponding to ς_i such that $\sum_{i=1}^{n} \omega_i = 1$. The q-ROFHWDMM operator is thus defined as

$$\text{q-ROFHWDMM}^P(\varsigma_1, \varsigma_2, \ldots, \varsigma_n) = \frac{1}{\sum_{j=1}^{n} p_j}\left(\bigotimes_{\pi \in S_n}\bigoplus_{j=1}^{n}\left(p_j \varsigma_{\pi(j)}^{nw_{\pi(j)}}\right)\right)^{\frac{1}{n!}}$$

where S_n is the symmetric group of degree n.

Theorem 2 *For any collection $\{\varsigma_1, \varsigma_2, \ldots, \varsigma_n\}$ of q-ROFNs, the aggregated value using q-ROFHWDMM operator is also a q-ROFN and it is defined as*

$$\text{q-ROFHWDMM}^P(\varsigma_1, \varsigma_2, \ldots, \varsigma_n) =$$

$$\left(\begin{array}{c}\left(\dfrac{\left(\left(\prod\limits_{\pi\in S_n}(\phi_2'+(\gamma^2-1)\varphi_2')\right)^{\frac{1}{n!}}+(\gamma^2-1)\left(\prod\limits_{\pi\in S_n}(\phi_2'-\varphi_2')\right)^{\frac{1}{n!}}\right)^{\frac{1}{\sum_{j=1}^{n}p_j}}-\left(\left(\prod\limits_{\pi\in S_n}(\phi_2'+(\gamma^2-1)\varphi_2')\right)^{\frac{1}{n!}}-\left(\prod\limits_{\pi\in S_n}(\phi_2'-\varphi_2')\right)^{\frac{1}{n!}}\right)^{\frac{1}{\sum_{j=1}^{n}p_j}}}{\left(\left(\prod\limits_{\pi\in S_n}(\phi_2'+(\gamma^2-1)\varphi_2')\right)^{\frac{1}{n!}}+(\gamma^2-1)\left(\prod\limits_{\pi\in S_n}(\phi_2'-\varphi_2')\right)^{\frac{1}{n!}}\right)^{\frac{1}{\sum_{j=1}^{n}p_j}}+(\gamma-1)\left(\left(\prod\limits_{\pi\in S_n}(\phi_2'+(\gamma^2-1)\varphi_2')\right)^{\frac{1}{n!}}-\left(\prod\limits_{\pi\in S_n}(\phi_2'-\varphi_2')\right)^{\frac{1}{n!}}\right)^{\frac{1}{\sum_{j=1}^{n}p_j}}}\right)^{1/q}, \\ \left(\dfrac{\gamma\left(\left(\prod\limits_{\pi\in S_n}(\psi_2'+(\gamma^2-1)\chi_2')\right)^{\frac{1}{n!}}-\left(\prod\limits_{\pi\in S_n}(\psi_2'-\chi_2')\right)^{\frac{1}{n!}}\right)^{\frac{1}{\sum_{j=1}^{n}p_j}}}{\left(\left(\prod\limits_{\pi\in S_n}(\psi_2'+(\gamma^2-1)\chi_2')\right)^{\frac{1}{n!}}+(\gamma^2-1)\left(\prod\limits_{\pi\in S_n}(\psi_2-\chi_2')\right)^{\frac{1}{n!}}\right)^{\frac{1}{\sum_{j=1}^{n}p_j}}+(\gamma-1)\left(\left(\prod\limits_{\pi\in S_n}(\psi_2'+(\gamma^2-1)\chi_2')\right)^{\frac{1}{n!}}-\left(\prod\limits_{\pi\in S_n}(\psi_2'-\chi_2')\right)^{\frac{1}{n!}}\right)^{\frac{1}{\sum_{j=1}^{n}p_j}}}\right)^{1/q}\end{array}\right)$$

where

$$\phi_2' = \prod_{j=1}^{n}\left(\left(1+(\gamma-1)(1-\mu_{\pi(j)}^q)\right)^{nw_{\pi(j)}} + (\gamma^2-1)\left(\mu_{\pi(j)}^q\right)^{nw_{\pi(j)}}\right)^{p_j},$$

$$\varphi_2' = \prod_{j=1}^{n}\left(\left(1+(\gamma-1)(1-\mu_{\pi(j)}^q)\right)^{nw_{\pi(j)}} - \left(\mu_{\pi(j)}^q\right)^{nw_{\pi(j)}}\right)^{p_j},$$

$$\psi_2' = \prod_{j=1}^{n}\left(\left(1+(\gamma-1)\nu_{\pi(j)}^q\right)^{nw_{\pi(j)}} + (\gamma^2-1)\left(1-\nu_{\pi(j)}^q\right)^{nw_{\pi(j)}}\right)^{p_j},$$

$$\chi_2' = \prod_{j=1}^{n}\left(\left(1+(\gamma-1)\nu_{\pi(j)}^q\right)^{nw_{\pi(j)}} - \left(1-\nu_{\pi(j)}^q\right)^{nw_{\pi(j)}}\right)^{p_j}.$$

Corollary 1 *The q-ROFHDMM is a specific case of the q-ROFHWDMM operator. That is, for $w = (1/n, 1/n, \ldots, 1/n)^T$, the q-ROFHWDMM operator reduces to q-ROFHDMM operator.*

The two fundamental properties, monotonicity and boundedness, of the q-ROFHWDMM operator are discussed hereafter.

Property 5 (Monotonicity) *If $\varsigma_i = (\mu_i, \nu_i)$ and $\varsigma_i' = (\mu_i', \nu_i')$ for $i = 1, 2, \ldots, n$ are any two collection of q-ROFNs s.t. $\mu_i \le \mu_i'$, $\nu_i \ge \nu_i'$ for all i, then*

$$\textit{q-ROFHWDMM}^P(\varsigma_1, \varsigma_2, \ldots, \varsigma_n) \le \textit{q-ROFHWDMM}^P(\varsigma_1', \varsigma_2', \ldots, \varsigma_n').$$

Property 6 (Boundedness) *For any collection $\varsigma_i = (\mu_i, \nu_i)$ for $i = 1, 2, \ldots, n$ of q-ROFNs, if $\varsigma^- = \left(\min\limits_{i=1}^{n}(\mu_i), \max\limits_{i=1}^{n}(\nu_i)\right)$ and $\varsigma^+ = \left(\max\limits_{i=1}^{n}(\mu_i), \min\limits_{i=1}^{n}(\nu_i)\right)$, then*

$$\varsigma^- \le \textit{q-ROFHWDMM}^P(\varsigma_1, \varsigma_2, \ldots, \varsigma_n) \le \varsigma^+.$$

4 Application of the Proposed AOs on MADM

4.1 MADM Method Based on the q-ROFHWDMM Operator

Now we'll develop a MADM method that uses the q-ROFHWDMM operator. To implement this, let us take $\Im = \{\Im_1, \Im_2, ..., \Im_m\}$ be the set of all feasible alternatives, which are being evaluated on the basis of n-attributes $\{\zeta_1, \zeta_2, ..., \zeta_n\}$ with the weight vector $\omega = \{\omega_1, \omega_2, ..., \omega_n\}$ such that $\omega_j \in [0, 1]$ and $\sum_{j=1}^{w} \omega_j = 1$. Let $\hbar = (\aleph_{ij})_{m \times n}$ be the decision matrix, where $\aleph_{ij} = (\mu_{ij}, \nu_{ij})$ is an assessment value (as q-ROFN) of an alternative $\Im_i$ with respect to the attribute ζ_j.

The step-by-step approach of this generalised orthopair fuzzy MADM method is given hereafter.

Step 1. Normalisation of $\hbar$:

Generally, two types of attributes are involved in any decision matrix: cost and benefit types. To consider these attributes simultaneously, we need to normalise the decision matrix as follows:

$$\aleph_{ij} = (\mu_{ij}, \nu_{ij}) = \begin{cases} (\mu_{ij}, \nu_{ij}), & \text{for benefit attributes } \zeta_j \\ (\nu_{ij}, \mu_{ij}), & \text{for cost attributes } \zeta_j \end{cases}$$

Step 2. Evaluate comprehensive values:

To get a comprehensive value $\aleph_i$ for each alternative $\Im_i$, apply the proposed q-ROFHWDMM operator which aggregates the assessment values $\aleph_{ij} (j = 1, 2, ..., n)$.

$$\aleph_i = \text{q-ROFHWDMM}(\aleph_{i1}, \aleph_{i2}, ..., \aleph_{in})$$

Step 3. Find the score and accuracy values:

First, compute the $S(\aleph_i)$ for each $\aleph_i (i = 1, 2, ..., m)$. Now if any two or more score values match, then calculate their accuracy values $A(\aleph_i)$ according to the Eqs. (2) and (3), respectively.

Step 4. Rank the alternatives:

Now use definition 4 to rank the alternatives $(\Im_i)$ and choose the most appealing one.

4.2 An Illustrative Example

Now, a practical MADM problem adopted from [8] is presented to illustrate the applicability of the developed MADM technique. The target of this MADM problem is to help an organisation install an ERP system. For that, five viable ERP systems have been chosen by the project team. $\Im_i (i = 1, 2, 3, 4, 5)$ i.e. 5-alternatives and 4-attributes $\zeta_j (j = 1, 2, 3, 4)$ that are (1) function and technology ζ_1; (2) strategic fitness ζ_2; (3) vendor's ability ζ_3; (4) vendor's reputation ζ_4 and $\omega = (0.2, 0.1, 0.3, 0.4)$

Table 1 Decision matrix ($\hbar$) taken from [8]

Alternative	Attributes			
	ζ_1	ζ_2	ζ_3	ζ_4
$\Im_1$	(0.5, 0.8)	(0.6, 0.3)	(0.3, 0.6)	(0.5, 0.7)
$\Im_2$	(0.7, 0.5)	(0.7, 0.2)	(0.7, 0.2)	(0.4, 0.5)
$\Im_3$	(0.6, 0.4)	(0.5, 0.7)	(0.5, 0.3)	(0.6, 0.3)
$\Im_4$	(0.8, 0.1)	(0.6, 0.3)	(0.3, 0.4)	(0.5, 0.6)
$\Im_5$	(0.6, 0.4)	(0.4, 0.8)	(0.7, 0.6)	(0.5, 0.8)

Table 2 Final results of all $\Im_i$

Alternatives	Comprehensive values	Score values	Ranking
$\Im_1$	(0.6118, 0.5381)	0.0732	4
$\Im_2$	(0.7275, 0.3030)	0.3572	1
$\Im_3$	(0.6210, 0.3863)	0.1818	3
$\Im_4$	(0.7033, 0.2781)	0.3264	2
$\Im_5$	(0.6240, 0.6010)	0.0258	5

denotes the weight vector of these qualities. The associated information of these five alternative with respect to four attributes is given in the form of a decision matrix $\hbar = (\aleph_{ij})_{5\times 4}$ of q-ROFNs as provided in the Table 1.

In order to achieve the most suitable alternative, we utilised the MADM method given in Sect. 4.1.

Step 1. Normalisation of $\hbar$:
Here, the given decision matrix ($\hbar$) does not need to be normalised, as all four ζ_j are benefit type.
Step 2. Evaluate comprehensive values:
Now apply q-ROFHWDMM operator and compute the comprehensive values $\aleph_i (i = 1, 2, 3, 4, 5)$ for all alternatives $\Im_i (i = 1, 2, 3, 4, 5)$ using decision matrix $\hbar$ (Table 1), for $q = 3$, $\gamma = 1$, and $P = (1, 1, 1, 1)$. The comprehensive values are presented in column 2 of Table 2.
Step 3. Find the score and accuracy values:
For each $\aleph_i (i = 1, 2, 3, 4, 5)$, compute score value $S(\aleph_i)$. Computed score values are presented in column 3 of Table 2.
Step 4. Ranking of alternatives:
Finally, based on the calculated $S(\aleph_i)$, rank the alternatives $\Im_i$ as discussed in step 4 of section 4.1 and result are presented in column 4 of Table 2. From Table 2, it's clear that alternative A_2 is the best alternative among possible potential ERP systems. The final choice of alternative may depend on the parameters' values q, γ, P and AOs

applied. Therefore, it is obvious to investigate the efficiency of the proposed method corresponding to the parameters' values selected and AOs used. Therefore, sections 4.3 and 4.4 discusses sensitivity analysis and comparative analysis, respectively.

4.3 Sensitivity Analysis

To investigate flexibility and capability of the proposed MADM method, a sensitivity analysis has been carried out by changing the parameter q, γ and then P one by one. The effects on the final result due to these variations are analysed and discussed hereafter.

Table 3 shows the variation in score values by assigning different integer values to $q \in [2, 10]$ and fixing the values of $\gamma = 1$ and $P = (1, 1, 1, 1)$. Similarly in Table 4, γ varies from 1 to 10; however, the other two parameters q and P are fixed as 3 and $(1, 1, 1, 1)$, respectively. From Tables 3 and 4, it is observable that, on increasing the value of parameters q (Table 3) and γ (Table 4), the score values and ranking results of some alternatives changes accordingly, which reflects the influence of these two parameters (q and γ) on the final decision. The parameter q not just provides the larger assessment space but also influences the final results. Similarly, the γ parameter makes the aggregation process more flexible and affects the final results. However, for the studied MADM problem, the best alternative obtained through all considered variations is unanimously $\Im_2$. Further, to examine the effect of interrelationship among attributes, different values of the parameter vector P were analysed on fixing the values of parameters q and γ as 3 and 1 respectively, and evaluated score values and ranking results are shown in Table 5. In this case, Table 5 shows that, on considering the interdependency of multiple attributes, the ranking results are slightly different from those in the case of no interaction. But the best alternative for all the considered variations of P of multiple interrelationships is $\Im_2$.

4.4 Comparative Analysis

To demonstrate the compatibility of the developed AOs, this section compares six existing AOs, q-ROFWA and q-ROFWG [6], q-ROFWBM [7], q-ROFGWHM and q-ROFWGHM [8], q-ROFWMSM [9], and one proposed AO (q-ROFHWDMM) under same q-ROFNs environment with $q = 3$. The q-ROFWA and q-ROFWG has no additional parameter other than q [6]. The q-ROFWBM operator takes into account the correlation between any two attributes [7], and its additional parameters are set to $s = 1, t = 1$. The selected values of their extra parameters for applying q-ROFGWHM and q-ROFWGHM operators are $\phi = 1,\ \varphi = 1$, and they assess the

Table 3 Results by varying q in q-ROFHWDMM operator

q	Score values ($S(\aleph_i)$)					Ranking orders
2	$S(\aleph_1) =$ 0.0663	$S(\aleph_2) =$ 0.4201	$S(\aleph_3) =$ 0.2260	$S(\aleph_4) =$ 0.3873	$S(\aleph_5) =$ 0.0315	$\Im_2 \succ \Im_4 \succ \Im_3 \succ \Im_1 \succ \Im_5$
3	$S(\aleph_1) =$ 0.0732	$S(\aleph_2) =$ 0.3572	$S(\aleph_3) =$ 0.1818	$S(\aleph_4) =$ 0.3264	$S(\aleph_5) =$ 0.0258	$\Im_2 \succ \Im_4 \succ \Im_3 \succ \Im_1 \succ \Im_5$
4	$S(\aleph_1) =$ 0.0749	$S(\aleph_2) =$ 0.2899	$S(\aleph_3) =$ 0.1362	$S(\aleph_4) =$ 0.2658	$S(\aleph_5) =$ 0.0226	$\Im_2 \succ \Im_4 \succ \Im_3 \succ \Im_1 \succ \Im_5$
5	$S(\aleph_1) =$ 0.0709	$S(\aleph_2) =$ 0.2326	$S(\aleph_3) =$ 0.0994	$S(\aleph_4) =$ 0.2150	$S(\aleph_5) =$ 0.0196	$\Im_2 \succ \Im_4 \succ \Im_3 \succ \Im_1 \succ \Im_5$
6	$S(\aleph_1) =$ 0.0636	$S(\aleph_2) =$ 0.1869	$S(\aleph_3) =$ 0.0720	$S(\aleph_4) =$ 0.1742	$S(\aleph_5) =$ 0.0164	$\Im_2 \succ \Im_4 \succ \Im_3 \succ \Im_1 \succ \Im_5$
7	$S(\aleph_1) =$ 0.0551	$S(\aleph_2) =$ 0.1511	$S(\aleph_3) =$ 0.0521	$S(\aleph_4) =$ 0.1416	$S(\aleph_5) =$ 0.0132	$\Im_2 \succ \Im_4 \succ \Im_1 \succ \Im_3 \succ \Im_5$
8	$S(\aleph_1) =$ 0.0467	$S(\aleph_2) =$ 0.1230	$S(\aleph_3) =$ 0.0379	$S(\aleph_4) =$ 0.1156	$S(\aleph_5) =$ 0.0103	$\Im_2 \succ \Im_4 \succ \Im_1 \succ \Im_3 \succ \Im_5$
9	$S(\aleph_1) =$ 0.0389	$S(\aleph_2) =$ 0.1009	$S(\aleph_3) =$ 0.0276	$S(\aleph_4) =$ 0.0945	$S(\aleph_5) =$ 0.0079	$\Im_2 \succ \Im_4 \succ \Im_1 \succ \Im_3 \succ \Im_5$
10	$S(\aleph_1) =$ 0.0322	$S(\aleph_2) =$ 0.0833	$S(\aleph_3) =$ 0.0202	$S(\aleph_4) =$ 0.0775	$S(\aleph_5) =$ 0.0059	$\Im_2 \succ \Im_4 \succ \Im_1 \succ \Im_3 \succ \Im_5$

correlation between any two attributes [8].The q-ROFWMSM operator takes into account interactions among any number of attributes [9], and its granularity parameter is set to $k = 2$, allowing it to consider correlation between two any attributes for that very same interactional behavior. To maintain the same operational behavior for the developed AO (q-ROFHWDMM) also, the selected values of γ and P are 1 and $(1, 1, 0, 0)$ respectively. Table 6 suggested that the best alternative and the worst alternative obtained from all the different operators under investigation are almost the same.

Table 4 Results by changing γ in q-ROFHWDMM operator

γ	Score values($S(\aleph_i)$)					Ranking orders
1	$S(\aleph_1) =$ 0.0732	$S(\aleph_2) =$ 0.3572	$S(\aleph_3) =$ 0.1818	$S(\aleph_4) =$ 0.3264	$S(\aleph_5) =$ 0.0258	$\Im_2 \succ \Im_4 \succ \Im_3 \succ \Im_1 \succ \Im_5$
2	$S(\aleph_1) =$ 0.0606	$S(\aleph_2) =$ 0.3533	$S(\aleph_3) =$ 0.1929	$S(\aleph_4) =$ 0.3184	$S(\aleph_5) =$ 0.0288	$\Im_2 \succ \Im_4 \succ \Im_3 \succ \Im_1 \succ \Im_5$
3	$S(\aleph_1) =$ 0.0508	$S(\aleph_2) =$ 0.3490	$S(\aleph_3) =$ 0.1968	$S(\aleph_4) =$ 0.3099	$S(\aleph_5) =$ 0.0299	$\Im_2 \succ \Im_4 \succ \Im_3 \succ \Im_1 \succ \Im_5$
4	$S(\aleph_1) =$ 0.0425	$S(\aleph_2) =$ 0.3451	$S(\aleph_3) =$ 0.1980	$S(\aleph_4) =$ 0.3021	$S(\aleph_5) =$ 0.0297	$\Im_2 \succ \Im_4 \succ \Im_3 \succ \Im_1 \succ \Im_5$
5	$S(\aleph_1) =$ 0.0353	$S(\aleph_2) =$ 0.3415	$S(\aleph_3) =$ 0.1981	$S(\aleph_4) =$ 0.2951	$S(\aleph_5) =$ 0.0288	$\Im_2 \succ \Im_4 \succ \Im_3 \succ \Im_1 \succ \Im_5$
6	$S(\aleph_1) =$ 0.0290	$S(\aleph_2) =$ 0.3384	$S(\aleph_3) =$ 0.1976	$S(\aleph_4) =$ 0.2889	$S(\aleph_5) =$ 0.0275	$\Im_2 \succ \Im_4 \succ \Im_3 \succ \Im_1 \succ \Im_5$
7	$S(\aleph_1) =$ 0.0234	$S(\aleph_2) =$ 0.3355	$S(\aleph_3) =$ 0.1970	$S(\aleph_4) =$ 0.2833	$S(\aleph_5) =$ 0.0261	$\Im_2 \succ \Im_4 \succ \Im_3 \succ \Im_5 \succ \Im_1$
8	$S(\aleph_1) =$ 0.0183	$S(\aleph_2) =$ 0.3329	$S(\aleph_3) =$ 0.1962	$S(\aleph_4) =$ 0.2782	$S(\aleph_5) =$ 0.0247	$\Im_2 \succ \Im_4 \succ \Im_3 \succ \Im_5 \succ \Im_1$
9	$S(\aleph_1) =$ 0.0137	$S(\aleph_2) =$ 0.3305	$S(\aleph_3) =$ 0.1954	$S(\aleph_4) =$ 0.2736	$S(\aleph_5) =$ 0.0233	$\Im_2 \succ \Im_4 \succ \Im_3 \succ \Im_5 \succ \Im_1$
10	$S(\aleph_1) =$ 0.0095	$S(\aleph_2) =$ 0.3284	$S(\aleph_3) =$ 0.1945	$S(\aleph_4) =$ 0.2694	$S(\aleph_5) =$ 0.0218	$\Im_2 \succ \Im_4 \succ \Im_3 \succ \Im_5 \succ \Im_1$

5 Conclusions

In the light of the interrelationship between multiple attributes in MADM problems, this paper proposes two novel AOs that are q-ROFHDMM and q-ROFHWDMM operators. These are Hamacher TN and TCN-inspired DMM operators under the q-ROFN environment. The advantage of combining Hamacher TN and TCN-inspired arithmetic procedures with DMM in proposed AOs is that they can capture not only the correlation between multiple attributes but also provide a flexible aggregation process due to γ and P in AOs. Some essential properties of these AOs are also given in the paper. The generality of the developed AOs is investigated through some special cases. Further, utilising the proposed AO (q-ROFHWDMM), a MADM approach

Table 5 Results by altering P in q-ROFHWDMM operator

Parameter vector(P)	Score values ($S(\aleph_i)$)					Ranking results
$(1, 0, 0, 0)$	$S(\aleph_1) = -0.2377$	$S(\aleph_2) = 0.0953$	$S(\aleph_3) = 0.1019$	$S(\aleph_4) = -0.0027$	$S(\aleph_5) = -0.1826$	$\Im_3 \succ \Im_2 \succ \Im_4 \succ \Im_5 \succ \Im_1$
$(2, 0, 0, 0)$	$S(\aleph_1) = -0.2813$	$S(\aleph_2) = 0.0467$	$S(\aleph_3) = 0.0815$	$S(\aleph_4) = -0.0710$	$S(\aleph_5) = -0.2462$	$\Im_3 \succ \Im_2 \succ \Im_4 \succ \Im_5 \succ \Im_1$
$(1, 1, 0, 0)$	$S(\aleph_1) = -0.1366$	$S(\aleph_2) = 0.2682$	$S(\aleph_3) = 0.1456$	$S(\aleph_4) = 0.1570$	$S(\aleph_5) = -0.0531$	$\Im_2 \succ \Im_4 \succ \Im_3 \succ \Im_5 \succ \Im_1$
$(1, 1, 1, 0)$	$S(\aleph_1) = -0.0516$	$S(\aleph_2) = 0.3236$	$S(\aleph_3) = 0.1688$	$S(\aleph_4) = 0.2814$	$S(\aleph_5) = -0.0139$	$\Im_2 \succ \Im_4 \succ \Im_3 \succ \Im_5 \succ \Im_1$
$(1, 1, 1, 1)$	$S(\aleph_1) = 0.0732$	$S(\aleph_2) = 0.3572$	$S(\aleph_3) = 0.1818$	$S(\aleph_4) = 0.3264$	$S(\aleph_5) = 0.0258$	$\Im_2 \succ \Im_4 \succ \Im_3 \succ \Im_1 \succ \Im_5$
$(2, 2, 2, 2)$	$S(\aleph_1) = 0.0732$	$S(\aleph_2) = 0.3572$	$S(\aleph_3) = 0.1818$	$S(\aleph_4) = 0.3264$	$S(\aleph_5) = 0.0258$	$\Im_2 \succ \Im_4 \succ \Im_3 \succ \Im_1 \succ \Im_5$
$(3, 3, 3, 3)$	$S(\aleph_1) = 0.0732$	$S(\aleph_2) = 0.3850$	$S(\aleph_3) = 0.1819$	$S(\aleph_4) = 0.3479$	$S(\aleph_5) = 0.0258$	$\Im_2 \succ \Im_4 \succ \Im_3 \succ \Im_1 \succ \Im_5$
$(4, 4, 4, 4)$	$S(\aleph_1) = 0.0732$	$S(\aleph_2) = 0.3850$	$S(\aleph_3) = 0.2395$	$S(\aleph_4) = 0.3479$	$S(\aleph_5) = 0.0258$	$\Im_2 \succ \Im_4 \succ \Im_3 \succ \Im_1 \succ \Im_5$
$(1, 2, 3, 4)$	$S(\aleph_1) = -0.0316$	$S(\aleph_2) = 0.3114$	$S(\aleph_3) = 0.1650$	$S(\aleph_4) = 0.2693$	$S(\aleph_5) = -0.0279$	$\Im_2 \succ \Im_4 \succ \Im_3 \succ \Im_5 \succ \Im_1$

has been developed. To show the applicability of the proposed approach, a MADM problem related to the selection of an ERP system has been solved. Sensitivity analysis for different variations and comparative analysis with six existing AOs have also been done to demonstrate the efficiency and compatibility of the proposed AOs. Our analysis and results conclude that the developed AOs are more flexible and general and can solve a wide range of real-life MADM problems. In future research, the proposed AOs may be further extended in various directions, including changing the uncertain environment, considering the heterogeneous relationship among attributes and so on.

Table 6 Score and ranking results for different AOs

AOs	Score values ($S(\aleph_i)$)					Ranking order
$q-ROFWA$ [6]	$S(\aleph_1) =$ −0.1443	$S(\aleph_2) =$ 0.2015	$S(\aleph_3) =$ 0.1394	$S(\aleph_4) =$ 0.1635	$S(\aleph_5) =$ −0.0515	$\Im_2 \succ \Im_4 \succ \Im_3 \succ \Im_5 \succ \Im_1$
$q-ROFWG$ [6]	$S(\aleph_1) =$ −0.2377	$S(\aleph_2) =$ 0.0953	$S(\aleph_3) =$ 0.1019	$S(\aleph_4) =$ −0.0027	$S(\aleph_5) =$ −0.1826	$\Im_3 \succ \Im_2 \succ \Im_4 \succ \Im_5 \succ \Im_1$
$q-ROFWBM^{1,1}$ [7]	$S(\aleph_1) =$ −0.6917	$S(\aleph_2) =$ −0.4263	$S(\aleph_3) =$ −0.4687	$S(\aleph_4) =$ −0.4372	$S(\aleph_5) =$ −0.6853	$\Im_2 \succ \Im_4 \succ \Im_3 \succ \Im_5 \succ \Im_1$
$q-ROFGWHM^{1,1}$ [8]	$S(\aleph_1) =$ −0.3070	$S(\aleph_2) =$ 0.0635	$S(\aleph_3) =$ 0.0412	$S(\aleph_4) =$ 0.0055	$S(\aleph_5) =$ −0.2345	$\Im_2 \succ \Im_4 \succ \Im_3 \succ \Im_5 \succ \Im_1$
$q-ROFWGHM^{1,1}$ [8]	$S(\aleph_1) =$ −0.0821	$S(\aleph_2) =$ 0.2208	$S(\aleph_3) =$ 0.2241	$S(\aleph_4) =$ 0.1228	$S(\aleph_5) =$ −0.0044	$\Im_3 \succ \Im_2 \succ \Im_4 \succ \Im_5 \succ \Im_1$
$q-ROFWMSM^{k=2}$ [9]	$S(\aleph_1) =$ 0.4898	$S(\aleph_2) =$ 0.6936	$S(\aleph_3) =$ 0.6421	$S(\aleph_4) =$ 0.6254	$S(\aleph_5) =$ 0.5812	$\Im_2 \succ \Im_3 \succ \Im_4 \succ \Im_5 \succ \Im_1$
$q-ROFHWDMM^{(1,1,0,0)}$	$S(\aleph_1) =$ −0.1366	$S(\aleph_2) =$ 0.2682	$S(\aleph_3) =$ 0.1456	$S(\aleph_4) =$ 0.1570	$S(\aleph_5) =$ −0.0531	$\Im_2 \succ \Im_4 \succ \Im_3 \succ \Im_5 \succ \Im_1$

References

1. Zadeh, L.A.: Fuzzy sets. Inf. Control **8**, 338–356 (1965)
2. Atanassov, K.T.: Intuitionistic fuzzy sets. Fuzzy Sets Syst. **20**, 87–96 (1986)
3. Yager, R.R.: Pythagorean fuzzy subsets. In: Proceeding of The Joint IFSA World Congress and NAFIPS Annual Meeting, Edmonton, Canada pp. 57–61 (2013)
4. Yager, R.R.: Generalized orthopair fuzzy sets. IEEE Trans. Fuzzy Syst. **25**(5), 1222–1230 (2017)
5. Zhenghai, A., Xu, Z., Yager, R.R., Ye, J.: q-Rung orthopair fuzzy integrals in the frame of continuous Archimedean t-norms and t-conorms and their application. IEEE Trans. Fuzzy Syst. **29**(5), 996–1007 (2020)
6. Liu, P., Wang, P.: Some q-rung orthopair fuzzy aggregation Operators and their applications to multiple-attribute decision making. Int. J. Intell. Syst. **32**(2), 259–280 (2018)
7. Liu, P., Liu, J.: Some q-rung orthopair fuzzy Bonferroni mean operators and their application to multi-attribute group decision making. Int. J. Intell. Syst. **33**(2), 315–347 (2018)
8. Wei, G., Gao, H., Wei, Y.: Some q-rung orthopair fuzzy Heronian mean operators in multiple attribute decision making. Int. J. Intell. Syst. **33**(7), 1426–1458 (2018)
9. Wei, G., Wei, C., Wang, J., Gao, H., Wei, Y.: Some q-rung orthopair fuzzy Maclaurin symmetric mean operators and their applications to potential evaluation of emerging technology commercialization. Int. J. Intell. Syst. **34**(1), 50–81 (2019)
10. Rawat, S.S., Komal.: Multiple attribute decision making based on q-rung orthopair fuzzy Hamacher Muirhead mean operators. Soft Comput. **26**, 2465–2487 (2022)
11. Muirhead, R.F.: Some methods applicable to identities and inequalities of symmetric algebraic functions of *n* letters. Proc. Edinb. Math. Soc. **21**(3), 144–162 (1902)
12. Qin, J., Liu, X.: 2-tuple linguistic Muirhead mean operators for multiple attribute group decision making and its application to supplier selection. Kybernetes **45**(1), 2–29 (2016)

13. Hamacher, H.: Uber logische verknunpfungenn unssharfer Aussagen und deren Zugenhorige Bewertungsfunktione Trappl, Klir, Riccardi (Eds.), Progress in Cybernatics and Systems Research, vol. 3, pp. 276–288 (1978)
14. Batyrshin, I., Kaynak, O.: Parametric classes of generalized conjunction and disjunction operations for fuzzy modeling. IEEE Trans. Fuzzy Syst. **7**(5), 586–596 (1999)
15. Huang, J.Y.: Intuitionistic fuzzy Hamacher aggregation operators and their application to multiple attribute decision making. J. Intell. Fuzzy Syst. **27**, 505–513 (2014)
16. Wu, S.J., Wei, G.W.: Pythagorean fuzzy Hamacher aggregation operators and their application to multiple attribute decision making. Int. J. Inf. Technol. Decis. Making **21**(3), 189–201 (2017)
17. Akram, M., Peng, X., Sattar, A.: A new decision-making model using complex intuitionistic fuzzy Hamacher aggregation operators. Soft. Comput. **25**, 7059–7086 (2021)
18. Darko, A.P., Liang, D.: Some q-rung orthopair fuzzy Hamacher aggregation operators and their application to multiple attribute group decision making with modified EDAS method. Eng. Appl. Artif. Intell. **87**, 103259 (2020)
19. Liu, P., Wang, P.: Multiple-attribute decision-making based on Archimedean Bonferroni operators of q-rung orthopair fuzzy numbers. IEEE Trans. Fuzzy Syst. **27**(5), 834–848 (2018)

Convective Instability in a Composite Nanofluid Layer Under Local Thermal Non-equilibrium

Anurag Srivastava and B. S. Bhadauria

Abstract Linear, as well as weakly non-linear, analyses have been done to understand the onset of convection and heat and mass transport in a composite nanofluid horizontal layer heated from below under LTNE (local thermal non-equilibrium) effect. Two different types of nanoparticles are assumed to be suspended in the base fluid. Both the nanoparticles and the base fluid are taken to be at different temperature, and therefore, three temperature model is used for LTNE. Thermal Rayleigh number is evaluated analytically using Galerkin's approach while non-linear analysis is done numerically. The effect of both top-heavy and bottom-heavy configurations of nanoparticles over convective instability is examined. It is found that the system is more stable in case of bottom-heavy configuration when compared to that of top-heavy case. Moreover, the effect of LTNE depends upon the concentration of nanoparticles significantly. A comparison between streamlines, isotherms and isohalines for both LTE (local thermal equilibrium) and LTNE cases is also presented.

Keywords Composite nanofluids · Local thermal non-equilibrium · Non-linear analysis · Free-free boundaries

Glossary

Latin Symbols

a	Horizontal wave number.
C_1, C_2	Nanoparticle volume fraction.
d	Dimensional layer depth.

A. Srivastava (✉) · B. S. Bhadauria
Department of Mathematics, Babasaheb Bhimrao Ambedkar University, Lucknow 226025, India
e-mail: anurag.091@gmail.com

R. K. Sharma et al. (eds.), *Frontiers in Industrial and Applied Mathematics*,
Springer Proceedings in Mathematics & Statistics 410,
https://doi.org/10.1007/978-981-19-7272-0_9

D_{B1}, D_{B2}	Brownian diffusion coefficients.
D_{T1}, D_{T2}	Thermophoretic diffusion coefficients.
g	Gravitational acceleration.
h	Dimensional interphase heat transfer coefficient.
p	Pressure.
t	Time.
T	Temperature.
v_1, v_2, v_3	Nanofluid velocity components.
(x, y, z)	Rectangular coordinates.

Greek Symbols

α	Proportionality factor.
κ	Effective thermal conductivity.
μ	Viscosity.
ρ	Density.
ρc	Volumetric heat capacity at constant pressure.
ψ	Dimensionfree stream function.

Subscripts and Superscripts

$(^\circ)$	Perturbation variable.
0	Reference value.
b	Basic value.
f	Liquid phase.
$p1, p2$	Particle phase.
u, l	Upper and lower walls respectively.
1, 2	Two types of nanoparticles.

1 Introduction

In modern times, the demand of industries is to develop such devices which are sophisticated and compact with high functionality. For example, in the electronic industries, the demand is to have smaller and smaller devices like mobiles, laptops, computers, etc., with more and more capacity. A similar problem is also faced in the automotive industry. In the run of developing smaller devices and improving the overall performance, cooling and enhanced heat transfer are the major issues. Earlier, extended solid surfaces and traditional fluids like water, ethylene glycol, oil, etc., were used for the purpose. But these methods were not very effective to reduce the size, as well as for enhanced heat transfer, because of the lower thermal conductivities of the traditional fluids. Later on, in order to remove these difficulties, the solid–liquid systems were used with micro-sized metal particles as suggested by Maxwell, but it also did not work because of the clogging and abrasion problems due to the extra large size of particles. The problem was ultimately resolved after the invention of nano-sized particles. Choi [15] was the first person who developed the suspension of these nano-scaled particles into some base fluid to form a new type of fluid which he named nanofluid. The problem of clogging and abrasion automatically vanished because the size of nanoparticles is quite closer to that of fluid molecules and this prevents nanoparticles to settle down under the effect of gravity. Eastman et al. [16] and Das et al. [17] reported an increment of around 15–40% in the effective thermal conductivity of fluid on the addition of a small amount of these nano-sized particles.

Two mathematical models are generally used to study the process of natural convection in nanofluids, viz., Khanafer-Vafai-Lightstone single-phase model (Khanafer et al. [20]) and Buongiorno two-phase model (Buongiorno [14]). In the single-phase approach, nanoparticles and base fluid are considered as a single homogeneous fluid, while in two-phase approach, nanoparticles and base fluid are considered as two different phases. The two-phase approach consists of a separate governing equation for nanoparticle volume fraction. Single-phase model provides less detail about each phase but it is computationally more efficient than the two-phase model which provides a better sightedness of the two phases. An exponential growth in research in the field of nanofluids can be observed in the last two decades. Tzou [43] investigated the onset of convection in a horizontal nanofluid layer heated from below, with the help of Buongiorno two-phase model. Nield and Kuznetsov [29] studied thermal instability in a porous medium layer saturated by a nanofluid and found that critical Rayleigh number can be altered by a considerable amount, on changing the basic distribution of nanoparticles as top heavy or bottom heavy. Kuznetsov and Nield [30] again investigated a similar problem using the Brinkman model. Agarwal et al. [1] used Darcy model to investigate thermal instability in a rotating anisotropic porous medium layer saturated by a nanofluid. Agarwal et al. [3] studied a similar problem of thermal instability using a Binary nanofluid. Agarwal and Bhadauria [4] studied convective heat transport by longitudinal rolls in dilute nanoliquids. Nield and Kuznetsov [13] presented a revised model of their earlier work for studying thermal instability in a porous medium layer saturated by a nanofluid, based on the zero flux boundary conditions. Kiran et al. [21] studied the effect of gravity modulation and internal heating on thermal convection in a nanofluid saturated porous medium. Siddheshwar and Lakshmi [38] investigated the classical problem of Darcy-Bénard convection for Newtonian liquids and Newtonian nanoliquids in cylindrical enclosures and cylindrical annuli. Kanchana et al. [19] studied the effect of gravity, boundary temperature and rotation modulations on Rayleigh-Bénard convection in twenty-eight nanoliquids.

In all the studies mentioned above, it is assumed that the different phases are in LTE (local thermal equilibrium), i.e., the temperature difference between the fluid and particle phases, as well as fluid and solid-matrix phases, is assumed to be negligible.

But the assumption of LTE was not very appropriate with the physical nature of the problem. Because of the significant differences among the thermal conductivities of fluid, nanoparticles and solid-matrix phases, a thermal lagging is created among the different phases, and therefore, the concept of LTNE (local thermal non-equilibrium) came into picture. Kuznetsov [24] used thermal non-equilibrium condition to study forced convection in porous media, Rees and Banu [33] used a two-field model for the separate modelling of the fluid and solid phase temperature fields in a fluid saturated porous medium to investigate the onset of Darcy-Bénard convection, Baytas and Pop [8] studied the effect of local thermal non-equilibrium on natural convection in a square porous cavity, Baytas [9] again performed a similar study with a heat generating solid phase non-Darcy porous medium, Rees and Pop [34] investigated the effect of LTNE in porous medium convection, Saeid [35] used LTNE model to investigate the problem of two-dimensional steady mixed convection in a vertical porous layer, numerically, Malashetty et al. [26] examined the stability of a horizontal fluid saturated sparsely packed porous layer under the assumption of local thermal non-equilibrium between the fluid and solid phases, Malashetty et al. [27] performed the similar study with anisotropic porous layer, Malashetty et al. [28] used thermal non-equilibrium model to examine double diffusive convection in a porous layer, Agarwal et al. [2] studied the effect of local thermal non-equilibrium on the linear and non-linear thermal instability in a nanofluid saturated rotating porous layer using Darcy model, Agarwal et al. [5] studied famous Rayleigh-Bénard convection in a nanofluid layer using a thermal non-equilibrium model and concluded that convection starts earlier for LTNE as compared to LTE case, Agarwal et al. [6] investigated thermal instability of a nanofluid layer under local thermal non-equilibrium, Siddheshwar and Siddabasappa [37] investigated the effect of local thermal non-equilibrium on onset of Brinkman-Bénard convection and on heat transport using rigid-rigid and free-free boundaries and found that the classical results hold even under the assumption of LTNE, Lagziri and Bezzazi [25] examined the effect of Robin's boundary condition in the Darcy–Rayleigh problem with LTNE model.

In view of the fact that composite nanoparticle has many more advantages over single-nanoparticle, the use of composite nanoparticle in place of single-nanoparticle to form composite nanoparticle-based composite nanofluids (hybrid nanofluids) is a great opportunity for scientists and researchers as suggested by Sarkar et al. [36] who presented an extensive review of hybrid/composite nanofluids. Nanocomposites hold such physiochemical characteristics which do not appear in the individual components. This improved and superior performance of nanocomposite-dispersed composite nanofluids, makes them a rapidly expanding research area. Both physical and chemical processes can be used for the synthesis of composite materials (Hanemann et al. [18] and Zhang [45]). The properties of composite nanofluid lie in between those of its constituent nanofluids (Suleiman et al. [7]). Gupta et al. [11] studied the effect of vertical magnetic field over the instability of binary nanofluids. Sharma et al. [39] used top- heavy distribution of nanoparticles to study binary nanofluid convection in a rotating porous layer. Sharma et al. [40] investigated the effect of externally impressed magnetic field over binary nanofluid convection in a porous medium. Sharma et al. [41] numerically investigated the convective instability in a rotating binary nanofluid porous layer. Kumar et al. [23] studied the cell formation in Rayleigh-Bénard convection using metallic and non-metallic nanofluids. Sharma et al. [42] investigated the instability in nanofluids using the LTNE effect and Hall currents. Gupta et al. [12] used blood as a base liquid to investigate the double diffusive instability in Casson binary nanoliquids. Bhadauria and Srivastava [10] studied the joint impact of internal heating and through-flow in a nanofluid saturated porous medium under LTNE.

Kumar and Awasthi [22] were the first to examine the thermal instability of composite nanofluids. In this work, we have investigated the convective instability of composite nanofluid using a thermal non-equilibrium approach. We have taken both top-heavy and bottom-heavy configurations of nanoparticle concentration for our analysis. It is noticed that the system behaves differently for smaller and higher concentrations of nanoparticles in top/bottom-heavy case. To the best of our knowledge, no study is done till date, investigating the effect of local thermal non-equilibrium on composite nanofluids. In view of the tremendous applications of composite nanofluids, as well as LTNE, in science and technology, we got motivated to work in this area.

2 Mathematical Formulation

Figure 1 describes the schematic model of the problem. A suspension of two different types of nanoparticles into some base liquid is considered to be filled in between two infinitely extended horizontal plates at z = 0 and z = d. Free-Free isothermal boundaries have been considered. The base fluid and both types of nanoparticles are considered to be at different temperatures. Using the approximation of Oberbeck–Boussinesq and the above assumptions, the governing equations are: (Agarwal et al. [5], Kumar and Awasthi [22])

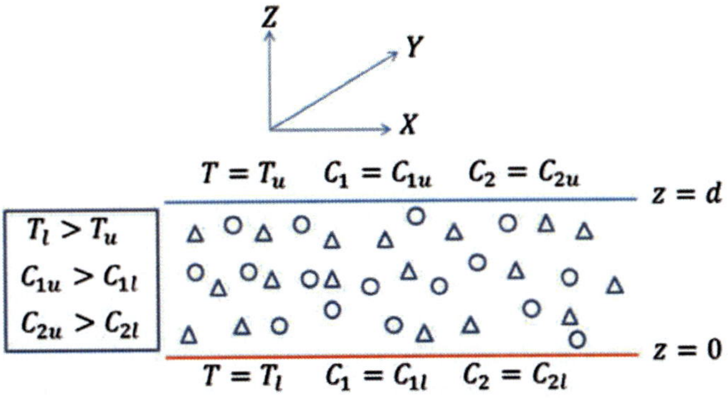

Fig. 1 Formal diagram

$$\nabla \cdot \mathbf{v} = 0, \tag{1}$$

$$\rho_f \left[\frac{\partial \mathbf{v}}{\partial t} + (\mathbf{v} \cdot \nabla)\mathbf{v} \right] = -\nabla p + \mu \nabla^2 \mathbf{v} + [C_1 \rho_{p1} + C_2 \rho_{p2} + \rho_f (1 - C_1 - C_2)(1 - \alpha(T_f - T_u))]\mathbf{g}, \tag{2}$$

$$(\rho c)_f \left[\frac{\partial T_f}{\partial t} + (\mathbf{v} \cdot \nabla) T_f \right] = \kappa_f \nabla^2 T_f + (\rho c)_{p1} \left[D_{B1} \nabla C_1 \cdot \nabla T_f + D_{T1} \frac{\nabla T_f \cdot \nabla T_f}{T_f} \right] + (\rho c)_{p2} \left[D_{B2} \nabla C_2 \cdot \nabla T_f + D_{T2} \frac{\nabla T_f \cdot \nabla T_f}{T_f} \right] + \frac{h_{p1}}{(1 - C_{1l})}(T_{p1} - T_f) + \frac{h_{p2}}{(1 - C_{2l})}(T_{p2} - T_f), \tag{3}$$

$$(\rho c)_{p1} \left[\frac{\partial T_{p1}}{\partial t} + (\mathbf{v} \cdot \nabla) T_{p1} \right] = \kappa_{p1} \nabla^2 T_{p1} + \frac{h_{p1}}{C_{1l}}(T_f - T_{p1}), \tag{4}$$

$$(\rho c)_{p2} \left[\frac{\partial T_{p2}}{\partial t} + (\mathbf{v} \cdot \nabla) T_{p2} \right] = \kappa_{p2} \nabla^2 T_{p2} + \frac{h_{p2}}{C_{2l}}(T_f - T_{p2}), \tag{5}$$

$$\frac{\partial C_1}{\partial t} + (\mathbf{v} \cdot \nabla) C_1 = D_{B1} \nabla^2 C_1 + \frac{D_{T1}}{T_u} \nabla^2 T_f, \tag{6}$$

$$\frac{\partial C_2}{\partial t} + (\mathbf{v} \cdot \nabla) C_2 = D_{B2} \nabla^2 C_2 + \frac{D_{T2}}{T_u} \nabla^2 T_f. \tag{7}$$

We assume that the temperature of fluid and the temperature and concentration of both the nanoparticles are constant at the boundaries, therefore, the boundary conditions are as follows (Agarwal et al. [5]):

$$\mathbf{v} = 0,\ T_f = T_l,\ T_{p1} = T_l,\ T_{p2} = T_l,\ C_1 = C_{1l},\ C_2 = C_{2l} \text{ at } z = 0, \tag{8}$$

$$\mathbf{v} = 0,\ T_f = T_u,\ T_{p1} = T_u,\ T_{p2} = T_u,\ C_1 = C_{1u},\ C_2 = C_{2u} \text{ at } z = \text{d}. \tag{9}$$

where, $C_{1u} > C_{1l}$, $C_{2u} > C_{2l}$ and $T_l > T_u$.

Now, to make the variables dimension-free, we use the following non-dimensional parameters:

$$(x^*, y^*, z^*) = (x, y, z)/d,\ (v_1^*, v_2^*, v_3^*) = (v_1, v_2, v_3)d/\alpha_f,\ t^* = t\alpha_f/d^2$$
$$p^* = \frac{pd^2}{\mu\alpha_f},\ C_1^* = \frac{C_1 - C_{1l}}{C_{1u} - C_{1l}},\ C_2^* = \frac{C_2 - C_{2l}}{C_{2u} - C_{2l}},\ T^* = \frac{T - T_u}{T_l - T_u}$$

where $\alpha_f = \dfrac{\kappa_f}{(\rho c)_f}$ is the nanofluid thermal diffusivity.

Making use of the above mentioned non-dimensional parameters in Eqs. (1)–(9) and leaving the asterisk, we get the following non-dimensional equations:

$$\nabla \cdot \mathbf{v} = 0, \tag{10}$$

$$\frac{1}{Pr}\left[\frac{\partial \mathbf{v}}{\partial t} + (\mathbf{v}\cdot\nabla)\mathbf{v}\right] = -\nabla p + \nabla^2 \mathbf{v} - C_1 Rn_1 \hat{e_z} - C_2 Rn_2 \hat{e_z} - Rm\hat{e_z} + RaT_f\hat{e_z}, \tag{11}$$

$$\begin{aligned}\frac{\partial T_f}{\partial t} + (\mathbf{v}\cdot\nabla)T_f = \nabla^2 T_f + \frac{N_{B1}}{Le_1}\nabla C_1\cdot\nabla T_f + \frac{N_{A1}N_{B1}}{Le_1}\nabla T_f\cdot\nabla T_f + \frac{N_{B2}}{Le_2}\nabla C_2\cdot\nabla T_f \\ + \frac{N_{A2}N_{B2}}{Le_2}\nabla T_f\cdot\nabla T_f + H_{p1}(T_{p1} - T_f) + H_{p2}(T_{p2} - T_f),\end{aligned} \tag{12}$$

$$\frac{\partial T_{p1}}{\partial t} + (\mathbf{v}\cdot\nabla)T_{p1} = \epsilon_1\nabla^2 T_{p1} + \gamma_1 H_{p1}(T_f - T_{p1}), \tag{13}$$

$$\frac{\partial T_{p2}}{\partial t} + (\mathbf{v}\cdot\nabla)T_{p2} = \epsilon_2\nabla^2 T_{p2} + \gamma_2 H_{p2}(T_f - T_{p2}), \tag{14}$$

$$\frac{\partial C_1}{\partial t} + (\mathbf{v}\cdot\nabla)C_1 = \frac{1}{Le_1}\nabla^2 C_1 + \frac{N_{A1}}{Le_1}\nabla^2 T_f, \tag{15}$$

$$\frac{\partial C_2}{\partial t} + (\mathbf{v}\cdot\nabla)C_2 = \frac{1}{Le_2}\nabla^2 C_2 + \frac{N_{A2}}{Le_2}\nabla^2 T_f, \tag{16}$$

$$\mathbf{v} = 0,\ T_f = 1,\ T_{p1} = 1,\ T_{p2} = 1,\ C_1 = 0,\ C_2 = 0 \text{ at } z = 0, \tag{17}$$

$$\mathbf{v} = 0,\ T_f = 0,\ T_{p1} = 0,\ T_{p2} = 0,\ C_1 = 1,\ C_2 = 1 \text{ at } z = 1. \tag{18}$$

Where, $Pr = \frac{\mu}{\rho_f\alpha_f}$ is the Prandtl number, $Rn_1 = \frac{(C_{1u}-C_{1l})(\rho_{p1}-\rho_f)gd^3}{\mu\alpha_f}$ and $Rn_2 = \frac{(C_{2u}-C_{2l})(\rho_{p2}-\rho_f)gd^3}{\mu\alpha_f}$ are the nanoparticle concentration Rayleigh numbers, $Rm = \frac{[\rho_{p1}C_{1l}+\rho_{p2}C_{2l}+\rho_f(1-C_{1l}-C_{2l})gd^3]}{\mu\alpha_f}$ is the basic density Rayleigh number, $Ra = \frac{\alpha(T_l-T_u)\rho_f gd^3}{\mu\alpha_f}$ is the thermal Rayleigh number, $N_{A1} = \frac{D_{T1}(T_l-T_u)}{D_{B1}T_u(C_{1u}-C_{1l})}$ and $N_{A2} = \frac{D_{T2}(T_l-T_u)}{D_{B2}T_u(C_{2u}-C_{2l})}$ are the modified diffusivity ratios, $N_{B1} = \frac{(\rho c)_{p1}(C_{1u}-C_{1l})}{(\rho c)_f}$ and $N_{B2} = \frac{(\rho c)_{p2}(C_{2u}-C_{2l})}{(\rho c)_f}$ are the modified particle-density increments, $Le_1 = \frac{\alpha_f}{D_{B1}}$ and $Le_2 = \frac{\alpha_f}{D_{B2}}$ are the Lewis numbers, $\epsilon_1 = \frac{\kappa_{p1}(\rho c)_f}{(\rho c)_{p1}\kappa_f}$ and $\epsilon_2 = \frac{\kappa_{p2}(\rho c)_f}{(\rho c)_{p2}\kappa_f}$ are thermal diffusivity ratios, $\gamma_1 = \frac{(1-C_{1l})(\rho c)_f}{C_{1l}(\rho c)_{p1}}$ and $\gamma_2 = \frac{(1-C_{2l})(\rho c)_f}{C_{2l}(\rho c)_{p2}}$ are the modified thermal capacity ratios, $H_{p1} = \frac{h_{p1}d^2}{(1-C_{1l})\kappa_f}$ and $H_{p2} = \frac{h_{p2}d^2}{(1-C_{2l})\kappa_f}$ are non-dimensional interphase heat transfer coefficients or the Nield numbers.

3 Basic State

At the basic state, it is assumed that all the physical quantities are time independent and function of z only. The conduction state is given by

$$\mathbf{v} = \mathbf{0},\ T_f = T_{fb}(z),\ T_{p1} = T_{p1b}(z),\ T_{p2} = T_{p2b}(z),\ C_1 = C_{1b}(z),\ C_2 = C_{2b}(z),\ p = p_b(z). \tag{19}$$

Using Eq. (19), in Eqs. (12)–(16), we have

$$0 = \frac{d^2T_{fb}}{dz^2} + \frac{N_{B1}}{Le_1}\frac{dC_{1b}}{dz}\frac{dT_{fb}}{dz} + \frac{N_{A1}N_{B1}}{Le_1}\frac{dT_{fb}}{dz}\frac{dT_{fb}}{dz} + \frac{N_{B2}}{Le_2}\frac{dC_{2b}}{dz}\frac{dT_{fb}}{dz} + \frac{N_{A2}N_{B2}}{Le_2}\frac{dT_{fb}}{dz}\frac{dT_{fb}}{dz} + H_{p1}(T_{p1b} - T_{fb}) + H_{p2}(T_{p2b} - T_{fb}), \tag{20}$$

$$0 = \epsilon_1\frac{d^2T_{p1b}}{dz^2} + \gamma_1 H_{p1}(T_{fb} - T_{p1b}), \tag{21}$$

$$0 = \epsilon_2\frac{d^2T_{p2b}}{dz^2} + \gamma_2 H_{p2}(T_{fb} - T_{p2b}), \tag{22}$$

$$0 = \frac{1}{Le_1}\frac{d^2C_{1b}}{dz^2} + \frac{N_{A1}}{Le_1}\frac{d^2T_{fb}}{dz^2}, \tag{23}$$

$$0 = \frac{1}{Le_2}\frac{d^2C_{2b}}{dz^2} + \frac{N_{A2}}{Le_2}\frac{d^2T_{fb}}{dz^2}. \tag{24}$$

Similarly, the boundary conditions (17) and (18) take the form:

$$T_{fb}(0) = T_{pb1}(0) = T_{pb2}(0) = 1 \text{ and } C_{1b}(0) = C_{2b}(0) = 0, \tag{25}$$

$$T_{fb}(1) = T_{pb1}(1) = T_{pb2}(1) = 0 \text{ and } C_{1b}(1) = C_{2b}(1) = 1. \tag{26}$$

In the basic state, it is assumed that both the nanoparticles and base fluid are in thermal equilibrium, i.e., there is no heat transfer between particle and fluid phases. Generally, in case of nanofluids, the order of Lewis number ranges between 10^4 – 10^7, while the values of N_{A1} and N_{A2} are always less than 10 (Buongiorno [14]). Applying these approximations, we have the following conduction state:

$$T_{fb}(z) = T_{pb1}(z) = T_{pb2}(z) = (1 - z), \ C_{1b}(z) = C_{2b}(z) = z. \tag{27}$$

4 Perturbed State

The parameters are now written in the manner: $\mathbf{v} = \mathbf{v}(0,0,0) + \mathbf{v}(\mathring{v}_1, \mathring{v}_2, \mathring{v}_3)$, $p = p_b + \mathring{p}$, $T_f = T_{fb} + \mathring{T}_f$, $T_{p1} = T_{p1b} + \mathring{T}_{p1}$, $T_{p2} = T_{p2b} + \mathring{T}_{p2}$, $C_1 = C_{1b} + \mathring{C}_1$, $C_2 = C_{2b} + \mathring{C}_2$ in order to impose small perturbations to the basic conduction state. Here the perturbed quantities are written under the ring ($^\circ$). The case of two-dimensional (X-Z) rolls is considered, for simplicity, which gives the liberty to take all the physical quantities to be independent of y. Also introducing the stream function ψ such that $v_1 = \frac{\partial \psi}{\partial z}$ and $v_3 = -\frac{\partial \psi}{\partial x}$. Substituting the new perturbed variables in Eqs. (10)–(16) and eliminating the pressure term and using the basic state solution (27), we have the following dimension-free reduced set of governing equations (after leaving the ring symbol ($^\circ$)):

$$\frac{1}{Pr}\frac{\partial}{\partial t}(\nabla^2\psi) = \nabla^4\psi + Rn_1\frac{\partial C_1}{\partial x} + Rn_2\frac{\partial C_2}{\partial x} - Ra\frac{\partial T_f}{\partial x} + \frac{1}{Pr}\frac{\partial(\psi, \nabla^2\psi)}{\partial(x, z)}, \tag{28}$$

$$\frac{\partial T_f}{\partial t} = -\frac{\partial \psi}{\partial x} + \nabla^2 T_f + H_{p1}(T_{p1} - T_f) + H_{p2}(T_{p2} - T_f) + \frac{\partial(\psi, T_f)}{\partial(x, z)}, \tag{29}$$

$$\frac{\partial T_{p1}}{\partial t} = -\frac{\partial \psi}{\partial x} + \epsilon_1\nabla^2 T_{p1} + \gamma_1 H_{p1}(T_f - T_{p1}) + \frac{\partial(\psi, T_{p1})}{\partial(x, z)}, \tag{30}$$

$$\frac{\partial T_{p2}}{\partial t} = -\frac{\partial \psi}{\partial x} + \epsilon_2\nabla^2 T_{p2} + \gamma_2 H_{p2}(T_f - T_{p2}) + \frac{\partial(\psi, T_{p2})}{\partial(x, z)}, \tag{31}$$

$$\frac{\partial C_1}{\partial t} = \frac{\partial \psi}{\partial x} + \frac{1}{Le_1}\nabla^2 C_1 + \frac{N_{A1}}{Le_1}\nabla^2 T_f + \frac{\partial(\psi, C_1)}{\partial(x, z)}, \tag{32}$$

$$\frac{\partial C_2}{\partial t} = \frac{\partial \psi}{\partial x} + \frac{1}{Le_2}\nabla^2 C_2 + \frac{N_{A2}}{Le_2}\nabla^2 T_f + \frac{\partial(\psi, C_2)}{\partial(x, z)}. \tag{33}$$

Both the boundaries at the top and bottom are considered to be free. The boundary conditions under this assumption are

$$C_1 = C_2 = \psi = \nabla^2\psi = T_f = T_{p1} = T_{p2} = 0 \text{ at z} = 0, 1. \tag{34}$$

5 Stability Analysis

5.1 *Linear Stability Analysis*

For stationary mode of convection, we seek the solutions of Eqs. (28)–(33) in the following form (Postelnicu and Rees [32]) so as to satisfy the boundary conditions (34):

$$(\psi, T_f, T_{p1}, T_{p2}, C_1, C_2) = [A\sin(ax), \{B, C, D, E, F\}\cos(ax)]\sin(\pi z) \tag{35}$$

here A, B, C, D, E and F are constants and 'a' denotes the wave number (horizontal).

Now substituting the values from Eq. (35) into the linearized format of Eqs. (28)–(33) and using the Galerkin's technique, we have

$$\begin{bmatrix} -\delta^4 & -aRa & 0 & 0 & aRn_1 & aRn_2 \\ a & (H_{p1} + H_{p2} + \delta^2) & -H_{p1} & -H_{p2} & 0 & 0 \\ a & -\gamma_1 H_{p1} & (\gamma_1 H_{p1} + \epsilon_1\delta^2) & 0 & 0 & 0 \\ a & -\gamma_2 H_{p2} & 0 & (\gamma_2 H_{p2} + \epsilon_2\delta^2) & 0 & 0 \\ -a & \frac{\delta^2 N_{A1}}{Le_1} & 0 & 0 & \frac{\delta^2}{Le_1} & 0 \\ -a & \frac{\delta^2 N_{A2}}{Le_2} & 0 & 0 & 0 & \frac{\delta^2}{Le_2} \end{bmatrix} \begin{bmatrix} A \\ B \\ C \\ D \\ E \\ F \end{bmatrix} = \begin{bmatrix} 0 \\ 0 \\ 0 \\ 0 \\ 0 \\ 0 \end{bmatrix} \tag{36}$$

The condition for the non-zero solution of the system (36) provides the expression for the Rayleigh number Ra.

$$\begin{aligned} Ra = {} & \frac{1}{a^2[H_{p1}H_{p2}(\gamma_1 + \gamma_2 + \gamma_1\gamma_2) + H_{p1}(1 + \gamma_1)\delta^2\epsilon_2 + \delta^2\epsilon_1(H_{p2}(1 + \gamma_2) + \delta^2\epsilon_2)]}\Big[\delta^6\{H_{p1}H_{P2}\gamma_2\epsilon_1 \\ & + \delta^4\epsilon_1\epsilon_2 + \delta^2(\gamma_1 + \epsilon_1)\epsilon_2 H_{p1} + H_{p1}H_{p2}\gamma_1(\gamma_2 + \epsilon_2) + \delta^2(\gamma_2 + \epsilon_2)\epsilon_1 H_{p2}\} - a^2\{(Le_1 Rn_1 + Le_2 Rn_2) \\ & (H_{p1}H_{P2}\gamma_2\epsilon_1 + \delta^4\epsilon_1\epsilon_2 + \delta^2(\gamma_1 + \epsilon_1)\epsilon_2 H_{p1} + H_{p1}H_{p2}\gamma_1(\gamma_2 + \epsilon_2) + \delta^2(\gamma_2 + \epsilon_2)\epsilon_1 H_{p2}) + N_{A1}Rn_1 \\ & (H_{p1}H_{p2}(\gamma_1 + \gamma_2 + \gamma_1\gamma_2) + H_{p1}(1 + \gamma_1)\delta^2\epsilon_2 + \delta^2\epsilon_1(H_{p2}(1 + \gamma_2) + \delta^2\epsilon_2)) + N_{A2}Rn_2(H_{p1}H_{p2}(\gamma_1 \\ & + \gamma_2 + \gamma_1\gamma_2) + H_{p1}(1 + \gamma_1)\delta^2\epsilon_2 + \delta^2\epsilon_1(H_{p2}(1 + \gamma_2) + \delta^2\epsilon_2))\}\Big] \end{aligned} \tag{37}$$

here, $\delta^2 = a^2 + \pi^2$

5.2 Non-linear Stability Analysis

For local non-linear stability analysis, the following Fourier series expressions are considered:

$$\psi = \sum_{m=1}^{\infty}\sum_{n=1}^{\infty} A_{mn}(t)\sin(m\pi ax)\sin(n\pi z), \tag{38}$$

$$T_f = \sum_{m=1}^{\infty}\sum_{n=1}^{\infty} B_{mn}(t)\cos(m\pi ax)\sin(n\pi z), \tag{39}$$

$$T_{p1} = \sum_{m=1}^{\infty}\sum_{n=1}^{\infty} C_{mn}(t)\cos(m\pi ax)\sin(n\pi z), \tag{40}$$

$$T_{p2} = \sum_{m=1}^{\infty}\sum_{n=1}^{\infty} D_{mn}(t)\cos(m\pi ax)\sin(n\pi z), \tag{41}$$

$$C_1 = \sum_{m=1}^{\infty}\sum_{n=1}^{\infty} E_{mn}(t)\cos(m\pi ax)\sin(n\pi z), \tag{42}$$

$$C_2 = \sum_{m=1}^{\infty}\sum_{n=1}^{\infty} F_{mn}(t)\cos(m\pi ax)\sin(n\pi z). \tag{43}$$

All these Fourier expressions contain infinite terms and taking all of them together would be very cumbersome. Therefore, for stream function, we consider (1,1) mode, while for other parameters, (1,1) and (0,2) modes have been taken (Agarwal et al. [3])

$$\psi = A_{11}(t)\,\sin(\pi ax)\sin(\pi z), \tag{44}$$

$$T_f = B_{11}(t)\,\cos(\pi ax)\sin(\pi z) + B_{02}(t)\sin(2\pi z), \tag{45}$$

$$T_{p1} = C_{11}(t)\,\cos(\pi ax)\sin(\pi z) + C_{02}(t)\sin(2\pi z), \tag{46}$$

$$T_{p2} = D_{11}(t)\,\cos(\pi ax)\sin(\pi z) + D_{02}(t)\sin(2\pi z), \tag{47}$$

$$C_1 = E_{11}(t)\,\cos(\pi ax)\sin(\pi z) + E_{02}(t)\sin(2\pi z), \tag{48}$$

$$C_2 = F_{11}(t)\,\cos(\pi ax)\sin(\pi z) + F_{02}(t)\sin(2\pi z). \tag{49}$$

here, $A_{11}, B_{11}, B_{02}, C_{11}, C_{02}, D_{11}, D_{02}, E_{11}, E_{02}, F_{11}$ and F_{02} are the time dependent functions, to be determined. Using Eqs. (44)–(49) into the Eqs. (28)–(33) and applying Galerkin's orthogonalization procedure, we have:

$$A'_{11}(t) = \frac{Pr}{\pi(1+a^2)}\left[-\pi^3(1+a^2)^2 A_{11} - aRaB_{11} + a(Rn_1E_{11} + Rn_2F_{11})\right], \quad (50)$$

$$B'_{11}(t) = -a\pi A_{11}(1+\pi B_{02}) - \{H_{p1} + H_{p2} + \pi^2(1+a^2)\}B_{11} + H_{p1}C_{11} + H_{p2}D_{11}, \quad (51)$$

$$B'_{02}(t) = -\{H_{p1} + H_{p2} + 4\pi^2\}B_{02} + \frac{a\pi^2}{2}A_{11}B_{11} + H_{p1}C_{02} + H_{p2}D_{02}, \quad (52)$$

$$C'_{11}(t) = \gamma_1 H_{p1}(B_{11} - C_{11}) - a\pi A_{11} - a\pi^2 A_{11}C_{02} - \epsilon_1\pi^2(1+a^2)C_{11}, \quad (53)$$

$$C'_{02}(t) = \gamma_1 H_{p1}B_{02} - (\gamma_1 H_{p1} + 4\epsilon_1\pi^2)C_{02} + \frac{a\pi^2}{2}A_{11}C_{11}, \quad (54)$$

$$D'_{11}(t) = \gamma_2 H_{p2}B_{11} - a\pi A_{11} - a\pi^2 A_{11}D_{02} - \{\gamma_2 H_{p2} + \epsilon_2\pi^2(1+a^2)\}D_{11}, \quad (55)$$

$$D'_{02}(t) = \gamma_2 H_{p2}B_{02} - (\gamma_2 H_{p2} + 4\epsilon_2\pi^2)D_{02} + \frac{a\pi^2}{2}A_{11}D_{11}, \quad (56)$$

$$E'_{11}(t) = \frac{a\pi Le_1 A_{11} - \pi^2(1+a^2)\{N_{A1}B_{11} + E_{11}\}}{Le_1} - a\pi^2 A_{11}E_{02}, \quad (57)$$

$$E'_{02}(t) = \frac{a\pi^2}{2}A_{11}E_{11} - \frac{4\pi^2(N_{A1}B_{02} + E_{02})}{Le_1}, \quad (58)$$

$$F'_{11}(t) = \frac{a\pi Le_2 A_{11} - \pi^2(1+a^2)\{N_{A2}B_{11} + F_{11}\}}{Le_2} - a\pi^2 A_{11}F_{02}, \quad (59)$$

$$F'_{02}(t) = \frac{a\pi^2}{2} A_{11} F_{11} - \frac{4\pi^2 (N_{A2} B_{02} + F_{02})}{Le_2}. \tag{60}$$

The above set of Eqs. (50)–(60) represents an autonomous system of ODE's which is difficult to solve analytically, and therefore, we move towards the numerical solutions of the same by using an inbuilt tool (NDSolve) of Mathematica.

6 Transport of Heat and Mass

The Nusselt number for fluid $Nu_f(t)$ is defined as (Agarwal et al. [3]):

$$Nu_f(t) = \frac{\text{Transport of heat by (conduction + convection)}}{\text{Transport of heat by conduction}} = 1 + \left[\frac{\int_0^{2/a} (\frac{\partial T_f}{\partial z}) dx}{\int_0^{2/a} (\frac{\partial T_{fb}}{\partial z}) dx} \right]_{z=0} \tag{61}$$

Using Eqs. (27) and (45) in Eq. (61), we get

$$Nu_f(t) = 1 - 2\pi B_{02}(t). \tag{62}$$

In a similar way, the Nusselt numbers for both the nanoparticles

$$Nu_{p1}(t) = 1 - 2\pi C_{02}(t) \text{ and } Nu_{p2}(t) = 1 - 2\pi D_{02}(t). \tag{63}$$

and the concentration Nusselt numbers $Nu_{C1}(t)$ and $Nu_{C2}(t)$ can be found to be

$$Nu_{C1}(t) = 1 + 2\pi E_{02}(t) + N_{A1}(1 - 2\pi B_{02}(t)), \tag{64}$$
$$Nu_{C2}(t) = 1 + 2\pi F_{02}(t) + N_{A2}(1 - 2\pi B_{02}(t)). \tag{65}$$

7 Results and Discussion

7.1 *Linear Stability Analysis*

In this paper, the effect of LTNE over composite nanofluid is studied. The expression for the thermal Rayleigh number obtained in Eq. (37) can be reduced for the LTE case by substituting $H_{p1} = H_{p2} = 0$ as follows:

$$Ra^{LTE} = \frac{\delta^6}{a^2} - Rn_1(Le_1 + N_{A1}) - Rn_2(Le_2 + N_{A2}). \tag{66}$$

Above expression for Ra^{LTE}, is exactly similar to that of obtained by Kumar and Awasthi [22] and can be further reduced for the case of simple nanofluid by taking either $Rn_1 = 0$ or $Rn_2 = 0$ in Eq. (66) to get $Ra^{LTE} = \frac{\delta^6}{a^2} - Rn_1(Le_1 + N_{A1})$ or $Ra^{LTE} = \frac{\delta^6}{a^2} - Rn_2(Le_2 + N_{A2})$. This expression for Ra^{LTE} for simple nanofluid bears a complete resemblance to that found by Nield and Kuznetsov [31]. It is notable that the modified diffusivity ratios N_{A1} and N_{A2} take positive values for top-heavy case and negative values for bottom-heavy case. But these positive or negative values of N_{A1} and N_{A2} do not much affect the final results because of the presence of Lewis number Le, and therefore, the investigation is done for $N_{A1} = N_{A2} = 5$, $Le_1 = Le_2 = 100$, $H_{p1} = H_{p2} = 10$, $\epsilon_1 = \epsilon_2 = 0.04$, $\gamma_1 = \gamma_2 = 5$ (Agarwal et al. [3], Kumar and Awasthi [22]) and for various values of Rn_1 and Rn_2 depending upon the top-heavy or bottom-heavy case. The positive values of Rn_1 and Rn_2 describe the top-heavy case, while the negative values describe the bottom-heavy case. The variation in critical Rayleigh number and critical wave number for positive and negative values of N_{A1} and N_{A2} in bottom-heavy case is shown in Table 1.

In Fig. 2a and b, the comparison between the onset of convection for LTE and LTNE cases is presented for top and bottom-heavy configurations of composite nanofluid, respectively. It can be observed that the onset of convection advances in both configurations for local thermal non-equilibrium between the fluid and nanoparticles. In LTNE case, there is a temperature difference, between the fluid and nanoparticle phases, which encourages the transfer of energy between them. This transfer of energy leads to enhancing the onset of convection. A similar conclusion can also be drawn from Table 1. Equation (66), suggests that for composite nanofluid, the onset of convection advances for the top-heavy case, while it gets delayed for the bottom-heavy case as far as the fluid and nanoparticles are in thermal equilibrium (LTE). But in the case of LTNE, the onset of convection advances for composite nanofluid as compared to that of normal nanofluid for both top and bottom-heavy configurations (Table 1).

Figure 3a depicts that for smaller concentration of nanoparticles ($Rn_1 = Rn_2 = -0.5$) in bottom-heavy case of LTNE, convection starts earlier for composite nanofluid as compared to that of normal nanofluid unlike the case of LTE, but for higher concentration of nanoparticles ($Rn_1 = Rn_2 = -2$), the convection starts delaying for composite nanofluid (Fig. 3b) and the effect of LTNE reduces.

Figure 4 represents the variation of critical Rayleigh number (Ra_c) along with the simultaneous variation of both the interphase heat transfer coefficients H_{p1} and H_{p2} for the fluid and particle phases. It can be seen that the behaviour of Ra_c is similar in both top-heavy and bottom-heavy cases. As H_{p1} or $H_{p2} \to 0$, there is no heat transfer between particles and fluid phases and therefor the critical Rayleigh number Ra_c is constant. Moreover, when H_{p1} or $H_{p2} \to \infty$, heat transfer between particles and fluid phases is too fast and therefore they again attain the state of thermal equilibrium, keeping the value of Ra_c still a constant. The maximum change (decrement) in the value of Ra_c can be seen for those value of $\log H_{p1}$ and $\log H_{p2}$ which lie near the halfway. The region, in which the change in the value of Ra_c is observed, may be

Table 1 Critical Rayleigh number (Ra_c) and critical wave number (a_c) for top-heavy and bottom-heavy configurations of nanofluid

Top heavy		
$Rn_1 = Rn_2 = 0.5, N_{A1} = N_{A2} = 5$	$\mathbf{a_c}$	$\mathbf{Ra_c}$
Normal nanofluid under LTE	2.22144	605.011
Normal nanofluid under LTNE	2.22050	508.759
Composite nanofluid under LTE	2.22144	552.511
Composite nanofluid under LTNE	2.21997	400.876
Bottom heavy		
$Rn_1 = Rn_2 = -0.5, N_{A1} = N_{A2} = 5$	$\mathbf{a_c}$	$\mathbf{Ra_c}$
Normal nanofluid under LTE	2.22144	710.011
Normal nanofluid under LTNE	2.22034	597.916
Composite nanofluid under LTE	2.22144	762.511
Composite nanofluid under LTNE	2.21944	556.479
Bottom heavy		
$Rn_1 = Rn_2 = -0.5, N_{A1} = N_{A2} = -5$	$\mathbf{a_c}$	$\mathbf{Ra_c}$
Normal nanofluid under LTE	2.22144	705.011
Normal nanofluid under LTNE	2.22034	592.916
Composite nanofluid under LTE	2.22144	752.511
Composite nanofluid under LTNE	2.21944	546.479

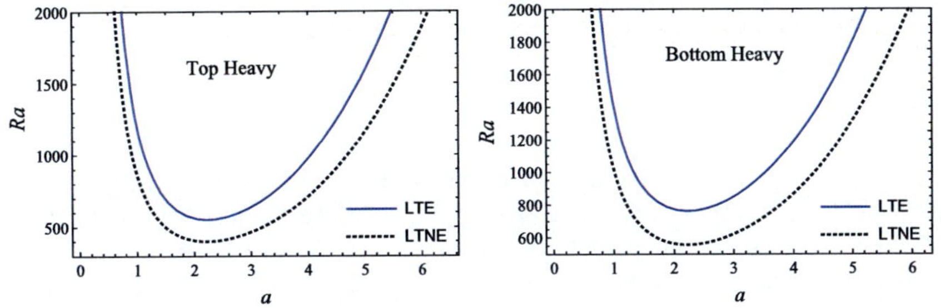

(a) $Rn_1 = Rn_2 = 0.5$ (b) $Rn_1 = Rn_2 = -0.5$

Fig. 2 Effect of LTNE over neutral stability curve

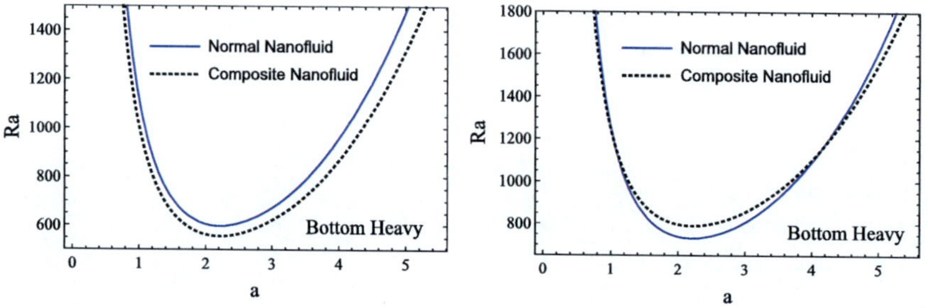

(a) $Rn_1 = Rn_2 = -0.5$ (b) $Rn_1 = Rn_2 = -2$

Fig. 3 Comparison between the onset of convection for normal and composite nanofluid under LTNE

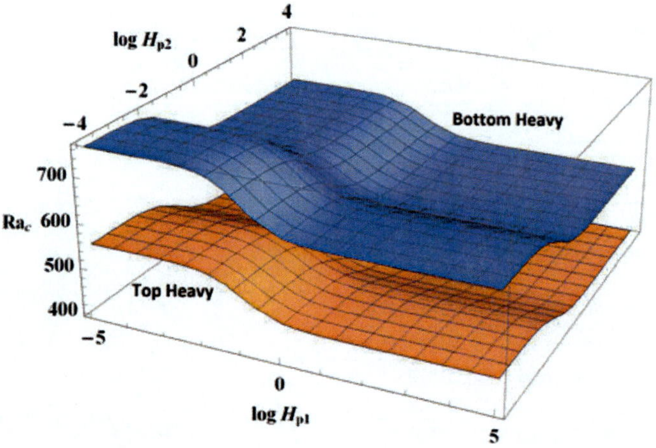

Fig. 4 Variation of critical Rayleigh number Ra_c with both interphase heat transfer coefficients, $Rn_1 = Rn_2 = 0.5$ for top heavy and $Rn_1 = Rn_2 = -0.5$ for bottom heavy

termed as the LTNE region because the major effect of LTNE can be seen only in this specific region.

Figure 5 describes the variation of critical wave number (a_c) along with the simultaneous variation of both the interphase heat transfer coefficients H_{p1} and H_{p2} for the fluid and particle phases. When H_{p1} or $H_{p2} \to 0$, the thermal field of the fluid is totally unaffected by the particle phase, while when H_{p1} or $H_{p2} \to \infty$, both fluid and nanoparticles attain uniform temperatures and start behaving like a single phase only. This is the reason why we don't observe any variation in the value of critical wave number (a_c) for the terminal values of H_{p1} and H_{p2}. The exact values of Ra_c and a_c are given in Table 2 for both top-heavy ($Rn_1 = Rn_2 = 0.5$) and bottom-heavy ($Rn_1 = Rn_2 = -0.5$) configurations. And it can be clearly noticed that the variation in the values of critical Rayleigh number (Ra_c) and critical wave number (a_c) can only be seen in the LTNE region.

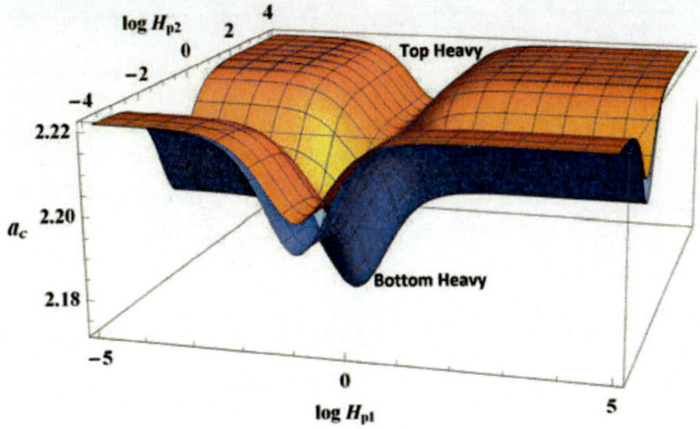

Fig. 5 Variation of critical wave number a_c with both interphase heat transfer coefficients, $Rn_1 = Rn_2 = 0.5$ for top heavy and $Rn_1 = Rn_2 = -0.5$ for bottom heavy

Table 2 Evaluation of Ra_c and a_c for different values of H_{p1} and H_{p2} under top-heavy and bottom-heavy case

			Top heavy		Bottom heavy	
	H_{p1}	H_{p2}	Ra_c	a_c	Ra_c	a_c
LTE region	10^{-5}	10^{-5}	552.4933	2.2214	762.4868	2.2214
	10^{-4}	10^{-4}	552.3308	2.2213	762.2660	2.2212
LTNE region	10^{-3}	10^{-3}	550.7245	2.2199	760.0833	2.2194
	10^{-2}	10^{-2}	536.3190	2.2089	740.4969	2.2045
	10^{-1}	10^{-1}	469.4622	2.1847	649.5693	2.1720
	10^{0}	10^{0}	411.5009	2.2091	570.9055	2.2048
	10^{1}	10^{1}	400.8765	2.2199	556.4793	2.2194
LTE region	10^{2}	10^{2}	399.7232	2.2213	554.9124	2.2212
	10^{3}	10^{3}	399.6069	2.2214	554.7544	2.2214
	10^{4}	10^{4}	399.5953	2.2214	554.7386	2.2214
	10^{5}	10^{5}	399.5941	2.2214	554.7370	2.2214

7.2 *Non-linear Stability Analysis*

The non-linear stability analysis is done based on the numerical solutions of ordinary differential Eqs. (50)–(60). The coefficient of heat transport for fluid (fluid-thermal Nusselt number $Nu_f(t)$), coefficient of heat transport for nanoparticles (particle-thermal Nusselt numbers $Nu_{p1}(t)$ and $Nu_{p2}(t)$) and the coefficient of nanoparticles concentration transport (concentration Nusselt numbers $Nu_{\phi 1}(t)$ and $Nu_{\phi 2}(t)$) are evaluated as a function of time 't' for $Pr = 5$, $Na_1 = Na_2 = 5$, $Le_1 = Le_2 = 100$, $H_{p1} = H_{p2} = 10$, $\epsilon_1 = \epsilon_2 = 0.04$, $\gamma_1 = \gamma_2 = 5$ (Agarwal et al. [3], Kumar and

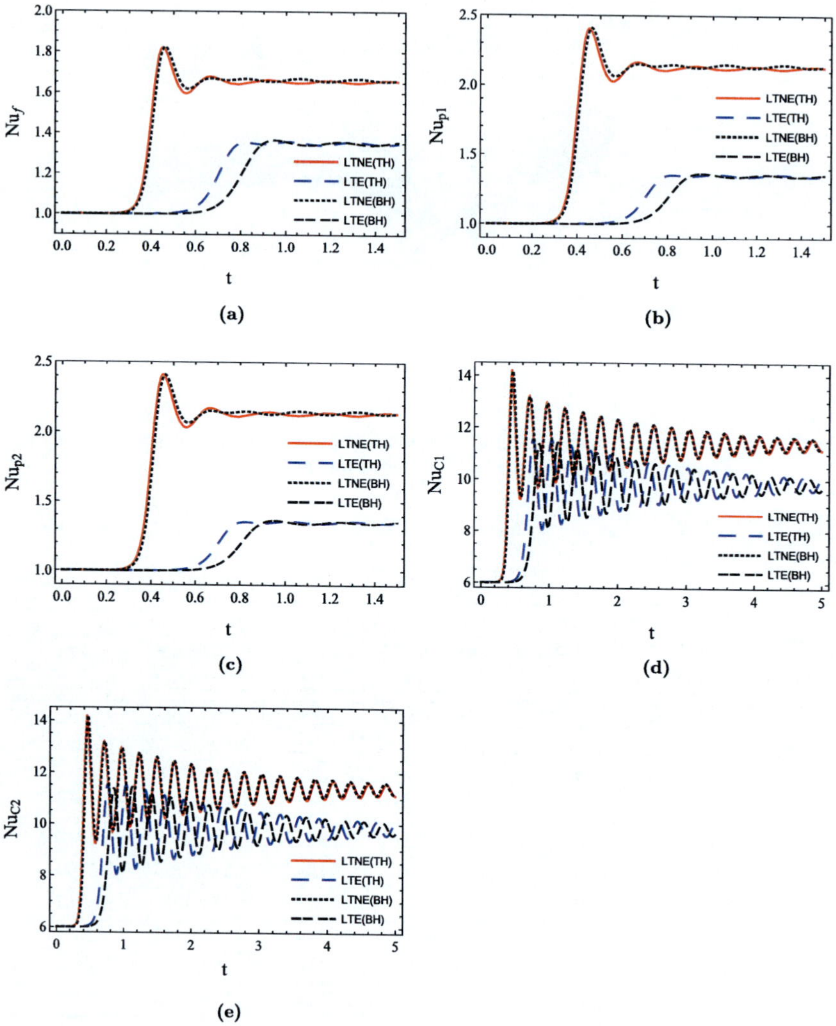

Fig. 6 $Rn_1 = Rn_2 = 5$ for top heavy and $Rn_1 = Rn_2 = -5$ for bottom heavy

Awasthi [22]). All the obtained results are presented graphically in Figs. 6 and 8. All these graphs of non-linear analysis have a common characteristic that can be discussed in three parts. In the first part, when t is close to 0, we observe a constant Nusselt number which is the indicative of conduction state. As time passes, a huge enhancement in heat, as well as mass transport can be seen in the second part, which is the indicative of convection taking place in the system. At last, in the third part, all the Nusselt numbers again tend to become constant followed by some oscillations. This indicates that the system has achieved a steady state.

Figure 6a–c represent the variation of thermal Nusselt number for fluid and particle phases. It can be clearly noticed that in case of LTNE, the heat transfer in both fluid and nanoparticles is higher than that of LTE. Moreover, the onset of convection also advances in the LTNE case for both top-heavy and bottom-heavy configurations. One more thing can be depicted in case of LTE, the onset of convection advances for top-heavy configuration as compared to that of bottom-heavy configuration, but we don't observe any such effect in the LTNE case. The heat transfer in nanoparticles is more than that of fluid phase, this increased heat transfer in nanoparticles is possibly due to their increased thermal conductivities. Figure 6d–e describe the variation of concentration Nusselt number with time for both nanoparticles, respectively. Mass transport also increases significantly in case of LTNE for both top-heavy and bottom-heavy configurations. This increased heat and mass transfer in the case of LTNE is possibly due to the extra energy transfer because of the temperature difference between the fluid and particle phases.

Figure 7 explains the heat and mass transport inside the system for top-heavy configuration under LTNE. We notice that convection starts earlier (Fig. 7a–c) in case of composite nanofluid as compared to that of normal nanofluid. Mass transport also starts slightly earlier in case of composite nanofluids for both the nanoparticles (Fig. 7d–e).

Figure 8 describes the transport of heat and mass inside the system for bottom-heavy configuration under LTNE. We notice that convection gets delayed (Fig. 8a–c) in case of composite nanofluid as compared to that of normal nanofluid, i.e., the system gets more stabilized for composite nanofluid in bottom-heavy case. This delay in onset of convection is probably due to the increased nanoparticle concentration near the bottom. Mass transport also gets delayed in case of composite nanofluids for both the nanoparticles (Fig. 8d–e).

In Fig. 9a, b and c, the streamlines, isotherms(fluid) and isohalines (for C_1 or C_2) have been shown, respectively, for t = 0.200, 0.225, 0.250, 0.275 and 0.300 under the case of local thermal equilibrium for composite nanofluid. It can be noticed that at t = 0.200, the magnitude of streamlines is very weak (Fig. 9c) and the isotherms are almost horizontal (Fig. 9a) indicating that initially the heat transfer is mainly due to conduction. Initially, the isohalines are also almost horizontal which depicts that mass transport is very slow. As time passes, the magnitude of streamlines gets stronger, which is indicative of convection taking place in the system. Isotherms are also getting curved which indicates the formation of convective cells. At t = 0.300, significant amount of mass transport can be seen in Fig. 9c. For higher values of time, the heat and mass transport become independent of time and the system achieves a steady state.

In Fig. 10a, b and c, the streamlines, isotherms(fluid) and isohalines (for C_1 or C_2) have been shown, respectively, for similar time as in Fig. 9, under the case of local thermal non-equilibrium for composite nanofluid. It can be easily analyzed that for the same time duration, the magnitude of streamlines is greater (Fig. 10a) than that of in LTE case, showing that the onset of convection advances in case of LTNE. The isotherms at t = 0.300 in Fig. 10b, describe that the transfer of heat is due to strict convection while for the same time t = 0.300 in LTE case, only a transition

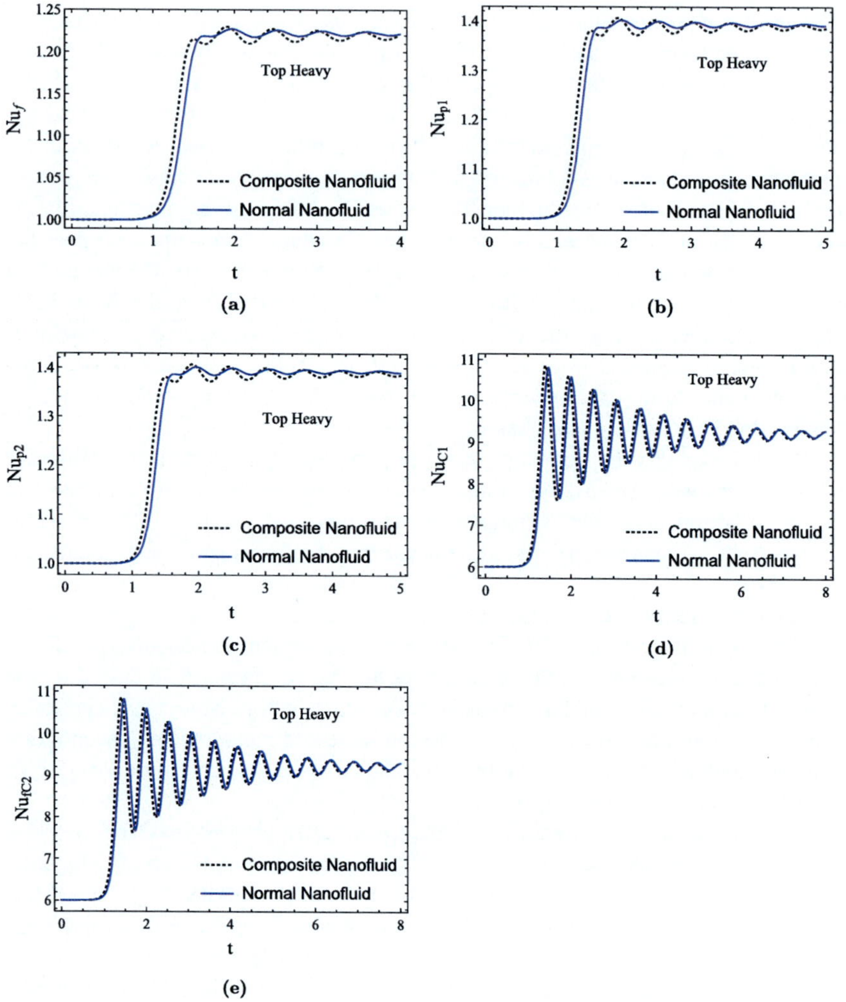

Fig. 7 $Rn_1 = Rn_2 = 5$ for composite nanofluid and $Rn_1 = 5$, $Rn_2 = 0$ for normal nanofluid

from conduction state to convection state can be visualized. More amount of mass transport can also be observed in the LTNE case (Fig. 10c) as compared to that of the LTE case.

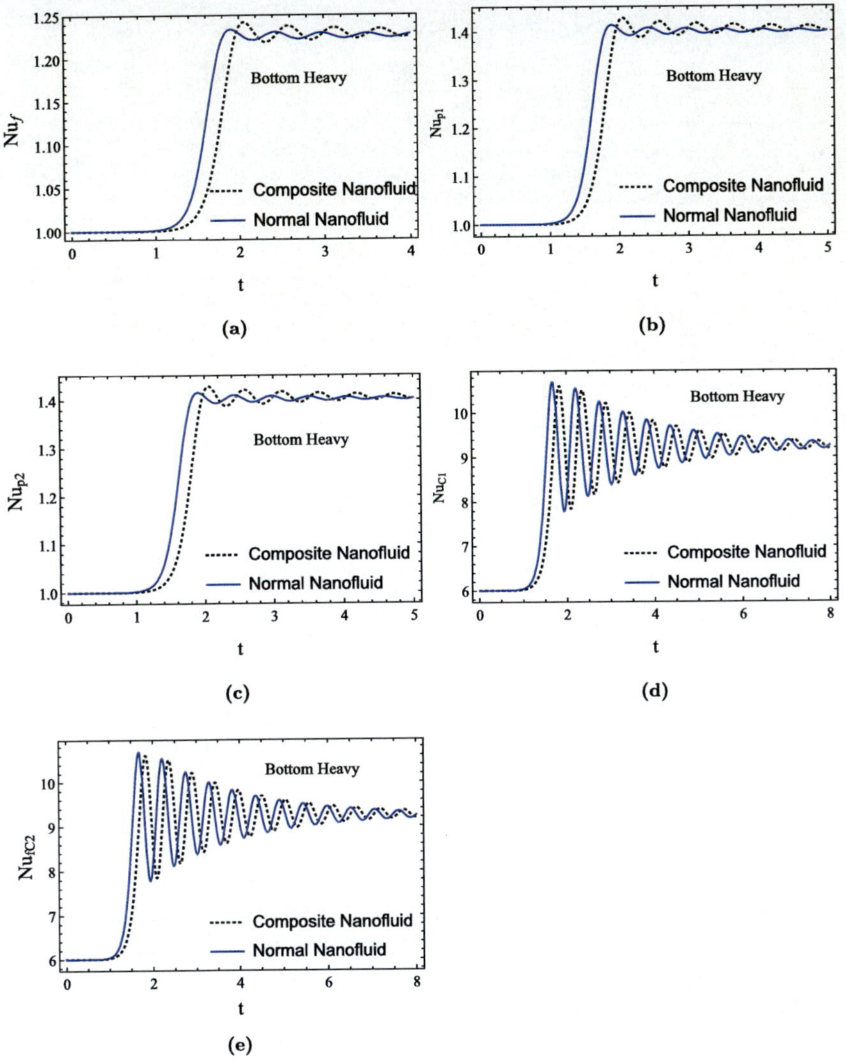

Fig. 8 $Rn_1 = Rn_2 = -5$ for composite nanofluid and $Rn_1 = -5$, $Rn_2 = 0$ for normal nanofluid

8 Conclusions

In order to investigate the effect of local thermal non-equilibrium over convective instability in a composite nanofluid layer, both linear and weakly non-linear stability analyses have been performed under free-free boundary conditions. All the results have been presented using graphs and tables. The prime conclusions are as follows:

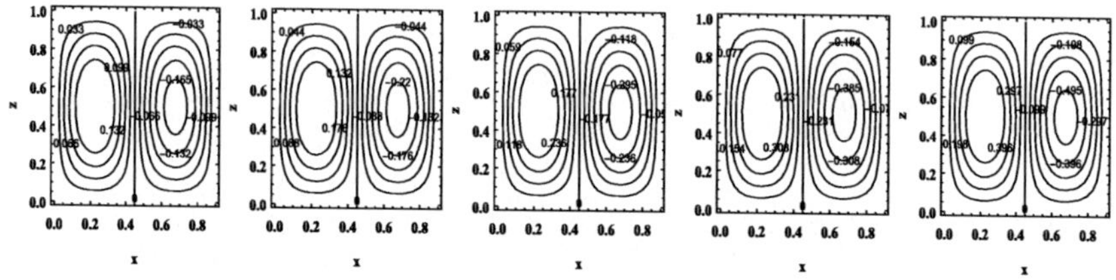

(a) Streamlines for t = 0.200, 0.225, 0.250, 0.275, 0.300

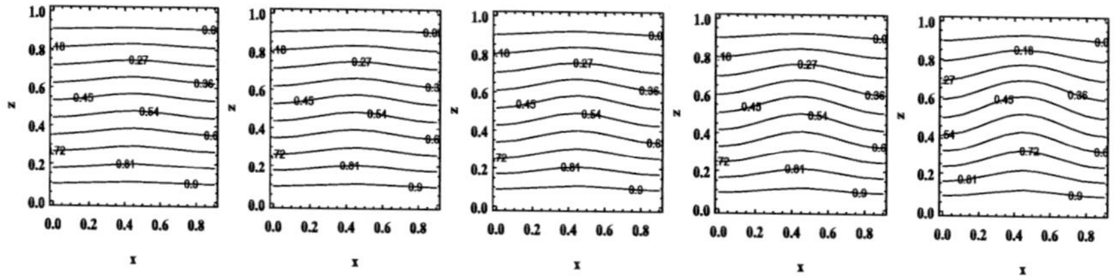

(b) Isotherms(fluid) for t = 0.200, 0.225, 0.250, 0.275, 0.300

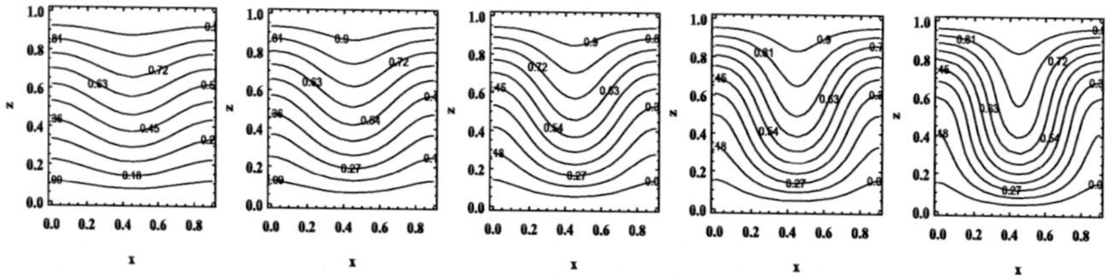

(c) Isohalines(C_1 or C_2) for t = 0.200, 0.225, 0.250, 0.275, 0.300

Fig. 9 Local thermal equilibrium

1. For smaller concentration of nanoparticles, the effect of LTNE is to advance the onset of convection as compared to LTE case for both top-heavy and bottom-heavy configurations.
2. For smaller concentration of nanoparticles, the effect of LTNE dominates over the delay in onset of convection for composite nanofluid in bottom-heavy case.
3. For higher concentration of nanoparticles in bottom-heavy case, the effect of LTNE reduces and onset of convection starts delaying for composite nanofluid as compared to normal nanofluid.
4. The heat transfer between particle and fluid phases takes place only for the intermedial values of both the interphase heat transfer coefficients, which affects the critical Rayleigh number and critical wave number in the LTNE region only.
5. The system is more stable in case of bottom-heavy configuration as compared to that of top-heavy case.
6. Heat and mass transport in fluid and both the nanoparticles in LTNE case is significantly greater than that of LTE case.

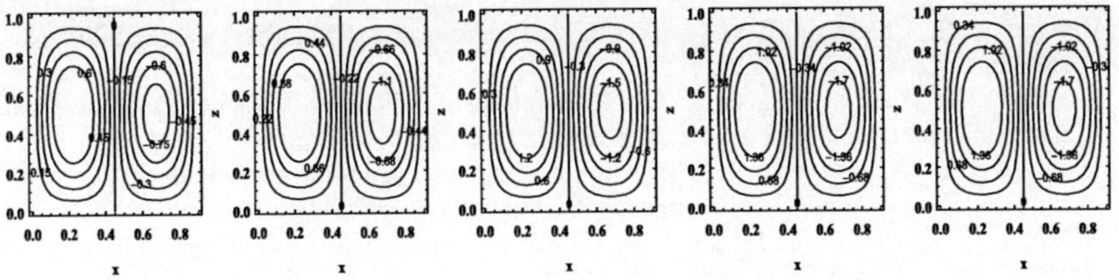

(a) Streamlines for t = 0.200, 0.225, 0.250, 0.275, 0.300

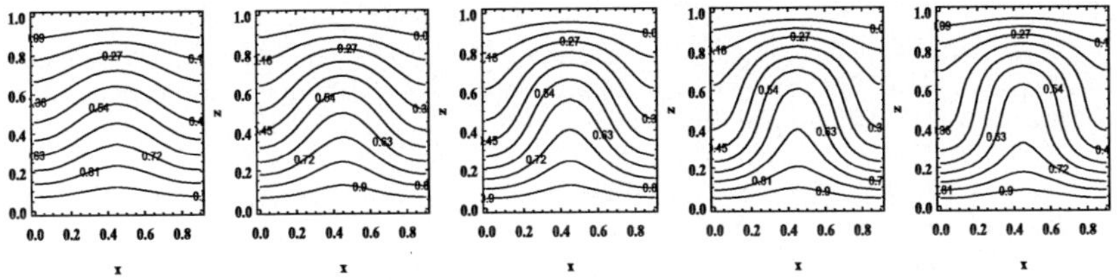

(b) Isotherms(fluid) for t = 0.200, 0.225, 0.250, 0.275, 0.300

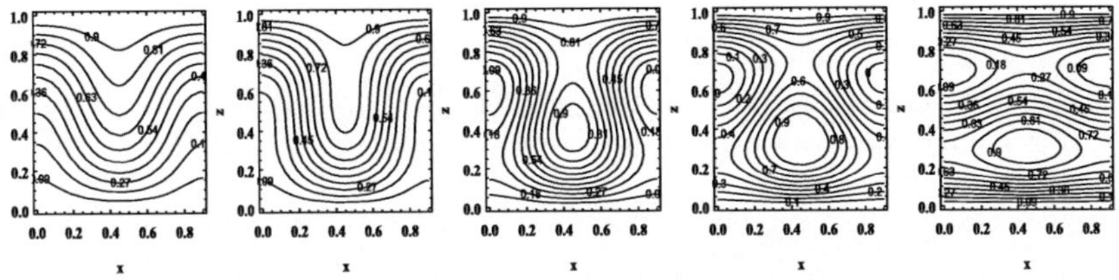

(c) Isohalines(C_1 or C_2) for t = 0.200, 0.225, 0.250, 0.275, 0.300

Fig. 10 Local thermal non-equilibrium

$Nu_f^{LTNE} > Nu_f^{LTE}$, $Nu_{p1}^{LTNE} > Nu_{p1}^{LTE}$, $Nu_{p2}^{LTNE} > Nu_{p2}^{LTE}$, $Nu_{\phi1}^{LTNE} > Nu_{\phi1}^{LTE}$, $Nu_{\phi2}^{LTNE} > Nu_{\phi2}^{LTE}$

7. In composite nanofluid top-heavy configuration, the onset of convection and mass transport advances, while for bottom- heavy configuration, the onset of convection and mass transport delays as compared to normal nanofluid.

References

1. Agarwal, S., Bhadauria, B.S., Siddheshwar, P.G.: Thermal instability of a nanofluid saturating a rotating anisotropic porous medium. STRPM **2**(1) (2011)
2. Agarwal, S., Bhadauria, B.S.: Natural convection in a nanofluid saturated rotating porous layer with thermal non-equilibrium model. Transp. Porous Med. **90**, 627–654 (2011)
3. Agarwal, S., Sacheti, N.C., Chandran, P., Bhadauria, B.S., Singh, A.K.: Non-linear convective transport in a binary nanofluid saturated porous layer. Transp. Porous Med. **93**, 29–49 (2012)

4. Agarwal, S., Bhadauria, B.S.: Convective heat transport by longitudinal rolls in dilute nanoliquids. J. Nanofluids **3**(4) (2014)
5. Agarwal, S., Rana, P., Bhadauria, B.S.: Rayleigh-Bénard convection in a nanofluid layer using a thermal non-equilibrium model. JHT **136**, 122501 (2014)
6. Agarwal, S., Bhadauria, B.S.: Thermal instability of a nanofluid layer under local thermal non-equilibrium. Nano Convergence (2015). https://doi.org/10.1186/s40580-014-0037-z
7. Akilu, S., Sharma, K.V., Baheta, A.V., Mamat, R.: A review of thermophysical properties of water based composite nanofluids. Renew. Sustain. Energy Rev. **66**, 654–678 (2016)
8. Baytas, A.C., Pop, I.: Free convection in a square porous cavity using a thermal non-equilibrium model. Int. J. Therm. Sci. **41**, 861–870 (2002)
9. Baytas, A.C.: Thermal non-equilibrium natural convection in a square enclosure filled with a heat generating solid phase non-Darcy porous medium. Int. J. Energy Res. **27**, 975–988 (2003)
10. Bhadauria, B.S., Srivastava, A.: Combined effect of internal heating and through-flow in a nanofluid saturated porous medium under local thermal nonequilibrium. J. Porous Media **25**(2), 75–95 (2022)
11. Gupta, U., Sharma, J., Sharma, V.: Instability of binary nanofluids with magnetic field. Appl. Math. Mech. **36**(6), 693–706 (2015)
12. Gupta, U., Sharma, J., Devi, M.: Double-diffusive instability of Casson nanofluids with numerical investigations for blood-based fluid. Eur. Phys. J. Spec. Top. **230**, 1435–1445 (2021)
13. Nield, D.A., Kuznetsov, A.V.: Thermal instability in a porous medium layer saturated by a nanofluid: a revised model. Int. J. Heat Mass Transf. **68**, 211–214 (2014)
14. Buongiorno, J.: Convective transport in nanofluids. ASME J. Heat Transfer **128**, 240–250 (2006)
15. Choi, S.: Enhancing thermal conductivity of fluids with nanoparticles. In: Signier, D.A., Wang, H.P. (eds.) Development and Applications of Non-Newtonian Flows, ASME FED, vol. 231/MD vol. 66, pp. 99–105 (1995)
16. Eastman, J.A., Choi, S.U.S., Li, S., Yu, W., Thompson, L.J.: Anomalously increased effective thermal conductivities of ethylene glycol-based nanofluids containing copper nanoparticles. Appl. Phys. Lett. **78**, 718–720 (2001)
17. Das, S.K., Putra, N., Thiesen, P., Roetzel, W.: Temperature dependence of thermal conductivity enhancement for nanofluids. ASME J. Heat Transf. **125**, 567–574 (2003)
18. Hanemann, T., Szabo, D.V.: Polymer-nanoparticle composites: from synthesis to modern applications. Materials **3**, 3468–517 (2010)
19. Kanchana, C., Siddheshwar, P.G., Zhao, Y.: Regulation of heat transfer in Rayleigh-Bénard convection in Newtonian liquids and Newtonian nanoliquids using gravity, boundary temperature and rotational modulations. J. Therm. Anal. Calorim. **142**, 1579–1600 (2020)
20. Khanafer, K., Vafai, K., Lightstone, M.: Buoyancy-driven heat transfer enhancement in a two-dimensional enclosure utilizing nanofluids. Int. J. Heat Mass Transf. **46**, 3639–3653 (2003)
21. Kiran, P., Bhadauria, B.S., Kumar, V.: Thermal convection in a nanofluid saturated porous medium with internal heating and gravity modulation. J. Nanofluids **5**, 1–12 (2016)
22. Kumar, V., Awasthi, M.K.: Thermal instability in a horizontal composite nano-liquid layer. SN Appl. Sci. **2**, 380 (2020)
23. Kumar, R., Sharma, J., Sood, J.: Rayleigh-Bénard cell formation of green synthesized nanoparticles of silver and selenium. Mater. Today: Proc. **28**, 1781–1787 (2020)
24. Kuznetsov, A.V.: Thermal non-equilibrium forced convection in porous media. In: Ingham, D.B., Pop, I. (eds.) Transport Phenomenon in Porous Media, pp. 103–130. Pergamon, Oxford (1998)
25. Lagziri, H., Bezzazi, M.: Robin boundary effects in the darcy-rayleigh problem with local thermal non-equilibrium model. Transport in Porous Media (2019). https://doi.org/10.1007/s11242-019-01301-2
26. Malashetty, M.S., Shivakumara, I.S., Sridhar, K.: The onset of Lapwood-Brinkman convection using a thermal nonequilibrium model. Int. J. Heat Mass Transf. **48**, 1155–1163 (2005)
27. Malashetty, M.S., Shivakumara, I.S., Sridhar, K.: The onset of convection in an anisotropic porous layer using a thermal non-equilibrium model. Transp. Porous Med. **60**, 199–215 (2005)

28. Malashetty, M.S., Swamy, M.S., Heera, R.: Double diffusive convection in a porous layer using a thermal non-equilibrium model. Int. J. Therm. Sci. **47**, 1131–1147 (2008)
29. Nield, D.A., Kuznetsov, A.V.: Thermal instability in a porous medium layer saturated by nonofluid. Int. J. Heat Mass Transf. **52**, 5796–5801 (2009)
30. Kuznetsov, A.V., Nield, D.A.: Thermal instability in a porous medium layer saturated by nonofluid: Brinkman Model. Transp. Porous Media **81**(3), 409–422 (2010)
31. Nield, D.A., Kuznetsov, A.V.: The effect of local thermal nonequilibrium on the onset of convection in a nanofluid. J. Heat Transf. 132/052405-1 (2010)
32. Postelnicu, A., Rees, D.A.S.: The onset of Darcy-Brinkman convection in a porous layer using a thermal nonequlibrium model-part I: stress-free boundaries. Int. J. Energy Res. **27**, 961–973 (2003)
33. Rees, D.A.S., Banu, N.: Onset of Darcy - Bénard convection using a thermal non-equilibrium model. Int. J. Heat Mass Transf. **45**, 2221–2228 (2002)
34. Rees, D.A.S., Pop, I.: Local thermal non-equilibrium in porous medium convection. In: Ingham, D.B., Pop, I. (eds.) Transport Phenomena in Porous Media, vol. III, pp. 147–173. Elsevier, Oxford (2005)
35. Saeid, N.H.: Analysis of mixed convection in a vertical porous layer using non-equilibrium model. Int. J. Heat Mass Transf. **47**, 5619–5627 (2004)
36. Sarkar, J., Ghosh, P., Adil, A.: A review on hybrid nano fluids: recent research, development and applications. Renew. Sustain. Energy Rev. **43**, 164–77 (2015)
37. Siddheshwar, P.G., Siddabasappa, C.: Linear and wealy nonlinear stability analyses of two-dimensional, steady Brinkman-Bénard convection using local thermal non-equilibrium model. Transp. Porous Med. (2017). https://doi.org/10.1007/s11242-017-0943-8
38. Siddheshwar, P.G., Lakshmi, K.M.: Darcy-Bénard convection of Newtonian liquids and Newtonian nanoliquids in cylindrical enclosures and cylindrical annuli. Phys. Fluids **31**, 084102 (2019). https://doi.org/10.1063/1.5109183
39. Sharma, J., Gupta, U.: Double-diffusive nanofluid convection in porous medium with rotation: Darcy-Brinkman model. Proc. Eng. **127**, 783–790 (2015)
40. Sharma, J., Gupta, U., Wanchoo, R.K.: Magneto binary nanofluid convection in porous medium. Int. J. Chem. Eng. **2016**, Article ID 9424036 (2016). https://doi.org/10.1155/2016/9424036
41. Sharma, J., Gupta, U., Wanchoo, R.K.: Numerical study on binary nanofluid convection in a rotating porous layer. Differ. Equ. Dyn. Syst. **25**(2), 239–249 (2017)
42. Sharma, J., Gupta, U.: Nanofluid convection under Hall currents and LTNE effects. Mater. Today: Proc. **26**(3), 3369–3377 (2020)
43. Tzou, D.Y.: Instability of nanofluids in natural convection. ASME J. Heat Transf. **130**(7), 072401 (2008). https://doi.org/10.1115/1.2908427
44. Tzou, D.Y.: Thermal instability of nanofluids in natural convection. Int. J. Heat Mass Transf. **51**(11–12), 2967–2979 (2008). https://doi.org/10.1016/j.ijheatmasstransfer.2007.09.014
45. Zhang, Q., Xu, Y., Wang, X., Yao, W.-T.: Recent advances in noble metal based composite nanocatalysts: colloidal synthesis, properties, and catalytic applications. Nanoscale **7**, 10559–83 (2015)

Investigation of Traffic Dynamics Considering Driver's Characteristics and Downstream Traffic Conditions

Nikita Madaan and Sapna Sharma

Abstract This paper aims to examine the impact of the driver's behavior with the downstream average flow on current traffic dynamics in the lattice hydrodynamic model. The influence of driver's behavior and downstream traffic conditions with different sites are examined theoretically with the help of linear stability. It is observed that traffic flow stability can be improved by incorporating both driver's behavior and the average flow of traffic downstream. Finally, numerical simulations show that present traffic dynamics may be improved by integrating the impacts of driver behavior and average downstream traffic conditions in order to alleviate traffic congestion. Also, it validates the theoretical findings.

Keywords Traffic flow · Lattice model · Downstream average flow · Driver's behavior

1 Introduction

Travel has now become a vital part of most people's daily lives. The rising economy and growing population have also increased congestion in metropolitan areas. To alleviate crowded road conditions while incurring minimal traffic expenditures, management agencies have prioritized transportation security and dependability. Since traffic congestion is rising, some scholars have used mathematics and physics to explain why it occurs and to anticipate how it will evolve through modeling and simulation.

In recent decades, a lot of research has been carried out to resolve the urban traffic issues. Multiple traffic flow models, such as microscopic [3, 4, 35–37] and macroscopic [1, 2, 5–9, 9–34] models have been created to better understand the intricate process of congested roadways. Macroscopic models represent the flow of

N. Madaan (✉) · S. Sharma
School of Mathematics (SOM), Thapar Institute of Engineering and Technology (TIET), Patiala 147004, India
e-mail: nikitamadaan1@gmail.com

R. K. Sharma et al. (eds.), *Frontiers in Industrial and Applied Mathematics*,
Springer Proceedings in Mathematics & Statistics 410,
https://doi.org/10.1007/978-981-19-7272-0_10

traffic by simulating the movement of liquids or gases and explore the overall average behavior of vehicles, whereas microscopic models are discrete models that simulate the individual behavior of vehicles.

Nagatani [5] created the fundamental one-lane unidirectional lattice hydrodynamic model (LHM) in 1998 by integrating characteristics of both microscopic and macroscopic models. This model allowed researchers to investigate the effect of real-world traffic conditions on traffic dynamics. Later, in real traffic flow, numerous different versions of Nagatani's lattice model were explored by investigating various aspects, including optimal current difference [13], driver's behavior [11], density difference effect [8] and, and so on. In addition, the lattice hydrodynamic one-lane unidirectional model is also expanded to include the curved road, two-lane, higher-dimensional lattice model in traffic systems [9, 16, 18–34].

In real-world traffic scenario, the intelligent transportation system (ITS) has been widely used in information and communication systems, making traffic information accessible to drivers in ITS environments more useful than ever before. In 1999, Nagatani [9] introduced a modified car-following model that includes interaction between the next-nearest-neighbor in front. Further, a car-following model is introduced by Kuang et al. [35] based on the effect of average headway. Later, Kuang et al. [34] modified the Zhu et al. model [36] by including the impact of average velocity as well as mean expected velocity field of forwarding vehicles in a vehicle to vehicle interaction. Subsequently, Chuan et al. [37] investigated the impact of multi-anticipation and also examined the influence of forwarding sites in the LHM. Later, Zhu et al. [17] developed a single-lane LHM that took into account the difference between optimal and real traffic flow, based on average density and prior traffic flow.

In regular traffic situations, driver characteristics (timid, aggressive, and normal) have a significant effect on traffic flow. Additionally, numerous studies [11, 12, 14, 15, 18, 19, 29] have been conducted to examine the impact of the behavior of drivers on traffic flow. According to studies, aggressive drivers create a strong impact on the stability of traffic flow, although timid drivers are found to have a negative influence on traffic flow stability. For this reason, it's more realistic to investigate traffic features in terms of the behavior of drivers.

The future era is of semi-automated vehicles. These vehicles partially depend on the information of the surroundings as well as downstream situations. The idea of this paper is to improve the traffic conditions by taking the driver's behavior with the average flow of front sites simultaneously. Therefore, the aim is to create a new lattice model that incorporates average flow on front sites and the behavior of drivers on current traffic conditions.

The following is the outline of the paper. The proposed lattice model, which incorporates the influence of downstream average flow and behavior of drivers on current traffic conditions, was described in Sect. 2 of this paper. Section 3 explains the proposed model's theoretical analysis. Section 4 contains the findings. Section 5 contains the conclusion.

2 Model

Nagatani [5] developed the basic LHM in 1998 to depict the density waves in traffic flow. The basic lattice model consists of two equations: a continuity equation and a flow evolution equation, as follows:

$$\partial_t \rho_j(t) + \rho_0(\rho_j(t)v_j(t) - \rho_{j-1}(t)v_{j-1}(t)) = 0, \tag{1}$$

$$\partial_t(\rho_j(t)v_j(t)) = a[\rho_0 V(\rho_{j+1}(t)) - \rho_j(t)v_j(t)]. \tag{2}$$

Here, ρ_j denotes the density and v_j presents the velocity, respectively at jth site on the one-dimensional lattice for time t. The average density is ρ_0, while a is the sensitivity of the drivers. $V(\rho_{j+1})$ is the Bando's [3, 4] optimal velocity function (OVF), given as

$$V(\rho) = \frac{v_{max}}{2}\left[\tanh\left(\frac{2}{\rho_0} - \frac{\rho}{\rho_0^2} - \frac{1}{\rho_c}\right) + \tanh\left(\frac{1}{\rho_c}\right)\right], \tag{3}$$

In Eq. (3), ρ_c and v_{max} denote the critical density and maximum velocity, respectively. In reality, drivers constantly analyze the state of the road ahead of them and attempt to adjust their vehicle's speed in response to the information received from ITS. Further, to explore the traffic situations more realistically, we propose a LHM to examine the driver's behavior while taking into account downstream average flow on forward sites. Thus, the continuity equation is the same in the new LHM, but the flow equation is reformed as

$$\begin{aligned}\partial_t(\rho_j v_j) = a\Big[\rho_0 V(\rho_{j+1}) - \rho_j v_j + \alpha(2p-1)\tau V'(\rho_{j+1})\partial_t \rho_{j+1}\Big] \\ + \lambda\Big[q_j^{avg} - \rho_j v_j\Big].\end{aligned} \tag{4}$$

The delay time is given by $\tau = 1/a$, and the anticipation coefficient is denoted by α in Eq. (4). The parameter $p \in [0, 1]$ demonstrates that how the behavior of the drivers impacts the traffic dynamics. Whenever $p < 0.5$, the driver exhibits the timid behavior; when $p = 0.5$, it exhibits normal behavior; and whenever $p > 0.5$, it exhibits aggressive behavior. Average flow difference is represented by λ and $q_j^{avg} = \frac{1}{n}\sum_{l=1}^{n}(\rho_{j+l}v_{j+l})$ is the average flow of the n forward sites.

After omitting v from Eqs. (1) and (4), the resulting density evolution equation be obtained as

$$\begin{aligned}\partial_t^2(\rho_j) + (\lambda + a)\partial_t \rho_j - \frac{\lambda}{n}\left(\sum_{l=1}^{n}\partial_t \rho_{j+l}\right) + a\alpha\rho_0^2\tau(2p-1)[V'(\rho_{j+1})\partial_t \rho_{j+1} \\ - V'(\rho_j)\partial_t \rho_j] + a\rho_0^2\Big(V(\rho_{j+1}) - V(\rho_j)\Big) = 0.\end{aligned} \tag{5}$$

In the new model, when $\alpha = 0$ or $p = 1/2$ with $n = 1$, it reduces to the Tian et al. model [7]. Furthermore, this model is identical to Nagatani's [5] model with $\alpha = 0$ or $p = 1/2$ and $\lambda = 0$.

3 Theoretical Analysis

To examine qualitative features of proposed LHM, we apply linear stability analysis. Consider a traffic flow with a constant density of ρ_0 and an optimal velocity of $V(\rho_0)$. As a result, traffic uniformity may be achieved by

$$\rho_j(t) = \rho_0, \quad v_j(t) = V(\rho_0), \tag{6}$$

where $V'(\rho_0) = \frac{dV(\rho)}{d\rho} |_{\rho=\rho_0}$. After adding a tiny fluctuation ($y_j(t)$) into the condition of smooth flow of traffic, i.e., $\rho_j(t) = \rho_0 + y_j(t)$ and using modified density in Eq. (5). Applying linearization, we obtain

$$\partial_t^2 y_j + (\lambda + a)\partial_t y_j - \frac{\lambda}{n} \sum_{l=1}^{n} (\partial_t y_{j+l}) + a\alpha\rho_0^2 V'(\rho_0)\tau(2p-1)(\partial_t y_{j+1} - \partial_t y_j)$$
$$+a\rho_0^2 V'(\rho_0)(y_{j+1} - y_j) = 0 \tag{7}$$

Now, in Eq. (7), we can describe the deviation $y_j(t)$ as an exponential function, i.e. $y_j(t) = exp(\iota\kappa j + \eta t)$, we get:

$$\eta^2 + (a+\lambda)\eta - \frac{\lambda}{n}\eta\Big(\sum_{l=1}^{n}(e^{ikl})\Big) + a\alpha\rho_0^2\tau(2p-1)V'(\rho_0)\eta(e^{\iota\kappa} - 1)$$
$$+ a\rho_0^2 V'(\rho_0)(e^{\iota\kappa} - 1) = 0. \tag{8}$$

On inserting $\eta = \eta_1(\iota\kappa) + \eta_2(\iota\kappa)^2 ...$ in Eq. (8), coefficients of $(\iota\kappa)$ and $(\iota\kappa)^2$ of the first and second order were obtained as follows:

$$\eta_1 = -\rho_0^2 V'(\rho_0), \tag{9}$$

$$\eta_2 = -\frac{\rho_0^2 V'(\rho_0)}{2} - \frac{(\rho_0^2 V'(\rho_0))^2}{a} - \frac{\alpha(\rho_0^2 V'(\rho_0))^2}{a} - \frac{\lambda\rho_0^2 V'(\rho_0)(n+1)}{2a}. \tag{10}$$

Homogeneous flow is uncertain for longer wavelength with $\eta_2 < 0$, but becomes stable with $\eta_2 > 0$. So, the neutral stability criterion is as follows:

$$a = -2\rho_0^2 V'(\rho_0)(1 - \alpha(2p-1)) - \lambda(n+1). \tag{11}$$

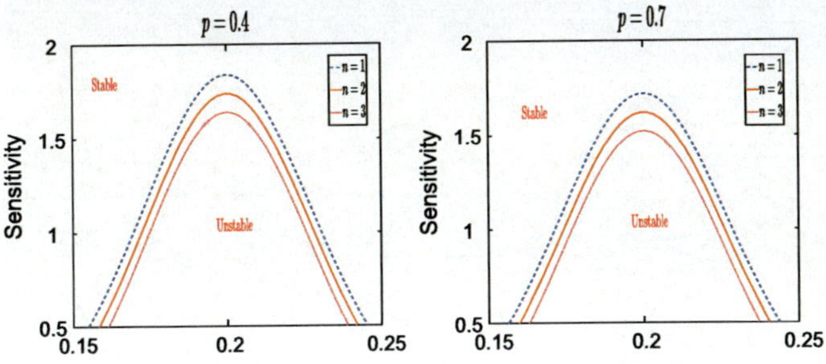

(a) (b)

Fig. 1 Phase plot for distinct number of forwarding sites n in parameter space (ρ, a), when $\alpha = 0.1$ and $\lambda = 0.1$ **a** $p = 0.4$, **b** $p = 0.7$

The following criterion can be used to stabilize a uniform flow:

$$a > -2\rho_0^2 V'(\rho_0)(1 - \alpha(2p - 1)) - \lambda(n + 1). \tag{12}$$

Figures 1a–b show the phase plot for distinct number of forwarding sites (n) and p, while all other parameters remain constant in the parameter space (ρ, a). Figure 1 illustrates the neutral stability curves. The apex of each curve reflects the crucial point (ρ_c, a_c) in the respective curves. In this manner, the phase plot is separated into stable and unstable regions. As seen in Fig. 1a, the amplitude of neutral stability curves reduces as the number of forwarded sites (n) increases when $p = 0.4$, implying that the stability of uniform traffic flow has been improved by using downstream average flow information. Also, it can be seen in Fig. 1b, that the sensitivity decreases as n increases with $p = 0.7$, indicating the widening of the stability region. This demonstrates that by considering both impacts simultaneously, i.e., downstream average flow and effect of behavior of drivers on traffic flow can help in strengthening the traffic flow stability.

4 Numerical Simulation

Numerical simulation with periodic boundary conditions is used to verify theoretical results. The following initial conditions are preferred:

$$\rho_j(0) = \begin{cases} \rho_0; & j \neq \frac{M}{2}, \frac{N}{2} - 1 \\ \rho_0 - \sigma; & j = \frac{N}{2} \\ \rho_0 + \sigma; & j = \frac{N}{2} - 1 \end{cases} \tag{13}$$

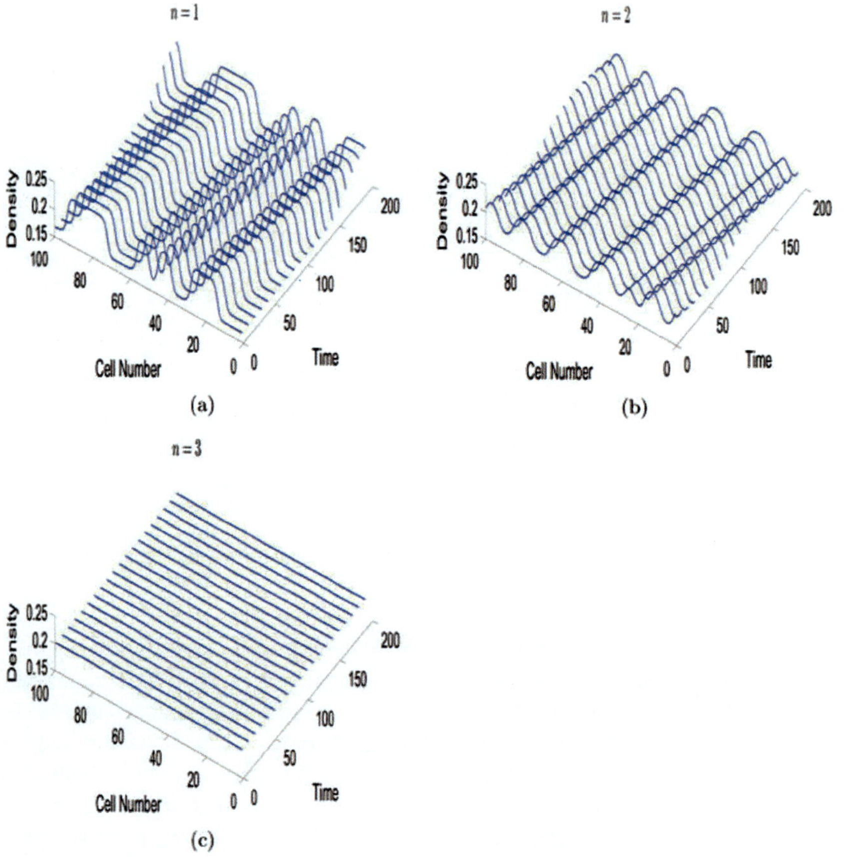

Fig. 2 When $a = 1.65$ and $p = 0.4$, the spatial-temporal evolution of density waves for distinct values of n

The associated variables are as follows: $\rho_0 = \rho_c = 0.2$, $\lambda = 0.1$, $v_{max} = 2$, $\alpha = 0.1$, and $t = 2 \times 10^4$ s. Here, $N = 100$ and $\sigma = 0.05$ represents the total number of sites and initial perturbation, respectively.

Figure 2 shows the space-time density wave for distinct values of forwarding sites (i.e., $n = 1, 2,$ and 3) at $t = 10^4$s, when $p = 0.4$, and $a = 1.65$. The density waves in the pattern of Fig. 2a, b are kink-antikink, since the stability requirement (Eq.(12)) is not met, and the flow goes from uniform to congested after the tiny disturbance. From the figures, one can observe that the kink-antikink density waves occurs for smaller values of n and propagates backwards. Further, when the value of n increases, stability region increases, especially for $n = 3$, the amplitude of density

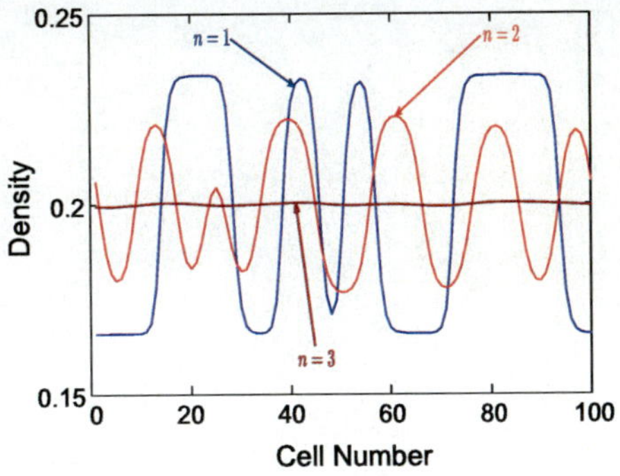

Fig. 3 Density patterns for distinct values of n at $t = 10000\,\text{s}$ with $a = 1.65$ and $p = 0.4$

wave vanishes completely. We found that if forward lattices are more than 3, then also it satisfies the stability condition. It indicates that traffic congestion can be reduced by having information about forward lattices.

The density pattern for distinct values of n with $p = 0.4$ shown in Fig. 3, which corresponds to Fig. 2. As n increases, the density wave's amplitude reduces, and finally, the flow goes into the homogeneous steady state for $n = 3$.

Figure 4 indicates the spatio-temporal density wave profiles for distinct values of forwarding sites (i.e., $n = 1, 2,$ and 3) at $t = 10^4$s, when $p = 0.7$, and $a = 1.5$. The density waves in the pattern of Fig. 4a, b demonstrate that an initial perturbation results in the kink-antikink solution propagating backward direction. When the instability criteria (Eq.(12)) is fulfilled, the flow transits from uniform to congested. The amplitude of density wave diminishes with the increase in n, however, as $n = 3$, the stability region increases.

The density pattern for distinct values of n with $p = 0.7$ shown in Fig. 5, which corresponds to Fig. 4. As n increases, the density wave's amplitude reduces, and for $n = 3$, the amplitude of density wave vanishes completely, which shows that the knowledge of prospective sites flow can help in minimizing the traffic congestion.

After examining all the simulation findings, we noticed that all the simulation results are completely similar to the theoretical results presented in the previous section. Also, in real traffic phenomenon, it is feasible for the drivers to adjust their speed, if they have adequate information about the forward traffic situation and then the traffic congestion reduces. All these results show that the information about the diver's behavior and the downstream average flow on forward sites is crucial in improving traffic flow stability.

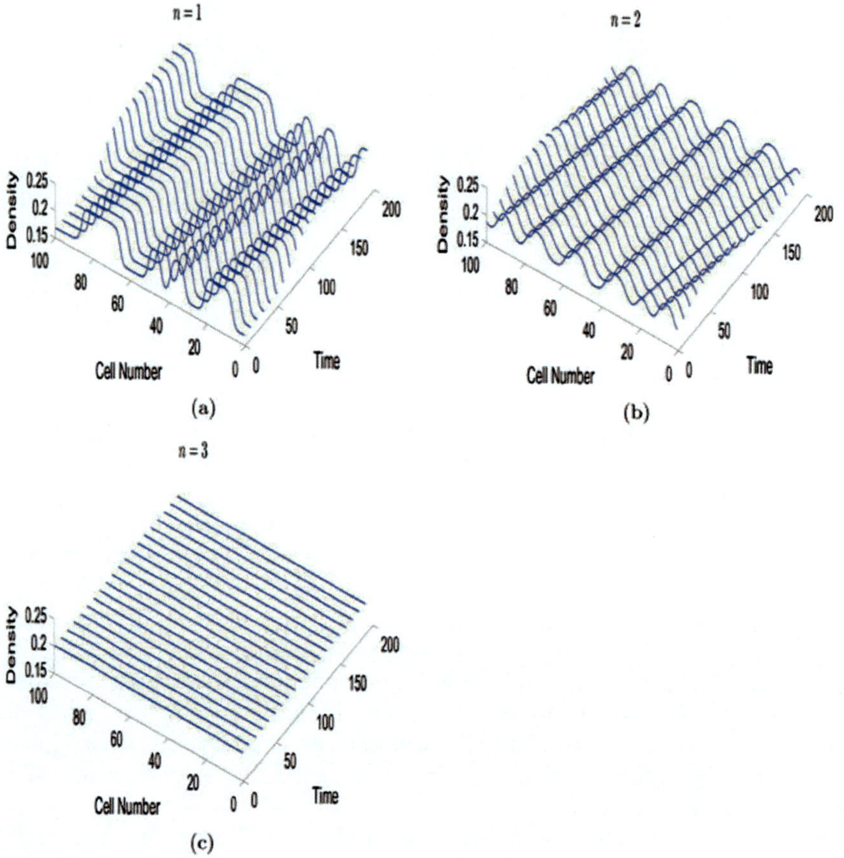

Fig. 4 When $a = 1.5$ and $p = 0.7$, the spatio-temporal evolution of density waves for distinct values of n

Fig. 5 Density patterns for distinct values of n at $t = 10000\,\text{s}$ with $a = 1.5$ and $p = 0.7$

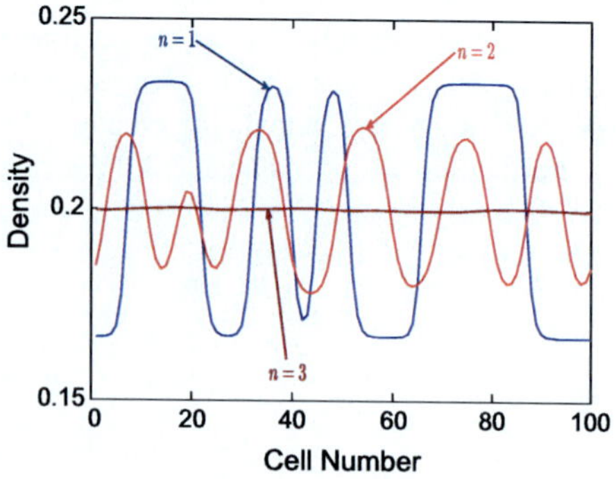

5 Conclusion

The current study presents a LHM for examining the influence of driver's behavior with downstream average flow on traffic dynamics. The stability condition of traffic dynamics is analyzed via theoretical analysis. From the study of the phase diagram, it is depicted that the new model considering the average flow of front sites with driver's behavior has a greater influence on reducing the traffic congestion than the basic lattice model. Furthermore, the numerical findings correspond well with the theoretical conclusions. Thus, it is prominent to make an aspect that the current traffic dynamics are influenced by the forward traffic information and it is favorable in reducing the traffic congestion.

References

1. Chowdhury, D., Santen, L., Schadschneider, A.: Statistical physics of vehicular traffic and some related systems. Phys. Rep. **329**(4–6), 199–329 (2000)
2. Nagatani, T.: The physics of traffic jams. Rep. Prog. Phys. **65**(9), 1331 (2002)
3. Bando, M., Hasebe, K., Nakayama, A., Shibata, A., Sugiyama, Y.: Structure stability of congestion in traffic dynamics. Jpn. J. Ind. Appl. Math. **11**(2), 203–223 (1994)
4. Bando, M., Hasebe, K., Nakayama, A., Shibata, A., Sugiyama, Y.: Dynamical model of traffic congestion and numerical simulation. Phys. Rev. E **51**(2), 1035 (1995)
5. Nagatani, T.: Modified KdV equation for jamming transition in the continuum models of traffic. Phys. A **261**(3), 599–607 (1998)
6. Ge, H., Cheng, R.: The backward looking effect in the lattice hydrodynamic model. Phys. A **387**(28), 6952–6958 (2008)
7. Jun-Fang, T., Bin, J., Xing-Gang, L., Zi-You, G.: Flow difference effect in the lattice hydrodynamic model. Chin. Phys. B **19**(4), 040303 (2010)
8. Tian, J.F., Yuan, Z.Z., Jia, B., Li, M.H., Jiang, G.H.: The stabilization effect of the density difference in the modified lattice hydrodynamic model of traffic flow. Phys. A: Stat. Mech. Appl. **391**, 4476–4482 (2012)
9. Nagatani, T.: Stabilization and enhancement of traffic flow by the next-nearest-neighbor interaction. Phys. Rev. E **60**, 6395–6401 (1999)
10. Gupta, A.K., Redhu, P.: Analyses of the driver's anticipation effect in a new lattice hydrodynamic traffic flow model with passing. Nonlinear Dyn. **76**(2), 1001–1011 (2014)
11. Li, X.Q., Fang, K.L., Peng, G.H.: A new lattice model of traffic flow with the consideration of the driver's aggressive characteristics. Phys. A **468**, 315–321 (2017)
12. Sharma, S.: Modeling and analyses of driver's characteristics in a traffic system with passing. Nonlinear Dyn. **86**(3), 2093–2104 (2016)
13. Tian, C., Sun, D., Zhang, M.: Nonlinear analysis of lattice model with consideration of optimal current difference. Commun. Nonlinear Sci. Numer. Simul. **16**(11), 4524–4529 (2011)
14. Kaur, R., Sharma, S.: Analysis of driver's characteristics on a curved road in a lattice model. Phys. A **471**, 59–67 (2017)
15. Madaan, N., Sharma, S.: Effects of multi-phase optimal velocity function on a lattice model accounting for driver's behavior. Int. J. Mod. Phys. B **33**(22), 1950248 (2019)
16. Kaur, D., Sharma, S.: The impact of the predictive effect on traffic dynamics in a lattice model with passing. Eur. Phys. J. B **93**(3), 1–10 (2020)
17. Zhu, C., Ling, S., Zhong, S., Liu, L.: A modified lattice model of traffic flow with the consideration of the downstream traffic condition. Mod. Phys. Lett. B **33**(02), 1950008 (2019)

18. Zhang, G., Sun, D., Liu, W., Zhao, M., Cheng, S.: Analysis of two-lane lattice hydrodynamic model with consideration of driver's characteristics. Phys. A **422**, 16–24 (2015)
19. Sharma, S.: Lattice hydrodynamic modeling of two-lane traffic flow with timid and aggressive driving behavior. Phys. A **421**, 401–411 (2015)
20. Nagatani, T.: Jamming transitions and the modified Korteweg'de Vries equation in a two-lane traffic flow. Phys. A **265**, 297–310 (1999)
21. Gupta, A.K., Redhu, P.: Analyses of driver's anticipation effect in sensing relative flux in a new lattice model for two-lane traffic system. Phys. A **392**(22), 5622–5632 (2013)
22. Wang, T., Gao, Z., Zhang, J., Zhao, X.: A new lattice hydrodynamic model for two-lane traffic with the consideration of density difference effect. Nonlinear Dyn. **75**, 27–34 (2014)
23. Peng, G.H., Kuang, H., Zhao, H., Qing, L.: Nonlinear analysis of a new lattice hydrodynamic model with the consideration of honk effect on flux for two-lane highway. Phys. A **515**, 93–101 (2019)
24. Tao, W., Zi-You, G., Xiao-Mei, Z., Jun-Fang, T.: Flow difference effect in the two-lane lattice hydrodynamic model. Chinese Phys. B **21**(7), 070507 (2012)
25. Sharma, S.: Effect of driver's anticipation in a new two-lane lattice model with the consideration of optimal current difference. Nonlinear Dyn. **81**(1–2), 991–1003 (2015)
26. Kaur, D., Sharma, S.: A new two-lane lattice model by considering predictive effect in traffic flow. Phys. A: Stat. Mech. Appl. **539**, 122913 (2020)
27. Gupta, A.K., Redhu, P.: Analysis of a modified two-lane lattice model by considering the density difference effect. Commun. Nonlinear Sci. Numer. Simul. **19**(5), 1600–1610 (2014)
28. Qi, X., Cheng, R., Ge, H.: Analysis of a novel two-lane lattice model with consideration of density integral and relative flow information. Eng. Comput. **37**, 2939–2955 (2020)
29. Qi, X., Ge, H., Cheng, R.: Analysis of a novel two-lane hydrodynamic lattice model accounting for Driver's aggressive effect and flow difference integral. Math. Probl. Eng. **2020** (2020). (https://doi.org/10.1155/2020/8258507)
30. Zhang, J., Xu, K., Li, S., Wang, T.: A new two-lane lattice hydrodynamic model with the introduction of driver's predictive effect. Phys. A: Stat. Mech. Appl. **551**, 124249 (2020)
31. Madaan, N., Sharma, S.: A lattice model accounting for multi-lane traffic system. Phys. A: Stat. Mech. Appl. **564**, 125446 (2021)
32. Kaur, D., Sharma, S.: Prior information affecting traffic dynamics in a two dimensional (2D) network. Eur. Phys. J. B **94**(9), 1–12 (2021)
33. Redhu, P., Gupta, A.K.: Effect of forward looking sites on a multi-phase lattice hydrodynamic model. Phys. A **445**, 150–160 (2016)
34. Kuang, H., Wang, M.T., Lu, F.H., Bai, K.Z., Li, X.L.: An extended car-following model considering multi-anticipative average velocity effect under V2V environment. Phys. A: Stat. Mech. Appl. **527**, 121268 (2019)
35. Kuang, H., Xu, Z.P., Li, X.L., Lo, S.M.: An extended car-following model accounting for the average headway effect in intelligent transportation system. Phys. A **471**, 778–787 (2017)
36. Zhu, W.X., Zhang, L.D.: A new car-following model for autonomous vehicles flow with mean expected velocity field. Phys. A **492**, 2154–2165 (2018)
37. Chuan, T., Di-Hua, S., Shu-Hong, Y.: A new lattice hydrodynamic traffic flow model with a consideration of multi-anticipation effect. Chinese Phys. B **20**(8), 088902 (2011)

Fractal Convolution Bessel Sequences on Rectangle

R. Pasupathi, M. A. Navascués, and A. K. B. Chand

Abstract Fractal functions provide a natural deterministic approximation of complex phenomena and also it has self-similarity. Recently, it has been recognized as an internal binary operation, called fractal convolution. In the present article, we obtain Bessel sequences of $L^2(\mathcal{I} \times \mathcal{J})$ composed of product of fractal convolutions, using the identification of $L^2(\mathcal{I} \times \mathcal{J})$ with the tensor product space $L^2(\mathcal{I}) \otimes L^2(\mathcal{J})$, where $\mathcal{I}$ and $\mathcal{J}$ are real compact intervals.

Keywords Fractals · Attractor · Fractal interpolation function · Convolution · Bessel sequences

1 Introduction

The most natural and scientific phenomena result in patterns that are highly complicated like snowflakes, electromagnetic waves, etc. These patterns are hard to be described by the usual Euclidean geometry as they don't resemble simple shapes with smooth margins. In his famous book "The Fractal Geometry of Nature", Mandelbrot [12] introduced the concept of fractal to capture non-uniformity in nature and in modeling a variety of phenomena in applied mathematics and engineering: image processing, bio-engineering, signal processing, turbulence, etc. (see, for instance, [3–6, 11, 15, 21, 25]). Fractal geometry plays a major role for modeling objects with infinite details in nature.

R. Pasupathi (✉) · A. K. B. Chand
Department of Mathematics, Indian Institute of Technology Madras, Chennai 600036, India
e-mail: pasupathi4074@gmail.com

A. K. B. Chand
e-mail: chand@iitm.ac.in

M. A. Navascués
Departmento de Matemática Aplicada, Escuela de Ingeniería y Arquitectura, Universidad de Zaragoza-50018, Zaragoza, Spain
e-mail: manavas@unizar.es

R. K. Sharma et al. (eds.), *Frontiers in Industrial and Applied Mathematics*,
Springer Proceedings in Mathematics & Statistics 410,
https://doi.org/10.1007/978-981-19-7272-0_11

Barnsley [1, 2] introduced the theory of Fractal Interpolation Function (FIF) using the concept of Iterated Function System (IFS) introduced by Hutchinson [9]. FIFs are defined as the fixed points of maps between spaces of functions. Some of the advantages of FIFs are : (i) FIFs retain self-similarity under magnification, (ii) for the suitable choice of scaling vectors, FIF can provide smooth or non-smooth approximations and one more remarkable advantage is that (iii) the graph of these approximations provide a non-integer dimension with respect to the Hausdorff magnitude of scaling factor.

FIF was introduced as a continuous function interpolating the prescribed data set and the graph of the fractal function is the attractor of a suitable IFS. Many authors developed the theory of FIFs, both in the univariate and multivariate settings (see for example, [13, 14, 24, 26]). For any continuous function f defined on a compact interval $\mathcal{I}$, we can get a continuous fractal function $f^{\omega}_{\Gamma,b}$ associated with the partition Γ of $\mathcal{I}$, base function b in $C(\mathcal{I})$ and scale vector ω, which interpolates the given function f on the nodes of Γ. The operator which takes $f \to f^{\omega}_{\Gamma,b}$ on $C(\mathcal{I})$ is called the fractal operator. The fractal operator can be extended to the Lebegue spaces $L^p(\mathcal{I})$, and in such manner, we construct fractal perturbation of p-integrable functions. Navascués and group [16, 17, 19, 23] have studied various properties of fractal functions by using the fractal operator. Latterly, the fractal function $f^{\omega}_{\Gamma,b}$ has been realized as the binary operation $f * b$ between the seed function f and the base function b in $L^p(\mathcal{I})$, where the partition and the scale vector are fixed [18, 20].

In this paper, we have constructed fractal convolution Bessel sequences of $L^2(\mathcal{I} \times \mathcal{J})$, by using the identification of $L^2(\mathcal{I} \times \mathcal{J})$ with the tensor product space $L^2(\mathcal{I}) \otimes L^2(\mathcal{J})$. Turning to the structure of our paper, in Sect. 2, we recall the concept of the tensor product of two Hilbert spaces. In Sect. 3, we give a brief outline of the fractal functions based on the notion of FIF, and by using the fractal function, we define fractal convolution in $L^p(\mathcal{I})$. Finally, in Sect. 4, we have constructed partial fractal convolution operators in $L^2(\mathcal{I})$ and $L^2(\mathcal{J})$ with fixing the null function, where $\mathcal{J}$ is also a real compact interval. From these operators and the identification of $L^2(\mathcal{I} \times \mathcal{J})$, we construct Bessel sequences of $L^2(\mathcal{I} \times \mathcal{J})$ composed of product of fractal functions.

2 Tensor Product of Hilbert Spaces

Let us recall the basic concepts of the tensor product of two Hilbert spaces $\mathcal{H}_1$ and $\mathcal{H}_2$ and tensor product of two linear bounded operators. For reference, the reader can see [8, 10].

Definition 1 Let $\mathcal{H}_1$ and $\mathcal{H}_2$ be Hilbert spaces on the field of real numbers. Then the tensor product of $\mathcal{H}_1$ and $\mathcal{H}_2$ is the collection $\mathcal{H}_1 \otimes \mathcal{H}_2$ defined by

$$\mathcal{H}_1 \otimes \mathcal{H}_2 := \{Q : \mathcal{H}_2 \to \mathcal{H}_1 : Q \text{ linear, bounded}, \sum_j \|Qe_j\|^2 < \infty\},$$

where $\{e_j\}$ is an orthonormal basis of $\mathcal{H}_2$.

If $\{l_i\}$ is an orthonormal basis of $\mathcal{H}_1$, then we have

$$\sum_j \|Qe_j\|^2 = \sum_j \sum_i |\langle Qe_j, l_i\rangle|^2 = \sum_i \sum_j |\langle e_j, Q^* l_i\rangle|^2 = \sum_i \|Q^* l_i\|^2,$$

where Q^* is the adjoint of Q. Then the sum $\sum_j \|Qe_j\|^2$ is independent of the basis chosen in $\mathcal{H}_2$. The space $\mathcal{H}_1 \otimes \mathcal{H}_2$ is a Hilbert space with the inner product

$$\langle Q, R\rangle = \sum_j \langle Qe_j, Re_j\rangle,$$

and this induced norm is

$$|||Q||| = (\sum_j \|Qe_j\|^2)^{1/2} = |||Q^*|||.$$

Now we define the tensor product of vectors $p \in \mathcal{H}_1$ and $q \in \mathcal{H}_2$ as the operator $p \otimes q : \mathcal{H}_2 \to \mathcal{H}_1$

$$(p \otimes q)(h) = \langle q, h\rangle p \qquad h \in \mathcal{H}_2. \tag{1}$$

and note that

(i) $\overline{span}(\{p \otimes q : p \in \mathcal{H}_1, q \in \mathcal{H}_2\}) = \mathcal{H}_1 \otimes \mathcal{H}_2$.
(ii) $\langle p \otimes q, \tilde{p} \otimes \tilde{q}\rangle = \langle p, \tilde{p}\rangle\langle q, \tilde{q}\rangle$.

We now consider the tensor product of bounded linear operators.

Definition 2 If T_1 and T_2 are bounded linear operators of $\mathcal{H}_1$ and $\mathcal{H}_2$, respectively, then define the tensor product of T_1 and T_2 by

$$T_1 \otimes T_2 : \mathcal{H}_1 \otimes \mathcal{H}_2 \to \mathcal{H}_1 \otimes \mathcal{H}_2$$

$$(T_1 \otimes T_2)P = T_1 P T_2^* \qquad P \in \mathcal{H}_1 \otimes \mathcal{H}_2.$$

Some basic properties of this tensor product are:

(a) $T_1 \otimes T_2$ is a linear bounded operator on $\mathcal{H}_1 \otimes \mathcal{H}_2$.
(b) $\|T_1 \otimes T_2\| = \|T_1\|\|T_2\|$, where $\|T_1 \otimes T_2\|$ is the operator norm of $T_1 \otimes T_2$ on $\mathcal{H}_1 \otimes \mathcal{H}_2$ with respect to $|||.|||$.
(c) $(T_1 \otimes T_2)(p \otimes q) = T_1(p) \otimes T_2(q)$.
(d) $(T_1 \otimes T_2)^{-1} = T_1^{-1} \otimes T_2^{-1}$.

Theorem 1 ([8] Theorem 7.16) *The identification of $p \otimes q$ with the function $p(x)q(y)$ extends uniquely to an isometric isomorphism of $L^2(\mathcal{I}) \otimes L^2(\mathcal{J})$ with $L^2(\mathcal{I} \times \mathcal{J})$ whose inverse identifies $P \in L^2(\mathcal{I} \times \mathcal{J})$ with the operator $s \to \int_{\mathcal{J}} P(., y)s(y)dy$, where $s \in L^2(\mathcal{J})$.*

If $p \in L^2(\mathcal{I})$ and $q \in L^2(\mathcal{J})$, we have

$$(p \otimes q)(s) = \langle q, s\rangle p = \int_{\mathcal{J}} p(.)q(y)s(y)dy.$$

The kernel of $p \otimes q$ is $p(x)q(y)$. Similarly, the operator $\sum_{i=1}^{n} p_i \otimes q_i$ has the kernel $\sum_{i=1}^{n} p_i(x)q_i(y)$.

Thus, by Theorem 1, the space $L^2(\mathcal{I} \times \mathcal{J})$ is identified with the tensor product space $L^2(\mathcal{I}) \otimes L^2(\mathcal{J})$.

3 Fractal Functions and Fractal Convolutions in $L^p(\mathcal{I})$

In this section, we provide a family of fractal functions in $L^p(\mathcal{I})$ (Lebegue spaces) in the notion of FIF. For details the reader is referred to [16, 17].

Let $\Gamma : \{\gamma_0, \gamma_1, \ldots, \gamma_M\}$ be a partition of the interval $\mathcal{I} = [\gamma_0, \gamma_M]$ such that $\gamma_0 < \gamma_1 < \cdots < \gamma_M$ and set $\mathcal{I}_l = [\gamma_{l-1}, \gamma_l]$, for all $l = 1, 2, \ldots, M$. The function $L_l : \mathcal{I} \to \mathcal{I}_l, l = 1, 2, \ldots, M$ is defined as

$$L_l(\gamma) = a_l\gamma + b_l,$$

where $$a_l = \frac{\gamma_l - \gamma_{l-1}}{\gamma_M - \gamma_0}, \qquad b_l = \frac{\gamma_M\gamma_{l-1} - \gamma_0\gamma_l}{\gamma_M - \gamma_0}.$$

Let us denote $L^p(\mathcal{I}), 1 \leq p < \infty$ as the collection of real-valued Lebegue integrable functions defined on $\mathcal{I}$ with respect to the L^p-norm defined by

$$\|k\|_p := \left(\int_{\mathcal{I}} |k(\gamma)|^p d\gamma\right)^{1/p}.$$

Let $f \in L^p(\mathcal{I})$ be the prescribed function, called the seed function or the germ function. Fix a function $b \in L^p(\mathcal{I})$, referred as the base function and finally choose the free variable $\omega = (\omega_l)_{l=1}^{M}$, ω_l is real for $l \in \{1, \ldots, M\}$ such that $\Omega := \max\{|\omega_l| : l = 1, 2, \ldots, M\} < 1$, which is called the scale vector.

We obtain FIF $f_{\Gamma,b}^{\omega} \in L^p(\mathcal{I})$ with respect to the germ function f, base function b, partition Γ, and the scale vector ω as the fixed point of the Read-Bajraktarević (RB) operator $T_{f,b,\Gamma}^{\omega} : L^p(\mathcal{I}) \to L^p(\mathcal{I})$ is defined by

$$T_{f,b,\Gamma}^{\omega} g(\gamma) = f(\gamma) + \omega_l (g - b) \circ L_l^{-1}(\gamma) \quad \forall\, \gamma \in \mathcal{I}_l,\ l = 1, 2, \ldots, M.$$

The magnitude of the scale vector gives the RB operator $T^{\omega}_{f,b,\Gamma}$ which is a contraction map on the Banach space $L^p(\mathcal{I})$. Consequently, the Banach contraction principle says $T^{\omega}_{f,b,\Gamma}$ has a unique fixed point $f^{\omega}_{\Gamma,b}$ (say) in $L^p(\mathcal{I})$. The FIF $f^{\omega}_{\Gamma,b}$ is called the fractal function, which is a self-referential function. And since $f^{\omega}_{\Gamma,b}$ is the fixed point of $T^{\omega}_{f,b,\Gamma}$ which satisfies the following functional equations:

$$f^{\omega}_{\Gamma,b}(\gamma) = f(\gamma) + \omega_l(f^{\omega}_{\Gamma,b} - b) \circ L_l^{-1}(\gamma) \quad \forall\, \gamma \in \mathcal{I}_l,\ l = 1, 2, \ldots, M, \tag{2}$$

this implies the following inequality:

$$\|f^{\omega}_{\Gamma,b} - f\| \le \frac{\Omega}{1-\Omega}\|f - b\|.$$

For any prescribed function f in $L^p(\mathcal{I})$, (2) produces a collection of self-referential functions $\{f^{\omega}_{\Gamma,b} : \omega \in (-1, 1)^M\}$ related with Γ, b. Observe that if the choosing scale vector is null, then $f^{\omega}_{\Gamma,b}$ coincides with the given function f. The elements Γ, b, and ω can be chosen appropriately so as to maintain or reform the properties of the given function f depending on our problem.

We define a binary operation $\mathcal{R} := \mathcal{R}_{\Gamma,\omega}$ on the space $L^p(\mathcal{I}) \times L^p(\mathcal{I})$, called the fractal convolution operator associated with the fixed partition Γ and the fixed scale vector ω by (cf. [18])

$$\mathcal{R}(f, b) := f^{\omega}_{\Gamma,b}.$$

In place of $\mathcal{R}(f, b)$, we write $f * b$. The binary operation $*$ on $L^p(\mathcal{I}) \times L^p(\mathcal{I})$ is called the fractal convolution of the functions f and b.

4 Partial Fractal Convolutions on Rectangle with Fixing Null Function

Consider the fractal convolution operators $\mathcal{R} = \mathcal{R}_{\Gamma,\omega}$ associated with the fixed partition Γ of the interval $\mathcal{I}$ and the fixed scale vector ω defined as

$$\mathcal{R} : L^2(\mathcal{I}) \times L^2(\mathcal{I}) \to L^2(\mathcal{I}) \quad \text{by} \quad (f, b) \to f *_1 b := f^{\omega}_{\Gamma,b}$$

and $\mathcal{S} = \mathcal{S}_{\tilde{\Gamma},\beta}$ associated with the fixed partition $\tilde{\Gamma}$ of the interval $\mathcal{J}$, and the fixed scale vector β defined as

$$\mathcal{S} : L^2(\mathcal{J}) \times L^2(\mathcal{J}) \to L^2(\mathcal{J}) \quad \text{by} \quad (g, \tilde{b}) \to g *_2 \tilde{b} := g^{\beta}_{\tilde{\Gamma},\tilde{b}}$$

where $f^{\omega}_{\Gamma,b}$ and $g^{\beta}_{\tilde{\Gamma},\tilde{b}}$ are fractal functions.

We also get partial fractal convolution operators on $L^2(\mathcal{I})$ by keeping one of the input coordinates of $\mathcal{R}$ being fixed. We define, for a fixed $f \in L^2(\mathcal{I})$, the operator

$$\mathcal{R}_f^1(b) = \mathcal{R}(f, b) : L^2(\mathcal{I}) \to L^2(\mathcal{I}) \quad \text{by} \quad b \to f *_1 b,$$

and for a fixed $b \in L^2(\mathcal{I})$

$$\mathcal{R}_b^2(f) = \mathcal{R}(f, b) : L^2(\mathcal{I}) \to L^2(\mathcal{I}) \quad \text{by} \quad f \to f *_1 b.$$

This is also called one-sided fractal convolutions. We call $\mathcal{R}_f^1$ as the f-left fractal convolution and $\mathcal{R}_b^2$ as the b-right fractal convolution of $\mathcal{R}$. Similarly, we define the one-sided fractal convolution operators on $L^2(\mathcal{J})$ of $\mathcal{S}$ as follows:

For a fixed $g \in L^2(\mathcal{J})$,

$$\mathcal{S}_g^1(\tilde{b}) = g *_2 \tilde{b} \quad \forall\, \tilde{b} \in L^2(\mathcal{J})$$

and for a fixed $\tilde{b} \in L^2(\mathcal{J})$,

$$\mathcal{S}_{\tilde{b}}^2(g) = g *_2 \tilde{b} \quad \forall\, g \in L^2(\mathcal{J}),.$$

By using (2), the one-sided fractal convolution operators satisfy the following inequalities (see [18] Theorem 3.3):

$$\|\mathcal{R}_f^1(b_1) - \mathcal{R}_f^1(b_2)\| \le \frac{\Omega}{1-\Omega}\|b_1 - b_2\|, \tag{3}$$

$$\|\mathcal{R}_b^2(f_1) - \mathcal{R}_b^2(f_2)\| \le \frac{1}{1-\Omega}\|f_1 - f_2\|, \tag{4}$$

for all $f, f_1, f_2, b, b_1, b_2 \in L^2(\mathcal{I})$, where $\Omega := \max_n\{|\omega_n|\}$ and $\|.\|$ is the L^2-norm (called Euclidean norm) and

$$\|\mathcal{S}_g^1(\tilde{b}_1) - \mathcal{S}_g^1(\tilde{b}_2)\| \le \frac{\tilde{\Omega}}{1-\tilde{\Omega}}\|\tilde{b}_1 - \tilde{b}_2\|, \tag{5}$$

$$\|\mathcal{S}_{\tilde{b}}^2(g_1) - \mathcal{S}_{\tilde{b}}^2(g_2)\| \le \frac{1}{1-\tilde{\Omega}}\|g_1 - g_2\|, \tag{6}$$

for all $g, g_1, g_2, \tilde{b}, \tilde{b}_1, \tilde{b}_2 \in L^2(\mathcal{J})$, where $\tilde{\Omega} := \max_m\{|\beta_m|\}$.

Remark 1 By using fractal convolution, we can get a connection between fractal interpolation theory and frame theory. Note that the fractal convolution operators $\mathcal{R}$ and $\mathcal{S}$ are linear. This gives that one-sided fractal convolution operators $\mathcal{R}_0^1$, $\mathcal{R}_0^2$, $\mathcal{S}_0^1$ and $\mathcal{S}_0^2$ are also linear.

Consider the one-sided fractal convolution operators $\mathcal{R}_0^1$, $\mathcal{R}_0^2$, $\mathcal{S}_0^1$ and $\mathcal{S}_0^2$ with the null function 0 being fixed.

$$\mathcal{R}_0^1(b) = 0 *_1 b, \qquad \mathcal{R}_0^2(f) = f *_1 0,$$

and

$$\mathcal{S}_0^1(\tilde{b}) = 0 *_2 \tilde{b}, \qquad \mathcal{S}_0^2(g) = g *_2 0.$$

From (3)–(6), we have $\mathcal{R}_0^1$ and $\mathcal{R}_0^2$ are bounded linear operators on $L^2(\mathcal{I})$ and $\mathcal{S}_0^1$ and $\mathcal{S}_0^2$ are bounded linear operators on $L^2(\mathcal{J})$ with

$$\|\mathcal{R}_0^1\| \leq \frac{\Omega}{1-\Omega}, \qquad \|\mathcal{R}_0^2\| \leq \frac{1}{1-\Omega}. \tag{7}$$

$$\|\mathcal{S}_0^1\| \leq \frac{\tilde{\Omega}}{1-\tilde{\Omega}}, \qquad \|\mathcal{S}_0^2\| \leq \frac{1}{1-\tilde{\Omega}}. \tag{8}$$

For the detailed exposition of the following definitions, the reader can refer to [7, 22].

Definition 3 A family of elements $\{\eta_n\}_{n=1}^{\infty}$ in a separable Hilbert space $\mathcal{H}$, is called a frame if there exist constants $\mathcal{A}, \mathcal{B} > 0$ such that

$$\mathcal{A}\|\eta\|^2 \leq \sum_{n=1}^{\infty} |\langle \eta, \eta_n \rangle|^2 \leq \mathcal{B}\|\eta\|^2 \qquad \forall\, \eta \in \mathcal{H}.$$

The constants $\mathcal{A}$ and $\mathcal{B}$ are called lower and upper frame bounds of $\{\eta_n\}_{n=1}^{\infty}$, respectively.

Definition 4 A family of elements $\{\eta_n\}_{n=1}^{\infty}$ in a separable Hilbert space $\mathcal{H}$, is called a Bessel sequence if there exists $\mathcal{B} > 0$ such that

$$\sum_{n=1}^{\infty} |\langle \eta, \eta_n \rangle|^2 \leq \mathcal{B}\|\eta\|^2 \qquad \forall\, \eta \in \mathcal{H}.$$

The constant $\mathcal{B}$ is called Bessel constant of $\{\eta_n\}_{n=1}^{\infty}$.

Proposition 1 ([10]) *If $\{\eta_n\}$ is a frame for $\mathcal{H}_1$ and $\{\lambda_m\}$ is a frame for $\mathcal{H}_2$, then $\{\eta_n \otimes \lambda_m\}$ is a frame for $\mathcal{H}_1 \otimes \mathcal{H}_2$.*

Lemma 1 *If $\{\eta_n\}$ is a Bessel sequence of $\mathcal{H}_1$ and $\{\lambda_m\}$ is a Bessel sequence of $\mathcal{H}_2$, then $\{\eta_n \otimes \lambda_m\}$ is a Bessel sequence of $\mathcal{H}_1 \otimes \mathcal{H}_2$.*

Proof We can conclude this lemma from the proof of the above proposition in [10].

Theorem 2 *If* $\{e_n\}$ *and* $\{r_m\}$ *are Bessel sequences of* $L^2(\mathcal{I})$ *and* $L^2(\mathcal{J})$ *respectively, then for* $L^2(\mathcal{I} \times \mathcal{J})$,

(i) $\{(0 *_1 e_n) \otimes (0 *_2 r_m)\}$ *is a Bessel sequence.*
(ii) $\{(0 *_1 e_n) \otimes (r_m *_2 0)\}$ *is a Bessel sequence.*
(iii) $\{(e_n *_1 0) \otimes (0 *_2 r_m)\}$ *is a Bessel sequence.*
(iv) $\{(e_n *_1 0) \otimes (r_m *_2 0)\}$ *is a Bessel sequence.*

Proof By Lemma 1, $\{e_n \otimes r_m\}$ is a Bessel sequence. From the properties (b) and (c) of the tensor product of the bounded linear operators, for any $Q \in L^2(\mathcal{I}) \otimes L^2(\mathcal{J})$, we have

$$\text{(i)} \quad \sum_{m,n} |\langle Q, (0 *_1 e_n) \otimes (0 *_2 r_m)\rangle|^2 = \sum_{m,n} |\langle Q, \mathcal{R}_0^1(e_n) \otimes \mathcal{S}_0^1(r_m)\rangle|^2$$

$$= \sum_{m,n} |\langle Q, (\mathcal{R}_0^1 \otimes \mathcal{S}_0^1)(e_n \otimes r_m)\rangle|^2$$

$$= \sum_{m,n} |\langle (\mathcal{R}_0^1 \otimes \mathcal{S}_0^1)^* Q, e_n \otimes r_m\rangle|^2$$

$$\leq \mathcal{B}|||(\mathcal{R}_0^1 \otimes \mathcal{S}_0^1)^* Q|||^2$$

$$\leq \mathcal{B}\|\mathcal{R}_0^1 \otimes \mathcal{S}_0^1\|^2 |||Q|||^2$$

$$= \mathcal{B}\|\mathcal{R}_0^1\|^2 \|\mathcal{S}_0^1\|^2 |||Q|||^2,$$

$$\text{(ii)} \quad \sum_{m,n} |\langle Q, (0 *_1 e_n) \otimes (r_m *_2 0)\rangle|^2 = \sum_{m,n} |\langle Q, (\mathcal{R}_0^1 \otimes \mathcal{S}_0^2)(e_n \otimes r_m)\rangle|^2$$

$$= \sum_{m,n} |\langle (\mathcal{R}_0^1 \otimes \mathcal{S}_0^2)^* Q, e_n \otimes r_m\rangle|^2$$

$$\leq \mathcal{B}|||(\mathcal{R}_0^1 \otimes \mathcal{S}_0^2)^* Q|||^2$$

$$\leq \mathcal{B}\|\mathcal{R}_0^1 \otimes \mathcal{S}_0^2\|^2 |||Q|||^2$$

$$= \mathcal{B}\|\mathcal{R}_0^1\|^2 \|\mathcal{S}_0^2\|^2 |||Q|||^2,$$

(iii) $$\sum_{m,n} |\langle \mathcal{Q}, (e_n *_1 0) \otimes (0 *_2 r_m)\rangle|^2 = \sum_{m,n} |\langle \mathcal{Q}, (\mathcal{R}_0^2 \otimes \mathcal{S}_0^1)(e_n \otimes r_m)\rangle|^2$$

$$= \sum_{m,n} |\langle (\mathcal{R}_0^2 \otimes \mathcal{S}_0^1)^* \mathcal{Q}, e_n \otimes r_m\rangle|^2$$

$$\leq \mathcal{B} |||(\mathcal{R}_0^2 \otimes \mathcal{S}_0^1)^* \mathcal{Q}|||^2$$

$$\leq \mathcal{B} \|\mathcal{R}_0^2 \otimes \mathcal{S}_0^1\|^2 |||\mathcal{Q}|||^2$$

$$= \mathcal{B} \|\mathcal{R}_0^2\|^2 \|\mathcal{S}_0^1\|^2 |||\mathcal{Q}|||^2,$$

(iv) $$\sum_{m,n} |\langle \mathcal{Q}, (e_n *_1 0) \otimes (r_m *_2 0)\rangle|^2 = \sum_{m,n} |\langle \mathcal{Q}, (\mathcal{R}_0^2 \otimes \mathcal{S}_0^2)(e_n \otimes r_m)\rangle|^2$$

$$= \sum_{m,n} |\langle (\mathcal{R}_0^2 \otimes \mathcal{S}_0^2)^* \mathcal{Q}, e_n \otimes r_m\rangle|^2$$

$$\leq \mathcal{B} |||(\mathcal{R}_0^2 \otimes \mathcal{S}_0^2)^* \mathcal{Q}|||^2$$

$$\leq \mathcal{B} \|\mathcal{R}_0^2 \otimes \mathcal{S}_0^2\|^2 |||\mathcal{Q}|||^2$$

$$= \mathcal{B} \|\mathcal{R}_0^2\|^2 \|\mathcal{S}_0^2\|^2 |||\mathcal{Q}|||^2,$$

where $\mathcal{B}$ is a Bessel constant of the sequence $\{e_n \otimes r_m\}$. Hence, proved.

5 Conclusion

In this article, we obtained a collection of fractal (self-referential) Bessel sequences of $L^2(\mathcal{I} \times \mathcal{J})$. Since the fractal functions and Bessel sequences own greater flexibility in order to choose good approximations of mappings separately, our obtained fractal Bessel sequences behave very nicely in order to approximate two-dimensional square-integrable maps whose domain is a rectangle. First, we considered partial fractal convolution (bounded linear) operators on both $L^2(\mathcal{I})$ and $L^2(\mathcal{J})$ with the null function being fixed. Consequently, by considering the bounded linear operators $\mathcal{R}_0^i \otimes \mathcal{S}_0^j$, $(i, j) \in \{1, 2\} \times \{1, 2\}$, we obtained Bessel sequences of $L^2(\mathcal{I} \times \mathcal{J})$, composed of product of fractal convolution maps on $L^2(\mathcal{I})$ and $L^2(\mathcal{J})$.

References

1. Barnsley, M.F.: Fractal Functions and Interpolation. Constr. Approx. **2**, 303–329 (1986)
2. Barnsley, M.F.: Fractals Everywhere. Academic, Boston (1988)
3. Barnsley, M.F., Hurd, L.P.: Fractal Image Compression. AK Peters Ltd, Wellesley (1993)
4. Chand, A.K.B., Vijender, N.: A new class of fractal interpolation surfaces based on functional values. Fractals **24**(1), 1650007, 1–17 (2016)
5. Chand, A.K.B., Vijender, N., Navascués, M.A.: Shape preservation of scientific data through rational fractal splines. Calcolo **51**, 329–362 (2014)
6. Chand, A.K.B., Viswanathan, P.: A constructive approach to cubic Hermite fractal interpolation function and its constrained aspects. BIT Numer. Math. **53**, 841–865 (2013)
7. Christensen, O.: Frames and Bases: An Introductory Course. Birkhauser, Boston (2008)
8. Folland, G.B.: A Course in Abstract Harmonic Analysis. CRC Press, Boca Raton (1995)
9. Hutchinson, J.: Fractals and self-similarity. Indiana Univ. Math. J. **30**, 713–747 (1981)
10. Khosravi, A., Asgari, M.S.: Frames and bases in tensor product of Hilbert spaces. Int. Math. J. **4**, 527–537 (2003)
11. Kumagai, Y.: Fractal structure of financial high frequency data. Fractals **10**(1), 13–18 (2002)
12. Mandelbrot, B.B.: The Fractal Geometry of Nature. Freeman, New York (1982)
13. Massopust, P.R.: Fractal Functions, Fractal Surfaces, and Wavelets. Academic Press Inc, San Diego (1994)
14. Massopust, P.R.: Interpolation and Approximation with Splines and Fractals. Oxford University Press, New York (2010)
15. Mazel, D.S., Hayes, M.H.: Using iterated function systems to model discrete sequences, U. IEEE Trans. Signal Process. **40**, 1724–1734 (1992)
16. Navascués, M.A.: Fractal polynomial interpolation. Z. Anal. Anwend. **25**(2), 401–418 (2005)
17. Navascués, M.A., Chand, A.K.B.: Fundamental sets of fractal functions. Acta Appl. Math. **100**, 247–261 (2008)
18. Navascués, M.A., Massopust, P.: Fractal convolution: a new operation between functions. Fract. Calc. Appl. Anal. **22**(3), 619–643 (2019)
19. Navascués, M.A., Mohapatra, R., Akhtar, M.N.: Fractal frames of functions on the rectangle. Fractal Fract. **42**(5) (2021)
20. Navascués, M.A., Viswanathan, P., Mohapatra, R.: Convolved fractal bases and frames. Adv. Oper. Theorem **42**(6) (2021)
21. Roy, A., Sujith, R.I.: Fractal dimension of premixed flames in intermittent turbulence. Combust. Flame **226**, 412–418 (2021)
22. Singer, I.: Bases in Banach Spaces I. Springer, New York (1970)
23. Viswanathan, P., Chand, A.K.B.: Fractal rational functions and their approximation properties. J. Approx. Theory **185**, 31–50 (2014)
24. Viswanathan, P., Chand, A.K.B., Agarwal, R.P.: Preserving convexity through rational cubic spline fractal interpolation function. J. Comput. Appl. Math. **263**, 262–276 (2014)
25. Vrscay, E.R.: Iterated function systems: Theory, applications and the inverse problem. In: Proceedings of the NATO Advanced Study Institute on Fractal Geometry, Montreal, July 1989, Kluwer Academic Publishers (1990)
26. Wang, H.Y., Yu, J.S.: Fractal interpolation functions with variable parameters and their analytical properties. J. Approx. Theory **175**, 1–18 (2013)

Uniform Approximation of Functions Belonging to $L[0, \infty)$-Space Using $C^\gamma.T$-Means of Fourier–Laguerre Series

Sachin Devaiya and Shailesh Kumar Srivastava

Abstract Recently, Singh and Saini [Uniform approximation in $L[0, \infty)$-space by Cesàro means of Fourier–Laguerre series. Proc. Natl. Acad. Sci., India, Sect. A Phys. Sci. (2021)] determined the degree of approximation of functions f belonging to $L[0, \infty)$ by Cesàro means of its Fourier–Laguerre series for any $x > 0$. In this paper, we obtain the error of approximation of functions $f \in L[0, \infty)$ using product mean $C^\gamma.T (\gamma \geq 1)$ of its Fourier–Laguerre series for any $x > 0$. Further, we also discuss some particular cases of $C^\gamma.T$-means.

Keywords $C^\gamma.T$-mean · Error of approximation · Fourier–Laguerre series

1 Introduction

The Fourier–Laguerre expansion of function $f \in L[0, \infty)$ is given by

$$f(x) \sim \sum_{n=0}^{\infty} a_n L_n^{(\alpha)}(x), \tag{1}$$

where $L_n^{(\alpha)}(x)$ is nth Laguerre polynomial of order $\alpha > -1$, is defined by the generating function

$$\sum_{n=0}^{\infty} L_n^{(\alpha)}(x)\, \omega^n = \frac{\exp(\frac{\omega x}{\omega - 1})}{(1 - \omega)^{1+\alpha}},$$

S. Devaiya (✉) · S. K. Srivastava
Department of Mathematics and Humanities, Sardar Vallabhbhai National Institute of Technology, Surat, Gujarat 395007, India
e-mail: sbdevaiya18695@gmail.com

R. K. Sharma et al. (eds.), *Frontiers in Industrial and Applied Mathematics*, Springer Proceedings in Mathematics & Statistics 410,
https://doi.org/10.1007/978-981-19-7272-0_12

and

$$a_n = \frac{n!\,\alpha!}{(n+\alpha)!\,\Gamma(\alpha+1)} \int_0^{\infty} x^{\alpha} \exp(-x)\, f(x)\, L_n^{(\alpha)}(x)dx. \tag{2}$$

It is pretended that integral (2) exists.

The kernel polynomial $J_k^{\alpha}(x, y)$, is given as

$$J_k^{\alpha}(x, y) = \sum_{m=0}^{k} \frac{L_m^{(\alpha)}(x)\, L_m^{(\alpha)}(y)}{\Gamma(\alpha+1)\binom{m+\alpha}{m}}. \tag{3}$$

$$J_k^{\alpha}(x, y) = \frac{k+1}{\Gamma(\alpha+1)\binom{k+\alpha}{\alpha}} \frac{L_k^{(\alpha)}(x)\, L_{k+1}^{(\alpha)}(y) - L_{k+1}^{(\alpha)}(x)\, L_k^{(\alpha)}(y)}{x-y}. \tag{4}$$

In a more appropriate form,

$$J_k^{\alpha}(x, y) = \frac{k+1}{\Gamma(\alpha+1)\binom{k+\alpha}{\alpha}} \frac{L_{k+1}^{(\alpha)}(x)\, L_{k+1}^{(\alpha-1)}(y) - L_{k+1}^{(\alpha-1)}(x)\, L_{k+1}^{(\alpha)}(y)}{x-y}, \tag{5}$$

and

$$J_k^{\alpha}(x, y) = \frac{k+1}{\Gamma(\alpha+1)\binom{k+\alpha}{\alpha}} \left(L_{k+1}^{(\alpha)}(x) \frac{L_{k+1}^{(\alpha-1)}(y) - L_{k+1}^{(\alpha-1)}(x)}{x-y} - L_{k+1}^{(\alpha-1)}(x) \frac{L_{k+1}^{(\alpha)}(y) - L_{k+1}^{(\alpha)}(x)}{x-y} \right). \tag{6}$$

For more details, one can see [15, pp. 101, 266].

The $(n+1)$th partial sum of the Fourier–Laguerre series of equation (1) is defined by

$$s_n(f; x) = \sum_{k=0}^{n} a_k\, L_k^{(\alpha)}(x),\ n \in \mathrm{N}_0. \tag{7}$$

Define

$$[t]_n(f; x) = \sum_{k=0}^{n} a_{n,k}\, s_k(f; x),\ n \in \mathrm{N}_0,$$

where $T \equiv (a_{n,k} \geq 0$ for every $n, k)$ is a lower triangular matrix such that $a_{n,-1} = 0$, $A_{n,k} = \sum_{k=r}^{n} a_{n,k}$ and $A_{n,0} = 1, n \in \mathrm{N}_0$. The Fourier–Laguerre series is called T-summable to s, if $[t]_n(f; x) \to s$ as $n \to \infty$.

If $a_{n,k} = \begin{cases} \frac{n!\,\gamma!}{(n+\gamma)!}\binom{n+\gamma-k-1}{\gamma-1}, & 0 \leq k \leq n, \\ 0, & k > n, \end{cases}$ then the matrix T converts to Cesàro matrix of order $\gamma \geq 1$ and denoted by C^{γ}. The Fourier-Laguerre series is called C^{γ}-summable to s_1, if $[C^{\gamma}]_n(f; x) \to s_1$ as $n \to \infty$.

The product of C^γ-summable with T-summable defines $C^\gamma.T$-summable. Thus, $C^\gamma.T$-summability of sequence $\{s_n(f;x)\}$ denoted by

$$[C^\gamma.T]_n(f;x) = \frac{n!\,\gamma!}{(n+\gamma)!}\sum_{v=0}^{n}\binom{n+\gamma-v-1}{\gamma-1}\sum_{k=0}^{v} a_{v,k}s_k(f;x). \tag{8}$$

If $[C^\gamma.T]_n(f;x) \to s_2$ as $n \to \infty$, then the Fourier–Laguerre series is called $C^\gamma.T$-summable to s_2. The regularity of T and C^γ methods implies the regularity of the $C^\gamma.T$ method.

The following cases are important and particular cases of $C^\gamma.T$ method:

1. If $a_{v,k} = \frac{1}{(v-k+1)\,log(v+1)}$, then $C^\gamma.T$ reduce to $C^\gamma.H$ or $(C,\gamma)(H,\frac{1}{v+1})$.
2. If $a_{v,k} = \frac{p_{v-k}}{P_v}$, where $P_v = \sum_{k=0}^{v} p_k \neq 0$, then $C^\gamma.T$ reduce to $C^\gamma.N_p$ or (C,γ) (N, p_v).
3. If $a_{v,k} = \frac{p_k}{P_v}$, then $C^\gamma.T$ reduce to $C^\gamma.\overline{N}_p$ or $(C,\gamma)(\overline{N}, p_v)$.
4. If $a_{v,k} = \frac{p_{v-k}q_k}{R_v}$, where $R_v = \sum_{k=0}^{v} p_k q_{v-k}$, then $C^\gamma.T$ reduce to $C^\gamma.N_{pq}$ or $(C,\gamma)(N,p,q)$.
5. If $a_{v,k} = \frac{1}{(1+q)^v}\binom{v}{k}q^{v-k}$, then $C^\gamma.T$ reduce to $C^\gamma.E^q$ or $(C,\gamma)(E,q)$.
6. If $a_{v,k} = \frac{1}{2^v}\binom{v}{k}$, then $C^\gamma.T$ reduce to $C^\gamma.E^1$ or $(C,\gamma)(E,1)$,

where p_v and q_v are a non-negative, monotonic, and non-increasing sequence of real constants.

If we take $\gamma = 1$ in the above cases, then we get $C^1.H$, $C^1.N_p$, $C^1.\overline{N}_p$, $C^1.N_{pq}$, $C^1.E^q$, and $(C,1)(E,1)$ are also particular cases of the $C^\gamma.T$ method.

Remark 1 We consider the matrix $a_{i,k} = \frac{1}{n^i}\binom{i}{k}(n-1)^{i-k}$, for $n \geq 2$, and the series $1 - 2n\sum_{i=1}^{\infty}(-2n+1)^{i-1}$, for $n \in \mathrm{N}$, then the ith partial sum of the series is given by $s_i = (-2n+1)^i$. It can be seen that the series is not T-summable and also not C^γ-summable (for $\gamma = 1$), but it is $C^\gamma.T$-summable (for $\gamma = 1$). We can observe that product summabilities are more effective than the single summability.

Many researchers have obtained the error of approximation of functions by various summability methods or operators; for instance, one can see [1–14]. In the last two decades, the error of approximation of functions f lies in $L[0,\infty)$ using different types of summability methods of its Fourier–Laguerre series became the area of interest for many investigators. The authors like Lal and Nigam [5], Nigam and Sharma [8], Sahani et al. [9], and Saini and Singh [10] have approximate functions using different types of summability methods such as (N,p,q), $(E,1)$, Nörlund means, and Hausdorff, respectively. On the other hand, Khatri and Mishra [3], Krasniqi [4], Mittal and Singh [7], and Sonker [14] have approximate functions using different types of product summability methods such as Harmonic–Euler, $(C,1)(E,q)$, Matrix–Euler mean, and $C^1.T$, respectively. But the authors mentioned above have approximated the function at a point $x = 0$. In 1976, Singh [11] gave an interesting result on the absolute $(C,1)$-summability of the series $\sum_{n=1}^{\infty}\frac{a_n L_n^{(\alpha)}(x)}{(\log(n+1))^{\epsilon+1}}$, where a_n is Fourier-Laguerre coefficient of $f \in L[0,\infty)$ with $x > 0$. The Laguerre functions form an orthogonal basis for $L_2[0,\infty)$-space, which successively defines the

Fourier–Laguerre series. It has also been shown that Laguerre's polynomial theory directly solves the problem of determining Fourier–Laguerre approximations for a large class of delay systems. Moreover, these findings are necessary for studying the regular order of identification as a standard method for identifying infinite-dimensional systems [6]. Recently, Singh and Saini [12] have approximate function f belonging to $L[0, \infty)$ by Cesàro means of the Fourier–Laguerre series of f for any $x > 0$. We also use the following notations:

$$\phi(x, y) = f(y) - f(x) \text{ and } \psi(x, u) = f(x \pm u) - f(x).$$

2 Main Results

We note that a lot of work has been done to approximate function $f \in L[0, \infty)$ using different types of summability methods of its Fourier–Laguerre series at a point $x = 0$, but very little work has been done for $x > 0$. Also, the importance of the product summability method, which is discussed in Remark 1 and particular cases of $C^{\gamma}.T$-means, are motivated us to study the problem of the error of approximation of functions f using $C^{\gamma}.T$-means of its Fourier–Laguerre series for $x > 0$. More precisely, we prove the following result:

Theorem *Let $T \equiv (a_{n,k})$ be a lower triangular regular matrix satisfy the following conditions:*

1. *$a_{n,k}$ be a non-negative and non-decreasing with respect to k, for $0 \leq k \leq n$,*
2. $\sum_{v=t}^{n} A_{v,v-t} = O(n+1), \; n \in \mathbb{N}_0$.

Then error of approximation of functions $f \in L[0, \infty)$-class by $C^{\gamma}.T$-means of its Fourier-Laguerre series at $x > 0$ by is given by

$$\|[C^{\gamma}.T]_n(f; x) - f(x)\| = o(\xi(n)), \tag{9}$$

where $\xi(t)$ is an increasing function (positive) of t such that $\xi(t) \to \infty$ as $t \to \infty$ and satisfies following conditions:

$$\Phi(t) = \int_t^{\epsilon} \frac{|\phi(x, y)|}{y^{1/4-\alpha/2}} dy = o\left(\xi\left(\frac{1}{t}\right)\right), \quad t \to 0, \tag{10}$$

$$\int_t^{\delta} \frac{|\psi(x, u)|}{u} du = o\left(\xi\left(\frac{1}{t}\right)\right), \quad t \to 0, \tag{11}$$

$$\int_n^{\infty} \frac{\exp(-y/2)\, |\phi(x, y)|}{y^{13/12-\alpha/2}} dy = o\left(\frac{\xi(n)}{n^{1/2}}\right), \quad n \to \infty, \tag{12}$$

where $\alpha \geq -1/2$, $\delta\,(>0)$ *is a fixed number and this holds uniformly for every fixed positive interval* $0 < \epsilon \leq x \leq \omega < \infty$.

Here, few lemmas are given, which are useful to prove our Theorem.

Lemma 1 *Let* ϵ *be a fixed positive constant and* α *be an arbitrary real number. Then*

$$L_n^{(\alpha)}(x) = \begin{cases} O(n^{\alpha}), & 0 \leq x \leq 1/n, \\ O(x^{-(2\alpha+1)/4}\, n^{(2\alpha-1)/4}), & 1/n \leq x \leq \epsilon, \end{cases} \quad \text{as } n \to \infty.$$

The proof is given in [15, pp. 177, Theorem 7.6.4].

Lemma 2 *Let* ρ *and* α *be arbitrary real numbers,* $0 < \eta < 4$ *and* $\omega > 0$. *Then*

$$max\ \exp(-x/2)x^{\rho}|L_n^{(\alpha)}(x)| = O(n^{Q}),$$

where

$$Q = \begin{cases} max\left(\rho - \frac{1}{2}, \frac{\alpha}{2} - \frac{1}{4}\right), & \omega \leq x \leq (4-\eta)n, \\ max\left(\rho - \frac{1}{3}, \frac{\alpha}{2} - \frac{1}{4}\right), & x > n. \end{cases}$$

The proof is given in [15, pp. 241, Theorem 8.91.7].

Lemma 3 *Let* ϵ *and* ω *be fixed positive constants and* α *be an arbitrary real number, then*

$$L_n^{(\alpha)}(x) = k(x)\, n^{\alpha/2-1/4} \cos(2\sqrt{nx} - (\alpha + 1/2)\pi/2) + O(n^{\alpha/2-3/4}),$$

where $k(x) = \frac{x^{-\alpha/2-1/4}\exp(x/2)}{\sqrt{\pi}}$ *and* $x \in [\epsilon, \omega]$.

The proof is given in [15, pp. 198, Theorem 8.22.1].

Lemma 4 *If* $x, y \in [1/n, \omega]$, *then*

$$\frac{L_n^{(\alpha)}(y) - L_n^{(\alpha)}(x)}{\sqrt{y} - \sqrt{x}} = k(y)\, n^{\alpha/2-1/4} \frac{\cos(2\sqrt{ny} + \lambda) - \cos(2\sqrt{nx} + \lambda)}{\sqrt{y} - \sqrt{x}} + x^{-\alpha/2-3/4}\, O(n^{\alpha/2-1/4}) + y^{-\alpha/2-3/4}\, O(n^{\alpha/2-1/4}),$$

where $\lambda = -(\alpha + 1/2)\pi/2$.

The proof is given in [15, pp. 237].

Lemma 5 *If condition (10) holds, then*

$$\int_0^t y^{\alpha}\, |\phi(x, y)|\, dy = o\left(t^{\alpha/2+1/4}\, \xi\left(\frac{1}{t}\right)\right).$$

The proof is given in [12, Lemma 5].

Lemma 6 *If condition (12) holds, then*

$$\int_{\omega}^{n} \exp(-y/2)\, y^{\alpha/2-3/4}\, |\phi(x, y)|\, dy = o(\xi(n)),$$

where ω (> 0) *is a fixed number and* $n \to \infty$.

The proof is given in [12, Lemma 6].

3 Proof of Theorem

We have

$$\begin{aligned} s_n(f; x) &= \sum_{k=0}^{n} a_k\, L_k^{(\alpha)}(x) \\ &= \sum_{k=0}^{n} \frac{\int_0^\infty \exp(-y)\, y^\alpha\, f(y)\, L_k^{(\alpha)}(y)\, L_k^{(\alpha)}(x)\, dy}{\Gamma(\alpha+1)\binom{k+\alpha}{\alpha}} \\ &= \frac{1}{\Gamma(\alpha+1)} \int_0^\infty \exp(-y)\, y^\alpha\, f(y) \sum_{k=0}^{n} \frac{L_k^{(\alpha)}(x)\, L_k^{(\alpha)}(y)}{\binom{k+\alpha}{\alpha}} dy \\ &= \int_0^\infty \exp(-y)\, y^\alpha\, f(y)\, J_n^\alpha(x, y)\, dy. \end{aligned} \tag{13}$$

Applying T-summability on Eq. (13), we get

$$\begin{aligned} [T]_n(f; x) &= \sum_{k=0}^{n} a_{n,k}\, s_k(x) \\ &= \sum_{k=0}^{n} a_{n,k} \int_0^\infty \exp(-y)\, y^\alpha\, f(y)\, J_k^\alpha(x, y)\, dy. \end{aligned} \tag{14}$$

Applying (C, γ)-summability on Eq. (14),

$$\begin{aligned} [C^\gamma.T]_n(f; x) = \frac{n!\,\gamma!}{(n+\gamma)!\,\Gamma(\alpha+1)} \sum_{v=0}^{n} \binom{n+\gamma-v-1}{\gamma-1} \sum_{k=0}^{v} a_{v,v-k} \times \\ \int_0^\infty \exp(-y)\, y^\alpha\, f(y)\, J_k^\alpha(x, y)\, dy, \end{aligned} \tag{15}$$

we have

$$[C^\gamma.T]_n(f; x) - f(x)$$

$$= \frac{n!\,\gamma!}{(n+\gamma)!}\sum_{v=0}^{n}\binom{n+\gamma-v-1}{\gamma-1}\sum_{k=0}^{v} a_{v,v-k}\int_0^{\infty}\exp(-y)\,y^{\alpha}\,[f(y)-f(x)]J_k^{\alpha}(x,y)\,dy$$

$$= \frac{n!\,\gamma!}{(n+\gamma)!}\sum_{v=0}^{n}\binom{n+\gamma-v-1}{\gamma-1}\sum_{k=0}^{v} a_{v,v-k}\int_0^{\infty}\exp(-y)y^{\alpha}\,\phi(x,y)\,J_k^{\alpha}(x,y)\,dy$$

$$= \frac{n!\,\gamma!}{(n+\gamma)!}\sum_{v=0}^{n}\binom{n+\gamma-v-1}{\gamma-1}\sum_{k=0}^{v} a_{v,v-k}\left[\int_0^{1/n}+\int_{1/n}^{\epsilon}+\int_{\epsilon}^{x-\delta}+\int_{x-\delta}^{x+\delta}+\int_{x+\delta}^{\omega}+\int_{\omega}^{n}+\int_{n}^{\infty}\right]\exp(-y)\,y^{\alpha}\,\phi(x,y)\,J_k^{\alpha}(x,y)\,dy$$

$$= \sum_{i=1}^{7} I_i. \tag{16}$$

Consider that x is restricted to a fixed positive number, then using Lemma 1, we have

$$\begin{aligned}|L_{k+1}^{(\alpha)}(x)| &= O(x^{-\alpha/2+1/4}k^{\alpha/2-1/4})\\ &= O(k^{\alpha/2-1/4}).\end{aligned} \tag{17}$$

Now, in Eq. (5) applying Lemma 1 for $0 \le y < 1/n$, we have

$$|J_k^{\alpha}(x,y)| = O(k^{1-\alpha}[k^{\alpha/2-1/4}k^{\alpha-1} + k^{\alpha/2-3/4}k^{\alpha}]). \tag{18}$$

Applying Eqs. (17), (18) and Lemma 5, we have

$$|I_1| \le \frac{n!\,\gamma!}{(n+\gamma)!}\sum_{v=0}^{n}\binom{n+\gamma-v-1}{\gamma-1}\sum_{k=0}^{v} a_{v,v-k}\int_0^{1/n} y^{\alpha}\,|\phi(x,y)|\,|J_k^{\alpha}(x,y)|\,dy$$

$$\begin{aligned}
&= \frac{n!\,\gamma!}{(n+\gamma)!}\sum_{v=0}^{n}\binom{n+\gamma-v-1}{\gamma-1}\sum_{k=0}^{v} a_{v,v-k}\,O\left(k^{1-\alpha}\int_0^{1/n} y^{\alpha}\,|\phi(x,y)|\left[k^{\alpha/2-1/4}k^{\alpha-1}+k^{\alpha/2-3/4}k^{\alpha}\right]dy\right)\\
&= \frac{n!\,\gamma!}{(n+\gamma)!}\sum_{v=0}^{n}\binom{n+\gamma-v-1}{\gamma-1}\sum_{k=0}^{v} a_{v,v-k}\,O\left(k^{\alpha/2+1/4}\int_0^{1/n} y^{\alpha}\,|\phi(x,y)|dy\right)\\
&= \frac{n!\,\gamma!}{(n+\gamma)!}\sum_{v=0}^{n}\binom{n+\gamma-v-1}{\gamma-1}\sum_{k=0}^{v} a_{v,v-k}\,o(\xi(n))\\
&= o\left(\xi(n)\,n^{-\gamma}\sum_{v=0}^{n}(n-v)^{\gamma-1}\right)\\
&= o(\xi(n)).
\end{aligned} \tag{19}$$

Now, in Eq. (5) applying Lemma 1 for $1/n \le y < \epsilon$, we have

$$|J_k^\alpha(x,y)| = O(k^{1-\alpha}[k^{\alpha/2-1/4}y^{-\alpha/2+1/4}k^{\alpha/2-3/4} + k^{\alpha/2-3/4}y^{-\alpha/2-1/4}k^{\alpha/2-1/4}]). \tag{20}$$

$$\begin{aligned}
|I_2| &\le \frac{n!\,\gamma!}{(n+\gamma)!}\sum_{v=0}^{n}\binom{n+\gamma-v-1}{\gamma-1}\sum_{k=0}^{v} a_{v,v-k}\int_{1/n}^{\epsilon} y^\alpha\,|\phi(x,y)|\,|J_k^\alpha(x,y)|\,dy\\
&= \frac{n!\,\gamma!}{(n+\gamma)!}\sum_{v=0}^{n}\binom{n+\gamma-v-1}{\gamma-1}\sum_{k=0}^{v} a_{v,v-k}\,O\left(k^{1-\alpha}\int_{1/n}^{\epsilon} y^\alpha\,|\phi(x,y)|\right.\\
&\qquad\left.[k^{\alpha/2-1/4}y^{-\alpha/2+1/4}k^{\alpha/2-3/4} + k^{\alpha/2-3/4}y^{-\alpha/2-1/4}k^{\alpha/2-1/4}]\,dy\right)\\
&= \frac{n!\,\gamma!}{(n+\gamma)!}\sum_{v=0}^{n}\binom{n+\gamma-v-1}{\gamma-1}\sum_{k=0}^{v} a_{v,v-k}O\left(\int_{1/n}^{\epsilon} y^{\alpha/2-1/4}\,|\phi(x,y)|dy\right)\\
&= \frac{n!\,\gamma!}{(n+\gamma)!}\sum_{v=0}^{n}\binom{n+\gamma-v-1}{\gamma-1}\sum_{k=0}^{v} a_{v,v-k}\,o(\xi(n))\\
&= o\left(\xi(n)\,n^{-\gamma}\sum_{v=0}^{n}(n-v)^{\gamma-1}\right)\\
&= o(\xi(n)),
\end{aligned} \tag{21}$$

in view of Eq. (20) and condition (10).

With the help Lemma 3 for $\epsilon \le y \le x-\delta$, we have

$$|L_{k+1}^{(\alpha-1)}(y)| = O\left(\frac{y^{-\alpha/2+1/4}\exp(y/2)}{\sqrt{\pi}}\,k^{\alpha/2-3/4}\cos(2\sqrt{ky}-(\alpha-1/2)\pi/2) + k^{\alpha/2-5/4}\right). \tag{22}$$

Applying formula (5), we have

$$\begin{aligned}
I_3 &= \frac{n!\,\gamma!}{(n+\gamma)!}\sum_{v=0}^{n}\binom{n+\gamma-v-1}{\gamma-1}\sum_{k=0}^{v} a_{v,v-k}\int_{\epsilon}^{x-\delta}\exp(-y)\,y^\alpha\\
&\qquad \phi(x,y)J_k^\alpha(x,y)dy\\
&= \frac{n!\,\gamma!}{(n+\gamma)!}\sum_{v=0}^{n}\binom{n+\gamma-v-1}{\gamma-1}\sum_{k=0}^{v} a_{v,v-k}O(k^{1-\alpha})\int_{\epsilon}^{x-\delta}\exp(-y)\,y^\alpha\,\phi(x,y)\\
&\qquad \frac{L_{k+1}^{(\alpha)}(x)\,L_{k+1}^{(\alpha-1)}(y) - L_{k+1}^{(\alpha-1)}(x)\,L_{k+1}^{(\alpha)}(y)}{x-y}dy\\
&= I_{31} + I_{32}.
\end{aligned} \tag{23}$$

Now, applying Lemma 3 and Eq. (22), we have

$$\begin{aligned}
I_{31} &= \frac{n!\,\gamma!}{(n+\gamma)!}\sum_{v=0}^{n}\binom{n+\gamma-v-1}{\gamma-1}\sum_{k=0}^{v}a_{v,v-k}O(k^{1-\alpha})\int_{\epsilon}^{x-\delta} \\
&\quad \frac{\exp(-y)\,y^{\alpha}\,|\phi(x,y)|}{x-y}O(k^{\alpha/2-1/4})\Bigg[\frac{y^{-\alpha/2+1/4}\exp(y/2)}{\sqrt{\pi}}k^{\alpha/2-3/4} \\
&\qquad \cos(2\sqrt{ky}-(\alpha-1/2)\pi/2)+O(k^{\alpha/2-5/4})\Bigg]dy \\
&= \frac{n!\,\gamma!}{(n+\gamma)!}\sum_{v=0}^{n}\binom{n+\gamma-v-1}{\gamma-1}\sum_{k=0}^{v}a_{v,v-k}\Bigg[O(1)\int_{\epsilon}^{x-\delta} \\
&\quad \frac{y^{\alpha/2+1/4}|\phi(x,y)|}{\exp(y/2)\,(x-y)}\cos(2\sqrt{ky}-(\alpha-1/2)\pi/2)dy+ \\
&\qquad O(k^{-1/2})\int_{\epsilon}^{x-\delta}\frac{\exp(-y)y^{\alpha}|\phi(x,y)|}{x-y}dy+o(1)\Bigg] \qquad (24) \\
&= \frac{n!\,\gamma!}{(n+\gamma)!}\sum_{v=0}^{n}\binom{n+\gamma-v-1}{\gamma-1}\sum_{k=0}^{v}a_{v,v-k}\,o(1) \\
&= o\left(n^{-\gamma}\sum_{v=0}^{n}(n-v)^{\gamma-1}\right) \\
&= o(1), \qquad (25)
\end{aligned}$$

in Eq. (24) using Riemann–Lebesgue theorem the first integral approaches to 0 and second integral approaches to 0 as $k\to\infty$.

Similarly,

$$|I_{32}| = o(1). \qquad (26)$$

Combining (23), (25), and (26), we have

$$|I_3| = o(1). \qquad (27)$$

Proceeding on the same lines

$$|I_5| = o(1). \qquad (28)$$

Applying Lemmas 3 and 4 in formula (6), we have

$$J_k^\alpha(x,y) = k^{1-\alpha} k^{\alpha/2-1/4} k^{\alpha/2-3/4} \frac{x^{-\alpha/2-1/4}\exp(x/2) y^{-\alpha/2-1/4}\exp(y/2)}{\pi(\sqrt{x}+\sqrt{y})}$$
$$\left[\sqrt{y}\cos(2\sqrt{kx}+\lambda)\frac{\sin(2\sqrt{ky}+\lambda)-\sin(2\sqrt{kx}+\lambda)}{\sqrt{y}-\sqrt{x}} - \sqrt{x}\right.$$
$$\left.\sin(2\sqrt{kx}+\lambda)\frac{\cos(2\sqrt{ky}+\lambda)-\cos(2\sqrt{kx}+\lambda)}{\sqrt{y}-\sqrt{x}} + O(1)\right]. \tag{29}$$

Here, the variables are confined to a fixed positive interval; so, the remainders in Lemmas 3 and 4 depend only on n (see [15, pp. 267]).

Following the calculation of [15, pp. 267], we have

$$J_k^\alpha(x,y) = \frac{1}{2}\sqrt{x}\left(\frac{\exp(x/2)\,x^{-\alpha/2-1/4}}{\sqrt{\pi}}\right)^2 y^{-1/2}\frac{\sin(2\sqrt{k}(\sqrt{y}-\sqrt{x}))}{(\sqrt{y}-\sqrt{x})} + O(1).$$

Thus, from the above equation, we have

$$\begin{aligned}
I_4 &= \frac{n!\gamma!}{(n+\gamma)!}\sum_{v=0}^{n}\binom{n+\gamma-v-1}{\gamma-1}\sum_{k=0}^{v} a_{v,v-k}\int_{x-\delta}^{x+\delta}\exp(-y)y^\alpha\phi(x,y)J_k^\alpha(x,y)dy\\
&= \frac{n!\gamma!}{(n+\gamma)!}\sum_{v=0}^{n}\binom{n+\gamma-v-1}{\gamma-1}\sum_{k=0}^{v} a_{v,v-k}\left[\frac{1}{2}\sqrt{x}\left(\frac{\exp(x/2)\,x^{-\alpha/2-1/4}}{\sqrt{\pi}}\right)^2\right.\\
&\quad\left.\int_{x-\delta}^{x+\delta}\exp(-y)\,y^{\alpha-1/2}\,\phi(x,y)\frac{\sin(2\sqrt{k}(\sqrt{y}-\sqrt{x}))}{\sqrt{y}-\sqrt{x}}dy + O(1)\right]\\
&= \frac{n!\gamma!}{(n+\gamma)!}\sum_{v=0}^{n}\binom{n+\gamma-v-1}{\gamma-1}\sum_{k=0}^{v} a_{v,v-k}O\left(\left[\int_{x-\delta}^{x-1/n} + \int_{x-1/n}^{x+1/n} + \right.\right.\\
&\quad\left.\left.\int_{x+1/n}^{x+\delta}\right]\exp(-y)\,y^{\alpha-1/2}\,\phi(x,y)\frac{\sin(2\sqrt{k}(\sqrt{y}-\sqrt{x}))}{\sqrt{y}-\sqrt{x}}dy\right) + O(1)\\
&= I_{41} + I_{42} + I_{43} + O(1).
\end{aligned} \tag{30}$$

Applying condition (11), we have

$$\begin{aligned}
|I_{41}| &= \frac{n!\,\gamma!}{(n+\gamma)!}\sum_{v=0}^{n}\binom{n+\gamma-v-1}{\gamma-1}\sum_{k=0}^{v}a_{v,v-k}O\left(\int_{x-\delta}^{x-1/n}\frac{|\phi(x,y)|(\sqrt{x}+\sqrt{y})}{|x-y|}dy\right)\\
&= \frac{n!\,\gamma!}{(n+\gamma)!}\sum_{v=0}^{n}\binom{n+\gamma-v-1}{\gamma-1}\sum_{k=0}^{v}a_{v,v-k}O\left(\int_{1/n}^{\delta}\frac{\psi(x,u)}{u}du\right)\\
&= o\left(\xi(n)\,n^{-\gamma}\sum_{v=0}^{n}(n-v)^{\gamma-1}\right)\\
&= o(\xi(n)). \qquad (31)
\end{aligned}$$

Proceeding on the same lines, we have

$$|I_{43}| = o(\xi(n)). \qquad (32)$$

Applying condition (11), we have

$$\begin{aligned}
|I_{42}| &= \frac{n!\,\gamma!}{(n+\gamma)!}\sum_{v=0}^{n}\binom{n+\gamma-v-1}{\gamma-1}\sum_{k=0}^{v}a_{v,v-k}O\left(\int_{x-1/n}^{x+1/n}|\phi(x,y)|\left|\frac{\sin(2\sqrt{k}(\sqrt{y}-\sqrt{x}))}{\sqrt{y}-\sqrt{x}}\right|dy\right)\\
&= \frac{n!\,\gamma!}{(n+\gamma)!}\sum_{v=0}^{n}\binom{n+\gamma-v-1}{\gamma-1}\sum_{k=0}^{v}a_{v,v-k}\,O\left(\sqrt{k}\int_{0}^{1/n}|\psi(x,u)|du\right)\\
&= \frac{n!\,\gamma!}{(n+\gamma)!}\sum_{v=0}^{n}\binom{n+\gamma-v-1}{\gamma-1}\sum_{k=0}^{v}a_{v,v-k}\,o\left(\frac{\sqrt{k}\,\xi(n)}{n}\right)\\
&= o\left(\xi(n)\,n^{-\gamma}\sum_{v=0}^{n}(n-v)^{\gamma-1}\right)\\
&= o(\xi(n)). \qquad (33)
\end{aligned}$$

Combining (30)–(33), we have

$$|I_4| = o(\xi(n)). \qquad (34)$$

Now, with the help of first part of Lemma 2 (for $\eta = 3$), we have

$$
\begin{aligned}
|I_6| &\le \frac{n!\,\gamma!}{(n+\gamma)!}\sum_{v=0}^{n}\binom{n+\gamma-v-1}{\gamma-1}\sum_{k=0}^{v}a_{v,v-k}\Bigg[O(k^{1-\alpha})\int_{\omega}^{n}\exp(-y)y^{\alpha-1}\\
&\qquad |\phi(x,y)||L_{k+1}^{(\alpha)}(x)||L_{k+1}^{(\alpha-1)}(y)|dy + O(k^{1-\alpha})\int_{\omega}^{n}\exp(-y)y^{\alpha-1}|\phi(x,y)|\\
&\qquad |L_{k+1}^{(\alpha-1)}(x)||L_{k+1}^{(\alpha)}(y)|dy\Bigg]\\
&= I_{61} + I_{62}.
\end{aligned} \tag{35}
$$

Applying Lemma 6, we have

$$
\begin{aligned}
|I_{61}| &= \frac{n!\,\gamma!}{(n+\gamma)!}\sum_{v=0}^{n}\binom{n+\gamma-v-1}{\gamma-1}\sum_{k=0}^{v}a_{v,v-k}O\Bigg(k^{-\alpha/2+3/4}\int_{\omega}^{n}\exp(-y/2)\\
&\qquad y^{\alpha/2-3/4}|\phi(x,y)|\exp(-y/2)\,y^{\alpha/2-1/4}|L_{k+1}^{(\alpha-1)}(y)|dy\Bigg)\\
&= \frac{n!\,\gamma!}{(n+\gamma)!}\sum_{v=0}^{n}\binom{n+\gamma-v-1}{\gamma-1}\sum_{k=0}^{v}a_{v,v-k}(k^{-\alpha/2+3/4}\,k^{\alpha/2-3/4})\\
&\qquad \int_{\omega}^{n}\exp(-y/2)\,y^{\alpha/2-3/4}\,|\phi(x,y)|dy\\
&= \frac{n!\,\gamma!}{(n+\gamma)!}\sum_{v=0}^{n}\binom{n+\gamma-v-1}{\gamma-1}\sum_{k=0}^{v}a_{v,v-k}\,(k^{-\alpha/2+3/4}\,k^{\alpha/2-3/4})\,o(\xi(n))\\
&= o\left(\xi(n)\,n^{-\gamma}\sum_{v=0}^{n}(n-v)^{\gamma-1}\right)\\
&= o(\xi(n)).
\end{aligned} \tag{36}
$$

Similarly, we can calculate

$$|I_{62}| = o(\xi(n)). \tag{37}$$

Combining (35)–(37), we have

$$|I_6| = o(\xi(n)). \tag{38}$$

Using second part of Lemma 2 for $n \le y < \infty$, we have

$$\exp(-x/2)\,x^{\alpha/2+1/12}\,|L_n^{(\alpha)}(x)| = O(n^{\alpha/2-1/4}). \tag{39}$$

Applying formula (4), we have

$$
\begin{aligned}
|I_7| &= \frac{n!\,\gamma!}{(n+\gamma)!}\sum_{v=0}^{n}\binom{n+\gamma-v-1}{\gamma-1}\sum_{k=0}^{v} a_{v,v-k}\,O(k^{1-\alpha}) \\
&\qquad \int_n^{\infty}\frac{\exp(-y)\,y^{\alpha}|\phi(x,y)|}{x-y}\{L_k^{(\alpha)}(x)\,L_{k+1}^{(\alpha)}(y)-L_{k+1}^{(\alpha)}(x)\,L_k^{(\alpha)}(y)\}dy \\
&= I_{71}+I_{72}.
\end{aligned}
\tag{40}
$$

Now, applying condition (12), and with the help of equation (39), we get

$$
\begin{aligned}
I_{71} &= \frac{n!\,\gamma!}{(n+\gamma)!}\sum_{v=0}^{n}\binom{n+\gamma-v-1}{\gamma-1}\sum_{k=0}^{v} a_{v,v-k}\,O(k^{1-\alpha}) \\
&\qquad \int_n^{\infty}\frac{\exp(-y)\,y^{\alpha}|\phi(x,y)|}{y-x}O(k^{\alpha/2-1/4})\exp(y/2)y^{-\alpha/2-1/12}k^{\alpha/2-1/4}dy \\
&= \frac{n!\,\gamma!}{(n+\gamma)!}\sum_{v=0}^{n}\binom{n+\gamma-v-1}{\gamma-1}\sum_{k=0}^{v} a_{v,v-k}\,O(k^{1/2})\int_n^{\infty}\exp(-y/2) \\
&\qquad\qquad y^{\alpha/2-13/12}|\phi(x,y)|dy \\
&= \frac{n!\,\gamma!}{(n+\gamma)!}\sum_{v=0}^{n}\binom{n+\gamma-v-1}{\gamma-1}\sum_{k=0}^{v} a_{v,v-k}\,o(\xi(n)) \\
&= o\left(\xi(n)\,n^{-\gamma}\sum_{v=0}^{n}(n-v)^{\gamma-1}\right) \\
&= o(\xi(n)).
\end{aligned}
\tag{41}
$$

Similarly, we can calculate

$$
|I_{72}| = o(\xi(n)). \tag{42}
$$

Combining (40)–(42), we have

$$
|I_7| = o(\xi(n)). \tag{43}
$$

Collecting (16), (19), (21), (27), (28), (34), (38), and (43), we have

$$
\left|[C^{\gamma}.T]_n(f;x)-f(x)\right| = o(\xi(n)).
$$

This completes the proof of Theorem. □

4 Corollaries

Here, few corollaries are given, which are derived from our Theorem.

Corollary 1 *If we take $a_{v,k}=\frac{1}{(v-k+1)\,log(v+1)}$ in Eq. (8), then $C^{\gamma}.T$ reduce to $C^{\gamma}.H$ or $(C,\gamma)(H,\frac{1}{v+1})$, then, for $f\in L[0,\infty)$, we have*

$$\left|[C^{\gamma}.H]_n(f;x) - f(x)\right| = o(\xi(n)).$$

Corollary 2 *If we take $a_{v,k} = \frac{p_{v-k}}{P_v}$, where $P_v = \sum_{k=0}^{v} p_k \neq 0$ in Eq. (8), then $C^{\gamma}.T$ reduce to $C^{\gamma}.N_p$ or $(C,\gamma)(N,p_v)$, then, for $f \in L[0,\infty)$, we have*

$$\left|[C^{\gamma}.N_p]_n(f;x) - f(x)\right| = o(\xi(n)).$$

Corollary 3 *If we take $a_{v,k} = \frac{p_k}{P_v}$ in Eq. (8), then $C^{\gamma}.T$ reduce to $C^{\gamma}.\overline{N}_p$ or $(C,\gamma)(\overline{N},p_v)$, then, for $f \in L[0,\infty)$, we have*

$$\left|[C^{\gamma}.\overline{N}_p]_n(f;x) - f(x)\right| = o(\xi(n)).$$

Corollary 4 *If we take $a_{v,k} = \frac{p_{v-k}\,q_k}{R_v}$, where $R_v = \sum_{k=0}^{v} p_k\, q_{v-k}$ in Eq. (8), then $C^{\gamma}.T$ reduce to $C^{\gamma}.N_{pq}$ or $(C,\gamma)(N,p,q)$, then, for $f \in L[0,\infty)$, we have*

$$\left|[C^{\gamma}.N_{pq}]_n(f;x) - f(x)\right| = o(\xi(n)).$$

Corollary 5 *If we take $a_{v,k} = \frac{1}{(1+q)^v}\binom{v}{k}q^{v-k}$ in Eq. (8), then $C^{\gamma}.T$ reduce to $C^{\gamma}.E^q$ or $(C,\gamma)(E,q)$, then, for $f \in L[0,\infty)$, we have*

$$\left|[C^{\gamma}.E^q]_n(f;x) - f(x)\right| = o(\xi(n)).$$

Corollary 6 *If we take $a_{v,k} = \frac{1}{2^v}\binom{v}{k}$ in Eq. (8), then $C^{\gamma}.T$ reduce to $C^{\gamma}.E^1$ or $(C,\gamma)(E,1)$, then, for $f \in L[0,\infty)$, we have*

$$\left|[C^{\gamma}.E^1]_n(f;x) - f(x)\right| = o(\xi(n)).$$

Remark 2 If we take $\gamma = 1$ in the above cases, then we get $C^1.H$, $C^1.N_p$, $C^1.\overline{N}_p$, $C^1.N_{pq}$, $C^1.E^q$, and $(C,1)(E,1)$ are also particular cases of the $C^{\gamma}.T$ method.

Acknowledgements This work was supported by the Council of Scientific and Industrial Research (CSIR), New Delhi, India [Award No.: 09/1007(0008)/2020-EMR-I], and Sardar Vallabhbhai National Institute of Technology, Surat-395007, Gujarat [Grant No.: Seed Money/2020-21/1481 dated: 08/12/2020].

References

1. Gairola, A.R., Singh, K.K., Mishra, L.N.: Degree of approximation by certain Durrmeyer type operators. Discontinuity Nonlinearity Complex **11**(2), 253–273 (2022)
2. Kajla, A., Mohiuddine, S.A., Alotaibi, A., Goyal, M., Singh, K.K.: Approximation by ϑ-Baskakov-Durrmeyer-type hybrid operators. Iran. J. Sci. Technol. Trans. A: Sci. **44**, 1111–1118 (2020). https://doi.org/10.1007/s40995-020-00914-3
3. Khatri, K., Mishra, V.N.: Approximation of functions belonging to $L[0,\infty)$ by product summability means of its Fourier-Laguerre series. Cogent Math. **3**(1), 1250854 (2016). https://doi.org/10.1080/23311835.2016.1250854

4. Krasniqi, X.Z.: On the degree of approximation of a function by (C, 1)(E, q) means of its Fourier-Laguerre series. Int. J. Anal. Appl. **1**(1), 33–39 (2013)
5. Lal, S., Nigam, H.K.: Degree of approximation by (N, p, q) summability means of the Fourier-Laguerre expansion. Tamkang J. Math. **32**(2), 143–150 (2001). https://doi.org/10.5556/j.tkjm.32.2001.357
6. Mäkilä, P.M.: Laguerre series approximation of infinite dimensional systems. Automatica (Oxf). **26**(6), 985–995 (1990). https://doi.org/10.1016/0005-1098(90)90083-T
7. Mittal, M.L., Singh, M.V.: Error estimation of functions by Fourier-Laguerre polynomials using Matrix-Euler operators. Int. J. Anal. **2015** (2015). https://doi.org/10.1155/2015/478345
8. Nigam, H.K., Sharma, A.: A study on degree of approximation by *(E, 1)* summability means of the Fourier-Laguerre expansion. Int. J. Math. Math. Sci. **2010** (2010). https://doi.org/10.1155/2010/351016
9. Sahani, S.K., Mishra, V.N., Pahari, N.P.: On the degree of approximation of a function by Nörlund means of its Fourier-Laguerre series. Nepal J. Math. Sci. **1**, 65–70 (2020). https://doi.org/10.3126/njmathsci.v1i0.34164
10. Saini, S., Singh, U.: Degree of approximation of $f \in L[0, \infty)$ by means of Fourier-Laguerre series. In: Agrawal, P.N., Mohapatra, R.N., Singh, U., Srivastava, H.M. (eds) Mathematical Analysis and its Applications. Springer Proceedings in Mathematics and Statistics, vol. 143, pp. 207–217. Springer India, New Delhi (2015). https://doi.org/10.1007/978-81-322-2485-3_16
11. Singh, T.: On the absolute summability factors of Fourier-Laguerre expansion. Indian J. Pure Appl. Math. **7**(9), 961–968 (1976)
12. Singh, U., Saini, S.: Uniform approximation in $L[0, \infty)$-space by Cesáro means of Fourier-Laguerre series. Proc. Natl. Acad. Sci., India, Sect. A Phys. Sci. (2021). https://doi.org/10.1007/s40010-021-00747-8
13. Sinha, T.A.K., Singh, K.K., Sharma, A.K.: On simultaneous approximation and combinations of Lupas type operators. Kragujevac J. Math. **48**(4), 619–627 (2024)
14. Sonker, S.: Approximation of functions by $(C^1.T)$ means of its Fourier-Laguerre series. Proc. ICMS-2014 **1**(1), 122–125 (2014)
15. Szegö, G.: Orthogonal Polynomials. American Mathematical Society, New York (1959)

Numerical Modelling and Experimental Validation of Mechanical Separation of Helminth Eggs for Wastewater Purification

M. Diederich, F. Gül, C. Özman, A. C. Benim, L. Ihringer, and D. Möller

Abstract Hydrodynamics of wastewater, which is contaminated with helminth eggs is computationally and experimentally investigated, for laboratory conditions and for a small sewage treatment plant. In the computational analysis, the flow is mathematically modelled within the framework of a Eulerian–Lagrangian framework, where the continuous water phase is treated by an Eulerian, and the discrete particle phase (helminth eggs) is treated by a Lagrangian formulation. For turbulent flows, the Shear Stress Transport model is used to model the turbulence of the continuous phase. The effect of the latter on the discrete phase is modelled by a discrete random walk model. In modelling the momentum exchange between the phases, a special emphasis is placed upon the accurate determination of the drag coefficient for the helminth eggs. For this purpose, flow around individual eggs is analysed and laboratory measurements of other authors are inspected. Before applying these results, measurements are performed on a small sewage treatment plant using surrogate spheres, for validating the remaining aspects of the Eulerian–Lagrangian hydrodynamics modelling. Subsequently, the operation of the small sewage plant is analysed for wastewater containing helminth eggs for its optimization.

Keywords Helminth eggs · Wastewater treatment · CFD

M. Diederich · F. Gül · C. Özman · A. C. Benim (✉)
Center of Flow Simulation, Düsseldorf University of Applied Sciences, Düsseldorf, Germany
e-mail: alicemal@prof-benim.com

L. Ihringer · D. Möller
Menk'sche GmbH, Monheim Am Rhein, Germany
e-mail: geschaeftsfuehrung@menksche.de

D. Möller
e-mail: d.moeller@aquato.de

R. K. Sharma et al. (eds.), *Frontiers in Industrial and Applied Mathematics*,
Springer Proceedings in Mathematics & Statistics 410,
https://doi.org/10.1007/978-981-19-7272-0_13

1 Introduction

A large population of the world does not have access to clean water. Microorganisms such as helminth eggs (HE) in water are causing death especially among children and people with weakened immune system. As a method of purifying wastewater (WW), the sedimentation technique (ST) is quite often utilized. For being able to utilize the ST in an efficient way, it is important to know the sinking speeds (SS) of the particulates. If a small-sized sewage treatment plant (SSTP) is of concern, the problem is more complicated: the times of residence are shorter. Moreover, flow turbulence that can occur can affect the motion of the particulates, additionally.

Different approaches for cleaning WW were previously presented by several authors [1–4]. The purpose of the current work is establishing a numerical simulation model to obtain the SS of HE, which is verified by comparisons to measurements. The procedure is subsequently to be applied to calculate the separation behaviour of a SSTP and for optimizing its performance. For achieving this goal, measurements and calculations are performed. For the calculations, an approach based on Computational Fluid Dynamics (CFD) methodology is preferred. The measurements are to be obtained on an SSTP.

In the past, the sedimentation phenomenon in various areas of application was numerically analysed by a number of different researchers [5, 6]. For a detailed analysis of the sedimentation by means of a CFD approach, the two-phase mixture prevailing in the WW shall first be described by adequate means. Here, an Eulerian–Lagrangian approach is adopted to this purpose, which will be explained in more detail below. In the past analyses of similar kind, the shapes of the particulates were always assumed to be spherical, without paying attention to the actual forms of them. Consequently, the law of Stokes [7] was always used for determining the SS, which has its validity for spherical forms. In reality, the geometries of HE differ from that of a sphere. Therefore, one cannot a priori assume that the Stokes law would deliver precise results for them. Therefore, in the current analysis, the emphasis is placed upon the precise calculation of the SS and deduction of more precise drag coefficient expressions for HE, depending on their individual shapes. As means of validation of the numerical approach, the measurements of Sengupta et al. [8] are employed as basis.

2 Outline of Computational Modelling

Under certain conditions, flows may be described by ordinary differential equations [9]. In many applications, like the present one, a description by the full Navier–Stokes equations is needed [10], which can be solved only by numerical methods. For the latter, various approaches are possible, including the Lattice Boltzmann Method [11]. The Finite Volume Method (FVM) [12] is the most popular one, and it is also applied presently.

Thus, for continuous phase, described in Eulerian frame, incompressible flow of water as described by Navier–Stokes equations is computationally modelled applying the FVM. Steady-state and unsteady calculations in two-dimensional and three-dimensional geometries are performed. In cases with free surface (air–water interface), the Volume of Fluid method is used [13]. The characteristics of the considered flows range from laminar [14], over transitional [15], to turbulent [16]. Therefore, for turbulent flows, the Shear Stress Transport model [17] is used as turbulence model within RANS or URANS framework [18], as it copes comparably well with transitional flows [15].

As far as the dispersed phase is considered, the relevant particle sizes may be considered within the range 50–100 μm. Although the Brownian motion was argued to play a role [8], it is currently assumed that this is relevant rather for nanoscales [19] but not for the present particle size range. Thus, only the macroscopic fluid drag and gravity/buoyancy forces are assumed to act externally on the particles [20]. Currently, a dilute disperse phase is assumed and particle–particle interactions are neglected. In describing particulate flows, the dispersed phase may also be described within an Eulerian framework [21, 22]. Alternatively a Lagrangian framework can be used, where trajectories of individual particles [23] are calculated. The Lagrangian formulation is currently adopted, as it is more convenient for present purposes. In case of turbulent flow of the continuous phase, its effect on the dispersed phase is modelled by the discrete random walk model [24].

In the numerical formulation, the velocity–pressure coupling is treated by the so-called SIMPLEC procedure [25]. The convective terms are discretized by a second-order accurate upwinding scheme [26]. Within the iterative procedure, the under-relaxation factors used for the pressure and velocity, 1.0 and 0.5, are used, respectively. For convergence, the threshold values for the scaled residuals are set to 10^{-9}, which are 10^6 times smaller than the default value of 10^{-3}. Grid independence is always ensured by preceding grid independence studies. The analysis is performed based on the general-purpose CFD software ANSYS Fluent 18.0 [27].

3 Investigations on Isolated Helminth Eggs

In this part, the flow around individual helminth eggs is investigated. The purpose is to determine the drag coefficient accurately [28], which gives the basis for the momentum transfer between the continuous water and discrete particle (helminth eggs) phases. In relationship with mechanical wastewater purification, the discussion is quite often carried out based on the so-called sink velocity, V [8], which delivers a more direct information for the special case of sinking particles in quiescent fluid. This quantity is closely related to the drag coefficient, C, which allows a more general description for non-quiescent fluid and, thus, a more convenient means for mathematical formulation.

In analysing similar flows [8], quite often the Stokes law [29] is used for determining the sink velocity (or drag coefficient), which is valid for a spherical particle shape and vanishingly small relative velocity between the flow and particle. The drag coefficient for sphere (C), as function of the relative Reynolds number (Re) according to Stokes law, is given below:

$$C = \frac{64}{\mathrm{Re}} \tag{1}$$

with

$$\mathrm{Re} = \frac{\rho\, \Delta u\, d}{\mu} \tag{2}$$

where Δu denotes the magnitude of speed difference between the fluid and particle, while ρ, μ and d denote the fluid density, viscosity and sphere diameter, respectively.

For arbitrary C, the following relationship can be derived for the drag coefficient and the sink velocity V, where g and $\Delta\rho$ denote the gravitational acceleration and the density difference between particle and fluid, respectively.

$$V = \sqrt{\frac{4}{3}\frac{1}{C}\frac{\Delta\rho}{\rho} g d} \tag{3}$$

In the common commercial CFD software, empirical expressions for the drag coefficient for sphere are normally implemented such as the following one [30] for $\mathrm{Re} \leq 1000$:

$$C = \frac{24}{\mathrm{Re}}\left(1 + 0.15\,\mathrm{Re}^{0.687}\right) \tag{4}$$

which approach the Stokes law for small relative Reynolds number.

A main issue in this respect is that the shapes of helminth eggs are not spherical, but the Stokes law as well as the further commonly used drag laws are given for a sphere. In order to cope with this fact, an approach that is normally adopted is to assume a "representative diameter" [8] for the non-spherical egg shape, and use the Stokes law for sphere, based on this diameter. However, an accurate result cannot a priori be assumed, by this approach. This point is addressed in the present part of the study.

Three types of helminth eggs are considered, namely, Oesophagostomum spp. (oes), Trichuris suis (tri) and Ascaris suum (asc), the generic shapes of which are qualitatively sketched in Fig. 1. Sengupta et al. [8] performed experiments on these egg types, in which they determined their sink velocities using an Owen tube [31], using populations of 500–600 eggs of each kind. In each group, the properties and the measured velocities show a scatter, of course. Here, the mean values are considered, unless otherwise is stated. As the eggs do not have spherical shapes, their length

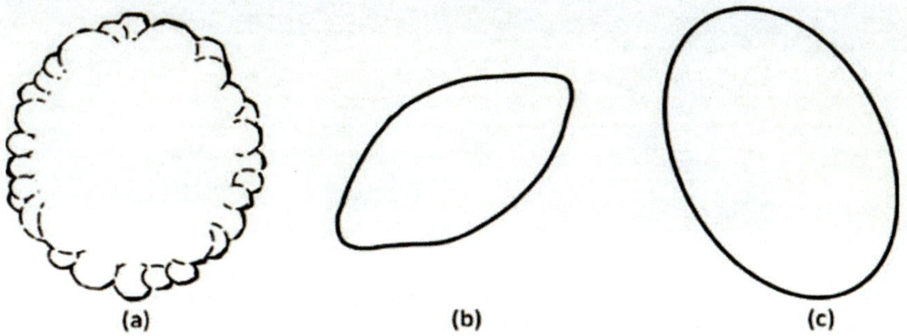

Fig. 1 Qualitative sketches of typical helminth eggs of different types, **a** Ascaris suum (asc), **b** Trichuris sui (tri), **c** Oesophagostomum spp. (oes)

Table 1 Geometry parameters and densities of helminth eggs [8]

Egg type	L (μm)	W (μm)	Density (kg/m^3)
asc	67.20	55.41	1120
tri	62.16	30.78	1100
oes	76.17	50.33	1070

(L) and width (W) are used [8] to define an equivalent diameter as their arithmetic average

$$d_E = \frac{L + W}{2} \tag{5}$$

Geometry parameters and densities of the considered helminth eggs that are measured in the experiments of Sengupta et al. [8] are provided in Table 1. Note that the measured parameters showed a scatter and the mean values are shown in the table.

In the simulations, the shapes (Fig. 1) are exactly considered. Since the surface of asc was very irregular and nearly arbitrary, it is considered with some idealization.

The sink velocities for the eggs are obtained indirectly, via drag coefficient (Eq. 3). The latter is obtained by means of steady-state calculations as function of Re. The expected range of Re is rather low. Thus seven Re are considered between 0.001 and 0.01. In agreement with Ref. [8], tap water is considered, with $\rho = 997.8$ kg/m^3, $\mu = 0.00094$ Pa · s.

The calculated flow fields for Re = 0.01, for a certain relative orientation of the eggs to the flow direction, are presented in Fig. 2, where the egg surface and the velocity vector fields (relative to egg) in two perpendicular sections are displayed for the three egg types. The low velocity region near the egg surface due to the no-slip condition, i.e. the thick boundary layer can be observed. The velocity variation near surface is stronger for asc, due to the surface structures (Fig. 2a).

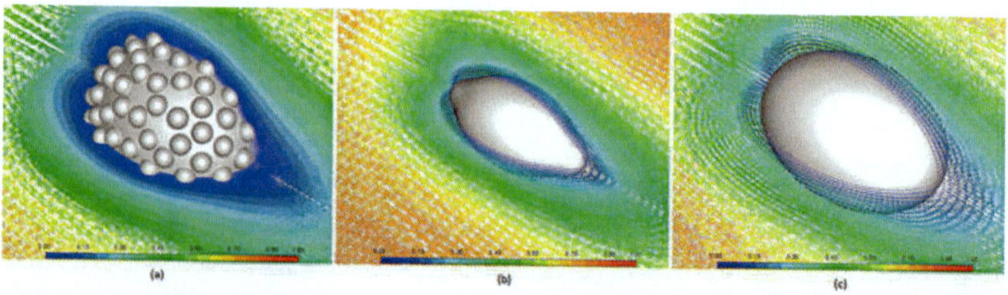

Fig. 2 Vectors of nondimensional velocity (relative to egg) in two planes, Re $= 0.01$, **a** asc, **b** tri, **c** oes

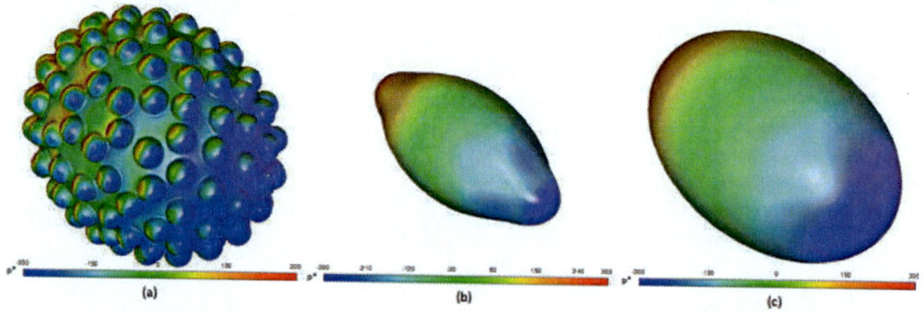

Fig. 3 Distribution of nondimensional gauge pressure on surface, Re $= 0.01$, **a** asc **b** tri, **c** oes

For the same case (Fig. 2), the predicted nondimensional gauge pressure distributions on the egg surface are displayed in Fig. 3, for the three egg types. One can see that there are similarities in large, for the three egg types, where a pressure difference between the upwind and downwind sides that contributes to drag force. As the surface pressure distribution is rather smooth for tri (Fig. 3b) and oes (Fig. 3c), a patterned distribution is observed for asc (Fig. 3a) due to the surface structure.

In addition to the pressure differential, a further contribution to the drag force is provided by the shear stress. The predicted nondimensional shear stress magnitude distributions for the three egg types, for the same flow configuration (Fig. 2), are displayed in Fig. 4. It is interesting to see that the shear stress patterns are qualitatively quite different between the egg types. For oes, a very homogeneous distribution is observed (Fig. 4c), whereas a more variable but a smooth distribution is observed for tri (Fig. 4b). For asc, a very inhomogeneous pattern is observed, where the tips of the protrusions experience locally high values, whereas very low values are observed for the remaining surfaces (Fig. 4a).

The predicted sink velocities for the three egg types are compared with the measured values by Sengupta et al. [8] (EXP) and those provided by the Stokes law, in Table 2. Please note that the experimental values are the averages of individual measurements varying within a range. In Figs. 2, 3 and 4, only one relative orientation between the flow direction and egg is displayed. Please note that the drag coefficients, which are used to obtain the sink velocities, are obtained by considering

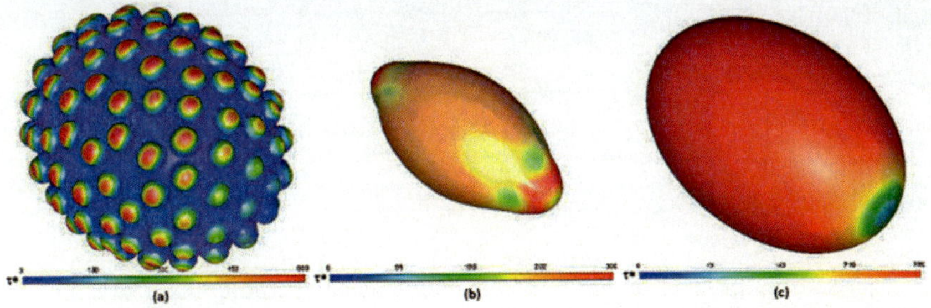

Fig. 4 Distribution of nondimensional shear stress magnitude on surface, Re = 0.01, **a** asc, **b** tri, **c** oes

Table 2 Predicted sink velocities compared with experiments and the Stokes law

Egg type	V (mm/s)		
	EXP	Stokes	Prediction
asc	0.06	0.27	0.24
tri	0.15	0.13	0.15
oes	0.13	0.17	0.17

a number of different relative orientations, taking their average to obtain a resultant/effective drag coefficient, since the relative orientation of the egg can arbitrarily change, in reality. One can see that the predicted values do not differ much from the values indicated by the Stokes law. For oes, the predicted value is the same as that of the Stokes law and overpredicts the experimental value. For tri, the predicted value agrees exactly with the experimental value, whereas the Stokes law shows an underprediction. The largest disagreement to the experiments is observed for asc. Here, both the Stokes law and predictions indicate a much higher sink velocity than the measured one.

This was rather unexpected, since the surface protrusions let one to expect high drag. However, it turns out that the surface protrusions cause a low overall drag. It seems that most parts of the surface are "protected" by the protrusions and experience a very low shear, whereas the high shear zones are restricted to small areas at the tips of the protrusions (Fig. 4a).

Based on the results of Table 2, for tri and oes, one can attest a reasonable accuracy to the predictions that is slightly better than that of the Stokes law, for tri. The disagreement observed for asc indicates that additional effects such as agglomeration might be playing a role, in reality, which are not considered in the predictions. However, one can still see that the CFD predictions provided an improvement against the Stokes law (Table 3), even it is rather small.

Since the present CFD predictions could not deliver very accurate results, especially for asc, probably due to the omission of agglomeration in the present model, drag coefficient correlations are derived, based directly on the experimental results. For this purpose, Oseen drag law for sphere [32] is taken as basis, which has a broader

Table 3 Derived correction factors for the drag law of different egg types

Egg type	Correction factor, f	
	Tap water	Wastewater
asc	4.3477	1.6610
tri	0.8593	1.4206
oes	1.3249	1.5203

range of validity towards higher Reynolds numbers compared to the Stokes flows, and the developed correlations are obtained by correcting the Oseen drag law, by a factor f, as indicated below:

$$C = f \times \frac{24}{\mathrm{Re}}\left(1 + \frac{3}{16}\mathrm{Re}\right) \tag{4}$$

The derived correction factors for different egg types are provided in Table 3. Using the wastewater data of Sengupta et al. [8], correction factors for wastewater are also developed. Using these correlations (Eq. 4, Table 3), CFD calculations are performed within the Eulerian–Lagrangian framework, where the particle trajectories are calculated in the Owen tube [31], like in the experiments of Sengupta et al. [8]. Please note that these CFD simulations are of different character than the previously discussed CFD calculations. In the previous results, individual eggs are simulated by considering the exact geometry, without needing a drag law, but for the purpose of deriving a drag law. In the CFD predictions that follow, egg geometries are not resolved but considered as particles that obey a certain drag law (the presently derived one, Eq. 4, Table 3), and based on this, particle trajectories are calculated. It shall also be noted that the developed drag coefficient correlations are implemented in the used software via User-Defined Functions (UDF). From the calculated particle trajectories in the Owen tube, based on the developed drag coefficients, the corresponding sink velocities are obtained.

The sink velocities, predicted in this manner, in comparison with experiments and the Stokes law are presented in Table 4. A very good agreement with the experimental values can be observed. Please note that the settling behaviours are quite different between tap water and wastewater. This can be due to different flocculation behaviours.

4 Investigation of the Wastewater Treatment Plant

The sewage treatment plant under investigation is shown in Fig. 5. The computational model of the plant, with the indication of water fill levels for a certain operation point, is depicted in Fig. 6.

The predicted distributions of the velocity magnitude in the first vessel, in the vertical middle plane are shown in Fig. 7. In both sub-figures (Fig. 7a, b), the velocity

Table 4 Predicted sink velocities using derived drag correlations compared with experiments

Egg type	V (mm/s)			
	Tap water		Wastewater	
	EXP	Lagrangian prediction by corrected drag law	EXP	Lagrangian prediction by corrected drag law
asc	0.06	0.06	0.16	0.16
tri	0.15	0.15	0.09	0.09
oes	0.13	0.13	0.11	0.11

Fig. 5 Small sewage treatment plant

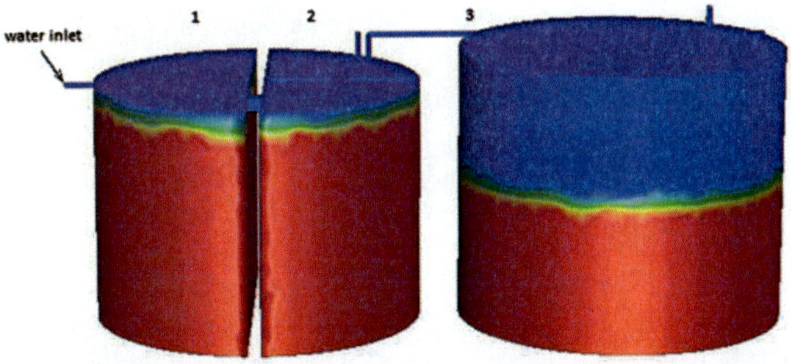

Fig. 6 Computational model of the small sewage plant with indication of water volume fraction for an operation condition (red: water, blue: air)

magnitude (V) is made nondimensional by the velocity at the inlet of the water intake pipe (V_0) during the filling process. The sub-figures show states, where the vessel is nearly filled out, shortly before closing the water intake (the same V_0 value is used in both sub-figures). Figure 7a corresponds to the instance short before the start

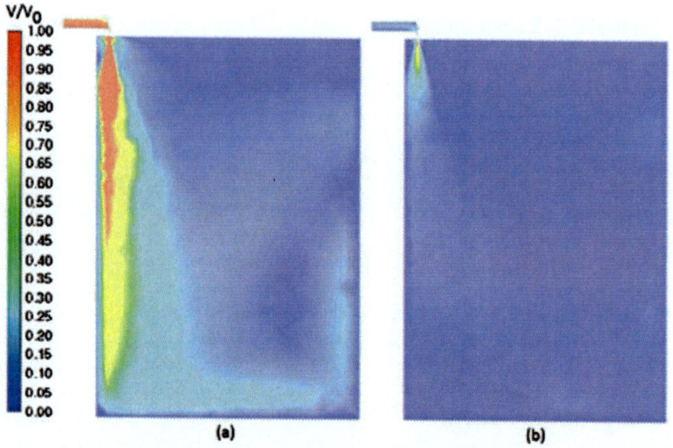

Fig. 7 Velocity magnitude distribution in middle plane of the first vessel, **a** State I, **b** State II

of closing the water inlet valve (State I). Figure 7b shows the state just before the valve is fully closed, where the discharge is nearly 10% of the nominal value (State II). In the figure, one can see that quite high velocities can occur at entrance of the falling water jet into the pool, which can be much higher than the adjusted discharge velocity (V_0) upstream the intake pipe, just because of the gravitational acceleration. This causes quite high velocities to occur in large regions of the pool (Fig. 7a).

Obviously, for the throttled state (Fig. 7b), the velocities are much smaller. Also here, the velocities at the jet entry are near V_0 (Fig. 7b).

The distribution of turbulence intensity (Tu) for the states shown in Fig. 7 is displayed in Fig. 8 (Tu is calculated by dividing the local fluctuational velocity u' by V_0, where u' is obtained from the turbulence kinetic energy). One can see that quite high levels of turbulence can occur, due to the high velocities, which decline, of course, by the velocity reduction (Fig. 8). The trajectories of about 300 helminth eggs injected through the water inlet at State I for a period of 20 s after injection are displayed in Fig. 9, where the colour indicated the dimensionless velocity magnitude (in the sense discussed for Fig. 7). In the figure, it can be observed, first, that the trajectories are dispersed due to the variations in the velocity field and turbulence.

Secondly, one can see that the eggs move with a much higher velocity than the sink velocity due to the fluid motion, which confirms that the formulation based on drag coefficient, which incorporates the special case of quiescent fluid, is principally more convenient than the formulation based on sink velocity in detailed analysis of such systems. Thirdly, it is observed that the differences between the egg types are not very much different. This is affected by the fact that the presented distribution is governed by the water flow with a much higher speed than the sink velocity. On the other hand, here drag coefficient correlations that are obtained for wastewater are used, which anyway do not imply substantial differences between the egg types (Table 3).

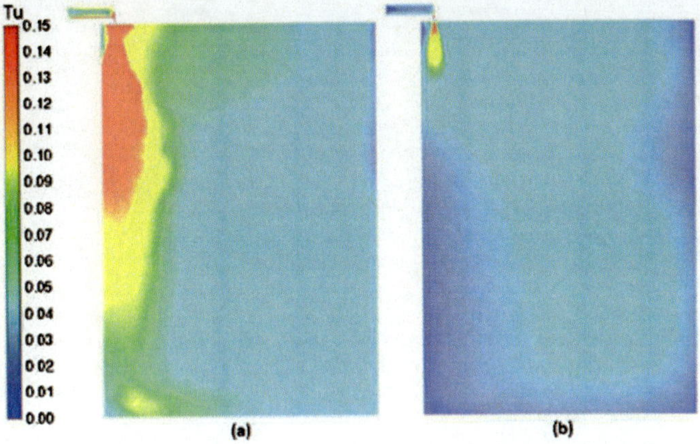

Fig. 8 Turbulence intensity distribution in middle plane of first vessel, **a** State I, **b** State II

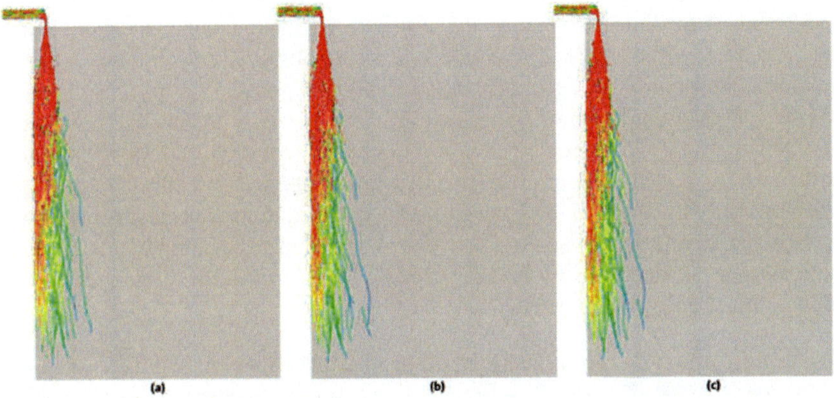

Fig. 9 Trajectories of eggs for a period of 20 s after injection (State I), **a** asc, **b** tri, **c** oes

The trajectories of the eggs injected at State II for a period of 100 s are depicted in Fig. 10. One can see that the distances travelled in 100 s for State I are much shorter than those within 20 s for State I, due to the much smaller flow velocities.

5 Conclusions

Hydrodynamics of wastewater, which is contaminated with helminth eggs is investigated, for laboratory conditions and for a small sewage treatment plant. The first part was devoted to the investigation of effect of the individual shapes of the eggs on the sink velocity. Here, the flow around the individual eggs of three different

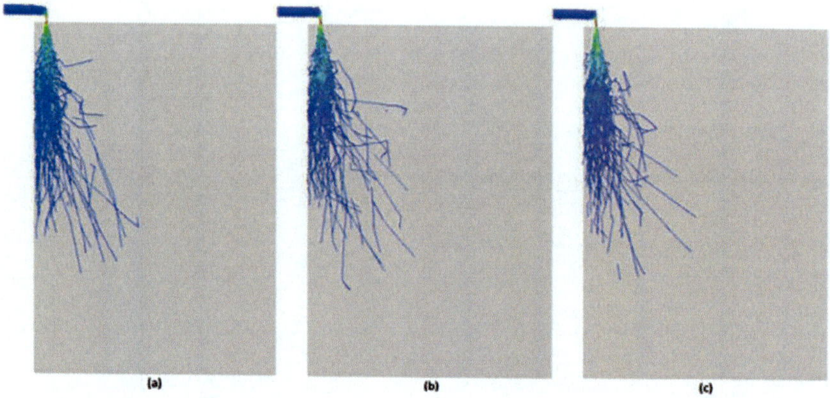

Fig. 10 Trajectories of eggs for a period of 100 s after injection (State II), **a** asc, **b** tri, **c** oes

helminth types is analysed, with the purpose of obtaining relationships for their drag coefficients that would also lead to the sink velocity information. The sink velocities predicted by the obtained drag coefficients are observed not to agree with the measured values very well especially for a certain egg type (asc), for tap water. Consequently, the measured sink velocities are directly utilized to obtain drag coefficient correlations for the three egg types. This is done for tap water and for wastewater, both, and the correlations are implemented in the applied software by means of User-Defined Function (UDF). Employing the correlations, the processes in a small sewage treatment plant are investigated. It is observed that rather high velocities and turbulence levels occur during the filling process that governs the distribution of the eggs in this phase.

Acknowledgements The authors would like to acknowledge the German Federal Environmental Foundation (Deutsche Bundesstiftung Umwelt, DBU), Osnabrück, Germany, for the funding support of this research project [Project number = 34847/01-23]. In particular, the continuous encouragement and support by Mr. Alexander Bonde and Mr. Franz-Peter Heidenreich are gratefully acknowledged.

References

1. Cornel, P., Kneidl, S.: Parasiten im Abwasser—Problematik und Lösungsansätze für die Wasserwiederverwendung. EXPOVAL, Statusseminar zum BMBF-Verbundprojekt EXPOVAL—Anpassung und Validierung deutscher Standards für Kläranlagen im Ausland, Hannover, 1–2 November 2015 (2015)
2. Cirelli, G.L., Consoli, S., Licciardello, F., Aiello, R., Giuffrida, F., Leonardi, C.: Treated municipal wastewater reuse in vegetable production. Agric. Water Manag. **104**, 163–170 (2012)
3. Cui, B., Luo, J., Jin, D., Jin, B., Zhuang, X., Nai, Z.: Investigating the bacterial community and ameobae population in rural domestic wastewater reclamation for irrigation. J. Environ. Sci. **70**, 97–105 (2018)

4. Ellis, K.V., Rodrigues, P.C.C., Gomez, C.L.: Parasite ova and cysts in waste stabilization ponds. Water Res. **27**(9), 1455–1460 (1993)
5. Rüther, N.: Computational fluid dynamics in fluvial sedimentation engineering. Dissertation, Norwegian Institute of Science and Technology, Trondheim (2006)
6. Shilton, A.N., Mara, D.D.: CFD modelling of baffles for optimizing tropical waste stabilization pond systems. Water Sci. Technol. **51**(12), 103–106 (2005)
7. Batchelor, G.K.: An Introduction to Fluid Dynamics. Cambridge University Press, Cambridge (1967)
8. Sengupta, M. E., Thamsborg, S. M., Andersen, T. J., Olsen, A., Dalsgaard, A.: Sedimentation of helminth eggs in water. Water Res. **45**(15), 4651–4660 (2011)
9. Ali, R., Farooq, A., Shahzad, A., Benim. A. C., Iqbal. A., Razzaq, M.: Computational approach on three-dimensional flow of couple-stress fluid with convective boundary conditions. Physica A **553**, 124056 (2020)
10. Bird, R.B., Stewart, W.E., Lightfoot, E.N.: Transport Phenomena. Wiley, NY (2002)
11. Aslan, E., Taymaz, I., Benim, A.C.: Investigation of LBM curved boundary treatments for unsteady flows. Eur. J. Mech. B/Fluids **51**, 68–74 (2015)
12. Moukalled, F., Mangani, L., Darwish, M.: The Finite Volume Method in Computational Fluid Dynamics, Springer, Berlin (2016)
13. Hirt, C.W., Nichols, B.D.: Volume of fluid (VOF) method for the dynamics of free boundaries. J. Comput. Phys. **1**(39), 201–225 (1981)
14. Bhattacharyya, S., Chattopadhyay, H., Benim, A.C.: Heat transfer enhancement of laminar flow of ethylene glycol through a square channel fitted with angular cut wavy strip. Procedia Eng. **157**, 19–28 (2016)
15. Benim, A.C., Cagan, M., Nahavandi, A., Pasqualotto, E.: RANS predictions of turbulent flow past a circular cylinder over the critical regime. In: Proceedings of the 5th IASME/WSEAS International Conference on Fluid Mechanics and Aerodynamics, Athens, Greece, 25–27 August 2007, pp. 232–237 (2007)
16. Benim, A.C., Nahavandi, A., Stopford, P.J., Syed, K.: URANS, LES and DES analysis of turbulent swirling flows in gas turbine combustors. WSEAS Trans. Fluid Mech. **1**(5), 465–472 (2006)
17. Menter, F.R.: Two-equation eddy-viscosity turbulence models for engineering applications. AIAA J. **32**(8), 1598–1605 (1994)
18. Pope, S.B.: Turbulent Flows. Cambridge University Press, Cambridge (2012)
19. Tahat, M.S., Benim, A.C.: Experimental analysis on thermophysical properties of Al_2O_3/CuO hybrid nano fluid with its effects on flat plate solar collector. Defect Diffus Forum **374**, 148–156 (2017)
20. Clift, R., Grace, J.R., Weber, M.E.: Bubbles, Drops, and Particles. Dover Publications, Mineola (2013)
21. Benim, A.C., Epple, B., Krohmer, B.: Modelling of pulverised coal combustion by a Eulerian-Eulerian two-phase flow formulation. Progr. Comput. Fluid Dyn. Int. J. **5**(6), 345–361 (2005)
22. Epple, B., Fiveland, W., Krohmer, B., Richards, G., Benim, A.C.: Assessment of two-phase flow models for the simulation of pulverized coal combustion. Int. J. Energy Clean Environ. **6**(3), 267–287 (2005)
23. Durst, F., Miloievic, D., Schönung, B.: Eulerian and Lagrangian predictions of particulate two-phase flows: a numerical study. Appl. Math. Model. **8**(2), 101–115 (1984)
24. Gosman, A.C., Ioannides, E.: Aspects of computer simulation of liquid-fuelled combustors. J. Energy **7**(6), 482–490 (1983)
25. Van Doormaal, J.P., Raithby, G.D.: Enhancements of the SIMPLE method for predicting incompressible fluid flows. Numer. Heat Transf. **7**, 147–163 (1984)
26. Barth, T.J., Jespersen, D.C.: The design and application of upwind schemes on unstructured meshes. AIAA Paper, 89-0366 (1989)
27. ANSYS Fluent 18.0, Theory Guide. www.ansys.com
28. Schlichting, H.: Boundary Layer Theory, 7th edn. McGraw-Hill, New York (1979)

29. Stokes, G. G.: On the effect of the internal friction of fluids on the motion of pendulums. Trans. Cambr. Phil. Soc. **9**, Pt. II, 8-106 (1851)
30. Wallis, G.B.: One-Dimensional Two-Phase Flow. McGraw-Hill, New York (1969)
31. Owen, M. W.: Determination of the Settling Velocities of Cohesive Muds. IT161, Hydraulics Research Station, Wallingford, England (1976)
32. Oseen, C.W.: Über die Stokessche Formel und über die verwandte Aufgabe in der Hydrodynamik. Arkiv för Mathematik, Astronomi och Fysik **6**(29), 1–20 (1910)

Heat Transfer and Second Law Analysis of Ag-Water Nanoliquid in a Non-Uniformly Heated Porous Annulus

H. A. Kumara Swamy, M. Sankar, N. Keerthi Reddy, and S. R. Sudheendra

Abstract In majority of industrial and engineering applications, enhanced heat transfer with minimum entropy production is the major concern. With several theoretical and experimental works, it has been found that replacing the traditional heat transfer liquids with nanoliquid is one of the reliable ways to enhance the thermal transport with minimum loss of system energy. In this regard, the current article deals with the convective nanoliquid flow and the associated thermal dissipation as well as entropy generation rates in a porous annular enclosure saturated nanoliquid. The vertical surface of interior and exterior cylinders is maintained with sinusoidal thermal conditions with different phase deviations, while the horizontal boundaries are thermally insulated. The governing physical equations are solved by implementing finite difference method (FDM). The variation in buoyant nanoliquid flow and the corresponding heat transport rates along with local and global entropy production rates are systematically examined. For the numerical simulations, a vast range of parameters such as the Rayleigh ($10^3 \leq \text{Ra} \leq 10^5$) and Darcy ($10^{-6} \leq \text{Da} \leq 10^{-2}$) numbers, phase deviation ($0 \leq \gamma \leq \pi$), and nanoparticle volume fraction ($0 \leq \phi \leq 0.05$) are considered in this analysis. The contributions of heat transfer entropy and fluid friction entropy to global entropy production in the geometry are determined through the Bejan number. The numerical results reveal the impact of various parameters on control of convective flow, heat transfer, and entropy generation rates. Further, the results are in excellent agreement with standard benchmark

H. A. K. Swamy (✉)
Department of Mathematics, CMR Institute of Technology, Bengaluru, India
e-mail: hakumarswamy96@gmail.com

M. Sankar
Department of General Requirements, University of Technology and Applied Sciences, Ibri, Sultanate of Oman

N. K. Reddy
Department of Mathematical Sciences, Ulsan National Institute of Science and Technology (UNIST), Ulsan, Republic of Korea

S. R. Sudheendra
Department of Mathematics, School of Engineering, Presidency University, Bengaluru, India

R. K. Sharma et al. (eds.), *Frontiers in Industrial and Applied Mathematics*, Springer Proceedings in Mathematics & Statistics 410,
https://doi.org/10.1007/978-981-19-7272-0_14

simulations. The predicted results could provide some vital information in choosing the proper choice of parameters to enhance the system efficiency.

Keywords Annulus · Entropy generation · Nanofluid · Sinusoidal heating · Porous medium

Abbreviations

Ar	Aspect ratio
D	Width of the annulus
Da	Darcy number
g	Acceleration due to gravity
H	Height of the annulus
k	Thermal conductivity
K	Permeability
Ra	Rayleigh number
T	Dimensionless temperature
α	Thermal diffusivity
β	Thermal expansion coefficient
γ	Phase deviation
θ	Dimensional temperature
μ	Dynamic viscosity
ρ	Density
φ	Porosity
ϕ	Nanoparticle volume fraction

1 Introduction

In view of various important applications on convective thermal transport which includes aeronautics, nuclear reactors, solar collectors, and heating and cooling devices, many experimental and theoretical investigations on buoyant movement of different fluids in various geometries have been investigated. Since several industrial applications characterize the convection heat transfer in a sealed annular region from two or more coaxial concentric cylinders, we considered this geometry in the present analysis. In many heat transfer industries, the thermal distribution may not be uniform. Due to this, many researchers have focused on the impact of nonuniform thermal distribution on heat transfer performance of the system. Buoyant convection with nonuniform thermal condition with dissimilar phase deviations on vertical boundaries has been analyzed by Deng and Chang [1] and concluded that maximum thermal transport produces with nonuniform thermal profile than constant thermal conditions. Kiran et al. [2] numerically analyzed the impact of nonuniform temperature distribution on flow strength and heat transport rate by considering different constraints. In various thermal transport applications, the vital drawback of utilizing the traditional fluids is their poor thermal conductivity. In view of this, several researchers have focused on enhancing the thermal transport system efficiency by upgrading the thermal conductivity of the liquids and this results to the invention of new type of liquids known as "nanoliquids". Choi and Eastman [3] made a pioneering attempt to study convection in a nanoparticle suspended fluid and concluded that the suspension of nanosize particles in base fluid increases the thermal conductivity and in turn enhances the heat transport rate. Earlier, Abouali and Falahatpisheh [4] made an attempt to analyze buoyant flow of Al_2O_3 nanoliquid in an annular enclosure. The impact of nanoparticle volume fraction on fluid flow and heat transfer rate has been numerically studied [5]. By considering nonuniform thermal conditions, Reddy et al. [6] studied the impact of hybrid nanoparticle concentration on fluid motion and heat dissipation rate in an annulus region. Recently, Sankar et al. [7] analyzed the effect of conductive solid wall on thermal dissipation rate of different nanofluids and found that Cu-H_2O nanofluid helps to dissipate maximum thermal energy.

For the design of several thermal transport equipment along with enhancing heat removal rate, entropy minimization is also a key parameter since the assessment of system efficiency can be estimated by the entropy production rate. Mejri et al. [8] numerically studied the influence of sinusoidal thermal profile on nanoliquid flow and entropy generation in a square cavity. Recently, the same study has been extended to annular enclosure by Sankar et al. [9]. The geometry saturated with porous medium shows vital change in flow strength and thermal performance than the nonporous geometry. With regard to this, several research works have been carried out to investigate the impact of porous medium and thermal performance of the system. Swamy et al. [10] analyzed the impact of geometric tilt angle on nanofluid flow and entropy in an annulus saturated porous medium. By utilizing Lattice Boltzmann Method, Ghasemi and Siavashi [11] reported that the particular choice of linear thermal distribution leads to enhance the thermal performance of the system. Later, Kashyap and Dass [12] analyzed entropy generated in a nanofluid-filled porous geometry subjected to various nonuniform thermal conditions. By considering the sedimentation of nanoparticles, Baghsaz et al. [13] studied heat transport and entropy generation of nanofluid-saturated porous cavity.

From a thorough and methodical scrutiny of literature, it is noticed that the thermal transport and irreversibility distribution of nanoliquid in a porous annular enclosure subjected to nonuniform thermal conditions has not been investigated and this motivates the current investigation. It is presumed that this investigation would provide some useful results/data to enhance the performance of thermal systems. In this work, the flow and thermal fields, thermal dissipation rate, entropy production, and Bejan number are predicted by varying the Rayleigh number, Darcy number, phase deviation, and volume fraction of nanoparticle.

2 Mathematical Statement

The system considered in this study is the annular enclosure formed from two coaxial vertical concentric cylinders with radii r_i and r_o of interior and exterior cylinders, respectively, as portrayed in Fig. 1. The vertical surfaces of inner and outer cylinders are maintained with sinusoidal thermal profiles with different phase deviations, while the top and bottom walls are maintained adiabatic [8]. The porous annular region is occupied with Ag-H_2O nanoliquid. The properties of nanoliquid are estimated using the correlations provided in [8, 11]. Thermo-physical properties of H_2O and Ag nanoparticle are taken from [11]. The vertical surfaces of interior and exterior cylinder are subjected to sinusoidal thermal distribution with different phase deviation, while the bottom and top surfaces are insulated. It is assumed that the H_2O and Ag nanoparticles are in thermal equilibrium, the thermal properties of porous matrix and nanoliquid are to be identical, fluid is incompressible, Newtonian. The fluid motion is considered to be two dimensional, laminar, unsteady, and axisymmetric. Also, Boussinesq approximation is adopted in this study. By imposing the above assumptions, the dimensional governing equations are as follows [5]:

$$\nabla.\vec{q} = 0 \tag{1}$$

$$\frac{\rho_{nf}}{\varphi}\left[\frac{\partial \vec{q}}{\partial t^*} + \frac{1}{\varphi}(\vec{q}.\nabla)\vec{q}\right] = -\nabla p + \frac{\mu_{nf}}{\varphi}\nabla^2\vec{q} - \frac{\mu_{nf}}{K}\vec{q} + (\rho\beta)_{nf}\,g(\theta - \theta_c) \tag{2}$$

$$\frac{\partial \theta}{\partial t^*} + (\vec{q}.\nabla)\theta = \alpha_{nf}\nabla^2\theta \tag{3}$$

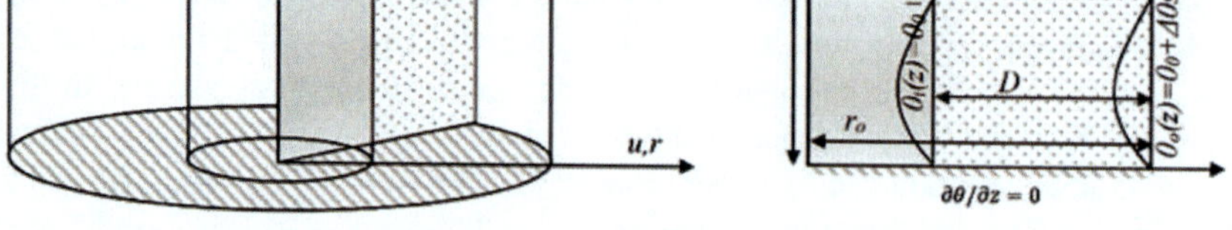

Fig. 1 Physical domain of the problem and its axisymmetric view

By using the non-dimensional variables [9], the dimensionless energy and vorticity stream function equations are as follows:

$$\frac{\partial T}{\partial t} + U\frac{\partial T}{\partial R} + W\frac{\partial T}{\partial Z} = \frac{\alpha_{nf}}{\alpha_f}\nabla^2 T \tag{4}$$

$$\frac{1}{\varphi}\left[\frac{\partial \zeta}{\partial t} + \frac{1}{\varphi}\left(U\frac{\partial \zeta}{\partial R} + W\frac{\partial \zeta}{\partial Z} - \frac{U\zeta}{R}\right)\right] = \frac{\mu_{nf}}{\rho_{nf}\alpha_f}\frac{1}{\varphi}\left[\nabla^2\zeta - \frac{\zeta}{R^2}\right] - \frac{\mu_{nf}}{\rho_{nf}\alpha_f}\frac{\zeta}{Da} - \frac{(\rho\beta)_{nf}}{\rho_{nf}\beta_f} Ra\, Pr\, \frac{\partial T}{\partial R} \tag{5}$$

$$\zeta = \frac{1}{R}\left[\frac{\partial^2\psi}{\partial R^2} - \frac{1}{R}\frac{\partial \psi}{\partial R} + \frac{\partial^2\psi}{\partial Z^2}\right] \tag{6}$$

Here, $U = \frac{1}{R}\frac{\partial \psi}{\partial Z}$, $W = -\frac{1}{R}\frac{\partial \psi}{\partial R}$ and $\nabla^2 = \frac{\partial^2}{\partial R^2} + \frac{1}{R}\frac{\partial}{\partial R} + \frac{\partial^2}{\partial Z^2}$

The total thermal transfer across the enclosure is defined as the sum of average Nusselt numbers calculated along the heating half of inner and outer surfaces and given by [8, 11].

$$\overline{Nu} = \left(-\frac{k_{nf}}{k_f}\right)\int_{\text{heating half}}\left[\left(\frac{\partial T}{\partial R}\right)_{R=\frac{1}{\lambda-1}} + \left(\frac{\partial T}{\partial R}\right)_{R=\frac{\lambda}{\lambda-1}}\right]dZ \tag{7}$$

3 Equation for Entropy Generation

Based on second law of thermodynamics and postulates made, the dimensionless form of local entropy generation due to heat transfer ($S_{l.T}$) and fluid friction ($S_{l.\psi}$) for fluid-saturated porous medium is written as

$$S_{l.T} = \frac{k_{nf}}{k_f}\left[\left(\frac{\partial T}{\partial R}\right)^2 + \left(\frac{\partial T}{\partial Z}\right)^2\right]$$

$$S_{l.\psi} = \Phi\frac{\mu_{nf}}{\mu_f}\left\{[U^2 + W^2] + Da\left[\begin{array}{l} 2\left\{\left(\frac{\partial U}{\partial R}\right)^2 + \left(\frac{\partial W}{\partial Z}\right)^2 + \left(\frac{U}{R}\right)^2\right\} \\ + \left(\frac{\partial U}{\partial Z} + \frac{\partial W}{\partial R}\right)^2 \end{array}\right]\right\} \tag{8}$$

Here, $\Phi = \frac{\mu_f}{k_f}\theta_0\left(\frac{\alpha_f}{\sqrt{K}\Delta\theta}\right)^2$ is known as irreversibility distribution ratio. The global entropy production in the geometry is given by the sum of heat and friction entropy, i.e., $S_{GEN} = S_{l.T} + S_{l.\psi}$. The total entropy production is calculated by integrating local entropy generation within the enclosure

$$S_{tot} = \frac{1}{V}\int\limits_{V} S_{GEN}\, dV = \frac{1}{V}\int\limits_{V} S_{l.T} + S_{l.\psi}\, dV$$

The above equation can be written as $S_{tot} = S_T + S_\psi$. The relative dominance of entropy production due to heat transfer and fluid friction is given by the parameter known as the Bejan number (Be_l) and is defined as

$$Be = \frac{1}{V}\int\limits_{V} \left(\frac{S_{l.T}}{S_{GEN}}\right) dV$$

If $Be < 0.5$, then S_ψ is dominant, if $Be > 0.5$, then S_T is dominant, and if $Be = 0.5$, it indicates that heat transfer entropy and friction entropy contribute equally.

4 Numerical Technique, Grid Sensitivity, and Validation

The governing PDEs are solved by adopting an implicit Finite Difference Method which gives algebraic tri-diagonal FD equations. The solutions of these equations are obtained using TDMA. Local entropy production due to individual components is obtained by solving Eq. 8 with a second-order central difference approximation. Finally, the average Nusselt number and total entropy generation are estimated by adopting, respectively, Simpson and Trapezoidal rules. The detailed discretization could be found in our previous work [10]. After performing the grid independency study with $Ra = 10^5$ $\gamma = \pi/2$, $Da = 10^{-2}$, and $\phi = 0.05$, we found that 161×161 is the suitable mesh size for this analysis. Table 1 provides the comparison between the average Nusselt number obtained by our code and those obtained by Abouali and Falahatpisheh [4] through the heat transfer correlations.

Table 1 Comparison of average Nusselt number of present study with Abouali and Falahatpishesh [4]

Ra $(Gr \times Pr)$	Abouali and Falahatpishesh [4]	Present study	Relative difference (%)	
6×10^3	2.8673	2.8843	0.59	$\phi = 0.0$
	2.7291	2.7401	0.40	$\phi = 0.02$
6×10^4	5.4385	5.4796	0.75	$\phi = 0.0$
	5.1762	5.1836	0.14	$\phi = 0.02$
6×10^5	10.3150	10.4015	0.83	$\phi = 0.0$
	9.8178	9.9132	0.96	$\phi = 0.02$

5 Discussion on Results

The prime focal point of this study is to detect the flow pattern, heat dissipation rate, and entropy production of nanoliquid filled in porous annular enclosure subjected to nonuniform thermal conditions. The numerical simulations have been performed for vast range of Rayleigh number (*Ra*), Darcy number (*Da*), phase deviation (γ), and nanoparticle volume fraction (ϕ). The influence of these parameters on natural convection and entropy production of nanoliquid-saturated porous annular enclosure has been analyzed.

The effect of *Ra* on streamlines, isotherms and entropy contours for water and nanoliquid is predicted in Fig. 2 by fixing other parameters as constant. At lower *Ra* (conduction dominant mode), streamlines exhibit one larger and three smaller eddies. Through thermal lines it is clear that the heat from the hot region is supplied to the cold region of same wall. Because of conductive mode, friction entropy is negligible due to this the entropy contours akins the thermal condition profile. The addition of ϕ has not altered the contour patterns significantly. Increase in *Ra* to 10^5 increases the flow strength, and as a result two eddies at the top merge and also increase the size of larger eddy. The isothermal and entropy contour pattern on interior and exterior walls are varying with rise in *Ra* which causes an appreciable change in thermal transport and entropy production with an increment in *Ra*.

Figure 3 deals with the streamlines, isotherms, and entropy generation of nanoliquid and water for $\gamma = 0$ and π. For $\gamma = 0$, the streamlines exhibit four vortices

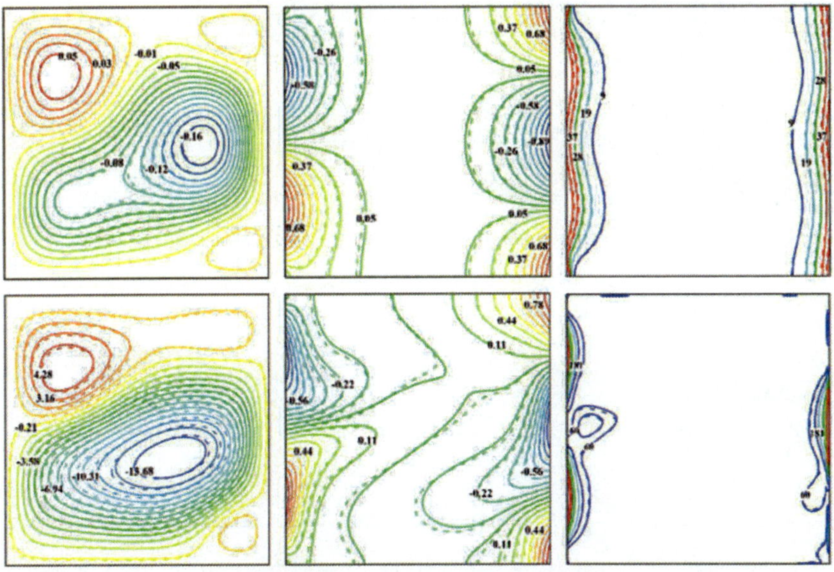

Fig. 2 Streamlines, isotherms, and entropy generation contours for $Ra = 10^3$ (top), $Ra = 10^5$ (bottom) at $\gamma = \pi/2$, $Da = 10^{-2}$, $\phi = 0$ (dotted line) and $\phi = 0.04$ (solid line)

which are parallel along midline. The left bottom and right top eddies are rotating clockwise while the other vortexes are in anti-clockwise direction. Since the thermal profile of both vertical walls is similar, the isotherms and entropy contours appear similar on left and right walls. For $\gamma = \pi$, the liquid movement takes place in two vortices. This reduction in number of eddies is due to the variation in the position of hot and cold region along the outer cylinder. It is interesting to note that the fluidity of both the eddies is similar. As the magnitude of γ increases, the thermal lines and entropy pattern along the interior wall have not been varied significantly; however, profound change along the exterior wall can be noticed indicating that change in γ affects the exterior boundary.

The impact of *Da* on nanoliquid flow, thermal and entropy distribution at $Ra = 10^5$ and $\gamma = 3\pi/4$ is depicted in Fig. 4. For both *Da* (10^{-6} and 10^{-2}), the nanoliquid movement takes place in two vortices where the bottom vortex is rotating in clockwise and other in counterclockwise direction. Though the flow takes place in same number of eddies, the fluidity at $Da = 10^{-6}$ is very much smaller than fluidity at $Da = 10^{-2}$. This is because of permeability difference. For $Da = 10^{-6}$, the permeability is low, due to this the thermal transfer takes place through conduction mode which can be observed through isotherms and also due to conduction dominance, and the entropy contours appear to be parallel along the vertical walls. As the permeability is enhanced (increase in *Da*), the resistance of liquid flow declines and leads to convective dominance. In this situation, significant change in isotherm and entropy

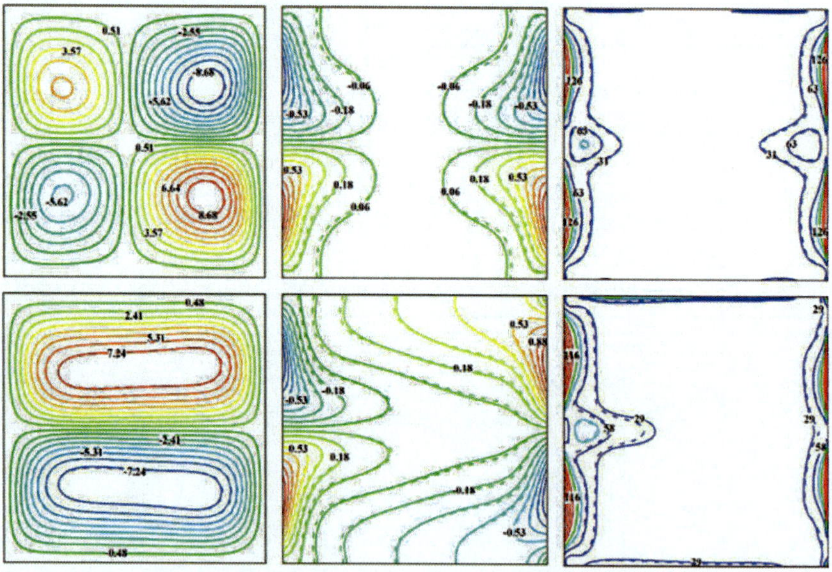

Fig. 3 Streamlines, isotherms, and entropy generation contours for $\gamma = 0$ (top), $\gamma = \pi$ (bottom) at $Ra = 10^5$, $Da = 10^{-2}$, $\phi = 0$ (dotted line) and $\phi = 0.04$ (solid line)

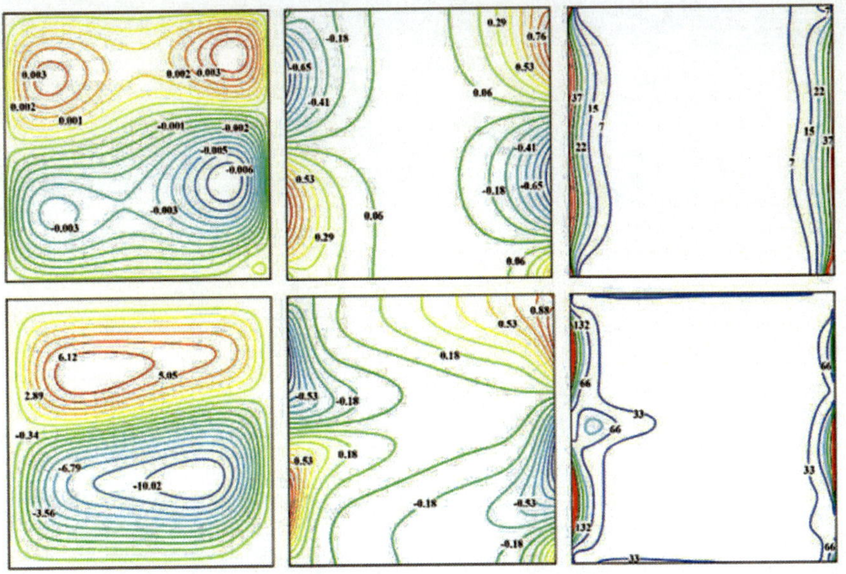

Fig. 4 Streamlines, isotherms, and S_{GEN} contours for $Da = 10^{-6}$ (top), $Da = 10^{-2}$ (bottom) at $\gamma = 3\pi/4$ $Ra = 10^5$, and $\phi = 0.04$

patterns can be noticed. Due to enhancement in fluidity, the friction entropy enhances and leads to increase in entropy production in the enclosure.

The influence of γ on thermal dissipation rate, total entropy generation, and Bejan number for various *Ra* and *Da* is predicted in Figs. 5 and 6, respectively, by fixing other parameters. It is noticed that sinusoidal thermal condition with $\gamma = \pi/2$ dissipated greater thermal energy during conduction dominant/week convection mode ($Ra \leq 10^4$ or $Da \leq 10^{-4}$); however, during strong convective mode ($Ra = 10^5$ or $Da = 10^{-2}$) $\gamma = 3\pi/4$ contributes to greater thermal energy dissipation. The same result has also been noticed in the study of Deng and Chang [1]. For $\gamma \leq 3\pi/4$, increase in *Ra* or *Da* enhances the average *Nu*. But for $\gamma = \pi$, average *Nu* slightly decreases during change in mode of heat transfer. It is noticed that irrespective of γ, enhancement of *Ra* or *Da* leads to enrichment in S_{tot}. For $Ra \leq 10^4$ or $Da \leq 10^{-4}$, generation of entropy is minimal with respect to every γ compared to $Ra = 10^5$ or $Da = 10^{-2}$. This is because, for $Ra \leq 10^4$ or $Da \leq 10^{-4}$, conduction mode is dominant and this does not cause friction entropy, while for higher *Ra* or *Da*, fluidity enhances and causes greater friction entropy which leads to greater S_{tot}. This can also be seen through *Be* plot that for conduction dominant/week convection mode *Be* is around 0.5 and for stronger convective flow $Be < 0.5$.

Figure 7 illustrates the impact of nanoparticle volume fraction and Rayleigh number on average *Nu*, S_{tot}, and *Be* at $Da = 10^{-2}$ and $\gamma = \pi/2$. As expected, enhancement of ϕ enhances the thermal conductivity of liquid which helps in maximizing thermal dissipation rate. Hence, on enhancing ϕ, global *Nu* enhances for all *Ra*. For

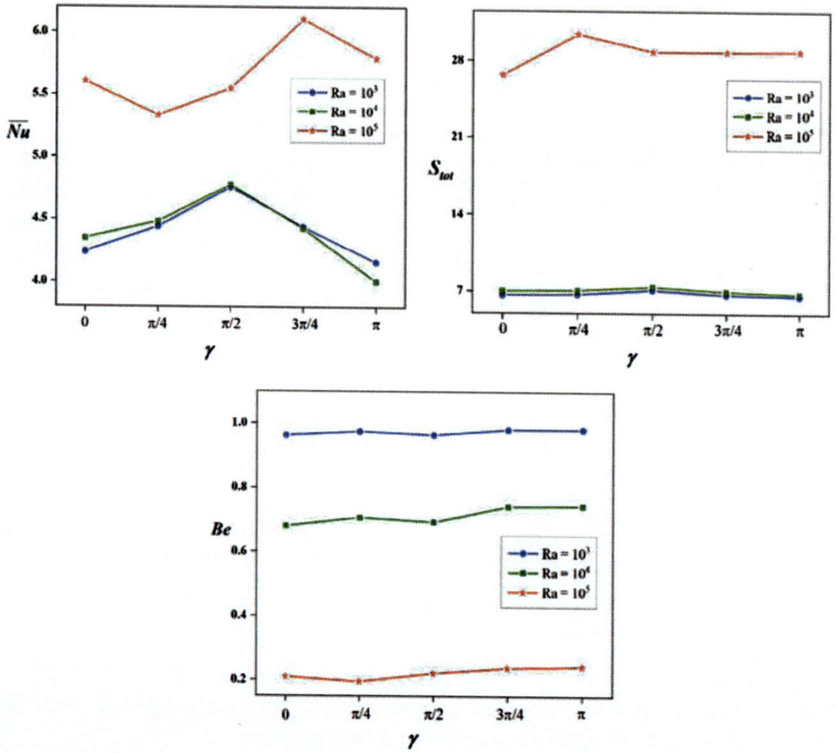

Fig. 5 Impact of *Ra* and γ on average Nu, S_{tot}, and *Be* at $Da = 10^{-2}$ and $\phi = 0.05$

$Ra = 10^3$ and 10^4, it can be noticed that no profound change in thermal dissipation rate and it is same for all nanoparticle volume fraction. As the fluidity is higher for $Ra = 10^5$, friction entropy generation will be higher and this enhances the total entropy generation compared to lower and moderate *Ra*. Though the magnitude of flow strength declines, the thermal conductivity enhances on enrichment of nanoparticle concentration. Due to this, heat transfer entropy increases and results in enhancement of S_{tot} with concentration of nanoparticle. The behavior of *Be* is similar as discussed earlier.

The influence of phase deviation and solid particle concentration on global Nusselt number, total entropy generation, and Bejan number has been illustrated in Fig. 8 for $Ra = 10^5$ and $Da = 10^{-2}$. As discussed earlier, enrichment of nanoparticle volume fraction leads to enhancement in heat dissipation rate due to increase in thermal conductivity. This holds good for all phase deviations. Among the three phase deviations ($\gamma = 0, \pi/2, \pi$) considered, it has been found that $\gamma = \pi$ dissipates greater heat compared to other phase deviations for all nanoparticle concentrations. The interesting fact to note is that the choice of parameters, i.e., $Ra = 10^5$, $Da = 10^{-2}$, and $\gamma = \pi/2$ is not suitable for better thermal performance of the system as it dissipates less

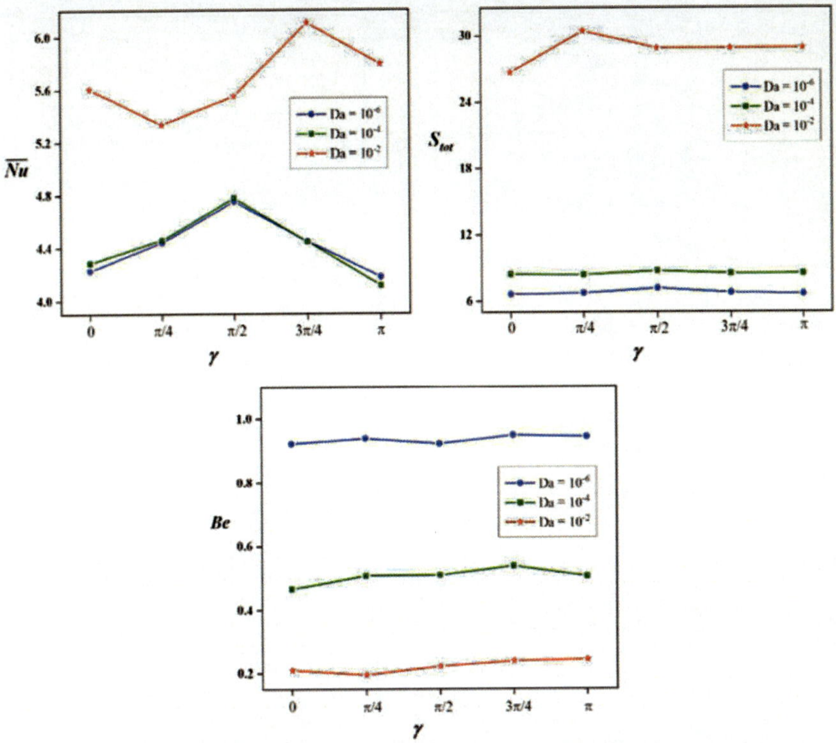

Fig. 6 Impact of *Da* and γ on average *Nu*, S_{tot}, and *Be* at $Ra = 10^5$ and $\phi = 0.05$

amount of heat with greater entropy. Since both *Ra* and *Da* are greater, the convective flow takes place with greater magnitude and leads to more friction entropy due to this $Be < 0.5$ in all the considered cases.

Figures 9 and 10 illustrate the impact of Darcy number for various Rayleigh number and nanoparticle volume fraction, respectively. During this study the phase deviation is maintained at π so that the thermal boundary conditions of both vertical walls will be exactly opposite. During conduction/weak convection mode, the change in heat transfer rate is not much significant but in stronger convective mode ($Ra = 10^5$ and $Da = 10^{-2}$) the heat transfer rate has been enhanced by 39.02%. But in the case of nanoparticle volume fraction, heat transport rate has been increased for both Darcy values and this may be due to an increase in thermal conductivity. Similar mechanism can be noticed in entropy generation also. The Bejan number is noticed to be much greater than 0.5 during conduction being dominant and less than 0.5 during convective being dominant indicating that during conduction mode S_T contributes more to S_{tot} while during convective mode S_ψ contributes more to S_{tot}.

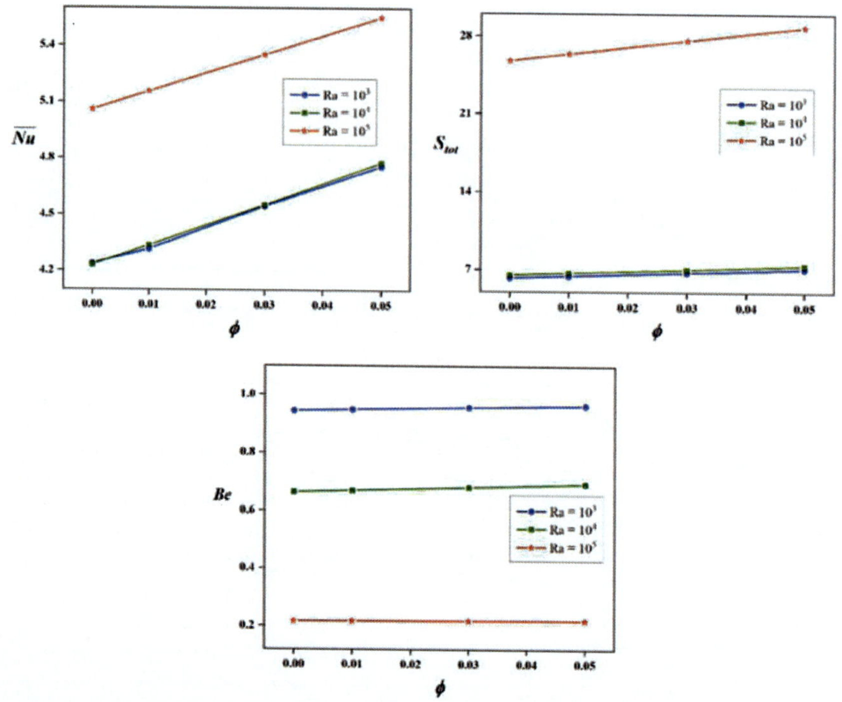

Fig. 7 Impact of Ra and ϕ on average Nu, S_{tot}, and Be at $Da = 10^{-2}$ and $\gamma = \pi/2$

6 Conclusions

A numerical analysis on heat transfer and second law of thermodynamics in a sinusoidally heated porous annulus has been performed. The various parameters adopted in this investigation are Rayleigh number, Darcy number, phase deviation, and nanoparticle volume fraction. Through the vast numerical simulations, it has been found that Ra, Da, and γ play vital role on flow movement pattern, fluidity, heat transport, and irreversibility distribution. During conduction being dominant ($\text{Ra} \leq 10^4$ or $\text{Da} \leq 10^{-4}$), maximum amount of thermal dissipation takes place at $\gamma = \pi/2$, whereas, during convective mode, $\gamma = 3\pi/4$ helps to dissipate high amount of thermal energy. It is also found that the porous enclosure with minimum permeability produces minimal entropy. It has also been observed that during convective-dominant flow, $\gamma = \pi/2$ decreases the efficiency of the thermal system, i.e., dissipates lower thermal energy with greater entropy generation.

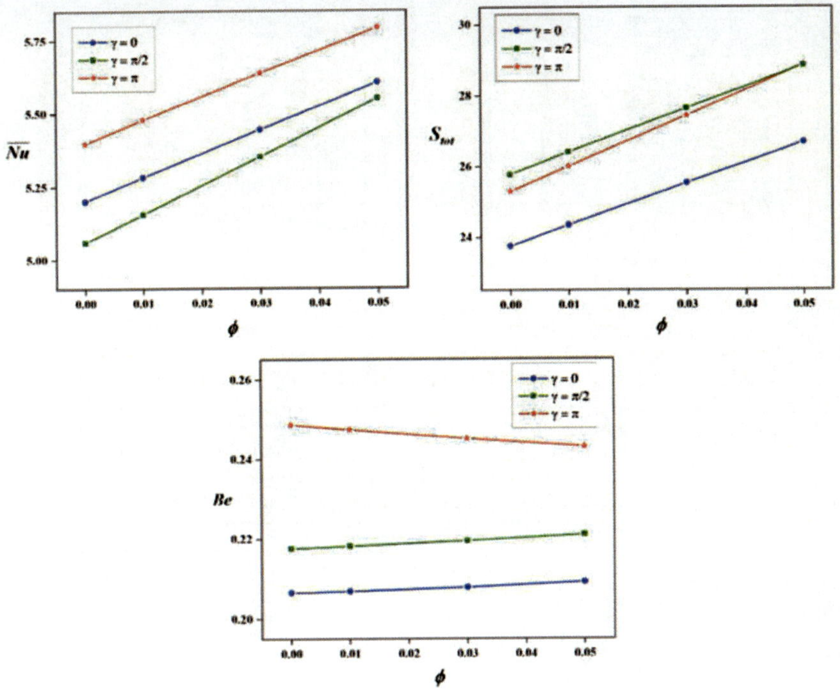

Fig. 8 Impact of γ and ϕ on average *Nu*, S_{tot}, and *Be* at $Da = 10^{-2}$ and $Ra = 10^5$

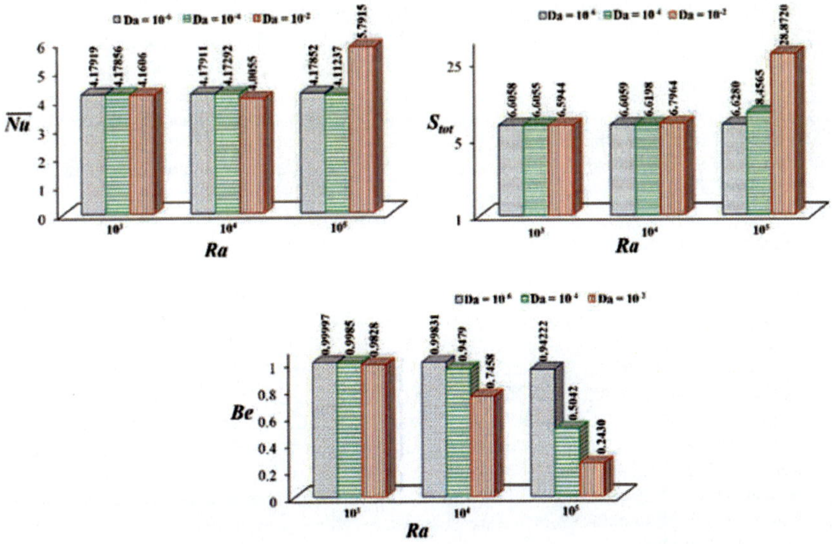

Fig. 9 Impact of *Ra* and *Da* on average *Nu*, S_{tot}, and *Be* at $\phi = 0.05$ and $\gamma = \pi/2$

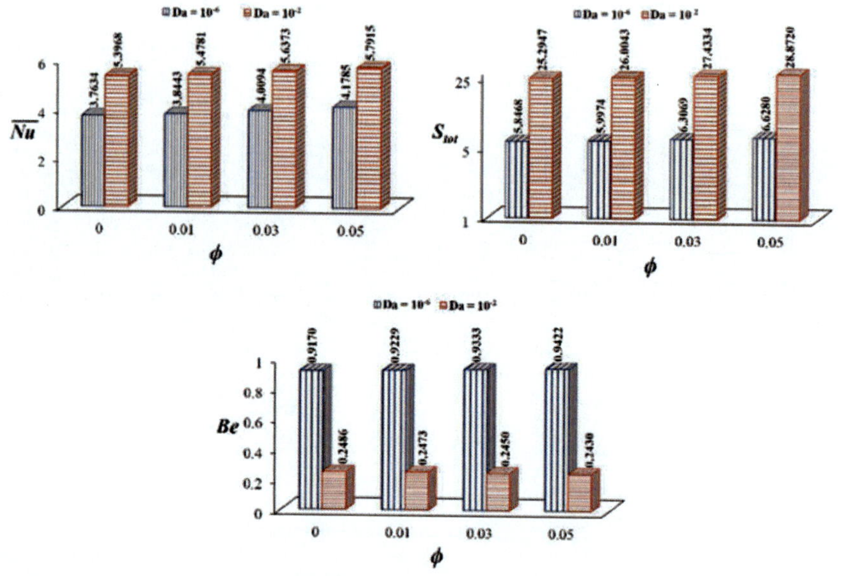

Fig. 10 Impact of *Da* and ϕ on average *Nu*, S_{tot}, and *Be* at $Ra = 10^5$ and $\gamma = \pi$

References

1. Deng, Q., Chang, J.: Natural convection in a rectangular enclosure with sinusoidal temperature distributions on both sidewalls. Numer. Heat Transf. A. **54**(5), 507–524 (2008)
2. Kiran, S., Sankar, M., Swamy, H.A.K., Makinde, O.D.: Unsteady buoyant convective flow and thermal transport analysis in a non-uniformly heated annular geometry. Comput. Therm. Sci. **14**(2), 1–17 (2022)
3. Choi, S.U., Eastman, J.A.: Enhancing thermal conductivity of fluids with nanoparticles. ASME Int Mech Eng Congr Expo. (1995)
4. Abouali, O., Falahatpisheh, A.: Numerical investigation of natural convection of Al_2O_3 nanofluid in vertical annuli. Heat Mass Transf. **46**, 15–23 (2009)
5. Mebarek-Oudina, F., Bessaih, R.: Numerical simulation of natural convection heat transfer of copper-water nanofluid in a vertical cylindrical annulus with heat source. Thermophys. Aeromech. **26**(3), 325–334 (2019)
6. Reddy, N.K., Swamy, H.A.K., Sankar, M.: Buoyant convective flow of different hybrid nanoliquids in a non-uniformly heated annulus. Eur. Phys. J. Spec. Top. **230**(5), 1213–1255 (2021)
7. Sankar, M., Reddy, N.K., Do, Y.: Conjugate buoyant convective transport of nanofluids in an enclosed annular geometry. Sci. Rep. **11**, 17122 (2021)
8. Mejri, I., Mahmoudi, A., Abbassi, M.A., Omri, A.: Magnetic field effect on entropy generation in a nanofluid-filled enclosure with sinusoidal heating on both side walls. Powder Technol. **266**, 340–353 (2014)
9. Sankar, M., Swamy, H.A.K., Do, Y., Altmeyer, S.: Thermal effects of nonuniform heating in a nanofluid-filled annulus: Buoyant transport versus entropy generation. Heat Transfer **51**(1), 1062–1091 (2022)
10. Swamy, H.A.K., Sankar, M., Reddy, N.K.: Analysis of entropy generation and energy transport of Cu-water nanoliquid in a tilted vertical porous annulus. Int. J. Appl. Comput. Math. **8**(1), 10 (2022)

11. Alsabery, A.I., Chamkha, A.J., Saleh, H., Hashim, I.: Natural convection flow of a nanofluid in an inclined square enclosure partially filled with a porous medium. Sci. Rep. **7**, 2357 (2017)
12. Kashyap, D., Dass, A.K.: Two-phase lattice Boltzmann simulation of natural convection in a Cu-water nanofluid-filled porous cavity: effects of thermal boundary conditions on heat transfer and entropy generation. Adv. Powder Technol. **29**(11), 2707–2724 (2018)
13. Baghsaz, S., Rezanejad, S., Moghimi, M.: Numerical investigation of transient natural convection and entropy generation analysis in a porous cavity filled with nanofluid considering nanoparticle sedimentation. J. Mol. Liq. **279**, 327–341 (2019)

Qualitative Analysis of Peer Influence Effects on Testing of Infectious Disease Model

Anjali and Manoj Kumar Singh

Abstract Outbreaks minimization has become the need of the hour. If we start clouding the list of infectious diseases, the list will point out that there is a rise in infectious diseases in the current era, and after the first case of any illness for knowing about its spread, testing methods are introduced. So testing plays a key role, and it is necessary to detect any infectious disease effects on humans. However, the peer influence effect of persons who have recovered from disease without visiting any doctor encourages other people to not get tested and take self-medication. They do not understand the need for tests and spread fake scenarios convincing others to follow them. The paper studies the impact of these individuals on the emergence of the disease by analyzing the mathematical model proposed in the situation, which is further analyzed and studied through simulation. The analysis section comprises local and global stability of the equilibrium points, primary reproduction number, and threshold analysis of the proposed model. Numerical simulation has provided a clear view of the qualitative analysis through the graphs and the plots.

Keywords Peer-effect · Testing · Infectious disease · Reproduction number · Lyapunov · Liénard Chipart criterion · Threshold analysis

1 Introduction

The outbreak is a word that symbolizes the observed scenario, which was unexpected and unusual. That is why the emergence of infectious disease has generally been termed an outbreak. Management of an outbreak requires a stepwise plan to handle the

Anjali (✉) · M. K. Singh
Faculty of Mathematics and Computing, Department of Mathematics and Statistics, Banasthali Vidyapith, 304022 Niwai, Rajasthan, India
e-mail: anjalipanwarepidemiology@gmail.com

R. K. Sharma et al. (eds.), *Frontiers in Industrial and Applied Mathematics*,
Springer Proceedings in Mathematics & Statistics 410,
https://doi.org/10.1007/978-981-19-7272-0_15

situation patiently and overcome it. The steps can be called as outbreak identification, investigation, defining the case, description of the outbreak, proposing and testing the possibilities, controlling steps, and communication. Laboratory testing and medical equipment invention have served us greatly. Many unusual happenings have been recorded during laboratory tests or screening tests, such as human avian flu, diarrhea, and many more [3]. In 2019, SARS-CoV-2 virus [31] said to have initials in Wuhan, China which infects from human to another [30] was also tested with RTPCR and antigen testing kits. This helped to keep the infected people under consideration. Open testing centers in large numbers was the initial step taken by the government.

The mathematical models are the best to detect the dynamics of the importance of testing and peer effect during infectious disease transmission. The mathematical models help us to determine all the concerned cases. In 1760, Daniel Bernoulli [1] proposed the mathematical model to know about the smallpox mortality rate at that time. To study more complex mathematical models [7, 20, 21, 25, 29, 32, 33] researchers are working hard. The analysis of the mathematical models is the efficient way to provide exact results and predictions.

After emergence, it is clear that testing plays an important role to diagnose the infection in the host, noting down the rise and working on other actions required to like what are the necessities. Different types of microbes are diagnosed in different ways. Peer effect/pressure [4] is influencing other people indirectly or directly by the peers. This effect can surely encourage or discourage persons. To be its part, an individual need not be in any organization or a group. It works in our society very actively. Name it find it can be said for the issue which is affected by this effect.

2 Mathematical Model Description

Communicable infectious disease, transferred from human to human when enters the environment, can infect persons as at that stage no one is considered to have immunity to virus (susceptible). On being infected, the person can infect others, and the persons who contact these individuals are considered prone to the disease (exposed). After testing methods are introduced, like RTPCR is for coronavirus. It helps to detect whether the person is infected or not. This can help one take medications and recover. While some of the deadly infections are also sometimes taken lightly by people, visiting for tests can be skipped. The mathematical model here is proposed to focus on the following questions that are

- What is the effect of people who have recovered without any tests and carry a mentality that virus is no big deal and can be recovered without any tests or following guidelines?
- Could these people increase the number of infections?

Parameters and variables of the model are decided according to the need of the situation. Variables include $S(t)$, $E(t)$, $I(t)$, $J(t)$, $R_i(t)$, and $R_j(t)$ which represents the susceptible class, exposed class, infected and tested class, infected and not tested class, recovered after being tested and lastly the focused class which will play an important role is recovered after not being tested class respectively. The parameters B and Θ are for the birth and death rate of humans. Also the parameters η and γ are transmissible infection multiple of $I(t)$ and $J(t)$ class respectively. The infection rate of $I(t)$ and $J(t)$ class from $E(t)$ class are α_i and α_j respectively. The recovery rate of $R_i(t)$ is β_i and of $R_j(t)$ is β_j. Lastly, ψ is the rate of peer effect of $R_j(t)$. The mathematical model with the help of the ordinary differential equation can be written as Eq. (1):

$$\begin{cases} \frac{dS(t)}{dt} = B - (\eta I + \gamma J)\frac{S}{N} - \Theta S, \\ \frac{dE(t)}{dt} = (\eta I + \gamma J)\frac{S}{N} - (\alpha_i + \alpha_j + \psi R_j)E - \Theta E, \\ \frac{dI(t)}{dt} = \alpha_i E - \beta_i I - \Theta I, \\ \frac{dJ(t)}{dt} = (\alpha_j + \psi R_j)E - \beta_j J - \Theta J, \\ \frac{dR_i(t)}{dt} = \beta_i I - \Theta R_i, \\ \frac{dR_j(t)}{dt} = \beta_j J - \Theta R_j. \end{cases} \tag{1}$$

The Model Assumptions for the mathematical model [4] that initially there is no testing kit available. There is no recovered individual initially so $R_i(0) = R_j(0) = 0$ also there is no reinfection in the environment. The total humans in the environment at the time t is $N(t)$ which is equal to the sum of $S(t)$, $E(t)$, $I(t)$, $J(t)$, $R_i(t)$, and $R_j(t)$ (Fig. 1).

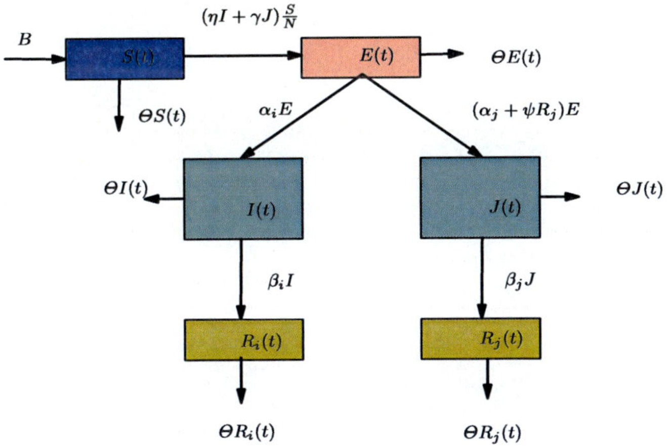

Fig. 1 Diagrammatic representation of set of Eq. (1)

3 Model Analysis

3.1 Positivity and Boundedness of Solution

According to the assumptions, the compartmental model (1) has all its associated parameters and variables non-negative. To show the positivity of the compartmental model solution, the particular seed conditions defined in theorem (1) are necessary.

Theorem 1 *Let the seed conditions be*

$$S_0 = (S(0), E(0), I(0), J(0), R_i(0), R_j(0)) \geq 0; \quad t \geq 0.$$

Then, $\overline{S} = (S(t), E(t), I(t), J(t), R_i(t), R_j(t)) > 0$ *where* $\overline{S}$ *is the bounded solution of the system of equations of model (1). That is*

$$\overline{\Pi} = \left\{ \overline{S} \in \mathbb{R}^6_+, \ 0 \leq N(t) \leq \max\left\{ N(0) + \frac{\Theta}{\mu} \right\} \right\}$$

is the positively invariant feasible region.

Proof Let $\bar{t} := \sup\{t > 0 \ : \ S(0) > 0\} \in [0, t]$ and considering the first equation of the system of Eq. (1)

$$\frac{dS}{dt} \geq -(\eta I + \gamma J)\frac{S}{N} - \Theta S. \tag{2}$$

From Eq. (2). We get

$$S(\bar{t}) \geq S(0) \exp\left[-\left\{ (\Theta)\bar{t} + \int_0^{\bar{t}} (\eta I(s) + \gamma J(s)) \frac{1}{N(s)} ds \right\} \right] > 0.$$

Similarly we can find that $E(\bar{t})$, $I(\bar{t})$, $J(\bar{t})$, $R_i(\bar{t})$ and $R_j(\bar{t})$ are positive. Adding all the equations of the system (1). We get

$$\frac{dN}{dt} = B - \Theta N. \tag{3}$$

From Eq. (3). We get

$$N(t) = \frac{B}{\Theta} + \left(N(0) - \frac{B}{\Theta} \right) \exp(-\Theta t), \tag{4}$$

which results to give $N(t) \to \frac{B}{\Theta}$, whenever $t \to \infty$. That is for $N(0) < \frac{B}{\Theta}$, $N(t)$ increases to $\frac{B}{\Theta}$ and for $N(0) > \frac{B}{\Theta}$, $N(t)$ decreases to $\frac{B}{\Theta}$. Therefore, we say that $N(t)$ is bounded above implying $\overline{S}$ is bounded above.

3.2 Local Stability

Let disease-free equilibrium point DFE be

$$DFE = (S(0), E(0), I(0), J(0), R_i(0), R_j(0)), \tag{5}$$

which is $(\frac{B}{\Theta}, 0, 0, 0, 0, 0)$ of the model (1). Here F is the infection matrix and V is the transmission matrix inspired from the compartmental model (1).

$$F = \begin{bmatrix} 0 & -\eta & -\gamma \\ 0 & 0 & 0 \\ 0 & 0 & 0 \end{bmatrix}, \; V = \begin{bmatrix} \alpha_i + \alpha_j + \Theta & 0 & 0 \\ -\alpha_i & \beta_i + \Theta & 0 \\ -\alpha_j & 0 & \beta_j + \Theta \end{bmatrix}, \tag{6}$$

which gives FV^{-1} as

$$\frac{1}{k_1 k_2 k_3} \begin{bmatrix} \alpha_i \eta k_3 + \alpha_j \gamma k_2 & \eta k_2 k_3 & \gamma k_1 k_2 \\ 0 & 0 & 0 \\ 0 & 0 & 0 \end{bmatrix},$$

where $k_1 = \alpha_i + \alpha_j + \Theta$, $k_2 = \beta_i + \Theta$ and $k_3 = \beta_j + \Theta$. Calculating by the next generation matrix [28], we get $\rho(FV^{-1}) = \Re_0$. Here ρ is spectral radius.

$$\Re_0 = \left(\frac{\alpha_i \eta k_3 + \alpha_j \gamma k_2}{k_1 k_2 k_3} \right) = \frac{\alpha_i \eta}{k_1 k_2} + \frac{\alpha_j \gamma}{k_1 k_3}. \tag{7}$$

Let $\Re_1 = \frac{\alpha_i \eta}{k_1 k_2}$ and $\Re_2 = \frac{\alpha_j \gamma}{k_1 k_3}$ spread of infection due to tested and infected class while infected and not tested class respectively.

Lemma 1 *The disease-free equilibrium point DFE of the mathematical model (1), that is (5), is locally asymptotically stable if* $\Re_0 < 1$ *and unstable whenever* $\Re_0 > 1$.

The $\Re_0$ is the threshold quantity giving an average number, one infected individual spreading the infection and a certain amount of susceptible individuals becoming infectious [28].

Theorem 2 *The system of equations of the model (1) is locally asymptotically stable at disease-free equilibrium point if reproduction number is less than one and unstable if not.*

Proof The Jacobian matrix at DFE for the proposed model (1) is

$$J_{DFE} = \begin{bmatrix} -\Theta & 0 & -\eta & -\gamma\psi & 0 & 0 \\ 0 & -(\alpha_i + \alpha_j + \Theta) & \eta & \gamma & 0 & 0 \\ 0 & \alpha_i & -(\beta_i + \Theta) & 0 & 0 & 0 \\ 0 & \alpha_j & 0 & -(\beta_j + \Theta) & 0 & 0 \\ 0 & 0 & \beta_i & 0 & -\Theta & 0 \\ 0 & 0 & 0 & \beta_j & 0 & -\Theta \end{bmatrix}.$$

Hence, the characteristic equation of J_{DFE} can be expressed as

$$(x + \Theta)^3 (x^3 + A_1 x^2 + A_2 x + A_3), \tag{8}$$

where $A_1 = 3\Theta + \alpha_i + \alpha_j + \beta_i + \beta_j$, $A_2 = 3\Theta^2 - \eta\alpha_i + 2\Theta\alpha_i - \gamma\alpha_j + 2\Theta\alpha_j + 2\Theta\beta_i + \alpha_i\beta_i + \alpha_j\beta_i + 2\Theta\beta_j + \alpha_i\beta_j + \alpha_j\beta_j + \beta_i\beta_j$, $A_3 = \Theta^3 - \eta\Theta\alpha_i + \Theta^2\alpha_i - \gamma\Theta\alpha_j + \Theta^2\alpha_j + \Theta^2\beta_i + \Theta\alpha_i\beta_i - \gamma\alpha_j\beta_i + \Theta\alpha_j\beta_i + \Theta^2\beta_j - \eta\alpha_i\beta_j + \Theta\alpha_i\beta_j + \Theta\alpha_j\beta_j + \Theta\beta_i\beta_j + \alpha_i\beta_i\beta_j + \alpha_j\beta_i\beta_j$.

$$x^3 + A_1 x^2 + A_2 x + A_3 = 0 \tag{9}$$

Equation (8) shows that the Jacobian matrix J_{DFE} has six roots from which three are real and negative and other three eigenvalues can be determined by solving Eq. (9).

To detect the sign of the real parts of the roots of Eq. (9), the Liénard Chipart [6] criterion has been taken into account. The necessary and sufficient condition of the Liénard Chipart criterion, if A_1, A_2, and $A_3 > 0$ of Eq. (9) then the real parts of the roots will be negative for the same.

We have

$$\begin{aligned} A_1 &= k_1 + k_2 + k_3, \\ A_2 &= k_1 k_2 (1 - \mathfrak{R}_1) + k_1 k_3 (1 - \mathfrak{R}_2) + k_2 k_3, \\ A_3 &= (1 - \mathfrak{R}_0) k_1 k_2 k - 3. \end{aligned}$$

Whenever $\mathfrak{R}_0 < 1$, the values of A_1, A_2, and A_3 are greater then zero. This leads that all the eigenvalues of J_{DFE} have the negative real part if $\mathfrak{R}_0 < 1$. This implies local asymptotic stability at DFE if $\mathfrak{R}_0 < 1$.

The section epidemiologically states, if $\mathfrak{R}_0 < 1$, the infection spread decreases. Whereas, if $\mathfrak{R}_0 > 1$, the spread increases.

3.3 *Endemic Equilibrium*

The proposed model (1), endemic equilibria could be obtained by equating each equation of the system to zero. Hence, Eq. (10) is obtained.

$$\begin{cases} S^* = \frac{B}{(\lambda^*+\Theta)}, \\ E^* = \frac{\lambda^* S^*}{(k_1+\psi R_j^*)}, \\ I^* = \frac{\alpha_i E^*}{k_2}, \\ J^* = \frac{(\alpha_j+\psi R_j^*)E^*}{k_3}, \\ R_i^* = \frac{\beta_i I^*}{\Theta}, \\ R_j^* = \frac{\beta_j J^*}{\Theta}, \end{cases} \tag{10}$$

where $\lambda^* = \frac{(\eta I^*+\gamma J^*)}{N^*}$.

$$a(\lambda^*)^2 + b\lambda^* - c = 0 \tag{11}$$

where $a = k_2\alpha_i(1 + \frac{\beta_i}{\Theta}) + k_2(\alpha_j + \psi R_j)(1 + \frac{\beta_j}{\Theta}) + k_2k_3,\ b = k_1k_2k_3 + \psi k_2k_3R_j$, and $c = \gamma\alpha_i k_3 + \gamma k_2(\alpha_j + \psi R_j)$. From Eq. (11), both the roots $\lambda^* = \frac{-b\pm\sqrt{b^2+4ac}}{2a}$ are real. In accordance with Descartes' rule of signs one must be negative and other is positive for which the endemic equilibria exist for the proposed model (1).

3.4 Global Stability

The section emphasizes the global dynamics of the endemic equilibria of the model (1). It discusses the impact of the primary reproduction number on global asymptotic stability [27].

Theorem 3 *The endemic equilibria of the model (1) has global asymptotic stability if* $\mathfrak{R}_0 > 1,\ \gamma = \Theta$ *and* $\beta_j = 0$ *where endemic equilibria is represented as* $\mathbb{E}^* = (S^*,\ E^*,\ I^*,\ J^*,\ R_i^*,\ R_j^*)$.

Proof Let the Lyapunov function [27] for the system (1) with positive undetermined constants c_2, c_3, c_4, and c_6 be

$$V = c_2E(t) + c_3I(t) + c_4J(t) + c_6R_j(t).$$

Thus, we have

$$\begin{aligned} \dot{V} &\le c_2(\eta I + \gamma J - (\alpha_i + \alpha_j + \psi R_j)E - \Theta E) + c_3(\alpha_i E - \beta_i I - \Theta I) \\ &\quad + c_4((\alpha_j + \psi R_j)E - \beta_j J - \Theta J) + c_6(\beta_j J - \Theta R_j) \\ &\le E(c_4\alpha_j + c_3\alpha_i - c_2(\alpha_i + \alpha_j + \Theta)) + I(c_2\eta - c_3(\beta_i + \Theta)) \\ &\quad + J(c_6\beta_j + c_2\gamma - c_4(\beta_j + \Theta)) + R_jE(-c_2\psi + c_4\psi) - c_6\Theta R_j \\ &\le E(c_4\alpha_j + c_3\alpha_i - c_2(\alpha_i + \alpha_j + \Theta)) + I(c_2\eta - c_3(\beta_i + \Theta)) \\ &\quad + J(c_6\beta_j + c_2\gamma - c_4(\beta_j + \Theta)) + R_jE(-c_2\psi + c_4\psi). \end{aligned}$$

Choosing $c_2 = 1, c_3 = \frac{\eta}{k_2}$, and $c_4 = \frac{\gamma}{k_3}$ with $\beta_j = 0$, and $\gamma = \Theta$ one can obtain that

$$\dot{V} \leq k_1(\mathfrak{R}_0 - 1)E \tag{12}$$

Thus from Eq. (12), $\frac{dV(t)}{dt} < 0$ whenever $\mathfrak{R}_0 < 1$.

From an epidemiological point of view, the above theorem states that, if $\psi = 0,\ \eta = 0$, the disease spreads in the population whenever $\mathfrak{R}_0 > 1$. Therefore in $\overline{\Pi}$, the endemic equilibrium point is said to have global asymptomatic stability by using LaSalle's invariant principle [19].

3.5 Threshold Analysis

The section focuses on the effect of an infected and not tested person on the transmission variability of the infection of the proposed model (1)-the partial derivative of primary reproduction number with respect to α_j parameter analyzes the threshold analysis.

Theorem 4 *The infected and not tested class on the exposed class has positive or negative population-level effect if* γ *less or greater than* γ^*, *where* $\gamma^* = \frac{(\beta_j+\Theta)\alpha_i\eta}{(\beta_i+\Theta)(\alpha_i+\Theta)}$.

Proof The basic reproduction number $\mathfrak{R}_0$ is differentiated partially with respect to α_j is

$$\frac{\partial \mathfrak{R}_0}{\partial \alpha_j} = \frac{\eta \alpha_i}{(\beta_i + \Theta)(\alpha_i + \alpha_j + \Theta)^2} - \frac{(\alpha_i + \Theta)\gamma}{(\alpha_i + \alpha_j + \Theta)^2(\beta_j + \Theta)}.$$

Let

$$\gamma^* = \frac{(\beta_j + \Theta)\alpha_i \eta}{(\beta_i + \Theta)(\alpha_i + \Theta)}.$$

One can easily conclude that $\frac{\partial \mathfrak{R}_0}{\partial \alpha_j} < 0$, if $\gamma < \gamma^*$ and $\frac{\partial \mathfrak{R}_0}{\partial \alpha_j} > 0$, if $\gamma > \gamma^*$.

Thus, the value of the primary reproduction number $\mathfrak{R}_0$ will depend on α_j and will be decreasing function when the infected and not tested persons do not exceed the threshold value γ^* and therefore, disease burden will reduce. Further, the primary reproduction number $\mathfrak{R}_0$ will be an increasing function of the parameter α_j when the infected and not tested persons exceed the threshold value γ^* and therefore, the disease will increase.

4 Numerical Simulation

The section focuses on the analytical findings of model (1), verified through numerical simulations with assumed parametric values (per day) $B = 2;\ \Theta = 0.1;\ \eta = 0.6;\ \gamma = 0.2;\ \alpha_i = 0.252;\ \alpha_j = 0.81;\ \psi = 0.2;\ \beta_i = 0.2;\ \beta_j = 0.8$ which results

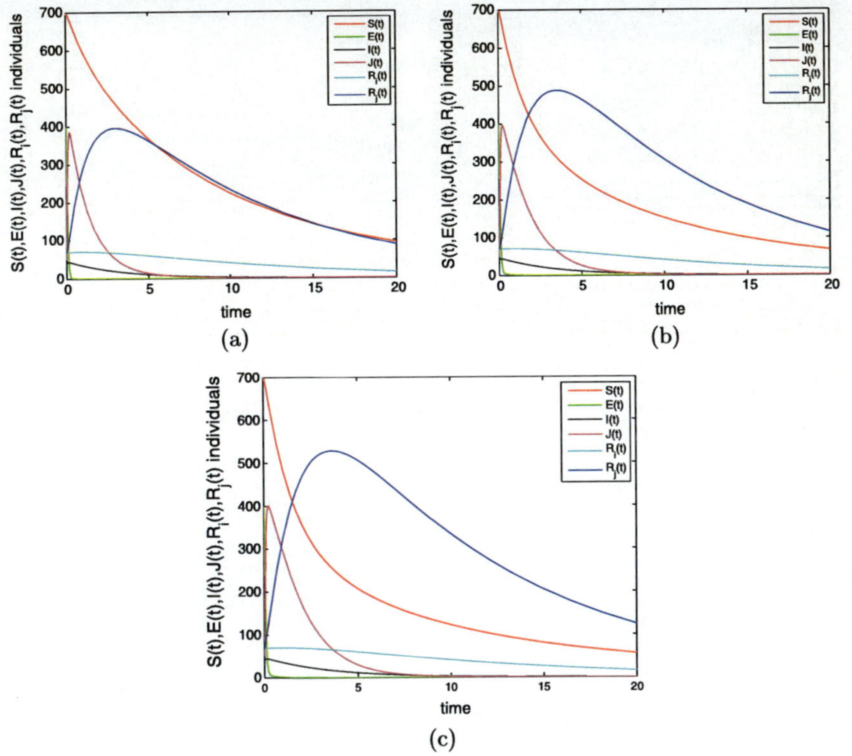

Fig. 2 Variation in the proposed model classes according to different values of γ changing the reproduction number **a** $\gamma = 0.2$ with $\mathfrak{R}_0 < 1$, **b** $\gamma = 0.7311$ with $\mathfrak{R}_0 = 1$, and **c** $\gamma = 0.95$ with $\mathfrak{R}_0 > 1$

to give the basic reproduction number as 0.6145. Hence the model (1) is local asymptotic stable at disease-free equilibrium point (Fig. 2a). Figure 2 depicts the variation in all classes (susceptible class, exposed class, infected class, infected and tested class, infected and not tested class, recovered after infected class and tested and recovered after infected and not tested class) according to different reproduction numbers (a) with $\mathfrak{R}_0 < 1$, (b) with $\mathfrak{R}_0 = 1$, and (c) with $\mathfrak{R}_0 > 1$. Figure 3 shows the variation in the total number of infected citizens and total number of recovered citizens in accordance with the basic reproduction number $\mathfrak{R}_0 < 1$, $= 1$ and > 1 and Fig. 4 represents the effective effect of γ on the basic reproduction which decreases with decrease in the value of γ with threshold value $\gamma^* = 0.7311$.

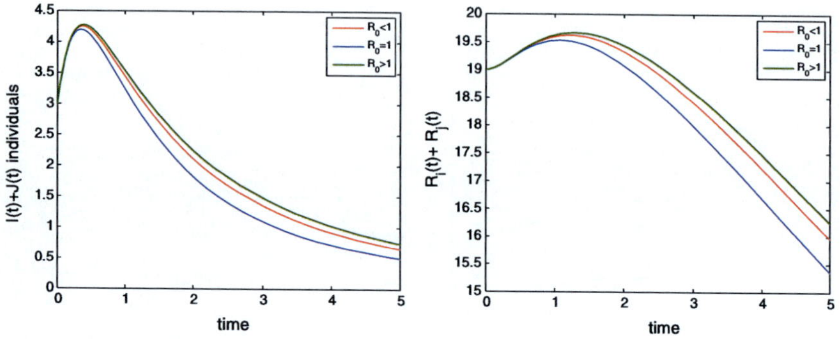

(a) (b)

Fig. 3 Variation in the **a** total infected citizens according **b** total recovered citizens in accordance with $\mathfrak{R}_0 < 1, = 1$ and > 1

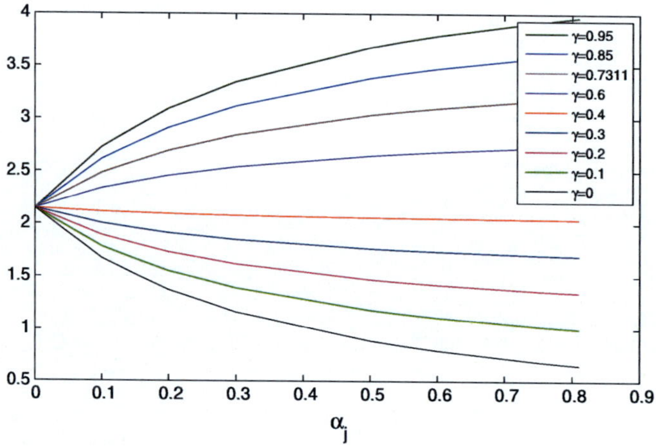

Fig. 4 The effect of infection rate(α_j) of $J(t)$ class from $E(t)$ on $\mathfrak{R}_0$

5 Results and Discussion

The effect of people in the class of recovered without being tested (R_j) or following any guidelines has been studied through the mathematical epidemiological model (1). Also, the increase in the number of infections and its importance with the testing class is elaborated through the qualitative analysis. The analysis helped to evaluate the importance of peer influence effects importance in society. The dynamics and behavior of the mathematical model have been theoretically and numerically determined. It concludes with the following epidemiological as well as mathematical results itemized as follows:

(*i*) The mathematical model has the system of equations resulting to provide non-negative and bounded solution for all $t > 0$, when the seed conditions are non-negative (Theorem 1) which proves that the model (1) is mathematical and epidemiological well-posed.

(*ii*) The model has local asymptotic stability at disease-free equilibria (DFE) if the primary reproduction number $\mathfrak{R}_0$ is less than unity (Theorem 2). Epidemiologically, for each case, each infected individual will infect less than an individual during the entire time of their infection period. The result leads to an infection decline and exhaustion. It concludes that the primary reproduction number can control the infection. This stops the disease from shaping into a pandemic or epidemic.

(*iii*) The mathematical model has an endemic equilibrium if the primary reproduction number $\mathfrak{R}_0$ is more significant than one. The local asymptotic stability and global asymptotic stability for the particular case when $\mathfrak{R}_0$ is more significant than one is established in (Theorem 3). Epidemiologically, the primary reproduction number is more significant than one, leading the infection to grow at a tremendous rate. Each infected individual infects more than a single individual during the entire infection period. It concludes that the disease invades the susceptible class at the time and leads to a pandemic or epidemic.

(*iv*) The primary reproduction number $\mathfrak{R}_0$ depends on the parameter α_j. The value of the primary reproduction number $\mathfrak{R}_0$ decreases when the class of infected and not tested persons do not exceed the value γ^* called the threshold value (Theorem 4). Therefore, we can see that the untested individuals can increase or decrease the infection in society at a certain rate.

6 Conclusion

The information opinion indicates the potential value of the study regarding the reporting issue of outbreak changes the thinking process of the citizens. The phase of the global crisis is unlike any other scenario in the past few years—people deaths, the spread of human suffering, and turning upside down of lives leads to different absurd pieces of information everywhere. However, this is much more than a health crisis. It is an ecological, economic, financial, human and social crisis. Different prevention strategies play different roles in society in decreasing the damages, and testing is one of them. The mindset of the individual can stop them from being tested due to the scenario. The peer influence effect can lead a society to a different path at that time.

References

1. Bernoulli, D.: Essai d'une nouvelle analyse de la mortalite causee par la petite verole. Mem. Math. Phys. Acad. Roy. Sci., Paris (1766)
2. Biswas, S.K., Ghosh, J.K., Sarkar, S., Ghosh, U.: COVID-19 Pandemic in India: A Mathematical Model Study. Springer Nature B.V (2020). https://doi.org/10.1007/s11071-020-05958-z
3. Brauer, F., Chavez, C.: Mathematical Models in Population Biology and Epidemiology. Springer, New York (2011). ISSN:978-1-4757-3516-1
4. Buonomo, B., Lacitignola, D.: Modeling peer influence effects on the spread of high-risk alcohol consumption behavior. Ricerche mat. **63**, 101–117 (2014). https://doi.org/10.1007/s11587-013-0167-3
5. Cani, J., Yakowitz, S., Blount, M.: The spread and quarantine of HIV infection in a prison system. Soc. Ind. Appl. Math. J. Appl. Math. **57**, 1510–1530 (1997). https://doi.org/10.1137/S0036139995283237
6. Daud, A.A.M.: A note on Lienard-Chipart criteria and its application to epidemic models. Math. Stat. **9**, 41–45 (2021). https://doi.org/10.13189/ms.2021.090107
7. Dhar, J., Sharma, A.: The role of the incubation period in a disease model. Appl. Math. e-Notes **9**, 146–153 (2009)
8. Erdem, M., Safan, M., Chavez, C.: Mathematical analysis of an SIQR influenza model with imperfect quarantine. Bull. Math. Biol. **79**, 1612–1636 (2017). https://doi.org/10.1007/s11538-017-0301-6
9. Esteva, L., Gumel, A.B., Vargas, C.: Qualitative study of transmission dynamics of drug-resistant malaria. Math. Comput. Model. **50**, 611–630 (2009). https://doi.org/10.1016/j.mcm.2009.02.012
10. Esteva, L., Vargas, C.: Influence of vertical and mechanical transmission on the dynamics of dengue disease. Math. Biosci. **167**, 51–64 (2000). https://doi.org/10.1016/s0025-5564(00)00024-9
11. Hale, J.K.: Ordinary Differential Equations. Wiley, New York (1969)
12. Hamer, W.H.: The Milroy lectures on epidemic disease in England-the evidence of variability and persistence of type. The Lancet **1**, 733–739 (1906)
13. Hethcote, H.W., Thieme, H.R.: Stability of the endemic equilibrium in epidemic models with subpopulations. Math. Biosci. **75**, 205–227 (1985). https://doi.org/10.1016/0025-5564(85)90038-0
14. Hethcote, H., Ma, Z., Shengbing, L.: Effects of quarantine in six endemic models for infectious diseases. Math. Biosci. **180**, 141–160 (2002). https://doi.org/10.1016/s0025-5564(02)00111-6
15. Hethcote, H.W., Ma, Z., Liao, S.B.: Effects of quarantine in six endemic models for infectious diseases. Math. Biosci. **180**, 141–160 (2002). https://doi.org/10.1016/s0025-5564(02)00111-6
16. Hyman, J.M., Li, J.: Modeling the effectiveness of isolation strategies in preventing STD epidemics. Soc. Ind. Appl. Math. J. Appl. Math. **58**, 912–925 (1998)
17. Khan, M.A., Atangana, A.: Modeling the dynamics of novel coronavirus (2019-nCov) with fractional derivative. Alex. Eng. J. **59**, 2379–2389 (2020). https://doi.org/10.1016/j.aej.2020.02.033
18. Lakshmikantham, V., Leela, M.S., Matynyuk, A.A.: Stability Analysis of Nonlinear Systems. Marcel Dekker Incorporated New York and Basel (1989)
19. La Salle, J.P.: The stability of dynamical systems. Soc. Ind. Appl. Math. (1976)
20. Lee, S.H., Kang, H., Song, H.S.: Effect of Individual self-protective behavior on epidemic spreading. J. Biol. Syst. **27**, 531–542 (2019). https://doi.org/10.1142/S0218339019500219
21. Misra, A.K., Rai, R.K., Takeuchi, Y.: Modeling the effect of time delay in budget allocation to control an epidemic through awareness. Int. J. Biomath. **11** (2018). https://doi.org/10.1142/S1793524518500274
22. Perko, L.: Differential Equations and Dynamical Systems. Springer, New York (1996). https://doi.org/10.1007/978-1-4684-0249-0

23. Ross, R.: The Prevention of Malaria, 2nd edn. John Murray, London (1911)
24. Safi, M.A., Gumelb, A.B.: Dynamics of a model with quarantine-adjusted incidence and quarantine of susceptible individuals. J. Math. Anal. Appl. **399**, 565–575 (2013). https://doi.org/10.1016/j.jmaa.2012.10.015
25. Sahu, G.P., Dhar, J.: Analysis of an SVEIS epidemic model with partial temporary immunity and saturation incidence rate. Appl. Math. Model. **36**, 908–923 (2012). https://doi.org/10.1016/j.apm.2011.07.044
26. Sahua, G.P., Dhar, J.: Dynamics of an SEQIHRS epidemic model with media coverage, quarantine and isolation in a community with pre-existing immunity. J. Math. Anal. Appl. **421**, 1651–1672 (2015). https://doi.org/10.1016/j.jmaa.2014.08.019
27. Ullah, S., Khan, M.A.: Modeling the impact of non-pharmaceutical interventions on the dynamics of novel coronavirus with optimal control analysis with a case study. Chaos, Solitons Fractals **139** (2020). https://doi.org/10.1016/j.chaos.2020.110075
28. Van den Driesschea, P., Wamough, J.: Reproduction number and sub-threshold endemic equilibria for compartment models of disease transmisson. Math. Biosci. **180**, 29–48 (2002). https://doi.org/10.1016/S0025-5564(02)00108-6
29. Wang, S., Xu, F., Rong, L.: Bistability analysis of an HIV model with immune response. Journal of Biological System **25**, 677–695 (2017). https://doi.org/10.1142/S021833901740006X
30. WHO. https://www.who.int/docs/default-source/coronaviruse/situation-reports/20200311-sitrep-51-covid-19.pdf?sfvrsn=1ba62e57-10
31. WHO. https://covid19.who.int/
32. Xing, Z., Cardona, C.J.: Pre-existing immunity to pandemic (H1N1) 2009. Emerg. Infect. Dis. **15**, 1847–1849 (2009). https://doi.org/10.3201/eid1511.090685
33. Zhou, X., Cui, J.: Analysis of stability and bifurcation for an SEIV epidemic model with vaccination and nonlinear incidence rate. Nonlinear Dyn. **63**, 639–653 (2011). https://doi.org/10.1007/S11071-010-9826-Z

B-Splines Collocation Approach to Simulate Secondary Dengue Virus (DENV) Infection Model with Diffusion

Rohit Goel, R. C. Mittal, and Neha Ahlawat

Abstract Dengue fever is a mosquito-borne viral infection caused by the dengue virus (DENV) found worldwide in tropical and sub-tropical urban and non-urban areas. Dengue viruses (DENV) spread through the bite of an infected *Ades* species mosquito. There is not available any specific treatment or cure for this DENV infection. The dynamics of the secondary dengue virus infection considering the spatial mobility of dengue virus particles and cells can better be studied and analyzed with reaction–diffusion mathematical models. A reaction–diffusion mathematical model consisting of five simultaneous nonlinear partial differential equations to characterize the dynamics of secondary Dengue infection is studied in this paper. The spatial mobility of the dengue particles and cells is considered in the model. A numerical simulation technique based on the cubic B-splines collocation is proposed to approximate the solution of the considered model.

Keywords Dengue virus infection · Cubic basis splines · Thomas algorithm · RK4 method

1 Introduction

Dengue fever caused by dengue virus (DENV) also known as breakbone fever is a viral disease prominent in tropical and sub-tropical regions that approximately affects 250 million people worldwide [1]. Dengue (DFH) hemorrhagic fever/dengue (DSS) shock syndrome [2, 3] are widespread among the four serologically [4] severe

R. Goel (✉) · R. C. Mittal · N. Ahlawat
Jaypee Institute of Information Technology, Noida, India
e-mail: rohitdd777@gmail.com

R. Goel
Deshbandhu College (University of Delhi), New Delhi, India

R. K. Sharma et al. (eds.), *Frontiers in Industrial and Applied Mathematics*, Springer Proceedings in Mathematics & Statistics 410,
https://doi.org/10.1007/978-981-19-7272-0_16

DENV syndromes. The antibody's immune responses and the CTLs are the two major constituents that destroy the DENV-infected cells [5, 6] and kill the DENV particles [7] respectively.

In the past and the recent years, the dynamics of within-host DENV primary infection [8–11] and pertaining to another stereotype secondary infection [12–15] has been studied through various mathematical models. However, no significant attempts have been made yet to describe diffusion-based dynamics of DENV infection. A mathematical model describing the global dynamics of the secondary DENV infection taking into consideration the diffusion impacts is proposed by Elaiw and Alofi [16]. The non-negativity, the boundedness of the solution, and the stability of the equilibrium points are analyzed and discussed in the extended model.

The reaction–diffusion mathematical models account to a wide variety of physical and dynamical phenomena occurring in day-to-day life [24, 25]. In the present paper, the solutions of the DENV infection model [16] have been found numerically using the cubic B-splines collocation method. The splines which are precisely the piecewise continuous polynomial functions [17] constitute an elegant framework for dealing with the discretization and the interpolation simulation problems. In this paper, cubic B-splines are collocated over the finite elements to approximate the spatial variables and its derivatives. The B-splines are preferred over the other traditional schemes for their inheritance of continuity and the small local support over the given partition of the domain. Mittal et al. [18–20] have proposed the collocation scheme to estimate solutions of various linear and nonlinear partial differential equations. The authors also used the proposed scheme for a larger dimension malaria infection reaction–diffusion model, M1 cancer virotherapy, COVID-19 infection, computer virus dynamics, and NPZ-SIR models and achieved the accurate solutions.

2 The DENV Reaction–Diffusion Model

Elaiw and Alofi [16] proposed the following DENV secondary infection diffusion model:

$$\frac{\partial K(u,t)}{\partial t} = d_K \Delta K(u,t) + \delta - \mu K(u,t)M(u,t) - \xi K(u,t)$$

$$\frac{\partial L(u,t)}{\partial t} = d_L \Delta L(u,t) + \mu K(u,t)M(u,t) - \rho L(u,t)$$

$$\begin{aligned}\frac{\partial M(u,t)}{\partial t} &= d_M \Delta M(u,t) + \tau L(u,t) - \eta M(u,t)\\ &- \omega_1 M(u,t)N(u,t) - \omega_2 M(u,t)P(u,t)\end{aligned}$$

$$\frac{\partial N(u,t)}{\partial t} = d_N \Delta N(u,t) + \lambda_1 M(u,t)N(u,t) - \alpha_1 N(u,t)$$

$$\frac{\partial P(u,t)}{\partial t} = d_P \Delta P(u,t) + \lambda_2 M(u,t)P(u,t) - \alpha_2 P(u,t) \tag{1}$$

for time $t > 0$ and position $u \in \Gamma$, where $K(u,t)$, $L(u,t)$, $M(u,t)$, $N(u,t)$, and $P(u,t)$, respectively, denote the concentrations of the target cells, DENV infected cells, DENV particles, heterologous antibodies formed from the primary DENV infection and the homologous antibodies formed from the secondary DENV infection, respectively. $\partial\Gamma$ is the smooth boundary of the bounded, connected, and the continuous domain $\Gamma \subset \mathbb{R}^m (m \geq 1)$. $\Delta = \frac{\partial^2}{\partial u^2}$ being the Laplacian operator and d_Λ denotes the diffusion coefficient of the component Λ. $\lambda_1 M(u,t)N(u,t)$ and $\lambda_2 M(u,t)P(u,t)$ are, respectively, the rates of activation of the two antibodies.

The model is associated with the non-negative, continuous, and biologically justified initial conditions

$$\begin{aligned} K(u,0) &= \psi_1(u), L(u,0) = \psi_2(u), M(u,0) = \psi_3(u), \\ N(u,0) &= \psi_4(u), P(u,0) = \psi_5(u) \end{aligned}$$

and the homogeneous Neumann boundary conditions representing a natural dispersal barrier and signifying that the cells and the viruses cannot cross the isolated boundary.

$$\frac{\partial K}{\partial \vec{n}} = \frac{\partial L}{\partial \vec{n}} = \frac{\partial M}{\partial \vec{n}} = \frac{\partial N}{\partial \vec{n}} = \frac{\partial P}{\partial \vec{n}} = 0; t > 0, u \in \partial\Gamma$$

where $\frac{\partial}{\partial \vec{n}}$ being the outward normal derivative on the boundary $\partial\Gamma$.

3 Mathematical Formulation

The solution domain $[a, b]$ is uniformly partitioned into a mesh of uniform step size length $h = u_{i+1} - u_i = \frac{(b-a)}{n}$ for $i = 0, 1, 2 \ldots.., (n-1)$ by the knots u_i where $i = 0, 1, 2 \ldots.., n$ where n is the number of grid points in the partition such that $a = u_0 < u_1 \ldots\ldots\ldots\ldots < u_n = b$.

The approximate solutions to find $K^n(u,t)$, $L^n(u,t)$, $M^n(u,t)$, $N^n(u,t)$, $P^n(u,t)$ takes the following form:

$$K^n(u,t) = \sum_{j=-1}^{n+1} \sigma_j{}^{(K)}(t) C_j(u), \quad a \leq u \leq b, t > 0 \tag{2}$$

$$L^n(u,t) = \sum_{j=-1}^{n+1} \sigma_j{}^{(L)}(t) C_j(u), \quad a \leq u \leq b, t > 0 \tag{3}$$

$$M^n(u,t) = \sum_{j=-1}^{n+1} \sigma_j{}^{(M)}(t) C_j(u), \quad a \le u \le b, t > 0 \tag{4}$$

$$N^n(u,t) = \sum_{j=-1}^{n+1} \sigma_j{}^{(N)}(t) C_j(u), \quad a \le u \le b, t > 0 \tag{5}$$

$$P^n(u,t) = \sum_{j=-1}^{n+1} \sigma_j{}^{(P)}(t) C_j(u), a \le u \le b, t > 0 \tag{6}$$

where $\sigma_j{}^{(i)}(t)$; $i = K, L, M, N, P$ are the time-dependent numbers. And $C_j(u)$ is the cubic B-spline basis function defined by

$$C_j(u) = \frac{1}{h^3} \begin{cases} (u - u_{j-2})^3 & u \in [u_{j-2}, u_{j-1}) \\ (u - u_{j-2})^3 - 4(u - u_{j-1})^3 & u \in [u_{j-1}, u_j) \\ (u_{j+2} - u)^3 - 4(u_{j+1} - u)^3 & u \in [u_j, u_{j+1}) \\ (u_{j+2} - u)^3 & u \in [u_{j+1}, u_{j+2}) \\ 0 & \text{otherwise} \end{cases} \tag{7}$$

where the functions $C_{-1}, C_0, C_1, \ldots\ldots\ldots C, C_N, C_{n+1}$ form a basis over the domain $a \le u \le b$. The values of the functions $C_j(u)$ and their two successive derivatives $C'_j(u)$, $C''_j(u)$ over the prescribed set of knots are given in Table 1.

Using the B-spline function (7) in the approximate solution function (2), the approximate values can be expressed in terms of time-dependent numbers $\sigma_j{}^{(k)}(t)$ as

$$\left.\begin{aligned} K_j &= \sigma_{j-1}{}^{(K)} + 4\sigma_j{}^{(K)} + \sigma_{j+1}{}^{(K)} \\ hK'_j &= 3(\sigma_{j+1}{}^{(K)} - \sigma_{j-1}{}^{(K)}) \\ h^2K''_j &= 6(\sigma_{j+1}{}^{(K)} - 2\sigma_j{}^{(K)} + \sigma_{j+1}{}^{(K)}) \end{aligned}\right\} \tag{8}$$

The respective values of the estimated solutions for the other four variables and their derivatives can be expressed in the similar manner.

Table 1 Values of cubic B-spline coefficients

	u_{j-2}	u_{j-1}	u_j	u_{j+1}	u_{j+2}
$C_j(u)$	0	1	4	1	0
$C'_j(u)$	0	−3/h	0	3/h	0
$C''_j(u)$	0	6/h^2	−12/h^2	6/h^2	0
$C_j(u)$	0	1	4	1	0

4 Treatment at Boundary Conditions

If $r_0(t)$ and $r_1(t)$ are the prescribed boundary conditions for $K(u, t)$, respectively,

$$\left(\frac{\partial K}{\partial u}\right)_{u=a} = r_0(t) \text{ and } \left(\frac{\partial K}{\partial u}\right)_{u=b} = r_1(t)$$

Then

$$K_u(u_0, t) = \sum_{j=-1}^{1} \sigma_j{}^{(K)} C'{}_j(u_0) = r_0(t)$$

$$K_u(u_n, t) = \sum_{j=n-1}^{n+1} \sigma_j{}^{(K)} C'{}_j(u_n) = r_1(t)$$

Using Table 1, we get

$$\sigma_1{}^{(K)} - \sigma_{-1}{}^{(K)} = \left(\frac{h}{3}\right) r_0(t)$$

$$\sigma_{n+1}{}^{(K)} - \sigma_{n-1}{}^{(K)} = \left(\frac{h}{3}\right) r_1(t)$$

so that

$$\sigma_{-1}{}^{(K)} = \sigma_1{}^{(K)} - \left(\frac{h}{3}\right) r_0(t)$$

$$\sigma_{n+1}{}^{(K)} = \sigma_{n-1}{}^{(K)} + \left(\frac{h}{3}\right) r_1(t)$$

Thus, the two-time dependent quantities falling outside the prescribed knots are determined. The remaining other variables can also be treated at the boundary conditions similarly.

5 Implementation of a RD Equation

The equation for $K(u, t)$ is given by

$$\frac{\partial K(u, t)}{\partial t} = d_K \Delta K(u, t) + \phi_1(K, L, M, N, P) \tag{9}$$

where d_K is diffusion coefficient and ϕ_1 corresponds to the reaction term.

Using (8) in the approximate solution expression (2) for $H^n(x,t)$, the given differential Eq. (9) results in the following system:

$$\dot{\sigma}_{j-1}^{(K)} + 4\dot{\sigma}_{j}^{(K)} + \dot{\sigma}_{j+1}^{(K)} = (\sigma_{j-1}{}^{(K)} - 2\sigma_j{}^{(K)} + \sigma_{j+1}{}^{(K)})d_K + \phi_{1j};\ 0 \le j \le n$$

ϕ_{1j} is the right-hand side quantity of H at the j-th node, respectively. Upon deporting $\dot{\sigma}_{-1}^{(K)}, \dot{\sigma}_{n+1}^{(K)}, \sigma_{-1}{}^{(K)}, \sigma_{n+1}{}^{(K)}$, the following system of differential equations is obtained:

$$A\dot{\widehat{\sigma}}^{(K)} = B_K\widehat{\sigma}^{(K)} + \widehat{\phi_1} \tag{10}$$

where $B_K = d_K.B$

$$A = \begin{bmatrix} 4 & 2 & & & & \\ 1 & 4 & 1 & & & \\ & \cdots & \cdots & \cdots & & \\ & \cdots & \cdots & \cdots & & \\ & & & 1 & 4 & 1 \\ & & & & 2 & 4 \end{bmatrix}_{(n+1)\times(n+1)} \quad B = \left(\frac{6}{h^2}\right)\begin{bmatrix} -2 & 2 & & & & \\ 1 & -2 & 1 & & & \\ & \cdots & \cdots & \cdots & & \\ & \cdots & \cdots & \cdots & & \\ & & & 1 & -2 & 1 \\ & & & & 2 & -2 \end{bmatrix}_{(n+1)\times(n+1)}$$

and

$$\dot{\widehat{\sigma}}^{(K)} = \begin{bmatrix} \dot{\sigma}_0^{(K)} \\ \dot{\sigma}_1^{(K)} \\ \cdots \\ \cdots \\ \cdots \\ \dot{\sigma}_{n-1}^{(K)} \\ \dot{\sigma}_n^{(K)} \end{bmatrix} \widehat{\phi_1} = \begin{bmatrix} \phi_{10} + \left(\frac{h}{3}\right)r_0(t) \\ \phi_{11} \\ \cdots \\ \cdots \\ \cdots \\ \phi_{1(n-1)} \\ \phi_{1n} - \left(\frac{h}{3}\right)r_1(t) \end{bmatrix}_{(n+1)\times 1}$$

Similar treatment can be applied in the corresponding time-dependent quantities for the other variables of the system.

6 Evaluation of Initial Vector $\widehat{\sigma^0}^{(K)}$

The initial vector $\widehat{\sigma^0}^{(K)}$ can be evaluated by using the given data of initial conditions and the boundary values of their derivatives.

$$K_x(a,0) = K_u(u_0,0) = r_0(0) \tag{11}$$

$$K(u_j,0) = \psi_1(u_j),\ j = 1,2\ldots\ldots,n-1 \tag{12}$$

$$K_u(b, 0) = K_u(u_n, 0) = r_1(0) \tag{13}$$

Similar expressions will be derived for the other remaining variables. Equations (11)–(13) on applying to (2) yield a $(n + 1) \times (n + 1)$ system of the form

$$A\widehat{\sigma^0}^{(K)} = \widehat{\phi_1^0} \tag{14}$$

where

$$\widehat{\sigma^0}^{(K)} = \begin{bmatrix} \sigma_{K0}^0 \\ \sigma_{K1}^0 \\ \cdots \\ \cdots \\ \cdots \\ \sigma_{K(n-1)}^0 \\ \sigma_{Kn}^0 \end{bmatrix}, \widehat{\phi_1^0} = \begin{bmatrix} \psi_1(u_0) + (\frac{h}{3})r_0(0) \\ \psi_1(u_1) \\ \cdots \\ \cdots \\ \cdots \\ \psi_1(u_{n-1}) \\ \psi_1(u_n) - (\frac{h}{3})r_1(0) \end{bmatrix}$$

and so on. This system being reduced to a tridiagonal system can finally be simplified by the well-known Thomas algorithm for tridiagonal systems [22, 23].

7 Method Implementation

The considered DENV infection model (1) can be rewritten in the standard reaction–diffusion form as

$$\frac{\partial K(u,t)}{\partial t} = d_K \Delta K(u,t) + \phi_1(K, L, M, N, P)$$

$$\frac{\partial L(u,t)}{\partial t} = d_L \Delta L(u,t) + \phi_2(K, L, M, N, P)$$

$$\frac{\partial M(u,t)}{\partial t} = d_M \Delta M(u,t) + \phi_3(K, L, M, N, P)$$

$$\frac{\partial N(u,t)}{\partial t} = d_N \Delta N(u,t) + \phi_4(K, L, M, N, P)$$

$$\frac{\partial P(u,t)}{\partial t} = d_P \Delta P(u,t) + \phi_5(K, L, M, N, P)$$

where each of the $d_i's$ are the respective diffusion coefficients and the corresponding terms are the diffusion terms and ϕ_i represents the reaction terms in each the corresponding reaction–diffusion differential equations.

Using the B-splines approximation for the above reaction–diffusion system, the following system of differential equations is obtained

$$\boldsymbol{M}\dot{\widehat{\boldsymbol{\sigma}}} = \boldsymbol{P}\widehat{\boldsymbol{\sigma}} + \boldsymbol{F}$$

where M and P are the $5(n+1)$ ordered block diagonal matrices.

$$\boldsymbol{M} = \begin{bmatrix} A & & & & \\ & A & & & \\ & & A & & \\ & & & A & \\ & & & & A \end{bmatrix}_{5(n+1)\times 5(n+1)} \qquad \boldsymbol{P} = \begin{bmatrix} B_K & & & & \\ & B_L & & & \\ & & B_M & & \\ & & & B_{LN} & \\ & & & & B_P \end{bmatrix}_{5(n+1)\times 5(n+1)}$$

Here $\dot{\widehat{\boldsymbol{\sigma}}}$

and **F** are, respectively, the column vectors.

$$\dot{\widehat{\boldsymbol{\sigma}}} = \begin{bmatrix} \dot{\widehat{\sigma}}^{(K)} \\ \dot{\widehat{\sigma}}^{(L)} \\ \dot{\widehat{\sigma}}^{(M)} \\ \dot{\widehat{\sigma}}^{(N)} \\ \dot{\widehat{\sigma}}^{(P)} \end{bmatrix}, \quad \mathbf{F} = \begin{bmatrix} \widehat{\phi_1} \\ \widehat{\phi_2} \\ \widehat{\phi_3} \\ \widehat{\phi_4} \\ \widehat{\phi_5} \end{bmatrix}_{5(n+1)\times 1}$$

The parameter vector $\dot{\widehat{\boldsymbol{\sigma}}}$ in the above system is determined at a given time level using the Thomas algorithm [21]. Then the approximate solutions at any desired time can be found by computing the time-dependent numbers in the estimated solution by using the RK4 method.

8 Numerical Simulations

Numerical calculations are being performed to authenticate and implement the proposed scheme. For this purpose, the spatial domain $\Gamma = [0, 2]$ is considered with step size of $\Delta u = 0.02$. For calculations, the time step size of $\Delta t = 0.1$ is selected. The following parameter values are taken: $\delta = 10$, $\xi = 0.01$, $\rho = 0.3$, $\tau = 5$, $\eta = 3$, $\omega_1 = 0.3$, $\omega_2 = 0.1$, $\alpha_1 = 0.1$, $\alpha_2 = 0.1$. All the remaining parameters are given in Table 2 taken according to the strategies as classified by the stability analysis.

Depending upon the convergence of the positive solutions of the model, the simulations are being performed under the four distinct strategies as given below.

Table 2 Variable parameters values for different strategies

Strategy	μ	λ_1	λ_2
I	0.00004	0.005	0.001
II	0.0004	0.005	0.001
III	0.004	0.05	0.001
IV	0.004	0.01	0.02

Subject to the global stability and the existence of the equilibrium points $\Omega_0, \Omega_1, \Omega_2, \Omega_3$ [16], there are considered four strategies for the computation of the simulated results. This paper is chiefly related with the proposed efficient numerical simulation technique to solve the DENV infection model. A detailed analysis of these threshold parameters and their biological aspects can be well studied in Ref. [16] and the available literature. The model is accompanied with the following set of initial condition:

$$\psi_1(u) = 500\big(1 + 0.5\cos^2(\pi u)\big)$$

$$\psi_2(u) = 30\big(1 + 0.5\cos^2(\pi u)\big)$$

$$\psi_3(u) = 4\big(1 + 0.5\cos^2(\pi u)\big)$$

$$\psi_4(u) = 2\big(1 + 0.5\cos^2(\pi u)\big)$$

$$\psi_5(u) = 2\big(1 + 0.5\cos^2(\pi u)\big)$$

9 Results and Discussions

The numerical simulations accomplished by the proposed scheme for the considered DENV secondary infection model [16] are summarized by performing simulations for the following four global stability dependent different strategies. The detailed biological significances of the results obtained can be well described by the biologists. As available in the literature, the parameters $\mathcal{R}_0, \mathcal{R}_1, \mathcal{R}_2$ are taken as follows:

$$\mathcal{R}_0 = \frac{\delta\tau\mu}{\xi\rho\eta}, \mathcal{R}_1 = \frac{\mathcal{R}_0}{1 + \frac{\mu\alpha_1}{\xi\lambda_1}}, \mathcal{R}_1 = \frac{\mathcal{R}_0}{1 + \frac{\mu\alpha_2}{\xi\lambda_2}}$$

Strategy I, concerns with the stability of the infection free equilibrium point $\Omega_0 = (\delta/\xi, 0, 0, 0, 0)$ signifying that the DENV is cleared.

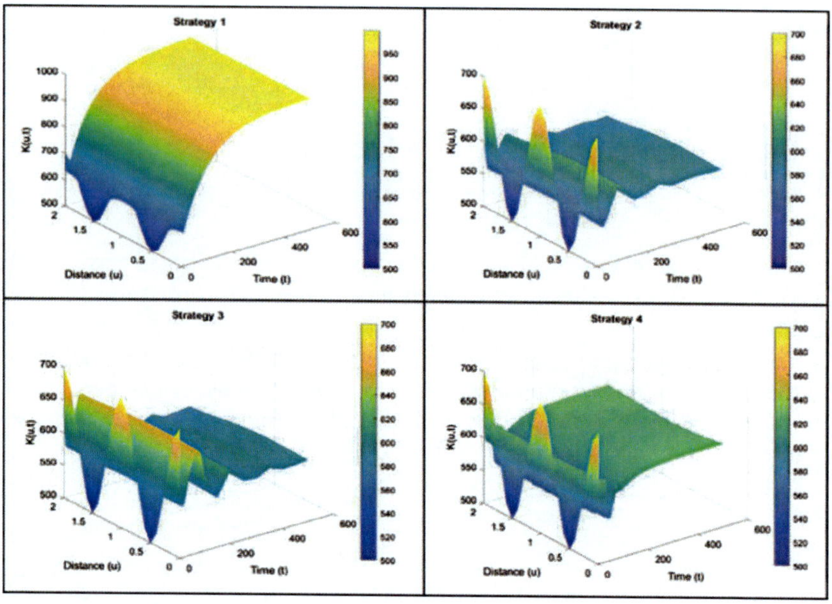

Fig. 1 Numerical simulations for target cells $K(u, t)$

Strategy II, concerns with the stability of the persistent DENV equilibrium point $\Omega_1 = (\frac{K_0}{\mathcal{R}_0}, \frac{\eta\xi(\mathcal{R}_0-1)}{\tau\mu}, \frac{\xi(\mathcal{R}_0-1)}{\mu}, 0, 0)$ signifying the state of persistent DENV infection without any effective immune body responses.

Strategy III, concerns with the stability of the equilibrium point $\Omega_2 = (\frac{\lambda_1\delta}{\xi\lambda_1+\mu\alpha_1}, \frac{\mu\delta\alpha_1}{\rho(\xi\lambda_1+\mu\alpha_1)}, \frac{\alpha_1}{\lambda_1}, \frac{\eta(\mathcal{R}_1-1)}{\omega_1}, 0)$ signifying the persistence of DENV infection with effective heterologous antibody immune responses.

Strategy IV, concerns with the stability of the equilibrium point $\Omega_3 = (\frac{\lambda_2\delta}{\xi\lambda_2+\mu\alpha_2}, \frac{\mu\delta\alpha_2}{\rho(\xi\lambda_2+\mu\alpha_2)}, \frac{\alpha_2}{\lambda_2}, 0, \frac{\eta(\mathcal{R}_2-1)}{\omega_2})$ signifying the persistence of DENV infection with homologous antibody immune responses.

The catholic stability and the existence of the equilibrium points signifies that the prescribed initial conditions do not govern the solutions in long run.

10 Conclusions

The considered DENV infection model is successfully simulated by the proposed collocation method of cubic B-splines. It is also observed that the effect of initial conditions does not run long. The results achieved are quite convincing and in sufficient agreement with those available in Ref. [16]. The method being easy to implement, reliable, and economical is appropriate for like reaction–diffusion mathematical models. The approach can be proved beneficiary for many biologists. Due to the

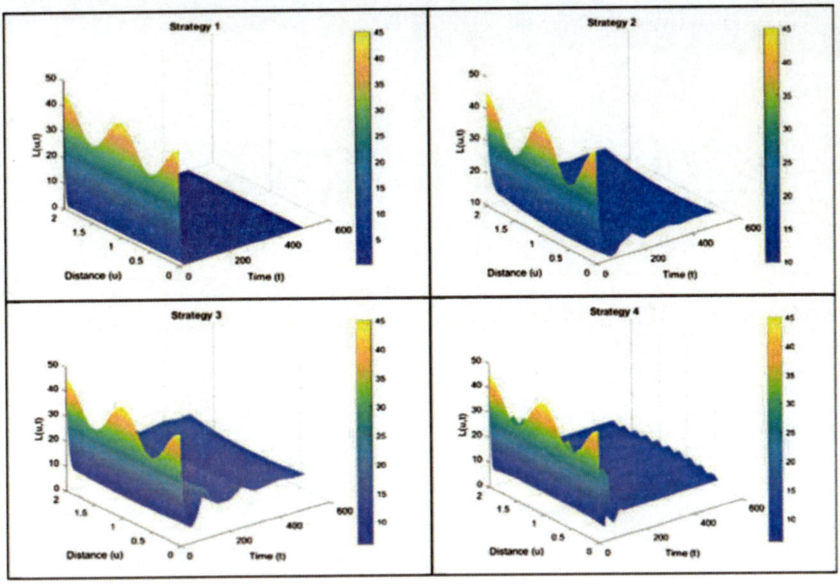

Fig. 2 Numerical simulations for DENV infected cells $L(u, t)$

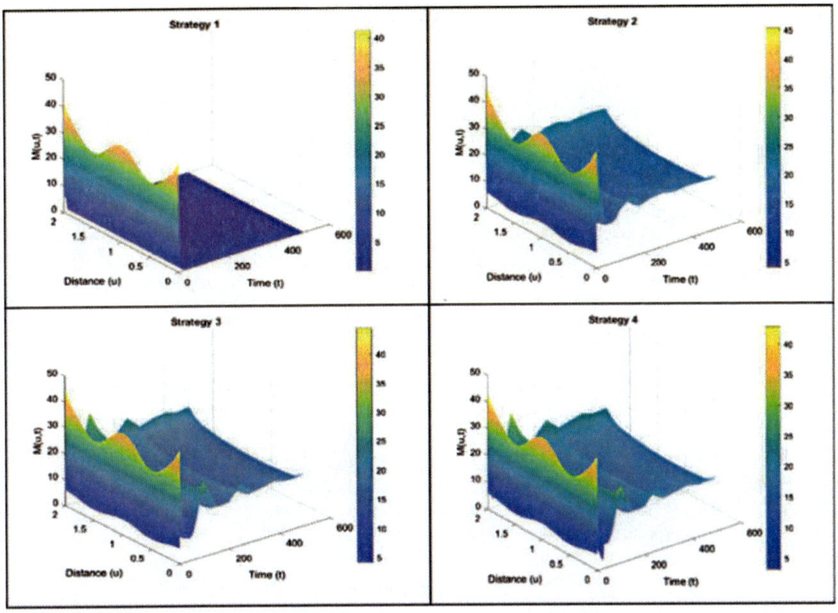

Fig. 3 Numerical simulations for DENV particles $M(u, t)$

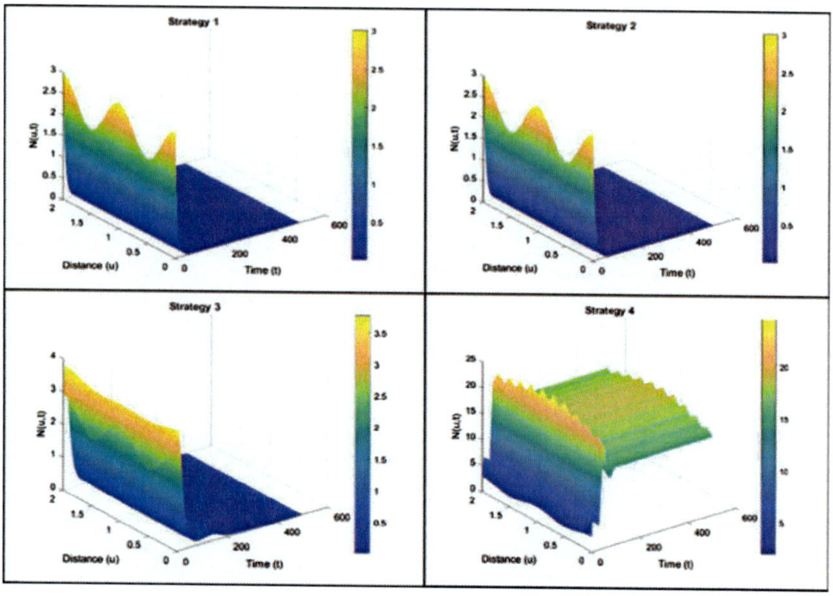

Fig. 4 Numerical simulations for heterologous antibodies $N(u, t)$

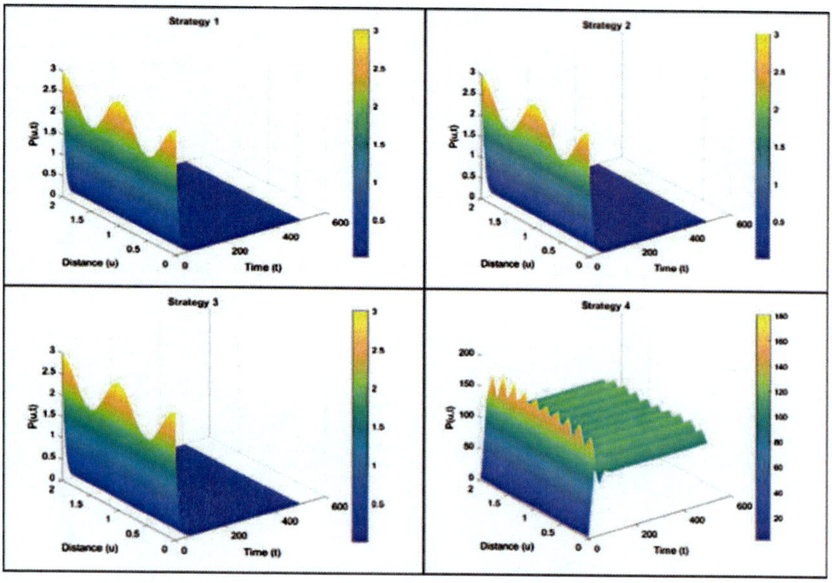

Fig. 5 Numerical simulations for homologous antibodies $P(u, t)$

complexity and the bigger dimensions of the model the proposed technique has prevalence over the traditional numerical simulation techniques. The proposed scheme is thus recommended as a burning alternative to deal a variety of similar mathematical models occurring in the field of medical sciences.

References

1. Obi, J.O., et al.: Current trends and limitations in dengue antiviral research. Trop. Med. Infect. Dis. **6**(4), 180 (2021)
2. Halstead, S.B.: Dengue. Lancet **370**(9599), 1644–1652 (2007)
3. Halstead, S.B.: The XXth century dengue pandemic: need for surveillance and research. World Health Stat. Q. **45**(2–3), 292–298 (1992)
4. Navarro-Sanchez, F., et al.: Innate immune responses to dengue virus. Arch. Med. Res. **36**(5), 425–435 (2005)
5. Kliks, S.C., et al.: Antibody-dependent enhancement of dengue virus growth in human monocytes as a risk factor for dengue haemorrhagic fever. Am. J. Trop. Med. Hyg. **40**(4), 444–451 (1989)
6. Willey, J.M., et al.: Microbiology. 7th edn. Mc-Graw-Hill, New York, NY, USA (2008)
7. Gibbons, R.V., Vaughn, D.W.: Dengue: an escalating problem. BMJ **324**(7353), 1563–1566 (2002)
8. Mishra, A., Gakkhar, S.: A micro-epidemic model for primary dengue infection. Commun. Nonlinear Sci. Numer. Simul. **47**, 426–437 (2017)
9. Sasmal, S.K., et al.: Mathematical modelling on t-cell mediated adaptive immunity in primary dengue infections. J. Theor. Biol. **429**, 229–240 (2017)
10. Murphy, B.R., Whitehead, S.S.: Immune response to dengue virus and prospects for a vaccine. Annu. Rev. Immunol. **29**(1), 587–619 (2011)
11. Perera, S., Perera, S.S.N.: Mathematical modeling and analysis of innate and humoral immune responses to dengue infections. Int, J. Biomath. **12**(7) (2019). Article ID 1950077
12. Ben-Shachar, R., Koelle, K.: Minimal within-host dengue models highlight the specific roles of the immune response in primary and secondary infections. J. R. Soc. Interface **12** (2015). Article ID 20140886
13. Comez, M.C., Yang, H.M.: A simple mathematical model to describe antibody dependent enhancement in heterologous secondary infection in dengue. Math. Med. Biol. **36**, 411–438 (2019)
14. Nikin-Beers, R., Ciupe, S.M.: Modelling original antigenic sin in dengue viral infection. Math. Med. Biol. **35**(2), 257–272 (2018)
15. Borisov, M., et al.: Modelling the host immune response to mature and immature dengue viruses. Bull. Math. Biol. **81**(12), 4951–4976 (2019)
16. Elaiw, A.M., Alofi, A.S.: Global dynamics of secondary infection with diffusion. J. Math. (2021). https://doi.org/10.1155/2021/5585175. Article ID 5585175
17. Unser, M.: Splines: a perfect fit for medical imaging. In: Proceedings of SPIE, The International Society for Optical Engineering, vol. 4684 (2002)
18. Mittal, R.C., Jain, R.K.: Numerical solutions of nonlinear Burger's equation with modified cubic B-splines collocation method. Appl. Math. Comput. **218**(15), 7839–7855 (2012)
19. Mittal, R.C., Jain, R.K.: Redefined cubic B-splines collocation method for solving convection-diffusion equations. Appl. Math. Model. **36**, 5555–5573 (2012)
20. Mittal, R.C., Jain, R.K.: Numerical solutions of non-linear Burger's equation with modified cubic B-splines collocation method. Appl. Math. Comput. (Elsevier) **218**(15), 7839–7855 (2012)
21. Cont, S.D., Boor, C.: Elementary Numerical Analysis: An Algorithmic Approach. McGraw-Hill Book Company

22. Martin, A., Boyd, I.: Variant of the Thomas Algorithm for opposite-bordered tri-diagonal systems of equations. Int. J. Num. Meth. Biomed. Eng. **26**(6), 752–759 (2010)
23. Hale, J.K., Lunel, S.M.V.: Introduction to Functional Differential Equations. Springer, New York (1993)
24. Henry, D.: Geometric Theory of Semi-linear Parabolic Equations. Springer, New York (1993)
25. Wang, L., Li, M.: Diffusion driven instability in reaction-diffusion systems. J. Math. Anal. Appl. **254**(1), 138–153 (2001)

Study of Heat and Mass Transfer in a Composite Nanofluid Layer

Awanish Kumar, B. S. Bhadauria, and Anurag Srivastava

Abstract A non-linear analysis is done to analyze the heat and mass transport in a composite nanofluid layer confined between two parallel horizontal plates, heated from below. The Nusselt number for temperature and nanoparticle concentrations is obtained as a function of time. It is observed that the suspension of two different nanoparticles in a base fluid significantly affects the heat and mass transport. We observe that the modified diffusivity ratios and the Lewis numbers for the first and second types of nanofluids only affect the mass transportation of the first and second types of nanofluids, respectively.

Keywords Non-linear theory · Composite nanofluids · Free–free boundaries

Nomenclature

Latin Symbols

D_{B_1}, D_{B_2}	Brownian Diffusion coefficients.
D_{T_1}, D_{T_2}	Thermophoretic diffusion coefficients.
Pr	Prandtl number.
L	Dimensional layer depth.
Le_1, Le_2	Lewis numbers.
N_{A1}, N_{A2}	Modified diffusivity ratios.
N_{B1}, N_{B2}	Modified particle-density increments.
p	Pressure.
g	Gravitational acceleration.

A. Kumar (✉) · B. S. Bhadauria · A. Srivastava
Department of Mathematics, Babasaheb Bhimrao Ambedkar University, Lucknow 226025, India
e-mail: awanish.425@gmail.com

R. K. Sharma et al. (eds.), *Frontiers in Industrial and Applied Mathematics*,
Springer Proceedings in Mathematics & Statistics 410,
https://doi.org/10.1007/978-981-19-7272-0_17

t	Time.
T	Temperature.
$\mathbf{V} = (u, v, w)$	Nanofluid velocity.

Greek Symbols

$\alpha_f = \kappa/\rho c$	Thermal diffusivity of the nanofluid.
κ	Thermal conductivity of the nanofluid.
β_T	Thermal volumetric coefficient.
μ	Dynamic viscosity.
ρ_{p_1}, ρ_{p_2}	Mass densities of nanoparticles.
ϕ_1, ϕ_2	Nanoparticle volume fractions.

1 Introduction

In order to enhance the poor thermal conductivity of liquids, Maxwell, suggested to add solid particles of high thermal conductivity into the liquids, more than a century ago. His idea was implemented with millimeter- or micrometer-sized particles but it was not very fruitful because of such extra-sized particles. The major issues with such particles were settling down under gravity, clogging, and abrasion. So, there was a search for particles smaller than micro-sized particles and this search ultimately ended with the invention of nanofluids (by Choi [1]) which are the fluids comprising a little amount of uniformly dispersed and suspended nanometer-sized particles in a base fluid. Around 15–40% increment in the thermal conductivity (Eastman et al. [2], Das et al. [3]) of the fluid is observed on adding a small amount of nanoparticles into the base fluid. Moreover, the size of nano-particles becomes quite closer to fluid molecules' size and this prevents nanoparticles to settle down under gravity.

Because of these important properties, nanofluids are widely used in various industries, especially in those processes where cooling is essentially required. Buongiorno [4] was the first to study convective transport in nanofluids in 2006. He noticed that other than base fluid velocity, Brownian diffusion and thermophoresis are mainly responsible for nanoparticles' absolute velocity in the absence of turbulent motion. Tzou [5, 6] used the Buongiorno model to study the onset of convection in a horizontal nanofluid layer heated from below. Nield and Kuznetsov [7–11] further analyzed the similar problem with porous media. After them, many researchers are still working in this field. Bhadauria et al. [12] described the non-linear study for bi-dimensional convection in a nanofluid-saturated porous medium.

Apart from the direct study of the onset of convection, heat, and mass transfer, various researchers showed their interest in the study of convective flows under the effect of various external modulations like thermal modulation, gravity modulation, magnetic field modulation, etc. These modulations have various practical applications in different industries. Venezian [13] was the first to introduce the effect of modulating the boundary temperatures. Later on, Umavathi [14] studied the thermal modulation in the case of nanofluids. Gresho and Sani [15] were the first to study the consequences of modulating gravitational field on Rayleigh–Bénard Convection. Bhadauria et al. [16] did the non-linear study of thermal instability under temperature/gravity modulation. Bhadauria et al. [17] studied the effect of gravity modulation and internal heating over convection in a nanofluid-saturated porous medium. Thomson [18] and Chandrasekhar [19] were the first to discuss the idea about magneto-convection. This has now become a huge area of research. Kiran et al. [20] recently published an article about magneto-convection under magnetic field modulation. Yadav [21] presented a numerical solution of the onset of buoyancy-driven nanofluid convective motion in an anisotropic porous medium layer with internal heating and variable gravity. Sakshath et al. [22] investigated the effect of horizontal pressure gradient on Rayleigh–Bénard convection of a Newtonian nanoliquid in a high porosity medium using a local thermal non-equilibrium model.

After various kind of modulations, a new type of nanofluid, known as composite nanofluid, has now become an advanced area of interest among the researchers in the recent years. A composite nanofluid is prepared by suspending two or more types of nanoparticles in a base fluid in order to get a stable and homogeneous mixture. The synthesis of such composite materials can be done either by chemical or physical processes (Hanemann and Szabo [23], Zhang et al. [24]). The characteristics of the composite nanofluids lie in between the properties of their constituents. The thermophysical properties of composite nanofluids can be altered to converge to the required heat transfer demands. An extensive review on composite nanofluids and their properties is given by Suleiman et al. [25]. Linear and nonlinear analysis in Hele-Shaw cell in the presence of through-flow and gravity modulation have been done by Bhadauria et al. [26].

The very first study of thermal instability for composite nanofluids is presented by Kumar and Awasthi [27] recently. They concluded that the maximum stability is achieved only when both kinds of nanoparticles are in the same ratio. To the best of our knowledge, no non-linear study on this topic is present in literature till date. This idea motivated us to present this study of heat and mass transfer in a composite nanofluid layer.

2 Mathematical Formulation

An infinitely extended horizontal layer, of composite nanoliquid in which two different types of nanoparticles are suspended homogeneously, restricted between $Z = 0$ and $Z = L$ has been considered. The upper plate at $Z = L$ is assumed to be at temperature T_0, while the lower plate is at slightly higher temperature $T_0 + \Delta T$ as shown in Fig. 1. The cartesian coordinate system has been used. Both nanoparticles and the base fluid are assumed to be in local thermal equilibrium. Boundaries are considered to be Free–Free and perfectly insulating. The linearization of equations is done using

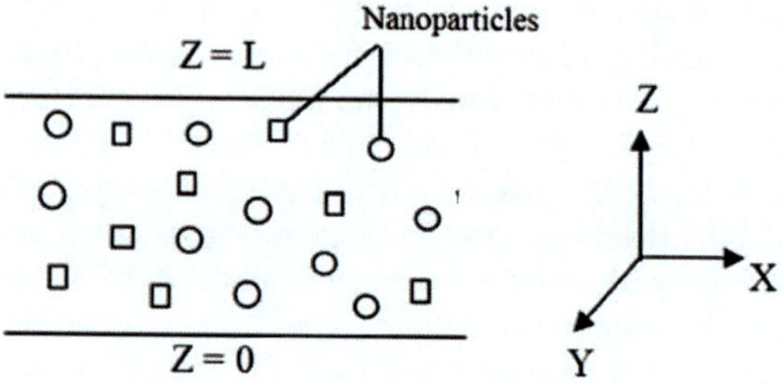

Fig. 1 Formal diagram

the Oberbeck–Boussinesq approximation. The governing equations of the system are as follows (Kumar and Awasthi [27]):

$$\nabla \cdot \mathbf{V} = 0 \tag{1}$$

$$\rho\left[\frac{\partial}{\partial t} + (\mathbf{V}\cdot\nabla)\right]\mathbf{V} = -\nabla p + \mu\nabla^2\mathbf{V} + [\phi_1\rho_{p_1} + \phi_2\rho_{p_2} + \rho(1-\phi_1-\phi_2)\{1-\beta_T(T-T_0)\}]\mathbf{g} \tag{2}$$

$$\rho c\left[\frac{\partial}{\partial t} + (\mathbf{V}\cdot\nabla)\right]T = \kappa\nabla^2 T + (\rho c)_{p_1}[D_{B_1}\nabla\phi_1\nabla T + \frac{D_{T_1}}{T_0}\nabla T\nabla T] + (\rho c)_{p_2}[D_{B_2}\nabla\phi_2\nabla T + \frac{D_{T_2}}{T_0}\nabla T\nabla T] \tag{3}$$

$$\left[\frac{\partial}{\partial t} + (\mathbf{V}\cdot\nabla)\right]\phi_1 = D_{B_1}\nabla^2\phi_1 + \left(\frac{D_{T_1}}{T_0}\right)\nabla^2 T \tag{4}$$

$$\left[\frac{\partial}{\partial t} + (\mathbf{V}\cdot\nabla)\right]\phi_2 = D_{B_2}\nabla^2\phi_2 + \left(\frac{D_{T_2}}{T_0}\right)\nabla^2 T \tag{5}$$

where $\nabla^2 \equiv \frac{\partial^2}{\partial x^2} + \frac{\partial^2}{\partial y^2} + \frac{\partial^2}{\partial z^2}$

At the boundaries, the volume fractions of nanoparticles are assumed to be constant. The boundary conditions under consideration are as follows:

$$\left.\begin{aligned} w = 0,\ \frac{\partial w}{\partial z} + \lambda_1 L\frac{\partial^2 w}{\partial z^2} = 0,\ \phi_1 = \phi_{10},\ \phi_2 = \phi_{20}\ \text{ at } z = 0 \\ w = 0,\ \frac{\partial w}{\partial z} - \lambda_2 L\frac{\partial^2 w}{\partial z^2} = 0,\ \phi_1 = \phi_{11},\ \phi_2 = \phi_{21}\ \text{ at } z = L \end{aligned}\right\} \tag{6}$$

where λ_1 and λ_2 take the value "0" and "∞" for rigid–rigid and free–free boundaries, respectively. Also $\phi_{11} > \phi_{10}$ and $\phi_{21} > \phi_{20}$. In order to non-dimensionalize the equations, we use the following substitutions:

$$\left.\begin{aligned} &(x, y, z) = L(x', y', z'),\quad (u, v, w) = (u', v', w')\frac{\alpha_f}{L}, \\ &t = \frac{L^2}{\alpha_f}t',\quad p = \frac{\mu\alpha_f}{L^2}p', \\ &T' = \frac{T - T_0}{\Delta T},\quad \phi'_{i(=1or2)} = \frac{\phi_i - \phi_{i0}}{\phi_{i1} - \phi_{i0}}. \end{aligned}\right\} \tag{7}$$

Making use of (7) into the Eqs. (1)–(6) and leaving the primes for simplicity, we obtain the following non-dimensional equations:

$$\nabla \cdot \mathbf{V} = 0 \tag{8}$$

$$\frac{1}{Pr}\left[\frac{\partial}{\partial t} + (\mathbf{V}\cdot\nabla)\right]\mathbf{V} = -\nabla p + \nabla^2\mathbf{V} + \mathbf{e_z}\left[RaT - Rn_1\phi_1 - Rn_2\phi_2 - Rm\right] \tag{9}$$

$$\left[\frac{\partial}{\partial t} + (\mathbf{V}\cdot\nabla)\right]T = \nabla^2 T + \left(\frac{N_{B1}}{Le_1}\right)\nabla\phi_1\cdot\nabla T + \left(\frac{N_{A1}N_{B1}}{Le_1}\right)\nabla T\cdot\nabla T + \left(\frac{N_{B2}}{Le_2}\right)\nabla\phi_2\cdot\nabla T + \left(\frac{N_{A2}N_{B2}}{Le_2}\right)\nabla T\cdot\nabla T \tag{10}$$

$$\left[\frac{\partial}{\partial t} + (\mathbf{V}\cdot\nabla)\right]\phi_1 = \left(\frac{1}{Le_1}\right)\nabla^2\phi_1 + \left(\frac{N_{A1}}{Le_1}\right)\nabla^2 T \tag{11}$$

$$\left[\frac{\partial}{\partial t} + (\mathbf{V}\cdot\nabla)\right]\phi_2 = \left(\frac{1}{Le_2}\right)\nabla^2\phi_2 + \left(\frac{N_{A2}}{Le_2}\right)\nabla^2 T \tag{12}$$

The dimensional-less boundary conditions are

$$\left.\begin{aligned} T = 1, w = \phi_1 = \phi_2 = 0, \frac{\partial w}{\partial z} + \lambda_1\frac{\partial^2 w}{\partial z^2} = 0 \text{ at } z = 0, \\ T = w = 0, \phi_1 = \phi_2 = 1, \frac{\partial w}{\partial z} - \lambda_2\frac{\partial^2 w}{\partial z^2} = 0 \text{ at } z = 1, \end{aligned}\right\} \tag{13}$$

where $Ra = \frac{\rho g\beta_T L^3\Delta T}{\mu\alpha_f}$ is the thermal Rayleigh number, $Rn_1 = \frac{(\rho_{p_1} - \rho)(\phi_{11} - \phi_{10})gd^3}{\mu\alpha_f}$ and $Rn_2 = \frac{(\rho_{p_2} - \rho)(\phi_{21} - \phi_{20})gd^3}{\mu\alpha_f}$ are the nanoparticle concentration Rayleigh numbers, $Rm = \frac{\{\rho_{p_1}\phi_{10} + \rho_{p_2}\phi_{20} + \rho(1 - \phi_{10} - \phi_{20})\}gd^3}{\mu\alpha_f}$ is the basic density Rayleigh number, $Pr = \frac{\mu}{\rho\alpha_f}$ is Prandtl number, $Le_1 = \frac{\alpha_f}{D_{B_1}}$ and $Le_2 = \frac{\alpha_f}{D_{B_2}}$ are the Lewis numbers, $N_{A1} = \frac{D_{T_1}\Delta T}{D_{B_1}T_0(\phi_{11} - \phi_{10})}$ and $N_{A2} = \frac{D_{T_2}\Delta T}{D_{B_2}T_0(\phi_{21} - \phi_{20})}$ are the modified diffusivity ratios, and $N_{B1} = (\rho c)_{p_1}\frac{\phi_{11} - \phi_{10}}{\rho c}$ and $N_{B2} = (\rho c)_{p_2}\frac{\phi_{21} - \phi_{20}}{\rho c}$ are the modified particle-density increments.

3 Conduction State

The temperature, pressure, and nanoparticle volume fractions are taken to be the functions of "z" only. The time-independent quiescent solution of Eqs. (8)–(12) is obtained under the following assumptions:

$$\mathbf{V} = \mathbf{0},\ T = T_b(z),\ p = p_b(z),\ \phi_1 = \phi_{1b}(z),\ \phi_2 = \phi_{2b}(z). \tag{14}$$

The desired conduction state is evaluated (Kumar and Awasthi [27]) as:

$$T_b(z) = 1 - z,\ \phi_{1b}(z) = z,\ \phi_{2b}(z) = z. \tag{15}$$

4 Perturbed State

We impose small perturbations on the conduction state:

$$\mathbf{V} = \tilde{\mathbf{V}},\ \ p = p_b + \tilde{p},\ \ T = T_b + \tilde{T},\ \ \phi_1 = \phi_{1b} + \tilde{\phi_1},\ \ \phi_2 = \phi_{2b} + \tilde{\phi_2}. \tag{16}$$

Using Eq. (16) in Eqs. (8)–(12) and assuming all the physical quantities to be free from "y", we get the following perturbed equations:

$$\nabla \cdot \tilde{\mathbf{V}} = 0 \tag{17}$$

$$\frac{1}{Pr}\left[\frac{\partial}{\partial t} + \left(\tilde{u}\frac{\partial}{\partial x} + \tilde{w}\frac{\partial}{\partial z}\right)\right]\tilde{\mathbf{V}} = -\nabla\tilde{p} + \nabla^2\tilde{\mathbf{V}} + \mathbf{e_z}\left[Ra\tilde{T} - Rn_1\tilde{\phi_1} - Rn_2\tilde{\phi_2}\right] \tag{18}$$

$$\frac{\partial \tilde{T}}{\partial t} - \tilde{w} + \left(\tilde{u}\frac{\partial}{\partial x} + \tilde{w}\frac{\partial}{\partial z}\right)\tilde{T} = \nabla^2\tilde{T} + \frac{N_{B1}}{Le_1}\left[\frac{\partial \tilde{T}}{\partial z} - \frac{\partial \tilde{\phi_1}}{\partial z}\right] + \frac{N_{B2}}{Le_2}\left[\frac{\partial \tilde{T}}{\partial z} - \frac{\partial \tilde{\phi_2}}{\partial z}\right]$$
$$-\frac{2N_{A1}N_{B1}}{Le_1}\frac{\partial \tilde{T}}{\partial z} - \frac{2N_{A2}N_{B2}}{Le_2}\frac{\partial \tilde{T}}{\partial z} \tag{19}$$

$$\frac{\partial \tilde{\phi_1}}{\partial t} + \tilde{w} + \left(\tilde{u}\frac{\partial}{\partial x} + \tilde{w}\frac{\partial}{\partial z}\right)\tilde{\phi_1} = \frac{1}{Le_1}\nabla^2\tilde{\phi_1} + \frac{N_{A1}}{Le_1}\nabla^2\tilde{T} \tag{20}$$

$$\frac{\partial \tilde{\phi_2}}{\partial t} + \tilde{w} + \left(\tilde{u}\frac{\partial}{\partial x} + \tilde{w}\frac{\partial}{\partial z}\right)\tilde{\phi_2} = \frac{1}{Le_2}\nabla^2\tilde{\phi_2} + \frac{N_{A2}}{Le_2}\nabla^2\tilde{T} \tag{21}$$

The corresponding perturbed boundary conditions are

$$\left.\begin{aligned} \tilde{T} = 0, \tilde{w} = \tilde{\phi_1} = \tilde{\phi_2} = 0, \frac{\partial \tilde{w}}{\partial z} + \lambda_1\frac{\partial^2 \tilde{w}}{\partial z^2} = 0 \text{ at } z = 0,\\ \tilde{T} = 0, \tilde{w} = \tilde{\phi_1} = \tilde{\phi_2} = 0, \frac{\partial \tilde{w}}{\partial z} - \lambda_2\frac{\partial^2 \tilde{w}}{\partial z^2} = 0 \text{ at } z = 1. \end{aligned}\right\} \tag{22}$$

where $\tilde{\mathbf{V}} = (\tilde{u}, \tilde{v}, \tilde{w})$.

Now eliminating the pressure term in Eq. (18), introducing the stream function ψ in Eqs. (18)–(21), and removing the tildes, we get the following transformed equations:

$$\frac{1}{Pr}\left[\frac{\partial}{\partial t}(\nabla^2\psi)\right] = \nabla^4\psi - Ra\frac{\partial T}{\partial x} + Rn_1\frac{\partial \phi_1}{\partial x} + Rn_2\frac{\partial \phi_2}{\partial x} + \frac{1}{Pr}\left[\frac{\partial(\psi, \nabla^2\psi)}{\partial(x, z)}\right] \tag{23}$$

$$\frac{\partial T}{\partial t} + \frac{\partial \psi}{\partial x} = \nabla^2 T + \frac{N_{B1}}{Le_1}\left[\frac{\partial T}{\partial z} - \frac{\partial \phi_1}{\partial z}\right] - \frac{2N_{A1}N_{B1}}{Le_1}\frac{\partial T}{\partial z} + \frac{N_{B2}}{Le_2}\left[\frac{\partial T}{\partial z} - \frac{\partial \phi_2}{\partial z}\right] - \frac{2N_{A2}N_{B2}}{Le_2}\frac{\partial T}{\partial z} + \frac{\partial(\psi, T)}{\partial(x, z)} \tag{24}$$

$$\frac{\partial \phi_1}{\partial t} - \frac{\partial \psi}{\partial x} = \frac{1}{Le_1}\nabla^2\phi_1 + \frac{N_{A1}}{Le_1}\nabla^2 T + \frac{\partial(\psi, \phi_1)}{\partial(x, z)} \tag{25}$$

$$\frac{\partial \phi_2}{\partial t} - \frac{\partial \psi}{\partial x} = \frac{1}{Le_2}\nabla^2\phi_2 + \frac{N_{A2}}{Le_2}\nabla^2 T + \frac{\partial(\psi, \phi_2)}{\partial(x, z)} \tag{26}$$

where $u = \frac{\partial \psi}{\partial z}$ and $\mathrm{w} = -\frac{\partial \psi}{\partial \mathrm{x}}$.

5 Non-linear Stability Analysis

A non-linear stability analysis is done using the below-mentioned truncated Fourier expressions (Bhadauria et al. [17]):

$$\psi = A_{11}(t)\sin(\mathrm{kx})\sin(\pi \mathrm{z}) \tag{27}$$

$$T = B_{11}(t)\cos(\mathrm{kx})\sin(\pi \mathrm{z}) + \mathrm{B}_{02}(\mathrm{t})\sin(2\pi \mathrm{z}) \tag{28}$$

$$\phi_1 = C_{11}(t)\cos(\mathrm{kx})\sin(\pi \mathrm{z}) + \mathrm{C}_{02}(\mathrm{t})\sin(2\pi \mathrm{z}) \tag{29}$$

$$\phi_2 = D_{11}(t)\cos(\mathrm{kx})\sin(\pi \mathrm{z}) + \mathrm{D}_{02}(\mathrm{t})\sin(2\pi \mathrm{z}) \tag{30}$$

All these expressions are taken in such a way to satisfy the free–free boundary conditions:

$$\psi = \nabla^2\psi = T = \phi_1 = \phi_2 = 0 \text{ at } \mathrm{z} = 0, 1, \tag{31}$$

where $A_{11}(t)$, $B_{11}(t)$, $B_{02}(t)$, $C_{11}(t)$, $C_{02}(t)$, $D_{11}(t)$ and $\mathrm{D}_{02}(\mathrm{t})$ are unknowns and the functions of "t".

Making use of Eqs. (27)–(30) into the Eqs. (23)–(26) and using the condition of orthogonality with the eigenfunctions, we obtain

$$A'_{11}(t) = Pr[-\delta^2 A_{11}(t) - \frac{\mathrm{k}}{\delta^2}\{RaB_{11}(t) - Rn_1C_{11}(t) - Rn_2D_{11}(t)\}] \quad (32)$$

$$B'_{11}(t) = -\mathrm{k}A_{11}(t) - \mathrm{k}\pi A_{11}(t)B_{02}(t) - \delta^2 B_{11}(t) \quad (33)$$

$$B'_{02}(t) = -4\pi^2 B_{02}(t) + \frac{\mathrm{k}\pi}{2}A_{11}(t)B_{11}(t) \quad (34)$$

$$C'_{11}(t) = \mathrm{k}[A_{11}(t) - \pi A_{11}(t)C_{02}(t)] - \frac{\delta^2}{Le_1}[N_{A1}B_{11}(t) + C_{11}(t)] \quad (35)$$

$$C'_{02}(t) = -\frac{4\pi^2}{Le_1}[N_{A1}B_{02}(t) + C_{02}(t)] + \frac{\mathrm{k}\pi}{2}A_{11}(t)C_{11}(t) \quad (36)$$

$$D'_{11}(t) = \mathrm{k}[A_{11}(t) - \pi A_{11}(t)D_{02}(t)] - \frac{\delta^2}{Le_2}[N_{A2}B_{11}(t) + D_{11}(t)] \quad (37)$$

$$D'_{02}(t) = -\frac{4\pi^2}{Le_2}[N_{A2}B_{02}(t) + D_{02}(t)] + \frac{\mathrm{k}\pi}{2}A_{11}(t)D_{11}(t) \quad (38)$$

where $\delta^2 = (\mathrm{k}^2 + \pi^2)$

The above autonomous simultaneous ODEs (32)–(38) are solved numerically using NDSolve of Mathematical under suitably chosen initial conditions.

6 Heat and Mass Transport

The heat transport Nusselt number, $Nu_T(t)$ is defined as

$$Nu_T(t) = \frac{\text{Heat transport by (conduction+convection)}}{\text{Heat transport by conduction}}$$

$$Nu_T(t) = 1 + \left[\frac{\int_0^{2\pi/\mathrm{k}} \left(\frac{\partial T}{\partial z}\right) dx}{\int_0^{2\pi/\mathrm{k}} \left(\frac{\partial T_b}{\partial z}\right) dx}\right]_{z=0} \quad (39)$$

On putting the values of T and $T_b(z)$ from Eqs. (28) and (15) into the Eq. (39), we have

$$Nu_T(t) = 1 - 2\pi B_{02}(t) \quad (40)$$

The nanoparticle concentration Nusselt number for the first type of nanoparticles, $Nu_{\phi 1}(t)$, can be defined as

$$Nu_{\phi 1}(t) = 1 + \left[\frac{\int_0^{2\pi/\mathrm{k}} (\frac{\partial \phi_1}{\partial z})dx}{\int_0^{2\pi/\mathrm{k}} (\frac{\partial \phi_{1b}}{\partial z})dx}\right]_{z=0} + N_{A1}\left\{1 + \left[\frac{\int_0^{2\pi/\mathrm{k}} (\frac{\partial T}{\partial z})dx}{\int_0^{2\pi/\mathrm{k}} (\frac{\partial T_b}{\partial z})dx}\right]_{z=0}\right\} \quad (41)$$

Making use of Eqs. (28), (29) and (15) into the Eq. (41), we get

$$Nu_{\phi 1}(t) = (1 + 2\pi C_{02}(t)) + N_{A1}(1 - 2\pi B_{02}(t)) \tag{42}$$

Similarly, we can find the nanoparticle concentration Nusselt number for the second type of nanoparticles, $Nu_{\phi 2}(t)$, as follows:

$$Nu_{\phi 2}(t) = (1 + 2\pi D_{02}(t)) + N_{A2}(1 - 2\pi B_{02}(t)) \tag{43}$$

7 Results and Discussion

In non-linear analysis, we study heat and mass transport in the system. By thermal Nusselt number and concentration Nusselt number, we study how heat and mass transport, respectively, happens inside the system. Here thermal Nusselt number and concentration Nusselt number are functions of time. The general parametric values are taken as $Le_1 = 100$, $Le_2 = 100$, $N_{A1} = 2$, $N_{A2} = 2$, $Rn_1 = 5$, $Rn_2 = 5$, $R_a = 5000$, and $\mathrm{k} = 2.22144$. We found a common thing in all observations that the graph of thermal Nusselt number and both concentration Nusselt numbers are horizontal for a short time initially which shows a conduction state. After some time, they start increasing which shows a convection state and also oscillate for some time and go to constant which denotes a steady state. In ordinary nanofluid, Bhadauria et al. [17] examined that modified particle density increments and Lewis number have no significant effect on heat transfer. Here we also found the similar result in composite nanofluid which is shown in Figs. 2, 3, 4, and 5. If the value of Prandtl number (Pr) is increased, we observe that heat transfer starts sooner by convection in comparison to the previous Prandtl number which is shown in Fig. 6. In the case of composite nanofluid, we observe that heat transfer by convection is delayed in comparison to ordinary nanofluid which is equivalent to the result of Kumar and Awasthi [27] and shown in Fig. 7. Kumar and Awasthi [27] compared the onset of convection between ordinary and composite nanofluids under the heavy top condition and found a delay in the onset of convection in composite nanofluids. In the case of the first nanoparticle concentration Nusselt number, the effect of N_{A1} enhances the mass transport and has no effect of N_{A2} on mass transport which is shown in Figs. 8, 9. In the case of the second nanoparticle concentration Nusselt number, N_{A1} has no effect and N_{A2} enhances the mass transport which contradicts the result of first nanoparticle concentration Nusselt number and shown in Figs. 16, 17. The above result is similar to the result for ordinary nanofluid, which is compared to the result of Bhadauria et al. [17].

If we increase the value of Le_1, we observe that the amplitude of oscillations of nanoparticle concentration Nusselt number for the first nanoparticle, i.e., $Nu_{\phi 1}$ slightly increases, while increment in Le_1 has no effect on nanoparticle concentration Nusselt number for the second nanoparticle, i.e., $Nu_{\phi 2}$ (Figs. 10, 18). Similarly, if we increase the value of Le_2, the amplitude of oscillations of $Nu_{\phi 2}$ is slight increased,

Fig. 2 Plot of Nu_T with t for varying N_{A1}

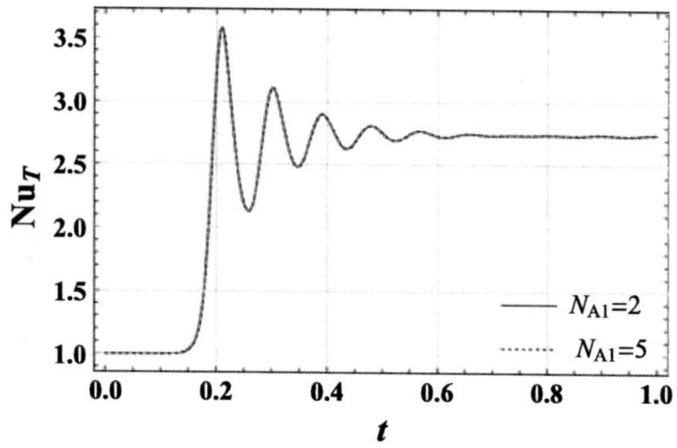

Fig. 3 Plot of Nu_T with t for varying N_{A2}

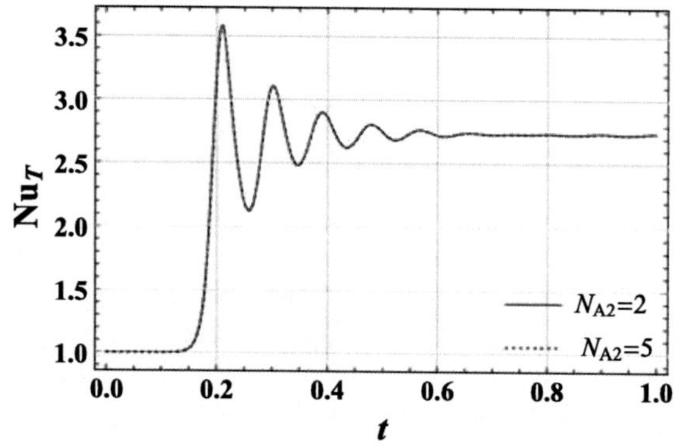

Fig. 4 Plot of Nu_T with t for varying Le_1

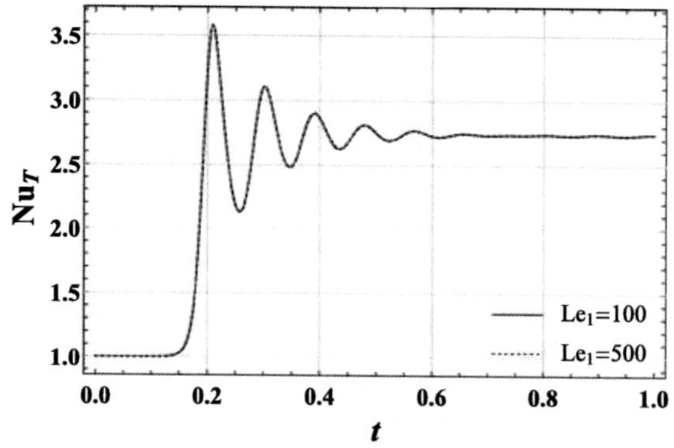

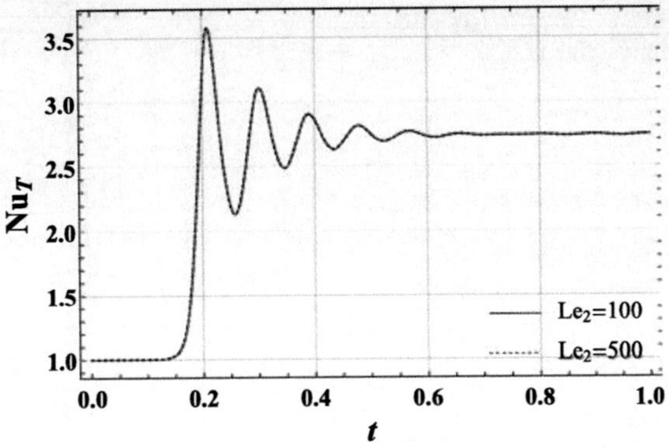

Fig. 5 Plot of Nu_T with t for varying Le_2

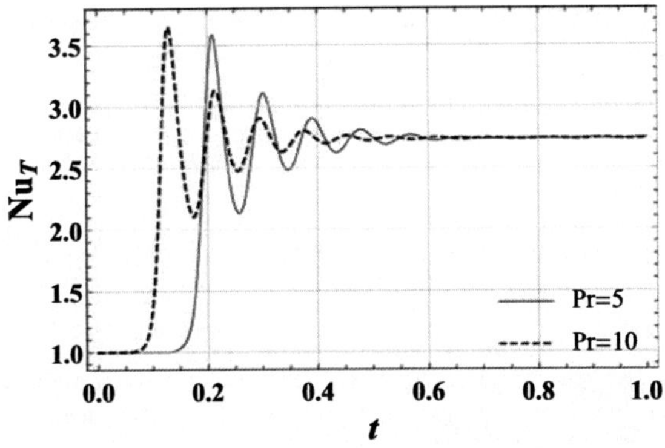

Fig. 6 Plot of Nu_T with t for varying Pr

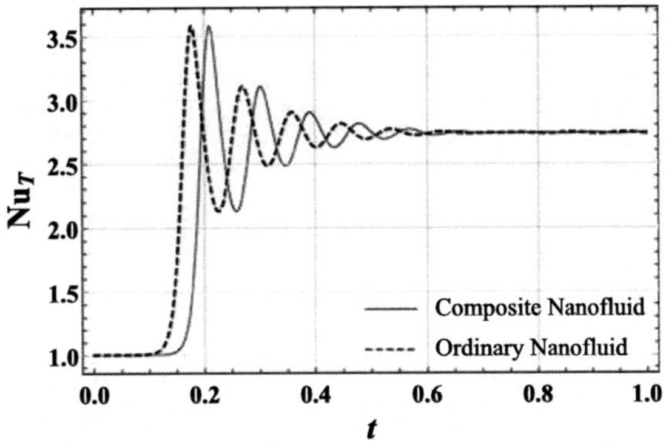

Fig. 7 Comparison of heat transfer in ordinary and composite nanofluid

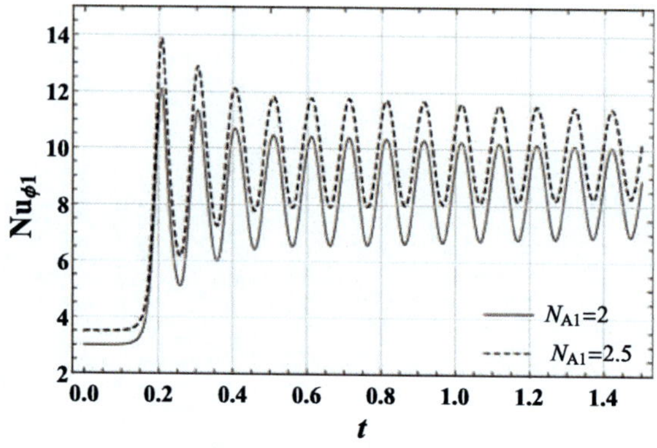

Fig. 8 Plot of $Nu_{\phi 1}$ with t for varying N_{A1}

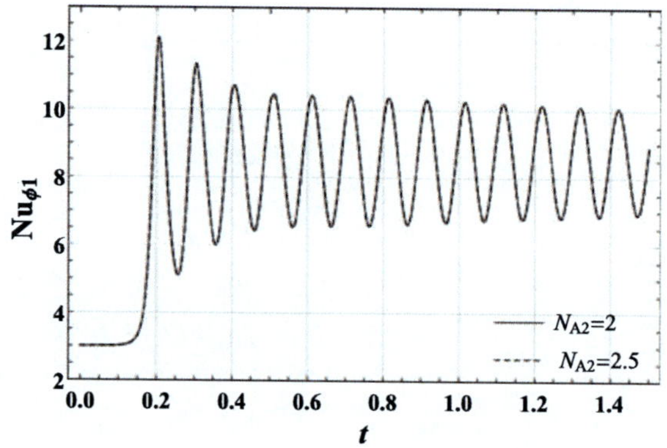

Fig. 9 Plot of $Nu_{\phi 1}$ with t for varying N_{A2}

while it has no effect over $Nu_{\phi 1}$ (Figs. 11, 19). Let us now discuss the effect of Prandtl number on mass transport in both cases. We found the same effect in both cases which is shown in Figs. 12, 20 and the result is same as the result of Bhadauria et al. [12]. If the ratio of Rn_1 and Rn_2 are different in composite nanofluid, then the mass transport by convection takes place sooner in comparison to the same ratio which is shown in Figs. 13, 14, 21 and 22. If nanoparticle concentration is top heavy then we found that there is a delay in the mass transport by convection in comparison to bottom heavy which is shown in Figs. 15, 23.

In Fig. 25a, b, the streamlines and isothermals have been shown, respectively, at conduction state for t = 0, 0.025, and 0.050. In Fig. 25a, we observe that the magnitude of streamlines is very weak for t = 0–0.050; therefore, the movement of fluid in the system is almost negligible, which means that heat transfer is only due to conduction. Figure 25b describes that the temperature of all the horizontal fluid layers is almost constant throughout the system, which indicates the conduction state. Figure 26a shows that as time "t" increases from 0.1 to 0.15, the magnitude of streamlines also increases slightly. It means that a small movement of fluid particles has started in the system and, therefore, the heat transfer is due to both conduction and convection, which indicates a transition from conduction to convection state.

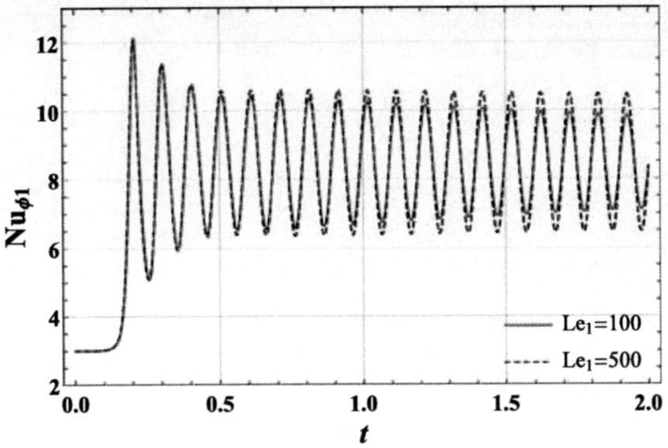

Fig. 10 Plot of $Nu_{\phi 1}$ with t for varying Le_1

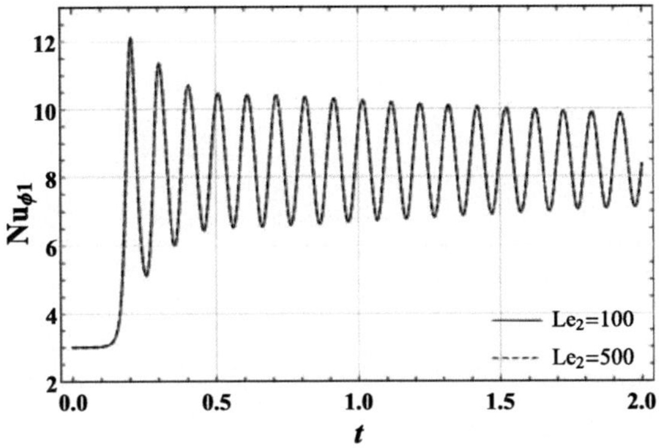

Fig. 11 Plot of $Nu_{\phi 1}$ with t for varying Le_2

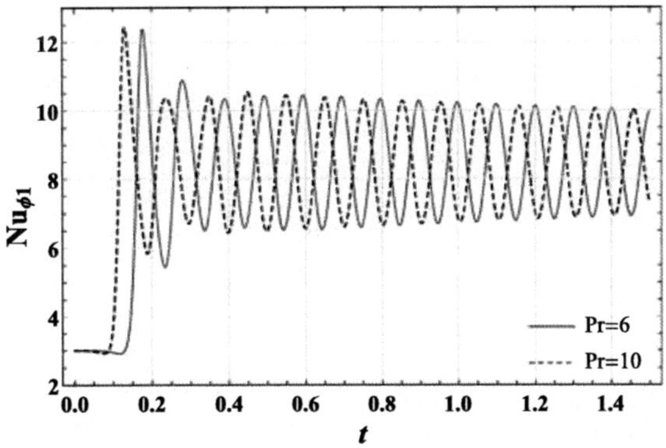

Fig. 12 Plot of $Nu_{\phi 1}$ with t for varying Pr

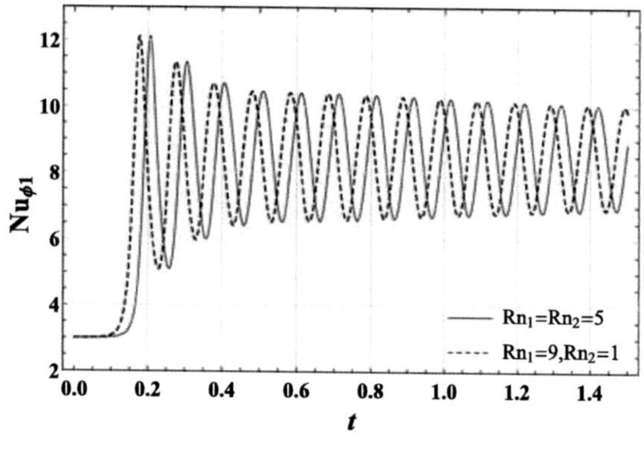

Fig. 13 Comparison of $Nu_{\phi 1}$ for same ratio ($Rn_1 = Rn_2$) and different ratio ($Rn_1 > Rn_2$)

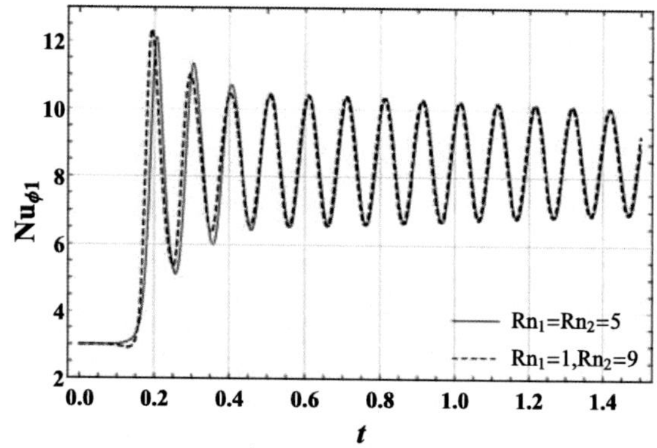

Fig. 14 Comparison of $Nu_{\phi 1}$ for same ratio ($Rn_1 = Rn_2$) and different ratio ($Rn_1 < Rn_2$)

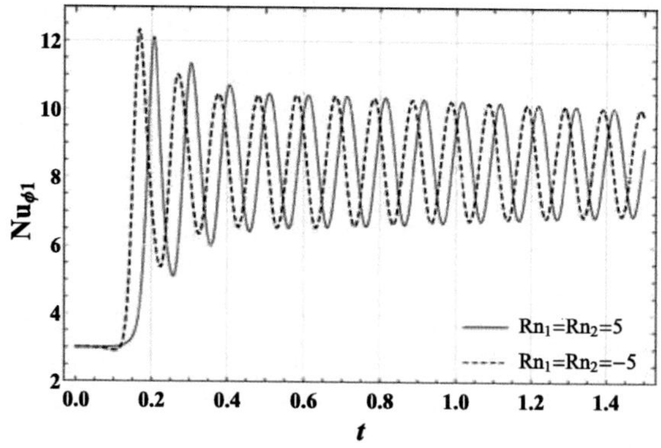

Fig. 15 Comparison of $Nu_{\phi 1}$ for top and bottom heavy configurations

Fig. 16 Plot of $Nu_{\phi 2}$ with t for varying N_{A1}

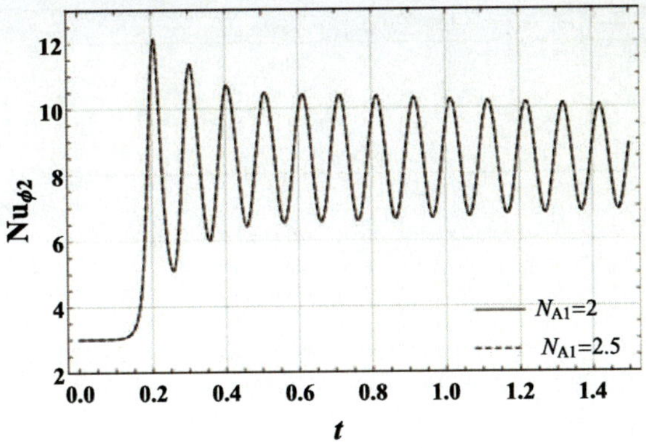

Fig. 17 Plot of $Nu_{\phi 2}$ with t for varying N_{A2}

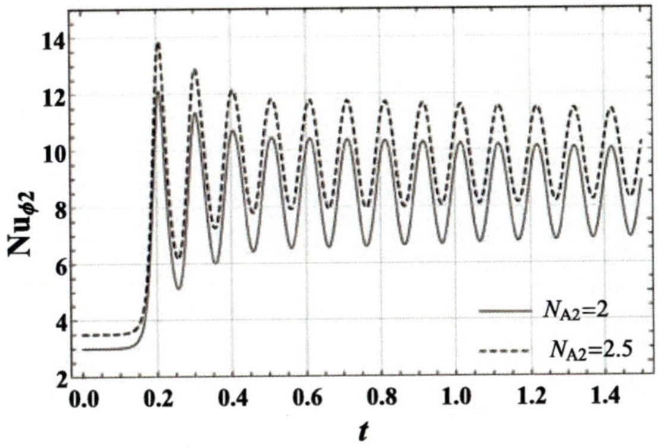

Figure 26b represents that the isothermals have started deforming from their original horizontal position as "t" increases from 0.1 to 0.15, which shows the very beginning stage of the formation of convection cells. Further, the magnitude of streamlines becomes stronger as time increases and in isothermals, fully developed convective cells can be seen with increasing time as presented in Fig. 27a, b, respectively. In Fig. 28a, b, there is no change in the magnitudes of streamlines and in the position of isothermals with increasing time, which shows that the system has achieved the steady state. In Fig. 29a, it can be noticed that the isohalines are parallel and horizontal, which means that the concentration of nanoparticles is constant with horizontal fluid layers and mass transport is almost negligible in the system for $t = 0–0.05$. With the passage of time, mass transport starts in the system as depicted by Fig. 29b. Mass transportation also achieves the steady state for higher values of time as shown by Fig. 24.

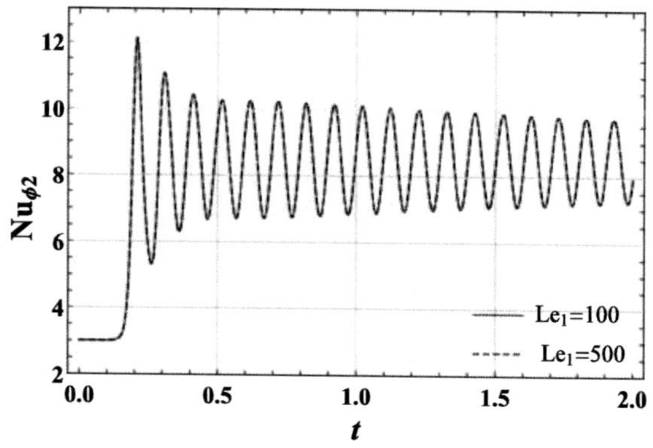

Fig. 18 Plot of $Nu_{\phi 2}$ with t for varying Le_1

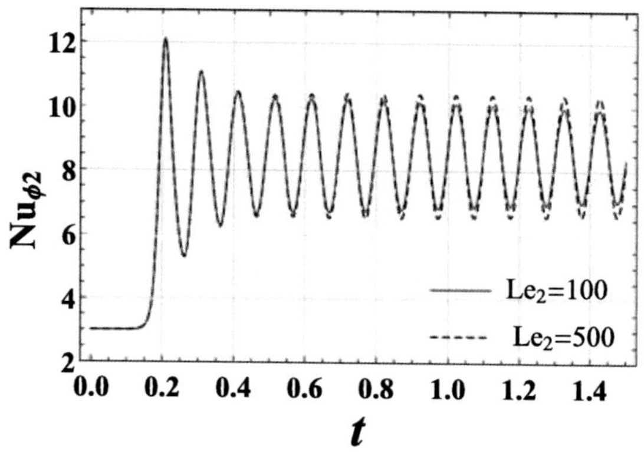

Fig. 19 Plot of $Nu_{\phi 2}$ with t for varying Le_2

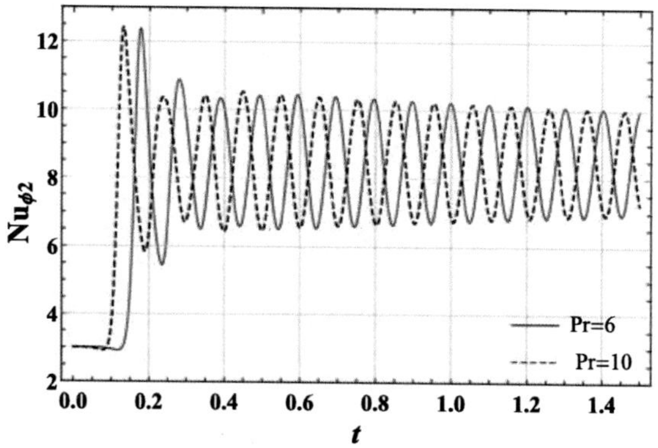

Fig. 20 Plot of $Nu_{\phi 2}$ with t for varying Pr

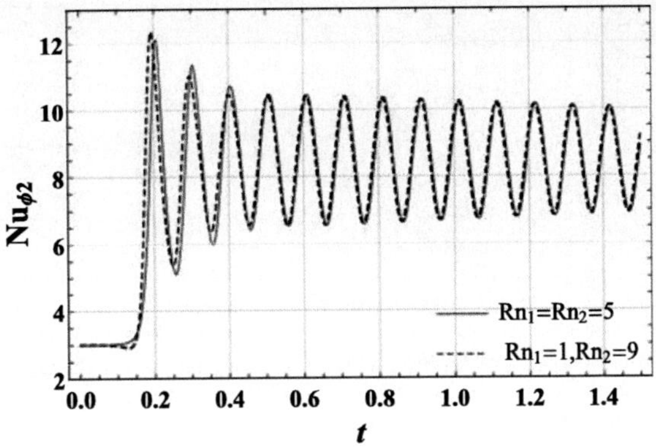

Fig. 21 Comparison of $Nu_{\phi 2}$ for same ratio ($Rn_1 = Rn_2$) and different ratio ($Rn_1 < Rn_2$)

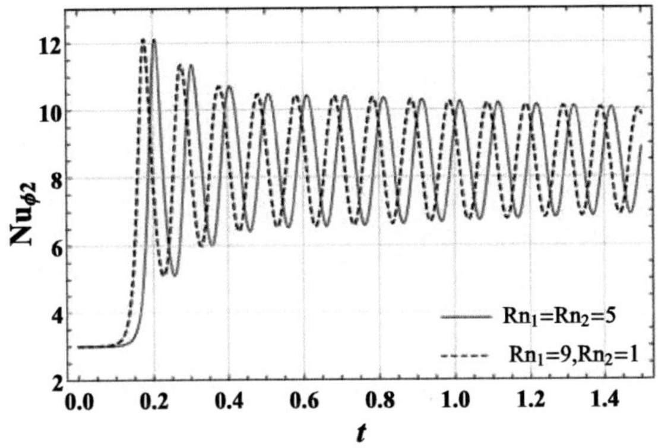

Fig. 22 Comparison of $Nu_{\phi 2}$ for same ratio ($Rn_1 = Rn_2$) and different ratio ($Rn_1 > Rn_2$)

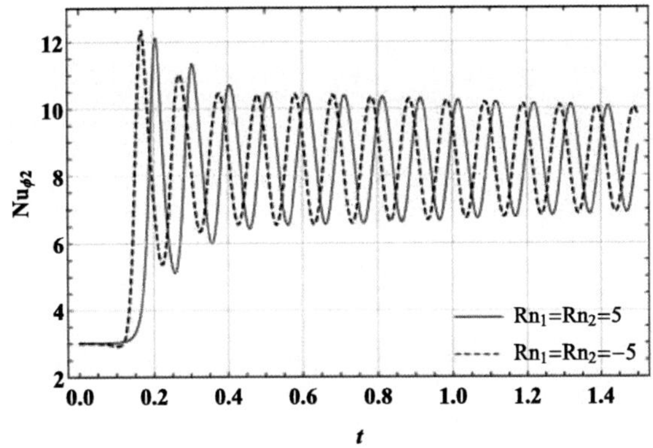

Fig. 23 Comparison of $Nu_{\phi 2}$ for top and bottom heavy configurations

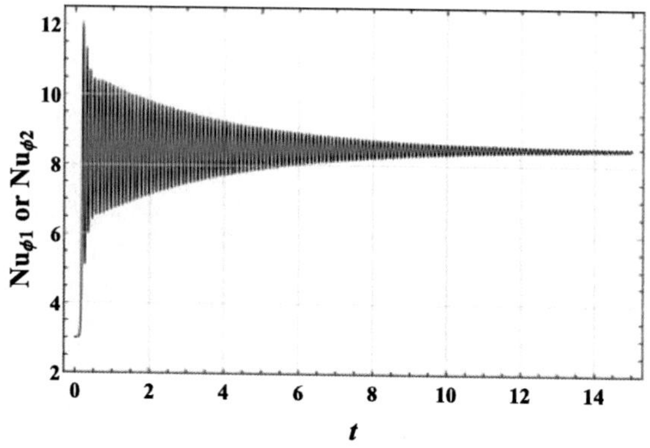

Fig. 24 Behavior of mass transport for higher value of time

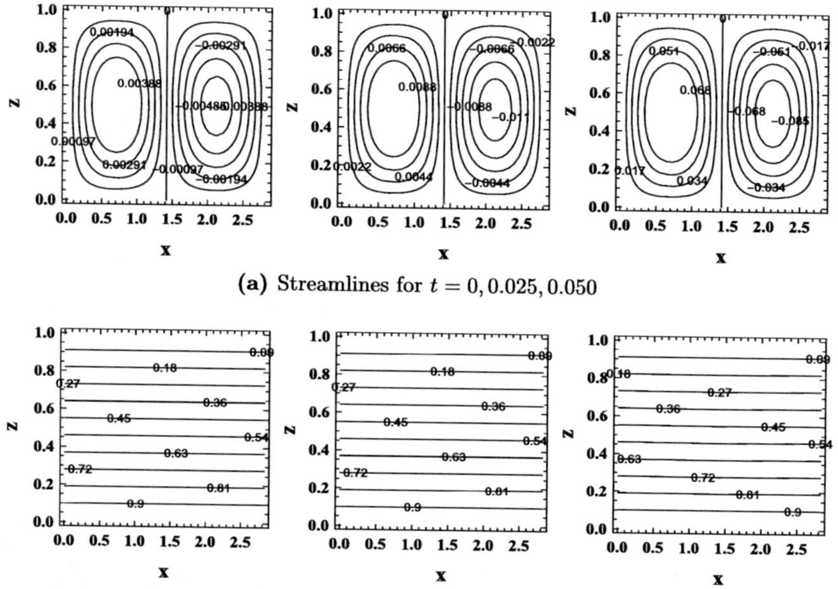

(a) Streamlines for $t = 0, 0.025, 0.050$

(b) Isothermals for $t = 0, 0.025, 0.050$

Fig. 25 Conduction state

8 Conclusions

We have investigated the heat and mass transport in a horizontal composite nanoliquid layer by performing a non-linear analysis. All the results have been presented graphically. These are the major conclusions:

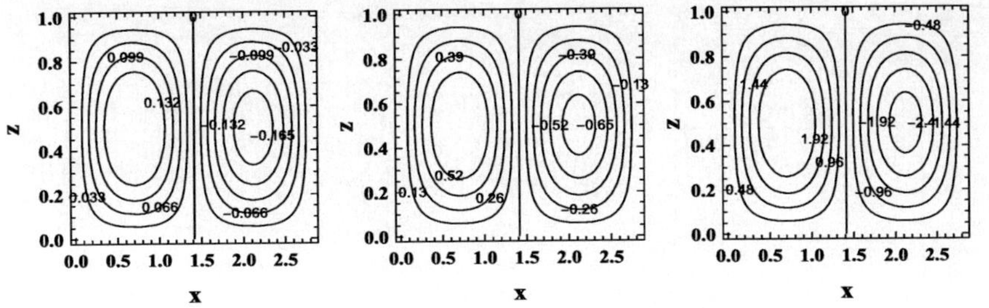

(a) Streamlines for $t = 0.1, 0.125, 0.150$

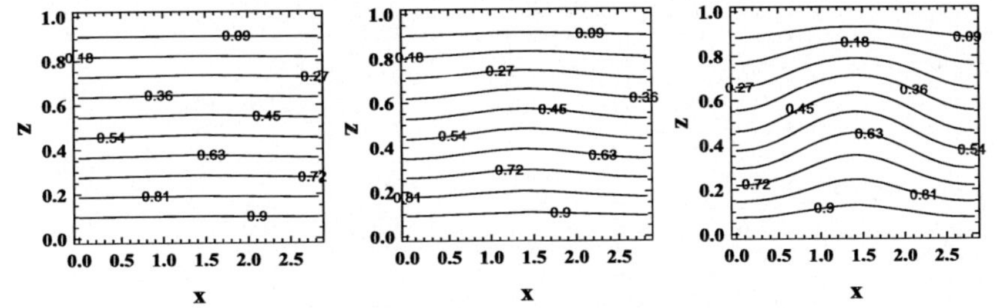

(b) Isothermals for $t = 0.1, 0.125, 0.150$

Fig. 26 Transition state (conduction to convection)

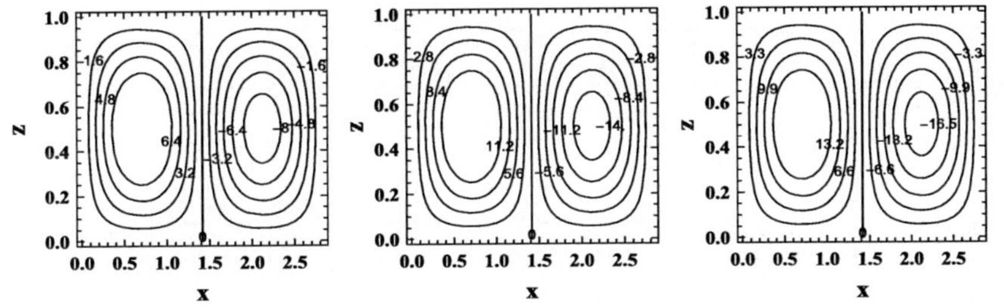

(a) Streamlines for $t = 0.175, 0.190, 0.205$

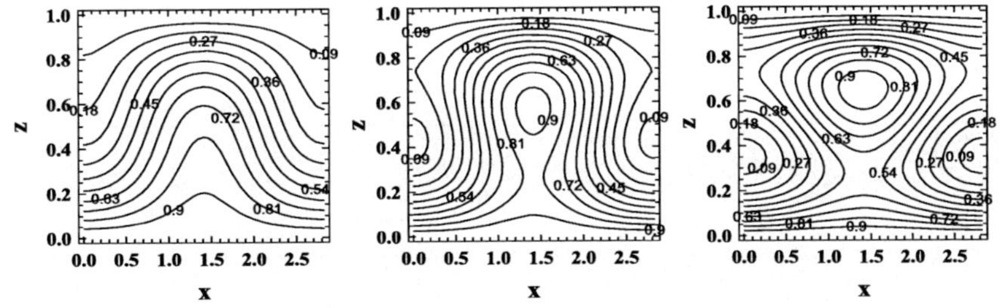

(b) Isothermals for $t = 0.175, 0.190, 0.205$

Fig. 27 Convection state

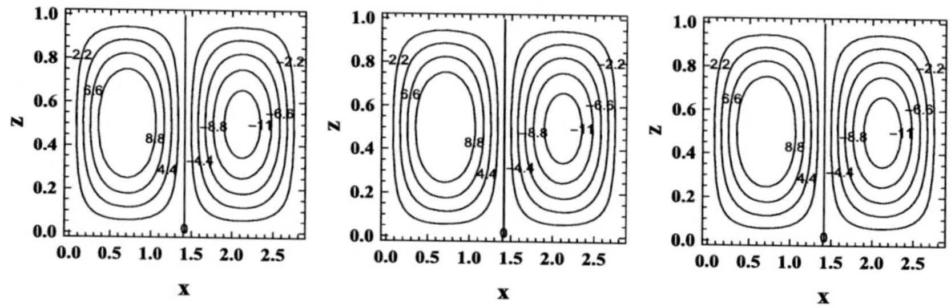

(a) Streamlines for $t = 1.025, 1.050, 1.075$

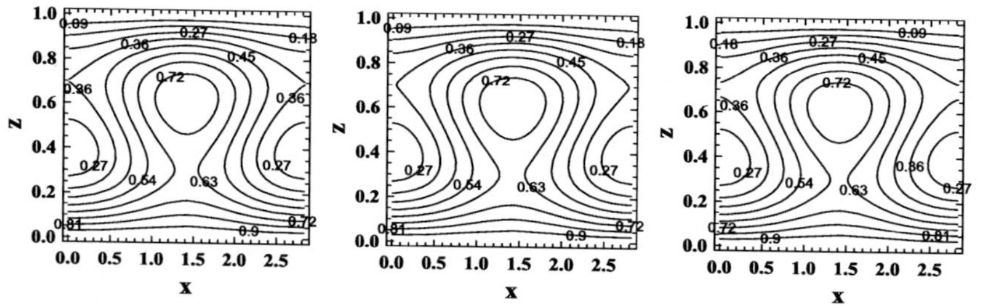

(b) Isothermals for $t = 1.025, 1.050, 1.075$

Fig. 28 Steady state

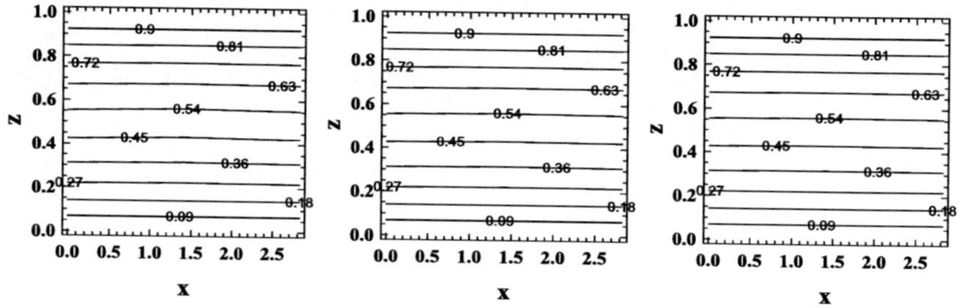

(a) Isohalines for $t = 0, 0.025, 0.050$

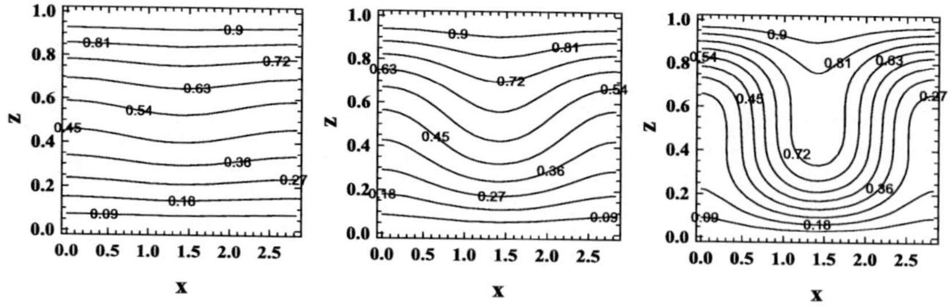

(b) Isohalines for $t = 0.1, 0.125, 0.150$

Fig. 29 Isohalines

1. We found same effect of modified particle density increments in composite nanofluid as compared to the ordinary nanofluid on heat transfer.
2. Lewis number has also same effect in composite nanofluid as compared to the ordinary nanofluid on heat and mass transfer.
3. We found that heat transfer by convection is delayed in composite nanofluid as compared to ordinary nanofluid.
4. Prandtl number has also same effect in composite nanofluid as compared to ordinary nanofluid on heat and mass transfer.
5. The effect of modified particle density increments on mass transport depends upon the nanoparticle concentration Nusselt number, i.e., the effect of N_{A1} is only on the first nanoparticle concentration Nusselt number and the effect of N_{A2} is only on the second nanoparticle concentration Nusselt number.
6. Le_1 has its effect only on the nanoparticle concentration Nusselt number for the first nanoparticle, i.e., $Nu_{\phi 1}$, while Le_2 has its effect only on the nanoparticle concentration Nusselt number for the second nanoparticle, i.e., $Nu_{\phi 2}$.

References

1. Choi, S.: Enhancing thermal conductivity of fluids with nanoparticles. In: Signier, D.A., Wang, H.P. (eds.) Development and Applications of Non-Newtonian flows, ASME FED, vol. 231/MD vol. 66, pp. 99–105 (1995)
2. Eastman, J.A., Choi, S, Li, S., Yu, W., Thompson, L.J.: Anomalously increased effective thermal conductivities of ethylene glycol-based nanofluids containing copper nanoparticles. Appl. Phys. Lett. **78**, 718–720 (2001)
3. Das, S.K., Putra, N., Thiesen, P., Roetzel, W.: Temperature dependence of thermal conductivity enhancement for nanofluids. ASME J. Heat Transf. **125**, 567–574 (2003)
4. Buongiorno, J.: Convective transport in nanofluids. ASME J. Heat Transf. **128**, 240–250 (2006)
5. Tzou, D.Y.: Instability of nanofluids in natural convection. ASME J. Heat Transf. **130**(7), 072401. https://doi.org/10.1115/1.2908427
6. Tzou, D.Y.: Thermal instability of nanofluids in natural convection. Int. J. Heat Mass Transf. **51**(11–12), 2967–2979 (2007). https://doi.org/10.1016/j.ijheatmasstransfer.09.014
7. Nield, D.A., Kuznetsov, A.V.: Thermal instability in a porous medium layer saturated by a nanofluid. Int. J. Heat Mass Transf. **52**, 5796–5801 (2009)
8. Kuznetsov, A.V., Nield, D.A.: Thermal instability in a porous medium saturated by a nanofluid: Brinkman model. Transp. Porous Media **81**, 409–422 (2010)
9. Kuznetsov, A.V., Nield, D.A.: The onset of double-diffusive nanofluid convection in a layer of a saturated porous medium. Transp. Porous Media **85**, 941–951 (2010)
10. Nield D.A., Kuznetsov A.V.: The onset of double-diffusive convection in a nanofluid layer. Int. J Heat Fluid Flow **32**, 771–776 (2011)
11. Nield, D.A., Kuznetsov, A.V.: The effect of vertical through flow on thermal instability in a porous medium layer saturated by nanofluid. Transp. Porous Media **87**, 765–775 (2011)
12. Bhadauria, B.S., Agarwal, S., Kumar, A.: Non-linear two-dimensional convection in a nanofluid saturated porous medium. Transp. Porous Media **90**, 605–625 (2011)
13. Venezian, G.: Effect of modulation on the onset of thermal convection. J. Fluid. Mech. **35**(243), 254 (1969)
14. Umavathi, J.C.: Effect of thermal modulation on the onset of convection in a porous medium layer saturated by a nanofluid. Transp. Porous Med. **98**, 59–79 (2013). https://doi.org/10.1007/s11242-013-0133-2

15. Gresho, P.M., Sani, R.L.: The effects of gravity modulation on the stability of a heated fluid layer. J. Fluid Mech. **40**(4), 783–806 (1970)
16. Bhadauria, B.S., Siddheshwar, P.G., Suthar, O.P.: Nonlinear thermal instability in a rotating viscous fluid layer under temperature/gravity modulation. J. Heat Transf. **134/102502-1** (2012)
17. Bhadauria, B.S., Kiran, P., Kumar, V.: Thermal convection in a nanofluid saturated porous medium with internal heating and gravity modulation. J. Nanofluids **5**, 1–12 (2016)
18. Thomson, W.: Thermal convection in a magnetic field. Phil. Mag. **42**(1417), 1432 (1951)
19. Chandrasekhar, S.: Hydrodynamic and Hydromagnetic stability. Oxford University Press, London (1961)
20. Kiran, P., Bhadauria, B.S., Narasimhulu, Y.: Oscillatory magneto-convection under magnetic field modulation. Alexandria Eng. J. **57**(1), 445–453 (2018)
21. Yadav, D.: Numerical solution of the onset of Buoyancy-driven nanofluid convective motion in an anisotropic porous medium layer with variable gravity and internal heating. Heat Transfer-Asian Res. **49**(3), 1170–1191 (2020)
22. Sakshath, T.N., Joshi, A.P.: Effect of horizontal pressure gradient on Rayleigh-Bénard convection of a Newtonian nanoliquid in a high porosity medium using a local thermal non-equilibrium model. Heat Transf. 1631–1657 (2021)
23. Hanemann, T., Szabo, D.V.: Polymer-nanoparticle composites: from synthesis to modern applications. Materials **3**, 3468–517 (2010)
24. Zhang, Q., Xu, Y., Wang, X., Yao, W.-T.: Recent advances in noble metal based composite nanocatalysts: colloidal synthesis, properties, and catalytic applications. Nanoscale **7**, 10559–83 (2015)
25. Suleiman, A., Sharma, K.V., Baheta, A.V., Mamat, R.: A review of thermophysical properties of water based composite nanofluids. Renew. Sustain. Energy Rev. **66**, 654–678 (2016)
26. Bhadauria, B.S., Kumar, A.: Throughflow and gravity modulation effect on thermal instability in a hele-shaw cell saturated by nanofluid. J. Porous Media **24**(6), 31–51 (2021)
27. Kumar, V., Awasthi, M.K.: Thermal instability in a horizontal composite nano-liquid layer. SN Appl. Sci. **2**, 380 (2020)

On the Existence and Stability Analysis for Ψ-Caputo Fractional Boundary Value Poblem

Bhagwat R. Yewale and Deepak B. Pachpatte

Abstract In this paper, we study the existence and uniqueness results of the solutions for non-linear boundary value problems involving Ψ-Caputo fractional derivative. Furthermore, we prove some stability results of the given problem. The tools used in the analysis are relies on Banach fixed point theorem and Ψ-fractional Gronwall inequality.

Keywords Fractional differential equations · Ψ-Caputo fractional derivative · Gronwall inequality · Stability · Fixed point theorem

1 Introduction

In this paper, we are concerned with the nonlinear fractional differential equations of the type

$$\mathfrak{D}_0^{\bar{\theta},\Psi}\upsilon(\mathfrak{t}) = \mathscr{G}\big(\mathfrak{t}, \upsilon(\mathfrak{t})\big), \text{ for all } \mathfrak{t} \in [0, \bar{\chi}] = \mathrm{I}, \tag{1}$$

$$\upsilon(0) + \mathrm{h}(\upsilon) = \upsilon_0, \upsilon(\bar{\chi}) = \upsilon_{\bar{\chi}}, \qquad \upsilon_0, \upsilon_{\bar{\chi}} \in \mathbb{R} \tag{2}$$

where $1 < \bar{\theta} < 2$, $\mathfrak{D}_0^{\bar{\theta},\Psi}$ is the Ψ-Caputo fractional derivative, $\mathscr{G} : [0, \bar{\chi}] \times \mathbb{R} \to \mathbb{R}$, $\mathrm{h} : \mathcal{C}(\mathrm{I}, \mathbb{R}) \times \mathbb{R} \to \mathbb{R}$ are nonlinear and continuous functions and $\upsilon \in \mathcal{C}(\mathrm{I}, \mathbb{R})$; $\mathcal{C}(\mathrm{I}, \mathbb{R})$ the space of continuous function from I to $\mathbb{R}$ with the supremum norm $\|.\|$.

Fractional order derivatives and integrals are more general cases of integer order derivatives and integrals as it provide arbitrary order derivatives and integration. It has been seen that many researchers have revealed the efficiency of fractional

B. R. Yewale (✉) · D. B. Pachpatte
Department of Mathematics, Dr. Babasaheb Ambedkar Marathwada University, Aurangabad 431004, Maharashtra, India
e-mail: yewale.bhagwat@gmail.com

R. K. Sharma et al. (eds.), *Frontiers in Industrial and Applied Mathematics*, Springer Proceedings in Mathematics & Statistics 410,
https://doi.org/10.1007/978-981-19-7272-0_18

differential equations (FDE$_s$) in the modelling of physical phenomena in different fields of science and engineering [3, 5, 12, 15, 16], which helped fractional calculus to become a very useful and attractive research field. In the literature, there are several approaches by which authors have defined numerous fractional differential and integral operators see [9]. One such class of fractional operators is an integration and differentiation of one function with respect to another function, referred to as, Ψ-Fractional calculus. For instance, Almeida [1], presented Ψ-Caputo fractional derivative which is modified version of Caputo derivative. In [17], authors established Ψ-Hilfer fractional derivative.

On the other hand, these Ψ-fractional operators have been utilized to perform a qualitative analysis of FDE$_s$. In particular, Almeida et al. [2], investigated the existence, uniqueness, continuous dependence and stability of the Ψ-Caputo FDE$_s$ with the help of Banach fixed point theorem. Kucche et al. [10], studied existence and uniqueness of Ψ-Hilfer FDE$_s$ with the help of Schauder's fixed point theorem as well as continuous dependence of the corresponding system have been studied by employing Weissinger theorem. Recently, Pachpatte [14] have used the Banach fixed point theorem to study the existence, uniqueness and stability of the Ψ-Hilfer partial FDE$_s$. In [20], Wahash et al. proved estimate and stability of the solution involving Ψ-Caputo derivative by using Ψ-Gronwall inequality. We mention here some recent studies that focus on the qualitative properties of Ψ-fractional differential equations [4, 6, 11, 18, 19, 21].

Motivated by above work, in this paper we discuss existence, uniqueness and stability of (1)–(2). In Sect. 2, we give some preliminaries. In Sect. 3, we prove existence and uniqueness of the solution of (1)–(2) in the view of Banach fixed point theorem. In Sect. 4, we present Stability analysis of (1)–(2). In Sect. 5, an illustrative example is given to demonstrate our results.

2 Preliminaries

Here, we provide some basic definitions and important results which are used throughout this work.

Definition 2.1 ([9]) Let $\bar{\theta} > 0$ and υ be an integrable function defined on I. Let $\Psi \in \mathcal{C}^1(\mathrm{I}, \mathbb{R})$ be an increasing function such that $\Psi'(\mathfrak{t}) \neq 0$, for all $\mathfrak{t} \in \mathrm{I}$. Then Ψ-Riemann Liouville fractional integral of υ of order $\bar{\theta}$ is defined as

$$\mathfrak{J}_{0+}^{\bar{\theta},\Psi}\upsilon(\mathfrak{t}) = \frac{1}{\Gamma(\bar{\theta})}\int_0^{\mathfrak{t}} \Psi'(\kappa)(\Psi(\mathfrak{t}) - \Psi(\kappa))^{\bar{\theta}-1}\upsilon(\kappa)d\kappa, \ \mathfrak{t} > 0. \tag{3}$$

Definition 2.2 ([1]) Let $\bar{\theta} > 0$ and $\Psi \in \mathcal{C}^n(\mathrm{I}, \mathbb{R})$, the Ψ-Caputo fractional derivative of a function $\upsilon \in \mathcal{C}^{n-1}(\mathrm{I}, \mathbb{R})$ of order $\bar{\theta}$ is defined as

$$\mathfrak{D}_{0+}^{\bar{\theta},\Psi}\upsilon(\mathfrak{t}) = \mathfrak{D}_{0+}^{\bar{\theta},\Psi}\left[\upsilon(\mathfrak{t}) - \sum_{m=0}^{n-1}\frac{\upsilon_{\Psi}^{[m]}(0)}{m!}(\Psi(\mathfrak{t}) - \Psi(0))^m\right], \tag{4}$$

where $n = \lceil\bar{\theta}\rceil + 1$ for $\bar{\theta} \notin \mathbb{N}$, $n = \bar{\theta}$ for $\bar{\theta} \in \mathbb{N}$.
and

$$\upsilon_{\Psi}^{[m]}(\mathfrak{t}) := \left(\frac{1}{\Psi'(\mathfrak{t})}\frac{d}{d\mathfrak{t}}\right)^m \vartheta(\mathfrak{t}).$$

Lemma 2.1 ([1]) *Let* $\bar{\theta} > 0$. *If* $\upsilon \in C^1(\mathrm{I}, \mathbb{R})$, *then*

$$\mathfrak{D}_{0+}^{\bar{\theta},\Psi}\mathfrak{J}_{0+}^{\bar{\theta},\Psi}\upsilon(\mathfrak{t}) = \upsilon(\mathfrak{t}),$$

and if $\upsilon \in C^n(\mathrm{I}, \mathbb{R})$, *then*

$$\mathfrak{J}_{0+}^{\bar{\theta},\Psi}\mathfrak{D}_{0+}^{\bar{\theta},\Psi}\upsilon(\mathfrak{t}) = \upsilon(\mathfrak{t}) - \sum_{m=0}^{n-1}\frac{\vartheta_{\Psi}^{[m]}(0)}{m!}(\Psi(\mathfrak{t}) - \Psi(0))^m. \tag{5}$$

Lemma 2.2 ([9]) *For* $\bar{\theta},\ \bar{\theta}_1 > 0$ *and* $\upsilon \in C^n(\mathrm{I})$, *we have*

$$\mathfrak{J}_{0+}^{\bar{\theta},\Psi}\mathfrak{J}_{0+}^{\bar{\theta}_1,\Psi}\upsilon(\mathfrak{t}) = \mathfrak{J}_{0+}^{\bar{\theta}+\bar{\theta}_1,\Psi}\upsilon(\mathfrak{t}), \quad \mathfrak{t} > 0. \tag{6}$$

Lemma 2.3 ([1]) *Let* $\bar{\theta} > 0$. *Then*

$$\mathfrak{D}_{0+}^{\bar{\theta},\Psi}(\Psi(\kappa) - \Psi(0))^k = 0, \text{ for all } k = 0, 1, 2, ..., n-1, \ n \in \mathbb{N}. \tag{7}$$

Lemma 2.4 ([8]) *Let X be a Banach space and* $B \subset X$ *be closed. If* $\zeta : B \to B$ *is a contraction mapping, then* ζ *has a fixed point in B.*

Lemma 2.5 *Let* $1 < \bar{\theta} < 2$ *and* $\mathscr{G} : \mathrm{I} \times \mathbb{R} \to \mathbb{R}$ *be a continuous function. Then the problem* (1)–(2) *is equivalent to*

$$\begin{aligned}\upsilon(\mathfrak{t}) =& \left(1 - \frac{\Psi(\mathfrak{t}) - \Psi(0)}{\Psi(\bar{\chi}) - \Psi(0)}\right)\upsilon_0 + \left(\frac{\Psi(\mathfrak{t}) - \Psi(0)}{\Psi(\bar{\chi}) - \Psi(0)} - 1\right)\mathrm{h}(\upsilon)\\ &+ \left(\frac{\Psi(\mathfrak{t}) - \Psi(0)}{\Psi(\bar{\chi}) - \Psi(0)}\right)(\upsilon_{\bar{\chi}} - \mathfrak{J}_0^{\bar{\theta},\Psi}\mathscr{G}(\bar{\chi}, \upsilon(\bar{\chi})) + \mathfrak{J}_0^{\bar{\theta},\Psi}\mathscr{G}(\mathfrak{t}, \upsilon(\mathfrak{t})).\end{aligned} \tag{8}$$

Proof Operating $\mathfrak{J}_0^{\bar{\theta},\Psi}$ on both the sides of (1) and using Lemma 2.1, we get

$$\upsilon(\mathfrak{t}) = c_0 + c_1(\Psi(\mathfrak{t}) - \Psi(0)) + \mathfrak{J}_0^{\bar{\theta},\Psi}\mathscr{G}(\mathfrak{t}, \upsilon(\mathfrak{t}))$$

Since $\upsilon(0) = \upsilon_0 - \mathrm{h}(\upsilon)$ and $\upsilon(\bar{\chi}) = \upsilon_{\bar{\chi}}$, we have

$$c_0 = \upsilon_0 - \mathrm{h}(\upsilon), \qquad c_1 = \frac{\upsilon_{\bar{\chi}} - \upsilon_0 + \mathrm{h}(\upsilon) - \mathfrak{J}_0^{\bar{\theta},\Psi}\mathcal{G}(\bar{\chi}, \upsilon(\bar{\chi}))}{\Psi(\bar{\chi}) - \Psi(0)}.$$

Then

$$\begin{aligned}\upsilon(\mathfrak{t}) =& \Big(1 - \frac{\Psi(\mathfrak{t}) - \Psi(0)}{\Psi(\bar{\chi}) - \Psi(0)}\Big)\upsilon_0 + \Big(\frac{\Psi(\mathfrak{t}) - \Psi(0)}{\Psi(\bar{\chi}) - \Psi(0)} - 1\Big)\mathrm{h}(\upsilon) \\ &+ \Big(\frac{\Psi(\mathfrak{t}) - \Psi(0)}{\Psi(\bar{\chi}) - \Psi(0)}\Big)(\upsilon_{\bar{\chi}} - \mathfrak{J}_0^{\bar{\theta},\Psi}\mathcal{G}(\bar{\chi}, \upsilon(\bar{\chi})) + \mathfrak{J}_0^{\bar{\theta},\Psi}\mathcal{G}(\mathfrak{t}, \upsilon(\mathfrak{t})). \end{aligned} \tag{9}$$

Conversely, suppose that υ satisfies (8). Then from (8), for $\mathfrak{t} = 0$ and $\mathfrak{t} = \bar{\chi}$, we obtain (2). Applying $\mathfrak{D}_{0+}^{\bar{\theta},\Psi}$ on both the sides of (8) and using Lemmas 2.1, 2.3, we get (1). □

3 Existence and Uniqueness

Theorem 3.1 *Let the function $\mathcal{G}$ and* h *satisfying:*
[H1]: there exists $\mathcal{W}_1 > 0$ and $0 < \mathcal{W}_2 < 1$ such that

$$|\mathcal{G}(\mathfrak{t}, \upsilon) - \mathcal{G}(\mathfrak{t}, \upsilon^*)| \leq \mathcal{W}_1 |\upsilon - \upsilon^*|,$$

and

$$|\mathrm{h}(\upsilon) - \mathrm{h}(\upsilon^*)| \leq \mathcal{W}_2 |\upsilon - \upsilon^*|.$$

If

$$\mathcal{W}_2 + 2\frac{(\Psi(\bar{\chi}) - \Psi(0))^{\bar{\theta}}}{\Gamma(\bar{\theta} + 1)}\mathcal{W}_1 < 1, \tag{10}$$

then (1)–(2) *has a unique solution.*

Proof Define $\mathcal{T} : C(\mathrm{I}, \mathbb{R}) \to C(\mathrm{I}, \mathbb{R})$ as follows:

$$\begin{aligned}(\mathcal{T}\upsilon)(\mathfrak{t}) =& \Big(1 - \frac{\Psi(\mathfrak{t}) - \Psi(0)}{\Psi(\bar{\chi}) - \Psi(0)}\Big)\upsilon_0 + \Big(\frac{\Psi(\mathfrak{t}) - \Psi(0)}{\Psi(\bar{\chi}) - \Psi(0)} - 1\Big)\mathrm{h}(\upsilon) \\ &+ \Big(\frac{\Psi(\mathfrak{t}) - \Psi(0)}{\Psi(\bar{\chi}) - \Psi(0)}\Big)(\upsilon_{\bar{\chi}} - \mathfrak{J}_0^{\bar{\theta},\Psi}\mathcal{G}(\bar{\chi}, \upsilon(\bar{\chi})) + \mathfrak{J}_0^{\bar{\theta},\Psi}\mathcal{G}(\mathfrak{t}, \upsilon(\mathfrak{t})). \end{aligned} \tag{11}$$

Then for $\upsilon, \upsilon^* \in C(\mathrm{I}, \mathbb{R})$, we have

$$
\begin{aligned}
|(\mathcal{T}\upsilon)(\mathfrak{t})-(\mathcal{T}\upsilon^*)(\mathfrak{t})| \leq & \Big(\frac{\Psi(\mathfrak{t})-\Psi(0)}{\Psi(\bar{\chi})-\Psi(0)}-1\Big)|\mathrm{h}(\upsilon)-\mathrm{h}(\upsilon^*)| \\
& +\Big(\frac{\Psi(\mathfrak{t})-\Psi(0)}{\Psi(\bar{\chi})-\Psi(0)}\Big)\mathfrak{J}_{0+}^{\bar{\theta},\Psi}|\mathcal{G}(\bar{\chi},\upsilon(\bar{\chi}))-\mathcal{G}(\bar{\chi},\upsilon^*(\bar{\chi}))| \\
& +\mathfrak{J}_{0+}^{\bar{\theta},\Psi}|\mathcal{G}(\mathfrak{t},\upsilon(\mathfrak{t}))-\mathcal{G}(\mathfrak{t},\upsilon^*(\mathrm{t}))| \\
\leq & \Big(\frac{\Psi(\mathfrak{t})-\Psi(0)}{\Psi(\bar{\chi})-\Psi(0)}-1\Big)\mathcal{W}_2|\upsilon-\upsilon^*| \\
& +\Big(\frac{\Psi(\mathfrak{t})-\Psi(0)}{\Psi(\bar{\chi})-\Psi(0)}\Big)\frac{\mathcal{W}_1}{\Gamma(\bar{\theta})}\int_0^{\bar{\chi}}\Psi'(\kappa)(\Psi(\bar{\chi})-\Psi(\kappa))^{\bar{\theta}-1}|\upsilon-\upsilon^*|d\kappa \\
& +\frac{\mathcal{W}_1}{\Gamma(\bar{\theta})}\int_0^{\mathfrak{t}}\Psi'(\kappa)(\Psi(\mathfrak{t})-\Psi(\kappa))^{\bar{\theta}-1}|\upsilon-\upsilon^*|d\kappa \\
\leq & \Big(\frac{\Psi(\mathfrak{t})-\Psi(0)}{\Psi(\bar{\chi})-\Psi(0)}\Big)\mathcal{W}_2\|\upsilon-\upsilon^*\|+\Big(\frac{\Psi(\mathfrak{t})-\Psi(0)}{\Psi(\bar{\chi})-\Psi(0)}\Big) \\
& \frac{(\Psi(\bar{\chi})-\Psi(0))^{\bar{\theta}}}{\Gamma(\bar{\theta}+1)}\mathcal{W}_1\|\upsilon-\upsilon^*\|+\frac{(\Psi(\mathfrak{t})-\Psi(0))^{\bar{\theta}}}{\Gamma(\bar{\theta}+1)}\mathcal{W}_1\|\upsilon-\upsilon^*\| \\
\leq & \Big(\mathcal{W}_2+2\frac{(\Psi(\bar{\chi})-\Psi(0)^{\bar{\theta}}}{\Gamma(\bar{\theta}+1)}\mathcal{W}_1\Big)\|\upsilon-\upsilon^*\|.
\end{aligned}
$$

In view of (10), $\mathcal{T}$ is contraction mapping. By Lemma 2.4, υ is a unique solution of the problem (1)–(2). □

4 Stability Analysis

In this section, by using Ψ-fractional Gronwall inequality, we analysis the Ulam-Hyers (UH), Generalized Ulam-Hyers (GHU), Ulam-Hyers-Rassias (UHR) and Generalized Ulam-Hyers-Rassias (GUHR) of the problem (1)–(2).

Let $\varepsilon > 0$ and $f : \mathrm{I} \to \mathbb{R}$ be a continuous function. We consider following inequalities:

$$
|\mathfrak{D}_{0+}^{\bar{\theta},\Psi}\omega(\mathfrak{t})-\mathcal{G}(\mathfrak{t},\omega(\mathfrak{t}))| \leq \varepsilon; \quad \mathfrak{t}\in[0,\bar{\chi}] \tag{12}
$$

and

$$
|\mathfrak{D}_{0+}^{\bar{\theta},\Psi}\omega(\mathfrak{t})-\mathcal{G}(\mathfrak{t},\omega(\mathfrak{t}))| \leq \varepsilon f(\mathfrak{t}); \quad \mathfrak{t}\in[0,\bar{\chi}]. \tag{13}
$$

Definition 4.1 The Eqs. (1)–(2) is said to be UH stable if there exists a real number $\delta > 0$ such that for each $\varepsilon > 0$ and for each solution $\omega \in \mathcal{C}(\mathrm{I},\mathbb{R})$ of the inequality (12), there exists a solution $\upsilon \in \mathcal{C}(\mathrm{I},\mathbb{R})$ satisfying

$$
\mathfrak{D}_{a}^{\bar{\theta},\Psi}\upsilon(\mathfrak{t})=\mathcal{G}\big(\mathfrak{t},\upsilon(\mathfrak{t})\big),\ \text{for all } \mathfrak{t}\in\mathrm{I},\ 1<\bar{\theta}<2, \tag{14}
$$

$$\upsilon(0) = \omega(0), \upsilon(\bar{\chi}) = \omega(\bar{\chi}) \tag{15}$$

with

$$|\omega(\mathfrak{t}) - \upsilon(\mathfrak{t})| \leq \delta\varepsilon, \quad \mathfrak{t} \in \mathrm{I}. \tag{16}$$

Definition 4.2 The Eqs. (1)–(2) is said to be GUH stable if there exists a continuous function $\varphi : \mathrm{I} \to \mathrm{I}$ with $\varphi(0) = 0$ such that for every $\varepsilon > 0$ and for each solution $\omega \in \mathcal{C}(\mathrm{I}, \mathbb{R})$ of (12), there exist a solution $\upsilon \in \mathcal{C}(\mathrm{I}, \mathbb{R})$ of (1)–(2) with

$$|\omega(\mathfrak{t}) - \upsilon(\mathfrak{t})| \leq \varphi(\varepsilon), \quad \mathfrak{t} \in \mathrm{I}. \tag{17}$$

Definition 4.3 The Eqs. (1)–(2) is said to be UHR stable with respect to the function f if there exists a real number $\delta > 0$ such that for every $\varepsilon > 0$ and for each solution $\omega \in \mathcal{C}(\mathrm{I}, \mathbb{R})$ of (13), there exist a solution $\upsilon \in \mathcal{C}(\mathrm{I}, \mathbb{R})$ of (1)–(2) with

$$|\omega(\mathfrak{t}) - \upsilon(\mathfrak{t})| \leq \delta\varepsilon f(\mathfrak{t}), \quad \mathfrak{t} \in \mathrm{I}. \tag{18}$$

Definition 4.4 The Eqs. (1)–(2) is GUHR stable with respect to the function f if there exists a real number $\delta > 0$ such that for each solution $\omega \in \mathcal{C}(\mathrm{I}, \mathbb{R})$ of (13), there exist a solution $\upsilon \in \mathcal{C}(\mathrm{I}, \mathbb{R})$ of (1)–(2) with

$$|\omega(\mathfrak{t}) - \upsilon(\mathfrak{t})| \leq \delta f(\mathfrak{t}), \quad \mathfrak{t} \in \mathrm{I}. \tag{19}$$

Remark 4.1 A function $\omega \in \mathcal{C}(\mathrm{I}, \mathbb{R})$ is a solution of (12) if and only if there exists a function $\mathrm{g} \in \mathcal{C}(\mathrm{I}, \mathbb{R})$ (where g depends on ω) such that

(1) $|\mathrm{g}(\mathfrak{t})| < \varepsilon$

(2) $\mathfrak{D}_{0+}^{\bar{\theta},\Psi}\omega(\mathfrak{t}) = \mathscr{G}(\mathfrak{t}, \omega(\mathfrak{t})) + \mathrm{g}(\mathfrak{t}), \ \mathfrak{t} \in \mathrm{I}.$

Remark 4.2 A function $\omega \in \mathcal{C}(\mathrm{I}, \mathbb{R})$ is a solution (13) if and only if there exists function $\mathrm{g}, f \in \mathcal{C}(\mathrm{I}, \mathbb{R})$ (where g depends on ω) such that

(1) $|\mathrm{g}(\mathfrak{t})| < \varepsilon f(\mathfrak{t})$

(2) $\mathfrak{D}_{0+}^{\bar{\theta},\Psi}\omega(\mathfrak{t}) = \mathscr{G}(\mathfrak{t}, \omega(\mathfrak{t})) + \mathrm{g}(\mathfrak{t}), \ \mathfrak{t} \in \mathrm{I}.$

Lemma 4.1 ([18]) *Ψ-Gronwall inequality:*
Assume that υ *and* u *are nonnegative integrable functions on* I. *Let* ρ *be a nonnegative continuous function on* I *such that* ρ *is nondecreasing. If*

$$\upsilon(\mathfrak{t}) \leq \mathrm{u}(\mathfrak{t}) + \rho(\mathfrak{t}) \int_0^{\mathfrak{t}} \Psi'(\kappa)(\Psi(\mathfrak{t}) - \Psi(\kappa))^{\bar{\theta}-1}\upsilon(\kappa)d\kappa, \tag{20}$$

then

$$\upsilon(\mathfrak{t}) \leq \mathrm{u}(\mathfrak{t}) \int_0^{\mathfrak{t}} \sum_{m=1}^{\infty} \frac{[\rho(\mathfrak{t})\Gamma(\bar{\theta})]^m}{\Gamma(\bar{\theta}m)} \Psi'(\kappa)(\Psi(\mathfrak{t}) - \Psi(\kappa))^{\bar{\theta}-1}\mathrm{u}(\kappa)d\kappa, \tag{21}$$

for $\mathfrak{t} \in \mathrm{I}$.

Remark 4.3 ([18]) Under the assumptions of Lemma 4.1, let $\upsilon(\mathfrak{t})$ be a nondecreasing function on I. Then we have

$$\upsilon(\mathfrak{t}) \leq \mathrm{u}(\mathfrak{t}) E_{\bar{\theta}}(\rho(\mathfrak{t})\Gamma(\bar{\theta})(\Psi(\mathfrak{t}) - \Psi(0))^{\bar{\theta}}),$$

where $E_{\bar{\theta}}(\mathfrak{t}) = \sum_{m=0}^{\infty} \frac{\mathfrak{t}^m}{\Gamma(\bar{\theta}+1)}$.

In the next theorem, we discuss the UH stability of the problem (1)–(2) with the help of Ψ-Gronwall inequality.

Theorem 4.1 *Suppose that [H1] hold and inequality* (12) *is satisfied, then the problem* (1)–(2) *is UH stable.*

Proof Let $\varepsilon > 0$. Assume that υ be a solution of (1)–(2). Then

$$\upsilon(\mathfrak{t}) = \Phi_{\upsilon} + \mathfrak{J}_0^{\bar{\theta},\Psi}\mathcal{G}(\mathfrak{t}, \upsilon(\mathfrak{t})), \tag{22}$$

where

$$\begin{aligned}\Phi_{\upsilon} =& \Big(1 - \frac{\Psi(\mathfrak{t}) - \Psi(0)}{\Psi(\bar{\chi}) - \Psi(0)}\Big)\upsilon_0 + \Big(\frac{\Psi(\mathfrak{t}) - \Psi(0)}{\Psi(\bar{\chi}) - \Psi(0)} - 1\Big)\mathrm{h}(\upsilon) \\ &+ \Big(\frac{\Psi(\mathfrak{t}) - \Psi(0)}{\Psi(\bar{\chi}) - \Psi(0)}\Big)(\upsilon_{\bar{\chi}} - \mathfrak{J}_0^{\bar{\theta},\Psi}\mathcal{G}(\bar{\chi}, \upsilon(\bar{\chi})).\end{aligned} \tag{23}$$

From (15), we can write

$$\upsilon(\mathfrak{t}) = \Phi_{\omega} + \mathfrak{J}_0^{\bar{\theta},\Psi}\mathcal{G}(\mathfrak{t}, \upsilon(\mathfrak{t})), \tag{24}$$

where

$$\begin{aligned}\Phi_{\omega} =& \Big(1 - \frac{\Psi(\mathfrak{t}) - \Psi(0)}{\Psi(\bar{\chi}) - \Psi(0)}\Big)\omega_0 + \Big(\frac{\Psi(\mathfrak{t}) - \Psi(0)}{\Psi(\bar{\chi}) - \Psi(0)} - 1\Big)\mathrm{h}(\omega) \\ &+ \Big(\frac{\Psi(\mathfrak{t}) - \Psi(0)}{\Psi(\bar{\chi}) - \Psi(0)}\Big)\omega_{\bar{\chi}} - \mathfrak{J}_0^{\bar{\theta},\Psi}\mathcal{G}(\bar{\chi}, \omega(\bar{\chi})).\end{aligned} \tag{25}$$

Since $\omega \in \mathcal{C}(\mathrm{I}, \mathbb{R})$ is a solution of inequality (12). By Remark 4.1, we have

$$|\mathfrak{D}_0^{\bar{\theta},\Psi}\omega(\mathfrak{t}) - \mathcal{G}\big(\mathfrak{t}, \omega(\mathfrak{t})\big)| \leq \varepsilon, \quad \text{for all } \mathfrak{t} \in \mathrm{I}. \tag{26}$$

Operating $\mathfrak{J}_0^{\bar{\theta},\Psi}$ on both the sides of (26), we obtain

$$|\omega(\mathfrak{t}) - \Phi_\omega - \frac{1}{\Gamma(\bar{\theta})}\int_0^{\mathfrak{t}} \Psi'(\kappa)(\Psi(\mathfrak{t}) - \Psi(\kappa))^{\bar{\theta}-1} \mathcal{G}(\kappa, \omega(\kappa)d\kappa| \leq \frac{(\Psi(\bar{\chi}) - \Psi(0))^{\bar{\theta}}}{\Gamma(\bar{\theta}+1)}\varepsilon. \tag{27}$$

By our assumption and from (24) and (27), we obtain

$$\begin{aligned}
|\omega(\mathfrak{t}) - \upsilon(\mathfrak{t})| = & \; |\omega(\mathfrak{t}) - \Phi_\omega - \frac{1}{\Gamma(\bar{\theta})}\int_0^{\mathfrak{t}} \Psi'(\kappa)(\Psi(\mathfrak{t}) - \Psi(\kappa))^{\bar{\theta}-1}\mathcal{G}(\kappa, \upsilon(\kappa))d\kappa| \\
\leq & \; |\omega(\mathfrak{t}) - \Phi_\omega - \frac{1}{\Gamma(\bar{\theta})}\int_0^{\mathfrak{t}} \Psi'(\kappa)(\Psi(\mathfrak{t}) - \Psi(\kappa))^{\bar{\theta}-1}\mathcal{G}(\kappa, \omega(\kappa))d\kappa| \\
& + \frac{1}{\Gamma(\bar{\theta})}\int_0^{\mathfrak{t}} \Psi'(\kappa)(\Psi(\mathfrak{t}) - \Psi(\kappa))^{\bar{\theta}-1}|\mathcal{G}(\kappa, \omega(\kappa)) - \mathcal{G}(\kappa, \upsilon(\kappa))|d\kappa \\
\leq & \; \frac{(\Psi(\bar{\chi}) - \Psi(0))^{\bar{\theta}}}{\Gamma(\bar{\theta}+1)}\varepsilon + \frac{\mathcal{W}_1}{\Gamma(\bar{\theta})}\int_0^{\mathfrak{t}} \Psi'(\kappa)(\Psi(\mathfrak{t}) - \Psi(\kappa))^{\bar{\theta}-1}|\omega(\kappa) - \upsilon(\kappa)|d\kappa. \quad (28)
\end{aligned}$$

Applying Lemma 4.1 to (28), we get

$$\begin{aligned}
& |\omega(\mathfrak{t}) - \upsilon(\mathfrak{t})| \\
\leq & \; \frac{(\Psi(\bar{\chi}) - \Psi(0))^{\bar{\theta}}}{\Gamma(\bar{\theta}+1)}\varepsilon\Big[1 + \int_0^{\mathfrak{t}} \sum_{m=1}^{\infty} \frac{\mathcal{W}_1^m}{\Gamma(\bar{\theta}m)}\Psi'(\kappa)(\Psi(\mathfrak{t}) - \Psi(\kappa))^{\bar{\theta}m-1}d\kappa\Big] \\
= & \; \frac{(\Psi(\bar{\chi}) - \Psi(0))^{\bar{\theta}}}{\Gamma(\bar{\theta}+1)}\varepsilon\Big[1 + \sum_{m=1}^{\infty} \frac{\mathcal{W}_1^m}{\Gamma(\bar{\theta}m)}\int_0^{\mathfrak{t}} \Psi'(\kappa)(\Psi(\mathfrak{t}) - \Psi(\kappa))^{\bar{\theta}m-1}d\kappa\Big] \\
\leq & \; \frac{(\Psi(\bar{\chi}) - \Psi(0))^{\bar{\theta}}}{\Gamma(\bar{\theta}+1)}\varepsilon\Big[1 + \sum_{m=1}^{\infty} \frac{\mathcal{W}_1^m}{\Gamma(\bar{\theta}m+1)}(\Psi(\bar{\chi})) - \Psi(0))^{\bar{\theta}m}\Big] \\
= & \; \frac{\varepsilon(\Psi(\bar{\chi})) - \Psi(0))^{\bar{\theta}}}{\Gamma(\bar{\theta}+1)}E_{\bar{\theta}}(\mathcal{W}_1(\Psi(\bar{\chi}) - \Psi(0))^{\bar{\theta}}). \quad (29)
\end{aligned}$$

Put

$$\delta = \frac{(\Psi(\bar{\chi}) - \Psi(0))^{\bar{\theta}}}{\Gamma(\bar{\theta}+1)}E_{\bar{\theta}}(\mathcal{W}_1(\Psi(\bar{\chi}) - \Psi(0))^{\bar{\theta}}). \tag{30}$$

Therefore

$$|\omega(\mathfrak{t}) - \upsilon(\mathfrak{t})| \leq \delta\varepsilon. \tag{31}$$

Hence, the problem (1)–(2) is UH stable. □

Theorem 4.2 *If there exists a function continuous function* $\varphi : \mathrm{I} \to \mathrm{I}$ *with* $\varphi(0) = 0$. *Then under the assumption of Theorem 4.1, the problem (1)–(2) is GUH stable*

Proof In a same fashion similar to Theorem 4.1, setting $\varphi(\varepsilon) = \delta\varepsilon$ with $\varphi(0) = 0$, we get

$$|\omega(\mathfrak{t}) - \upsilon(\mathfrak{t})| \leq \varphi(\varepsilon). \tag{32}$$

□

In order to prove UHR and GUHR stability, the following hypothesis must be satisfied:
[H2]: There exist an increasing function $f \in \mathcal{C}(\mathrm{I}, \mathbb{R})$ and $\gamma > 0$ such that

$$\mathfrak{J}_{0+}^{\bar{\theta},\Psi} f(\mathfrak{t}) \leq \gamma f(\mathfrak{t}), \quad \mathfrak{t} \in \mathrm{I}.$$

Lemma 4.2 *Let* $\varepsilon > 0$ *and* $\omega(\mathfrak{t}) \in \mathcal{C}(\mathrm{I}, \mathbb{R})$ *be a solution* (13). *Then*

$$\left|\omega(\mathfrak{t}) - \Phi_\omega - \mathfrak{J}_{0+}^{\bar{\theta},\Psi}\mathcal{G}(\mathfrak{t}, \omega(\mathfrak{t}))\right| \leq \varepsilon\gamma f(\mathfrak{t}). \tag{33}$$

Proof By Remark 4.2, g, $f \in \mathcal{C}(\mathrm{I}, \mathbb{R})$ such that

$$|\mathfrak{D}_{0+}^{\bar{\theta},\Psi}\omega(\mathfrak{t}) - \mathcal{G}(\mathfrak{t}, \omega(\mathfrak{t}))| = |\mathrm{g}(\mathfrak{t})| \leq \varepsilon f(\mathfrak{t}). \tag{34}$$

Operating $\mathfrak{J}_{0+}^{\bar{\theta},\Psi}$ and using the hypothesis [H2], we deduce that

$$|\omega(\mathfrak{t}) - \Phi_\omega - \mathfrak{J}_{0+}^{\bar{\theta},\Psi}\mathcal{G}(\mathfrak{t}, \omega(\mathfrak{t}))| \leq \varepsilon\mathfrak{J}_{0+}^{\bar{\theta},\Psi} f(\mathfrak{t}) \leq \gamma\varepsilon f(\mathfrak{t}). \tag{35}$$

□

Theorem 4.3 *Let* $\varepsilon > 0$ *and* $\omega \in \mathcal{C}(J, \mathbb{R})$ *be a solution* (13) *and* $\mathscr{W}_1\gamma \neq 1$, *then* (1)–(2) *is UHR stable.*

Proof Let $\upsilon(\mathfrak{t})$ be a solution of (1)–(2) and using $\Phi_\upsilon = \Phi_\omega$ Then

$$\upsilon(\mathfrak{t}) = \Phi_\omega + \mathfrak{J}_{0+}^{\bar{\theta},\Psi}\mathcal{G}(\mathfrak{t}, \omega(\mathfrak{t})). \tag{36}$$

By hypothesis [H1] and Lemma 4.2, we get

$$\begin{aligned}
|\omega(\mathfrak{t}) - \upsilon(\mathfrak{t})| \leq\ & \left|\omega(\mathfrak{t}) - \Phi_\omega - \frac{1}{\Gamma(\bar{\theta})}\int_0^{\mathfrak{t}} \Psi'(\kappa)(\Psi(\mathfrak{t}) - \Psi(\kappa))^{\bar{\theta}-1}\mathcal{G}(\kappa, \omega(\kappa))\right| d\kappa \\
& + \frac{1}{\Gamma(\bar{\theta})}\int_0^{\mathfrak{t}} \Psi'(\kappa)(\Psi(\mathfrak{t}) - \Psi(\kappa))^{\bar{\theta}-1}|\mathcal{G}(\kappa, \omega(\kappa)) - \mathcal{G}(\kappa, \upsilon(\kappa))| d\kappa \\
\leq\ & \gamma\varepsilon f(\mathfrak{t}) + \frac{\mathscr{W}_1}{\Gamma(\bar{\theta})}\int_0^{\mathfrak{t}} \Psi'(\kappa)(\Psi(\mathfrak{t}) - \Psi(\kappa))^{\bar{\theta}-1}|\omega(\kappa) - \upsilon(\kappa)| d\kappa.
\end{aligned} \tag{37}$$

Applying Lemma 4.1 to (37) and using hypothesis [H2], we obtain

$$
\begin{aligned}
|\omega(\mathfrak{t})-\upsilon(\mathfrak{t})| \leq\ & \gamma\varepsilon f(\mathfrak{t})+\gamma\varepsilon\int_0^{\mathfrak{t}}\sum_{k=1}^{\infty}\frac{\mathscr{W}_1^k}{\Gamma(\bar{\theta}m)}\Psi'(\kappa)(\Psi(\mathfrak{t})-\Psi(\kappa))^{\bar{\theta}k-1}f(\kappa)d\kappa \\
=\ & \gamma\varepsilon f(\mathfrak{t})+\gamma\varepsilon\Big[\int_0^{\mathfrak{t}}\frac{\mathscr{W}_1}{\Gamma(\bar{\theta})}\Psi'(\kappa)(\Psi(\mathfrak{t})-\Psi(\kappa))^{\bar{\theta}-1}f(\kappa)d\kappa \\
+\ & \int_0^{\mathfrak{t}}\frac{\mathscr{W}_1^2}{\Gamma(2\bar{\theta})}\Psi'(\kappa)(\Psi(\mathfrak{t})-\Psi(\kappa))^{2\bar{\theta}-1}f(\kappa)d\kappa+....\Big] \\
=\ & \gamma\varepsilon f(\mathfrak{t})+\gamma\varepsilon\big[\mathscr{W}_1\mathfrak{J}_{0+}^{\bar{\theta},\Psi}f(\mathfrak{t})+\mathscr{W}_1^2\mathfrak{J}_{0+}^{2\bar{\theta},\Psi}f(\mathfrak{t})+....\big] \\
\leq\ & \gamma\varepsilon f(\mathfrak{t})+\gamma\varepsilon\big[\mathscr{W}_1\gamma f(\mathfrak{t})+(\mathscr{W}_1\gamma)^2 f(\mathfrak{t})+....\big] \\
=\ & \gamma\varepsilon f(\mathfrak{t})\sum_{k=0}^{\infty}(\mathscr{W}_1\gamma)^k \\
=\ & \frac{\gamma}{1-\mathscr{W}_1\gamma}\varepsilon f(\mathfrak{t}).
\end{aligned}
\tag{38}
$$

Setting

$$\delta=\frac{\gamma}{1-\mathscr{W}_1\gamma}. \tag{39}$$

From (38) and (39), we have

$$|\omega(\mathfrak{t})-\upsilon(\mathfrak{t})|\leq\delta\varepsilon\rho(\mathfrak{t}).$$

□

Theorem 4.4 *Under the assumption of Theorem 4.3, problem (1)–(2) is GUHR stable.*

Proof In a same fashion similar to Theorem 4.3, setting $\varepsilon=1$, we get

$$|\omega(\mathfrak{t})-\upsilon(\mathfrak{t})|\leq\delta f(\mathfrak{t}). \tag{40}$$

□

5 Example

Example 5.1 Consider the following fractional differential equation involving Ψ-Caputo derivative

$$\mathfrak{D}_0^{\frac{3}{2},\Psi}\upsilon(\mathfrak{t})=\mathfrak{t}+\frac{1}{6}sin\upsilon(\mathfrak{t}),\ \text{for all } \mathfrak{t}\in[0,1], \tag{41}$$

$$\upsilon(0) + \frac{1}{4}\upsilon(\frac{1}{3}) = 0, \upsilon(1) = \frac{1}{2}. \tag{42}$$

Here, $\bar{\theta} = \frac{3}{2}$, $\mathscr{G}\big(\mathfrak{t}, \upsilon(\mathfrak{t})\big) = \mathfrak{t} + \frac{1}{6}sin\upsilon(\mathfrak{t})$, $\mathrm{h}(\upsilon) = \frac{1}{4}\upsilon(\frac{1}{3})$. Then for $\mathfrak{t} \in [0, 1]$,

$$|\mathscr{G}(\mathfrak{t}, \upsilon) - \mathscr{G}(\mathfrak{t}, \upsilon^*)| \leq \frac{1}{6}|\upsilon - \upsilon^*| \text{ and } |\mathrm{h}(\upsilon) - \mathrm{h}(\upsilon^*)| \leq \frac{1}{4}|\upsilon - \upsilon^*|.$$

Therefore $\mathscr{W}_1 = \frac{1}{6}$ and $\mathscr{W}_2 = \frac{1}{4}$. For $\Psi(\mathfrak{t}) = \mathfrak{t}$, we have

$$\mathscr{W}_2 + 2\frac{(\Psi(\bar{\chi}) - \Psi(0))^{\bar{\theta}}}{\Gamma(\bar{\theta}+1)}\mathscr{W}_1 = \frac{1}{4} + \frac{2(1-0)^{\frac{3}{2}}}{6\Gamma(\frac{5}{2})} = \frac{1}{4} + \frac{4}{9\sqrt{\pi}} < 1.$$

Hence, all the conditions of Theorem 3.1 are satisfied. Thus, by the Theorem 3.1, problem (41)–(42) has unique solution.

6 Concluding Remark

In this research work, the existence and uniqueness of the proposed system have been successfully examined using Banach fixed point theorem under some specific assumptions and conditions. Along with the existence and uniqueness, we established stability results such as UH, GUH, UHR and GUHR in the sense of Ψ-Gronwall inequality. It should be noted that, for different values of Ψ, the Ψ-Caputo fractional derivative reduces to many classical fractional operators such as Caputo [9], Caputo-Hadamard [7], Caputo-Erdélyi-Kober [13] fractional derivative. Thus, we believe that the results derived in this article are general in character and contributes in the theory of fractional differential equations.

References

1. Almeida, R.: A Caputo fractional derivative of a function with respect to another function. Commun. Nonlinear Sci. Numer. Simul. **44**, 460–481 (2017). https://doi.org/10.1016/j.cnsns.2016.09.006
2. Almeida, R., Malinowska, A.B., Monteiro, M.T.: Fractional differential equations with a Caputo derivative with respect to a kernel function and their applications. Math. Meth. Appl. Sci. **41**, 336–352 (2017). https://doi.org/10.1002/mma.4617
3. Baskonus, H.M., Sánchez Ruiz, L.M., Ciancio, A.: A new challenging arising in engineering problems with fractional and integer order. Fractal Fract. **5**(2), 35 (2021). https://doi.org/10.3390/fractalfract5020035
4. Derbazi, C., Baitiche, Z., Benchohra, M., N′Guérékata, G. M.: Existence, uniqueness, approximation of solutions and $E\alpha$-Ulam stability results for a class of nonlinear fractional differential equations involving ψ-Caputo derivative with initial conditions. Math. Morav. **25**(1), 1-30(2021). https://doi.org/10.5937/MatMor2101001D

5. Debnath, L.: Recent applications of fractional calculus to science and engineering, Int. J. Math. Math. Sci. **2003**, Article ID 753601, 3413–3442 (2003). https://doi.org/10.1155/S0161171203301486
6. Douriah, S., Foukrach, D., Benchohra, M., Graef, J.: Existence and uniqueness of periodic solutions for some nonlinear fractional pantograph differential equations with ψ-Caputo derivative. Arab. J. Math. (2021). https://doi.org/10.1007/s40065-021-00343-z
7. Gambo, Y.Y., Jarad, F., Baleanu, D., Abdeljawad, T.: On Caputo modification of the Hadamard fractional derivatives. Adv. Differ. Equ., Art. no. 10 (2014). https://doi.org/10.1186/1687-1847-2014-10
8. Granas, A., Dugundji, J.: Fixed Point Theory. Springer, New York (2003). https://doi.org/10.1007/978-0-387-21593-8
9. Kilbas, A.A., Srivastava, H.M., Trujillo, J.J.: Theory and Applications of Fractional Differential Equations. Elsevier, Amsterdam (2006)
10. Kucche, K.D., Mali, A.D., Sousa, J.V.D.C.: On the nonlinear ψ-Hilfer fractional differential equations. Comput. Appl. Math. **38**, 73 (2019). https://doi.org/10.1007/s40314-019-0833-5
11. Kucche, K.D., Kharade, J.P., Sousa, J.V.D.C.: On the nonlinear impulsive ψ-Hilfer fractional differential equations. Math. Model. Anal. **25**(2), 642–660 (2020). https://doi.org/10.3846/mma.2020.11445
12. Kumar, D., Singh, J.: Fractional Calculus in Medical and Health Science. CRC Press, New York (2020)
13. Luchko, Y., Trujillo, J.J.: Caputo-type modification of the Erdélyi-Kober fractional derivative. Fract. Calc. Appl. Anal. **10**(3), 249–267 (2007)
14. Pachpatte, D.B.: Existence and stability of some nonlinear ψ-Hilfer partial fractional differential equation, Part. differ. Equ. Apl. Math. **3** (2021). https://doi.org/10.1016/j.padiff.2021.100032
15. Pandey, P., Chu,Y.-M., Gómez-Aguilar, J.F., Jahanshahi, H., Aly, A.A.: A novel fractional mathematical model of COVID-19 epidemic considering quarantine and latent time. Results Phys. **26** (2021). https://doi.org/10.1016/j.rinp.2021.104286
16. Srivastava, H.M., Dubey, R.S., Jain, M.: A study of the fractional order mathematical model of dibeties and it's resulting complications. Math. Methods Appl. Sci. **42**(13), 4570–4583 (2019). https://doi.org/10.1002/mma.5681
17. Sousa, J.V.D.C., De Oliveira, E.C.: On the ψ-Hilfer fractional derivative. Commun. Nonlinear Sci. Numer. Simul. **60**, 72–91 (2018). https://doi.org/10.1016/j.cnsns.2018.01.005
18. Sousa, J.V.D.C., De Oliveira, E.C.: A Gronwall inequality and the Cauchy type problem by means of ψ-Hilfer operator. Differ. Equ. Appl. **11**(1), 87–106 (2019). https://doi.org/10.7153/dea-2019-11-02
19. Sousa, J.V.D.C., Kucche, K.D., De Oliveira, E.C.: On the Ulam-Hyers stabilities of the solutions of ψ-Hilfer fractional differential equation with abstract volterra operator. Math. Methods Appl. Sci. **42**(12), 3021–3032 (2019). https://doi.org/10.1002/mma.5562
20. Wahash, H.A., Panchal, S.K., Abdo, M.S.: Existence and stability of a nonlinear fractional differential equation involving a ψ-Caputo operator. ATNAA **4**(4), 266–278 (2020). https://doi.org/10.31197/atnaa.664534
21. Wahash, H.A., Abdo, M.S., Panchal, S.K.: Existence and Ulam-Hyers stability of the implicit fractional boundary value problem with ψ-Caputo fractional derivative. JAMCM **19**(1), 89–101 (2020). https://doi.org/10.17512/jamcm.2020.1.08

Alternative Crack-Tip Enrichment Functions for X-FEM in Arbitrary Polarized Piezoelectric Media

Rajalaxmi Rath and Kuldeep Sharma

Abstract In this paper, a new approach is proposed to study the fracture mechanics problems in 2-D arbitrary polarized piezoelectric media using X-FEM. The existing six-fold crack-tip enrichment functions defined for the generalized case of poling and alignment of crack in piezoelectric media are re-defined here by considering the localized solution of crack-tip field based on Lekhnitskii's formalism in the transformed coordinate system obtained from material axes to crack-axes, whereas the existing crack-tip enrichment functions were developed other way round. Using the proposed enrichment functions, some benchmark problems such as center cracks, edge cracks, double-edge cracks, and macro–micro-collinear cracks have been studied under arbitrary poling direction, plain strain, and impermeable crack-face conditions. An excellent agreement of normalized intensity factors (IFs) has been obtained for all the cases with the results of existing six-fold enrichment functions.

Keywords Crack-tip enrichment · Intensity factors · Lekhnitskii's formalism · Piezoelectric · X-FEM

1 Introduction

In this smart material's era, with rapid change in smart technology, the use of intellectual devices increases from our domestic appliances to spacecraft equipment and many other electromechanical devices. Piezoelectric material is one of the smart materials which has the best inherent quality that converts electrical energy to mechanical energy and vice versa. Due to this electromechanical coupling effect, these materials have been broadly used as sensors, ultrasonic generators, actuators, and transducers in many sophisticated, electromechanical devices of submarines, aeronautics, medical appliances, etc. However, such materials are mechanically brittle and electrically ductile in nature. That's why, whenever in manufacturing process

R. Rath · K. Sharma (✉)
Nit Uttarakhand, Srinagar, Garhwal, Uttarakhand 246174, India
e-mail: ksharma@nituk.ac.in

R. K. Sharma et al. (eds.), *Frontiers in Industrial and Applied Mathematics*,
Springer Proceedings in Mathematics & Statistics 410,
https://doi.org/10.1007/978-981-19-7272-0_19

or in service time, while they undergo high electromechanical loadings, the crack persists. In course of time, the developed crack could reach the critical limit and damage the structural integrity which degrades the performance of the material. Therefore, for many decades, it has been a major issue to analyze the pre-exist crack to predict the strength and efficiency of these materials for better performance and utilization.

Since the last decades, various analytical and numerical techniques have been implemented for the study of cracks in piezoelectric materials under different kinds of loadings and environments. X-FEM as one of the prominent numerical techniques for the study of fracture mechanics problems has also been developed and implemented for the study of static and propagating cracks in piezoelectric materials. Considering Sosa [1] and William's eigenfunction approach, Bechet and his coworkers [2] developed the six linearly independent crack-tip enrichment functions to incorporate the near-tip solution in the X-FEM framework. They developed these functions for a generalized case of crack and poling direction and studied the crack problems subjected to impermeable crack-face conditions. Bharagva and Sharma [3] also proposed the crack-tip enrichment functions for piezoelectric materials independent of Bechet et al. [2] but for poling axis perpendicular to crack-axis and studied the two-collinear cracks problem. Thereafter, many researchers implemented [4, 5] or extended these basis functions for studying the various kinds of problems such as transient, fatigue loading, multiple cracks, sub-interface crack, thermal loadings, semipermeable crack-face conditions, etc., in piezoelectric materials.

Applying Leikhntski's technique, Xu [6] extended Sosa's approach [1] for the study of crack and branch crack problems in arbitrary poling direction. Considering Xu's [6] approach and the work of Bechet et al. [2] in piezoelectric media, authors have proposed here an alternative approach for the development of crack-tip enrichment functions in piezoelectric materials under the generalized case of poling direction and developed six enrichment functions independent of the existing ones.

2 Fundamental Equations in Piezoelectric Media

Due to the coupling nature of piezoelectric material the interrelation between stresses and electrical displacements in terms of strain and electrical field components are represented as the constitutive equations. These are defined as

$$\sigma_{ij} = C_{ijkp}\epsilon_{kp} - e_{ijk}E_k \tag{1}$$

$$D_i = e_{ijk}\epsilon_{jk} + \tilde{\kappa}_{ij}E_j \tag{2}$$

where σ_{ij}, ϵ_{kp}, D_i, and E_p represent Cauchy stress tensor, mechanical strain tensor, electric displacement vector, and electric field vector, respectively. Additionally, C_{ijkp} and e_{ijk} represent elastic and piezoelectric constants and $\tilde{\kappa_{ij}}$ are dielectric permittivity constants.

The kinematic relations between strain tensor ϵ_{ij} and displacement vector u_i as well as electric field vector E_i, and scalar electric potential ϕ_i are given by

$$\epsilon_{ij} = \frac{1}{2}(u_{i,j} + u_{j,i}) \tag{3}$$

$$E_i = -\phi,_i \tag{4}$$

The Cauchy stress tensor and the electric displacement vector in the absence of body forces and charges satisfy the equilibrium equations as

$$\sigma_{ij,j} = 0 \tag{5}$$

$$D_{i,j} = 0. \tag{6}$$

2.1 Boundary Conditions

Considering a linear piezoelectric domain with a crack under plain strain conditions, a piezoelectric boundary value problem is stated as

$$\sigma_{ij}n_j = t_i^* \; on \; \Gamma_a; \; D_j n_j = -\omega^* \; on \; \Gamma_a, \tag{7}$$

$$u_j = u_j^* \; on \; \Gamma_u; \; \phi = \phi^* \; on \; \Gamma_\phi. \tag{8}$$

Here, t^* and ω^* represent stress and charge on the surface of the boundary, respectively.

2.2 Crack-Face Boundary Conditions

The analysis of fracture in linear piezoelectric media requires the information on medium present inside the crack faces. Researchers have classified three different crack-face boundary conditions: impermeable, permeable, and semipermeable. For the present study, impermeable crack-face conditions have been considered which are mathematically defined as below:

$$\sigma_{ij}n_j = 0 \; and \; D_j n_j = 0 \; on \; \Gamma_C. \tag{9}$$

3 X-FEM-Based Approximate Solution

X-FEM is one of the elegant numerical techniques to study the crack problems as it models the crack(s) geometry independent of the mesh and avoids re-meshing when the crack propagates. Using the concept of partition of unity, linearly independent functions which approximate the near tip or local solution incorporated into the FEM approximations for the elements of a particular region of interest. In the study of cracks, the Heaviside function is used to represent the discontinuity in the displacement or the primary variable(s) across the crack faces and the crack-tip enrichment functions for representing the near-tip asymptotic solutions. In piezoelectric material, the X-FEM-based approximate displacement and electric potential functions are defined as

$$\begin{aligned} u^h(x, y) &= \sum_{I \in N} N_I(x, y) u_I + \sum_{I \in N^{\widetilde{cr}}} N_I(x, y)(H(f^h(x, y)) - H(f_I))\hat{a}_I \\ &+ \sum_{I \in \widetilde{N^{Tip}}} N_I(x, y) \sum_{k=1}^{k=6} (F^k(r, \theta, \mu_k^{re}, \mu_k^{im}) - F^k(x_I, y_I, \mu_k^{re}, \mu_k^{im}))\hat{b}_I^k \end{aligned} \tag{10}$$

$$\begin{aligned} \phi^h(x, y) &= \sum_{I \in N} N_I(x, y) \phi_I + \sum_{I \in N^{\widetilde{cr}}} N_I(x, y)(H(f^h(x, y)) - H(f_I))\hat{c}_I \\ &+ \sum_{I \in \widetilde{N^{Tip}}} N_I(x, y) \sum_{k=1}^{k=6} (F^k(r, \theta, \mu_k^{re}, \mu_k^{im}) - F^k(x_I, y_I, \mu_k^{re}, \mu_k^{im}))\hat{d}_I^k \end{aligned} \tag{11}$$

where $H(f(x, y))$ is a Heaviside step function.

$$H\left(z^*\right) = \begin{cases} -1 & \text{if } z^* < 0 \\ 1 & \text{if } z^* > 0 \end{cases} \tag{12}$$

$f(x)$ represents an Level set function. $N^{\widetilde{cr}}$ and $\widetilde{N^{Tip}}$ denotes the set of enriched nodes associated with crack surfaces and crack tips, respectively. $N_i(x)$ represents the shape functions associated with node i. u_i and ϕ_i are the vectors of nodal degrees of freedom (DOF) containing the nodal displacements and electric potentials. Furthermore, $\hat{a}_I$, $\hat{b}_I^k$ and $\hat{c}_I$, $\hat{d}_I^k$ are the additional enriched DOFs in the elements containing the crack. $F^k(r, \theta, , \mu_k^{re}, \mu_k^{im})$ are the near crack-tip enrichment functions. The split and tip nodes enriched by the use of level set functions are depicted in Fig. 1.

Substituting the approximate solutions into the weak form and assembling the element-wise solution, the global system of equations can be written as follows:

$$Ku = f \tag{13}$$

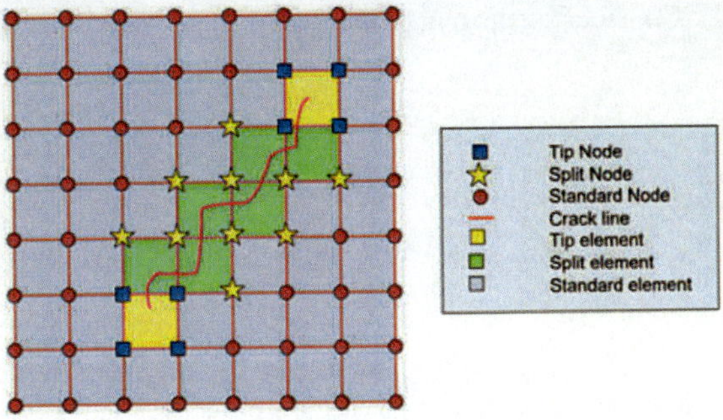

Fig. 1 Schematic diagram for enriched nodes selected for crack path description

4 Existing and Alternative Enrichment Functions

In this section, the existing and proposed enrichment functions for arbitrary polarized piezoelectric materials have been discussed in detail.

4.1 Existing Crack-Tip Enrichment Functions

Applying Sosa's approach [1], Bechet et al. [2] firstly obtained the near tip solution for transversely isotropic piezoelectric materials with respect to the material axes x_1 and x_2 (poling direction) as shown in Fig. 2 and then expand the obtained solution in terms of Laurent-like series expansion of the form:

$$U(x_1, x_2) = \sum_{m} \sum_{i=1}^{i=6} G_i(\gamma_m)(x_1 + \mu_i x_2)^{\gamma_m+2} \tag{14}$$

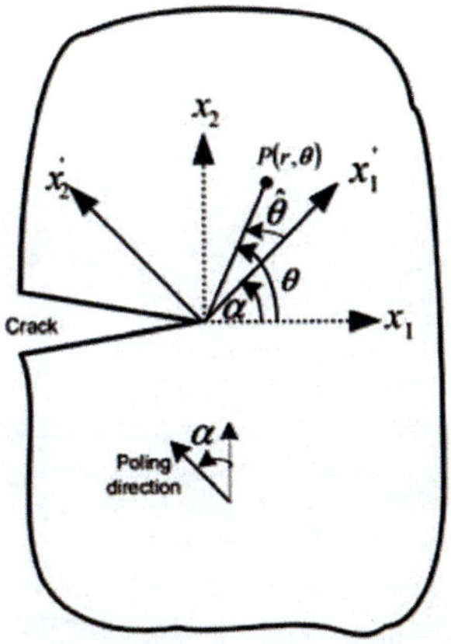

Fig. 2 Arbitrary polarized cracked piezoelectric media

Considering the real property of $U(x_1, x_2)$, it is expressed in polar form as follows:

$$U(x_1, x_2) = U(r, \theta) = \sum_{i=1}^{i=3} G_i(\gamma_m)(rcos(\theta) + \mu_i rsin(\theta))^{\gamma_m+2} + \sum_{i=1}^{i=3} \overline{G_i(\gamma_m)}(rcos(\theta) + \overline{\mu_i} rsin(\theta))^{\gamma_m+2} \quad (15)$$

Further, to define the generalized solution for arbitrary poling direction (makes an angle α w.r.t x_2 axis), they [2] simply expressed the solution (15) w.r.t the material axes, i.e., x_1' and x_2' (arbitrary poling direction) implies the angle θ is replace by an angle $\hat{\theta} = \theta - \alpha$ in (15). The generalized solution for arbitrary poling direction defined by Bechet et al. [2] is

$$U(x_1, x_2) = U(r, \theta) = \sum_{i=1}^{i=3} G_i(\gamma)(rcos(\theta - \alpha) + \mu_i rsin(\theta - \alpha))^{\gamma+2} + \sum_{i=1}^{i=3} \overline{G_i(\gamma)}(rcos(\theta - \alpha) + \overline{\mu_i} rsin(\theta - \alpha))^{\gamma+2} \quad (16)$$

Hence, the six linearly independent crack-tip enrichment functions were devised for the implementation of X-FEM in piezoelectric media as follows:

$$F(r, \theta, \mu_i^{(re)}, \mu_i^{(im)}) = \{\sqrt{r} p_1(\hat{\theta}), \sqrt{r} p_2(\hat{\theta}), \sqrt{r} p_3(\hat{\theta}), \sqrt{r} p_4(\hat{\theta}), \sqrt{r} p_5(\hat{\theta}), \sqrt{r} p_6(\hat{\theta})\} \quad (17)$$

where

$$p_m\left(\hat{\theta}\right) = \begin{cases} \rho\left(\hat{\theta}, \mu_i^{(re)}, \mu_i^{(im)}\right) \cos\left(\frac{\psi\left(\hat{\theta}, \mu_i^{(re)}, \mu_i^{(im)}\right)}{2}\right) & \text{if } \mu_i^{(im)} > 0 \\ \rho\left(\hat{\theta}, \mu_i^{(re)}, \mu_i^{(im)}\right) \sin\left(\frac{\psi\left(\hat{\theta}, \mu_i^{(re)}, \mu_i^{(im)}\right)}{2}\right) & \text{if } \mu_i^{(im)} \leq 0 \end{cases} \quad (18)$$

$$\psi = \frac{\pi}{2} + \pi \, int\left(\frac{\hat{\theta}}{\pi}\right) - arctan\left(\frac{\cos\left(\hat{\theta} - \pi \, int\left(\frac{\hat{\theta}}{\pi}\right)\right) + \mu_i^{re} \sin\left(\hat{\theta} - \pi \, int\left(\frac{\hat{\theta}}{\pi}\right)\right)}{\mid \mu_i^{(im)} \mid \sin\left(\hat{\theta} - \pi \, int\left(\frac{\hat{\theta}}{\pi}\right)\right)}\right) \quad (19)$$

$$\rho = \frac{1}{\sqrt{2}} \sqrt[4]{\left(\mu_i^{(re)}\right)^2 + \left(\mu_i^{(im)}\right)^2 + \mu_i^{(re)} \sin 2\hat{\theta} - \left[\left(\mu_i^{(re)}\right)^2 + \left(\mu_i^{(im)}\right)^2 - 1\right] \cos 2\hat{\theta}} \quad (20)$$

and $\mu_i = \mu_i^{re} + i\mu_i^{im}$ are the six root of the characteristic equations [2, 3].

4.2 Proposed/Alternative Crack-Tip Enrichment Functions

In the proposed approach, the piezoelectric material is considered as transversely isotropic w.r.t the axes x_1' and x_2' (arbitrary poling direction) in place of axes x_1 and x_2 which is similar to Xu [6] approach. Then, the constitutive equations for piezoelectric materials in $x_1'x_2'x_3'$ system can be expressed as

$$\sigma' = [C][\epsilon'] - [e]^T[E'];\ D' = [e][\epsilon'] + [\kappa][E'] \tag{21}$$

the material axis by an angle α in the clockwise direction. After transformation, the constitutive equations in $x_1x_2x_3$ coordinate system are of the form:

$$\sigma = [C^*][\epsilon] - [e^*]^T[E];\ D = [e^*][\epsilon] + [\kappa^*][E] \tag{22}$$

Also for plane strain case, these relations in $x_1 - x_2$ system can be written as

$$\begin{Bmatrix} \epsilon_{11} \\ \epsilon_{22} \\ 2\epsilon_{12} \end{Bmatrix} = \begin{pmatrix} a_{11}^* & a_{12}^* & a_{13}^* \\ a_{21}^* & a_{22}^* & a_{23}^* \\ a_{31}^* & a_{32}^* & a_{33}^* \end{pmatrix} \begin{Bmatrix} \sigma_{11} \\ \sigma_{22} \\ \sigma_{12} \end{Bmatrix} + \begin{pmatrix} b_{11}^* & b_{21}^* \\ b_{21}^* & b_{22}^* \\ b_{31}^* & b_{32}^* \end{pmatrix} \begin{Bmatrix} D_1 \\ D_2 \end{Bmatrix} \tag{23}$$

$$\begin{Bmatrix} E_1 \\ E_2 \end{Bmatrix} = -\begin{pmatrix} b_{11}^* & b_{12}^* & b_{13}^* \\ b_{21}^* & b_{22}^* & b_{23}^* \end{pmatrix} \begin{Bmatrix} \sigma_{11} \\ \sigma_{22} \\ \sigma_{12} \end{Bmatrix} + \begin{pmatrix} d_{11}^* & d_{12}^* \\ d_{21}^* & d_{22}^* \end{pmatrix} \begin{Bmatrix} D_1 \\ D_2 \end{Bmatrix} \tag{24}$$

where a_{ij}^*,b_{ij}^* and d_{ij}^* are the material constants in transformed axes, i.e., in the $x_1 - x_2$ coordinate system.

Further, by applying Lekhnitskii's formalism approach [1, 2, 6], one can obtain the generalized solution for arbitrary poling direction. Now to develop the ansatz for the proposed approach, the methodology of Bechet et al. [2] has been applied. Accordingly, the solution evaluated for generalized case of poling is defined as

$$\begin{aligned} U^*(x_1, x_2) = U^*(r, \theta) &= \sum_{i=1}^{i=3} E_i(\gamma)(rcos(\theta) + \mu_i^* rsin(\theta))^{\gamma+2} \\ &+ \sum_{i=1}^{i=3} \overline{E_i(\gamma)(rcos(\theta) + \mu_i^* rsin(\theta))^{\gamma+2}} \end{aligned} \tag{25}$$

where μ_i^* is an eigenvalue of the characteristic equation in the transformed system with positive imaginary part and depending upon the polarization angle α.

Since the solution is obtained here after transforming the material axes along the crack-axis, the solution in polar form depends on θ and not on $\theta - \alpha$. Also, in this approach, the effect of polarization has been observed in the evaluated solution but only in terms of material eigenvalues (μ_i^* and $\overline{\mu_i^*}$) and not as a function of $\theta - \alpha$ as in case of Bechet et al. [2] approach. Similar to Bechet et al. [2], six linearly independent functions have been developed corresponding to three independent singular eigenfunctions (at $\gamma = \frac{-1}{2}$) after imposing the homogeneous crack-face boundary conditions. The developed alternative six linearly independent functions are as follows:

$$F^*(r, \theta, \mu^{*(re)}, \mu^{*(im)}) = \{\sqrt{r}p_1^*(\theta), \sqrt{r}p_2^*(\theta), \sqrt{r}p_3^*(\theta), \sqrt{r}p_4^*(\theta), \sqrt{r}p_5^*(\theta), \sqrt{r}p_6^*(\theta)\} \tag{26}$$

where

$$p_m^*(\theta) = \begin{cases} \rho^*\left(\theta, \mu_i^{*(re)}, \mu_i^{*(im)}\right) \cos\left(\frac{\psi^*\left(\theta, \mu_i^{*(re)}, \mu_i^{*(im)}\right)}{2}\right) & \text{if } \mu_i^{*(im)} > 0 \\ \rho^*\left(\theta, \mu_i^{*(re)}, \mu_i^{*(im)}\right) \sin\left(\frac{\psi^*\left(\theta, \mu_i^{*(re)}, \mu_i^{*(im)}\right)}{2}\right) & \text{if } \mu_i^{*(im)} \leq 0 \end{cases} \tag{27}$$

$$\begin{aligned} \psi^* &= \frac{\pi}{2} + \pi\, int\left(\frac{\theta}{\pi}\right) \\ &- arctan\left(\frac{\cos\left(\theta - \pi\, int\left(\frac{\theta}{\pi}\right)\right) + \mu_i^{*(re)} \sin\left(\theta - \pi\, int\left(\frac{\theta}{\pi}\right)\right)}{\mid \mu_i^{*(im)} \mid \sin\left(\theta - \pi\, int\left(\frac{\theta}{\pi}\right)\right)}\right) \end{aligned} \tag{28}$$

$$\rho^* = \frac{1}{\sqrt{2}} \sqrt[4]{\left(\mu_i^{*(re)}\right)^2 + \left(\mu_i^{*(im)}\right)^2 + \mu_i^{*(re)} \sin 2\theta - \left[\left(\mu_i^{*(re)}\right)^2 + \left(\mu_i^{*(im)}\right)^2 - 1\right] \cos 2\theta} \tag{29}$$

5 Results and Discussion

In this section, comparative analysis has been presented for the results of intensity factors obtained using existing and proposed crack-tip enrichment functions subjected to arbitrary poling direction ($\alpha = 30°$). The benchmark problems of fracture mechanics such as center cracks, edge cracks, double-edge cracks, and major–minor cracks are considered for the analysis under variations in mechanical loadings, electrical loadings, and poling direction. The geometries of the specimens considered for analysis are shown in Fig. 3. The piezoelectric material $BaTio_3$ has been taken for numerical studies and its material constants are defined in Table 1. The plain strain, impermeable crack-face conditions, and linear rectangular elements have been taken here for X-FEM analysis. The IFs have been evaluated under electromechanical loadings (if not specified, $\sigma_{yy}^{\infty} = 40\,\text{MPa}$ and $D_y^{\infty} = 0.02\,\text{c/m}^2$) using the interaction integral approach as explained in [3]. The details of the geometric parameters and the number of elements for each problem are presented in Table 2.

Figure 4 shows the variations in normalized mode-I mechanical stress intensity factor (K_I^*) and electrical displacement intensity factor (K_{IV}^*) w.r.t variations in mechanical loading, electrical loading, and polarization angle, respectively. Here,

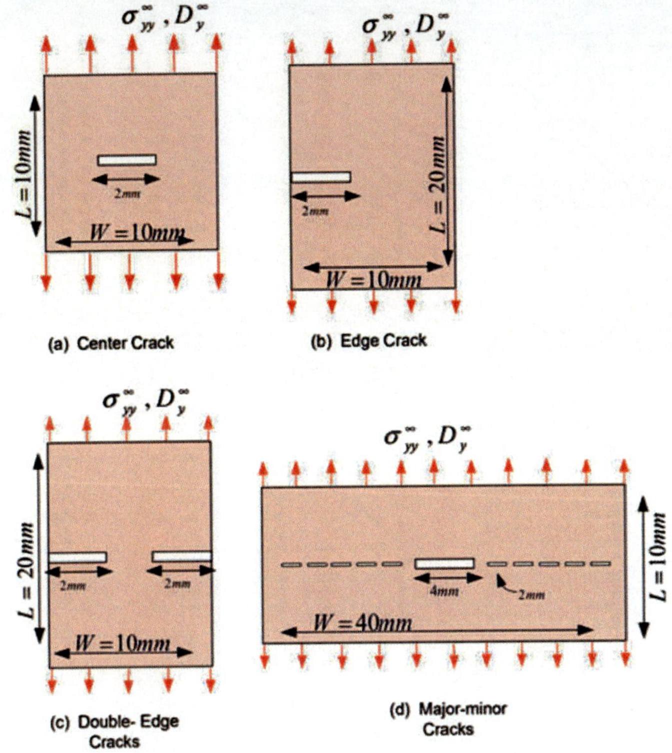

Fig. 3 Geometries of the cracked piezoelectric specimens considered for study

Table 1 Material properties for BaTio$_3$

Elastic constants	Piezoelectric constants	Permittivity
$c_{11} = 16.6$	$e_{31} = -4.4$	$\tilde{\kappa}_{11} = 14.343$
$c_{12} = 7.66$	$e_{33} = 1.6$	$\tilde{\kappa}_{33} = 16.823$
$c_{13} = 7.75$	$e_{15} = 11.6$	$\tilde{\kappa}_{11} = 14.343$
$c_{44} = 4.29$		
$c_{33} = 16.2$		

Table 2 Geometry and loading parameters used for analysis

Problem	Dimensions (in mm)	No. of elements
Center crack	$L = W = 10, a = 1$	99×99
Edge crack	$W = 10, L = 20, a = 2$	99×199
Double-edge cracks	$W = 10, L = 20, a = 2$	99×199
Major–minor cracks	$W = 40, L = 10, c = 2,$	399×99
	$a = 1, d = 1$	

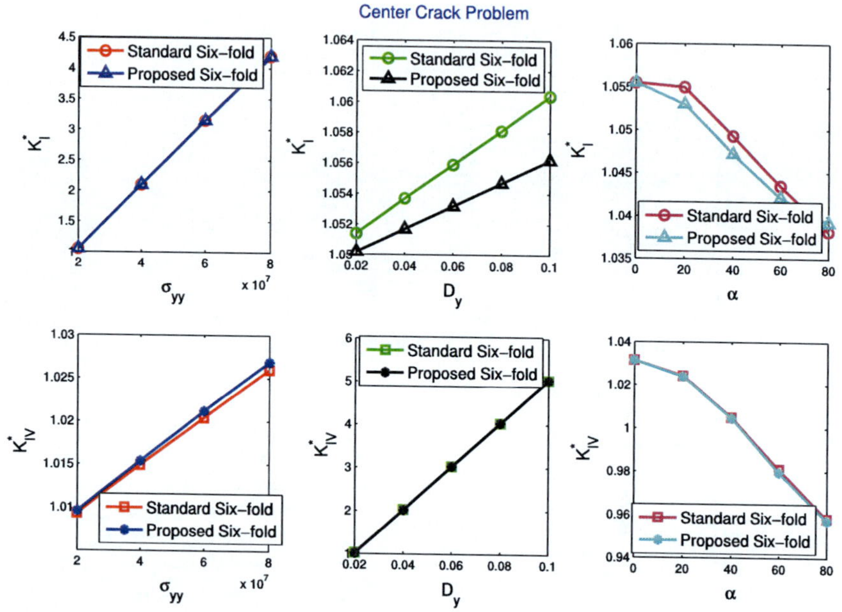

Fig. 4 Variations in normalized IFs with increasing mechanical loading, electrical loading, and polarization angle for center crack problem

the K_I^* and K_{IV}^* are evaluated w.r.t to the analytical results of IFs corresponding to the applied loadings but, for variations in mechanical loading, $K_I^* = \frac{K_I}{20\times10^6\times\sqrt{\pi a}}$ and, for variations in electrical loading, $K_{IV}^* = \frac{K_I}{0.02\times\sqrt{\pi a}}$. It has been observed that the results of IFs obtained using proposed enrichment functions are in good agreement with the results of existing crack-tip enrichment functions [2]. The effects on IFs w.r.t the variations in mechanical loadings, electrical loadings, and poling direction are observed, and the behavior obtained here is similar to the established results.

Similarly, Figs. 5, 6, and 7 represent the numerical studies of the IFs w.r.t variations in mechanical loading, electrical loading, and polarization angle for edge crack, double-edge crack, and major–minor collinear crack problems, respectively. These figures also demonstrate the efficacy of the proposed enrichment functions as in all the cases the maximum error found in both the normalizing IFs is less than 0.6%. Moreover, from the present numerical results, the effects of loadings, poling direction, position of crack/cracks in the specimen, number of cracks, and interaction of collinear cracks can be observed. One can also find that these effects are similar to the results available in literature [7] for impermeable crack-face conditions.

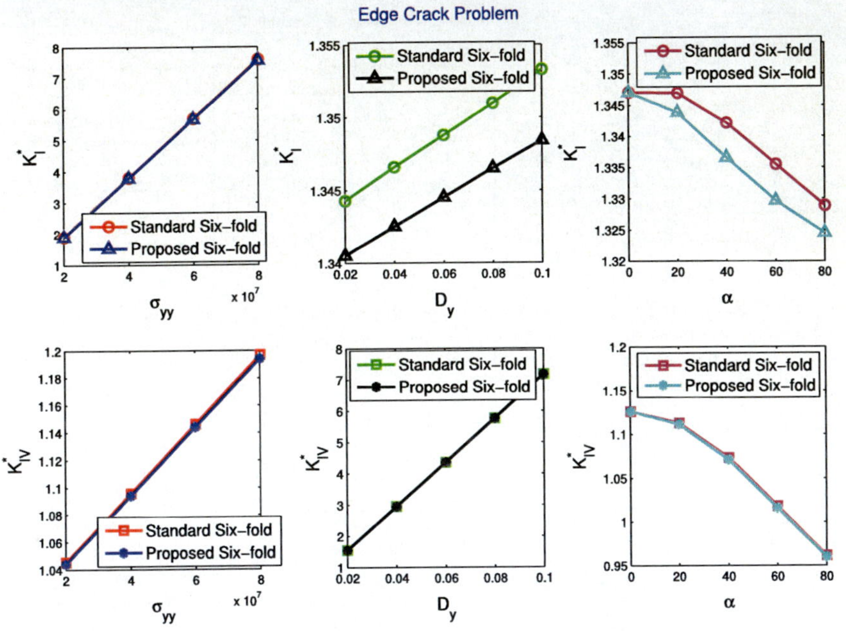

Fig. 5 Variations in normalized IFs with increasing mechanical loading, electrical loading, and polarization angle for edge crack problem

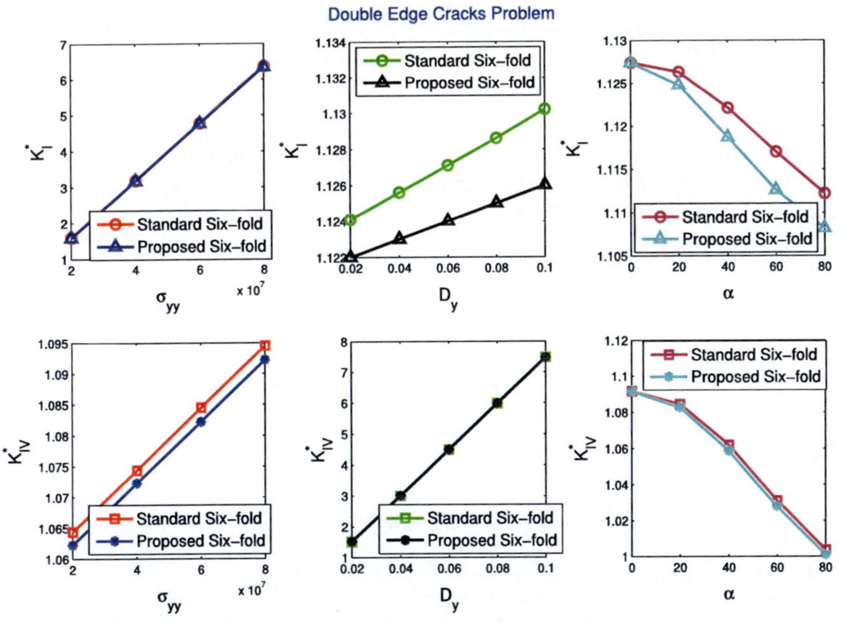

Fig. 6 Variations in normalized IFs with increasing mechanical loading, electrical loading, and polarization angle for double-edge crack problem

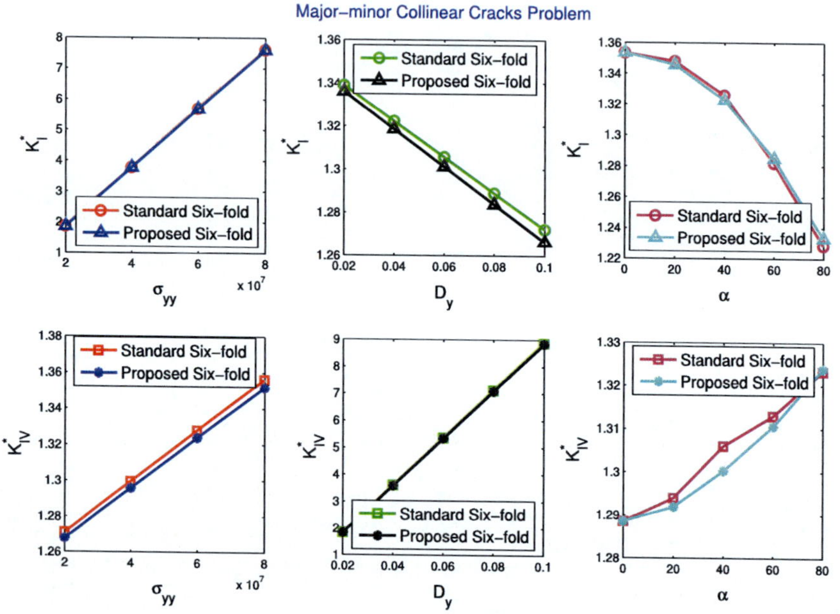

Fig. 7 Variations in normalized IFs with increasing mechanical loading, electrical loading and polarization angle for major–minor collinear crack problem

6 Conclusion

In the present work, an alternative approach is proposed to develop the six linearly independent crack-tip enrichment functions for X-FEM analysis in arbitrary polarized piezoelectric material. The results of IFs obtained by the proposed approach are found in good agreement with the results of existing enrichment functions for arbitrary polarized center crack, edge crack, double-edge crack, and major–minor collinear crack problems. Hence, we conclude that the proposed crack-tip enrichment functions could be considered as alternative crack-tip enrichment functions for the X-FEM-based fracture mechanics study in arbitrary polarized piezoelectric media.

References

1. Sosa, H.: Plane problems in piezoelectric media with defects. Int. J. Solids Struct. **28**, 491–505 (1991)
2. Bechet, E., Scherzer, M., Kuna, M.: Application of the X-FEM to the fracture of piezoelectric materials. Int. J. Numer. Meth. Eng. **77**, 1535–1565 (2009)
3. Bhargava, R.R., Sharma, K.: X-FEM simulation for two-unequal-collinear cracks in 2-D finite piezoelectric specimen. Int. J. Mech. Mater. Des. **8**, 129–148 (2012)

4. Sharma, K., Bui, T.Q., Zhang, Ch., Bhargava, R.R.: Analysis of a subinterface crack in piezoelectric bimaterials with the extended finite element method. Engg. Fract. Mech. **104**, 114–139 (2013)
5. Mishra, R.K.: A review on fracture mechanics in piezoelectric structures. Mater. Today.: Proced. **5**, 5407–5413 (2018)
6. Xu, X.-L., Rajapakse, R.K.N.D.: A theoretical studies of branched cracks in piezoelectrics. Acta Mater. **48**, 1865–1882 (2000)
7. Sharma, K., Bui, T.Q., Singh, S.: Numerical distributed dislocation modeling of multiple cracks in piezoelectric media considering different crack-face boundary conditions and finite size effects. Strength. Fract. Complex. **10**, 49–72 (2017)

Convergence Analysis of a Layer Resolving Numerical Technique for a Class of Coupled System of Singularly Perturbed Parabolic Convection-Diffusion Equations Having an Interface

S. Chandra Sekhara Rao and Abhay Kumar Chaturvedi

Abstract In this article, we consider a time-dependent weakly coupled system of $m(\geq 2)$ singularly perturbed convection-diffusion equations in the domain $G := \Omega \times S$ that has an interface $\Gamma_d := \{(d, t) : t \in S\}$, $d \in \Omega := (0, \ 1)$ and $S := (0, \ T]$. The source terms in the system of equations have discontinuities along Γ_d. Also, the second-order term of each equation is multiplied by a small positive parameter. These parameters can be arbitrarily small and different in magnitude due to which overlapping boundary and interior layers appear in the solution. An appropriate Shishkin mesh is used to discretize the domain. At the mesh points that are not on the interface line, the problem is discretized using an upwind central difference scheme. For the mesh points on the interface line, a particular upwind central difference scheme is used. An appropriate decomposition of exact and numerical solutions is made to analyze the parameters-uniform convergence of the considered numerical scheme. The numerical approximations yielded by this scheme are parameters-uniformly convergent of first-order in time and almost first-order in space concerning the perturbation parameters. Numerical results are presented to validate the theoretical results.

Keywords Singular perturbation · Parabolic problems · Weakly coupled system · Interior layers · Boundary layers · Parameter-uniform convergence · Shishkin mesh

S. C. S. Rao (✉) · A. K. Chaturvedi
Department of Mathematics, Indian Institute of Technology Delhi, Hauz Khas, New Delhi 110016, India
e-mail: sscsr@maths.iitd.ac.in

R. K. Sharma et al. (eds.), *Frontiers in Industrial and Applied Mathematics*, Springer Proceedings in Mathematics & Statistics 410,
https://doi.org/10.1007/978-981-19-7272-0_20

1 Introduction

Let the domains $G_1 := \Omega_1 \times S$ and $G_2 := \Omega_2 \times S$, where $\Omega_1 := (0,\ d)$ and $\Omega_2 := (d,\ 1)$. The considered singularly perturbed initial-boundary value problem is to find $\boldsymbol{u} \in C(\overline{G})^m \cap C^4(G_1 \cup G_2)^m$ such that

$$\boldsymbol{L}\boldsymbol{u} \equiv \frac{\partial \boldsymbol{u}}{\partial t} - \boldsymbol{E}\frac{\partial^2 \boldsymbol{u}}{\partial x^2} - \boldsymbol{B}\frac{\partial \boldsymbol{u}}{\partial x} + \boldsymbol{A}\boldsymbol{u} = \boldsymbol{f} \quad \text{in} \quad G_1 \cup G_2, \tag{1a}$$

$$\boldsymbol{u}(0,t) = \boldsymbol{p}(t), \quad \boldsymbol{u}(1,t) = \boldsymbol{q}(t) \quad \text{on} \quad S, \quad \text{and} \quad \boldsymbol{u}(x,0) = \boldsymbol{r}(x) \quad \text{on} \quad \Omega, \tag{1b}$$

and $\boldsymbol{u} = (u_1, \ldots, u_m)^T$ satisfies the following interface conditions

$$[\boldsymbol{u}]\,(d,t) := \boldsymbol{u}(d+,t) - \boldsymbol{u}(d-,t) = \boldsymbol{0}, \qquad -\boldsymbol{E}\left[\!\left[\frac{\partial \boldsymbol{u}}{\partial x}\right]\!\right](d,t) = \boldsymbol{0} \quad \text{along} \quad \Gamma_d, \tag{1c}$$

where $\boldsymbol{E} = \text{diag}(\varepsilon_1, \ldots, \varepsilon_m)$ with $0 < \varepsilon_1 \leq \ldots \leq \varepsilon_m \leq 1$, $\boldsymbol{B} = \text{diag}(b_1, \ldots, b_m)$ and the coupling matrix $\boldsymbol{A} = (a_{ij})_{m\times m}$, and $\boldsymbol{f} = (f_1, \ldots, f_m)^T$. Assume for each $(x,t) \in \overline{G}$, the matrices $\boldsymbol{A}$ and $\boldsymbol{B}$ satisfy

$$a_{ij}(x,t) \leq 0,\ i \neq j, \quad \sum_{k=1}^{m} a_{ik}(x,t) \geq 0, \quad \text{and} \quad b_i(x,t) \geq \beta_i > 0, \quad 1 \leq i, j \leq m. \tag{2}$$

Let $\beta = \min_i \beta_i$. Further, it is assumed that the source term f_i, $1 \leq i \leq m$, is sufficiently smooth on $\overline{G} \setminus \Gamma_d$, and have finite jump discontinuities in the spatial-variable only along the interface Γ_d. The widths of these layers depend on perturbation parameters $\varepsilon_1, \ldots, \varepsilon_m$. The boundary layers appearing in the kth component of the solution to the system (1a), at the vicinity of the boundary $x = 0$, have a width of $\mathcal{O}(\varepsilon_k \ln \frac{1}{\varepsilon_k})$, for each $1 \leq k \leq m$ [9, 11]. Accordingly, we assumed that $d \neq \mathcal{O}(\varepsilon_k^r \ln \frac{1}{\varepsilon_k})$, for any $1 \leq k \leq m$ and $r \geq 1$, that is, the point d is sufficiently away from the boundary layer regions; so the associated interior layers in the solution's component do not interact/overlap with the boundary layers. A scalar singularly perturbed convection-reaction-diffusion initial-boundary value problem with discontinuous coefficients (or discontinuous inhomogeneous term) is studied in [5, 13], where the boundary/initial conditions are sufficiently smooth and satisfy the compatibility conditions at the corners. A convection-diffusion initial-boundary value problem that has an interior layer in the initial condition is investigated in [6]. A considerably large amount of work is already available for the numerical aspect of singularly perturbed initial-boundary/boundary value linear convection-diffusion problems having only boundary layers in their solutions (see [3, 4, 7, 8, 10] and the references therein). In [15–19], singularly perturbed linear system of steady and unsteady reaction-diffusion equations and initial value problems with a interior layers are considered. However, we know of very few articles that deal with the time-dependent convection-diffusion problems having boundary and interior layers. Shishkin et al. considered a class of singularly perturbed convection-diffusion problems with discontinuous coefficients and source term across the interface [22]; in this article, an almost parameter-uniform convergence is proved. In [1], the author has considered a time-dependent problem in a composite domain, in one part of the domain the problem is parabolic reaction-diffusion type and in the other part of the domain the problem is parabolic convection-diffusion type; therein, an inverse-monotone finite volume method on a condensed Shishkin meshes is used for discretization, and an almost second-order parameter-uniform convergence convergence in the spatial is proved. In [5, 13], numerical techniques for some singularly perturbed time-dependent problems with discontinuous coefficients are designed, and their parameter-uniform convergence is investigated. A class of singularly perturbed parabolic convection-diffusion problems exhibiting strong interior layers is considered in [12]; therein, the authors proved the parameter–uniform convergence of the method in the discrete maximum norm. In [14], a finite-difference upwind scheme is constructed for a two-dimensional singularly perturbed convection-reaction-diffusion problem using an appropriate mesh fitted to the interior and boundary layers and the scheme is proved to be an almost first-order parameter-uniformly convergent. In [20], a parameters-uniform numerical method for a time-dependent weakly coupled system of two convection-diffusion equations that has a discontinuity, along the line Γ_d, in the source term is considered; therein,it is proved that the numerical method is parameters-uniformly convergent of almost first-order in space and first-order in time concerning both perturbation parameters.

However, we know no article dealing with the numerical analysis of the system of $m(> 2)$ parabolic convection-diffusion problem with an interface. This article aims to design and analyze a parameter-uniform numerical method for a class of coupled system of $m(\geq 2)$ parabolic convection-diffusion equations with an interface of the type (1). The considered problem has discontinuities in the source term across the interface Γ_d, and the magnitude of the diffusion parameters are different and can be arbitrarily small. Therefore, interacting and overlapping boundary layers near $x = 0$ and weak interior layers to the right of the interface Γ_d appear in the solution. The presence of interior and boundary layers in the solution reduces the desired order of accuracy in any numerical technique applied to the problem. We discretize the problem using a special finite difference scheme on an appropriate Shishkin mesh that is condensed in the layer regions. We decompose the exact solution into regular and layer components and derive some proper bounds on the solution and its derivatives. Using these bounds and an appropriate decomposition of the numerical solution, we prove parameter-uniform convergence for the approximation generated by the special finite difference scheme in a discrete maximum norm.

The article is arranged in the following manner. The properties of the continuous solution, a maximum principle for the operator L and stability results are discussed in Sect. 2. In Sect. 2, a decomposition of the solution into regular and singular components is also given to acquire sharper bounds on the solution and its derivatives. In Sect. 3, discretization of the problem, which is appropriate to the layers, is given.

In Sect. 4, the parameters-uniformly convergence of the scheme is proved. Outcome of numerical experiments are demonstrated in Sect. 5, and conclusions are included in Sect. 6.

Notations. Throughout C with or without a subscript and $\boldsymbol{C} = (C_1, \ldots, C_m)^T$ denote a generic positive constant and constant vector, respectively, which are independent of $\varepsilon_1, \ldots, \varepsilon_m$ and the mesh parameters. $\|\cdot\|_S$ denotes the maximum norm, where S is a closed and bounded set.

2 Properties of the Continuous Solution

We assume that the data $\boldsymbol{p}$, $\boldsymbol{q}$ and $\boldsymbol{r}$ are sufficiently smooth on the corresponding boundary of the domain $\overline{G}$ and satisfy the following compatibility conditions.

At the corners $(0, 0)$ and $(1, 0)$:

$$\boldsymbol{r}(0) = \boldsymbol{p}(0), \quad \boldsymbol{r}(1) = \boldsymbol{q}(0), \tag{3a}$$

and

$$\boldsymbol{p}'(0) = \boldsymbol{E}\boldsymbol{r}''(0) + \boldsymbol{B}(0)\boldsymbol{r}'(0) - \boldsymbol{A}(0,0)\boldsymbol{r}(0) + \boldsymbol{f}(0,0), \tag{3b}$$

$$\boldsymbol{q}'(0) = \boldsymbol{E}\boldsymbol{r}''(1) + \boldsymbol{B}(1)\boldsymbol{r}'(1) - \boldsymbol{A}(1,0)\boldsymbol{r}(1) + \boldsymbol{f}(1,0). \tag{3c}$$

At the point $(d, 0)$:

$$[\boldsymbol{r}]\,(d) := \boldsymbol{r}(d+) - \boldsymbol{r}(d-) = \boldsymbol{0}, \qquad -\boldsymbol{E}\left[\boldsymbol{r}'\right](d) = \boldsymbol{0}, \tag{3d}$$

$$-\boldsymbol{E}\left[\boldsymbol{r}''\right](d) - \boldsymbol{B}\left[\boldsymbol{r}'\right](d) - \left[\boldsymbol{f}\right](d,0) = \boldsymbol{0}. \tag{3e}$$

Under the above assumptions, the problem (1) has a solution $\boldsymbol{u} \in C(\overline{G})^m \cap C^{1+\gamma}(G)^m \cap C^{3+\gamma}(G_1 \cup G_2)^m$, $0 < \gamma \leq 1$ [20]. Here onwards, without the loss of any generality, we assume that the boundary and initial conditions (1b) are homogeneous, that is, and $\boldsymbol{p} \equiv \boldsymbol{q} \equiv \boldsymbol{0}$ on $[0,\ T]$ and $\boldsymbol{r} \equiv \boldsymbol{0}$ on $\overline{\Omega}$.

The differential operator $\boldsymbol{L}$ defined in (1a) satisfies the following maximum principle which can be proved analogusly as in [20].

Theorem 1 (Maximum Principle) *Suppose a map $\boldsymbol{\Psi} \in C(\overline{G})^m \cap C^2(G_1 \cup G_2)^m$ satisfies $\boldsymbol{\Psi} \geq \boldsymbol{0}$ on ∂G, $\left[\frac{\partial \boldsymbol{\Psi}}{\partial x}\right](d,t) \leq \boldsymbol{0}$ along Γ_d, and $\boldsymbol{L\Psi} \geq \boldsymbol{0}$ in $G_1 \cup G_2$, then $\boldsymbol{\Psi} \geq \boldsymbol{0}$ in $\overline{G}$.*

The stability result is a direct consequence of Theorem 1 which is given as follows.

Corollary 1 *Suppose $\boldsymbol{u}$ solves (1). Then*

$$\|\boldsymbol{u}\|_{\overline{G}} \leq \max\left\{\|\boldsymbol{u}\|_{\partial G},\ \frac{1}{\beta}\|\boldsymbol{f}\|_{\overline{G}}\right\}.$$

To establish parameters-uniform convergence of the numerical scheme depicted in Sect. 3, we give some bounds on the solution and its derivatives in the following Theorem.

Theorem 2 *Suppose* $\boldsymbol{u}$ *is a solution of* (1) *satisfying* $\boldsymbol{u} \in C(\overline{G})^m \cap C^{1+\gamma}(G)^m \cap C^{2+\gamma}(G_1 \cup G_2)^m$, $0 < \gamma \leq 1$. *Let the integers* k, m *satisfy* $0 \leq l \leq 2$, $0 \leq k + l \leq 2$. *Then for each* $(x, t) \in G_1 \cup G_2$

$$\left|\frac{\partial^{k+l} u_i}{\partial x^k \partial t^l}(x, t)\right| \leq C\varepsilon_i^{-k}, \quad \left|\frac{\partial^3 u_i}{\partial x^3}(x, t)\right| \leq C\varepsilon_i^{-1}(\varepsilon_i^{-2} + \sum_{k=1,\ k\neq i}^{m} \varepsilon_k^{-1}), \quad 1 \leq i \leq m.$$

Proof For the case $k = 0,\ l = 0$, the proof follows from Corollary 1. For the other cases, the proof follows using the idea given in [21, Part II, Sect. 2.2] and [2, 20]. □

The reduced solution $\boldsymbol{u}_0$ corresponding to (1) is a solution of the reduced problem defined as follows:

$$\frac{\partial \boldsymbol{u}_0}{\partial t} - \boldsymbol{B}\frac{\partial \boldsymbol{u}_0}{\partial x} + \boldsymbol{A}\boldsymbol{u}_0 = \boldsymbol{f} \ \text{ in } G_1 \cup G_2, \tag{4a}$$

$$\boldsymbol{u}_0(1, t) = \boldsymbol{0} \ \text{ on } S, \ \ \boldsymbol{u}_0(x, 0) = \boldsymbol{0} \ \text{ on } \Omega, \ \text{ and } \ [\boldsymbol{u}_0]\,(d, t) = \boldsymbol{0} \ \text{ along } \Gamma_d. \tag{4b}$$

The bounds given in Theorem 2 are not sharp enough to analyze parameters-uniform convergence of the numerical scheme depicted in Sect. 3. To obtain sharper bounds, $\boldsymbol{u}$ is decomposed into the sum $\boldsymbol{u} = \boldsymbol{v} + \boldsymbol{w}$, where $\boldsymbol{v} = (v_1, \ldots, v_m)^T$ and $\boldsymbol{w} = (w_1, \ldots, w_m)^T$ are regular and singular components, respectively.

The component $\boldsymbol{v}$ is a solution to the problem:

$$\boldsymbol{L}\boldsymbol{v} = \boldsymbol{f} \ \text{ in } G_1 \cup G_2, \tag{5a}$$

$$\boldsymbol{v}(0, t) = \boldsymbol{u}_0(0, t) \ \ \boldsymbol{v}(1, t) = \boldsymbol{0} \ \text{ on } S, \ \boldsymbol{v}(x, 0) = \boldsymbol{0} \ \text{ on } \Omega, \ \text{ and} \tag{5b}$$

$$[\boldsymbol{v}]\,(d, t) = \boldsymbol{0}, \ \boldsymbol{E}\left[\!\!\left[\frac{\partial \boldsymbol{v}}{\partial x}\right]\!\!\right](d, t) = \left[\!\!\left[\frac{\partial \boldsymbol{u}_0}{\partial x}\right]\!\!\right](d, t) \ \text{ along } \Gamma_d. \tag{5c}$$

The component $\boldsymbol{w}$ is a solution to the problem:

$$\boldsymbol{L}\boldsymbol{w} = \boldsymbol{0} \ \text{ in } G_1 \cup G_2, \tag{6a}$$

$$\boldsymbol{w}(0, t) = -\boldsymbol{v}(0, t), \ \boldsymbol{w}(1, t) = \boldsymbol{0} \ \text{ in } S, \ \boldsymbol{w}(x, 0) = \boldsymbol{0} \ \text{ on } \Omega, \ \text{ and} \tag{6b}$$

$$[\boldsymbol{w}]\,(d, t) = \boldsymbol{0}, \ \ \left[\!\!\left[\frac{\partial \boldsymbol{w}}{\partial x}\right]\!\!\right](d, t) = -\left[\!\!\left[\frac{\partial \boldsymbol{v}}{\partial x}\right]\!\!\right](d, t) + \boldsymbol{r}(t) \ \text{ along } \Gamma_d. \tag{6c}$$

We again decompose $\boldsymbol{v}$ into the sum

$$\boldsymbol{v} = \boldsymbol{v}_0 + \left(\prod_{i=1}^{m} \varepsilon_i\right) \boldsymbol{v}_1 + \left(\prod_{i=1}^{m} \varepsilon_i^2\right) \boldsymbol{v}_2, \tag{7}$$

where $\boldsymbol{v}_k = (v_{k1}, \ldots, v_{km})^T$, $k = 0, 1, 2$.

$\boldsymbol{v}_0$ is defined to be a solution to the problem:

$$\frac{\partial \boldsymbol{v}_0}{\partial t} - \boldsymbol{B}\frac{\partial \boldsymbol{v}_0}{\partial x} + \boldsymbol{A}\boldsymbol{v}_0 = \boldsymbol{f} \text{ in } G_1 \cup G_2, \tag{8a}$$

$$\boldsymbol{v}_0(1,t) = \boldsymbol{0} \text{ on } S, \quad \boldsymbol{v}_0(x,0) = \boldsymbol{0} \text{ on } \overline{\Omega}, \quad [\boldsymbol{v}_0](d,t) = \boldsymbol{0} \text{ along } \Gamma_d. \tag{8b}$$

$\boldsymbol{v}_1$ is defined to be a solution to the problem:

$$\frac{\partial \boldsymbol{v}_1}{\partial t} - \boldsymbol{B}\frac{\partial \boldsymbol{v}_1}{\partial x} + \boldsymbol{A}\boldsymbol{v}_1 = \boldsymbol{E}^{-1}\frac{\partial^2 \boldsymbol{v}_0}{\partial x^2} \text{ in } G_1 \cup G_2, \tag{9a}$$

$$\boldsymbol{v}_1(1,t) = \boldsymbol{0} \text{ on } S, \quad \boldsymbol{v}_1(x,0) = \boldsymbol{0} \text{ on } \overline{\Omega}, \quad [\boldsymbol{v}_1](d,t) = \boldsymbol{0} \text{ along } \Gamma_d. \tag{9b}$$

$\boldsymbol{v}_2$ is defined to be a solution to the problem:

$$\boldsymbol{L}\boldsymbol{v}_2 = \boldsymbol{E}^{-1}\frac{\partial^2 \boldsymbol{v}_1}{\partial x^2} \text{ in } G_1 \cup G_2, \tag{10a}$$

$$\boldsymbol{v}_2(0,t) = \boldsymbol{u}_0(0,t) - \boldsymbol{v}_0(0,t) - \boldsymbol{v}_1(0,t), \quad \boldsymbol{v}_2(1,t) = \boldsymbol{0} \text{ on } S, \quad \boldsymbol{v}_2(x,0) = \boldsymbol{0} \text{ on } \overline{\Omega}, \tag{10b}$$

$$[\boldsymbol{v}_2](d,t) = \boldsymbol{0} \text{ and } \left[\!\left[\frac{\partial \boldsymbol{v}_2}{\partial x}\right]\!\right](d,t) = \left(\left[\!\left[\frac{\partial \boldsymbol{u}_0}{\partial x}\right]\!\right] - \left[\!\left[\frac{\partial \boldsymbol{v}_0}{\partial x}\right]\!\right] - \left[\!\left[\frac{\partial \boldsymbol{v}_1}{\partial x}\right]\!\right]\right)(d,t) \text{ along } \Gamma_d. \tag{10c}$$

Again, decompose $\boldsymbol{w}$ into the sum $\boldsymbol{w} = \boldsymbol{w}_1 + \boldsymbol{w}_2$, where the boundary layer component $\boldsymbol{w}_1 = (w_{11}, \ldots, w_{1m})^T$ is the solution to the problem:

$$\boldsymbol{L}\boldsymbol{w}_1 = \boldsymbol{0} \text{ in } G, \tag{11a}$$

$$\boldsymbol{w}_1(0,t) = -\boldsymbol{v}(0,t), \quad \boldsymbol{w}_1(1,t) = \boldsymbol{0} \text{ on } S, \tag{11b}$$

$$\text{and } \boldsymbol{w}_1(x,0) = \boldsymbol{0} \text{ on } \overline{\Omega}, \tag{11c}$$

and the interior layer component $\boldsymbol{w}_2 = (w_{21}, \ldots, w_{2m})^T$ is the solution to the problem:

$$\boldsymbol{L}\boldsymbol{w}_2 = \boldsymbol{0} \text{ in } G_1 \cup G_2, \tag{12a}$$

$$\boldsymbol{w}_2(0,t) = \boldsymbol{w}_2(1,t) = \boldsymbol{0} \text{ on } S, \quad \boldsymbol{w}_2(x,0) = \boldsymbol{0} \text{ on } \overline{\Omega}, \tag{12b}$$

$$[\boldsymbol{w}_2](d,t) = \boldsymbol{0} \text{ and } \left[\!\left[\frac{\partial \boldsymbol{w}_2}{\partial x}\right]\!\right](d,t) = -\left[\!\left[\frac{\partial \boldsymbol{v}}{\partial x}\right]\!\right](d,t) \text{ along } \Gamma_d. \tag{12c}$$

Using (7)–(10), the following Lemma can be proved.

Lemma 1 *The regular component* $\boldsymbol{v}$ *satisfies the following bounds*

$$\left|\frac{\partial^{k+l}\boldsymbol{v}}{\partial x^k \partial t^l}(x,t)\right| \le \boldsymbol{C}, \quad (x,t) \in G_1 \cup G_2, \quad 0 \le l \le 2, \quad 0 \le k+l \le 3.$$

Now using (11), the bounds on the boundary layer component $\boldsymbol{w}_1$ are given in the following Lemma.

Lemma 2 *For any $(x,t) \in G$, components of $\mathbf{w}_1$ and their derivatives satisfy the following bounds*

$$\left|w_{1i}(x,t)\right| \le C \exp\left(-\frac{\alpha x}{\varepsilon_i}\right),\ 1 \le i \le m,$$

$$\left|\frac{\partial^{k+l} w_{1i}}{\partial x^k \partial t^l}(x,t)\right| \le C \sum_{j=i}^{m} \varepsilon_j^{-k} \exp\left(-\frac{\alpha x}{\varepsilon_j}\right),\ 0 \le l \le 2,\ 0 \le k+l \le 2,\ 1 \le i \le m,$$

$$\left|\frac{\partial^3 w_{11}}{\partial x^3}(x,t)\right| \le C \sum_{j=1}^{m} \varepsilon_j^{-3} \exp\left(-\frac{\alpha x}{\varepsilon_j}\right),\ \left|\frac{\partial^3 w_{1i}}{\partial x^3}(x,t)\right| \le C\varepsilon_i^{-1} \sum_{j=i}^{m} \varepsilon_j^{-2} \exp\left(-\frac{\alpha x}{\varepsilon_j}\right),$$

for $2 \le i \le m$. Now using (12) and Lemma 2, the bounds on the interior layer component $\boldsymbol{w}_2$ are given in the following Lemma.

Lemma 3 *For any $(x,t) \in G$, components of $\mathbf{w}_2$ satisfy the following bounds*

$$|\mathbf{w}_2(x,t)| \le \varepsilon_m C, \qquad \left|\frac{\partial^l \mathbf{w}_2}{\partial t^l}(x,t)\right| \le C,$$

$$\left|\frac{\partial^{k+l} w_{2i}}{\partial x^k \partial t^l}(x,t)\right| \le \begin{cases} C \sum_{j=i}^{m} \varepsilon_j^{1-k} \exp\left(-\frac{\alpha x}{\varepsilon_j}\right), & (x,t) \in G_1, \\ C \sum_{j=i}^{m} \varepsilon_j^{1-k} \exp\left(-\frac{\alpha (x-d)}{\varepsilon_j}\right), & (x,t) \in G_2, \end{cases}$$

and $$\left|\frac{\partial^3 w_{2i}}{\partial x^3}(x,t)\right| \le \begin{cases} C\varepsilon_i^{-1} \sum_{j=1}^{m} \varepsilon_j^{-1} \exp\left(-\frac{\alpha x}{\varepsilon_j}\right), & (x,t) \in G_1, \\ C\varepsilon_i^{-1} \sum_{j=1}^{m} \varepsilon_j^{-1} \exp\left(-\frac{\alpha (x-d)}{\varepsilon_j}\right), & (x,t) \in G_2. \end{cases}$$

for $l = 1, 2$, $0 \le k + l \le 3$ and $1 \le i \le m$.

3 Discretization of the Problem

3.1 The Mesh

Let $\overline{\Omega}^N : 0 = x_0 < x_1 < \ldots < x_N = 1$ be a partition of $\overline{\Omega}$ and $\overline{S}^M : 0 = t_0 < t_1 < \ldots < t_M = T$ be a partition of $\overline{S}$. Define $\overline{G}^{N,M} := \overline{\Omega}^N \times \overline{S}^M$ to be discretization of G. We consider a uniform mesh in time-variable, that is, $\overline{S}^M = \{t_k = k\Delta t,\ 0 \le k \le M,\ \Delta t = T/M\}$. However, in spatial-variable, we construct a variant of piecewise-uniform Shishkin mesh $\overline{\Omega}^N$. Define the transition parameters

$$\sigma_m^L := \min\left\{\frac{d}{2}, \varepsilon_m\alpha_0 \ln N\right\}, \quad \sigma_m^R := \min\left\{\frac{(1-d)}{2}, \varepsilon_m\alpha_0 \ln N\right\},$$

$$\sigma_l^L := \min\left\{\frac{\sigma_{l+1}^L}{2}, \varepsilon_l\alpha_0 \ln N\right\}, \quad \sigma_l^R := \min\left\{\frac{\sigma_{l+1}^R}{2}, \varepsilon_l\alpha_0 \ln N\right\},$$

for $l = m-1, m-2, \ldots, 1$, where $\alpha_0 \geq (1/\beta)$. For the construction of piecewise-uniform Shishkin mesh, we assume N to be a multiple of $4m$. Let $\sigma_0^L = \sigma_0^R = 0$. We divide $\overline{\Omega}_1$ into $m+1$ sub-intervals $\cup_{l=1}^m [\sigma_{l-1}^L, \sigma_l^L] \cup [\sigma_m^L, d]$. Further, for $1 \leq l \leq m$, $[\sigma_{l-1}^L, \sigma_l^L]$ is divided into $N/4m$ equidistant elements and $[\sigma_m^L, d]$ is divided into $N/4$ equidistant elements. Similarly, $\overline{\Omega}_2$ is partition into $m+1$ sub-intervals $\cup_{l=1}^m [d+\sigma_{l-1}^R, d+\sigma_l^R] \cup [d+\sigma_m^R, 1]$, and for $1 \leq l \leq m$, $[d+\sigma_{l-1}^R, d+\sigma_l^R]$ is partition into $N/4m$ equidistant elements and $[d+\sigma_m^R, 1]$ is divided into $N/4$ equidistant elements. The nodal points so constructed are denoted by $\overline{\Omega}_1^N := \{x_i\}_{i=0}^{\frac{N}{2}}$ and $\overline{\Omega}_2^N := \{x_i\}_{i=\frac{N}{2}}^N$, respectively, and $\overline{\Omega}^N = \overline{\Omega}_1^N \cup \overline{\Omega}_2^N$. Let the ith mesh size $h_i = x_i - x_{i-1}$ and $\overline{h}_i = (h_i + h_{i+1})/2$. Let $\partial G^{N,M} := \overline{G}^{N,M} \cap \partial G$ and $G^{N,M} := \overline{G}^{N,M} \setminus \partial G^{N,M}$.

3.2 The Discrete Problem

Define the discrete operator $\boldsymbol{L}^{N,M}$ as follows: for any mesh function $\boldsymbol{U}$ on $\overline{G}^{N,M}$

$$\boldsymbol{L}^{N,M}\boldsymbol{U} := D_t^-\boldsymbol{U} - \boldsymbol{E}\delta_x^2\boldsymbol{U} - \boldsymbol{B}D_x^+\boldsymbol{U} + \boldsymbol{A}\boldsymbol{U} \text{ for } (x_i, t_k) \in G^{N,M}, \tag{13}$$

where

$$\delta_x^2 Z(x_i, t_k) = \frac{(D_x^+ Z(x_i, t_k) - D_x^- Z(x_i, t_k))}{\overline{h}_i}, \quad D_x^+ Z(x_i, t_k) = \frac{Z(x_{i+1}, t_k) - Z(x_i, t_k)}{h_{i+1}},$$
$$D_x^- Z(x_i, t_k) = \frac{Z(x_i, t_k) - Z(x_{i-1}, t_k)}{h_i}, \quad D_t^- Z(x_i, t_k) = \frac{Z(x_i, t_k) - Z(x_i, t_{k-1})}{\Delta t}.$$

The discrete analog of (1) is to find a mesh function $\boldsymbol{U}$ such that for $(x_i, t_k) \in \overline{G}^{N,M}$ such that

$$\boldsymbol{L}^{N,M}\boldsymbol{U}(x_i, t_k) = \overline{\boldsymbol{f}}(x_i, t_k) \text{ for } (x_i, t_k) \in G^{N,M} \tag{14a}$$

subject to

$$\boldsymbol{U}(x_0, t_k) = \boldsymbol{p}(t_k), \ \boldsymbol{U}(x_N, t_k) = \boldsymbol{q}(t_k), \ \forall t_k \in S^M, \text{ and } \boldsymbol{U}(x_i, t_0) = \boldsymbol{r}(x_i), \ \forall x_i \in \Omega^N, \tag{14b}$$

where

$$\overline{f}(x_i,t_k) := \begin{cases} \dfrac{h_i f(x_i-,t_k) + h_{i+1} f(x_i+,t_k)}{2\overline{h}_i}, & (x_i,t_k) \in \Gamma_d, \\ f(x_i,t_k), & \text{otherwise.} \end{cases}$$

The discrete operator $\boldsymbol{L}^{N,M}$ satisfies the following discrete maximum principle.

Lemma 4 (Discrete Maximum Principle) *Suppose a mesh function* $\boldsymbol{Z}$ *satisfies* $\boldsymbol{Z}(x_i,t_k) \geq \boldsymbol{0}$ *for* $(x_i,t_k) \in \partial G^{N,M}$, $\boldsymbol{L}^{N,M}\boldsymbol{Z}(x_i,t_k) \geq \boldsymbol{0}$ *for* $(x_i,t_k) \in G^{N,M}$. *Then* $\boldsymbol{Z}(x_i,t_k) \geq \boldsymbol{0}$ *for all* $(x_i,t_k) \in \overline{G}^{N,M}$.

Proof This Lemma can be proved using similar arguments given in [20]. □

Corollary 2 *Let the mesh function* $\boldsymbol{U}$ *be the solution of (14), then*

$$\|\boldsymbol{U}\|_{\overline{G}^{N,M}} \leq C \max\{\|\boldsymbol{U}\|_{\partial G^{N,M}}, \ \|\boldsymbol{L}^{N,M}\boldsymbol{U}\|_{G^{N,M}}\},$$

where $\|\cdot\|_{\overline{G}^{N,M}}$ *denotes the discrete maximum norm.*

Proof The proof follows using a suitable barrier function and Lemma 4. □

4 Convergence Analysis

Decomposed the discrete solution $\boldsymbol{U}$ into the sum $\boldsymbol{U} = \boldsymbol{V} + \boldsymbol{W}$, where the mesh function $\boldsymbol{V}$ is a solution to the following problem:

$$\boldsymbol{L}^{N,M}\boldsymbol{V}(x_i,t_k) = \boldsymbol{f}(x_i,t_k), \qquad \text{for all } (x_i,t_k) \in G^{N,M} \setminus \Gamma_d, \tag{15a}$$

$$\boldsymbol{V}(0,t_k) = \boldsymbol{v}(0,t_k), \quad \boldsymbol{V}(d,t_k) = \boldsymbol{v}(d,t_k), \quad \boldsymbol{V}(1,t_k) = \boldsymbol{v}(1,t_k), \quad \boldsymbol{V}(x_i,0) = \boldsymbol{v}(x_i,0), \tag{15b}$$

and the function $\boldsymbol{W}$ is a solution to the following problem:

$$\boldsymbol{L}^{N,M}\boldsymbol{W}(x_i,t_k) = \boldsymbol{0}, \quad \text{for all } (x_i,t_k) \in G^{N,M} \setminus \{(d,t_k) : t_k \in S^M\}, \tag{16a}$$

$$\left.\begin{aligned} \boldsymbol{W}(0,t_k) &= \boldsymbol{w}(0,t_k), \quad \boldsymbol{W}(1,t_k) = \boldsymbol{w}(1,t_k), \\ \boldsymbol{W}(x_i,0) &= \boldsymbol{w}(x_i,0), \quad [D_x\boldsymbol{W}](d,t_k) = -[D_x\boldsymbol{V}](d,t_k), \end{aligned}\right\} \tag{16b}$$

where the jump $[D_x\boldsymbol{Z}](d,t_k) := D_x^+\boldsymbol{Z}(d,t_k) - D_x^-\boldsymbol{Z}(d,t_k)$.

Further, decompose $\boldsymbol{W}$ as $\boldsymbol{W} = \boldsymbol{W}_1 + \boldsymbol{W}_2$, where the mesh function $\boldsymbol{W}_1$ is a solution to the following problem:

$$\boldsymbol{L}^{N,M}\boldsymbol{W}_1(x_i,t_k) = \boldsymbol{0}, \quad \text{for all } (x_i,t_k) \in G^{N,M}, \tag{17a}$$

$$\boldsymbol{W}_1(0,t_k) = \boldsymbol{w}(0,t_k), \quad \boldsymbol{W}_1(1,t_k) = \boldsymbol{0}, \quad \boldsymbol{W}_1(x_i,0) = \boldsymbol{w}_1(x_i,0), \tag{17b}$$

and the mesh function $\boldsymbol{W}_2$ is a solution to the following problem:

$$\boldsymbol{L}^{N,M}\boldsymbol{W}_2(x_i,t_k)=\boldsymbol{0},\quad \text{for all } (x_i,t_k)\in G^{N,M}\setminus\Gamma_d, \tag{18a}$$

$$\left.\begin{array}{l}\boldsymbol{W}_2(0,t_k)=\boldsymbol{0}=\boldsymbol{W}_2(1,t_k),\quad \boldsymbol{W}_2(x_i,0)=\boldsymbol{w}_2(x_i,0),\\ \left[D_x\boldsymbol{W}_2\right](d,t_k)=-\left[D_x\boldsymbol{V}\right](d,t_k)-\left[D_x\boldsymbol{W}_1\right](d,t_k).\end{array}\right\} \tag{18b}$$

Using Taylor's expansion of any function $\boldsymbol{\Phi}$ having sufficient regularity, $i=1,\ldots,N/2-1,N/2+1,\ldots,N$ and $j=1,\ldots,M$, we have the following bounds:

$$\left|\left(\frac{\partial}{\partial t}-D_t^-\right)\boldsymbol{\Phi}(x_i,t_k)\right|\le C(t_k-t_{j-1})\max_{s\in[t_{j-1},t_k]}\left|\frac{\partial^2\boldsymbol{\Phi}}{\partial t^2}(x_i,s)\right|, \tag{19a}$$

$$\left|\left(\frac{\partial^2}{\partial x^2}-\delta_x^2\right)\boldsymbol{\Phi}(x_i,t_k)\right|\le C\max_{s\in[x_{i-1},x_{i+1}]}\left|\frac{\partial^2\boldsymbol{\Phi}}{\partial x^2}(s,t_k)\right|, \tag{19b}$$

$$\left|\left(\frac{\partial^2}{\partial x^2}-\delta_x^2\right)\boldsymbol{\Phi}(x_i,t_k)\right|\le C(h_{i+1}-h_i)\max_{s\in[x_{i-1},x_{i+1}]}\left|\frac{\partial^3\boldsymbol{\Phi}}{\partial x^3}(s,t_k)\right|, \tag{19c}$$

$$\left|\left(\frac{\partial^2}{\partial x^2}-\delta_x^2\right)\boldsymbol{\Phi}(x_i,t_k)\right|\le C(h_{i+1}-h_i)^2\max_{s\in[x_{i-1},x_{i+1}]}\left|\frac{\partial^4\boldsymbol{\Phi}}{\partial x^4}(s,t_k)\right|, \tag{19d}$$

$$\left|\left(\frac{\partial}{\partial x}-D_x^+\right)\boldsymbol{\Phi}(x_i,t_k)\right|\le Ch_{i+1}\max_{s\in[x_i,x_{i+1}]}\left|\frac{\partial^2\boldsymbol{\Phi}}{\partial x^2}(s,t_k)\right|. \tag{19e}$$

Along the interface: $(x_{N/2},t_k)\in\Gamma_d$.

We approximate $-\boldsymbol{E}\left[\!\left[\frac{\partial\boldsymbol{u}}{\partial x}\right]\!\right](x_{\frac{N}{2}},t_k)$ using the simple first-order approximation $-\frac{\boldsymbol{E}}{\bar{h}_{\frac{N}{2}}}(D_x^+\boldsymbol{U}-D_x^-\boldsymbol{U})(x_{\frac{N}{2}},t_k)$.

Using $[\boldsymbol{u}](x_{\frac{N}{2}},t_k)=\boldsymbol{0}$ and $-\boldsymbol{E}\left[\!\left[\frac{\partial\boldsymbol{u}}{\partial x}\right]\!\right](x_{\frac{N}{2}},t_k)=\boldsymbol{0}$ along the interface Γ_d, we obtain

$$\begin{aligned}&-\frac{\boldsymbol{E}}{\bar{h}_{\frac{N}{2}}}(D_x^+\boldsymbol{U}-D_x^-\boldsymbol{U})(x_{\frac{N}{2}},t_k)\\ &=\frac{1}{2\bar{h}_{\frac{N}{2}}}\left(-h_{\frac{N}{2}+1}\boldsymbol{E}\frac{\partial^2\boldsymbol{u}}{\partial x^2}(x_{\frac{N}{2}}+,t_k)-h_{\frac{N}{2}}\boldsymbol{E}\frac{\partial^2\boldsymbol{u}}{\partial x^2}(x_{\frac{N}{2}}-,t_k)\right)+\boldsymbol{E}\left(\mathcal{O}(h^2_{\frac{N}{2}+1}+h^2_{\frac{N}{2}})\right).\end{aligned}$$

Using (1a) in the above equation, we have

$$
\begin{aligned}
&-\frac{\boldsymbol{E}}{\overline{h}_{\frac{N}{2}}}(D_x^+\boldsymbol{U}-D_x^-\boldsymbol{U})(x_{\frac{N}{2}},t_k)\\
&=\frac{1}{2\overline{h}_{\frac{N}{2}}}\left(h_{\frac{N}{2}+1}\left(-\frac{\partial \boldsymbol{u}}{\partial t}+\boldsymbol{B}\frac{\partial \boldsymbol{u}}{\partial x}-\boldsymbol{A}\boldsymbol{u}+\boldsymbol{f}\right)(x_{\frac{N}{2}}+,t_k)\right.\\
&\left.+h_{\frac{N}{2}}\left(-\frac{\partial \boldsymbol{u}}{\partial t}+\boldsymbol{B}\frac{\partial \boldsymbol{u}}{\partial x}-\boldsymbol{A}\boldsymbol{u}+\boldsymbol{f}\right)(x_{\frac{N}{2}}-,t_k)\right)+\boldsymbol{E}\Big(\mathcal{O}(h_{i+1}^2+h_{\frac{N}{2}}^2)\Big)\\
&=\left(-\frac{\partial \boldsymbol{u}}{\partial t}+\boldsymbol{B}\frac{\partial \boldsymbol{u}}{\partial x}-\boldsymbol{A}\boldsymbol{u}+\overline{\boldsymbol{f}}\right)(x_{\frac{N}{2}},t_k)+\boldsymbol{E}\Big(\mathcal{O}(h_{\frac{N}{2}+1}^2+h_{\frac{N}{2}}^2)\Big).
\end{aligned}
$$

This gives the scheme

$$
\boldsymbol{L}^{N,M}\boldsymbol{U}(x_{\frac{N}{2}},t_k)=\overline{\boldsymbol{f}}(x_{\frac{N}{2}},t_k)+\frac{\boldsymbol{E}}{\overline{h}_{\frac{N}{2}}}\Big(\mathcal{O}(h_{\frac{N}{2}+1}^2+h_{\frac{N}{2}}^2)\Big)++\mathcal{O}(h_{\frac{N}{2}+1})+\mathcal{O}(h_{\frac{N}{2}+1})+\mathcal{O}(\Delta t).
$$

Therefore, using Taylor's expansion in the left and right neighborhood of the interface Γ_d, we obtain

$$
\begin{aligned}
&\left|\boldsymbol{L}^{N,M}(\boldsymbol{u}-\boldsymbol{U})(x_{\frac{N}{2}},t_k)\right|\\
&\le C\Big(\Delta t\max_{s\in[t_{j-1},t_k]}\left|\frac{\partial^2\boldsymbol{u}}{\partial t^2}(x_i,s)\right|+h_{\frac{N}{2}+1}\max_{s\in[x_{\frac{N}{2}},x_{\frac{N}{2}+1}]}\left|\frac{\partial^2\boldsymbol{u}}{\partial x^2}(s+,t_k)\right|\\
&+\frac{h_{\frac{N}{2}}^2}{\overline{h}_{\frac{N}{2}}}\boldsymbol{E}\max_{s\in[x_{\frac{N}{2}-1},x_{\frac{N}{2}}]}\left|\frac{\partial^3\boldsymbol{u}}{\partial x^3}(s-,t_k)\right|+\frac{h_{\frac{N}{2}}^2}{\overline{h}_{\frac{N}{2}}}\boldsymbol{E}\max_{s\in[x_{\frac{N}{2}},x_{\frac{N}{2}+1}]}\left|\frac{\partial^3\boldsymbol{u}}{\partial x^3}(s+,t_k)\right|\Big).
\end{aligned}\tag{20}
$$

Using the decompositions of $\boldsymbol{U}$ and $\boldsymbol{u}$ and the bounds given in Lemma 1, the truncation errors satisfy the following estimates.

Lemma 5 *Let $\boldsymbol{v}$ and $\boldsymbol{V}$ be the solution to the problems (5) and (15), respectively. Then*

$$
\left|\boldsymbol{L}^{N,M}(\boldsymbol{v}-\boldsymbol{V})(x_i,t_k)\right|\le(N^{-1}\ln N+M^{-1})\boldsymbol{C},\quad \textit{for}\ (x_i,t_k)\in\overline{G}^{N,M}.
$$

Furthermore,

$$
\|\boldsymbol{v}-\boldsymbol{V}\|_{\overline{G}^{N,M}}\le C(N^{-1}\ln N+M^{-1}).
$$

Proof If $(x_i,t_k)\in\overline{G}^{N,M}$ but $(x_i,t_k)\notin\Gamma_d$, then classical technique can be applied to prove the Lemma (see for example [20]). Along the interface, that is, $(x_i,t_k)\in\Gamma_d$, we use the following technique.

Using the decompositions of exact and discrete solutions $\boldsymbol{u}$ and $\boldsymbol{U}$, respectively, (16) and the bounds in (20), we derive the following estimates.

$$\left|\boldsymbol{L}^{N,M}(\boldsymbol{v}-\boldsymbol{V})(x_{\frac{N}{2}},t_k)\right|$$
$$\leq C\Big(\Delta t \max_{s\in[t_{j-1},t_k]}\left|\frac{\partial^2 \boldsymbol{v}}{\partial t^2}(x_i,s)\right| + h_{\frac{N}{2}+1}\max_{s\in[x_{\frac{N}{2}},x_{\frac{N}{2}+1}]}\left|\frac{\partial^2 \boldsymbol{v}}{\partial x^2}(s+,t_k)\right|$$
$$+\frac{h^2_{\frac{N}{2}}}{\bar{h}_{\frac{N}{2}}}\boldsymbol{E}\max_{s\in[x_{\frac{N}{2}-1},x_{\frac{N}{2}}]}\left|\frac{\partial^3 \boldsymbol{v}}{\partial x^3}(s-,t_k)\right| + \frac{h^2_{\frac{N}{2}}}{\bar{h}_{\frac{N}{2}}}\boldsymbol{E}\max_{s\in[x_{\frac{N}{2}},x_{\frac{N}{2}+1}]}\left|\frac{\partial^3 \boldsymbol{v}}{\partial x^3}(s+,t_k)\right|\Big).$$

Using the bounds given in Lemma 1, it can be easily obtained that

$$\left|\boldsymbol{L}^{N,M}(\boldsymbol{v}-\boldsymbol{V})(x_{\frac{N}{2}},t_k)\right| \leq (N^{-1}\ln N + M^{-1})\boldsymbol{C}.$$

Define the mesh function $\Psi^{\pm}(x_i,t_k)$ as

$$\Psi^{\pm}(x_i,t_k) := \begin{cases}(N^{-1}\ln N + M^{-1})(d-x_i)\boldsymbol{C} \pm (\boldsymbol{V}-\boldsymbol{v})(x_i,t_k) & \text{for } (x_i,t_k)\in \overline{G}_1^N,\\ (N^{-1}\ln N + M^{-1})(1-x_i)\boldsymbol{C} \pm (\boldsymbol{V}-\boldsymbol{v})(x_i,t_k) & \text{for } (x_i,t_k)\in \overline{G}_2^N.\end{cases}$$

Using Lemma 4 and the barrier function $\Psi^{\pm}$, we conclude that the error in the regular component satisfies

$$|(\boldsymbol{V}-\boldsymbol{v})(x_i,t_k)| \leq (N^{-1}\ln N + M^{-1})\boldsymbol{C}, \quad \text{for } (x_i,t_k)\in \overline{G}^{N,M},$$

and hence

$$\|(\boldsymbol{V}-\boldsymbol{v}\|_{\overline{G}^{N,M}} \leq C(N^{-1}\ln N + M^{-1}).$$

□

Lemma 6 *Let $\boldsymbol{w}$ and $\boldsymbol{W}$ be the solution to the problems (6) and (16), respectively. Then*

$$\left|\boldsymbol{L}^{N,M}(\boldsymbol{w}-\boldsymbol{W})(x_i,t_k)\right| \leq \begin{cases}(N^{-1}\ln N + M^{-1})\boldsymbol{C}, & \text{for } (x_i,t_k)\notin \Gamma_d,\\ \left(\dfrac{N^{-1}(\ln N)^2}{1+\varepsilon_1\ln N} + M^{-1}\right)\boldsymbol{C}, & \text{for } (x_i,t_k)\in \Gamma_d.\end{cases}$$

Furthermore,

$$\|\boldsymbol{w}-\boldsymbol{W}\|_{\overline{G}^{N,M}} \leq C(N^{-1}+M^{-1}).$$

Proof If $(x_i,t_k)\in \overline{G}^{N,M}$ but $(x_i,t_k)\notin \Gamma_d$, then classical technique can be applied to prove the Lemma (see for example [20]). Along the interface, that is, $(x_i,t_k)\in \Gamma_d$, we use the following technique.

Using the decomposition of exact and discrete solutions $\boldsymbol{u}$ and $\boldsymbol{U}$, respectively, (16) and the bounds in (20), we derive the following estimates.

$$\left|L^{N,M}(\boldsymbol{w}-\boldsymbol{W})(x_{\frac{N}{2}},t_k)\right|$$
$$\leq C\Big(\Delta t \max_{s\in[t_{j-1},t_k]}\left|\frac{\partial^2\boldsymbol{w}}{\partial t^2}(x_i,s)\right| + h_{\frac{N}{2}+1}\max_{s\in[x_{\frac{N}{2}},x_{\frac{N}{2}+1}]}\left|\frac{\partial^2\boldsymbol{w}}{\partial x^2}(s+,t_k)\right|$$
$$+\frac{h^2_{\frac{N}{2}}}{\bar{h}_{\frac{N}{2}}}\boldsymbol{E}\max_{s\in[x_{\frac{N}{2}-1},x_{\frac{N}{2}}]}\left|\frac{\partial^3\boldsymbol{w}}{\partial x^3}(s-,t_k)\right| + \frac{h^2_{\frac{N}{2}+1}}{\bar{h}_{\frac{N}{2}}}\boldsymbol{E}\max_{s\in[x_{\frac{N}{2}},x_{\frac{N}{2}+1}]}\left|\frac{\partial^3\boldsymbol{w}}{\partial x^3}(s+,t_k)\right|\Big).$$

Using Theorem 2, Lemmas 2 and 3, it holds that $\left|\frac{\partial^2\boldsymbol{w}}{\partial t^2}(x_{\frac{N}{2}},s)\right|\leq C$ and hence we obtain

$$\Delta t \max_{s\in[t_{j-1},t_k]}\left|\frac{\partial^2\boldsymbol{w}}{\partial t^2}(x_{\frac{N}{2}},s)\right|\leq M^{-1}\boldsymbol{C}. \tag{21}$$

Since $h_{\frac{N}{2}+1}=\dfrac{8\varepsilon_1\ln N}{\alpha N}$, $h_{\frac{N}{2}}\leq CN^{-1}$ and $\boldsymbol{w}=\boldsymbol{w}_1+\boldsymbol{w}_2$, using the bounds given in Lemmas 2 and 3, we have

$$\varepsilon_l\frac{h^2_{\frac{N}{2}+1}}{\bar{h}_{\frac{N}{2}}}\max_{x\in[x_{\frac{N}{2}},x_{\frac{N}{2}+1}]}\left|\frac{\partial^3 w_l}{\partial x^3}(x+,t_k)\right| = \frac{2\varepsilon_l h^2_{\frac{N}{2}+1}}{h_{\frac{N}{2}+1}+h_{\frac{N}{2}}}\max_{x\in[x_{\frac{N}{2}},x_{\frac{N}{2}+1}]}\left|\frac{\partial^3 w_l}{\partial x^3}(x+,t_k)\right|$$
$$\leq\frac{CN^{-1}(\ln N)^2}{1+\varepsilon_1\ln N}. \tag{22}$$

Now, using the bounds given in Lemmas 2 and 3, and the arguments $\dfrac{\exp\left(\frac{-\alpha x}{\varepsilon_l}\right)}{\varepsilon_l^j}\leq C,\ \ j=1,2,3,\ 1\leq l\leq m$, we obtain

$$\varepsilon_l\frac{h^2_{\frac{N}{2}}}{\bar{h}_{\frac{N}{2}}}\max_{x\in[x_{\frac{N}{2}-1},x_{\frac{N}{2}}]}\left|\frac{\partial^3 w_{1l}}{\partial x^3}(x-,t_k)\right|\leq\frac{CN^{-1}}{1+e_1\ln N},$$

and

$$\varepsilon_l\frac{h^2_{\frac{N}{2}}}{\bar{h}_{\frac{N}{2}}}\max_{x\in[x_{\frac{N}{2}-1},x_{\frac{N}{2}}]}\left|\frac{\partial^3 w_{2l}}{\partial x^3}(x-,t_k)\right|\leq\frac{CN^{-1}}{1+\varepsilon_1\ln N}.$$

Combining the above estimates, we have

$$\varepsilon_l\frac{h^2_{\frac{N}{2}}}{\bar{h}_{\frac{N}{2}}}\max_{x\in[x_{\frac{N}{2}-1},d]}\left|\frac{\partial^3 w_l}{\partial x^3}(x-,t_k)\right|\leq\frac{CN^{-1}}{1+\varepsilon_1\ln N}. \tag{23}$$

In a similar manner, it also holds that

$$\varepsilon_l \frac{h^2_{\frac{N}{2}}}{\bar{h}_{\frac{N}{2}}} \max_{x\in[x_{\frac{N}{2}},\, x_{\frac{N}{2}+1}]} \left|\frac{\partial^2 w_l}{\partial x^2}(x+, t_k)\right| \le \frac{CN^{-1}}{1+\varepsilon_1 \ln N}. \tag{24}$$

Using (21)–(24), we have

$$\left|\boldsymbol{L}^{N,M}(\boldsymbol{w}-\boldsymbol{W})(x_{\frac{N}{2}}, t_k)\right| \le \left(\frac{N^{-1}(\ln N)^2}{1+\varepsilon_1 \ln N} + M^{-1}\right)\boldsymbol{C}.$$

Now, consider the following barrier function $\boldsymbol{\Phi} = (\phi_1, \ldots, \phi_m)^T$ defined by

$$\boldsymbol{\Phi}(x_i, t_k) := (N^{-1}\ln N + M^{-1})(1-x_i)\boldsymbol{C} + \begin{cases} (N^{-1}\ln N + M^{-1})\boldsymbol{C}, & 0 \le i < \frac{N}{2}, \\ \left(\sum_{l=1}^{m} S_{\varepsilon_l,i} N^{-1}\ln N + M^{-1}\right)\boldsymbol{C}, & \frac{N}{2} \le i \le N, \end{cases}$$

where $\boldsymbol{C}$ is sufficiently large and the mesh function $S_{\varepsilon_l,i}$, $1 \le l \le m$, is given by

$$S_{\varepsilon_l,i} := \prod_{j=i+1}^{N} \left(1 + \frac{\alpha h_j}{2\varepsilon_l}\right)^{-1}, \quad \text{for } \frac{N}{2} + 1 \le i \le N-1,$$

with $S_{\varepsilon_l,\frac{N}{2}} := 1$, $S_{\varepsilon_l,N} := C$.

Using the barrier function $\boldsymbol{\Phi}$ and Lemma 4, we obtain the following result

$$\|\boldsymbol{w}-\boldsymbol{W}\|_{\overline{G}^{N,M}} \le (N^{-1}\ln N + M^{-1})\boldsymbol{C}.$$ □

We complete this section with the following main result.

Theorem 3 *Let $\boldsymbol{u}$ and $\boldsymbol{U}$ be the solution to the problems (1) and (14), respectively. Then*

$$\|\boldsymbol{U}-\boldsymbol{u}\|_{\overline{G}^{N,M}} \le C(N^{-1}\ln N + M^{-1}).$$

Proof The proof of the theorem follows, using the decompositions of $\boldsymbol{U}$ and $\boldsymbol{u}$ into regular and singular components, triangular inequality, and the bounds in Lemmas 5 and 6. □

5 Numerical Experiments

We use the following test example to verify the theoretical conclusions and to examine the error estimates numerically for various values of mesh and perturbation parameters. We estimate the maximum point-wise errors and corresponding orders of parameters-uniform convergence for this test example. The outcomes of numerical experiments are highlighted in the tables.

Example *We consider the following weakly coupled system of four convection-diffusion equations of type* (1)*, where the domain* $G = \Omega \times S, \Omega = (0,\ 1), S = (0,\ 1]$*, the reaction coefficient*

$$A = \begin{pmatrix} 4 & -\exp(-(x+t^2)) & -\exp(-(x+t^2)) & -\exp(-(x+t^2)) \\ -\exp(-(x+t^2)) & 5 & -\exp(-(x+t^2)) & -\exp(-(x+t^2)) \\ -\exp(-(x+t^2)) & -\exp(-(x+t^2)) & 6 & -\exp(-(x+t^2)) \\ -\exp(-(x+t^2)) & -\exp(-(x+t^2)) & -\exp(-(x+t^2)) & 7 \end{pmatrix},$$

the convection coefficient

$$\boldsymbol{B} = (\mathrm{diag}\left((1+4x\exp(x)),\ (1+5x\exp(x)),\ (1+2x\exp(x)),\ (1+6x\exp(x))\right))^{-1},$$

$d = 0.5$*, and the source term* $\boldsymbol{f} = (f_1,\ f_2,\ f_3,\ f_4)^T$ *is*

$$f_1(x,t) = \begin{cases} 1, & \text{if } x < 0.5, \\ 1.5, & \text{if } x \geq 0.5, \end{cases} \quad f_2(x,t) = \begin{cases} 2, & \text{if } x < 0.5, \\ 2.5, & \text{if } x \geq 0.5. \end{cases}$$

$$f_3(x,t) = \begin{cases} 3, & \text{if } x < 0.5, \\ 3.5, & \text{if } x \geq 0.5, \end{cases} \quad f_4(x,t) = \begin{cases} 4, & \text{if } x < 0.5, \\ 4.5, & \text{if } x \geq 0.5. \end{cases}$$

Also, the subdomains $G_1 = (0,\ 0.5) \times (0,\ 1]$ *and* $G_2 = (0.5,\ 1) \times (0,\ 1]$*.*

We compute the parameters-uniform errors in the numerical solution and the orders of convergence for the above example using the numerical technique described in Sect. 3. We employ the double mesh principle to estimate errors in the numerical solution as we do not have the exact solution to the above example. The double mesh $\hat{G}^{N,M} := \hat{\Omega}^N \times \hat{S}^M$, where $\hat{\Omega}^N$ is obtained by creating a new mesh points in the middle of each pair of consecutive mesh points $x_{i-1}, x_i \in \overline{\Omega}^N$, $i = 1, \ldots, N$, and $\hat{S}^M := S^{2M}$ (see [20]).

For numerical experiments, we choose the parameter ε_1 is from the set $\mathcal{E}_1 := \{2^{-4j} : j = 0, 1, \ldots, 7\}$ and for each $\varepsilon_l \in \mathcal{E}_l$, $1 \leq l \leq m-1$, we choose the parameter ε_{l+1} from the set $\mathcal{E}_{l+1} := \{2^{-4j} : j = 0, 1, \ldots, 7 \text{ and } 2^{-4j} \geq \varepsilon_l\}$. For various values of parameters N, M and for different choices of parameter $\varepsilon_l \in \mathcal{E}_l$, $1 \leq l \leq m$, we compute the maximum point-wise errors using the expression $D^{N,M}_{\varepsilon_1,\ldots,\varepsilon_m} := \|\hat{\boldsymbol{U}}^{N,M} - \boldsymbol{U}^{N,M}\|_{\overline{G}^{N,M}}$, where $\boldsymbol{U}^{N,M}$ is the numerical solution of (14) using the mesh $\overline{G}^{N,M}$ and $\hat{\boldsymbol{U}}^{N,M}$ is the numerical solution of (14) using the double mesh $\hat{G}^{N,M}$. Next, for each $\varepsilon_l \in \mathcal{E}_l$, the $(\varepsilon_{l+1}, \ldots, \varepsilon_m)$-uniform errors are computed using the expression $D^{N,M}_{\varepsilon_1,\ldots,\varepsilon_l} := \max_{\varepsilon_l \in \mathcal{E}_l,\ l+1 \leq l \leq m} D^{N,M}_{\varepsilon_1,\ldots,\varepsilon_m}$, and parameters-uniform errors are computed using the expression $D^{N,M} := \max_{\varepsilon_l \in \mathcal{E}_l,\ 1 \leq l \leq m} D^{N,M}_{\varepsilon_1,\ldots,\varepsilon_m}$.

Theoretically, the numerical scheme is proved to be first-order parameters-uniformly convergent in time and almost first-order parameters-uniformly conver-

Table 1 The $(\varepsilon_2, \varepsilon_3, \varepsilon_4)$-uniform errors $D_{\varepsilon_1}^{N,M}$, parameters-uniform errors $D^{N,M}$, and spatial orders of parameters-uniform convergence ρ^N for the Example

	N = 128	N = 256	N = 512	N = 1024	N = 2048	N = 4096
2^0	3.51E-03	1.82E-03	9.29E-04	4.69E-04	9.05E-05	4.34E-05
2^{-4}	5.04E-03	2.65E-03	1.36E-03	6.87E-04	1.27E-04	5.05E-04
2^{-8}	3.13E-02	1.96E-02	1.11E-02	5.97E-03	2.10E-03	1.45E-03
2^{-12}	4.45E-02	3.41E-02	2.33E-02	1.50E-02	9.51E-03	5.43E-03
2^{-16}	5.78E-02	4.55E-02	3.08E-02	2.01E-02	1.19E-02	6.58E-03
2^{-20}	6.08E-02	4.74E-02	3.22E-02	2.09E-02	1.22E-02	6.83E-03
2^{-24}	6.27E-02	4.85E-02	3.32E-02	2.14E-02	1.22E-02	6.83E-03
2^{-28}	6.27E-02	4.85E-02	3.32E-02	2.14E-02	1.22E-02	6.83E-03
$D^{N,M}$	6.27E-02	4.85E-02	3.32E-02	2.14E-02	1.22E-02	6.83E-03
ρ^N	0.46	0.66	0.75	0.94	0.96	

gent in space. To verify these outcomes numerically, we estimate the parameters-uniform errors $D^{N,M}$ for the different values of the parameters N and M. Due to different discretization in space and time variables, two types of errors are contributed in the numerical solution by the parameters N and M depending on their numerical values. Accordingly, in the numerical experiments, we attempt to offset the two errors by choosing the parameters N and M effectively while computing the orders of parameters-uniform convergence. To calculate the orders of parameters-uniform convergence in the spatial direction, for a given value of the mesh parameter M in the time direction, we pick the value of the mesh parameter N in the spatial direction such that $N^{-1} \ln N \geq M^{-1}$. Thus, we use the formula $\rho^N := \left(\ln D^{N,M} - \ln D^{2N,M}\right) / (\ln(2 \ln N) - \ln(\ln 2N))$ to compute parameter-uniform convergence orders in the spatial direction. While computing the orders of parameters-uniform convergence in the time direction, for a given value of the mesh parameter N in the spatial direction, we pick the value of the mesh parameter M in the time direction such that $N^{-1} \ln N \leq M^{-1}$. Hence, we use the formula $\eta^M := \left(\ln D^{N,M} - \ln D^{N,2M}\right) / \ln 2$ to compute parameter-uniform convergence orders in the time direction.

We conducted some numerical experiments on the above-given Example with $\alpha_0 = 1 + 6\exp(1)$. The outcomes of numerical experiments employing the numerical technique defined in Section 3 are presented in Tables 1, 2 and 3. For different values of the mesh parameters N, M ($N = M$ for Tables 1 and 3) and each $\varepsilon_l \in \mathcal{E}_k$, $l = 1, \ldots, 4$, the $(\varepsilon_2, \varepsilon_3, \varepsilon_4)$-uniform errors $D_{\varepsilon_1}^{N,M}$ are demonstrated in these tables. The second row from the last in Tables 1, 2 and 3 depicts the parameters-uniform errors $D^{N,M}$. The last row of Tables 1 and 3 corresponds to the orders of parameters-uniform convergence in the spatial direction. In contrast, the last row of Table 2 is associated with the orders of parameters-uniform convergence in the time direction.

Figure 1a portrays the log-log plots of $(N^{-1} \ln N + M^{-1})$ and $D^{N,M}$ on the y-axis versus N on the x-axis using the data from Table 1. Similarly, Fig. 1b portrays the

Table 2 The $(\varepsilon_2, \varepsilon_3, \varepsilon_4)$-uniform errors $D_{\varepsilon_1}^{N,M}$, parameters-uniform errors $D^{N,M}$, and temporal orders of parameters-uniform convergence η^M for the Example

ε_1	N = 176 M = 32	N = 512 M = 64	N = 1280 M = 128	N = 3760 M = 256	N = 8096 M = 512
2^0	1.14E-02	6.52E-03	3.51E-03	1.82E-03	9.30E-04
2^{-4}	1.21E-02	6.47E-03	3.48E-03	1.81E-03	9.42E-04
2^{-8}	3.54E-02	1.56E-02	6.95E-03	2.65E-03	1.27E-03
2^{-12}	5.91E-02	4.86E-02	3.84E-02	1.93E-02	8.13E-03
2^{-16}	6.13E-02	4.93E-02	3.87E-02	2.11E-02	9.66E-03
2^{-20}	7.43E-02	5.81E-02	4.30E-02	2.14E-02	1.03E-02
2^{-24}	8.44E-02	6.85E-02	4.35E-02	2.17E-02	1.08E-02
2^{-28}	8.44E-02	6.85E-02	4.35E-02	2.17E-02	1.08E-02
$D^{N,M}$	8.44E-02	6.85E-02	4.35E-02	2.17E-02	1.08E-02
η^M	0.30	0.66	1.00	1.01	

Table 3 The $(\varepsilon_2, \varepsilon_3, \varepsilon_4)$-uniform errors $D_{\varepsilon_1}^{N,M}$, parameters-uniform errors $D^{N,M}$, and spatial orders of parameters-uniform convergence ρ^N by using f-values rather than $\overline{f}$ along Γ_d for the Example

ε_1	N = 128	N = 256	N = 512	N = 1024	N = 2048	N = 4096
2^0	3.60E-03	1.87E-03	3.64E-04	2.83E-04	8.87E-05	6.95E-05
2^{-4}	5.10E-03	2.68E-03	1.43E-03	6.22E-04	4.56E-04	2.43E-04
2^{-8}	3.94E-02	2.47E-02	5.01E-03	4.34E-03	2.38E-03	1.01E-03
2^{-12}	6.01E-02	5.48E-02	6.81E-03	5.16E-03	3.55E-03	3.07E-03
2^{-16}	6.31E-02	5.64E-02	7.12E-03	5.23E-03	3.97E-03	3.28E-03
2^{-20}	6.40E-02	5.98E-02	7.32E-03	6.17E-03	4.23E-03	3.78E-03
2^{-24}	6.41E-02	6.44E-02	7.63E-03	6.28E-03	4.36E-03	3.74E-03
2^{-28}	6.47E-02	5.98E-02	7.78E-03	6.33E-03	4.44E-03	3.77E-03
$D^{N,M}$	6.47E-02	6.44E-02	7.78E-03	6.33E-03	4.44E-03	3.78E-03
ρ^N	0.01	3.67	0.35	0.59	0.27	

log-log plots of $(N^{-1} \ln N + M^{-1})$ and $D^{N,M}$ on the y-axis versus M on the x-axis using Table 2. The first curve from the top in Fig. 1a is associated with theoretical parameters-uniform error estimates$(N^{-1} \ln N + M^{-1})$. In contrast, the second curve in Fig. 1a is related to the parameters-uniform error estimates $D^{N,M}$. In Fig. 1b, the first curve from the top is associated with the theoretical parameters-uniform error estimates $(N^{-1} \ln N + M^{-1})$. In contrast, the second curve in Fig. 1b relates to the parameters-uniform error estimates $D^{N,M}$. The slope of these curves represents the order of parameters-uniform convergence for their corresponding errors. These curves also indicate that the analogous estimated errors decrease with the increase of mesh sizes.

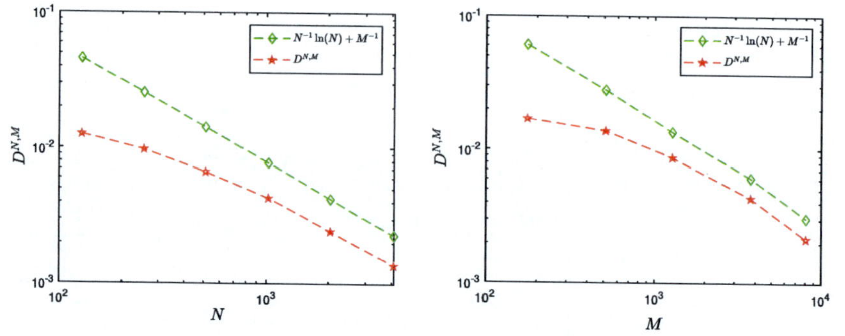

(a) $(N^{-1}\ln N+M^{-1})$, $D^{N,M}$ on y-axis versus N on x-axis.

(b) $(N^{-1}\ln N+M^{-1})$, $D^{N,M}$ on y-axis versus M on x-axis.

Fig. 1 Log-log plots of parameters-uniform errors using the data from Tables 1 and 2, respectively

Along the interface Γ_d, the proposed scheme (14) uses the value $\overline{f}$ rather than the general choice f. Theoretically, this is very crucial to establish almost first-order parameters-uniform convergence of the discrete scheme in the spatial variable. To corroborate in practice, we perform some numerical experiments on the Example data corresponding to the choice of the values of f (that is, $f(x_i, t_k)$ rather than $\overline{f}(x_i, t_k)$) in (14). The outcomes of the numerical experiments are demonstrated in Table 3. Comparing the results of Table 3 with Table 1, we notice the decrease in the orders of parameters-uniform convergence and increase in maximum errors for the larger N in Table 3. Consequently, the value of $\overline{f}$ is essential for theoretical and practical purposes in developing the discrete scheme (14) to attain almost first-order parameters-uniform convergence in the spatial direction.

6 Conclusions

A general weakly coupled system of $m(\geq 2)$ linear parabolic convection-diffusion equations is considered in the regime of singular perturbation problems. The considered system has discontinuities in the source term along the interface Γ_d. The discretization of the domain has been obtained using appropriate piecewise uniform Shishkin mesh, which is condensed in the layer region. For the nodal points not on the interface, the problem is discretized using an upwind central difference scheme. However, the problem is discretized using a special upwind central difference scheme for the nodal points on the interface. A decomposition of solution technique is used for the exact solution and its numerical analog concerning the parameters-uniform convergence analysis of the discrete scheme. Some appropriate bounds on the exact solution and its derivatives have been given. It is proved that the scheme is almost first-order in space and first-order in time parameters-uniformly convergent. We implemented the proposed scheme on a test example to verify our theoretical results. The test example given in Section 5 involves jump discontinuity in the source term, demonstrating that the method preserves its theoretically proven accuracy.

Acknowledgements The authors acknowledge the IIT Delhi HPC facility for computational resources. This research work is supported by the Science and Engineering Research Board (SERB) under the Project No. MTR/2019/000614.

References

1. Brayanov, I.A.: Numerical solution of a mixed singularly perturbed parabolic-elliptic problem. J. Math. Anal. Appl. **320**(1), 361–380 (2006). https://doi.org/10.1016/j.jmaa.2005.06.098
2. Cen, Z.: Parameter-uniform finite difference scheme for a system of coupled singularly perturbed convection-diffusion equations. Int. J. Comput. Math. **82**(2), 177–192 (2005). https://doi.org/10.1080/00207160420003 01798
3. Clavero, C., Gracia, J.L., Jorge, J.C.: High-order numerical methods for one-dimensional parabolic singularly perturbed problems with regular layers. Numer. Methods Partial Differ. Equ. **21**(1), 148–169 (2005). https://doi.org/10.1002/num.20030
4. Clavero, C., Gracia, J.L., Stynes, M.: A simpler analysis of a hybrid numerical method for time-dependent convection-diffusion problems. J. Comput. Appl. Math. **235**(17), 5240–5248 (2011). https://doi.org/10.1016/j.cam.2011.05.025
5. Dunne, R.K., O'Riordan, E.: Interior layers arising in linear singularly perturbed differential equations with discontinuous coefficients. In: Proceedings of the Fourth International Conference on Finite Difference Methods: Theory and Applications, pp. 29–38. Lozenetz, Bulgaria (2006)
6. Gracia, J.L., O'Riordan, E.: A singularly perturbed convection-diffusion problem with a moving interior layer. Int. J. Numer. Anal. Model. **9**(4), 823–843 (2012)
7. Kadalbajoo, M.K., Awasthi, A.: A parameter uniform difference scheme for singularly perturbed parabolic problem in one space dimension. Appl. Math. Comput. **183**(1), 42–60 (2006). https://doi.org/10.1016/j.amc.2006.05.023
8. Kumar, D.: An implicit scheme for singularly perturbed parabolic problem with retarded terms arising in computational neuroscience. Numer. Methods Partial Differ. Equ. **34**(6), 1933–1952 (2018). https://doi.org/10.1002/num.22269
9. Linß, T.: Layer-Adapted Meshes for Reaction-convection-diffusion Problems, Lecture Notes in Mathematics, vol. 1985. Springer, Berlin (2010). https://doi.org/10.1007/978-3-642-05134-0
10. Linß, T.: Layer-Adapted Meshes for Reaction-Convection-Diffusion Problems. Lecture Notes in Mathematics, vol. 1985. Springer, Berlin (2010)
11. Miller, J.J.H., O'Riordan, E., Shishkin, G.I.: Fitted Numerical Methods for Singular Perturbation Problems. World Scientific Publishing Co. Pte. Ltd., Hackensack, NJ, revised edn. (2012). https://doi.org/10.1142/9789814390743, error estimates in the maximum norm for linear problems in one and two dimensions
12. Mukherjee, K., Natesan, S.: Optimal error estimate of upwind scheme on Shishkin-type meshes for singularly perturbed parabolic problems with discontinuous convection coefficients. BIT **51**(2), 289–315 (2011). https://doi.org/10.1007/s10543-010-0292-2
13. O'Riordan, E., Shishkin, G.I.: Singularly perturbed parabolic problems with non-smooth data. In: Proceedings of the International Conference on Boundary and Interior Layers—Computational and Asymptotic Methods (BAIL 2002), vol. 166, pp. 233–245 (2004). https://doi.org/10.1016/j.cam.2003.09.025
14. Rao, S.C.S., Chaturvedi, A.K.: Parameter-uniform numerical method for a two-dimensional singularly perturbed convection-reaction-diffusion problem with interior and boundary layers. Math. Comput. Simul. **187**, 656–686 (2021). https://doi.org/10.1016/j.matcom.2021.03.016

15. Rao, S.C.S., Chawla, S.: Numerical solution for a coupled system of singularly perturbed initial value problems with discontinuous source term. In: Mathematical Analysis and Its Applications. Springer Proceedings in Mathematics and Statistics, vol. 143, pp. 753–764. Springer, New Delhi (2015). https://doi.org/10.1007/978-81-322-2485-3_60
16. Rao, S.C.S., Chawla, S.: Second order uniformly convergent numerical method for a coupled system of singularly perturbed reaction-diffusion problems with discontinuous source term. In: Boundary and Interior Layers, Computational and Asymptotic Methods—BAIL 2014. Lecture Notes in Computational Science and Engineering, vol. 108, pp. 233–244. Springer, Cham (2015)
17. Rao, S.C.S., Chawla, S.: Numerical solution of singularly perturbed linear parabolic system with discontinuous source term. Appl. Numer. Math. **127**, 249–265 (2018). https://doi.org/10.1016/j.apnum.2018.01.006
18. Rao, S.C.S., Chawla, S.: The error analysis of finite difference approximation for a system of singularly perturbed semilinear reaction-diffusion equations with discontinuous source term. In: Finite Difference Methods. Lecture Notes in Computational Science, vol. 11386, pp. 175–184. Springer, Cham (2019)
19. Rao, S.C.S., Chawla, S.: Parameter-uniform convergence of a numerical method for a coupled system of singularly perturbed semilinear reaction-diffusion equations with boundary and interior layers. J. Comput. Appl. Math. **352**, 223–239 (2019). https://doi.org/10.1016/j.cam.2018.11.021
20. Rao, S.C.S., Chawla, S., Chaturvedi, A.K.: Numerical analysis for a class of coupled system of singularly perturbed time-dependent convection-diffusion equations with a discontinuous source term. Numer. Methods Partial Differ. Eq. Accepted, 1–31 (2021). https://doi.org/10.1002/num.22845
21. Roos, H.G., Stynes, M., Tobiska, L.: Robust Numerical Methods for Singularly Perturbed Differential Equations. Springer Series in Computational Mathematics, vol. 24, 2nd edn. Springer, Berlin (2008). convection-diffusion-reaction and flow problems
22. Shishkin, G.I., Shishkina, L.P., Hemker, P.W.: A class of singularly perturbed convection-diffusion problems with a moving interior layer. An a posteriori adaptive mesh technique. Comput. Methods Appl. Math. **4**(1), 105–127 (2004). https://doi.org/10.2478/cmam-2004-0007

Filtering in Time-Dependent Problems

P. Megha and G. Chandhini

Abstract Spectral methods are efficient, robust and highly accurate methods in numerical analysis. When it comes to approximating a discontinuous function with spectral methods, it produces spurious oscillations at the point of discontinuity, which is called Gibbs' phenomenon. Gibbs' phenomenon reduces the spectral accuracy of the method globally. Filtering is a widely used method to prevent the oscillations due to Gibbs' phenomenon by which the accuracy of the spectral methods is regained up to an extent. In this work, we study the effects of various filters in time-dependent problems and do a comparison of numerical results.

Keywords Filter · Spectral method · Time-dependent problem

1 Introduction

Spectral methods are the numerical approaches of representing the solution of an equation as a truncated series of base functions. If the base functions used in the series are Fourier functions, it is called Fourier spectral methods, and if the base functions are polynomial functions, it is called polynomial spectral methods. The most commonly used polynomial spectral methods are Chebyshev spectral methods, Legendre spectral methods, Hermite spectral methods, etc. In this paper, we concentrate on Fourier spectral methods.

Spectral methods are significant in the sense that high order accuracy can be obtained with a lesser number of terms. Also, they enjoy the spectral rate of con-

P. Megha (✉) · G. Chandhini
Department of Mathematical and Computational Sciences, National Institute of Technology Karnataka, Mangalore, Karnataka, India
e-mail: meghap11394@gmail.com

R. K. Sharma et al. (eds.), *Frontiers in Industrial and Applied Mathematics*, Springer Proceedings in Mathematics & Statistics 410,
https://doi.org/10.1007/978-981-19-7272-0_21

vergence, i.e., if u is a function having s bounded derivatives and let $\mathcal{S}_{\mathcal{N}}u(x) = \sum_{|k|\leq \frac{N}{2}} \hat{u}_k e^{ikx}$ be the truncated Fourier series of the function u, then

$$|\mathcal{S}_{\mathcal{N}}u(x) - u(x)| \leq Const \parallel u \parallel_{C_s} N^{1-s}.$$

Whenever u is infinitely smooth, exponential accuracy is obtained.

The problem arises when there is discontinuity. The presence of discontinuity induces spurious oscillations at the point of discontinuity, which not only affects the convergence rate at the point of discontinuity, but also it reduces the convergence rate to linear order at the points away from the discontinuity. So the spectral accuracy enjoyed by spectral methods is lost globally. There have been many attempts to overcome this major issue with partial success.

Gottlieb et al. [1] developed a two-parameter family of mollifies with which they have convolved the function and this has improved convergence at the points away from the discontinuity. An essentially non-oscillatory Fourier spectral method [2] has been proposed by Cai et al., for hyperbolic differential equations having piecewise analytic functions as solutions. The modification involves appending a non-smooth function to the Fourier basis, and filters are considered away from the discontinuity. This has increased the convergence of the Fourier approximation by one order; however, the method has smoothened the solution at the point of discontinuity too. Vandeven [10] has introduced a class of filters that does not require the prior knowledge of the position of the discontinuity and produces exponential accuracy, nevertheless only for points away from the discontinuity. Shu and Wong [8] have compared the Fourier solutions of nonlinear conservation law problems post-processed using the Gegenbauer polynomial method [5] as well as filtering with vanishing viscosity approach. The Gegenbauer reconstruction has the drawback of round-off error and it suffers from the Runge phenomenon, hence not a robust method. A robust Gibbs complimentary basis is developed by Gelb et al. [4]. The standard Fourier Pade approximation is extended for functions with jumps by [3], which reduced the Gibbs overshoot by 2.5% and is pictorially accurate globally. Inspired by the Gegenbauer reconstruction method, an inverse polynomial reconstruction method is proposed by Shizgal et al. [7] for the Fourier series. The numerical results showed faster convergence for this new method compared to the Gegenbauer reconstruction.

In the present work, we attempt to make a comparison among various filters for linear advection equations with discontinuous solutions.

2 Filtering

Definition 1 Any C^∞ even function σ, whose support is $[-1, 1]$ and such that $\sigma(0) = 1$, is called **filter**.

Examples of filters

1. Lanczos, $\sigma_1(x) = \dfrac{1}{\pi x}\sin(\pi x)$.
2. Raised cosine, $\sigma_2(x) = (1 + \cos(\pi x))/2$.
3. Sharpened raised cosine, $\sigma_3(x) = \sigma_2^4(35 - 84\sigma_2 + 70\sigma_2^2 - 20\sigma_2^3)$.
4. Exponential filter, $\sigma_4(x) = \exp^{-\alpha^p x}$, $\alpha > 0$, p is the order of the filter.

The process,

$$\mathcal{S}_N^\sigma u(x) = \sum_{k=0}^{N} \sigma(k/N)\hat{u}_k \zeta_k(x),$$

with $\zeta_k(x)$ being either e^{ikx} or $\phi_k(x)$, is called **filtering**. The physical space (time-space) analogous to this process is called **mollification**, which includes convolving a unit mass, compact support non-negative kernel $\Phi(x)$, called **mollifier**, with the approximation series, i.e.,

$$\Phi * \mathcal{S}_{\mathcal{N}} u(x) = \int_\Omega \Phi(y)\mathcal{S}_{\mathcal{N}} u(x-y)dy$$

.

For instance, in the case of Fourier series, filtering is defined by

$$\mathcal{S}_N^\sigma u(x) = \sum_{|k|\le \frac{N}{2}} \sigma\left(\frac{k}{N/2}\right)\hat{u}_k e^{ikx},$$

corresponding to a filter $\sigma(x)$. Also,

$$\mathcal{S}_N^\sigma u(x) = \Phi * \mathcal{S}_{\mathcal{N}} u(x) = \int_0^{2\pi} \Phi(y) \sum_{|k|\le \frac{N}{2}} \hat{u}_k e^{ik(x-y)} dy,$$

where

$$\Phi(y) = \sum_{k=-\infty}^{k=\infty} \sigma(k, N)e^{ikx},$$

and for a mollification $\Phi * \mathcal{S}_{\mathcal{N}} u$, defined in the physical space, its associated filter samples $\sigma(k, N) = \hat{\Phi}_k$ are the Fourier coefficients of the mollifier Φ.

A milestone work on filters for reducing Gibbs' phenomenon was done by Vandeven [10] by developing a modal filter.

Theorem 1 ([10]) *Let u be a 2π periodic function such that $\exists$ an integer α and real numbers $(c_m)_{m=1}^{\alpha+1}$, with $0 \le c_1 \le c_2 \le \dots \le c_\alpha < 2\pi$ and $c_{\alpha+1} = c_1 + 2\pi$. Further, for any integer m with $1 \le m \le \alpha$, there exists an open set $\Omega_m \in C$ which contains $[c_m, c_{m+1}]$ and a function $v_m : \Omega_m \to C$ such that v_m is holomorphic on Ω_m, $\forall\, x \in (c_m, c_{m+1})$, $v_m(x) = u(x)$. Let $\epsilon > 0$ be a real number. Suppose that the filter is defined by*

$$\sigma(x) = 1 - \frac{(2p-1)!}{(p-1)!^2}\int_0^x (t(1-t))^{p-1}dt$$

with $p = c\left(\frac{N}{2}\right)^{\epsilon/4}$, *where* c *is a positive constant independent of* N. *Then the following estimate holds:*

$$\sup_{x\in\mathbb{R},\ d(x)>\left(\frac{N}{2}\right)^{-1+\epsilon}} |u(x) - S_N^{\sigma}u(x)| \leq \left(\frac{N}{2}\right)^{\beta} (CN^{-\epsilon/2})^{\left(\frac{N}{2}\right)^{\epsilon/4}}$$

for a positive constant C *independent of* N *and a positive constant* β *independent of both* u *and* N, *where*

$$S_N^{\sigma}u(x) = \sum_{|k|\leq\frac{N}{2}} \sigma(k/N)\hat{u}_k e^{ikx} \tag{1}$$

and $d(x) = inf\{|x - (c_m + 2k\pi) : 1 \leq m \leq \alpha,\ k \in \mathbb{Z}\}$.

Remark 1 Mostly filtering and mollification treats the Gibbs phenomenon as noise and hence it smoothens the discontinuity part of the function even though this improves the convergence rate significantly. Actually, Gibbs phenomenon contains enough information to reconstruct the function.

A breakthrough work on filtering was done by Tadmor and Tanner [9] by developing an adaptive filter incorporating the position of discontinuity. Their result can be stated as follows.

Theorem 2 ([9]) *Given the Fourier projection* $S_N u$ *of a piecewise analytic function* u, *we consider a* $C_0^{\infty}[-1\ 1]$ *filter* $\sigma(\zeta)$, *such that* σ *has* G_{α}*-regularity and that it is accurate of order* p *in the sense of satisfying the moments condition,* $\sigma^n(0) = \delta_{n0},\quad n = 0, 1, ..., p-1$. *Then,*

$$|u(x) - S_N^{\sigma}u(x)| \leq Const(1 + Nd(x))e^{-\alpha(\eta N d(x))^{1/\alpha}},$$

where we set the adaptive order $p(x) = (\eta N d(x))^{1/\alpha}$, *depending on the distance function* $d(x) = dist(x, singsupp\ u)$ *and constant* η *is dictated by the specific Gevrey and piecewise-analyticity properties of* σ *and* u.

3 Numerical Experiments

Example 1

$$u_t = -2\pi u_x,\quad -1 \leq x \leq 1,\ t > 0\ \ u(0,t) = u(2\pi, t) \tag{2}$$

$$u(x,0) = \begin{cases} x, 0 \leq x \leq \pi \\ x - 2\pi, \pi < x <= 2\pi \end{cases}$$

Table 1 Absolute error using various filters at time T = 10 of Example 1. The discontinuity point is π and time step, h = 0.0001

Points	Unfiltered	Raised cosine	Lanczos	Sharpened raised cosine
0.1567	0.0079	0.0006	0.0006	0.0006
0.3917	0.0006	0.0006	0.0006	0.0006
1.0968	0.0050	0.0006	0.0006	0.0006
1.7236	0.0125	0.0006	0.0006	0.0006
2.8204	0.0422	0.0008	0.0003	0.0006
3.0868	0.2083	0.0267	0.0109	0.3743
3.1024	0.0845	0.0170	0.0740	0.3068
3.1181	0.5620	0.6079	0.3301	0.4821
3.1338	1.1060	2.0863	1.9072	2.0791
3.1494	1.3742	2.2354	2.0794	2.2310
3.1651	0.5485	0.6923	0.4085	0.5817
3.1808	0.0232	0.0358	0.0724	0.2755
3.4471	0.0050	0.0007	0.0006	0.0006
3.9172	0.0113	0.0006	0.0008	0.0006
4.7006	0.0070	0.0006	0.0007	0.0006
5.3274	0.0039	0.0006	0.0006	0.0006
5.6408	0.0060	0.0006	0.0006	0.0006
6.2675	0.0061	0.0006	0.0007	0.0006

is solved using the Fourier-Galerkin method.

Table 1 shows the absolute values of the error using different filters for Example 1. The time integration is done by the Runge-Kutta fourth-order (RK4) method using time step $h = 0.0001$. The values show that the error has reduced at the points away from the discontinuity when the filters are used. But at the point of discontinuity, filtering does not have any effects even though the solution graphs in Fig. 1 show that filtering helps to diminish Gibbs' phenomenon in the closer neighbourhood. Figure 2 shows the error graph of the corresponding filters. In the closer vicinity of the discontinuity, accuracy has not been improved uniformly even with filtering. As we move away from the discontinuity raised cosine and Lanczos filters are consistently better than sharpened raised cosine filters.

A new filtering method based on the sigmoidal transformation was developed by Yun et al. [11]. This filter is a generalization of the existing Lanczos filter. The present Sidi-Lanczos-type sigmoidal filter is referred to as Sidi-LSF. The function is given by

$$\sigma_m(x) = \frac{2^{1-m}\sqrt{\pi}\gamma(m)\gamma(\frac{m+1}{2})}{\gamma(m/2)\gamma(\frac{m+1}{2} - x)\gamma(\frac{m+1}{2} + x)}.$$

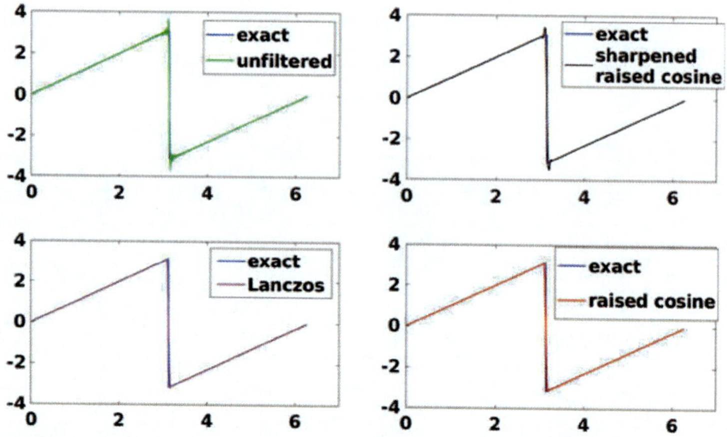

Fig. 1 Filtered solution at T = 10 using various filters and $N = 128$ (Example 1)

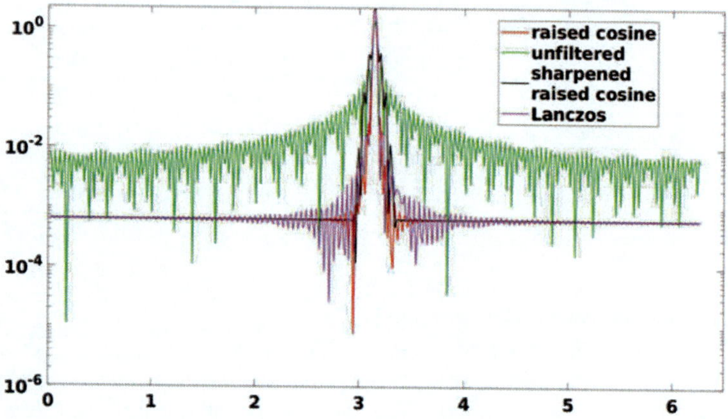

Fig. 2 Error graph of the solution at T = 10 is using various filters and $N = 128$ (Example 1)

Authors [11] observed that by increasing the parameter 'm' overall accuracy has been improved in the neighbourhood of the discontinuity points. Using the following example, we have made a comparison study between traditional raised cosine and Lanczos ('$m = 1$') filters with the other values of 'm'(= 5 & 10).

Example 2

$$u_t = -u_x, \quad -1 \le x \le 1, \; t > 0 \tag{3}$$

$$u(x,0) = \begin{cases} 1, -1 \le x \le -0.5 \\ \sin(\pi(x+0.5)), -0.5 \le x \le 0.5 \\ 1, 0.5 < x \le 1. \end{cases}$$

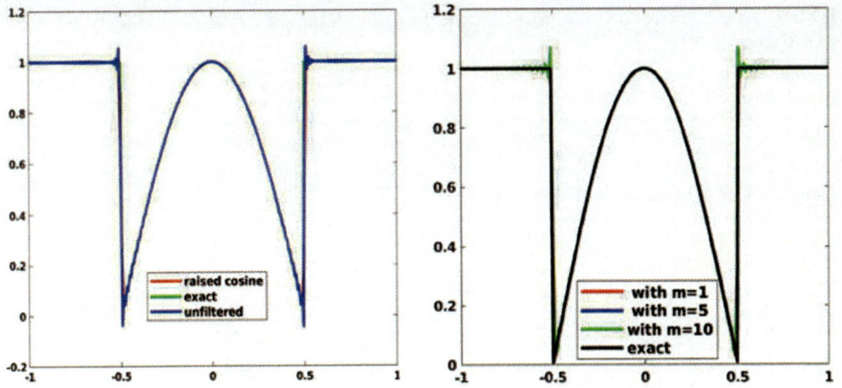

(a) Raised cosine filter (b) Lanczos average filter

Fig. 3 Filtered solution at the time T = 10 and $N = 128$ (Example 2)

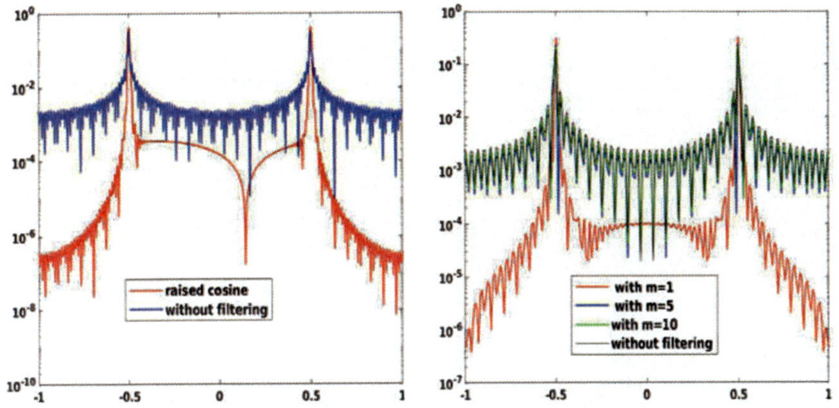

(a) Raised cosine filter (b) Lanczos average filter

Fig. 4 Error graph of the filtered solution at the time step T = 10 and $N = 128$ (Example 2)

The Example 2 is another first-order linear hyperbolic equation, whose solution has two discontinuous points, namely -0.5 and 0.5. Fourier-Galerkin with RK4 is used to obtain the solution of Example 2. Figure 3 gives a comparison of the solution obtained using raised cosine and Sidi-Lanczos-type sigmoidal filter (Sidi-LSF) while Fig. 4 provides corresponding error plots. However, it is observed that overall accuracy has not been improved as 'm' increases. Identifying the optimal value of 'm' is open. Error at various points that are given in Table 2 also shows that accuracy has not been improved consistently for the filtered solution.

We have also made another attempt to extend the adaptive filter,

Table 2 Absolute error using various filters at T = 10 of Example 2. The discontinuity points are –0.5 and 0.5. h = 0.0001

Points	Unfiltered	Raised cosine	Sidi-LSF m = 1	Sidi-LSF m = 5	Sidi-LSF m = 10
-0.9799	0.0004	0.0000	0.0000	0.0003	0.0004
-0.7544	0.0008	0.0000	0.0000	0.0004	0.0005
-0.5138	0.0354	0.0036	0.0107	0.0179	0.0250
-0.5088	0.0812	0.0617	0.0246	0.0546	0.0682
-0.5038	0.0683	0.2628	0.2324	0.1214	0.1002
-0.4987	0.3548	0.4318	0.4103	0.3622	0.3530
-0.4937	0.0655	0.1543	0.1037	0.0201	0.0430
-0.4035	0.0039	0.0004	0.0001	0.0027	0.0032
-0.0025	0.0012	0.0002	0.0001	0.0009	0.0011
0.2481	0.0027	0.0001	0.0001	0.0017	0.0021
0.3484	0.0011	0.0002	0.0001	0.0011	0.0013
0.4737	0.0128	0.0011	0.0020	0.0108	0.0124
0.4987	0.3307	0.4196	0.4103	0.3622	0.3530
0.5038	0.0851	0.2737	0.2324	0.1214	0.1002
0.5088	0.0839	0.0669	0.0246	0.0546	0.0682
0.5138	0.0323	0.0029	0.0107	0.0179	0.0250
0.7494	0.0034	0.0000	0.0000	0.0023	0.0028
0.8496	0.0019	0.0000	0.0000	0.0013	0.0016

$$\sigma_p(x) = \begin{cases} \exp\left(cx^p\right)/(x^2-1), & |x| < 1 \\ 0, & |x| \geq 1 \end{cases}$$

developed by Tadmor and Tanner [9] to linear hyperbolic problem. The parameter 'p' is chosen adaptively by incorporating the position of the discontinuity in the initial data. To illustrate, we have improved the Fourier-Galerkin solutions of Examples 1 and 2 using the adaptive filter. Figures 5 and 6 are the improved solutions and the corresponding error graphs of these examples. Table 3 compares an adaptive filtered solution with an unfiltered one.

It is observed that the adaptive filtered solution also does not improve the values near the discontinuity. However, comparing Figs. 2 and 5b, we can see that adaptive filtering on Fourier approximation has dramatically improved the accuracy at the points away from the discontinuity.

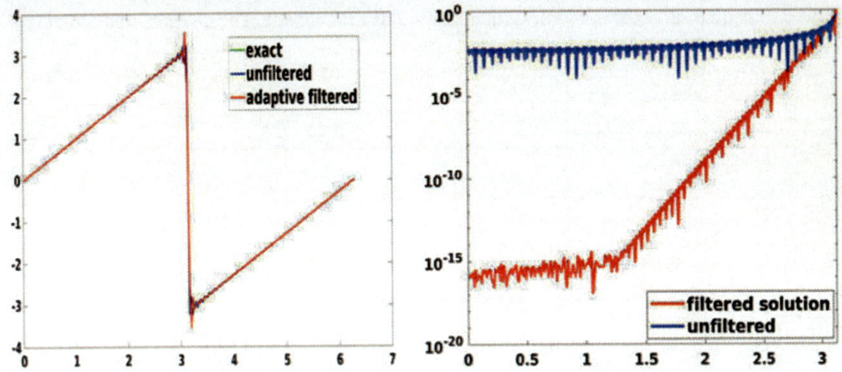

(a) Adaptive filtered solution

(b) Error graph of the adaptive filtered solution

Fig. 5 Filtered solution using adaptive filter at T = 10 and $N = 128$ (Example 1)

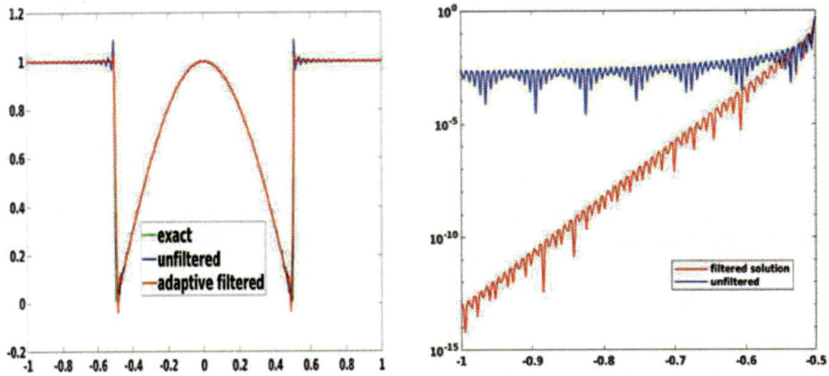

(a) Adaptive filtered solution

(b) Error graph of the adaptive filtered solution (interval is (-1,-0.5)

Fig. 6 Filtered solution at the time step T = 10 and $N = 128$ (Example 2)

4 Conclusion

Over the past many years, different filters have been developed for various problems having non-smooth or discontinuous solutions. The numerical results obtained in the last section show that a filter that is apt for a particular problem need not be suitable for another model problem. Also, filters providing good accuracy at points away from discontinuities are not appropriate at the discontinuous points. The adaptive filter developed by Tadmor and Tanner [9] was a breakthrough in this concept. It is observed by Kanevsky et al. [6] that merely applying filtering in time-dependent problems in each time step may reduce the convergence rate. As a result, they have

Table 3 Absolute error using adaptive filter at T = 10 of Example 1. The discontinuous point is π

Points	Unfiltered	Adaptive filtered
0.0781	0.0045	0.0×10^{-15}
0.3126	0.0049	0.0×10^{-15}
0.4689	0.0042	0.0×10^{-15}
0.7815	0.0009	0.0×10^{-15}
1.094	0.0064	0.0×10^{-15}
1.4067	0.0902	80.0×10^{-15}
1.563	0.0023	95.0×10^{-15}
1.719	0.0101	23.3×10^{-13}
1.875	0.0102	78.5×10^{-12}
2.032	0.0051	21.6×10^{-10}
2.344	0.0063	19.9×10^{-8}
2.970	0.0902	35.1×10^{-3}
3.126	0.0769	60.2×10^{-2}
3.1416	3.1416	3.1416

developed an idempotent filter from an exponential filter for a nozzle flow problem satisfying certain conditions. However, the choice of various parameters involved in the proposed idempotent filter for various problems is a challenge. Hence, our conclusion is that the choice of filters depends upon the particular problem at hand and the discontinuity position. Thus, the choice of filters for a class of problems is still an unresolved area that has a wide scope for future works.

References

1. Abarbanel, S., Gottlieb, D., Tadmor, E.: Spectral Methods for Discontinuous Problems, Technical Report, Institute for Computer Applications in Science and Engineering, Hampton, Virginia (1985)
2. Cai, W., Gottlieb, D., Shu, C.W.: Essentially nonoscillatory spectral Fourier methods for shock wave calculations. Math. Comput. **52**, 389–410 (1989)
3. Driscoll, T.A., Fornberg, B.: A Padé-based algorithm for overcoming the Gibbs phenomenon. Numer. Algorithms **26**, 77–92 (2001)
4. Gelb, A., Tanner, J.: Robust reprojection methods for the resolution of the Gibbs phenomenon. Appl. Comput. Harmon. Anal. **20**, 3–25 (2006)
5. Gottlieb, D., Shu, C.W., Solomonoff, A., Vandeven, H.: On the Gibbs phenomenon I: recovering exponential accuracy from the Fourier partial sum of a nonperiodic analytic function. J. Comput. Appl. Math. **43**, 81–98 (1992)
6. Kanevsky, A., Carpenter, M.H.: Idempotent filtering in spectral and spectral element methods. J. Comput. Phys. 41–58 (2006)
7. Shizgal, B.D., Jung, J.H.: Towards the resolution of the Gibbs phenomena. J. Comput. Appl. Math. **161**, 41–65 (2003)

8. Shu, C.W., Wong, P.S.: A note on the accuracy of spectral method applied to nonlinear conservation laws. J. Sci. Comput. **10**(3), 357–369 (1995)
9. Tadmor, E., Tanner, J.: Adaptive filters for piecewise smooth spectral data. IMA J. Numer. Anal. 635–647 (2005)
10. Vandevan, H.: Family of spectral methods for discontinuous problems. J. Sci. Comput. **6**, 159–192 (1991)
11. Yun, B.I., Rim, K.S.: Construction of Lanczos type filters for the Fourier series approximation. Appl. Numer. Math. **59**, 280–300 (2009)

Heat Transfer Model for Silk Finishing Calender

Neelam Gupta and Neel Kanth

Abstract Calendering is a finishing process used in many process industries like paper, textile and leather where the web passes through two or more rotating cylindrical bowls in touch with an aim to get special effects like smoothness, gloss and uniform flattening of the thin sheet. The key factor in the calendering process for getting desired results is pressure and temperature. Several unappealing elements such as damage to fabric and strength reduction of the fabric arise if pressure and temperature increase in excess. Temperature gradient calendering is used to overcome these undesirable factors. In this paper, the influence of parameters like cylindrical bowl temperature, dwell time and thermal diffusivity on the temperature of the fabric in the stiffness direction of the web inside the calender nip has been discussed for temperature gradient calenders of the textile industry using the heat balance integral method.

Keywords Calendering · Heat conduction · Nip mechanics · Nip width · Thermal diffusivity · Integral method

1 Introduction

At the final stage of fabric manufacturing, the calendering finishing process is used in the textile industry in which the fabric passes through nips formed by two or more rotating cylindrical bowls in touch at high pressure and temperature. The fabric runs through cylindrical bowls at different speeds depending on the fabric quality required. These cylindrical bowls are hard or soft which describes the type of calender. In hard nip calendering, all the cylindrical bowls are hard, while there are alternate hard and

N. Gupta (✉) · N. Kanth
Jaypee University of Information Technology, Waknaghat, Solan, Himachal Pradesh, India
e-mail: shah.neelam28@yahoo.in

R. K. Sharma et al. (eds.), *Frontiers in Industrial and Applied Mathematics*,
Springer Proceedings in Mathematics & Statistics 410,
https://doi.org/10.1007/978-981-19-7272-0_22

soft cylindrical bowls in the case of soft nip calendering. Nowadays, soft nip calenders like rolling calender, silk finishing calender, friction calender and schreiner calender are used in the textile industry because of better finishing results as compared to hard nip calenders [1–3].

Silk finishing calenders are soft nip calenders having a combination of alternating hard and soft cylindrical bowls. The basic difference in these calenders is in the material composition of soft cylindrical bowls. Silk finishing calender uses a special type of soft cylindrical bowl having less elastic modulus as compared to soft cylindrical bowls used in other soft calenders. Silk finishing calendering can process all types of fabrics, but it is used often for high-content cotton and silk fabric[4, 5].

The major difference between the hard nip and soft nip calenders is the dwell time between the cylindrical bowls and fabric as dwell time directly depends upon nip width and inversely on calender speed. Depending upon the type of calender, nip mechanics models are used to evaluate the nip width. Dwell time, temperature and pressure are the basic process parameters of calendering operation. Design and type of cylindrical bowls are another important design parameters in calendering system [1, 4].

Heat is conducted for a very short time from a heated cylindrical bowl to the fabric. With rise in dwell time, more heat is conducted to the fabric from the cylindrical bowl which enhances the gloss and smoothness of the fabric. In hard nip calender, both the cylindrical bowls are heated and may or may not be at similar temperatures, while in silk finishing calender, the hard cylindrical bowl is at high temperature and the soft cylindrical bowl is at the room temperature [1–6].

The appropriate stiffness, smoothness and gloss of the fabric are achieved by controlling the pressure in the nips or through temperature adjustment of heated cylindrical bowls. Pressure has the largest influence on stiffness and air resistance, whereas temperature has the largest influence on gloss and roughness of the fabric. Higher temperature, higher load and lower speed all increase gloss. With rise in pressure, surface smoothness can be enhanced by rising pressure, but sometimes damage to the fibre bond takes place by an excess rise in pressure which leads to fabric strength reduction. So for better finishing of the fabric, pressure should be increased to a particular limit.

Temperature gradient calendering (TGC) is used for removing several unappealing elements. In TGC, there are alternating hard and soft cylindrical bowls in which hard cylindrical bowl is heated up to a temperature of 300 °C and soft cylindrical bowl is at room temperature. So the fibres present on the surface of the fabric which is in touch with the heated cylindrical bowl get deformed permanently, while the fibres on the other side and up to the mid of the fabric do not get deformed due to which the surface properties of the fabric are developed while maintaining the bulk and strength properties [7, 8].

In this paper, analysis of heat conduction from cylindrical bowls to fabric passing through silk finishing calender nip is done for temperature gradient calender using one-dimensional heat conduction equation which has been solved using heat balance integral method.

2 Heat Conduction Model For Silk Finishing Calender for Semi-Infinite Medium

Three-dimensional transient state heat conduction equation is given by [9–19]

$$\frac{\partial \varphi}{\partial t} = \alpha[(\frac{\partial^2 \varphi}{\partial x^2})_{yz} + (\frac{\partial^2 \varphi}{\partial y^2})_{xz} + (\frac{\partial^2 \varphi}{\partial z^2})_{xy}] + \frac{q_v}{c_p \rho} \tag{1}$$

where x, y, z are "space coordinates", q_v is "heat generation term", t is "time", φ is "temperature", ρ is "density of the material", c_p is "specific heat"and α is "thermal diffusivity".

Also, thermal diffusivity is defined as

$$\alpha = \frac{\kappa}{\rho c_p}$$

where κ is "thermal conductivity".

After ignoring the heat generation term $q_v = 0$ and approximating the above equation in x direction only, Eq. (1) changes to

$$\frac{\partial \varphi}{\partial t} = \alpha \frac{\partial^2 \varphi}{\partial x^2} \tag{2}$$

For finding the temperature distribution inside the fabric in stiffness direction, solution of equation (2) under different initial and boundary conditions can be used depending upon the type of calender.

In TGC, temperature of the cold cylindrical bowl has no contribution to temperature distribution because the time of contact between the heated cylindrical bowl and fabric is very short. So heat transfer during this procedure is considered as "transient heat conduction into a semi-infinite medium". In this medium, fabric is bounded by the plane $x = 0$ and for the other side, x tends to infinity in the positive direction. In this case, the face $x = d$ has been moved to $x \longrightarrow \infty$. This case is applied where heating or cooling affects the surface of a body for a very short period of time.

The I.C. is taken as

$$\varphi(x, 0) = \varphi_0 \tag{3}$$

and B.C. are taken as

$$\left.\begin{array}{ll} \varphi(0, t) & = \varphi_h \\ \lim_{x \to \infty} \varphi(x, t) & = \varphi_0 \end{array}\right\} \tag{4}$$

where φ_0 is the initial fabric temperature and φ_h is the temperature of the hot cylindrical bowl.

3 Solution of Heat Conduction Model For Silk Finishing Calender

Introducing dimensionless variables

$$\eta = \frac{\varphi - \varphi_0}{\varphi_h - \varphi_0} \tag{5}$$

$$\tau = \alpha t \tag{6}$$

The heat equation, I.C. and B.C. get transformed to the following forms:

$$\frac{\partial \eta}{\partial \tau} = \frac{\partial^2 \eta}{\partial x^2} \tag{7}$$

with I.C.

$$\eta(x, 0) = 0 \tag{8}$$

and B.C.

$$\left.\begin{aligned} \eta(0, \tau) &= 1 \\ \lim_{x \to \infty} \eta(x, \tau) &= 0 \end{aligned}\right\} \tag{9}$$

Equation (7) is solved under initial and boundary conditions given by Eqs. (8) and (9) using the heat balance integral method. Various heat and phase change problems are solved using the heat balance integral method. "Heat balance integral method is a simple approximate technique originally developed for analysing thermal problems". This method was first described by Goodman [20–23].

Define $\delta(t)$, i.e. distance over which the temperature changes are felt at time t. Integrate equation (5) from $x = 0$ to $x = \delta(t)$ which gives

$$\left(\frac{\partial \varphi}{\partial x}\right)_{x=\delta(t)} - \left(\frac{\partial \varphi}{\partial x}\right)_{x=0} = \frac{1}{\alpha} \int_0^{\delta(t)} \frac{\partial \varphi}{\partial t} dx \tag{10}$$

The right-hand side integral is performed by applying the rule of differentiation under the integral sign, hence

$$\left(\frac{\partial \varphi}{\partial x}\right)_{x=\delta(t)} - \left(\frac{\partial \varphi}{\partial x}\right)_{x=0} = \frac{1}{\alpha}\left[\frac{d}{dt} \int_0^{\delta(t)} \varphi dx - \varphi_{x=\delta} \frac{d\delta}{dt}\right] \tag{11}$$

But $\left(\frac{\partial \varphi}{\partial x}\right)_{x=\delta} = 0$ and $\varphi = \varphi_0$ at $x = \delta$.

Let

$$\theta = \int_0^{\delta(t)} \varphi dx \tag{12}$$

Hence, Eq. (5) becomes

$$-\alpha\left(\frac{\partial\varphi}{\partial x}\right)_{x=0} = \frac{d}{dt}(\theta - \varphi_0\delta) \tag{13}$$

$$-\kappa\left(\frac{\partial\varphi}{\partial x}\right)_{x=0} = \rho C_p\frac{d}{dt}(\theta - \varphi_0\delta) \tag{14}$$

Rate of input of energy at face $x = 0$ at any time t= Rate of energy of the sensible heat of the heated layer of stiffness $\delta(t)$.

The B.C. at ∞ with

$$\varphi(\delta(t), t) = \frac{\partial\eta}{\partial x}(\delta(t), t) = 0 \tag{15}$$

where δ is "sufficiently far from the boundary such that the boundary temperature has a negligible effect".

Therefore,

$$\left.\begin{array}{ll} \varphi & = \varphi_h \ at \ x = 0 \\ \varphi & = \varphi_0 \ at \ x = \delta \\ \frac{\partial\varphi}{\partial x} & = 0 \ at \ x = \delta \end{array}\right\} \tag{16}$$

Another condition can be obtained by evaluating the differential equation at $x = \delta(t)$, where $\varphi = \varphi_h$=constant.

Therefore,

$$\left.\begin{array}{ll} \frac{\partial\varphi}{\partial t} & = 0 \ at \ x = \delta \\ \frac{\partial^2\varphi}{\partial x^2} & = 0 \ at \ x = \delta \end{array}\right\} \tag{17}$$

Using the above conditions, the solution for $G(x, t)$ is an appropriate approximate polynomial [20]:

$$\eta(x, t) = \left(1 - \frac{x}{\delta}\right)^m \tag{18}$$

On integrating over $x \in [0, \delta]$ and substituting for $G(x, t)$ using Eq. (18) and then again on integration, it leads to [20]

$$\delta = \sqrt{2m(m + 1)t} \tag{19}$$

Taking seventh order polynomial, using Eqs. (18) and (19), the resulting solution is given by

$$\varphi(x,t) = \varphi_0 + (\varphi_h - \varphi_0)\left[1 - 7\left(\frac{x}{\delta}\right) + 21\left(\frac{x}{\delta}\right)^2 - 35\left(\frac{x}{\delta}\right)^3 + 35\left(\frac{x}{\delta}\right)^4 - 21\left(\frac{x}{\delta}\right)^5 + 7\left(\frac{x}{\delta}\right)^6 - 21\left(\frac{x}{\delta}\right)^7\right] \tag{20}$$

with

$$\delta = \sqrt{112\alpha t} \tag{21}$$

4 Simulation of Heat Conduction Model

Influence of cylindrical bowl temperature on the temperature of the fabric in stiffness direction, influence of dwell time and influence of thermal diffusivity have been examined using the mathematical model given by Eq. (20) for single nip silk finishing calender. Cylindrical bowl temperatures (BT) are taken in the range from 180 °–270 °C, and the initial temperature of fabric is taken as 75 °C. Data used for simulation is given in Table 1.

Table 1 Simulation parameters

Calendering parameters	Simulation details
Composition	Alternate hard and soft bowls
Hard cylindrical bowl material	Cylinders having covering of chilled cast iron
Soft cylindrical bowl material	Cylinders having covering of soft material
Speed (m/min)	400 – 900
Linear load (kN/m)	120 – 480
Hot cylindrical bowl temperature (°C)	180 – 420
Hard Cover Stiffness (m)	0.105
Nip width (m)	0.0065
Soft material stiffness (m)	0.05
Diameter of cylindrical bowl (m)	0.36, 0.5
Specific heat ($J/Kg.K$)	1900
Thermal conductivity ($W/m.K$)	0.17
Density (Kg/m^3)	920
Dwell time (s)	0.008
Thermal diffusivity (m^2/s)	9.6×10^{-8}

5 Results and Discussion

5.1 *Influence of Cylindrical Bowl Temperature on Temperature of Fabric in Stiffness Direction*

Influence of cylindrical bowl temperature on the temperature of the fabric in stiffness direction inside the nip of silk finishing calender has been investigated from Eq. (20) as presented in Fig. 1, and calculated results are given in Table 2.

Results given in tables and figures clearly indicate that the side of the fabric which is in touch with the heated cylindrical bowl is at a very high temperature as compared to the side of the fabric which is in touch with the non-heated cylindrical bowl. Hence, temperature of the fabric decreases with rise in web depth, and while moving towards the mid of the fabric there is a very negligible influence of bowl temperature in stiffness direction. Also, results show that with a rise in cylindrical bowl temperature average fabric temperature rises.

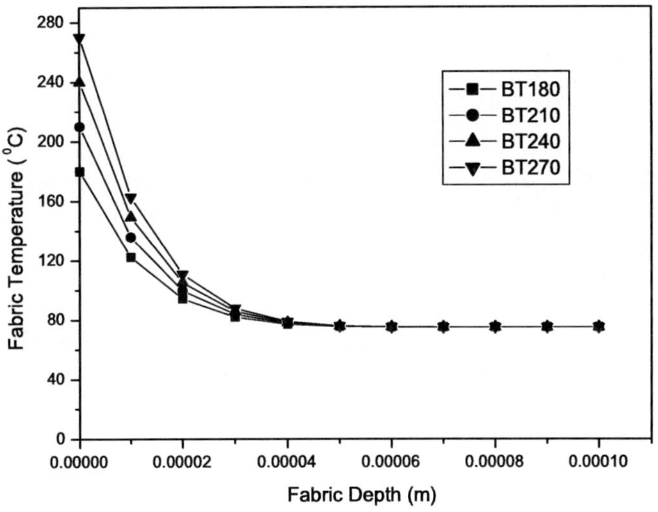

Fig. 1 Influence of cylindrical bowl temperature of silk finishing calender on temperature of fabric in stiffness direction

Table 2 Influence of cylindrical bowl temperature of silk finishing calender on temperature of fabric in stiffness direction

Fabric Depth (m)	Contact cylindrical bowl temperature (°C)			
	BT180	BT210	BT240	BT270
0	180	210	240	270
0.00001	122.244	135.742	149.241	162.739
0.00002	94.1768	99.6559	105.135	110.614
0.00003	81.8109	83.7569	85.7029	87.648
0.00004	77.0203	77.5976	78.1748	78.752
0.00005	75.4638	75.5963	75.7289	75.8614
0.00006	75.0718	75.0923	75.1128	75.1334
0.00007	75.0056	75.0072	75.0088	75.0104
0.00008	75.0001	75.0001	75.0002	75.0002
0.00009	75	75	75	75
0.0001	75	75	75	75
Average temperature	91.4358	96.1317	100.8277	105.5235

5.2 *Influence of Dwell Time*

The influence of dwell time has been found on the temperature at the mid part of the fabric in stiffness direction for the case of silk finishing calender from Eq. (20) as presented in Fig. 2, and calculated results are given in Table 3.

Results given in tables and figures clearly indicate that with rise in dwell time, fabric spends more time inside the calender nip due to which heat penetrates up to the centre of the fabric from the side which is in touch with the heated cylindrical bowl. Hence with rise in dwell time, temperature of the fabric rises.

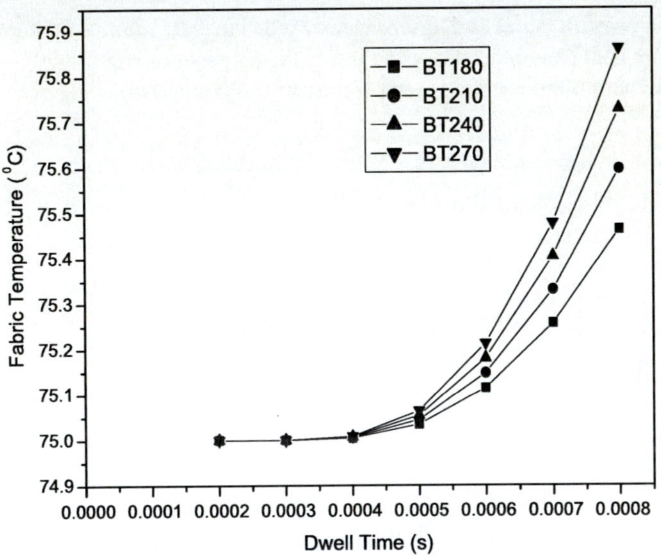

Fig. 2 Influence of dwell time at mid stiffness of the fabric (0.00005 m) at various temperatures in silk finishing calender

Table 3 Influence of dwell time at mid stiffness of the fabric (0.00005 m) at various temperatures in silk finishing calender

Dwell Time (s)	Contact cylindrical bowl temperature (°C)			
	BT180	BT210	BT240	BT270
0.0002	75	75	75	75
0.0003	75	75	75.0001	75.0001
0.0004	75.0045	75.0058	75.007	75.0083
0.0005	75.0346	75.0445	75.0543	75.0642
0.0006	75.1147	75.1474	75.1802	75.213
0.0007	75.2572	75.3307	75.4042	75.4777
0.0008	75.4638	75.5963	75.7289	75.8614

5.3 *Influence of Thermal Diffusivity on Temperature of Fabric in Stiffness Direction*

Influence of thermal diffusivity on the temperature of the fabric in stiffness direction has been investigated for silk finishing calendering with the heated cylindrical bowl temperature 180 °C using Eq. (20) as presented in Fig. 3, and calculated results are given in Table 4.

Results given in tables and figures clearly indicate that with rise in thermal diffusivity, more heat penetrates the fabric inside the calender nip up to the centre of the fabric and therefore the average temperature of the fabric rises.

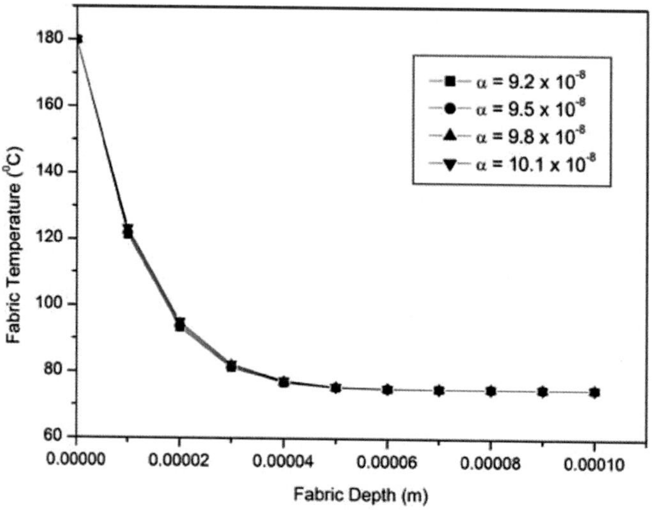

Fig. 3 Influence of thermal diffusivity of silk finishing calender on temperature of the fabric in stiffness direction

Table 4 Influence of thermal diffusivity on temperature of fabric in stiffness direction in silk finishing calender

Fabric Depth (m)	Thermal diffusivity (m^2/s)			
	9.2×10^{-8}	9.5×10^{-8}	9.8×10^{-8}	10.1×10^{-8}
0	180	180	180	180
0.00001	121.391	122.035	122.656	123.255
0.00002	93.397	93.9839	94.5586	95.1213
0.00003	81.3355	81.6922	82.0482	82.4033
0.00004	76.8006	76.9647	77.1329	77.305
0.00005	75.388	75.4442	75.5042	75.5678
0.00006	75.0542	75.0671	75.0818	75.0983
0.00007	75.0035	75.005	75.007	75.0094
0.00008	75	75.0001	75.0002	75.0003
0.00009	75	75	75	75
0.0001	75	75	75	75
Average Temperature	91.2154	91.3811	91.5445	91.7055

6 Conclusion

The analytical approximate solution of one-dimensional heat conduction equation is obtained using heat balance integral method. This also reflects the remarkable applicability of the heat balance integral method to solve the heat conduction model used in the textile industry. This model gives an evolutionary advantage and helps to predict the temperature of the fabric at distinct web depths in the stiffness direction inside the nip. In the case of the silk finishing calender, heat penetrates the outer surface of the fabric which is in touch with the heated cylindrical bowl. Therefore, in silk finishing calender, heat is not conducted to the mid of the fabric. Therefore, silk finishing calender gives more lustre and less dense fabric. To rise gloss and decrease roughness, higher temperature, higher load and lower speed are held accountable. Also, a decrease in the speed of the fabric passing through the calender nip results in the rise of dwell time which in turn rises the average temperature of the fabric. Desired fabric quality may not be obtained if calendering speed should not be increased beyond a certain limit. More pressure and heat are transported to the fabric in a stiffness direction which output the required quality of the fabric. So pressure and temperature influence should be balanced to get optimized results.

References

1. Gupta. N., Kanth. N.: Analysis of nip mechanics model for rolling calender used in textile industry. J. Serbian Soc. Comput. Mech. **12**(2), 39–52 (2018)
2. Gupta. N., Kanth. N.: Study of heat conduction inside rolling calender nip for different cylindrical bowl temperatures. J. Phys.: Conf. Ser. **1276**(1), 012044 1–9 (2019)
3. Litvinov. V., Farnood. R.: Modeling of the compression of coated papers in a soft rolling nip. J. Mater. Sci. **45**(1), 216–226 (2010)
4. Bhat, G.S., Jangala, P.K., Spruiell, J.E.: Thermal bonding of polypropylene nonwovens: effect of bonding variables on the structure and properties of the fabrics. J. Appl. Polym. Sci. **92**(6), 3593–3600 (2004)
5. Kanth. N., Ray. A.K., Dang. R.: Effect of design and process parameters on nip width of soft calendering. Int. J. Comput. Methods Eng. Sci. Mech. **17**(4), 247–252 (2016)
6. Gerstner. P., Paltakari. J., Gani. P.A.C.: Measurement and modelling of heat transfer in paper coating structure. J. Mater. Sci. **44**(2), 483–491 (2009)
7. Gratton. M.F., Hamel. J., McDonald. J.D.: Temperature-gradient calendering: From the laboratory to commercial reality. Pulp Paper Canada Ontario **98**, 62–71 (1997)
8. Holmstad. R., Kure. K.A., Chinga. G., Gregersen. Ø.W.: Effect of temperature gradient multi-nip calendering on the structure of SC paper. Nordic Pulp Paper Res. J. **19**(4), 489–494 (2004)
9. Hestmo. R.H., Lamvik. M.: Heat transfer during calendering of paper. J. Pulp Paper Sci. **28**(4), 128–135 (2002)
10. Gupta. N., Kanth. N.: Study of heat flow in a rod using homotopy analysis method and homotopy perturbation method. AIP Conf. Proc. 2019. **2061**(1), 020013 1–8 (2019)
11. Carslaw. H.S., Jaeger J.C.: Conduction of Heat in Solids. Oxford Science Publications (1959)
12. Kerekes. R.J.: Heat transfer in calendering. Trans. PPMC **5**(3), TR66–76 (1979)
13. Keller. S.: Heat transfer in a calender nip. J. Paper Sci. **20**(1), J33–J37 (1994)

14. Samula. S., Katoja. J.A., Niskanen. K.: Heat transfer to paper in a hot nip. Nordic Pulp Paper Res. J. **14**(4), 273–278 (1999)
15. Gupta. N., Kanth. N.: Analytical approximate solution of heat conduction equation using new homotopy perturbation method. Matrix Sci. Math. **3**(2), 01-07 (2019)
16. Gupta. N., Kanth. N.: Analysis of heat conduction inside the calender nip used in textile industry. AIP Conf. Proc. **2214**(1), 020008 (2020)
17. Gupta. N., Kanth. N.: Application of Perturbation theory in heat flow analysis. Collect. Papers Chaos Theory Appl. **173** (2021)
18. Gupta. N., Kanth. N.: Numerical solution of diffusion equation using method of lines. Indian J. Ind. Appl. Math. **10**(2), 194–203 (2019)
19. Gupta. N., Kanth. N.: A comparative study of new homotopy perturbation method and finite difference method for solving unsteady heat conduction equation. J. Serbian Soc. Comput. Mech. **15**(1), 98–109 (2021)
20. Mitchell, S.L., Myers, T.G.: Application of heat balance integral methods to one dimensional phase change problems. Int. J. Differ. Equ. **2012**, 1–22 (2012)
21. Langford. D.: The heat balance integral method. Int. J. Heat Mass Trans. 16(12), 2424–2428 (1973)
22. Baudouy. B.: Integral method for transient He II heat transfer in a semi-infinite domain. AIP Conf. Proc. **613**(1), 1349–1355 (2002)
23. Kot, V.A.: Integral method of boundary characteristics: the Dirichlet condition. Princ. Heat Trans. Res. **47**(10), 927–944 (2016)

A Multi-Criteria Decision Approach using Divergence Measures for Selection of the Best COVID-19 Vaccine

H. D. Arora, Anjali Naithani, and Aakanksha

Abstract COVID-19 is a worldwide health threat that has resulted in a significant number of deaths and complicated healthcare management issues. To prevent the COVID-19 pandemic, there is a need to choose a safe and most effective vaccine. Several Multi-criteria Decision-Making (MADM) techniques and approaches have been selected to choose the optimal probable options. The purpose of this article is to deliver divergence measures for fuzzy sets. To validate these measures, some of the properties were also proved. The Multi-criteria Decision-Making method is employed to rank and hence select the best vaccine out of available alternatives. The proposed research allows the ranking of different vaccines based on specified criteria in a fuzzy environment to aid in the selection process. The results suggest that the proposed model provides a realistic way to select the best vaccine from the vaccines available. A case study on the selection of the best COVID-19 vaccine and its experimental results using fuzzy sets are discussed.

Keywords TOPSIS · Multi-criteria decision-making · Triangular fuzzy sets

1 Introduction

The fuzzy set theory proposed by Zadeh in 1965 is a beneficial tool for solving problems in vague environments. Zadeh's fuzzy sets are intended to produce an analogue of crisp set theory in the field of uncertain conditions. Zadeh created a fuzzy set theory that may be used in incidents requiring ambiguity, vagueness, uncertainty

H. D. Arora · A. Naithani (✉) · Aakanksha
Department of Mathematics, Amity Institute of Applied Sciences, Amity University Uttar Pradesh, Noida, India
e-mail: anathani@amity.edu

H. D. Arora
e-mail: hdarora@amity.edu

Aakanksha
e-mail: aakanksha5@s.amity.edu

R. K. Sharma et al. (eds.), *Frontiers in Industrial and Applied Mathematics*, Springer Proceedings in Mathematics & Statistics 410,
https://doi.org/10.1007/978-981-19-7272-0_23

and hazy judgements. Theoretically, fuzzy sets might be used as a foundation for an expansion of mathematical concepts such as probability, topology and so forth whose traditional counterparts are based on the subject of crisp set theory. A Fuzzy Set primarily defines the degree to which a particular element belongs to a given set.

Fuzzy numbers were employed to get better outcomes in situations involving decision-making and evaluations. Fuzzy numbers which are an extended version of real numbers have their own features that may be linked to number theory. To make a connection between number theory and fuzzy numbers, triangular fuzzy numbers were introduced which mirror Pythagorean triples. Triangular Fuzzy Numbers (TFN) have been used to describe ambiguous and partial data in assessing risk, partial calls and knowledge-based systems.

Multi-criteria Decision-Making (MCDM) is a data science field that assesses multiple competing factors in decision-making. In domains where selecting the optimal solution is exceedingly complicated, the multi-criteria decision-making delivers robust decision taking. During the previous several years, Multi-criteria Decision-Making has had a tremendous amount of applications. Its relevance has risen considerably in a number of application sectors, especially when new techniques arise and current ones adapt. Multi-criteria Decision-Making is often utilized in a variety of fields, such as earth science, power generation, sustainability management, numerical methods and others. This study proposes a supplement to the fuzzy MCDM technique, in which the ranking of alternatives versus characteristics, as well as the weights of all criteria, are evaluated in semantic results calculated by Fuzzy numbers. Several academics in the field of linguistic modeling [4, 5] and fuzzy linguistic modeling [6] have presented the MCDM model in a fuzzy environment. Triantaphyllou et al. [7] gave Multi-criteria Decision-Making an Operations research approach. Harrera et al. [8] used a fuzzy set technique to provide a linguistic methodology for group decision-making. Kacprzyk et al. [9] propose fuzzy logic with linguistic expressions for group decision-making. Liu et al. [10] proposed a strategy for resolving fuzzy MADM issues with triangular Fuzzy Numbers depending on the connection number. For tackling multi attribute decision-making issues with given criterion weights, Wang and Gong [11] proposed a Set Pair Analysis-Based decision-making approach. Zhao and Zhang [12] presented the Set Pair Analysis-Based Triangular Fuzzy number MADM approach to handle difficulties with Multi Attribute Decision-Making when both characteristic weight and value are Triangular Fuzzy Numbers. To analyze the ambiguous MADM issue, Huang and Luo [13] proposed an index weight measure based on TFN. Moreover, Seikh et al. [14] gave Generalized triangular fuzzy numbers in an Intuitionistic fuzzy environment, and Sudha and Jayalalitha [15] defined Fuzzy triangular numbers in Sierpinski Triangle and Right-Angle Triangle. Also, Gani [16] proposed a new operation on Triangular Fuzzy Numbers for solving the fuzzy LPP.

The Technique for Order of Preference by Similarity to Ideal Solution (TOPSIS) is an MCDM approach established by Yoon and Hwang [17], which was updated by Yoon [18] and further by Hwang et al. [19]. The TOPSIS method is founded on the principle that the preferred choice should have the smallest Euclidean Distance from Positive Ideal Solution (PIS) and the greatest Euclidean Distance from Negative

Ideal Solution (NIS) [20]. The TOPSIS technique was used by several studies to investigate the MADM methodology [21]. TOPSIS method was employed for polar fuzzy linguistic [22, 24], environmental management [25, 26], supplier section [27] and several other realistic scenarios.

The COVID-19 pandemic is a worldwide health threat that has resulted in many deaths. In order to prevent further casualties, there is a need to choose the best vaccine when all the criteria are considered simultaneously. The criteria taken in this paper are taken from https://www.who.int/ [28]. India Today [29, 30] provided the data for the availability of different vaccines, and the data for the price of different vaccines [31, 32], their after-effects [33] and their efficacy [34] has been collected from Times of India [31, 34].

The following is how the entire article is structured: the second section discusses various fundamental definitions related to Fuzzy Sets, Triangular Fuzzy Numbers and Distance Measures. In the third section, a fuzzy TOPSIS algorithm is suggested as well as a case study to select the best COVID-19 vaccine is discussed and vaccines are ranked accordingly. Finally, Sect. 4 presents the paper's conclusion.

2 Preliminaries

The theoretical foundation of fuzzy sets suggested by Zadeh [35] and Zimmerman [36, 37] is covered in this section. The following is an overview of the fuzzy set concept.

Definition 2.1 [35]. The Fuzzy Set ***A*** in Y is described by the membership function:

$$\boldsymbol{A} = \{\langle y, \mu_A(y)\rangle | y \in \Upsilon\} \tag{1}$$

where $\mu_A(y)$: Y→ [0, 1] is the measure of the degree of belongingness of participation of an element $y \in$ Y in ***A***.

Definition 2.2 [38]. Let $A = [e, f, g, h]$ be any real Fuzzy Number, thus its membership function is as follows:

$$\mu_A(x) = \begin{cases} \mu_M^L(x) & e \le x \le f \\ 1 & f \le x \le g \\ \mu_M^U(x), & g \le x \le h \\ 0 & \text{otherwise} \end{cases} \tag{2}$$

where $\mu_M^L(x)$ and $\mu_M^U(x)$ are lower and the upper Membership Functions of the Fuzzy Number ***A***, respectively, and $p = -\infty$, or $p = q$, or $q = r$, or $r = s$, or $s = +\infty$.

Definition 2.3 [36, 37]. A Triangular Fuzzy Number (TFN) ***A*** is a Fuzzy Number with piece-wise linear membership function $\mu_A(x)$ described by

$$\mu_A(x) = \begin{cases} \frac{x-u}{v-u} & u \le x \le v \\ \frac{w-x}{w-v} & v \le x \le w \\ 0 & \text{otherwise} \end{cases} \tag{3}$$

which is represented as (u, v, w).

Definition 2.4 [39]. Let $P = (u, v, w)$ and $Q = (x, y, z)$ be any two TFNs. Then the Distance Measure function $D(P, Q)$ can be defined as

$$D(P, Q) = \sqrt{\frac{1}{3}\{(x-u)^2 + (y-v)^2 + (z-w)^2\}}. \tag{4}$$

3 Suggested Fuzzy TOPSIS Algorithm

Due to its capacity to examine several attributes concurrently, Multi Attribute Decision-Making (MADM) has appeared to be a promising technique to solve problems with inadequate or vague data. This section discusses the MADM issue in the fuzzy domain. A feasible procedure is made available to deal with MADM issues in a fuzzy environment. We know that each decision matrix in the MADM method has four main components: (a) Criteria, (b) Alternative, (c) Weights and (d) assessment value of alternatives in relation to the criteria. The method of the proposed technique will then be applied to the selection of the best COVID-19 vaccine.

The procedure proposed to solve the MADM issue in a fuzzy environment is explained by the following steps:

Step 1: Gather the decision maker's subjective opinion on the relevance of the weights.

Step 2: Compute the Fuzzy significant coefficients or weights founded on the decision maker's subjective judgements utilizing the table of linguistic variables and their accompanying Triangular Fuzzy Weights.

Step 3: Structure the normalized Decision Matrix.

Step 4: Create the Fuzzy Weighted Decision Matrix by multiplying normalized decision matrix by their corresponding fuzzy weights.

Step 5: Calculate the Fuzzy Positive Ideal Solution and the Fuzzy Negative Ideal Solution.

Step 6: Calculate the Euclidean Distance of all alternatives from fuzzy positive and negative ideal solutions.

Step 7: Determine the fuzzy closeness coefficient.

Step 8: Sort the alternatives corresponding to their closeness coefficients and chose the foremost option.

3.1 Case Study

Coronavirus disease (COVID-19) is an infectious illness caused by Severe Acute Respiratory Syndrome Coronavirus 2 (SARS-CoV-2). The symptoms of coronavirus range from none to life-threatening. COVID-19 can make anyone sick and cause them to become terminally sick or die at any time. To prevent these severe effects, vaccination is done in every country including India. The Oxford AstraZeneca vaccine, created by the Serum Institute of India (SII) under the title "Covishield" and BBV152 (Covaxin), a vaccine created by Bharat Biotech in conjunction with the National Institute of Virology and the Indian Council of Medical Research, was approved by the DCGI in January 2021. The DCGI authorized the Russian Sputnik V vaccine, which has been tested in India by Dr. Reddy's Laboratories, in April 2021. In late June 2021, DCGI authorized the Moderna vaccine for emergency use. The various criteria to choose the best COVID-19 vaccine are taken from https://www.who.int/publications/m/item/criteria-for-covid-19-vaccine-prioritization. The data for Covaxin, Covishield and Sputnik is taken for the Indian population, whereas the data for the Moderna vaccine is taken by considering the worldwide population as the jabs of the Moderna vaccine are given in India only in case of emergency. Data including various criteria Efficacy (C1), Availability (C2), Price (C3) and After Effect (C4) is given in the Table 1.

Based on the data given in the Table 1, the vaccines need to be ranked and the selection of the best COVID-19 vaccine needs to be done. The initial stage in MADM is to categorize the situation under consideration using benefit and cost criteria. Benefit criteria are those that are intended to have higher values, whereas cost criteria are those that are intended to have lower values. In the case study considered here, Efficacy (C1) and Availability (C2) are the criteria of benefit, and Price (C3) and After Effect (C4) are the criteria of cost. To proceed further, a 7-point scale of Triangular Fuzzy Numbers, as given in Table 2, must be chosen.

Step 1: Let there be four decision makers, DM1, DM2, DM3 and DM4 who will decide the best COVID-19 vaccine among the alternatives present. Table 3 given depicts the decision maker's choices in terms of linguistic factors as follows.

Step 2: Fuzzy weights are computed and given below, based on the subjective opinion of decision makers.

Table 1 Data set in the form of decision matrix

Alternative/criteria	Efficacy	Availability	Price	After Effect
Covaxin	81	55	1410	0.04
Covishield	90	75	780	0.03
Sputnik	91	15.6	1145	0.002
Moderna	95	0.75	800	0.00004

Table 2 Linguistic variables and their corresponding triangular fuzzy weights

Importance	Fuzzy weights
Very Low (VL)	(0, 0, 0.1)
Low (L)	(0, 0.1, 0.3)
Fairly Low (FL)	(0.1, 0.3, 0.5)
Medium (M)	(0.3, 0.5, 0.7)
Fairly High (FH)	(0.5, 0.7, 0.9)
High (H)	(0.7, 0.9, 1)
Very High (VH)	(0.9, 1, 1)

Table 3 Rating by decision makers on linguistic scale

Criteria/decision maker	DM1	DM2	DM3	DM4
Efficacy	H	FH	VH	VH
Availability	FH	H	M	FH
Price	M	FL	VL	FL
After effect	H	VH	FH	H

Step 3: Taking into account the highest, middle and lower values of the four ratings from Table 4, the aggregated fuzzy weights are generated as follows (Table 5).

Step 4: Multiply the Normalized Decision Matrix by its associated Fuzzy Weights to get the Fuzzy weighted Normalized Decision Matrix, as stated in the formula:

$$V = X \times W$$

Table 4 Conversion of linguistic rating of decision makers into fuzzy rating

Criteria/decision maker	DM1	DM2	DM3	DM4
C1	(0.7 0.9 1)	(0.5 0.7 0.9)	(0.9 1 1)	(0.9 1 1)
C2	(0.5 0.7 0.9)	(0.7 0.9 1)	(0.3 0.5 0.7)	(0.5 0.7 0.9)
C3	(0.3 0.5 0.7)	(0.1 0.3 0.5)	(0 0 0.1)	(0.1 0.3 0.5)
C4	(0.7 0.9 1)	(0.9 1 1)	(0.5 0.7 0.9)	(0.7 0.9 1)

Table 5 Aggregated fuzzy rating

Criteria/fuzzy weights	L fuzzy weight	M fuzzy weight	U fuzzy weight
Efficacy (C1)	0.75	0.90	0.98
Availability (C2)	0.50	0.70	0.88
Price (C3)	0.13	0.28	0.45
After effect (C4)	0.70	0.88	0.98

Table 6 Fuzzy weighted normalized decision matrix

	Efficacy			Availability		
Covaxin	0.3398	0.4077	0.4417	0.2916	0.4082	0.5103
Covishield	0.3775	0.4530	0.4908	0.3976	0.5567	0.6959
Sputnik	0.3817	0.4581	0.4962	0.0827	0.1158	0.1447
Moderna	0.3985	0.4782	0.5181	0.0040	0.0056	0.0070
	Price			After effect		
Covaxin	0.0826	0.1818	0.2975	0.5596	0.6994	0.7794
Covishield	0.0457	0.0457	0.1006	0.4197	0.5246	0.5845
Sputnik	0.0671	0.0671	0.1477	0.0280	0.0350	0.0390
Moderna	0.0469	0.0469	0.1032	0.0006	0.0007	0.0008

where V = v_{ij} (i = 1,…,4 and j = 1, 2, 3,…., 12) is normalized matrix, X = x_{ij} (i = 1,…,4 and j = 1,….., 4) is the decision matrix and W = w_{ij} (I = 1,….., 4, j = 1, 2, 3) are the aggregated fuzzy weights (Table 6).

Step 5: Using the following formulae, the fuzzy positive ideal solution (FPIS) A^{k+} and fuzzy negative ideal solution (NPIS) A^{k-} are calculated:

$$A^{k+} = \{r_1^{k+}, r_2^{k+}, \ldots, r_n^{k+}\} = \{(\max(r_{ij}^k)/j \in I), (\min(r_{ij}^k)/j \in J)\} \quad (5)$$

$$A^{k+} = \{r_1^{k-}, r_2^{k-}, \ldots, r_n^{k-}\} = \{(\min(r_{ij}^k)/j \in I), (\max(r_{ij}^k)/j \in J)\} \quad (6)$$

where I and J represent the criterion of benefit and criterion of cost, respectively.

Table 7 shows the results of the calculations.

Step 6: Separation measures S_i^+, S_i^- and the Euclidean Distance [39] $D(A_i, A^+), D(A_i, A^-)$ of each alternative from FPIS and FNIS have been determined using Formulae (7) and (8) and are provided in Tables 8 and 9.

$S_i^+ = \sum_{i=1}^n D(A_i, A^+)$, where

Table 7 Positive and negative ideal solution for each criterion

	Efficacy (C1)			Availability (C2)		
	Lower	Middle	Upper	Lower	Middle	Upper
A^+	0.3985	0.4782	0.5181	0.3976	0.5567	0.6959
A^-	0.3398	0.4077	0.4417	0.0040	0.0056	0.0070
	Price (C3)			After effect (C4)		
A^+	0.0457	0.1006	0.1646	0.0006	0.0007	0.0008
A^-	0.0826	0.1818	0.2975	0.5596	0.6994	0.7794

Table 8 Separation measures for FPIS for each criterion

		C1	C2	C3	C4	S_i^+
For FPIS	Covaxin	0.0689	0.1502	0.0924	0.6848	0.9964
	Covishield	0	0	0.5134	0.5134	0.5381
	Sputnik	0.4462	0.0536	0.0036	0.0336	0.5530
	Moderna	0.5578	0.0029	0.0029	0	0.5607

Table 9 Separation measures for FNIS for each criterion

		C1	C2	C3	C4	S_i^-
For FNIS	Covaxin	0	0.4075	0	0	0.4075
	Covishield	0.0443	0.5578	0.0924	0.1714	0.8659
	Sputnik	0.0492	0.1116	0.0389	0.6512	0.8509
	Moderna	0.0689	0	0.0895	0.6848	0.8432

$$D(A_i, A^+) = \sqrt{\frac{1}{3}\left\{(a_1 - b^+)^2 + \left(a_2 - b_2^+\right)^2 + \left(a_3 - b_3^+\right)^2\right\}} \quad \forall_i = 1, 2, 3, 4 \quad (7)$$

and

$S_i^- = \sum_{i=1}^{n} D(A_i, A^-)$, where

$$D(A_i, A^-) = \sqrt{\frac{1}{3}\left\{(a_1 - b^-)^2 + \left(a_2 - b_2^-\right)^2 + \left(a_3 - b_3^-\right)^2\right\}} \quad \forall_i = 1, 2, 3, 4 \quad (8)$$

Step 7: Equation (9) was used to get the closeness coefficient (R_i) for each evaluated alternative.

$$R_i = \frac{D(A_i, A^-)}{D(A_i, A^+) + D(A_i, A^-)} = \frac{S_i^-}{S_i^+ + S_i^-} \quad \text{where } 0 \leq R_i \leq 1, \quad i = 1, 2, 3, 4 \quad (9)$$

As stated in Table 10, the rankings were done in decreasing order of magnitude.

Table 10 Ranking result obtained from TOPSIS approach

	S_i^+	S_i^-	R_i	Rank
Covaxin	0.9964	0.4075	0.2903	4
Covishield	0.5381	0.8659	0.6168	1
Sputnik	0.5530	0.8509	0.6061	2
Moderna	0.5607	0.8432	0.6006	3

3.2 *Sensitivity Analysis*

Originally, the decision makers were given equal importance while ranking the different alternatives. However, there are instances where the decision maker's opinions are prioritized differently. In this section, such scenarios have been examined.

Different priorities, β_i, have been allotted to the four decision makers, where $\beta_i > 0, i = 1, 2, 3, 4$ and $\sum_{i=1}^{4} \beta_i = 1$. The distance measures D_r^+, D_r^- and the closeness coefficient (R_i) have been calculated using Eqs. (10), (11) and (12) and are introduced in Table 11.

$$D_r^+ = \sum_{r=1}^{s} \beta_i S_r^+ \tag{10}$$

$$D_r^- = \sum_{r=1}^{s} \beta_i S_r^- \tag{11}$$

$$\text{Also,} \quad R_i = \frac{v_r^-}{v_r^+ + v_r^-} \quad \text{where } 0 \leq R_i \leq 1, \quad i = 1, 2, 3, 4 \tag{12}$$

The results of the suggested technique remained the same when different priorities were assigned to the judgements of decision makers and Covishield stood out to be the best vaccine against COVID-19 in all circumstances, hence proving the validity and dependability of the suggested technique.

4 Conclusion

In this paper, a novel technique to solve issues involving Multi-criteria Decision-Making was proposed and the same was applied in order to select the best COVID-19 vaccine. The selection was done by considering different criteria and a team of experts. Then we ranked various vaccines with the help of the TOPSIS approach, also the selection for the best vaccine was done by assigning priorities to different criteria. Eventually, it was found that Covishield is the best vaccine out of the available alternatives. Despite the Multi-Criteria domain, this approach supports decision makers in producing unbiased and systematic judgements. In the long term, this study can be used in various Multi-criteria Decision-Making procedures and could help in the analysis of various vague situations.

Table 11 Aggregated closeness coefficient and ranking of each alternative

Vaccines	Distance measure		R_i	Rank	Best vaccine
	D_r^+	D_r^-			
(a) Case 1: $\beta_1 = 0.4$, $\beta_2 = 0.3$, $\beta_3 = 0.2$ and $\beta_4 = 0.1$					
Covaxin	0.1596	0.1223	0.4337	3	Covishield
Covishield	0.0612	0.2207	0.7829	1	
Sputnik	0.1558	0.1261	0.4472	2	
Moderna	0.1679	0.1139	0.4043	4	
(b) Case 2: $\beta_1 = 0.35$, $\beta_2 = 0.25$, $\beta_3 = 0.23$ and $\beta_4 = 0.17$					
Covaxin	0.1994	0.1019	0.3382	4	Covishield
Covishield	0.0959	0.2053	0.6817	1	
Sputnik	0.1365	0.1648	0.5470	2	
Moderna	0.1401	0.1611	0.5349	3	
(c) Case 3: $\beta_1 = 0.3$, $\beta_2 = 0.28$, $\beta_3 = 0.27$ and $\beta_4 = 0.17$					
Covaxin	0.2041	0.1141	0.3586	4	Covishield
Covishield	0.0947	0.2236	0.7025	1	
Sputnik	0.1510	0.1672	0.5254	2	
Moderna	0.1570	0.1613	0.5067	3	
(d) Case 4: $\beta_1 = 0.33$, $\beta_2 = 0.29$, $\beta_3 = 0.2$ and $\beta_4 = 0.18$					
Covaxin	0.2081	0.1182	0.3622	4	Covishield
Covishield	0.1005	0.2257	0.6918	1	
Sputnik	0.1527	0.1736	0.5321	2	
Moderna	0.1623	0.1639	0.5024	3	

References

1. Luca, A., Termini, S.: A definition of a nonprobabilistic entropy in the setting of fuzzy sets theory, **20**(5), 301–312 (1972)
2. Anand, M., Bharatraj, J.: Theory of triangular fuzzy number. In: National Conference on Advanced Trends in Mathematics, pp. 80–83. Thiruvalluvar University (2014). ISBN: 978 93 85126 14 7
3. Zhang, X., Ma, W., Chen, L.: New similarity of triangular fuzzy number and its application. Sci. World J. (2014). https://doi.org/10.1155/2014/215047
4. Bordogna, G., Fedrizzi, M., Pasi, G.: A linguistic modeling of consensus in group decision making based on OWA operators. IEEE Trans. Syst. Man Cybern. **27**(1), 126–132 (1997)
5. Chen, S.J., Hwang, C.L.: Fuzzy Multiple Attribute Decision Making. Springer, New York (1992)
6. Fodor, J.C., Roubens, M.: Fuzzy Preference Modelling and Multicriteria Decision Support. Kluwer Academic Publisher, Dordrecht (1994)
7. Triantaphyllou, E., Shu, B., Sanchez, S., Ray, T.: Multi-Criteria Decision Making: An Operations Research Approach. Wiley **15**, 175–186 (1998)
8. Herrera, F., Herrera, E., Viedma, Verdegay, J. L.: A linguistic decision process in group decision making. Group Decis. Negotiation **5**, 165–176 (1996)

9. Kacprzyk, J., Fedrizzi, M., Nurmi, H.: Group decision making and consensus under fuzzy preferences and fuzzy majority. Fuzzy Sets Syst. **49**(1), 21–31 (1992)
10. Liu, X.M., Zhao, K.Q., Wang, C.B.: New multiple attribute decision-making model with triangular fuzzy numbers based on connection numbers. Syst. Eng. Electron. **31**, 2399–2403 (2009)
11. Wang, J.Q., Gong, L.: Interval probability stochastic multi-criteria decision-making approach based on set pair analysis. Control Decis. **24**, 1877–1880 (2009)
12. Zhao, Y., Zhang, L.: Application of the set-pair analysis connection number in decision making of black-start vague set. CAAI Trans. Intell. Syst. **9**, 632–640 (2014)
13. Huang, Z.L., Luo, J.: Possibility degree relation method for triangular fuzzy number-based uncertain multi-attribute decision making. Control Decis. **30**, 1365–1371 (2015)
14. Seikh, M.R., Nayak, P.K., Pal, M.: Generalized triangular fuzzy numbers in intuitionistic fuzzy environment. Int. J. Eng. Res. Dev. **5**(1), 08–13 (2012)
15. Sudha, T., Jayalalitha, G.: Fuzzy triangular numbers in - Sierpinski triangle and right angle triangle. J. Phys. (2020). https://doi.org/10.1088/1742-6596/1597/1/012022
16. Gani, A.N.: A new operation on triangular fuzzy number for solving fuzzy linear programming problem, **6**(12), 525–532 (2012)
17. Hwang, C.L., Yoon, K.: Multiple Attribute Decision Making : Methods and Applications, vol. 40, pp. 721–727. Springer, New York (2004)
18. Yoon, K.: A reconciliation among discrete compromise situations. J. Oper. Res. Soc. **38**(3), 277–286 (1987)
19. Hwang, C.L., Lai, Y.J., Liu, T.Y.: A new approach for multiple objective decision making. Comput. Oper. Res. **20**(8), 889–899 (1993)
20. Assari, A., Mahesh, T., Assari, E.: Role of public participation in sustainability of historical city: usage of TOPSIS method. Indian J. Sci. Technol. **5**(3), 2289–2294 (2012)
21. Hwang, C. L., Yoon, K.: Multiple Objective Decision Making—Methods and Applications: A State-of-the-Art Survey. Lecture Notes in Economics and Mathematical Systems. Springer, New York (1981)
22. Adeel, A., Akram, M., Koam, A.N.A.: Group decision making based on mm-polar fuzzy linguistic TOPSIS method. Symmetry **11**(6), 735 (2019)
23. Akram, M., Adeel, A.: TOPSIS approach for MAGDM based on interval-valued hesitant fuzzy NN-soft environment. Int. J. Fuzzy Syst. **21**(3), 993–1009 (2019)
24. Akram, M., Shumaiza, Smarandache, F.: Decision making with bipolar neutrosophic TOPSIS and bipolar neutrosophic. ELECTRE-I. Axioms **7**(2), 33 (2018)
25. Ananda, J., Herath, G.: Analysis of forest policy using multi-attribute value theory. In: Herath, G., Prato, T. (eds.) Using Multi-criteria Decision Analysis in Natural Resource Management, pp. 11–40. Ashgate Publishing Ltd., Hampshire, (2006)
26. Askarifar, K., Motaffef, Z., Aazaami, S.: An investment development framework in Iran's seashores using TOPSIS and best-worst multi-criteria decision-making methods. Decis. Sci. Lett. **7**(1), 55–64 (2018)
27. Boran, F.E., Genc, S., Kurt, M., Akay, D.: A multi-criteria intuitionistic fuzzy group decision making for supplier selection with TOPSIS method. Expert Syst. Appl. **36**, 11363–11368 (2009)
28. WHO criteria page. https://www.who.int/publications/m/item/criteria-for-covid-19-vaccine-prioritization. Accessed 26 Oct 2021
29. COVID-19 vaccine homepage. https://www.indiatoday.in/coronavirus-outbreak/vaccine-updates/story/all-you-need-to-know-about-8-covid-vaccines-likely-to-be-given-in-india-in-2021-1802668-2021-05-14. Accessed 10 Oct 2021
30. Moderna vaccine availability page. https://www.indiatoday.in/coronavirus-outbreak/vaccine-updates/story/india-moderna-vaccine-covax-programme-who-1830179-2021-07-20. Accessed 9 Oct 2021
31. Blog. https://timesofindia.indiatimes.com/india/at-rs700-rs1500-price-of-covid-vaccine-in-indias-private-sector-among-costliest/articleshow/82509814.cms. Accessed 10 Oct 2021
32. COVID-19 vaccine price page. https://timesofindia.indiatimes.com/india/centre-caps-vaccine-prices-covishield-at-rs-780-covaxin-rs-1410/articleshow/83343406.cms. Accessed 10 Oct 2021.

33. COVID-19 vaccine after effects. https://timesofindia.indiatimes.com/india/only-2-4-infections-per-10k-found-in-those-vaccinated-with-two-doses-icmr/articleshow/82188882.cms. Accessed 9 Oct 2021
34. Vaccine efficacy page. https://pharmeasy.in/blog/covaxin-vs-covishield-a-detailed-comparison/. Accessed 9 Oct 2021
35. Zadeh, L.A.: Fuzzy sets. Inform Control **8**, 338–356 (1965)
36. Zimmermann, H.J.: Fuzzy Set, Decision Making and Expert System. Kluwer, Boston (1987)
37. Zimmermann, H.J.: Fuzzy Set Theory—And Its Application, 2nd edn. Kluwer, Boston (1991)
38. Chu, T.C., Lin. Y.C.: An interval arithmetic based fuzzy TOPSIS model. Exp. Syst. Appl. **36**, 10870–10876 (2009)
39. Chen, T.Y., Tsao, C.Y.: The interval-valued fuzzy TOPSIS method and experimental analysis. Fuzzy Sets Syst. **159**(11), 1410–1428 (2008)
40. Vaccine development. https://en.wikipedia.org/wiki/COVID-19_pandemic_in_India#Vaccine_development_and_production. Accessed 10 Oct 2021

Magnetohydrodynamic Mixed Convection Flow in a Vertical Channel Filled with Porous Media

Nidhi Singh and Manish K. Khandelwal

Abstract We report a linear instability mechanism of MHD mixed convection flow in a porous medium channel under a transverse magnetic field. The stability results are reported for an electrically conducting water-based electrolytes fluid. The governing equations are solved by a Chebyshev spectral collocation method. The linear disturbance equations formed a generalized eigenvalue problem. The results show that the basic flow contains the inflection point. The linear stability analysis shows that the growth of the disturbance reduces by increasing the strength of the magnetic field and decreasing the media permeability of the porous medium flow. The linear stability boundaries show that the relatively higher strength of the applied magnetic field stabilizes the flow, whereas an increase in the media permeability destabilizes the basic flow.

Keywords MHD flow · Mixed convection · Porous medium

1 Introduction

The phenomenon of mixed convection occurs due to thermal buoyancy force as well as external pressure gradient. The mixed convection flows through porous medium have been examined extensively due to wide applications in the electronic industry. The porous medium is an excellent candidate to enhance the heat in many heat transfer applications. The stability of mixed convection flow in vertical geometries has been the object of great interest in several applications namely heat exchangers, nuclear

N. Singh (✉) · M. K. Khandelwal
Department of Mathematics, Indira Gandhi National Tribal University, Amarkantak, MP 484887, India

M. K. Khandelwal
e-mail: khandelwal@igntu.ac.in; manish@iiitdm.ac.in

M. K. Khandelwal
Department of Mathematics, Indian Institute of Information Technology, Design and Manufacturing, Kancheepuram, Chennai 600127, India

R. K. Sharma et al. (eds.), *Frontiers in Industrial and Applied Mathematics*, Springer Proceedings in Mathematics & Statistics 410,
https://doi.org/10.1007/978-981-19-7272-0_24

reactors, solar collectors, electronic equipment. The researchers have already studied the hydrodynamic instability properties of mixed convection flow in the vertical configurations [1–6] under different types of heating conditions. The flow of an electrically conducting fluid through a porous medium under an applied magnetic field has received considerable attention in many technological and laboratory flows. The study of magnetohydrodynamic (MHD) flow of electrically conducting fluid through a porous medium is essential in many MHD-related applications such as blankets (e.g., dual-coolant lead–lithium (DCLL) blanket) for thermonuclear reactors, MHD generators, crystal growth, and stirring of melts in the metallurgical industry, and electronic devices [7, 8]. The interaction of electrically conducting fluid with the magnetic field through inter-connected porous medium flow gives a significant heat enhancement in the above-mentioned applications.

The understanding of hydrodynamic stability analysis of MHD porous media flow is a fundamental interest in many applications. The present paper focuses on the stability analysis of parallel mixed convection flow in a linearly heated vertical porous medium channel with a transverse magnetic field. The flow instabilities may appear by many factors, particularly by a magnetic field, heat transfer, and medium permeability of the porous medium. The present study will provide the basic concept of the flow instability mechanism of an electrically conducting fluid through a porous medium in the presence of a magnetic field. There are some studies relevant to the present investigation in a vertical configuration. We summarize some important conclusions, which may support the present investigation.

The instability mechanism of non-magnetic fully developed mixed convection flow in a vertical channel filled with a porous medium is well established in the open literature using linear and weakly nonlinear stability analysis [9–15]. In these studies, two different types of state the local thermal equilibrium (LTE) state and local thermal non-equilibrium (LTNE) state are used for energy equations to examine the instability characteristics of porous medium flow. The effect of media permeability [9], the influence of Prandtl number [10], impact of different models [11], the impact of inter-phase heat transfer coefficient [12] for porous medium flow are discussed in terms of stability analysis. These porous medium parallel mixed convection channel flow studies show that flow is most unstable under two-dimensional. The kinetic energy balance mainly gives three different instability types: buoyant, shear, and thermal-shear (mixed) instability for assisted flow and Rayleigh–Taylor instability for buoyancy-opposed flow.

The MHD flow through porous media is investigated very little in the open literature. However, few important theoretical and experimental studies on porous medium flow in the presence of a magnetic field in different geometries are available in the literature [7, 16–22]. Wallace et al. [16] have investigated an experimental study for the flow of mercury in porous media (sandstone) under a magnetic field. They have shown that the flow rate of mercury through the porous media under the magnetic field does not change. Later, Rudraiah et al. [22] have performed a theoretical and numerical study of Hartmann flow to validate experimentally obtained results of Wallace et al. [16]. Using multiple scale expansions, Geindreau and Auriault [7] have examined the macroscopic description of seepage in porous media under magnetic field.

In contrast to instability properties of porous medium flow under magnetic field, the instability mechanism for viscous-medium flow in a vertical channel under applied magnetic field is discussed rigorously. For example, a survey of existing published literature: related to natural convection [23–25], in connection with forced convection [26–28], and in connection with mixed convection [29–32] focuses on the instability mechanism of parallel channel flow under magnetic field. In these studies, the magnetic field, in general, stabilizes the basic flow. However, in the mixed convection flow, thermal buoyancy force destabilizes the MHD flow.

The above literature review shows that the hydrodynamic stability characteristics of mixed convection MHD flow in a vertical channel filled with a porous medium is not considered yet in our best knowledge. Therefore, we aim to discuss the stability properties of mixed convection MHD porous medium flow using linear stability analysis in a vertical channel. The present study will provide a new research development in many MHD applications.

2 Governing Equations

We consider an incompressible MHD flow in a long vertical channel filled with a porous medium. The width of the channel is *2L*. A uniform transverse magnetic field of strength $\boldsymbol{B_0}$ is applied perpendicularly to the direction of the flow, as shown in Fig. 1. Buoyancy force and an external pressure gradient drive the flow under a uniform magnetic field. A linearly varying temperature is considered on the walls as $T_w = T_0 + Cz$ where T_0 denotes the reference temperature and C is a positive constant. The schematic of flow configuration is displayed in Fig. 1. Thermo-physical properties of the fluid are assumed constant except density in the buoyancy force term. The Boussinesq approximation is used for the density variation. In the present study, we have adopted volume-averaged Navier–Stokes (VANS) equations for transporting porous medium [33] to analyze the flow instability. The non-dimensional governing equation of the present problem is given by

$$\nabla . \boldsymbol{V} = 0 \tag{1}$$

$$\frac{1}{\varepsilon}\frac{\partial \boldsymbol{V}}{\partial t} + \frac{1}{\varepsilon^2}(\boldsymbol{V}.\nabla)\boldsymbol{V} + F\boldsymbol{V}|\boldsymbol{V}| = -\nabla p + \frac{\lambda}{Re}\nabla^2\boldsymbol{V} - \frac{1}{DaRe}\boldsymbol{V} + \frac{Ra}{Re}\theta\overrightarrow{e_z} + \frac{Ha^2}{Re}(\boldsymbol{j}\times\overrightarrow{e_x}) \tag{2}$$

$$\sigma\frac{\partial\theta}{\partial t} + V.\nabla\theta = \frac{1}{RePr}\left(\nabla^2\theta - w\right) \tag{3}$$

$$\boldsymbol{j} = -\nabla\phi + (\boldsymbol{V}\times\overrightarrow{e_x}) \tag{4}$$

$$\nabla . \boldsymbol{j} = 0 \tag{5}$$

Fig. 1 Physical problem and coordinate system

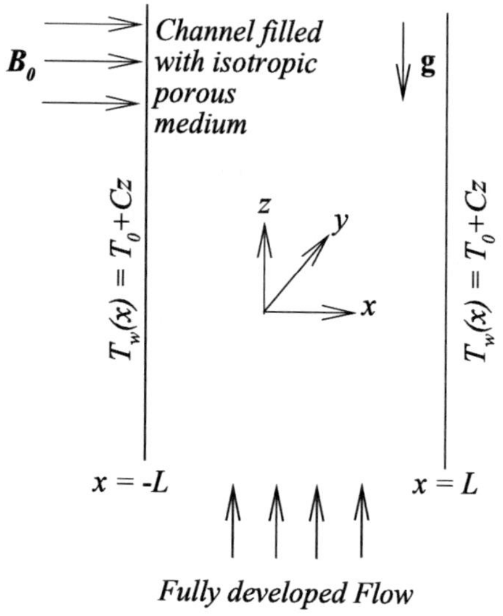

where $\vec{e_x}$, $\vec{e_z}$, σ, and ε are the unit vector in the x-direction and z-direction, the ratio of the volumetric heat capacities of the fluid and medium and porosity of the medium, respectively. Following non-dimensional quantities are used to non-dimensionalized the above governing equations

$$(x, y, z) = \frac{(x^*, y^*, z^*)}{L}, \boldsymbol{V} = \frac{\boldsymbol{V}^*}{\overline{W_0}}, P = \frac{p^*}{\rho_0 \overline{W_0}^2}, t = \frac{t^* \overline{W_0}}{L}, \theta = \frac{T - T_W}{CLRePr},$$

$$\phi = \frac{\varnothing}{L\boldsymbol{B}_0\overline{W_0}}, \boldsymbol{j} = \frac{\boldsymbol{j}^*}{\sigma_1 \boldsymbol{B}_0 \overline{W_0}} \quad (6)$$

where $\boldsymbol{V}, P, t, \theta, \boldsymbol{j}, \phi$ are the dimensionless velocity vector, pressure, time, temperature, current density, and electrical potential, respectively. The following non-dimensional parameters are appeared in the present problem: Rayleigh number ($Ra = gCL^4\beta_T/\nu\alpha$), Reynolds number ($Re = \overline{W_0}L/\nu$), Prandtl number ($Pr = \nu/\alpha$), Darcy number ($Da = K/L^2$), Forchheimer number ($F = C_F L/|K|^{1/2}$), viscosity ratio ($\lambda = \overline{\mu}/\mu_f$), interaction parameter ($N = \sigma_1 L\boldsymbol{B_0}^2/\rho_0\overline{W_0}$), and Hartmann number ($Ha = \sqrt{NRe}$). Furthermore, $\overline{W_0}$ is average base velocity, ρ_0 is reference fluid density, α is the thermal diffusivity, β_T is the thermal expansion coefficient, ν is the kinematic viscosity, g acceleration due to gravity, K the permeability of the porous medium, C_F is form drag coefficient, $\overline{\mu}$ is coefficient of effective viscosity, μ_f is the fluid viscosity, and σ_1 is the electrical conductivity. Note that the value of σ and λ is taken 1 for the present investigation.

2.1 Basic Flow

To investigate the instability mechanism of the mixed convective flow, first we derive the basic flow that is steady state, unidirectional, and fully developed. Using these conditions, the governing Eqs. (1)–(5) reduces into the following ordinary differential equations

$$\lambda\frac{d^2W_0}{dx^2} - \frac{1}{Da}W_0 - ReF|W_0|W_0 - Ha^2W_0 + Ra\Theta_0 = Re\frac{dP_0}{dz} \tag{7}$$

$$\frac{d^2\Theta_0}{dx^2} = W_0 \tag{8}$$

The boundary conditions for the basic flow at the channel walls are

$$W_0 = \Theta_0 = 0 \text{ at } x = \pm 1 \tag{9}$$

where W_0, Θ_0, and P_0 are the basic state velocity, temperature, and pressure, respectively.

2.2 Linear Stability Analysis

The classical normal mode analysis [34] is considered to examine the linear stability analysis of the above MHD mixed convection basic flow. The infinitesimal disturbance is imposed on the basic flow. Thus the velocity, temperature, and pressure field can be written as

$$(u, v, w, \theta, P) = \left(u', v', W_0(x) + w', \Theta_0(x) + \theta', P_0(z) + p'\right) \tag{10}$$

In the above equation, primed quantities denote infinitesimal disturbance to the corresponding field variable. The infinitesimal disturbance can be written in the form of traveling waves [34]

$$X'(x, y, z, t) = \widehat{X}(x)e^{i(\alpha z+\beta y-\alpha ct)} \tag{11}$$

where X' denotes field variables, α and β are wavenumbers in the z and y directions, respectively. $c = c_r + ic_i$ represents the complex wave speed. The behavior of disturbance (growth/decay) depends upon the sign of c_i. The flow is unstable, neutrally stable, or stable accordingly as $c_i > 0$, $c_i = 0$, or $c_i < 0$, respectively. The linear disturbance equations for above basic flow are given as

$$i\alpha\widehat{w}+i\beta\widehat{v}+\frac{d\widehat{u}}{dx}=0 \tag{12}$$

$$-\frac{1}{\varepsilon^2}i\alpha W_0\widehat{u}-\frac{d\widehat{p}}{dx}+\frac{\lambda}{Re}\left(\frac{d^2\widehat{u}}{dx^2}-(\alpha^2+\beta^2)\widehat{u}\right)-\frac{\widehat{u}}{DaRe}-F|W_0|\widehat{u}=-\frac{1}{\epsilon}i\alpha c\widehat{u} \tag{13}$$

$$-\frac{1}{\varepsilon^2}i\alpha W_0\widehat{v}-i\beta\widehat{p}+\frac{\lambda}{Re}\left(\frac{d^2\widehat{v}}{dx^2}-(\alpha^2+\beta^2)\widehat{v}\right)-\frac{\widehat{v}}{DaRe}-F|W_0|\widehat{v}$$
$$-N(i\alpha\widehat{\phi}+\widehat{v})=-\frac{1}{\varepsilon}i\alpha c\widehat{v} \tag{14}$$

$$-\frac{1}{\epsilon^2}i\alpha W_0\widehat{w}-i\alpha\widehat{p}+\frac{\lambda}{Re}\left(\frac{d^2\widehat{w}}{dx^2}-(\alpha^2+\beta^2)\widehat{w}\right)-\frac{1}{\epsilon^2}\frac{dW_0}{dx}\widehat{u}-\frac{\widehat{w}}{DaRe}$$
$$-2F|W_0|\widehat{w}+N\left(i\beta\widehat{\phi}-\widehat{w}\right)+\frac{Ra}{Re}\widehat{\theta}=-\frac{1}{\varepsilon}i\alpha c\widehat{w} \tag{15}$$

$$-i\alpha W_0\widehat{\theta}+\frac{1}{RePr}\left(\frac{d^2\widehat{\theta}}{dx^2}-(\alpha^2+\beta^2)\widehat{\theta}-\widehat{w}\right)-\frac{d\Theta_0}{dx}\widehat{u}=-\sigma i\alpha c\widehat{\theta} \tag{16}$$

$$\left[\frac{d^2\widehat{\phi}}{dx^2}-(\alpha^2+\beta^2)\widehat{\phi}\right]-i\beta\widehat{w}+i\alpha\widehat{v}=0 \tag{17}$$

Finally, Eqs. (12)–(17) constitute a generalized eigenvalue problem.

2.3 Numerical Method

In order to determine the numerical solution of Eqs. (7) and (8) (basic flow equations) and (12)–(17) (linear disturbance equations) along with boundary conditions, a high accurate spectral collocation method has been used. The Chebyshev polynomial is used as a basis set for spectral collocation method. The equations have been discretized along the x-direction at Gauss–Lobatto points. These are the extreme points of an M-degree Chebyshev polynomial and given by

$$x_j=cos\left(\frac{\pi j}{M}\right),\ j=0,1,2,\ldots,M$$

The main emphasis in spectral collocation method is to construct differential operator which is given by

Table 1 Comparison between published and present results under special case $Ha = 0$, $Da = 10^{12}$, $\varepsilon = 1$

Re	*Pr*	Published result [1]		Present	
		Ra_c	α_c	Ra_c	α_c
100	0.7	41.65	0.875	41.646	0.873
1000	0.7	30.26	1.355	30.263	1.355
100	7.0	15.73	0.24	15.738	0.237
1000	7.0	15.60	0.024	15.605	0.024
100	100	8.61	0.108	8.614	0.108
1000	100	8.6	0.011	8.597	0.011

$$D_{jk}^{(1)} = \begin{cases} \frac{c_j(-1)^{k+j}}{c_k\left(x_j - x_k\right)}, & j \neq k \\ \frac{x_j}{2\left(1 - x_j^2\right)}, & 1 \leq j = k \leq N - 1 \\ \frac{2N^2+1}{6}, & j = k = 0 \\ -\frac{2N^2+1}{6}, & j = k = N \end{cases}$$

The other higher order derivative can be obtained from lower order derivative by differentiating them. In this process, the differentiation operator takes the role of the derivative. Using spectral method discretization scheme, equations are transformed into a generalized eigenvalue problem of the form

$$AX = cBX \tag{18}$$

where X represents the eigenvector of the field variable, c is an eigenvalue. The square matrices A and B represent the coefficients of linear disturbance equations. The details of the considered numerical method with implementation procedure can be seen in the reference [32]. The eigenvalues and eigenvectors of the eigenvalue problem (18) are calculated by MATLAB software. The obtained results are compared with published results [1] under some special cases. The results calculated by our numerical code are in good match with the published one (Table 1).

3 Results and Discussion

In the present section, we discuss the results of mixed convection flow of electrically conducting fluid in a vertical channel filled with porous medium. The present problem is governed by six non-dimensional parameters, namely Darcy number (*Da*), Reynolds number (*Re*), Prandtl number (*Pr*), Rayleigh number (*Ra*), Hartmann number (*Ha*), and Forchheimer number (*F*). The main emphasis is considered on basic flow, disturbance growth rate profiles, and stability boundaries under a weak to a moderate value of the magnetic field. The stability results are determined for electrically conducting water-based electrolytes fluid, whose Prandtl number (Pr) is 7.01 [24]. The analysis is considered for high permeable porous medium flow. Therefore, the porosity of the medium is taken at 0.9. The Reynolds number is fixed at $Re = 1000$ for the present investigation. The value of F is calculated in terms of C_F and Da, i.e., $F = C_F/\sqrt{Da}$, where C_F is fixed at 0.006.

First, we examine the basic flow results to examine the impact of magnetic field and media permeability of the MHD porous medium flow. Figure 2 shows a variation of the basic velocity under different magnetic field strengths for $Da = 10^{-2}$. The basic velocity profiles contain the point of inflection near the channel walls. It is observed that point of inflection in the velocity profile smooth out slowly on increasing the value of magnetic field parameter Ha, and the velocity profile becomes flattened. The maximum velocity occurs near the channel walls. The impact of the media permeability in terms of the Darcy number under magnetic field on basic velocity is investigated in Fig. 3. The high permeable flow provokes a clear point of inflection in the velocity profile. The velocity profile is relatively more flattened under a low permeability case (see for $Da = 10^{-2}$). We have also examined the impact of thermal buoyancy force under magnetic field on basic velocity and temperature in Fig. 4a–b. It is observed that increasing the thermal buoyancy force in terms of Rayleigh number invites the point of inflection in the basic velocity profile. The magnitude of the basic velocity near channel walls increases on increasing the strength of the thermal buoyancy force. The increase in the value of Ra also results in increase in the magnitude of basic temperature (see Fig. 4b).

The above basic flow analysis indicates that the basic velocity profiles contain the inflection point, which could increase the flow's instability. Based on this analysis, we can predict that the instability of the flow decreases on increasing the value of Hartmann number, i.e., applied magnetic field has a tendency to stabilize the flow. However, increased media permeability acts in the reverse way of the magnetic field. We have examined the linear stability properties to confirm the observations of basic flow.

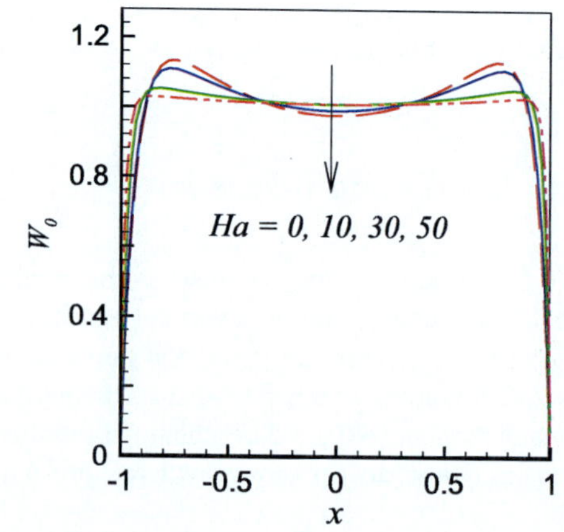

Fig. 2 Basic velocity profile for $Da = 10^{-2}$, $Ra = 150$, and different values of Ha

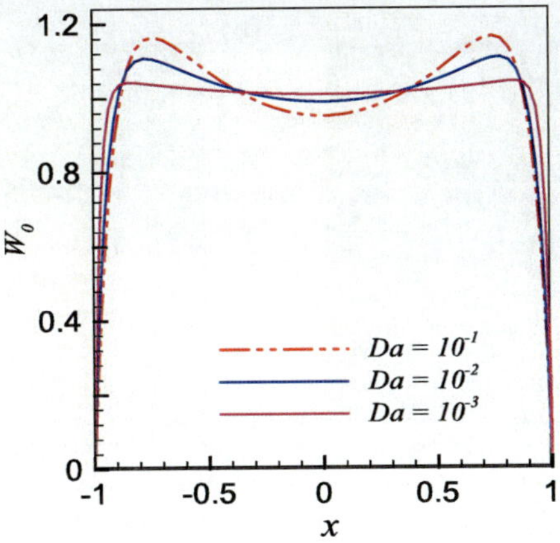

Fig. 3 Basic velocity profile for $Ha = 10$, $Ra = 150$, and different values of Da

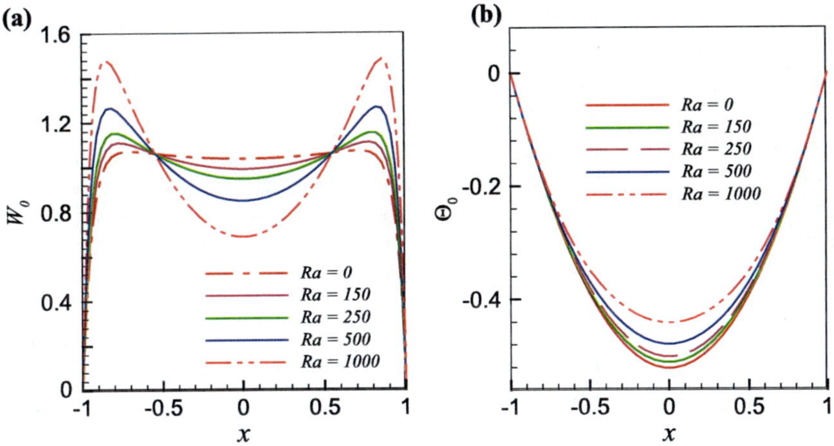

Fig. 4 **a** Basic velocity profile. **b** Basic temperature profile for $Ha = 10$, $Da = 10^{-2}$, and different values of Ra

We have tested several numerical tests for different parameter sets to know the least stable mode of linear stability. It is found that MHD porous medium flow is least stable under two-dimensional mode. Therefore, we have examined present linear stability results for spanwise wavenumber $\beta = 0$. The disturbance growth of the most unstable mode is one of the important features in the instability of the flow. To know the instability behavior of mixed convection MHD flow of electrically conducting water-based electrolytes fluid, we plot the disturbance growth rate contours in (Ha, α)-plane for $Da = 10^{-2}$ and $Da = 10^{-3}$. The positive (negative) value of the growth rate indicates the unstable (stable) nature. The disturbance growth rate contours show

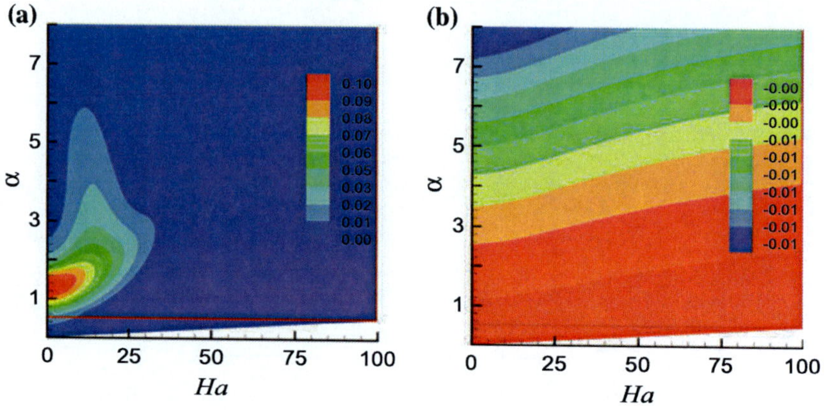

Fig. 5 Contour of the growth rate of the most unstable mode in (Ha, α)-plane at $Ra = 1000$ **a** $Da = 10^{-2}$ and **b** $Da = 10^{-3}$

that growth of the disturbance reduces on enhancing the strength of the magnetic field, i.e., flow instability reduces on enhancing the value of Hartmann number. Figure 5a shows an unstable zone for weak magnetic field strength, but for relatively high magnetic field strength, there is no unstable zone. We have also observed the decrease in the media permeability gives a more stable flow. Figure 5b shows a complete stable flow for the same parameters as Fig. 5a. The analysis shows that the instability of the flow grows by increasing the media permeability of the porous medium. The qualitative behavior of the growth rate analysis helps to examine the linear instability boundaries. The linear stability results are calculated in terms critical value of the Rayleigh number.

The instability boundaries for three different values of Darcy number ($Da = 10^{-1}$, 10^{-2}, and 10^{-3}) in (Ha, Ra_c)-plane is plotted in Fig. 6 to understand the influence of magnetic field on instability boundaries. The critical value of the Rayleigh number increases on increasing the strength of the magnetic field. For the higher value of Ha, the generated Lorentz force in the flow stabilizes the basic flow. On the other hand, increasing the Darcy number's value reduces the critical value of Rayleigh number, i.e., the flow stability reduces by increasing the media permeability of the porous medium flow.

To gain further insight about the characteristics of the instability mechanism, we have plotted eigenfunctions of a disturbance at a critical level for different values of magnetic field parameter, i.e., Hartmann number in Fig. 7 for $Da = 10^{-2}$. The disturbance fluctuation reduces by increasing the value of Ha. The magnitude of temperature disturbance is larger in comparison to the velocity disturbance eigenfunction for all considered values of the Ha.

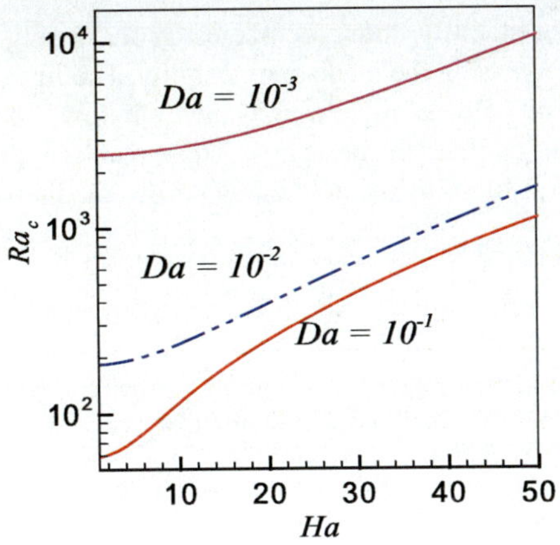

Fig. 6 Variation of critical *Ra* as a function of *Ha* for different values of *Da*

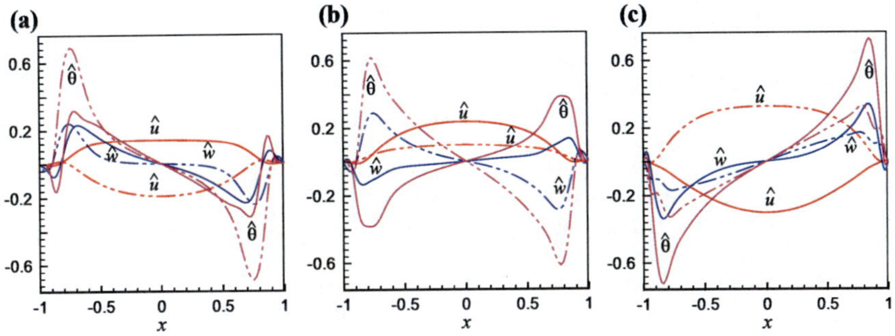

Fig. 7 Eigenfunctions ("—" real part, "–..–" imaginary part) of u (red), w (blue), and θ (purple) on linear stability critical point at $Da = 10^{-2}$ for **a** $Ha = 10$, **b** $Ha = 20$, and **c** $Ha = 50$

4 Conclusion

In this paper, we have studied the instability mechanism of MHD mixed convection flow in a vertical channel filled with the porous medium through a linear stability analysis. The present study results are examined for electrically conducting water-based electrolytes fluid at a fixed value of $Re = 1000$. A high permeable porous medium flow situation is considered. In the present study, we have examined basic flow characteristics, growth rate, and linear stability boundaries under a wide range of magnetic fields. The basic flow and linear disturbance equations are solved by Chebyshev spectral collocation method. The considered flow is least stable under two-dimensional disturbance. The enhancement in the media permeability and thermal buoyancy force could give rise to the inflection point in the velocity profile. The relatively strong magnetic field makes the velocity profile flatten. The growth rate profile shows that the stability domain enlarges by increasing the strength of the magnetic field and decreasing the media permeability. The linear stability conforms to the applied magnetic field stabilizes porous medium flow, whereas increase in the media permeability destabilizes the basic flow. These results of the present study may serve as a piece of fruitful information in many porous medium MHD applications.

References

1. Chen, Y.C., Chung, J.N.: The linear stability of mixed convection in a vertical channel flow. J. Fluid Mech. **325**, 29–51 (1996)
2. Chen, Y.C., Chung, J.N.: Stability of mixed convection in a differentially heated vertical channel. J. Heat Transf. **120**, 127–132 (1998)
3. Chen, Y.C., Chung, J.N.: A direct numerical simulation of k and h-type flow transition phenomenon in a heated vertical channel. Phys. Fluids **14**, 3327–3346 (2002)
4. Khandelwal, M.K., Bera, P.: Weakly nonlinear stability analysis of non-isothermal Poiseuille flow in a vertical channel. Phys. Fluids **27**, 064103-1-24 (2015)
5. Su, Y.C., Chung, J.N.: Linear stability analysis of mixed convection flow in a vertical pipe. J. Fluid Mech. **422**, 141–166 (2000)
6. Yao, L.S., Rogers, B.B.: The linear stability of mixed convection in a vertical annulus. J. Fluid Mech. **201**, 279–298 (1989)
7. Geindreau, C., Auriault, J.L.: Magnetohydrodynamic flows in porous media. J. Fluid Mech. **466**, 343–363 (2002)
8. Mc Whirter, J., Crawford, M., Klein, D., Sanders, T.: Modal for inertialess magnetohydrodynamic flow in packed beds. Fus. Technol. **34**, 187–197 (1998)
9. Bera, P., Khalili, A.: Stability of mixed convection in an anisotropic porous channel. Phys. Fluids **14**, 1617–1630 (2002)
10. Bera, P., Khalili, A.: Influence of Prandtl number on stability of mixed convective flow in a vertical channel filled with a porous medium. Phys. Fluids **18**, 124103 (2006)
11. Kumar, J., Bera, P., Khalili, A.: Influence of inertia and drag terms on the stability of mixed convection in a vertical porous-medium channel. Int. J. Heat Mass Transf. **53**, 5261–5271 (2010)
12. Bera, P., Khandelwal, M.K.: A thermal non-equilibrium perspective on instability mechanism of nonisothermal poiseuille flow in a vertical porous medium channel. Int. J. Ther. Sci. **105**, 159–173 (2016)
13. Khandelwal, M.K., Bera, P.: A thermal non equilibrium perspective on mixed convection in a vertical channel, Int. J. Ther. Sci. **56**, 23–34 (2012)
14. Sharma, A.K., Khandelwal, M.K., Bera, P.: Finite amplitude analysis of non-isothermal parallel flow in a vertical channel filled with a highly permeable porous medium. J. Fluid Mech. **857**, 469–507 (2018)
15. Khandelwal, M.K., Sharma, A.K., Bera, P.: Instability of mixed convection in a differentially heated channel filled with porous medium: a finite amplitude analysis. Phys. Fluids **33**, 024109 (2021)
16. Wallace, W.E., Pierce, C.I., Swayer, W.: Experiments on the flow of mercury in porous media in a transverse magnetic field. Technical Report, TN 23, U7, No. 7259. US Bureau of Mines
17. Raptis, A., Perdikis, C.: Magnetohydrodynamics effects on mass transfer flow through porous medium. Astrophys. Space Sci. **113**, 53–58 (1985)
18. Ram, G., Mishra, R.S.: Unsteady flow through magnetohydrodynamic porous media. Indian J. Pure Appl. Maths **8**, 637–647 (1977)
19. Tawil, M.A.E., Kamel, M.H.: MHD flow under stochastic porous media. Energy Conserv. Manag. **35**, 991–997 (1994)

20. Yih, K.A.: The effect of uniform suction/blowing on heat transfer of magnetohydrodynamic Hiemenz flow through porous media. Acta Mech. **130**, 147–158 (1998)
21. Prescott, P.J., Incropera, F.P.: Magnetically damped convection during solidification of a binary metal alloy. J. Heat Transf. **115**, 302–310 (1993)
22. Rudriah, N., Ramaiah, B.K., Rajasekhar, B.M.: Hartmann flow over a permeable bed. Intl J. Eng. Sci. **13**, 1–24 (1975)
23. Takashima, M.: The stability of natural convection in a vertical layer of electrically conducting fluid in the presence of a transverse magnetic field. Fluid Dyn. Res. **14**, 121–134 (1994)
24. Hudoba, A., Molokov, S.: Linear stability of buoyant convective flow in a vertical channel with internal heat sources and a transverse magnetic field. Phys. Fluids **28**(114103), 1–19 (2016)
25. Kolyshkin, A.A.: On the stability convection generated by internal heat sources in a magnetic field. Can. J. Phys. **66**, 990–993 (1988)
26. Stuart, J.T.: On the stability of viscous flow between parallel planes in the presence of co-planner magnetic field. Proc. Royal Soc. Lond. A **221**,189-XX (1954)
27. Lock, R.C.: The stability of the flow of an electrically conducting fluid between parallel plane under transverse magnetic field. Proc. R. Soc. A **233**, 105–125 (1955)
28. Hunt, J.C.R.: On the stability of parallel flows with parallel magnetic fields. Proc. R. Soc. A **293**, 342–358 (1966)
29. Satake, S., Sone, K., Furumi, K., Kunugi, T.: Direct numerical simulation of turbulent mixed convection in a vertical channel in a wall normal magnetic field. Fusion Eng. Des. **87**, 798–802 (2012)
30. Saleh, H., Hashim, I.: Flow reversal of fully-developed mixed MHD convection in vertical channels. Chin. Phys. Lett. **27**, 024401–024403 (2010)
31. Shankar, B.M., Kumar, J., Shivakumara, I.S.: Magnetohydrodynamic instability of mixed convection in a differentially heated vertical channel. Eur. Phys. J. Plus **134**(53), 1–12 (2019)
32. Singh, N., Khandelwal, M.K., Yu, P.: Instability of mixed convection flow in a differentially heated channel under a magnetic field with internal heating. Phys. Fluids **33**, 094102 (2021)
33. Nield, D.A., Bejan, A.: Convection in Porous Media, 4th edn. Springer, New York (2013)
34. Drazin, P.G., Reid, W.H.: Hydrodynamic Stability. Cambridge University Press, Cambridge (2004)

Group Action on Fuzzy Ideals of Near Rings

Asma Ali, Ram Prakash Sharma, and Arshad Zishan

Abstract In this paper, we introduce the group action on a near ring $\mathcal{N}$ and with it we study group action on fuzzy ideals of $\mathcal{N}$, $\mathcal{G}$-invariant fuzzy ideals, finite products of fuzzy ideals, and $\mathcal{G}$-primeness of fuzzy ideals of $\mathcal{N}$.

Keywords Fuzzy ideals · Prime fuzzy ideals · $\mathcal{G}$-invariant fuzzy ideals · $\mathcal{G}$-prime fuzzy ideals

2010 Mathematics Subject Classification. 16N60 · 16W25 · 16Y30

1 Introduction

A set $\mathcal{N}$ with two binary operations '+' and '·' is known as left near ring if (i) $(\mathcal{N}, +)$ is a group (not necessarily abelian), (ii) $(\mathcal{N}, \cdot)$ is a semigroup, (iii) $\alpha(\beta + \gamma) = \alpha \cdot \beta + \alpha \cdot \gamma \; \forall \; \alpha, \beta$ and γ in $\mathcal{N}$. Analogously, $\mathcal{N}$ is said to be a right near ring if $\mathcal{N}$ satisfies $(iii)'$ $(\beta + \gamma)\alpha = \beta \cdot \alpha + \gamma \cdot \alpha \; \forall \; \alpha, \beta$ and γ in $\mathcal{N}$. A near ring $\mathcal{N}$ with $0x = 0$, $\forall x \in \mathcal{N}$, is known as zero symmetric if $0x = 0$, (left distributively yields that $x0 = 0$).Throughout the paper, $\mathcal{N}$ represents a zero symmetric left near ring; for simplicity, we call it a near ring. An ideal of near ring $(\mathcal{N}, +, \cdot)$ is a subset $\mathcal{M}$ of $\mathcal{N}$ such that (i) $(\mathcal{M}, +) \lhd (\mathcal{N}, +)$, (ii)$\mathcal{N}\mathcal{M} \subset \mathcal{M}$, (iii) $(n_1 + m)n_2 - n_1 n_2 \in \mathcal{M} \; \forall$ $m \in \mathcal{M}$ and $n_1, n_2 \in \mathcal{N}$. Note that if $\mathcal{M}$ fulfils (i) and (ii), it's referred to as a left ideal of $\mathcal{N}$. It is termed a right ideal of $\mathcal{N}$ if $\mathcal{M}$ satisfies (i) and (iii). A mapping $\phi : \mathcal{N} \to \mathcal{N}'$ from near ring $\mathcal{N}$ to near ring $\mathcal{N}'$ is said to be a homomorphism if (i)

A. Ali · A. Zishan (✉)
Department of Mathematics, Aligarh Muslim University, Aligarh, India
e-mail: arshadzeeshan1@gmail.com

R. P. Sharma
Department of Mathematics, Himachal Pradesh University, Shimla, India

R. K. Sharma et al. (eds.), *Frontiers in Industrial and Applied Mathematics*,
Springer Proceedings in Mathematics & Statistics 410,
https://doi.org/10.1007/978-981-19-7272-0_25

$\phi(\alpha+\beta) = \phi(\alpha) + \phi(\beta)$ (ii) $\phi(\alpha\beta) = \phi(\alpha)\phi(\beta)\,\forall\,\alpha$ and $\beta \in \mathcal{N}$. A homomorphism $\phi : \mathcal{N} \to \mathcal{N}$ which is bijective is said to be an automorphism on $\mathcal{N}$. The set of all automorphism of $\mathcal{N}$ denoted by $Aut(\mathcal{N})$ forms a group under the operation of composition of mappings.

The study of group actions on rings led to the establishment of the Galois theory for rings. Lorenz and Passman [12], Montgomery [14], and others researched the skew grouping approach in the context of the Galois theory, as well as the groupring and fixed ring. The link between the $\mathcal{G}$-prime ideals of $\mathcal{R}$ and the prime ideals of skew groupring $\mathcal{RG}$ was identified by Lorenz and Passman [12]. Montgomery [14] investigated the relationship between the prime ideals of $\mathcal{R}$ and $\mathcal{R}^{\mathcal{G}}$, leading him to broaden the scope of the action of a group to spec$\mathcal{R}$.

Fuzzy sets were introduced independently by L.A. Zadeh and Dieter Klaua in 1965 as an extension of the classical notion of set. Liu [11] studied fuzzy ideals of a ring and many researchers [4, 6, 7, 20] extended the concepts.The concept of fuzzy ideals and related features have been applied to a variety of fields, including semigroups, [8–10, 18, 19], distributive lattice [2], BCK-algebras [16], and near rings [22]. Kim and Kim [5] defined the exact analogue of fuzzy ideals for near rings.

Sharma and Sharma [19] recently investigated the action of group on the fuzzy ideals of the ring $\mathcal{R}$ and found a relationship between the $\mathcal{G}$-prime fuzzy ideals of $\mathcal{R}$ and the prime fuzzy ideals of $\mathcal{R}$. We define the action of group on a near ring $\mathcal{N}$ and investigate the action of group on fuzzy ideals and $\mathcal{G}$-invariant fuzzy ideals of $\mathcal{N}$, finite products of fuzzy ideals, and $\mathcal{G}$-primeness of fuzzy ideals of $\mathcal{N}$. As a result, we extend Sharma and Sharma's conclusions to near ring $\mathcal{N}$.

2 Preliminaries

Definition 1 ([22]) If $\mathcal{N}$ is a near ring, then a fuzzy set $\tilde{F}$ in $\mathcal{N}$ is a set of ordered pair $\tilde{F} = \{(n, \eta_{\tilde{F}}(n)) | n \in \mathcal{N}\}$, $\eta_{\tilde{F}}(n)$ is called membership function.

Definition 2 ([22]) Let η and μ be two fuzzy subsets of a near ring $\mathcal{N}$. Then $\eta \cap \mu$ and $\eta \circ \mu$ are defined as follows:

$$\eta \cap \mu(m) = min\{\eta(m), \mu(m)\}.$$

And product $\eta \circ \mu$ is defined by

$$\eta \circ \mu(m) = \begin{cases} \sup\limits_{m=m_1m_2} \{\min(\eta(m_1), \mu(m_2))\} & \text{if } m = m_1m_2 \\ 0 & \text{if } m \neq m_1m_2. \end{cases} \tag{1}$$

Definition 3 ([22]) Let $(\mathcal{G}, +)$ be a group and η be a fuzzy subset of $\mathcal{G}$. Then η is fuzzy subgroup if
(i) $\eta(g_1 + g_2) \geq min(\eta(g_1), \eta(g_2))$, $\forall\, g_1, g_2$ in $\mathcal{G}$,
(ii) $\eta(g) = \eta(-g)$, $\forall\, g$ in $\mathcal{G}$.

Definition 4 ([22]) A fuzzy subset η of a near ring $\mathcal{N}$ is said to be a fuzzy subnear ring of $\mathcal{N}$ if η is a fuzzy subgroup of $\mathcal{N}$ with respect to the addition '+' and is a fuzzy groupoid with respect to the multiplication '.', i.e.,

(i) $\eta(x - y) \geq min(\eta(x), \eta(y))$ and (ii) $\eta(xy) \geq min(\eta(x), \eta(y))$ $\forall\, x, y \in \mathcal{N}$.

Definition 5 ([22]) A fuzzy subset η of a near ring $\mathcal{N}$ is said to be a fuzzy ideal of $\mathcal{N}$ if η satisfies following conditions:

(i) η is fuzzy subnear ring,
(ii) η is normal fuzzy subgroup with respect to '+',
(iii) $\eta(rs) \geq \eta(s)$; for all r,s in $\mathcal{N}$,
(iv) $\eta((r + t)s - rs) \geq \eta(t)$; $\forall\, r, s$ and t in $\mathcal{N}$.

If η satisfies (i),(ii), and (iii), then it is called a fuzzy left ideal of $\mathcal{N}$. If η satisfies (i),(ii), and (iv), then it is called a fuzzy right ideal of $\mathcal{N}$.

Definition 6 ([1]) Let $\mathcal{G}$ be a group and $\mathcal{Z}$ be a set. Then $\mathcal{G}$ is said to act on $\mathcal{Z}$ if there is a mapping $\phi : \mathcal{G} \times \mathcal{Z} \to \mathcal{Z}$, with $\phi(a, z)$ written $a * z$, such that
$(i)\ a * (b * z) = (ab) * z, \quad \forall a, b \in \mathcal{G}, z \in \mathcal{Z}$.
$(ii)\ e * z = z. \quad e \in \mathcal{G}, z \in \mathcal{Z}$. The mapping ϕ is called the action of $\mathcal{G}$ on $\mathcal{Z}$, and $\mathcal{Z}$ is said to be a $\mathcal{G}$-set.

Definition 7 ([1]) Let $\mathcal{G}$ be a group acting on a set $\mathcal{Z}$, and let $z \in \mathcal{Z}$. Then the set

$$\mathcal{G}z = \{az | a \in \mathcal{G}\}$$

is called the orbit of $\mathcal{Z}$ in $\mathcal{G}$.

Proposition 1 *Let $\mathcal{N}$ be a near ring and $\mathcal{G} = Aut(\mathcal{N})$, group of all automorphism of $\mathcal{N}$.Then $\mathcal{G}$ acts on $\mathcal{N}$ via following map*

$$\phi : \mathcal{G} \times \mathcal{N} \to \mathcal{N} \textit{which is defined by } \phi(h, a) = h(a) \textit{ or say } \ h * a = h(a).$$

Proof Take (h_1, a_1) *and* (h_2, a_2) such that

$$(h_1, a_1) = (h_2, a_2).$$

This implies that $h_1 = h_2$ *and* $a_1 = a_2$. Thus, we have

$$h_1(a_1) = h_2(a_1)$$

or

$$\phi(h_1, a_1) = \phi(h_2, a_2).$$

Hence, ϕ is well defined. Furthermore, we show that ϕ is the action of $\mathcal{G}$ on $\mathcal{N}$. Take any $g_1, g_2 \in \mathcal{G}$ *and* $b \in \mathcal{N}$. Then

$$g_1 * (g_2 * b) = g_1 * (g_2(b)) = g_1(g_2(b)) \tag{2}$$

$$(g_1 \circ g_2) * b = (g_1 \circ g_2)(b) = g_1(g_2(b)). \tag{3}$$

From (2) and (3), we get

$$(g_1 \circ g_2) * b = g_1 * (g_2 * b).$$

Also, we have

$$e * x = x.$$

Hence, ϕ is the action of $\mathcal{G}$ on $\mathcal{N}$.

Motivated by the definition of the group action of a finite group on fuzzy ideals of a ring [19], we define a $\mathcal{G}$–fuzzy ideal of $\mathcal{N}$ as follows:

Definition 8 Let $\mathcal{G}$ be a group. Then fuzzy set η of $\mathcal{N}$ is a $\mathcal{G}$–set or $\mathcal{G}$ act on η if

$$\eta^g(r) = \eta(r^g), \quad g \in \mathcal{G}$$

where r^g denotes g acts on r, $r \in \mathcal{N}$.

Example 1 Let $\mathcal{N} = \{0, 1, 2\}$ be a set. Then under following two binary operations $\mathcal{N}$ forms a zero symmetric near ring:

+	0	1	2
0	0	1	2
1	1	2	0
2	2	0	1

·	0	1	2
0	0	0	0
1	0	1	2
2	0	1	2

$$Aut(\mathcal{N}) = \{f | f : \mathcal{N} \to \mathcal{N} \ is \ isomorphism\}.$$

There are only two automorphisms (i) identity map and (ii) the map g defined as follows:

g(0)=0, g(1)=2, and g(2)=1.

We know that $Aut(\mathcal{N})$ forms a group. Define a map $\lambda : \mathcal{N} \to [0, 1]$ by

$$\lambda(a) = \begin{cases} 0.9 & a = 0 \\ 0.8 & a = 1, 2. \end{cases}$$

λ is a fuzzy ideal. By Definition 8, $\lambda^g : \mathcal{N} \to [0, 1]$ is defined as $\lambda^g(r) = \lambda(r^g)$, i.e.,

$$\begin{aligned} \lambda^g(0) &= \lambda(0^g) = \lambda(0) = 0.9 \\ \lambda^g(1) &= \lambda(1^g) = \lambda(2) = 0.8 \\ \lambda^g(2) &= \lambda(2^g) = \lambda(1) = 0.8. \end{aligned}$$

This implies that

$$\lambda^g = \{(0, 0.9), (1, 0.8), (2, 0.8)\} \quad \text{and} \tag{4}$$

$$\lambda^e = \lambda = \{(0, 0.9), (1, 0.8), (2, 0.8)\}. \tag{5}$$

This shows that λ^g is a fuzzy ideal of $\mathcal{N}$, since $\lambda = \lambda^g$.

3 Prime Fuzzy Ideals

Definition 9 ([19]) Let $\mathcal{Q}$ be a fuzzy ideal of $\mathcal{N}$. Then $\mathcal{Q}$ is said to be a prime ideal in $\mathcal{N}$ if $\mathcal{Q}$ is not a constant function and for any fuzzy ideals η and μ in $\mathcal{N}$, $\eta \circ \mu \subset \mathcal{Q}$ implies that either $\eta \subset \mathcal{Q}$ or $\mu \subset \mathcal{Q}$.

Example 2 Take $Z_4 = \{0, 1, 2, 3\}$ the zero symmetric left near ring under binary operations addition modulo 4 and for any $a, b \in Z_4$ multiplication is defined as

$$a \cdot b = \begin{cases} b & a \neq 0 \\ 0 & a = 0. \end{cases}$$

Define two maps $\eta_1, \eta_2 : Z_4 \to [0, 1]$ by $\eta_1(z_1) = \begin{cases} 0.9 & z_1 = 0 \\ 0.8 & z_1 \neq 0, \end{cases}$ and $\eta_2(z_2) = 0.9$ for all $z_1, z_2 \in Z_4$. It shows that $\eta_1 \circ \eta_2 \subseteq \eta_1$ and $\eta_1 \subseteq \eta_1$ but $\eta_2 \not\subset \eta_1$. As η_1 is non-constant function so η_1 is a prime fuzzy ideal.

Proposition 2 *If η is a fuzzy ideal of $\mathcal{N}$, then η^g is a fuzzy ideal of $\mathcal{N}$. Moreover, primeness of η as a fuzzy ideal implies the primeness of fuzzy ideal η^g of $\mathcal{N}$.*

Proof Assume that η is a fuzzy ideal of $\mathcal{N}$. Then we show that η^g is also a prime fuzzy ideal of $\mathcal{N}$, i.e., we will show that η^g satisfies following conditions:

Let $r, s \in \mathcal{N}$. Since η is a fuzzy ideal of $\mathcal{N}$, then we have

$$\eta^g(r - s) = \eta(r - s)^g = \eta(r^g - s^g) \geq min(\eta(r^g), \eta(s^g)),$$

i.e.,

$$\eta^g(r-s) \geq min(\eta^g(r), \eta^g(s)) \tag{6}$$

and

$$\eta^g(rs) = \eta(rs)^g = \eta(r^g s^g) \geq min(\eta(r^g), \eta(s^g)), \tag{7}$$

i.e.,

$$\eta^g(rs) \geq min(\eta(r^g), \eta(s^g)). \tag{8}$$

Equations (6) and (7) imply that η^g is a fuzzy subnear ring of $\mathcal{N}$.
Again $r, s \in \mathcal{N}$ and η is fuzzy ideal of $\mathcal{N}$, we have

$$\eta^g(r+s) = \eta(r+s)^g = \eta(r^g + s^g) \geq min(\eta(r^g), \eta(s^g)),$$

i.e.,

$$\eta^g(r+s) \geq min(\eta(r^g), \eta(s^g)). \tag{9}$$

Applying ([5], Lemma 2.3), we obtain

$$\eta^g(r) = \eta(r^g) = \eta(-r^g) = \eta^g(-r).$$

Also,

$$\eta^g(r) = \eta(r^g) = \eta(s^g + r^g - s^g) = \eta(s + r - s)^g,$$

i.e.,

$$\eta^g(r) = \eta^g(s + r - s). \tag{10}$$

Since η^g satisfies all conditions of normal subgroup, η^g is a normal fuzzy subgroup of $(\mathcal{N}, +)$. For $r, s \in \mathcal{N}$, we have

$$\eta^g(rs) = \eta(rs)^g = \eta(r^g s^g) \geq \eta(s^g),$$

i.e.,

$$\eta^g(rs) \geq \eta^g(s). \tag{11}$$

This implies that η^g is a fuzzy left ideal of $\mathcal{N}$. Now, for r, s *and* $t \in \mathcal{N}$, we have

$$\eta^g((r+t)s - rs) = \eta((r^g + t^g)s^g - r^g s^g) \geq \eta(t^g),$$

i.e.,

$$\eta^g((r+t)s - rs) \geq \eta^g(t). \tag{12}$$

This implies that η is a right fuzzy ideal. Thus, η is a fuzzy ideal(left fuzzy ideal as well as right fuzzy ideal) of $\mathcal{N}$.

Now we prove that η^g is a prime fuzzy ideal of $\mathcal{N}$. Let $\mathcal{A}$ and $\mathcal{B}$ be two fuzzy ideals of $\mathcal{N}$ such that $\mathcal{A} \circ \mathcal{B} \subset \eta^g$. Then $\mathcal{A}^{g^{-1}}$ and $\mathcal{B}^{g^{-1}}$ are also fuzzy ideals of $\mathcal{N}$, since $g^{-1} \in \mathcal{G}$ and as proved in η^g, we claim that $\mathcal{A}^{g^{-1}} \circ \mathcal{B}^{g^{-1}} \subset \eta$. Let $n \in \mathcal{N}$ and

$$\begin{aligned}
(\mathcal{A}^{g^{-1}} \circ \mathcal{B}^{g^{-1}})(n) &= \sup_{n=n_1n_2} \{\min(\mathcal{A}^{g^{-1}}(n_1), \mathcal{B}^{g^{-1}}(n_2))\} \\
&= \sup_{n^{g^{-1}}=n_1^{g^{-1}} n_2^{g^{-1}}} \left\{\min(\mathcal{A}(n_1^{g^{-1}}), \mathcal{B}(n_2^{g^{-1}}))\right\} \\
&= (\mathcal{A} \circ \mathcal{B})(n^{g^{-1}}) \\
&\leq \eta^g(n^{g^{-1}}) = \eta((n^{g^{-1}})^g) \\
&= \eta(n).
\end{aligned}$$

So, $\mathcal{A}^{g^{-1}} \circ \mathcal{B}^{g^{-1}} \subset \eta$. Since η is a prime fuzzy ideal, then we have $\mathcal{A}^{g^{-1}} \subset \eta$ or $\mathcal{B}^{g^{-1}} \subset \eta$. Suppose that $\mathcal{A}^{g^{-1}} \subset \eta$. Then for all $n \in \mathcal{N}$, we have

$$\mathcal{A}(n) = \mathcal{A}((n^g)^{g^{-1}}) = \mathcal{A}^{g^{-1}}(n^g) \leq \eta(n^g) = \eta^g(n).$$

Thus $\mathcal{A} \subset \eta^g$. This implies that η^g is a prime fuzzy ideal of $\mathcal{N}$.

Now we define a $\mathcal{G}$-invariant fuzzy ideal of a near ring.

Definition 10 A fuzzy ideal η of $\mathcal{N}$ is called a $\mathcal{G}$-invariant fuzzy ideal of $\mathcal{N}$ if and only if

$$\eta^g(r) = \eta(r^g) \geq \eta(r), \forall\, g \in \mathcal{G}, r \in \mathcal{N}.$$

Or

$$\eta(r) = \eta((r^g)^{g^{-1}}) \geq \eta(r^g).$$

Example 3 Let $\mathcal{X}$ be a near ring. Then

$$N = \left\{ \begin{pmatrix} x & 0 \\ 0 & y \end{pmatrix} \middle| \; x, y, \; 0 \in X \right\}$$

is near ring with regard to matrix addition and matrix multiplication. Let

$$I = \left\{ \begin{pmatrix} 0 & 0 \\ 0 & y \end{pmatrix} \middle| \; y, 0 \in X \right\}.$$

Then $\mathcal{I}$ is a fuzzy ideal of $\mathcal{N}$. Define a map $\eta : \mathcal{N} \to [0, 1]$ by

$$\eta(z) = \begin{cases} 0.9 & z = 0 \\ 0.8 & z \neq 0 \end{cases} .$$

Consider

$$\mathcal{G}(\subseteq Aut(\mathcal{N})) = \{f \mid f : \mathcal{N} \to \mathcal{N} \ is\ an\ isomorphism\}.$$

There are only two automorphisms that are identity map and the map $g : \mathcal{N} \to \mathcal{N}$ defined by

$$g\begin{pmatrix} x & 0 \\ 0 & y \end{pmatrix} = \begin{pmatrix} y & 0 \\ 0 & x \end{pmatrix}.$$

Since $\eta^g(r) = \eta(r^g) = \eta(r) \ \ for \ \ all \ \ g \in \mathcal{G} \ \ and \ \ r \in \mathcal{N}$, we get η is $\mathcal{G}-$invariant fuzzy ideal in $\mathcal{N}$.

Theorem 1 *Let η be a fuzzy ideal of $\mathcal{N}$ and $\eta^{\mathcal{G}} = \bigcap_{g \in \mathcal{G}} \eta^g$. Then $\eta^{\mathcal{G}}(r) = min\{\eta(r^g), g \in \mathcal{G}\}$. Moreover, fuzzy ideal η contains largest $\mathcal{G}$-invariant fuzzy ideal $\eta^{\mathcal{G}}$ of $\mathcal{N}$.*

Proof Assume that

$$\begin{aligned} \eta^{\mathcal{G}}(s) &= \bigcap_{k \in \mathcal{G}} \eta^k \\ &= min\{\eta^k(s), \ \ k \in \mathcal{G}\} = min\{\eta(s^k), \ \ k \in \mathcal{G}\}. \end{aligned}$$

We prove that $\eta^{\mathcal{G}}$ is a fuzzy ideal of $\mathcal{N}$.

Let $r, s \in \mathcal{N}$. Then

$$\begin{aligned} \eta^{\mathcal{G}}(r - s) &= min\{\eta(r - s)^g, g \in \mathcal{G}\} \\ &= min\{\eta(r^g - s^g), g \in \mathcal{G}\} \\ &= min\{min(\eta(r^g), \eta(s^g)), g \in \mathcal{G}\}. \end{aligned}$$

Since η is a fuzzy ideal, we have

$$\begin{aligned} \eta^{\mathcal{G}}(r - s) &\geq min\{min(\eta(r^g), g \in \mathcal{G}), min(\eta(s^g), g \in \mathcal{G})\} \\ &= min\{\eta^{\mathcal{G}}(r), \eta^{\mathcal{G}}(s)\}. \end{aligned}$$

This implies that

$$\eta^{\mathcal{G}}(r - s) \geq \{\eta^{\mathcal{G}}(r), \eta^{\mathcal{G}}(s)\}. \tag{13}$$

Also for any $r, s \in \mathcal{N}$

$$\begin{aligned}\eta^{\mathcal{G}}(rs) &= min\{\eta(rs)^g, g \in \mathcal{G}\}\\ &= min\{\eta(r^g s^g), g \in \mathcal{G}\}\\ &= min\{min(\eta(r^g), \eta(s^g)),\ g \in \mathcal{G}\}.\end{aligned}$$

Since η is a fuzzy ideal of $\mathcal{N}$, we have

$$\begin{aligned}\eta^{\mathcal{G}}(rs) &\geq min\{min(\eta(r^g), g \in \mathcal{G}), min(\eta(s^g), g \in \mathcal{G})\}\\ &= min\{\mu^G(r), \mu^G(s)\}.\end{aligned}$$

Thus,

$$\eta^{\mathcal{G}}(rs) \geq \{\eta^{\mathcal{G}}(r), \eta^{\mathcal{G}}(s)\}. \tag{14}$$

$$\begin{aligned}\eta^{\mathcal{G}}(s + r - s) &= min\{\eta(s + r - s)^g,\ g \in \mathcal{G}\}\\ &= min\{\eta(s^g + r^g - s^g), g \in \mathcal{G}\}\\ &= min\{\eta(r^g), g \in \mathcal{G}\}\\ &= \eta^{\mathcal{G}}(r).\end{aligned}$$

Therefore,

$$\eta^{\mathcal{G}}(s + r - s) = \eta^{\mathcal{G}}(r). \tag{15}$$

Now,

$$\begin{aligned}\eta^{\mathcal{G}}(rs) &= min\{\eta(rs)^g, g \in \mathcal{G}\}\\ &= min\{\eta(r^g s^g), g \in \mathcal{G}\}.\end{aligned}$$

Again since η is fuzzy ideal, we can write for $r, s \in \mathcal{N}$

$$\begin{aligned}\eta^{\mathcal{G}}(rs) &\geq min\{\eta(s^g), g \in \mathcal{G}\}.\\ &= \eta^{\mathcal{G}}(s),\end{aligned}$$

i.e.,

$$\eta^{\mathcal{G}}(rs) \geq \eta^{\mathcal{G}}(s) \tag{16}$$

$$\begin{aligned}\eta^G((r+t)s-rs) &= min\{\eta((r+t)s-rs)^g, g\in\mathcal{G}\}\\ &= min\{\eta((r+t)^g s^g - r^g s^g), g\in\mathcal{G}\}\\ &= min\{\eta((r^g+t^g)s^g - r^g s^g), g\in\mathcal{G}\}\\ &\geq min\{\eta(t^g), g\in\mathcal{G}\}.\\ &= \eta^{\mathcal{G}}(t)\end{aligned}$$

$$\eta^{\mathcal{G}}((r+t)s-rs) \geq \eta^{\mathcal{G}}(t). \tag{17}$$

Since η^G is the left and right fuzzy ideals of $\mathcal{N}$, then $\eta^{\mathcal{G}}$ is the fuzzy ideal of $\mathcal{N}$. It is still necessary to show that it is a $\mathcal{G}$-invariant fuzzy ideal of $\mathcal{N}$.

$$\begin{aligned}\eta^{\mathcal{G}}(r^g) &= \min\{\eta((r^g)^k), k\in\mathcal{G}\}\\ &= \min\{\eta(r^{gk}), k\in\mathcal{G}\}\\ &= \min\{\eta(r^{g'}), g'\in\mathcal{G}\}\\ &= \eta^{\mathcal{G}}(r).\end{aligned}$$

Now we prove that η^G is the largest. Assume that μ is any $\mathcal{G}$-invariant fuzzy ideal of $\mathcal{N}$ such that $\mu\subseteq\eta$. Then for any $g\in\mathcal{G}$

$$\mu(r^g) = \mu(r) \leq \eta(r).$$

Also,

$$\mu(r^g) = \mu(r) = \mu((r^g)^{g^{-1}}) \leq \eta(r^g).$$

This implies that

$$\mu(r) \leq min\{\eta(r^g), g\in\mathcal{G}\} = \eta^{\mathcal{G}}(r).$$

Thus,

$$\mu\subseteq\eta^{\mathcal{G}}.$$

Hence, $\eta^{\mathcal{G}}$ contained in η as the largest $\mathcal{G}$-invariant fuzzy ideal of $\mathcal{N}$.

Remark 1 If a fuzzy ideal η of $\mathcal{N}$ satisfies $\eta=\eta^{\mathcal{G}}$. Then η is called as $\mathcal{G}$-invariant fuzzy ideal of $\mathcal{N}$ and vice versa.

4 Union of Fuzzy Ideals of Near Ring

The following example demonstrates that the union of fuzzy ideals of a near ring $\mathcal{N}$ need not be a fuzzy ideal in $\mathcal{N}$.

Example 4 Let $\mathcal{Q}$ be a near ring. Then

$$\mathcal{N} = \left\{ \begin{pmatrix} 0 & p \\ 0 & q \end{pmatrix} \middle| \; p, q \;\; 0 \in \mathcal{Q} \right\}$$

is a near ring with regard to matrix addition and matrix multiplication. Let

$$\mathcal{I}_1 = \left\{ \begin{pmatrix} 0 & p \\ 0 & 0 \end{pmatrix} \middle| \; p, \; 0 \in \mathcal{Q} \right\}$$

and

$$\mathcal{I}_2 = \left\{ \begin{pmatrix} 0 & 0 \\ 0 & q \end{pmatrix} \middle| \; q, \; 0 \in \mathcal{Q} \right\}.$$

We can check that $\mathcal{I}_1$ and $\mathcal{I}_2$ are ideals of $\mathcal{N}$. Define maps

$$\eta_1 : \mathcal{N} \to [0, 1] \quad \textit{and} \quad \eta_2 : \mathcal{N} \to [0, 1]$$

by

$$\eta_1(x) = \begin{cases} 0.5 & x \in \mathcal{I}_1 \\ 0 & x \notin \mathcal{I}_1 \end{cases}$$

and

$$\eta_2(x) = \begin{cases} 0.6, & x \in \mathcal{I}_2 \\ 0, & x \notin \mathcal{I}_2. \end{cases}$$

Then η_1 and η_2 are fuzzy ideals of $\mathcal{N}$. However

$$(\eta_1 \cup \eta_2)(x) = \begin{cases} max\{0.5, 0.6\}, & x \in \mathcal{I}_1 \cup \mathcal{I}_2 \\ 0, & x \notin \mathcal{I}_1 \cup \mathcal{I}_2 \end{cases}$$

is not a fuzzy ideal of $\mathcal{N}$, since for $m = \begin{pmatrix} 0 & p \\ 0 & 0 \end{pmatrix} n = \begin{pmatrix} 0 & 0 \\ 0 & q \end{pmatrix}$, $m - n = \begin{pmatrix} 0 & p \\ 0 & -q \end{pmatrix} \notin \mathcal{I}_1 \cup \mathcal{I}_2$. We see that $\eta_1 \cup \eta_2(m - n) = 0$, $\eta_1 \cup \eta_2(m) = 0.6$, and $\eta_1 \cup \eta_2(n) = 0.5$. Thus,

$$\begin{aligned} \eta_1 \cup \eta_2(m - n) = 0 &\not\geq \max\{\eta_1 \cup \eta_2(m), \eta_1 \cup \eta_2(n)\} \\ &\not\geq \max\{0.6, 0.5\} \\ &\not\geq 0.6. \end{aligned}$$

Hence, $\eta_1 \cup \eta_2$ is not a fuzzy ideal of $\mathcal{N}$.

Proposition 3 *Let $\mathcal{C} = \{\eta_k\}$ be a chain of fuzzy ideals of $\mathcal{N}$. Then for any $m, n \in \mathcal{N}$*

$$min(\sup_k\{\eta_k(m)\}, \sup_k\{\eta_k(n)\}) = \sup_k\{min(\eta_k(m), \eta_k(n))\}.$$

Proof We can easily see that

$$\sup_k\{\min(\eta_k(m), \eta_k(n))\} \leq \min(\sup_k\{\eta_k(m)\}, \sup_k\{\eta_k(n)\}).$$

Now, assume that

$$\sup_k\{\min(\eta_k(m), \eta_k(n))\} = I.$$

And

$$I < \min(\sup_k\{\eta_k(m)\}, \sup_k\{\eta_k(n)\}).$$

Then

$$\sup_k\{\eta_k(m)\} > I, \quad or \quad \sup_k\{\eta_k(n)\} > I.$$

η_r and η_s exist in such a way that

$$\eta_r(m) > I, \quad \& \quad \eta_s(n) > I$$

or

$$\eta_r(m) > I \geq \min(\eta_r(m), \eta_r(n)) \tag{18}$$

and

$$\eta_r(n) > I \geq \min(\eta_s(m), \eta_s(n)). \tag{19}$$

Since, $\eta_r, \eta_s \in \mathcal{C}$, so without loss of generality, we may assume that $\eta_r \subseteq \eta_s$ and $\eta_s(n) \geq \eta_s(m)$ Therefore, from (18) and (19), we get

$$I < \eta_r(m) \leq \eta_s(m) = \min(\eta_s(m), \eta_s(n)).$$

This contradicts the fact that

$$I = \sup_k\{\min(\eta_k(m), \eta_k(n))\}.$$

Hence,

$$min(\sup_k\{\eta_k(m)\}, \sup_k\{\eta_k(n)\}) = \sup_k\{min(\eta_k(m), \eta_k(n))\}.$$

Corollary 1 *Assume that $C = \{\eta_k\}$ is a chain of fuzzy ideals of $\mathcal{N}$. Then for each $x_1, x_2, ..., x_m \in \mathcal{N}$,*

$$min(\sup_k\{\eta_k(x_1)\}, \sup_k\{\eta_k(x_2)\}, ..., \sup_k\{\eta_k(x_m)\}) = \sup_k\{min(\eta_k(x_1), \eta_k(x_2), ..., \eta_k(x_m))\}.$$

Theorem 1 *Let $C = \{\eta_k\}$ be a chain of fuzzy ideals of $\mathcal{N}$. Then $\bigcup_k \eta_k$ is a fuzzy ideal of $\mathcal{N}$.*

Proof Let $r, s \in \mathcal{N}$, and η_k be a fuzzy ideal of $\mathcal{N}$, where k is a natural number. Then

$$\begin{aligned}(\bigcup_k \eta_k)(r-s) &= \sup_k(\eta_k(r-s)) \\ &\geq \sup_k\{\min(\eta_k(r), \eta_k(s))\}.\end{aligned}$$

Using Corollary 1, we get

$$(\bigcup_k)(r-s) \geq \min\{\sup_k(\eta_k(r)), \sup_k(\eta_k(s))\},$$

i.e.,

$$(\bigcup_k \eta_k)(r-s) \geq \min\{(\bigcup_k \eta_k)(r), (\bigcup_k \eta_k)(s)\}. \tag{20}$$

Also,

$$\begin{aligned}(\bigcup_k \eta_k)(rs) &= \sup_k(\bigcup_k(rs)) \\ &\geq \sup_k\{\min(\eta_k(r), \eta_r(s))\}.\end{aligned}$$

Again from Corollary 1, we have

$$(\bigcup_k \eta_k)(rs) \geq \min\{\sup_k(\eta_k(r)), \sup_k(\eta_k(s))\},$$

i.e.,

$$(\bigcup_k \eta_k)(rs) \geq \min\{(\bigcup_k \eta_k)(r), (\bigcup_k \eta_k)(s)\}. \tag{21}$$

Now

$$(\bigcup_k \eta_k)(s + r - s) = \sup_k(\eta_k(s + r - s))$$
$$= \sup_k\{\eta_k(r)\}.$$

Since η_k is a fuzzy ideal in $\mathcal{N}$, we obtain

$$(\bigcup_k \eta_k)(s + r - s) = (\bigcup_k \eta_k)(r),$$

i.e.,

$$(\bigcup_k \eta_k)(s + r - s) = (\bigcup_k \eta_k)(r). \tag{22}$$

$$(\bigcup_k \eta_k)(rs) = \sup_k(\eta_k(rs))$$
$$\geq \sup_k\{\eta_k(s)\}.$$

Again using the fact that η_k is fuzzy ideal, we get

$$(\bigcup_k \eta_k)(rs) \geq (\bigcup_k \eta_k)(s) \tag{23}$$

$$(\bigcup_k \eta_k)((r + t)s - rs) = \sup_k(\eta_k((r + t)s - rs))$$
$$\geq \sup_k\{\eta_k(t)\}.$$

Also,

$$(\bigcup_k \eta_k)((r + t)s - rs) \geq (\bigcup_k \eta_k)(t). \tag{24}$$

Hence, $(\bigcup_k \eta_k)$ is a fuzzy ideal of $\mathcal{N}$.

5 G-Prime Fuzzy Ideals of a Near Ring

Motivated by the definition of $\mathcal{G}$-prime fuzzy ideals of the rings [19], we define $\mathcal{G}-$prime fuzzy ideals in a near ring as follows.

Definition 11 Let the fuzzy ideal η of $\mathcal{N}$ be $\mathcal{G}$-invariant and non-constant. If $\mu \circ \lambda \subseteq \eta$ implies that either $\mu \subseteq \eta$ or $\lambda \subseteq \eta$ for any two $\mathcal{G}$-invariant fuzzy ideals μ and λ of $\mathcal{N}$, then η is a $\mathcal{G}$-prime fuzzy ideal.

Example 5 Take $Z_3 = \{0, 1, 2\}$ which is a zero symmetric left near ring under binary operations addition modulo 3 and for any $r, s \in Z_3$ multiplication is defined as follows:

$$r \cdot s = \begin{cases} s & r \neq 0 \\ 0 & r = 0. \end{cases}$$

$$Aut(Z_3) = \{f | f : Z_3 \to Z_3 \ is \ isomorphism\}.$$

We can check that there are only two automorphisms on Z_3; one is the identity map and the other is the map g defined by

g(0)=0, g(1)=2 and g(2)=1.

$Aut(Z_3)$ forms a group under the composition of mappings. Now we define two maps $\eta_1, \eta_2 : Z_3 \to [0, 1]$ by $\eta_1(r) = \begin{cases} 0.9 & r = 0 \\ 0.8 & r \neq 0, \end{cases}$ and $\eta_2(s) = 0.9$ for all $r, s \in Z_3$. By Definition 8, $\eta_1^g : Z_3 \to [0, 1]$ is defined as $\eta_1^g(r) = \eta_1(r^g)$, i.e.,

$$\begin{aligned} \eta_1^g(0) &= \eta_1(0^g) = \eta_1(0) = 0.9 \\ \eta_1^g(1) &= \eta_1(1^g) = \eta_1(2) = 0.8 \\ \eta_1^g(2) &= \eta_1(2^g) = \eta_1(1) = 0.8. \end{aligned}$$

This implies that

$$\eta_1^g = \{(0, 0.9), (1, 0.8), (2, 0.8)\} \tag{25}$$

and

$$\eta_1^e = \eta_1 = \{(0, 0.9), (1, 0.8), (2, 0.8)\}. \tag{26}$$

Also, we can see that η_2 is a $\mathcal{G}$-invariant fuzzy ideal of Z_3. Since $\eta_1 \circ \eta_2 \subseteq \eta_1$ and $\eta_1 \subseteq \eta_1$ but $\eta_2 \not\subseteq \eta_1$, so it follows that η_1 is $\mathcal{G}$-prime fuzzy ideal as η_1 is non-constant function.

The following proposition is an extension of Lemma 2.6 of [22] in case of near rings:

Proposition 4 *If $\mathcal{N}$ is near ring and $\lambda_1, \lambda_2, ..., \lambda_n$ are fuzzy ideals of $\mathcal{N}$, then*

$$\lambda_1 \circ \lambda_2 \circ \cdots \circ \lambda_n \subset \lambda_1 \bigcap \lambda_2 \bigcap \cdots \bigcap \lambda_n.$$

Proof Let $\lambda_1 \circ \lambda_2 \circ \cdots \circ \lambda_n(x) = 0$. Then, there is nothing to demonstrate. Otherwise

$$\lambda_1 \circ \lambda_2 \circ \cdots \circ \lambda_n(x) = \sup_{x=x_1x_2\cdots x_n} \{\min(\lambda_1(x_1), \lambda_2(x_2), ..., \lambda_n(x_n))\}.$$

Since λ_i is a fuzzy ideal of $\mathcal{N}$, we get

$$\lambda_i((x+z)y - xy) \geq \lambda_i(z).$$

Since $\mathcal{N}$ is zero symmetric, we have

$$\begin{aligned}\lambda_1(x) = \lambda_1(x_1x_2\cdots x_n) &= \lambda_1((0+x_1)x_2\cdots x_n - 0\cdot x_1x_2\cdots x_n).\\ &\geq \lambda_1(x_1),\end{aligned}$$

i.e.,

$$\lambda_1(x) \geq \lambda_1(x_1).$$

Also, λ_2 is a fuzzy ideal; hence,

$$\begin{aligned}\lambda_2(x) = \lambda_2(x_1x_2\cdots x_n) \geq \lambda_2(x_2x_3\cdots x_n) &= \lambda_2((0+x_2)x_3\cdots x_n - 0\cdot x_2x_3\cdots x_n).\\ &\geq \lambda_2(x_2),\end{aligned}$$

i.e.,

$$\lambda_2(x) \geq \lambda_2(x_2).$$

In a similar manner, we can prove that

$$\lambda_3(x) \geq \lambda_3(x_3),$$

$$\lambda_4(x) \geq \lambda_4(x_4),$$

$$\cdots$$

$$\cdots$$

$$\dots$$

$$\lambda_{n-1}(x) \geq \lambda_{n-1}(x_{n-1}).$$

Since λ_n is a fuzzy ideal in $\mathcal{N}$, we get

$$\lambda_n(x) \geq \lambda_n(x_n).$$

Therefore,

$$\lambda_1 \circ \lambda_2 \circ \cdots \circ \lambda_n(x) = \min(\lambda_1(x_1), \lambda_2(x_2), \ldots., \lambda_n(x_n))$$

or

$$\underset{1 \leq i \leq n}{\circ} \lambda_i(x) \leq (\bigcap_{1 \leq i \leq n} \lambda_i)(x)$$

or

$$\underset{1 \leq i \leq n}{\circ} \lambda_i \subset \bigcap_{1 \leq i \leq n} \lambda_i.$$

Now we will prove the main result.

Theorem 2 *If η is a prime fuzzy ideal of $\mathcal{N}$. Then $\eta^{\mathcal{G}}$ is a $\mathcal{G}$-prime fuzzy ideal of $\mathcal{N}$. Conversely, if λ is a $\mathcal{G}$-prime fuzzy ideal of $\mathcal{N}$, then there exists a prime fuzzy ideal η of $\mathcal{N}$ such that $\eta^{\mathcal{G}} = \lambda$, η is unique up to its $\mathcal{G}$-orbit.*

Proof Assume that η is a prime fuzzy ideal of $\mathcal{N}$ and $\mathcal{P}$, $\mathcal{Q}$ are two $\mathcal{G}$-invariant fuzzy ideals of $\mathcal{N}$ such that $\mathcal{P} \circ \mathcal{Q} \subseteq \eta^{\mathcal{G}}$. Since $\eta^{\mathcal{G}}$ is the largest $\mathcal{G}$-invariant fuzzy ideal contained in η, then $\mathcal{P} \circ \mathcal{Q} \subseteq \eta$. Also primeness of η implies that either $\mathcal{P} \subseteq \eta$ or $\mathcal{Q} \subseteq \eta$. Therefore, by Theorem 1 either $\mathcal{P} \subseteq \eta^{\mathcal{G}}$ or $\mathcal{Q} \subseteq \eta^{\mathcal{G}}$. Thus, $\eta^{\mathcal{G}}$ is a $\mathcal{G}$-prime fuzzy ideal.

Conversely, suppose that λ is a $\mathcal{G}$-prime fuzzy ideal of $\mathcal{N}$ and consider

$$\mathcal{S} = \{\eta, \text{a fuzzy ideal of N} \mid \ \eta^{\mathcal{G}} \subseteq \lambda\}.$$

Before using Zorn's lemma on $\mathcal{S}$ to get the maximal element(i.e., maximal ideal), we have to show that if $\mathcal{C} = \{\eta_k\} \subset \mathcal{S}$ is a chain in $\mathcal{S}$, then $\bigcup_k \eta_k \in \mathcal{S}$.

Now, from Theorem 1, $\bigcup_k \eta_k$ is a fuzzy ideal of $\mathcal{N}$. Since $\eta_k \in \mathcal{S}$, we get $\eta_k^{\mathcal{G}} \subseteq \lambda$, and we can take any $r \in \mathcal{N}$ and $\eta_k \in \mathcal{C}$ such that

$$\eta_k^g(r) = \eta_k(r^g) \quad \text{and} \quad \eta_k^g \subseteq \lambda.$$

Then

$$\eta_k(r^g) = \eta_k^g(r) \leq \lambda(r),$$

or

$$\min(\eta_k(r^g), g \in \mathcal{G}) \leq \lambda(r).$$

This implies that

$$\sup\{\min(\eta_k(r^g), g \in \mathcal{G})\} \leq \lambda(r). \tag{27}$$

Since $\mathcal{G}$ is finite, by Corollary 1, we obtain

$$\min\{\sup(\eta_k(r^g), g \in \mathcal{G})\} = \sup_k\{\min(\eta_k(r^g), g \in \mathcal{G})\}. \tag{28}$$

From (27) and (28), we have

$$\min\{\sup_k(\eta_k(r^g), g \in \mathcal{G})\} \leq \lambda(r)$$

or

$$\min\{(\bigcup_k \eta_k)(r^g), g \in \mathcal{G}\} \leq \lambda(r).$$

Now by Theorem 1, we get

$$(\bigcup_k \eta_k)^{\mathcal{G}}(r) \leq \lambda(r).$$

Thus, we obtain

$$(\bigcup_k \eta_k)^{\mathcal{G}} \subseteq \lambda.$$

This shows that $(\bigcup_k \eta_l) \in \mathcal{S}$, i.e., $\mathcal{S}$ has upper bound. Now we use Zorn's lemma on $\mathcal{S}$ to choose a maximal fuzzy ideal say η. Let $\mathcal{P}$, $\mathcal{Q}$ be fuzzy ideals of $\mathcal{N}$ such that $\mathcal{P} \circ \mathcal{Q} \subseteq \eta$. Then

$$(\mathcal{P} \circ \mathcal{Q})^{\mathcal{G}} \subseteq \eta^{\mathcal{G}} \subseteq \lambda. \tag{29}$$

Since $\mathcal{P}^{\mathcal{G}}$ and $\mathcal{Q}^{\mathcal{G}}$ are the largest fuzzy ideals contained in $\mathcal{P}$ and $\mathcal{Q}$, respectively.

Now we prove that $\mathcal{P}^{\mathcal{G}} \circ \mathcal{Q}^{\mathcal{G}} \subseteq \mathcal{P} \circ \mathcal{Q}$ is a $\mathcal{G}$-invariant,

$$\begin{aligned}(\mathcal{P}^{\mathcal{G}} \circ \mathcal{Q}^{\mathcal{G}})(r^{g}) &= \sup_{r^{g}=ab} \{\min(\mathcal{P}^{\mathcal{G}}(a), \mathcal{Q}^{\mathcal{G}}(b))\} \\ &= \sup_{r=a^{g^{-1}}b^{g^{-1}}} \{\min(\mathcal{P}^{\mathcal{G}}(a^{g^{-1}}), \mathcal{Q}^{\mathcal{G}}(b^{g^{-1}}))\} \\ &= \mathcal{P}^{\mathcal{G}} \circ \mathcal{Q}^{\mathcal{G}}(r).\end{aligned}$$

Hence, by Theorem 1, $(\mathcal{P}^{\mathcal{G}} \circ \mathcal{Q}^{\mathcal{G}}) \subseteq (\mathcal{P} \circ \mathcal{Q})^{\mathcal{G}} \subseteq \lambda$. Since λ is $\mathcal{G}$-prime, then we have either $\mathcal{P}^{\mathcal{G}} \subseteq \lambda$ or $\mathcal{Q}^{\mathcal{G}} \subseteq \lambda$. By maximality of η either $\mathcal{P} \subseteq \eta$ or $\mathcal{Q} \subseteq \eta$. This implies that η is prime fuzzy ideal of $\mathcal{N}$. As $\lambda^{\mathcal{G}} = \lambda$, we have $\lambda \in \mathcal{S}$. But maximality of η gives that $\lambda \subseteq \eta$. Since λ and $\eta^{\mathcal{G}}$ are $\mathcal{G}$-invariant ideal and $\eta^{\mathcal{G}}$ is largest in η, we get

$$\lambda \subseteq \eta^{\mathcal{G}}. \tag{30}$$

Thus, from (29) and (30), we obtain

$$\eta^{\mathcal{G}} = \lambda.$$

Let there exist another prime fuzzy ideal σ of $\mathcal{N}$ such that $\sigma^{\mathcal{G}} = \lambda$. Then

$$\bigcap_{g \in \mathcal{G}} \eta^{g} = \eta^{\mathcal{G}} = \sigma^{G} \subseteq \sigma.$$

Since $\mathcal{G}$ is finite, so from Proposition 4, we get

$$\mathop{\circ}_{g \in \mathcal{G}} \eta^{g} \subseteq \bigcap_{g \in \mathcal{G}} \eta^{g}.$$

Or for any $h(\neq g) \in \mathcal{G}$, we have

$$\eta^{h} \circ (\bigcap_{\substack{g \in \mathcal{G} \\ g \neq h}} \eta^{g}) \subseteq \bigcap_{g \in \mathcal{G}} \eta^{g} \subseteq \sigma.$$

By fuzzy primeness either $\eta^{h} \subseteq \sigma$ or $\bigcap_{\substack{g \in \mathcal{G} \\ g \neq h}} \eta^{g} \subseteq \sigma$. If $\eta^{h} \subseteq \sigma$, then $\eta \subseteq \sigma^{h^{-1}}$ and maximality of η with $(\sigma^{h^{-1}})^{\mathcal{G}} \subseteq \lambda$ implies that

$$\eta = \sigma^{h^{-1}}. \tag{31}$$

On the other hand, if $\eta^{h} \nsubseteq \sigma$, we get $\bigcap_{\substack{g \in \mathcal{G} \\ g \neq h}} \eta^{g} \subseteq \sigma$. Thus, there exists some $(h \neq) g \in \mathcal{G}$ such that $\eta^{g} \subseteq \sigma$ and hence $\eta \subseteq \sigma^{g^{-1}}$. Again maximality of η with $(\sigma^{g^{-1}})^{\mathcal{G}} \subseteq \lambda$ yields that

$$\eta = \sigma^{g^{-1}}. \tag{32}$$

Equations (31) and (32) show that η is unique up to its $\mathcal{G}$-orbit.

Conclusion: In the future, we plan to study partial group action (the existence of $g * (h * x)$ implies the existence of $(gh) * x$, but not necessarily conversely) on fuzzy ideals of near rings. The theorems that we prove are the following which are generalizations of Theorems 1 and 2.

Open Problem 1. Can we establish relation between $\mathcal{G}-$invariant fuzzy ideal and largest $\mathcal{G}-$invariant fuzzy ideal of $\mathcal{N}$ under partial group action?

Open Problem 2. Can we investigate relationship between primeness and $\mathcal{G}$-primeness of fuzzy ideal if a group $\mathcal{G}$ partially acts on a fuzzy ideal?

Acknowledgements The authors are extremely thankful to the referees for their valuable comments and suggestions.

References

1. Bhattacharya, P.B., Jain, S.K., Nagpaul, S.R.: Basic Abstract Algebra. Cambridge University Press(2006)
2. Bo, Y., Wangming, W.: Fuzzy ideals on distributive lattice. Fuzzy Sets Syst. **35**, 231–240 (1990). https://doi.org/10.1016/0165-0114(90)90196-D
3. Clay, J.R.: Nearrings. Geneses and Applications. Oxford, New York (1992)
4. Dixit, V.N., Kumar, R., Ajal, N.: On Fuzzy rings. Fuzzy Sets Syst. **49**, 205–213 (1992). https://doi.org/10.1016/0165-0114(92)90325-X
5. Kim, S.D., Kim, H.S.: On Fuzzy Ideals of Near-Rings. Bull. Korean Math. Soc. **33**, 593–601 (1996). https://www.koreascience.or.kr/article/JAKO199611919482456.page
6. Kumar, R.: Certain fuzzy ideals of rings redefined. Fuzzy Sets Syst. 251–260 (1992). https://doi.org/10.1016/0165-0114(92)90138-T
7. Kumar, R.: Fuzzy irreducible ideals in rings. Fuzzy Sets Syst. **42**, 369–379 (1992). https://doi.org/10.1016/0165-0114(91)90116-8
8. Kuroki, N.: Fuzzy bi-ideals in semigroups. Comment. Math. Univ. St. Pauli **28**, 17–21(1979). https://doi.org/10.14992/00010265
9. Kuroki, N.: On fuzzy ideals and fuzzy bi-ideals in semigroups. Fuzzy Sets Syst. **5**, 203–215 (1981). https://doi.org/10.1016/0165-0114(81)90018-X
10. Kuroki, N.: Fuzzy semiprime ideals in semigroups. Fuzzy Sets Syst. **8**, 71–79 (1981). https://doi.org/10.1016/0165-0114(82)90031-8
11. Liu, W.: Fuzzy invariant subgroups and fuzzy ideals. Fuzzy Sets Syst. **8**, 133–139 (1982). https://doi.org/10.1016/0165-0114(82)90003-3
12. Lorenz, M., Passman, D.S.: Prime ideals in crossed products of finite groups: Israel J. Math. **33**(2) 89–132 (1979). https://doi.org/10.1007/BF02760553
13. McLean, R.G., Kummer, H.: Fuzzy ideals in semigroups. Fuzzy Sets Syst. **48**, 137–140 (1992). https://doi.org/10.1016/0165-0114(92)90258-6
14. Montgomery, S.: Fixed Rings of Finite Automorphism Groups of Associative Rings. Springer, Berlin (1980)
15. Mukherjee, T.K., Sen, M.K.: On fuzzy ideals of a ring I. Fuzzy Sets Syst. **21**, 99–104(1987). https://doi.org/10.1016/0165-0114(87)90155-2
16. Ougen, X.: Fuzzy BCK-algebras. Math. Japonica **36**, 935–942 (1991)
17. Pilz, G.: Near-Rings: North-Holland Publishing Company, Amsterdam (1983)

18. Rosenfeld, A.: Fuzzy groups. J. Math. Anal. Appl. **35**, 512–517 (1971)
19. Sharma, R.P., Sharma, S.: Group action on fuzzy ideals. Commun. Algebra 4207–4220 (1998). https://doi.org/10.1080/00927879808826406
20. Yue, Z.: Prime L-fuzzy ideals and primary L-fuzzy ideals. Fuzzy Sets Syst. **27**, 345–350 (1988). https://doi.org/10.1016/0165-0114(88)90060-7
21. Zadeh, L.A.: Fuzzy sets. Inf. Control **8**, 338–353 (1965). https://doi.org/10.1016/S0019-9958(65)90241-X
22. Zaid, S.A.: On fuzzy subnear-rings and ideals. Fuzzy Sets Syst. **44**, 139–146 (1989). https://doi.org/10.1016/0165-0114(91)90039-S

Effect of Viscosity on the Spherical Shock Wave Propagation in a Dusty Gas with Radiation Heat Flux and Exponentially Varying Density

Ravilisetty Revathi, Dunna Narsimhulu, and Addepalli Ramu

Abstract This paper investigates the effect of viscosity on the propagation of spherical shock waves in a dusty gas with a radiation heat flux and a density that grows exponentially. It is assumed that the dusty gas is a blend of fine solid particles and ideal gas. In a perfect gas, solid particles are uniformly distributed. To obtain several significant shock propagation properties, the solid particles are treated as a pseudo-fluid, and the mixture's heat conduction is neglected. The flow's equilibrium conditions are expected to be maintained in an optically thick gray gas model, and radiation is assumed to be of the diffusion type. The effects of modifying the viscosity parameter and time are explored, and non-similar solutions are found. The formal solution is determined by assuming that the shock wave's velocity is variable and its total energy is not constant.

Keywords Dusty gas · Radiation heat flux · Shock waves · Viscosity

1 Introduction

Numerous authors have investigated the propagation of shock waves in a medium with exponentially changing density [1–5]. Radiation's repercussions have not been

R. Revathi (✉)
School of Technology, Woxsen University, Kamkhole, Hyderabad, India
e-mail: revathiravilisetty@gmail.com

D. Narsimhulu
Department of Statistics and Applied Mathematics, Central University of Tamil Nadu, Thiruvarur, Tamil Nadu, India

A. Ramu
Department of Mathematics, Birla Institute of Technology and Science- Pilani, Hyderabad Campus, Hyderabad, India

R. K. Sharma et al. (eds.), *Frontiers in Industrial and Applied Mathematics*,
Springer Proceedings in Mathematics & Statistics 410,
https://doi.org/10.1007/978-981-19-7272-0_26

taken into consideration by these authors. Several researchers [6–9] have developed similar or non-similar solutions for the propagation of a shock wave with radiation heat transfer effects in an exponential medium. The propagation of a strong shock wave through a material whose density changes with distance from the point of the explosion was examined by [10, 11].

In many disciplines of science and engineering, the study of shock waves propagation in a dusty gas is significant due to their vast range of applications(see [12–14]). Pai et al. [15] have obtained a similarity solution for the propagation of a shock wave in dusty gas with constant density. Vishwakarma [16] then explored the propagation of shock waves with exponentially varying density in a dusty gas using a non-similarity method. Singh and Vishwakarma [17] explored shock wave propagation with exponentially changing density and radiation heat flux in a dusty gas. The consequences of viscosity have not been considered by these authors.

In the thin transition zone through which the gas travels from its initial state of thermodynamic equilibrium to its final, also equilibrium state, flow variables such as pressure, density, and particle velocity rapidly change. The shock front is the thermodynamic equilibrium inside this region, and it can be significantly affected. As a result, dissipative processes due to viscosity must be considered when analyzing shock wave propagation behind the shock front. Rankine [18], Rayleigh [19], and Taylor [20] explored the dissipative processes caused by viscosity and thermal conduction in the beginning. Henderson et al. [21] investigated the effects of thermal conductivity and viscosity on shock waves in argon. Simeonides [22] investigated the influence of viscousness on hypersonic flow. In a compressible gas, Huang et al. [23] explored viscous shock waves. It's worth noting that many previous research has remained focused on viscous shocks in a perfect gas. Nevertheless, it is widely recognized that the viscosity in a non-ideal gas, as compared to that in an ideal gas, plays a major role in the characterization of shocks.

To the best of the authors' knowledge, no research on the effects of viscosity on shock wave propagation has yet been reported. For this purpose, in the current work, we develop a non-similar solution taking viscosity into account for the propagation of a shock wave.

It is believed that the dusty gas is gray and opaque and that the shock is isothermal. Radiation energy and pressure are thought to be insignificant in comparison to material energy and pressure, hence only the radiation flux is taken into account. The non-linear dissipative mechanism due to viscosity q is assumed to be negligibly small, except in the neighborhood of the shock, and is taken as the function of flow variables and their derivatives as in von Neumann and Richtmyer [24]. Small solid particles are treated as a pseudo-fluid to accomplish some fundamental shock propagation properties, with the heat conduction of the mixture considered to be minimal and the flow field preserving the equilibrium flow state (see [25]).

2 Basic Equations and Boundary Conditions

The governing equations incorporating the viscosity term proposed, for the spherically symmetric, one-dimensional unsteady flow with radiation heat flux in a dusty gas, can be written as [9, 16, 24]

$$\begin{aligned} &\frac{\partial u}{\partial t}+u\frac{\partial u}{\partial r}+\frac{1}{\rho}\left(\frac{\partial p}{\partial r}+\frac{\partial q}{\partial r}\right)=0 \\ &\frac{\partial \rho}{\partial t}+u\frac{\partial \rho}{\partial r}+\rho\frac{\partial u}{\partial r}+\frac{2\rho u}{r}=0 \\ &\frac{\partial e_m}{\partial t}+u\frac{\partial e_m}{\partial r}-\frac{p+q}{\rho^2}\left(\frac{\partial \rho}{\partial t}+u\frac{\partial \rho}{\partial r}\right)+\frac{1}{\rho r^2}\frac{\partial}{\partial r}\left(Fr^2\right)=0 \end{aligned} \tag{1}$$

where t and r are the independent time and space coordinates, respectively, p—pressure of the mixture, u—radial direction flow velocity, ρ—density of the mixture, e_m—internal energy per unit mass of the mixture, F—radiation heat flux, and q—artificial viscosity.

The expression for artificial viscosity q is given by ([26] also see the references within)

$$q=\frac{1}{2}K^2\rho r^2\frac{\partial u}{\partial r}\left(|\frac{\partial u}{\partial r}|-\frac{\partial u}{\partial r}\right) \tag{2}$$

where K is a constant parameter that can be modified easily in any numerical experiment.

Using Rosseland's diffusion approximation and assuming local thermodynamic equilibrium, we have

$$F=-\frac{c\mu}{3}\frac{\partial}{\partial r}\left(aT^4\right) \tag{3}$$

where c—velocity of light, μ—mean free path of radiation, and $ac/4$—Stefan-Boltzmann constant.

The mean free path of radiation μ which is a function of absolute temperature T and density ρ is given by Wang [27] as

$$\mu=\mu_0\rho^{\alpha^\star}T^{\beta^\star} \tag{4}$$

where $\alpha^\star$, $\beta^\star$ are constants.

The dusty gas equation of state is as follows: (Pai [28])

$$p=\frac{1-K_p}{1-Z}\rho R^\star T \tag{5}$$

where $R^\star$—gas constant, Z—volume fraction of solid particles in the mixture, and K_p—mass concentration of solid particles.

Z and K_p are related as

$$K_p = \frac{Z\rho_{sp}}{\rho} \tag{6}$$

where ρ_{sp}—solid particle species density. For an equilibrium flow, K_p is constant throughout the flow.

The mixture's internal energy e_m can be expressed as follows:

$$e_m = \left[K_p C_{sp} + (1 - K_p)C_v\right]T = C_{vm}T \tag{7}$$

where C_{vm}—mixture's specific heat at constant volume, C_v—specific heat of a gas at a constant volume, and C_{sp}—specific heat of solid particles. The specific heat at constant pressure process is

$$C_{pm} = K_p C_{sp} + (1 - K_p)C_p \tag{8}$$

where C_p—specific heat of the gas at constant pressure process.

The ratio of the specific heats of the mixture is given by (see [28])

$$\Gamma = \frac{\gamma\left(1 + \frac{\sigma\beta'}{\gamma}\right)}{1 + \sigma\beta'} \tag{9}$$

where

$$\gamma = \frac{C_p}{C_v} \qquad \sigma = \frac{K_p}{1 - K_p} \qquad \beta' = \frac{C_{sp}}{C_v}. \tag{10}$$

Therefore, the internal energy e_m is given by

$$e_m = \frac{p(1 - Z)}{\rho(\Gamma - 1)}. \tag{11}$$

The propagation of a spherical shock wave into a resting medium with a small constant counter pressure is investigated. Further, the medium's initial density is assumed to follow the exponential law

$$\rho = Ae^{\alpha r} \tag{12}$$

where A, α are the constants which are positive.

The jump conditions across the shock are as follows:

$$u_2 = (1 - \beta)\,V$$

$$\rho_2 = \frac{\rho_1}{\beta}$$

$$p_2 = (1 - Z_1 - \frac{4K^2(1-\beta)^2}{\beta e^{\alpha r}})\rho_1 V^2$$

$$F_2 = (1-\beta)\left[\frac{(1+\Gamma)\beta + (1-\Gamma) - 2Z_1}{2(\Gamma - 1)} - \frac{1 - Z_1}{(\Gamma - 1)M_e^2} + \frac{4K^2(1-\beta)(Z_1 - \beta)}{\beta e^{\alpha r}(\Gamma - 1)}\right]\rho_1 V^3$$

$$z_2 = \frac{z_1}{\beta}$$

$$q_2 = \frac{4K^2(1-\beta)^2 \rho_1 V^2}{\beta e^{\alpha r}} \tag{13}$$

where R is the distance between the shock front and the point of symmetry, $U = \frac{dR}{dt}$ is the shock velocity, suffices "1" and "2" are the values just ahead and behind the shock, and $F_1 = 0$ (see [7]). Also, the expression for β is given by

$$\beta = Z_1 + \frac{1 - Z_1}{\Gamma M_e^2} \tag{14}$$

where

$$M_e^2 = \frac{V^2}{a_1^2} \qquad a_1^2 = \frac{\Gamma P_1}{\rho_1(1 - Z_1)} \tag{15}$$

M_e stands for the shock-Mach number, which refers to the sound speed a_1 in the dusty gas.

In general, Z_1, the solid particles' volume fraction at the initial state is not constant. However, because solid particles have a much higher density than gas (Miura and Glass [13]), the volume occupied by solid particles is extremely small, and Z_1 can be assumed to be a small constant. Z_1 is expressed as (Naidu et al. [29])

$$Z_1 = \frac{K_p}{G(1 - K_p) + K_p}. \tag{16}$$

Here, G is the solid particles density divided by the initial gas density.

Let the solution to Eqs. (1), (2), and (3) be of the form

$$u = t^{-1}U(\eta),\quad \rho = t^{\gamma^\star}D(\eta),\quad p = t^{\gamma^\star - 2}P(\eta),\quad F = t^{\gamma^\star - 3}Q(\eta),\quad q = t^{\gamma^\star - 2}S(\eta) \tag{17}$$

where

$$\eta = te^{\delta^\star r} \quad \delta^\star \neq 0 \tag{18}$$

and $\gamma^\star$, $\delta^\star$ are the constants that will be determined subsequently. The shock surface is chosen to be

$$\eta_0 = constant \tag{19}$$

as a result of which the velocity is given by

$$V = -\frac{1}{\delta^{\star} t}. \tag{20}$$

As a result, it is self-evident that $\delta^{\star} < 0$. In the form of (17), the solutions of the equations (1),(2), and (3) are compatible with the shock conditions only if

$$\alpha^{\star} = 1, \qquad \beta^{\star} = -\frac{5}{2}, \qquad \gamma^{\star} = 2, \qquad \delta^{\star} = -\frac{\alpha}{2}. \tag{21}$$

The Mach number M_e of the shock is given by

$$M_e^2 = \frac{V^2}{a_1^2} = -\frac{4(1 - Z_1)A}{\Gamma p_1 \alpha^2 \eta_0} = constant$$

For a very strong shock, as M_e is a constant, and p_1 is of order zero, we assume that the shock holds its enormous strength over a long period of time. As a result, the solutions in the following section are valid whenever $t > \tau$ until Z_1 stays small, where τ is the duration of the initial impulse. It can be obtained from Eqs. (20) and (21) that

$$R = \frac{2}{\alpha} log \frac{t}{\tau}. \tag{22}$$

3 Solution

By solving equations (1), (2), and (3), the flow variables in the flow field behind the shock front will be obtained. Equations (17), (20), and (21) provide us

$$\begin{aligned} \frac{\partial u}{\partial t} &= u\delta^{\star} V - V\frac{\partial u}{\partial r} & \qquad \frac{\partial \rho}{\partial t} &= V\rho\alpha - V\frac{\partial \rho}{\partial r} \\ \frac{\partial p}{\partial t} &= -V\frac{\partial p}{\partial r} & \qquad \frac{\partial q}{\partial t} &= -V\frac{\partial q}{\partial r}. \end{aligned} \tag{23}$$

Using the above Eq. (23) and considering the transformations

$$\begin{aligned} r' &= \frac{r}{R} & \qquad u' &= \frac{u}{V} & \qquad p' &= \frac{p}{p_2} \\ \rho' &= \frac{\rho}{\rho_2} & \qquad F' &= \frac{F}{F_2} & \qquad q' &= \frac{q}{q_2} \end{aligned} \tag{24}$$

in basic equations (1), (2), and (3), we get

$$\frac{d\rho'}{dr'} = \frac{\rho'}{1-u'}\left[2log\frac{t}{\tau} + \frac{du'}{dr'} + \frac{2u'}{r'}\right] \tag{25}$$

$$\frac{dp'}{dr'} = \frac{\rho'}{(1-Z_1)\beta}\left[(1-u')\frac{du'}{dr'} + u'log\frac{t}{\tau}\right] - \frac{\Theta}{1-Z_1-\Theta}\frac{dq'}{dr'} \tag{26}$$

$$\frac{du'}{dr'} = \frac{2(1-\beta)}{r'}\sqrt{\frac{q'}{\rho'(t/\tau)^{2r'}}} \tag{27}$$

$$\begin{aligned}\frac{dF'}{dr'} = &\frac{1-Z_1-\Theta}{(1-\beta)\left[\frac{(1+\Gamma)\beta+(1-\Gamma)-2Z_1}{2} - \frac{1-Z_1}{M_e^2} + \frac{\Theta(Z_1-\beta)}{(1-\beta)}\right]}\\ &\left\{\left[\frac{(\beta-Z_1\rho')(1-u')^2\rho'}{(1-Z_1)\beta^2} - \Gamma p'\right]\frac{du'}{dr'}\right.\\ &+\frac{(1-u')(\beta-Z_1\rho')\rho'u'log\frac{t}{\tau}}{(1-Z_1)\beta^2} - \frac{2\Gamma p'u'}{r'}\\ &-\frac{(\beta-Z_1\rho')(1-u')\rho'}{(1-Z_1)\beta^2}\frac{\Theta}{1-Z_1-\Theta}\frac{dq'}{dr'}\\ &\left.-\frac{\Theta}{1-Z_1-\Theta}q'(\Gamma-1)(1-u')\left[\frac{2log\frac{t}{\tau}}{u'-1} + \frac{1}{\rho'}\frac{d\rho'}{dr'}\right]\right\} - \frac{2F'}{r'}\end{aligned} \tag{28}$$

$$\begin{aligned}\frac{dq'}{dr'} = &\frac{1-Z_1-\Theta}{\Theta\beta(\beta-Z_1\rho')(1-u')(1-Z_1)}\\ &\left\{\frac{F'(1-Z_1)\beta(1-u')\sqrt{\rho'}log\frac{t}{\tau}}{NL\sqrt{p'}\sqrt{\beta-Z_1\rho'}} + (1-u')(\beta-Z_1\rho')\rho'u'log\frac{t}{\tau}\right.\\ &-2p'\beta^2(1-Z_1)log\frac{t}{\tau} - \frac{2p'u'\beta^2(1-Z_1)}{r'}\\ &\left.\left[p'\beta^2(1-Z_1) - (\beta-Z_1\rho')\rho'(1-u')^2\right]\frac{du'}{dr'}\right\}\end{aligned} \tag{29}$$

where

$$N = \frac{4ac\mu_0\alpha}{3\sqrt{R^{\star}}^3}$$

is a non-dimensional radiation parameter and

$$L = \frac{(\Gamma - 1)\sqrt{(1 - Z_1 - \Theta)^3}}{2(1-\beta)\left[\frac{(1+\Gamma)\beta + (1-\Gamma) - 2Z_1}{2} - \frac{1 - Z_1}{M_e^2} + \frac{\Theta(Z_1 - \beta)}{(1-\beta)}\right]\beta\sqrt{(1-K_p)^3}}$$

and

$$\Theta = \frac{4K^2(1-\beta)^2}{\beta(t/\tau)^{2r'}}.$$

The shock conditions get the following form in terms of dimensionless variables

$$r' = 1, \quad p' = 1, \quad \rho' = 1, \quad F' = 1, \quad q' = 1, \quad u' = 1 - \beta. \tag{30}$$

The solution to our problem is given by Eqs. (25)–(29) along with the boundary conditions (30). Due to the fact that the motion behind the shock can be calculated only when a certain time is supplied, the resulting solution is non-similar.

4 Results and Discussion

From the shock front $r' = 1$, we begin the numerical integration of Eqs. (25)–(29) along with boundary conditions (30) and work our way inwards to acquire the solutions. At specified instants when $t/\tau = 2$ or 4, distributions of flow variables $\rho' = \frac{\rho}{\rho_2}$, $p' = \frac{p}{p_2}$, $u' = \frac{u}{u_2}$ are obtained. For the sake of numerical integration, values of $K_p, \gamma, G, M_e^2, N, \chi$, and K are assumed to be $K_p = 0, 0.2, 0.4$; $\gamma = 1.4$; $G = 10, 50$ (see [15]), $M_e^2 = 20$, $N = 10$ (see [9]), $\chi = 1$ (see [13]), and $K = 0, 0.0349, 0.349$ (see [26]).

Figures 1, 2, and 3 depict the variation of ρ', p', and u' with reduced distance r' for varied values of viscosity parameter K at different times t/τ for fixed values of N, K_p, and G.

Density ρ' and pressure p' decline as we travel inwards from the shock front, as seen in Figs. 1 and 2. Further, from Fig. 3, it can be seen that the reduced flow velocity u' increases when $K = 0$, whereas, it decreases when $K \neq 0$. In the presence of viscosity, the nature of reduced flow variables(concave upwards) is in contrast with the case of no viscosity $K = 0$(concave downwards). It can be further observed that when $K = 0$, the values of ρ', p', and u' tend to be the same with the values of Singh and Vishwakarma [17] work.

An increase in the viscosity parameter K causes the density ρ', the pressure p', and the velocity u' to decrease as well as the slopes of ρ' and p' to decrease at any point in the flow behind the shock front. An increase in time t/τ causes density ρ', pressure p' to decline, and the flow velocity u' to rise.

Figures 4, 5, and 6 illustrate the reduced flow variables variation with reduced distance for varied values of solid particle mass concentration K_p and the ratio of solid particle density to initial gas density G for given values of N, K, and t/τ.

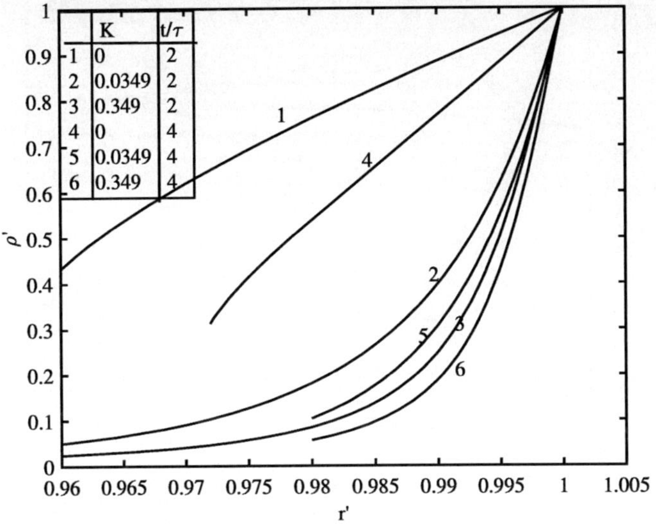

Fig. 1 Reduced density ρ' variation behind the shock front when $N = 10$, $G = 50$, and $K_p = 0.2$

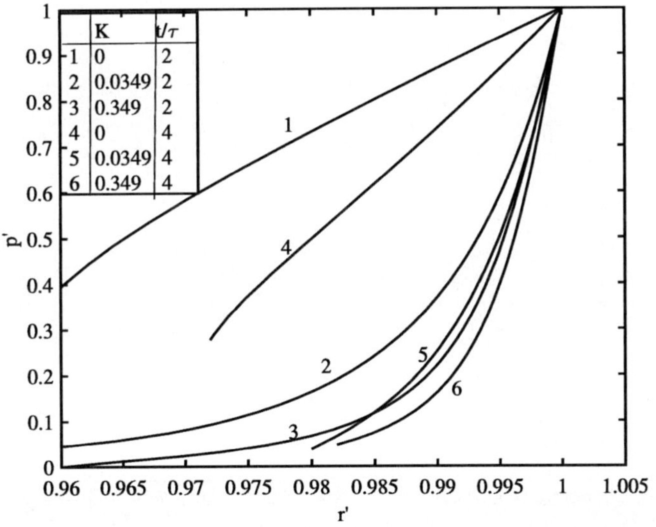

Fig. 2 Reduced pressure p' variation behind the shock front when $N = 10$, $G = 50$, and $K_p = 0.2$

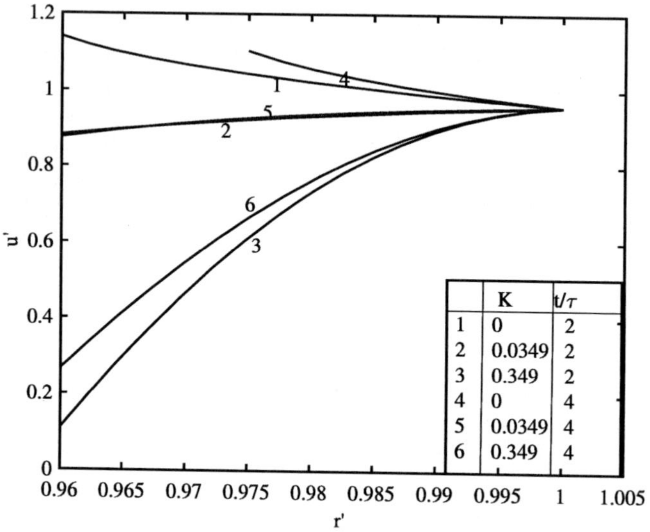

Fig. 3 Reduced flow velocity u' variation behind the shock front when $N = 10$, $G = 50$, and $K_p = 0.2$

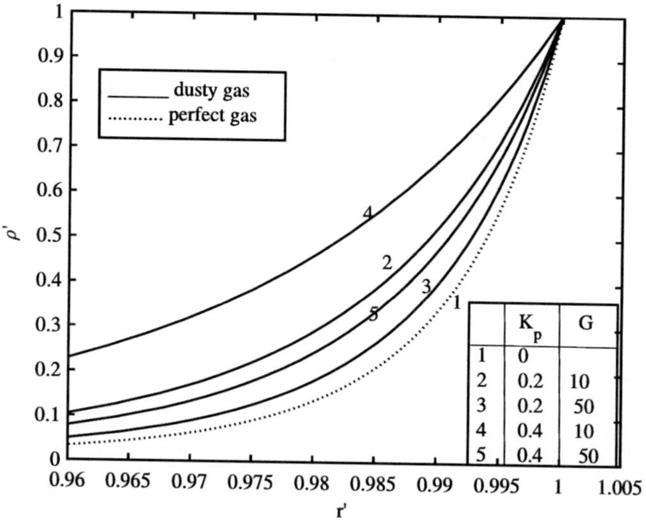

Fig. 4 Reduced density ρ' variation behind the shock front when $N = 10$, $K = 0.349$, and $t/\tau = 2$

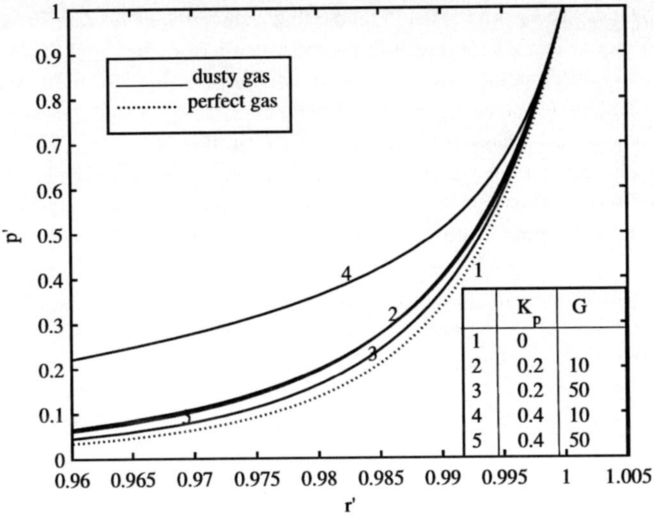

Fig. 5 Reduced pressure p' variation behind the shock front when $N = 10$, $K = 0.349$, and $t/\tau = 2$

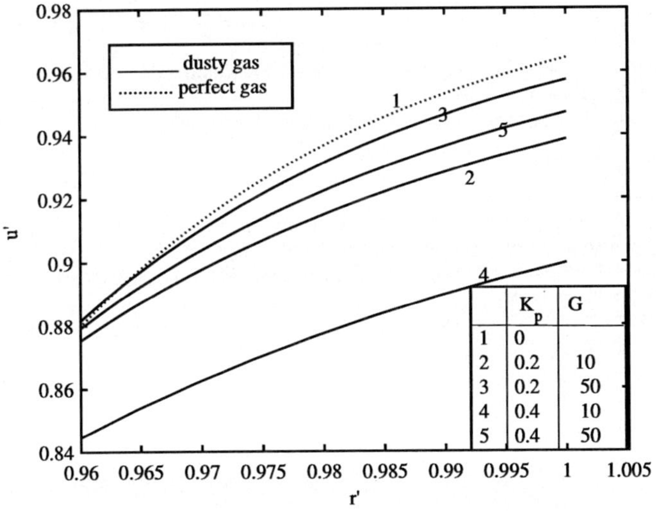

Fig. 6 Reduced flow velocity u' variation behind the shock front when $N = 10$, $K = 0.349$, and $t/\tau = 2$

It can be observed from Figs. 4, 5, and 6 that reduced density ρ', reduced pressure p', and reduced velocity decrease as one moves inside from the shock front. Whereas, the nature of reduced velocity u' (concave downwards) is opposite to those of density ρ' and pressure p' (concave upwards). When shock travels through a dusty gas, the values of density ρ', pressure p' increase; however, the shock speed u' drops when compared to a perfect gas ($K_p = 0$) or a dusty gas with a larger K_p. The existence of solid particles in dusty gas is accountable for this phenomenological behavior.

For given K, N, and G values, increasing the mass concentration of solid particles K_p increases ρ', p' and decreases u', as well, increases the slopes of density, pressure profiles. Also, an increase in the ratio of solid particle density to initial gas density G for fixed values of N, K, and K_p results in the decrease of density, pressure and an increase in fluid velocity.

References

1. Grover, R., Hardy, J.W.: The propagation of shocks in exponentially decreasing atmospheres. Astrophys. J. **143**, 48 (1966)
2. Hayes, W.D.: Self-similar strong shocks in an exponential medium. J. Fluid Mech. **32**(2), 305–315 (1968)
3. Deb Ray, G., Bhowmick, J.B.: Propagation of cylindrical and spherical explosion waves in an exponential medium. Defence Sci. J. **24**, 9–12 (1974)
4. Laumbach, D.D., Probstein, R.F.: A point explosion in a cold exponential atmosphere part i. J. Fluid Mech. **35**(1), 53–75 (1969)
5. Verma, B.G., Vishwakarma, J.P.: Axially symmetric explosion in magnetogasdynamics. Astrophys. Space Sci. **69**(1), 177–188 (1980)
6. Laumbach, D.D., Probstein, R.F.: A point explosion in a cold exponential atmosphere part ii, radiating flow. J. Fluid Mech. **40**(1), 833–858 (1970)
7. Laumbach, D.D., Probstein, R.F.: Self-similar strong shocks with radiation in a decreasing exponential atmosphere. Phys. Fluids **13**(5), 1178–1183 (1970)
8. Bhowmick, J.B.: An exact analytical solution in radiation gas dynamics. Astrophys. Space Sci. **74**(2), 481–485 (1981)
9. Singh, J.B., Srivastava, S.K.: Propagation of spherical shock waves in an exponential medium with radiation heat flux. Astrophys. Space Sci. **88**(2), 277–282 (1982)
10. Christer, A.H., Helliwell, J.B.: Cylindrical shock and detonation waves in magnetogasdynamics. J. Fluid Mech. **39**(4), 705–725 (1969)
11. Verma, B.G.: On a cylindrical blast ware propagating in a conducting gas. Zeitschrift fur angewandte Mathematik und Physik ZAMP **21**(1), 119–124 (1970)
12. Higashino, F., Suzuki, T.: The effect of particles on blast waves in a dusty gas. Zeitschrift für Naturforschung A **35**(12), 1330–1336 (1980)
13. Miura, H., Israel Glass, I.: On the passage of a shock wave through a dusty-gas layer. Proc. R. Soc. Lond. A. Math. Phys. Sci. **385**(1788), 85–105 (1983)
14. Popel, S.I., Gisko, A.A.: Charged dust and shock phenomena in the solar system. Nonlinear Process. Geophys. **13**(2), 223–229 (2006)
15. Pai, S.I., Menon, S., Fan, Z.Q.: Similarity solutions of a strong shock wave propagation in a mixture of a gas and dusty particles. Int. J. Eng. Sci. **18**(12), 1365–1373 (1980)
16. Vishwakarma, J.P.: Propagation of shock waves in a dusty gas with exponentially varying density. Eur. Phys. J. B-Condensed Matter Complex Syst. **16**(2), 369–372 (2000)
17. Singh, K.K., Vishwakarma, J.P.: Propagation of spherical shock waves in a dusty gas with radiation heat-flux. J. Theor. Appl. Mech. **45**(4), 801–817 (2007)

18. Macquorn Rankine, W.J.: Xv on the thermodynamic theory of waves of finite longitudinal disturbance. Philos. Trans. R. Soc. Lond. **160**, 277–288 (1870)
19. Rayleigh, L.: Aerial plane waves of finite amplitude. Proc. R. Soc. Lond. Ser. A, Contain. Pap. Math. Phys. Charact. **84**(570), 247–284 (1910)
20. Ingram Taylor, G.: The conditions necessary for discontinuous motion in gases. Proc. R. Soc. Lond. Ser. A, Contain. Pap. Math. Phys. Charact. **84**(571), 371–377 (1910)
21. Henderson, L.F., Crutchfield, W.Y., Virgona, R.J.: The effects of thermal conductivity and viscosity of argon on shock waves diffracting over rigid ramps. J. Fluid Mech. **331**, 1–36 (1997)
22. Simeonides, G.: Generalized reference enthalpy formulations and simulation of viscous effects in hypersonic flow. Shock Waves **8**(3), 161–172 (1998)
23. Huang, F., Matsumura, A., Shi, X.: Viscous shock wave and boundary layer solution to an inflow problem for compressible viscous gas. Commun. Math. Phys. **239**(1), 261–285 (2003)
24. Von Neumann, J., Richtmyer, R.D.: A method for the numerical calculation of hydrodynamic shocks. J. Appl. Phys. **21**(3), 232–237 (1950)
25. Suzuki, T., Ohyagi, S., Higashino, F., Takano, A.: The propagation of reacting blast waves through inert particle clouds. Acta Astronaut. **3**(7–8), 517–529 (1976)
26. Narsimhulu, D., Addepalli, R., Satpathi Dipak, K.: Similarity solution of spherical shock waves-effect of viscosity. Proyecciones (Antofagasta) **35**(1), 11–31 (2016)
27. Wang, K.C.: Approximate solution of a plane radiating "piston problem.Ã¢â‚¬Â Phys. Fluids **9**(10), 1922–1928 (1966)
28. Pai, S.: Two phase flows. Vol. 3, Vieweg Tracts in Pure and Applied Physics, pp. 56–80 (1977)
29. Narasimhulu Naidu, G., Venkatanandam, K., Ranga Rao, M.P.: Approximate analytical solutions for self-similar flows of a dusty gas with variable energy. Int. J. Eng. Sci. **23**(1), 39–49 (1985)

On the Stability of a Heated Inclined Fluid Layer with Gravity Modulation

Manisha Arora and Renu Bajaj

Abstract The effect of sinusoidal gravity modulation on the stability of natural convection in an inclined viscous fluid layer is studied using the energy stability theory. The variation of the critical value of the control parameter, the Rayleigh number, below which the basic flow is stable is discussed with the modulation parameters and the inclination of the fluid layer. An uncertain stability region is observed between the linear and the nonlinear marginal curves.

Keywords Energy method · Gravity modulation · Inclined fluid layer · Nonlinear stability

1 Introduction

The hydrodynamical stability of natural convection in an inclined fluid layer [10, 14] subjected to temperature gradient has been an interesting problem among researchers due to its non-zero basic flow. Researchers always attempt to control the rate of heat transfer across the fluid layer. Applying periodically modulated driving force is one of the methods to control the heat transfer rate. This hydrodynamical stability problem has direct applications in various fields such as material processing, nuclear science, large-scale convection problems, engineering, astrophysics, and geophysics.

The effect of time periodic gravity modulation on the stability of the basic flow of an infinite viscous fluid layer has been studied by various researchers [4, 5, 7–9, 11, 12]. Chen and Chen [7] have discussed the effect of gravity modulation on the linear stability of thermal convection in a vertical fluid layer. The mode of instability is found to depend on the Prandtl number of the fluid. The effect of the modulation parameters on the thermosolutal convection has been discussed by Bajaj [5] in magnetic fluids

M. Arora (✉) · R. Bajaj
Department of Mathematics, Panjab University, Chandigarh, India
e-mail: arora12794@gmail.com

R. Bajaj
e-mail: rbajaj@pu.ac.in

R. K. Sharma et al. (eds.), *Frontiers in Industrial and Applied Mathematics*,
Springer Proceedings in Mathematics & Statistics 410,
https://doi.org/10.1007/978-981-19-7272-0_27

using Floquet theory. Recently, Saravanan and Meenasaranya [15] have studied the energy stability of porous convection driven by periodically modulated boundary temperatures in the presence of magnetic field.

Linear instability analysis provides marginal boundary in the space of governing parameters above which the basic flow is unstable. Below the marginal curve, the basic flow is stable against perturbations of infinitesimally small magnitude. To determine the stability of the basic flow against finite perturbations, the energy stability analysis [16] is used.

Homsy [11] has discussed the stability characteristics in a horizontal fluid layer with temperature and gravity modulation using the energy method. Kaloni and Qiao [12] have discussed the nonlinear stability in a horizontal fluid layer with variable gravity force and inclined temperature gradient. Arora et al. [1, 3] have studied the nonlinear stability of a heated inclined fluid layer for the Prandtl numbers equal to 0.71 and 7.56. The region of uncertain stability has been found in the parametric space. The effect of internal heating on the stability of natural convection in an inclined fluid layer is discussed by Arora and Bajaj [2] using the energy method.

In this paper, we have studied the effect of time periodically varying gravity on the stability of the basic flow of a heated inclined fluid layer. The energy method is used to find the stability of the flow against arbitrary perturbations.

The outline of the paper is as follows: the mathematical formulation of the problem and the basic state are given in Sect. 2. The stability of the hydrodynamical system is discussed through the linear instability analysis and the energy stability analysis in Sect. 3. The Euler-Lagrange equations are derived and solved by using the shooting method. The critical value of the control parameter is obtained and the results are discussed in Sect. 4.

2 Governing Equations and the Basic State

Consider a viscous, incompressible flow of a fluid of uniform density ρ. The fluid layer is inclined at an angle ϕ to the horizontal and confined between two rigid, thermally conducting planes. The Cartesian coordinate system is assumed to be fixed in the fluid layer such that x-axis is normal and y-axis is along the fluid layer. The Boussinesq approximation [6], which makes density term constant in all the terms except in the body force term, is assumed. The variation of density with temperature gradient is given by the following relation

$$\rho = \rho_0[1 - \alpha(T - T_0)],$$

where α is the coefficient of volume expansion, ρ_0 is the density of the fluid at reference temperature T_0. The fluid flow is subjected to time dependent gravity modulation. The gravitational field is varying time periodically and is given by $\mathbf{g} = (g + \epsilon_0 \sin(\omega_0 t))(-\cos\phi, -\sin\phi, 0)$. Here, g is the mean value of acceleration due to gravity, ϵ_0 is the amplitude of the modulation and ω_0 is the frequency of the modulation. The governing equations are made dimensionless using the characteristic length d, time $\frac{d^2}{\kappa}$, temperature $\frac{T-T_0}{T_1-T_2}$, pressure $\frac{\kappa^2 \rho_0}{d^2}$ and modulation frequency $\frac{\kappa}{d^2}$. The governing equations representing the conservation of mass, linear momentum and energy in the dimensionless form are given by

$$\nabla \cdot \mathbf{v} = 0, \tag{1}$$

$$\begin{aligned}\frac{\partial \mathbf{v}}{\partial t} + \mathbf{v} \cdot \nabla \mathbf{v} = &- \nabla(\mathcal{P} + \frac{d^3 g}{\kappa^2}\{x \cos\phi + y \sin\phi\}) + \mathrm{Pr}\nabla^2 \mathbf{v} \\ &+ \mathrm{RaPr}T(1 + \epsilon \sin(\omega t))(\cos\phi \hat{i} + \sin\phi \hat{j}),\end{aligned} \tag{2}$$

$$\frac{\partial T}{\partial t} + \mathbf{v} \cdot \nabla T = \nabla^2 T, \tag{3}$$

The boundary conditions for the velocity and the temperature are

$$\mathbf{v}|_{x=\pm\frac{1}{2}} = \mathbf{0};\ T|_{x=\frac{1}{2}} = -\frac{1}{2};\ T|_{x=-\frac{1}{2}} = \frac{1}{2}. \tag{4}$$

In the above equations, $\mathbf{v}$ is the velocity, T the temperature, $\mathcal{P}$ the pressure, κ the thermal diffusivity, and ν the kinematic coefficient of viscosity of the fluid. All the quantities are in dimensionless form in the above equations. The dimensionless parameters governing the fluid flow are the Rayleigh number $\mathrm{Ra} = \frac{g\alpha d^3(T_1-T_2)}{\kappa\nu}$, the Prandtl number $\mathrm{Pr} = \frac{\nu}{\kappa}$, the amplitude of the modulation $\epsilon(= \epsilon_0/g)$ and the frequency of modulation ω.

2.1 The Basic State

By solving the system (1)–(4) analytically, we obtain the following basic state,

$$T_B(x) = -x, \tag{5}$$

$$\mathbf{v}_B(x, t) = (0, V_B(x, t), 0), \tag{6}$$

$$\mathcal{P}_B = -\frac{d^3 g}{\kappa^2}(x \cos\phi + y \sin\phi), \tag{7}$$

where

$$\begin{aligned}V_B(x, t) = &\mathrm{Ra} \sin\phi\Big(\frac{x^3}{6} - \frac{x}{24}\Big) + \mathrm{Ra} \sin\phi \frac{\epsilon \mathrm{Pr}}{\omega} x \sin(\omega t), \\ &- \mathrm{Ra} \sin\phi \frac{\epsilon \mathrm{Pr}}{2\omega}\Big(\frac{W_1(x)\sin(\omega t) + W_2(x)\cos(\omega t)}{\sinh^2(a)\cos^2(a) + \sin^2(a)\cosh^2(a)}\Big).\end{aligned} \tag{8}$$

The functions $W_1(x)$ and $W_2(x)$ are given by

$$W_1(x) = \cos(a)\sinh(a)\cos(2ax)\sinh(2ax) + \sin(a)\cosh(a)\sin(2ax)\cosh(2ax),$$
$$W_2(x) = \cos(a)\sinh(a)\sin(2ax)\cosh(2ax) - \sin(a)\cosh(a)\cos(2ax)\sinh(2ax),$$

where $a = \sqrt{\frac{\omega}{8\mathrm{Pr}}}$. The basic state is calculated with the assumption that the net flux across the fluid layer is zero, i.e.,

$$\int_{-\frac{1}{2}}^{\frac{1}{2}} V_B(x,t) = 0$$

Modulation of gravity has no effect on the basic temperature (5). The basic velocity is a periodic function of t with time period $\frac{2\pi}{\omega}$.

3 Stability Analysis

The basic state of the fluid system (5)–(8) is perturbed at an instant of time t to analyze its stability at that instant. Arbitrary disturbances of finite magnitude are imposed on the basic state of velocity, temperature and pressure. The perturbed quantities are $\mathbf{v}_B(x,t) + \mathbf{u}$, $T_B(x) + \theta$ and $\mathcal{P}_B(x,y) + p$, which would satisfy the governing equations of the system (1)–(4). The resulting perturbation equations are

$$\nabla \cdot \mathbf{u} = 0, \tag{9}$$

$$\begin{aligned}\frac{\partial \mathbf{u}}{\partial t} + \mathbf{v}_B \cdot \nabla \mathbf{u} + \mathbf{u} \cdot \nabla \mathbf{v}_B + \mathbf{u} \cdot \nabla \mathbf{u} = -\nabla p + \mathrm{Pr}\nabla^2 \mathbf{u} \\ + \mathrm{RaPr}\theta(1 + \epsilon \sin(\omega t))(\cos\phi \hat{i} + \sin\phi \hat{j}),\end{aligned} \tag{10}$$

$$\frac{\partial \theta}{\partial t} + \mathbf{u} \cdot \nabla \theta + u\frac{\partial T_B}{\partial x} + V_B\frac{\partial \theta}{\partial y} = \nabla^2 \theta. \tag{11}$$

The perturbations must get vanished at the rigid boundaries, which results in the following boundary conditions.

$$(u, v, w, \theta) = (0, 0, 0, 0) \text{ at } x = \mp\frac{1}{2}. \tag{12}$$

3.1 Linear Stability Analysis

To analyze the linear stability of the system, the disturbances are assumed to be of infinitesimally small amplitude and therefore the higher order terms in the equations (9)–(11) are ignored.

At the onset of instability, the fluid layer is assumed to get divided into cells, periodic along the layer ($y-$ axis) and transverse to the layer ($z-$ axis). Let V be the volume of one such periodic cell. The normal mode analysis is applied and the solution form is taken as

$$(u, v, w, \theta, p) = (\tilde{u}(x,t), \tilde{v}(x,t), \tilde{w}(x,t), \tilde{\theta}(x,t), \tilde{p}(x,t)) \exp\{\iota(k_1 y + k_2 z + \delta t)\} + \text{c.c.} \tag{13}$$

Here, k_1 and k_2 represent the wavenumbers in y and z direction, respectively, such that $k_1^2 + k_2^2 = a_1^2$ and $\delta = \delta_1 + \iota\delta_2$. The resulting system of ordinary differential equations is solved by using the shooting method [3]. The value of the control parameter Rayleigh number is determined by considering the fixed value of other parameters ϕ, Pr, ω, ϵ, t, k_1, k_2, and δ and is obtained numerically by minimizing its value with respect to the wavenumbers as

$$\tilde{Ra}_L(t) = \min_{k_1,\, k_2} Ra(\phi, \Pr, \omega, \epsilon, t, k_1, k_2) \tag{14}$$

The Rayleigh number is observed to be periodic with time period $\frac{2\pi}{\omega}$. Therefore, the critical value of the Rayleigh number is obtained as

$$Ra_L = \min_{t\in[0,\frac{2\pi}{\omega}]} \min_{k_1,\, k_2} Ra(\phi, \Pr, \omega, \epsilon, t, k_1, k_2) \tag{15}$$

The corresponding value of t, k_1, k_2 are denoted by t_L, k_{1L}, k_{2L}. $a_{1L} = \sqrt{k_{1L}^2 + k_{2L}^2}$. The basic state is stable with respect to infinitesimally small perturbations for Ra $<$ Ra$_L$.

3.2 Energy Stability Analysis

To study the stability with respect to arbitrary perturbations, the energy method [16] is used which provides a sufficient condition for the global stability of the basic flow. ∂V denotes the boundary of the periodic cell in the fluid layer at the onset of instability. Let $\|.\|$ denotes $L^2(V)$ norm derived from the inner product

$$\langle fg \rangle = \int_V fg \, dV.$$

The energy function $E(t)$ for the perturbations is defined as

$$E(t) = \frac{1}{2}||\mathbf{u}||^2 + \frac{\gamma}{2}||\theta||^2, \tag{16}$$

where γ is a coupling parameter. On taking L^2 product of Eq. (10) with $\mathbf{u}$ and Eq. (11) with θ, the following equations are obtained by using the continuity equation (5), Gauss Divergence theorem and the boundary conditions (12).

$$\frac{1}{2}\frac{d}{dt}||\mathbf{u}||^2 = -\langle uv\frac{\partial V_B}{\partial x}\rangle + \text{RaPr}(1+\epsilon\sin(\omega t))\{\cos\phi\langle\theta u\rangle + \sin\phi\langle\theta v\rangle\} - \text{Pr}||\nabla\mathbf{u}||^2, \tag{17}$$

$$\frac{1}{2}\frac{d}{dt}||\theta||^2 = \langle\theta u\rangle - ||\nabla\theta||^2. \tag{18}$$

From Eq. (16), the rate of change of energy function $E(t)$ is obtained as

$$\frac{dE}{dt} = \frac{1}{2}\frac{d}{dt}||\mathbf{u}||^2 + \frac{\gamma}{2}\frac{d}{dt}||\theta||^2 = \mathcal{I} - \mathcal{D}, \tag{19}$$

where

$$\begin{aligned}\mathcal{I} &= -\langle uv\frac{\partial V_B}{\partial x}\rangle + \text{RaPr}(1+\epsilon\sin(\omega t))\{\cos\phi\langle\theta u\rangle + \sin\phi\langle\theta v\rangle\} + \gamma\langle\theta u\rangle,\\ \mathcal{D} &= \text{Pr}||\nabla\mathbf{u}||^2 + \gamma||\nabla\theta||^2.\end{aligned} \tag{20}$$

We claim that $\frac{\mathcal{I}}{\mathcal{D}}$ is bounded over the space $\mathcal{H}$, where $\mathcal{H}$ is the space of all admissible solutions satisfying the perturbation equations (9)–(12). From the basic state of velocity, we have

$$\left|\frac{\partial V_B(x,t)}{\partial x}\right| \le \sup_{x,t}\left|\frac{\partial V_B(x,t)}{\partial x}\right| \le |\text{Ra}|\ \Omega(\epsilon,\omega),$$

where

$$\begin{aligned}\Omega(\epsilon,\omega) &= \frac{1}{12} + \frac{\epsilon\text{Pr}}{\omega}\left[1 + \frac{4a}{B}\{|\sinh(a)| + |\cosh(a)|\}\right],\\ B &= \sinh^2(a)\cos^2(a) + \sin^2(a)\cosh^2(a).\end{aligned}$$

By using the Cauchy-Schwarz inequality, the Arithmetic-Geometric mean inequality and the Poincaré inequality, we get

$$|\mathcal{I}| \le \frac{1}{K^2}\left(\frac{|\text{Ra}|}{\text{Pr}}\left(\frac{1}{12} + \Omega(\epsilon,\omega)\right) + |\text{Ra}|\sqrt{\frac{\text{Pr}}{\gamma}} + \frac{1}{2}\sqrt{\frac{\gamma}{\text{Pr}}}\right)\mathcal{D}. \tag{21}$$

Hence, $\frac{\mathcal{I}}{\mathcal{D}}$ is bounded over $\mathcal{H}$.

Now, using (21) and the Poincaré inequality [16] in (18),

$$\frac{dE}{dt} = \mathcal{I} - \mathcal{D} = -\mathcal{D}\left(1 - \frac{\mathcal{I}}{\mathcal{D}}\right) \le -\mathcal{D}(1-r) \tag{22}$$

where $r = \sup_{\mathcal{H}}\left(\frac{\mathcal{I}}{\mathcal{D}}\right)$. Using the Poincaré inequality [16], we get

$$2hK^2E(t) \leq \mathcal{D}$$

which implies,

$$\frac{dE}{dt} \leq -2h(1-r)K^2E(t)$$

where $h = \min\{1, \mathrm{Pr}\}$ and $K = \mathrm{diam}(V)$ is the Poincaré constant.

If $0 < r \leq 1$, then $\frac{dE(t)}{dt} \leq 0$. Thus all the perturbations decay with time for $r \in (0, 1)$. Thus, the periodic basic state is stable with respect to arbitrary disturbances.

The variation of $\left(\frac{\mathcal{I}}{\mathcal{D}}\right)$ is zero for $r = 1$, which is the maximum value of r for the stability of the basic state. This results in the following Euler-Lagrange equations:

$$\frac{\partial V_B}{\partial x}v - (\mathrm{RaPr}\cos\phi(1+\epsilon\sin(\omega t)) + \gamma)\theta - 2\mathrm{Pr}\nabla^2 u + \frac{\partial \lambda}{\partial x} = 0, \tag{23}$$

$$\frac{\partial V_B}{\partial x}u - \mathrm{RaPr}\sin\phi(1+\epsilon\sin(\omega t))\theta - 2\mathrm{Pr}\nabla^2 v + \frac{\partial \lambda}{\partial y} = 0, \tag{24}$$

$$-2\mathrm{Pr}\nabla^2 w + \frac{\partial \lambda}{\partial z} = 0, \tag{25}$$

$$-\mathrm{RaPr}(u\cos\phi + v\sin\phi)(1+\epsilon\sin(\omega t)) - \gamma u - 2\gamma\nabla^2\theta = 0, \tag{26}$$

$$\frac{\partial u}{\partial x} + \frac{\partial v}{\partial y} + \frac{\partial w}{\partial z} = 0, \tag{27}$$

where λ denotes the Lagrange multiplier associated with the equation of continuity (9). To solve the Euler-Lagrange equations, the following solution form is considered.

$$(u, v, w, \theta, \lambda) = (\hat{u}(x,t), \hat{v}(x,t), \hat{w}(x,t), \hat{\theta}(x,t), \hat{\lambda}(x,t))\exp\{\iota(l_1 y + l_2 z)\} + \text{c.c.}, \tag{28}$$

where l_1 and l_2 denote the wavenumbers in y and z direction, respectively. This results in the following system.

$$\frac{\partial V_B}{\partial x}\hat{v} - (\mathrm{RaPr}\cos\phi(1+\epsilon\sin(\omega t)) + \gamma)\hat{\theta} - 2\mathrm{Pr}(D^2 - a^2)\hat{u} + D\hat{\lambda} = 0, \tag{29}$$

$$\frac{\partial V_B}{\partial x}\hat{u} - \mathrm{RaPr}\sin\phi(1+\epsilon\sin(\omega t))\hat{\theta} - 2\mathrm{Pr}(D^2 - a^2)\hat{v} + \iota l_1\hat{\lambda} = 0, \tag{30}$$

$$-2\mathrm{Pr}(D^2 - a^2)\hat{w} + \iota l_2\hat{\lambda} = 0, \tag{31}$$

$$-\mathrm{RaPr}(\hat{u}\cos\phi + \hat{v}\sin\phi)(1+\epsilon\sin(\omega t)) - \gamma\hat{u} - 2\gamma(D^2 - a^2)\hat{\theta} = 0, \tag{32}$$

$$D\hat{u} + \iota l_1\hat{v} + \iota l_2\hat{w} = 0. \tag{33}$$

Here, $a_2^2 = l_1^2 + l_2^2$. The boundary conditions are given by

$$(\hat{u}, \hat{v}, \hat{w}, \hat{\theta}) = (0, 0, 0, 0) \text{ at } x = \mp\frac{1}{2}. \quad (34)$$

The above system (29)–(33) of the ordinary differential equations with the boundary conditions (34) is solved by using the shooting method [3, 13]. Define

$$\tilde{Ra}_M(t) = \max_{\lambda} \min_{l_1, l_2} Ra(\phi, \Pr, \omega, \epsilon, t, l_1, l_2, \lambda) \quad (35)$$

Here, t treated as a parameter, refers to the instant of time at which the perturbations are imposed on the basic state. $\tilde{Ra}_M(t)$ is found to be a periodic function of t with period $\frac{2\pi}{\omega}$. The critical value of the Rayleigh number is defined as

$$Ra_M = \min_{t \in [0, \frac{2\pi}{\omega}]} \tilde{Ra}_M(t) \quad (36)$$

The values of l_1, l_2, a_2, t and γ at which Ra_M is attained, are denoted by l_{1M}, l_{2M}, a_{2M}, t_M and γ_M, respectively. We have observed that the optimal value of Ra occurs at $l_{1M} = 0$, hence $a_{2M} = l_{2M}$.

4 Results and Discussion

We have obtained numerically the critical Rayleigh numbers, Ra_L corresponding to (15) (linear stability) and corresponding to (36) (energy stability). The value of the Prandtl number is fixed as 0.71, which corresponds to the Prandtl number of air.

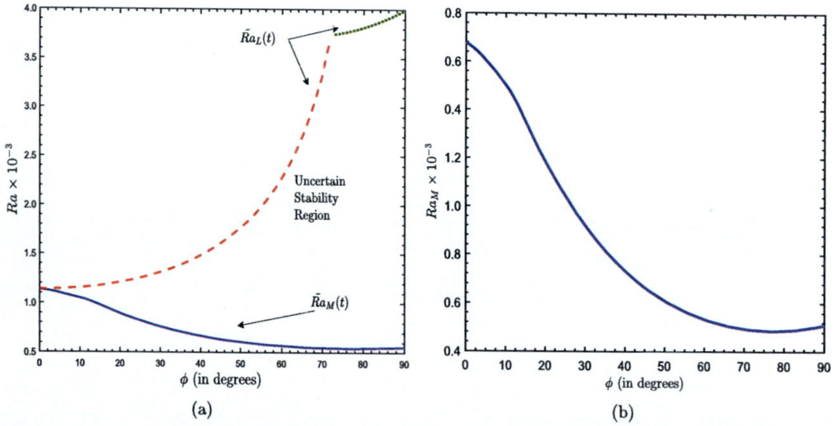

Fig. 1 For $\Pr = 0.71$, $\epsilon = 0.5$ and $\omega = 10$, **a** variation of $\tilde{\text{Ra}}_L(t)$ and $\tilde{\text{Ra}}_M(t)$ in (ϕ, Ra) plane at fixed values $t^* = \frac{t\omega}{2\pi} = 0.5$, **b** variation of Ra_M with the angle of inclination ϕ of the layer

Figure 1a depicts the variation of $\tilde{Ra}_L(t)$ and $\tilde{Ra}_M(t)$ with the angle ϕ of inclination of the fluid layer at fixed values of $t^*(=\frac{t\omega}{2\pi}) = 0.5$, $\omega = 10$ and $\epsilon = 0.5$. Here, $t^* = 0.5$ is the value of t^* at which $\tilde{Ra}_L(t)$ and $\tilde{Ra}_M(t)$ are minimum for a horizontal fluid layer. Dashed and dotted lines represent the variation of $\tilde{Ra}_L(t)$ and solid line shows the variation corresponding to $\tilde{Ra}_M(t)$. For $0 < \phi < 73°$, the instability occurs as longitudinal stationary mode and for $73° \leq \phi \leq 90°$, the preferred mode of instability is transverse stationary mode. The mode of instability changes from longitudinal stationary mode to transverse stationary mode at $\phi = 72°$ in the absence of modulation, i.e., $\epsilon = 0$ (see [3]). The value of $\tilde{Ra}_L(t)$ increases with the angle ϕ for $\phi \in [0, 90°]$. Also, $\tilde{Ra}_M(t)$ decreases with ϕ upto certain angle of inclination where it attains its minimum and then increases slightly with further increase in ϕ. Figure 1b shows the variation of Ra_M with the angle ϕ of inclination of fluid layer. The minimum value of Ra_M (from (36)) is attained at $\phi = \phi_c = 77°$.

To study the effect of the amplitude ϵ of modulation on the stability of the basic flow, the variation of the critical values of Ra_L and Ra_M with ϵ is represented in Fig. 2 for $\phi = 0,\ 10°$ and $90°$. The basic state is globally stable when $Ra \leq Ra_M$, it is linearly stable when $Ra \leq Ra_L$. We cannot predict the stability of the basic state in the region between the linear instability boundary and the global stability boundary (also termed as nonlinear stability boundary). This region is called uncertain stability region. Subcritical instabilities may be present in this region.

In Fig. 2, for $\phi = 0$, the linear and the nonlinear boundaries coincide. This is due to the fact that the linear operator of perturbation equations is self-adjoint for $\phi = 0$ [16]. We get distinct linear and nonlinear boundaries for $\phi = 10°$ and $90°$, which results in the existence of the uncertain stability region as shown in Fig. 2b and 2c. The values of Ra_L and Ra_M both decrease with increase in ϵ.

The variation of Ra_L and Ra_M with the frequency of the sinusoidal gravity modulation is shown in Fig. 3. For $\epsilon = 0.5$ and $\phi = 0$, the values of Ra_L and Ra_M are equal, i.e., $Ra_L = Ra_M = 1138.51$ for all values of ω. Figure 3a depicts that for $\phi = 10°$, Ra_L is equal to 1156.07 for all values of ω while Ra_M increases with increase in the value of ω. For $\omega \geq 20$, the variation of Ra_M is very small with increase in ω. We have checked numerically that for $\epsilon = 0.5$, Ra_L does not vary with ω for $\phi \in (0, 72°)$ but its value increases with ω for $\phi > 72°$. For $\phi = 90°$, the linear stability boundary and the nonlinear stability boundary in (ω, Ra) plane are shown in Fig. 3b. The value of both Ra_L and Ra_M increases with increase in ω. We have observed that the rate of increase of Ra_L is relatively large as compared to that of Ra_M.

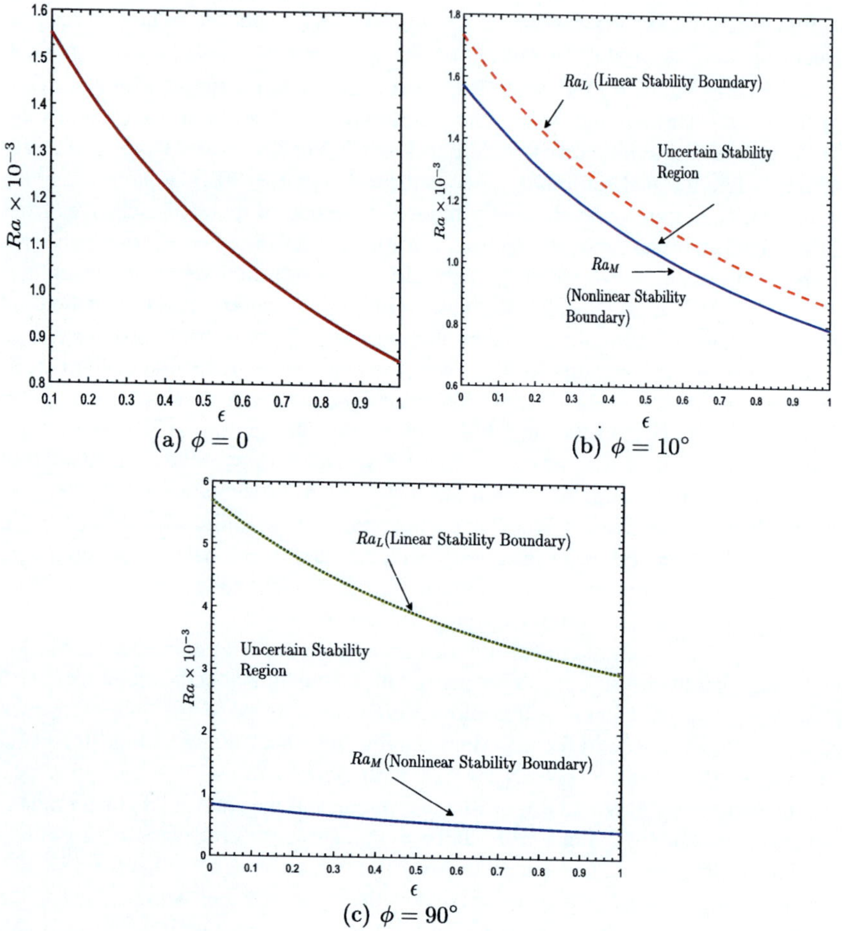

Fig. 2 At $\omega = 10$, the variation of Ra_L and Ra_M with the amplitude ϵ of modulation for **a** $\phi = 0$, **b** $\phi = 10°$ and **c** $\phi = 90°$

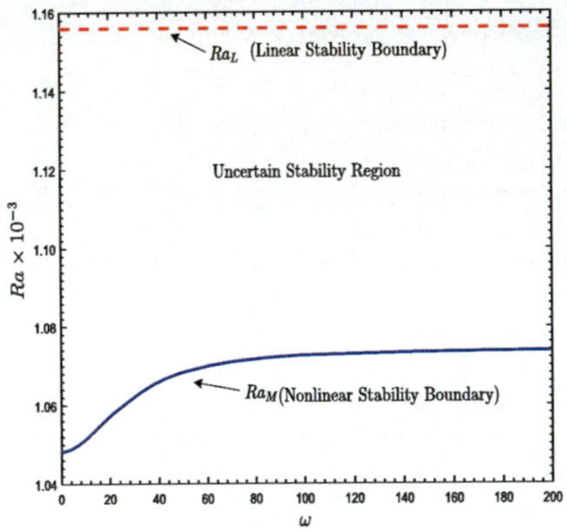

(a) $\phi = 10°$

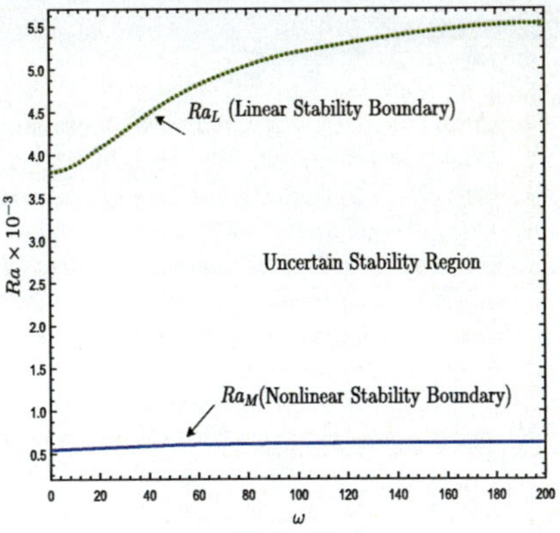

(b) $\phi = 90°$

Fig. 3 At $\epsilon = 0.5$, variation of Ra_L and Ra_M in (ω, Ra) plane for **a** $\phi = 10°$, **b** $\phi = 90°$

5 Conclusions

The stability of natural convection in an inclined fluid layer is studied under the effect of sinusoidal gravity modulation. We have obtained analytically the basic state. The basic velocity oscillates time periodically with the driving frequency. The global stability results are obtained by using the energy method. In the present analysis, all the results are obtained for $\mathrm{Pr} = 0.71$. We have obtained the following conclusions:

- A sufficient condition for the stability of the basic flow is obtained using energy method.
- The critical value of the control parameter depends upon the instant of time t at which the basic flow is perturbed.
- Using energy method, $\phi = 77°$ is found be the least stable configuration with respect to arbitrary perturbations of inclined fluid layer for fixed values of $\omega = 10$ and $\epsilon = 0.5$.
- Uncertain stability regions are observed in (ϕ, Ra), (ϵ, Ra) and (ω, Ra) parametric space.
- At fixed values of ϕ and ω, we have observed decrement in the values of Ra_L and Ra_M with increase in the amplitude ϵ of gravity modulation.
- With increase in the value of the frequency ω of the modulation, the critical value of Rayleigh number Ra_M increases. However, the increase is significant upto $\omega = 20$ only.

Acknowledgements Financial aid from the University Grants Commission (UGC), New Delhi through grant number F.16-6(DEC.2016)2017(NET)(403462) to Manisha Arora in the form of Senior Research Fellowship is gratefully acknowledged.

References

1. Arora, M., Bajaj, R.: Stability of transient natural convection in impulsively heated inclined fluid layer. Fluid Dyn. Res. **52**(5), 055501 (2020)
2. Arora, M., Bajaj, R.: Global stability of natural convection in internally heated inclined fluid layer. J. Eng. Math. **128**(1), 7 (2021)
3. Arora, M., Singh, J., Bajaj, R.: Nonlinear stability of natural convection in an inclined fluid layer. Int. J. Appl. Comput. Math **6**, 21 (2020)
4. Bajaj, R.: Thermo-magnetic convection in ferrofluids with gravity-modulation. Indian J. Eng. Mater. Sci. **10**, 282–291 (2003)
5. Bajaj, R.: Thermodiffusive magneto convection in ferrofluids with two-frequency gravity modulation. J. Magn. Magn. Mater. **288**, 483–494 (2005)
6. Chandrasekhar, S.: Hydrodynamic and Hydromagnetic Stability. Oxford University Press, Oxford (1961)
7. Chen, W.Y., Chen, C.: Effect of gravity modulation on the stability of convection in a vertical slot. J. Fluid Mech. **395**, 327–344 (1999)
8. Farooq, A., Homsy, G.: Linear and nonlinear dynamics of a differentially heated slot under gravity modulation. J. Fluid Mech. **313**, 1–38 (1996)
9. Gresho, P.M., Sani, R.L.: The effects of gravity modulation on the stability of a heated fluid layer. J. Fluid Mech. **40**(4), 783–806 (1970)
10. Hart, J.E.: Stability of the flow in a differentially heated inclined box. J. Fluid Mech. **47**(3), 547–576 (1971)
11. Homsy, G.M.: Global stability of time-dependent flows. part 2. modulated fluid layers. J. Fluid Mech. **62**(2), 387–403 (1974)
12. Kaloni, P., Qiao, Z.: Non-linear convection in a porous medium with inclined temperature gradient and variable gravity effects. Int. J. Heat Mass Transf. **44**(8), 1585–1591 (2001)
13. Kincaid, D., Kincaid, D.R., Cheney, E.W.: Numerical analysis: mathematics of scientific computing, vol. 2, American Mathematical Society (2009)
14. Lappa, M.: Thermal Convection: Patterns, Evolution, and Stability. Wiley (2010)
15. Saravanan, S., Meenasaranya, M.: Energy stability of modulation driven porous convection with magnetic field. Meccanica **56**, 2777–2788 (2021)
16. Straughan, B.: The energy method, stability, and nonlinear convection, vol. 91, Springer, Berlin (1992)

Dynamical Study of an Epidemiological Model with Harvesting and Infection in Prey Population

Smriti Chandra Srivastava and Nilesh Kumar Thakur

Abstract The analysis of prey–predator in an eco-epidemiological system has become the major concern of scientific research in the field of mathematics and disease dynamical studies. Our studies concern with three-tier species model system when infection is spreading among the prey populations. The impact of infection affecting population dynamics is more complicated studies in natural dynamics. Therefore, we investigate an eco-epidemiological model system's local and global stability around the biologically feasible equilibrium point. In order to analyze the local and global stability of the model system, we perform a detailed numerical experiments. We analyze the resulting model through various mathematical characteristics like boundedness, global stability, local stability, and bifurcation. We further investigate time evaluation, phase portraits, and bifurcation diagrams and results show the complexity of the eco-epidemiological system. The analytical results are verified through simulations.

Keywords Bifurcation · Eco-epidemic · Harvesting · Prey-predator · Stability

1 Introduction

The study of prey-predator systems dynamics is one of the predominant research area in mathematical ecology, epidemiology, and marine system. Many researchers have studied the marine ecological complexities that arise in the ocean. Because of the universality and importance of the aquatic, the prey-predator model may help to investigate the complexities and salvaging marine animals. In nature, millions of fish die every year with different causes in the aquatic dynamical system. Overexploitation of biological resources and overfishing are responsible for the removable

S. C. Srivastava (✉) · N. K. Thakur
National Institute of Technology Raipur, Chhatisgarh 492010, India
e-mail: ssmriti.srivastava@gmail.com

N. K. Thakur
e-mail: nkthakur.maths@nitrr.ac.in

R. K. Sharma et al. (eds.), *Frontiers in Industrial and Applied Mathematics*,
Springer Proceedings in Mathematics & Statistics 410,
https://doi.org/10.1007/978-981-19-7272-0_28

of marine resources, so the harvesting policy may play a significant role in the fishery and agriculture sectors to save species at the risk of extinction. According to the United Nations (UN) food and agriculture organization, the fishery is a sustainable livelihood, and the recent coverage shows approximately 39 million people are engaged in the fishery, and 20.5 million people are involved in aquaculture. As a result of fish exports in 2018, approximately USD 164 billion worth of revenue was earned from the global seafood industry, compared to 93 million tonnes in 2017 [1]. However, it comes with several challenges. Ecological and economic studies has been focused on the issue of reasonable harvesting policy for a long time because it has proved to be an exciting challenge. Disease in the fish [2] is a severe problem in fish production and revenue earning. The increasing aquatic fish diseases are believed to be caused by an unbalanced environment like climate change, global warming, and industrial pollutants [3, 4].

We may also include the economic perspective in fishery that contains both the susceptible as well as infected fish. According to a fact, fish with the disease are more likely to be caught in fisheries. Therefore, it is essential to analyze the fishery models mathematically in economic and ecological aspects in the presence of infection. The literature show that much remarkable works on eco-epidemiolocal modelling have considered the ecology and economic aspects [5–8]. Lv et al. [9] examined the influence of harvesting on a phytoplankton-zooplankton system. They determined that excessive harvesting could destroy population viability, while proper harvesting ensures the population's survival. Pei et al. [10] developed a model of planktonic harvesting with two species of zooplankton. Lafferty et al. [11] conducted a study on the infected killifish (Fundulus parvipinnes) and found these fishes more vulnerable to the bird predators as they tend to stay closer to the sea surface. Upadhyay et al. [12] studied the emergence of spatial patterns and developed a spatial model considering the Tilapia and Pelican population in a damaged diffusive eco-epidemiological system. The results suggested that removing infected Tilapia at regular time intervals and controlling salinity can help restore the system, providing a conservation strategy point of view. Further, Upadhyay et al. [13] studied an ecosystem of the Salton Sea in the crisis and suggested that it is essential to plan for a transition at the ecosystem level so that human and bird residents living adjacent to the shrinking and salinization sea are not adversely affected. Recently, Pal et al. [14] studied the eco-epidemiological problem by combining harvesting in the prey-predator model with prey refuge and predator switching phenomena. Motivated by the above research, we have developed a modified model by introducing harvesting efforts in prey populations and intra-specific competition among predators[12].

There are the following sections in the paper: in Sect. 2, we describe the model formulation and its parameters in detail. in Sect. 3, we study the boundedness, equilibrium, local stability and global stability for the proposed model. Furthermore, in Sect. 4, we carry out simulation and its discussion to verify the analytical findings and in Sect. 5, we include a summary of the paper with result interpretation and discussion at the end of the paper.

2 Model Formulation

We present an eco-epidemiological system where the fish population is infected by viruses. Viruses may lead to two categories of fish populations: susceptible fish and infected fish. Therefore, the total fish population at time t can be expressed as $N(t) = S(t) + I(t)$, where $N(t)$ is the fish population, $S(t)$ is the susceptible fish population, and $I(t)$ is the infected fish population. It is assumed that only susceptible fish populations can reproduce, but the infected fish populations die before reproducing. However, the infected fish populations still contribute to the susceptible S in population growth towards their carrying capacity. The transmission of viruses from infected fish species to susceptible fish is described by Holling type II function response. Spread of disease is assumed in the prey population only that is not inherited genetically. The relationship between (S, I, P) is defined by the Holling type II function response. Viruses transmit among fish species at the rate of β while C_1 is a half-saturation constant. Birds predate the infected fish population with the rate of ω_2 and predate the susceptible fish population with the rate of ω_1 while C_2 is the half-saturation constant. The present model is based on the following assumptions: (a) bird populations are forced to compete for food (intra-specific competition) when there is a shortage of fish, so bird population competed for food at the rate of ϵ (intra-specific coefficient), and (b) harvesting efforts are implemented in the fish population to eradicate the infection from the dynamics and stabilize it. The catchability coefficient represents q_1 and q_2 for susceptible and infected fish, respectively, along with the harvesting effort E introduced in both susceptible and infected populations. Although virus-induced fish are more vulnerable and easy to pick therefore we assume that $q_2 > q_1$. With this assumption the model system is as follows:

$$
\begin{aligned}
\frac{dS}{dt} &= rS\left(1 - \frac{S+I}{k}\right) - \frac{\beta SI}{S+I+C_1} - \frac{\omega_1 SP}{S+C_2} - q_1 ES,\\
\frac{dI}{dt} &= \frac{\beta SI}{S+I+C_1} - \frac{\omega_2 IP}{I+C_2} - \delta I - q_2 EI,,\\
\frac{dP}{dt} &= \frac{\omega_3 SP}{S+C_2} + \frac{\omega_4 IP}{I+C_2} - eP - \epsilon P^2,
\end{aligned}
\tag{1}
$$

with initial conditions $S(0) > 0$, $I(0) > 0$ and $P(0) > 0$.

3 Stability Analysis

In this section, we investigate the boundedness, existence criteria and the linear stability analysis of the model system (1).

Theorem 1 *Let $E < \frac{r}{q_1}$, the system (1) has non-negative and unique solution uniformly bounded by octant*

$$\phi = \left\{(S, I, P) \in T_+^3;\ S(t) + I(t) + \frac{\omega_1}{\omega_3} P(t) = \frac{k}{4r\eta}(\eta + r - q_1 E)^2 + \in, \in > 0\right\},$$

where, $\eta \le min\ (\delta + q_2 E, e)$ and $\omega_2\omega_3 > \omega_1\omega_4$.

The proof of this theorem can be obtained by simple calculation and hence omitted. To evaluate the local behaviour of model system (1), variational matrix for the model system (1) to understand whether the model (1) is locally stable at each equilibrium point $L(S, I, P)$ can be written as follows

$$J(L) = \begin{pmatrix} S\frac{\partial h_1}{\partial S} + h_1 & S\frac{\partial h_1}{\partial I} & S\frac{\partial h_1}{\partial Z} \\ I\frac{\partial h_2}{\partial S} & I\frac{\partial h_2}{\partial I} + h_2 & I\frac{\partial h_2}{\partial Z} \\ P\frac{\partial h_3}{\partial S} & P\frac{\partial h_3}{\partial I} & P\frac{\partial h_3}{\partial Z} + h_3 \end{pmatrix},$$

$$\frac{\partial h_1}{\partial S} = -\frac{r}{k} + \frac{\beta I}{(S + I + C_1)^2} + \frac{\omega_1 P}{(S + C_2)^2}, \frac{\partial h_1}{\partial I} = -\frac{r}{k} - \frac{\beta (S + C_1)}{(S + I + C_1)^2},$$
$$\frac{\partial h_1}{\partial P} = -\frac{\omega_1}{(S + C_2)}, \frac{\partial h_2}{\partial S} = \frac{\beta (I + C_1)}{(S + I + C_1)^2}, \frac{\partial h_2}{\partial I} = -\frac{\beta S}{(S + I + C_1)^2} + \frac{\omega_2 P}{(I + C_2)},$$
$$\frac{\partial h_2}{\partial P} = -\frac{\omega_2}{(I + C_2)}, \frac{\partial h_3}{\partial S} = \frac{\omega_3 C_2}{(S + C_2)^2}, \frac{\partial h_3}{\partial I} = \frac{\omega_4 C_2}{(I + C_2)^2}, \frac{\partial h_3}{\partial P} = -\epsilon.$$

There are five equilibrium points that exist.

(i) The trivial equilibrium point $L_0(0, 0, 0)$ always exists and corresponding eigenvalues of $J(L_0)$ are $(r - q_1 E, -(\delta + q_2 E), -e)$. Clearly the system is stable manifold in IP direction. If the system holds $r \le q_1 E$, then the system is stable around $L_0 = (0, 0, 0)$, otherwise, unstable or saddle.

(ii) The infected and predator-free equilibrium point $L_1(S, 0, 0)$, where $S = \frac{k(r - q_1 E)}{r}$ exist if the condition $r > q_1 E$ hold. Eigenvalues of $J(L_1)$ are $\left(-(r + q_1 E), -(\delta + q_2 E) + \frac{\beta S}{S + C_1}, -e + \frac{w_3 S}{S + C_2}\right)$. Therefore, L_1 is locally asymptotically stable provided

$$\frac{\beta k}{(k + C_1)(\delta + q_2 E)} \le 1, \tag{2}$$

and

$$\frac{w_2 k}{(k + C_2)\, e} \le 1, \tag{3}$$

and L_1 is a saddle point if at least one condition (2) and (3) hold.

(iii) The predator-free equilibrium point $L_2(S', I', 0)$, where

$$I' = \frac{S'(\beta - \phi) - C_1\phi}{\phi}, \&S' = \frac{\phi(r(k + C_1) - k(\beta - \phi)) + \sqrt{\gamma}}{2\beta r}, \tag{4}$$

where $\phi = q_2 E + \delta$ and $\gamma = k(\beta - \phi) - r(C_1 + k) + 4rkC_1(\beta - q_1 E)$ exist if the following conditions $S' \geq \frac{C_1\phi}{\beta-\phi}$ and $\beta \geq \phi$ hold.

The Jacobian matrix with respect to the point L_2 is

$$J(L_2) = \begin{pmatrix} \frac{\beta S'I'}{(S'+I'+C_1)^2} - \frac{rS'}{k} & -\frac{rS'}{k} - \frac{\beta S'(S'+C_1)}{(S'+I'+C_1)^2} & -\frac{w_1 S'}{(S'+C_2)} \\ \frac{\beta I'(I'+C_1)}{(S'+I'+C_1)^2} & -\frac{\beta S'I'}{(S'+I'+C_1)^2} & \frac{-w_2 I'}{I'+C_2} \\ 0 & 0 & \frac{\omega_3 S'}{S'+C_2} + \frac{\omega_4 I'}{I'+C_2} - e \end{pmatrix}.$$

Clearly, one eigenvalue corresponding to $J(L_2)$ is given as $\frac{\omega_3 S'}{S'+C_2} + \frac{\omega_4 I'}{I'+C_2} - e$. The eigenvalue is negative or positive, depending on following nature:

$$\frac{\omega_3 S'}{S' + C_2} + \frac{\omega_4 I'}{I' + C_2} < e, \tag{5}$$

or,

$$\frac{\omega_3 S'}{S' + C_2} + \frac{\omega_4 I'}{I' + C_2} > e. \tag{6}$$

The other eigenvalues are root of sub matrix

$$J(L^{21}) = \begin{pmatrix} \frac{\beta S'I'}{(S'+I'+C_1)^2} - \frac{rS'}{k} & -\frac{rS'}{k} - \frac{\beta S'(S'+C_1)}{(S'+I'+C_1)^2} \\ \frac{\beta I'(I'+C_1)}{(S'+I'+C_1)^2} & -\frac{\beta S'I'}{(S'+I'+C_1)^2} \end{pmatrix}, \tag{7}$$

The eigenvalues of submatrix $J(L^{21})$ have negative value if $tr(J(L^{21})) < 0$ and $det(J(L^{21})) > 0$. Therefore,

$$tr(J(L^{21})) = -\frac{rS'}{k} < 0, \tag{8}$$

and

$$det(J(L^{21})) = \frac{\beta S'I'(r(S' + I' + \beta)^2 + \beta C_1 k)}{(S' + I' + C_1)^3} > 0. \tag{9}$$

Thus, the predator-free equilibrium point L_2 is locally asymptotically stable if conditions Eqs. (5), (8) and (9) hold.

(iv) The infection-free equilibrium point $L_3(\bar{S}, 0, \bar{P})$, where

$$\bar{S} = \frac{C_2\left(e+\epsilon\bar{P}\right)}{\omega_3-\left(e+\epsilon\bar{P}\right)}, \& \bar{P} = \frac{\left(k\left(r-q_1E\right)-r\bar{S}\right)\left(S+C_2\right)}{\omega_1 K}, \tag{10}$$

exist if the conditions $w_3 \leq \left(e+\epsilon\bar{P}\right)$ and $k \leq \frac{r\bar{S}}{r-q_1E}$ hold.
The Jacobian matrix with respect to the point L_3 is

$$J(L_3) = \begin{pmatrix} r\left(1-\frac{2\bar{S}}{k}\right)-\frac{\omega_1\bar{P}C_2}{(\bar{S}+C_2)^2}-q_1E & -\frac{r\bar{S}}{k}-\frac{\beta\bar{S}(C_1-\bar{S})}{(\bar{S}+C_1)^2} & -\frac{\omega_1\bar{S}}{(\bar{S}+C_2)} \\ 0 & \frac{\beta\bar{S}(\bar{S}+C_1)}{(\bar{S}+C_1)^2}-\frac{w_4\bar{P}}{C_2}-(\delta+q_1E) & 0 \\ \frac{C_2w_3\bar{P}}{(\bar{S}+C_2)^2} & -\frac{w_4\bar{P}}{C_2} & 0 \end{pmatrix}.$$

Clearly, one eigenvalue corresponding to L_3 is given as $\frac{\beta\bar{S}(\bar{S}+C_1)^2}{S+C_1}-\frac{\omega_2\bar{P}}{C_2}-(q_1E+\delta)$. The eigenvalue is negative or positive, depending on following nature:

$$\frac{\beta\bar{S}\left(\bar{S}+C_1\right)^2}{\bar{S}+C_1} > \frac{\omega_2\bar{P}}{C_2}+(q_1E+\delta), \tag{11}$$

or,

$$\frac{\beta\bar{S}\left(\bar{S}+C_1\right)^2}{\bar{S}+C_1} < \frac{\omega_2\bar{P}}{C_2}+(q_1E+\delta). \tag{12}$$

The other eigenvalues are root of sub matrix

$$J(L^{31}) = \begin{pmatrix} \frac{r(k-2\bar{S})}{k}-\frac{\omega_1\bar{P}C_2}{(\bar{S}+C_2)^2}-q_1E & -\frac{\omega_1\bar{S}}{(\bar{S}+C_2)} \\ \frac{C_2\omega_3\bar{P}}{(C_2+\bar{S})^2} & 0 \end{pmatrix}. \tag{13}$$

The eigenvalues of submatrix $J(L^{31})$ have negative value if $tr(J(L^{31})) < 0$ and $det(J(L^{31})) > 0$. Therefore,

$$tr(J(L^{31})) = r\left(1-\frac{2\bar{S}}{k}\right)-\frac{\omega_1C_2\bar{P}}{\left(\bar{S}+C_2\right)^2} < 0, \tag{14}$$

and

$$det(J(L^{21})) = \frac{\omega_1\omega_3C_2\bar{P}\bar{S}}{\left(\bar{S}+C_2\right)^4} > 0. \tag{15}$$

Thus, the infection-free equilibrium point L_3 is locally asymptotically stable if conditions Eqs. (12), (14) and (15) hold.

(v) The non-trivial equilibrium point $L_4(S^*, I^*, P^*)$ has been obtained from model system (1). From straightforward calculation

$$P^* = \frac{1}{\epsilon}\left(\frac{\omega_3 S^*}{S^* + C_2} + \frac{\omega_4 I^*}{I^* + C_2} - e\right). \tag{16}$$

From model system (1), we have

$$F(S^*) = S^{*2} + AS^* + B = 0, \tag{17}$$

where

$$A = \frac{(I + C_2)(C_2(\epsilon\beta - e\omega_2) - (\delta I + q_2 E))}{(I + C_2)(\epsilon\beta - e\omega_2 + (\delta I + q_2 E)(I + C_2)) - \omega_2\omega_4}$$
$$- \frac{\left(I + C_1(\omega_2(\omega_4 I - eC_2) + (I + C_2)e\omega) - (\delta I + q_2 E)(I + C_2)^2\right) - \omega_2\omega_4}{(I + C_2)(\epsilon\beta - e\omega_2 + (\delta I + q_2 E)(I + C_2)) - \omega_2\omega_4},$$
$$B = \frac{C_2(I + C_1)(I + C_2(\delta I + q_2 E(I + C_2) - e\omega_2) - \omega_2(\omega_4 I + \omega_3))}{(I + C_2)(\epsilon\beta - e\omega_2 + (\delta I + q_2 E)(I + C_2)) - \omega_2\omega_4}.$$

Roots of Eq. (17) can be represented as $S^* = \frac{-A + \sqrt{A^2 - 4AB}}{2}$ that consist atleast one positive root in the following cases

(i) $A < 0$ and $B < 0$,
(ii) $A < 0$, $B > 0$ and $A^2 - 4B > 0$,
(iii) $A > 0$ and $B < 0$.

In a same manner, we can obtain the roots of $F(I^*)$

$$F(I^*) = I^{*2} + A_1 I^* + B_1 = 0, \tag{18}$$

where,

$$A_1 = \frac{\omega_1 P^* k + 2rS^* + C_1 r + \beta k + k(q_1 E - r)}{r(S^* + C_2)},$$
$$B_1 = \frac{\left(S^* + C_1\left(\omega_1 P^* k + (S^* + C_2)(S^* - k)r\right)\right) + S^* q_1 Ek\left(S^* + C_1 + 2C_2\right)}{r(S^* + C_2)}. \tag{19}$$

Roots of Eq. (18) can be represented as $I^* = \frac{-A_1 + \sqrt{A_1^2 - 4A_1 B_1}}{2}$ that consist atleast one positive root in the following cases

(i) $A_1 < 0$ and $B_1 < 0$,
(ii) $A_1 < 0$, $B_1 > 0$ and $A_1^2 - 4B_1 > 0$,
(iii) $A_1 > 0$ and $B_1 < 0$.

The variational matrix along $L_4(S^*, I^*, P^*)$ is given by

$$J(L^*) = \begin{pmatrix} h_{11} & h_{12} & \varrho_{13} \\ h_{21} & h_{22} & h_{23} \\ h_{31} & h_{32} & h_{33} \end{pmatrix},$$

$$h_{11} = -\frac{rs}{k} + \frac{\beta S^* I^*}{(S^* + I^* + C_1)^2} + \frac{\omega_1 S^* P^*}{(S^* + C_2)^2}, h_{12} = -\frac{rS^*}{k} - \frac{(S^* + C_1) S^* \beta}{(S^* + I^* + C_1)^2},$$
$$h_{13} = -\frac{\omega_1 P^* S^*}{(S^* + C_2)^2}, h_{21} = \frac{\beta I^* (I + C_1)}{(S^* + I^* + C_1)^2},$$
$$h_{22} = -\frac{\beta S^* I^*}{(S^* + I^* + C_1)^2} + \frac{\omega_2 P^* I^*}{(I^* + C_2)^2}, h_{23} = -\frac{\omega_2 I^*}{(I^* + C_2)}, h_{31} = \frac{\omega_3 C_2 P^*}{(S^* + C_2)^2},$$
$$h_{32} = \frac{\omega_4 C_2 P^*}{(I + C_2)^2}, h_{33} = -2\epsilon P^* + \frac{\omega_3 S^*}{(S + C_2)} + \frac{\omega_4 I^*}{(I^* + C_2)} - e.$$

The characteristics equation of $L_4(S^*, I^*, Z^*)$ is given by

$$V^3 + A_1 V^2 + A_2 V + A_3 = 0,$$

where

$$A_1 = -(h_{11} + h_{22} + h_{33}),$$
$$A_2 = (h_{11}h_{22} - h_{12}h_{21}) + (h_{22}h_{33} - h_{23}h_{32}) + (h_{11}h_{33} - h_{13}h_{31}),$$
$$A_3 = (h_{13}h_{22} - h_{12}h_{23}) h_{31} + (h_{11}h_{23} - h_{13}h_{21}) h_{32} + (h_{12}h_{21} - h_{11}h_{22}) h_{33},$$

Theorem 2 *Assume that the $L_4\,(S^*, I^*, P^*)$ is positive equilibrium point of the system (1). This point is locally asymptotically stable when $A_1 > 0$, $A_2 > 0$ and $A_1 A_2 - A_3 > 0$ are satisfied.*

The proof is simple and can be derived from the Routh-Hurwitz criterion.

Theorem 3 *Assume that the positive equilibrium point $L_4(S^*, I^*, P^*)$ is locally asymptotically stable of the model system,then it is a globally stable in the interior of the positive octant (i.e., int R^3_+) provided that*

$$\frac{3r}{2k} + \frac{\beta (S^* + C_1)}{2\rho_{11}} > \frac{w_1 P^*}{2\rho_{22}} + \frac{\beta (k_1 + 2) I^* + k_1 C_1}{2\rho_{11}}, \tag{20}$$

$$\frac{r}{2kk_1} + \frac{\beta (k_1 (2S^* + C_1) + I)}{2k_1\rho_{11}} > \frac{k_1 w_2 P^*}{2\rho_{22}} + \frac{\beta (S^* + C_1)}{2k_1\rho_{11}}, \tag{21}$$

where $\rho_{11} = (S^* + I^* + C_1)(S + I + C_1)$ and $\rho_{22} = (S^* + C_2)(S + C_2)$.

Proof Consider the following positive definite Lyapunov function about the equilibrium point

$$\vartheta\left(S, I, P\right)=\left(S-S^{*}-S^{*}\ln\frac{S}{S^{*}}\right)+k_{1}\left(I-I^{*}-I^{*}\ln\frac{I}{I^{*}}\right)+k_{2}\left(P-P^{*}-P^{*}\ln\frac{P}{P^{*}}\right). \tag{22}$$

Differentiating ϑ with respect to time t along the solution of the system (1), we obtain

$$\frac{d\vartheta}{dt}=\frac{d\vartheta_{1}}{dt}+\frac{d\vartheta_{2}}{dt}+\frac{d\vartheta_{3}}{dt},$$

$$\begin{aligned}\frac{d\vartheta}{dt}&=-L_{11}(S-S^{*})^{2}+L_{12}\left(S-S^{*}\right)\left(I-I^{*}\right)-k_{1}L_{22}(I-I^{*})^{2}-k_{2}L_{33}\left(I-I^{*}\right)^{2}\\&\quad+L_{23}\left(P-P^{*}\right)\left(I-I^{*}\right)+L_{31}\left(S-S^{*}\right)\left(P-P^{*}\right),\end{aligned}$$

where,

$$\begin{aligned}L_{11}&=\frac{r}{k}-\frac{\beta I}{\left(S^{*}+I^{*}+C_{1}\right)\left(S^{*}+I^{*}+C_{1}\right)}+\frac{w_{1}P^{*}}{\left(S^{*}+C_{2}\right)\left(S+C_{2}\right)},\\L_{22}&=\frac{\beta S^{*}}{\left(S^{*}+I^{*}+C_{1}\right)\left(S+I+C_{1}\right)}-\frac{w_{2}P^{*}}{\left(I^{*}+C_{2}\right)\left(I+C_{2}\right)},\\L_{33}&=\epsilon\left(P-P^{*}\right)^{2}, L_{12}=-\frac{r}{k}-\frac{\beta\left(k_{1}\left(I+C_{1}\right)-\left(S^{*}+C_{1}\right)\right)}{\left(S^{*}+I^{*}+C_{1}\right)\left(S+I+C_{1}\right)},\\L_{23}&=\frac{k_{2}w_{4}C_{2}-k_{1}w_{2}\left(I^{*}+C_{1}\right)}{\left(I^{*}+C_{2}\right)\left(I+C_{2}\right)}, L_{31}=\frac{k_{2}w_{3}C_{2}-w_{1}\left(S^{*}+C_{2}\right)}{\left(S^{*}+C_{2}\right)\left(S+C_{2}\right)}.\end{aligned}$$

Sufficient conditions for $\frac{d\vartheta}{dt}$ to be negative definite required that conditions (20) and (21) hold. This proves the result.

4 Numerical Simulation

Numerical simulation has been carried out using Matlab to validate the results obtained analytically.

$$\begin{aligned}&r=2.1, k=100, \beta=1.931, C_{1}=10, C_{2}=10, q_{1}=0.6, q_{2}=0.8, E=0.002,\\&a=0.218, e=1.2, \epsilon=0.004, \omega_{1}=1.02, \omega_{2}=0.1, \omega_{3}=2, \omega_{4}=1.65.\end{aligned} \tag{23}$$

The set of parameters are taken from [12]. Two parameters, carrying capacity and intra-specific coefficients, have an enormous impact on disease dynamics and exhibit diverse behaviour within the eco-epidemiological system (1).

In Fig. 1 we can observe the local stability of the system (1) around the interior equilibrium point. As depicted in Fig. 1, system exhibits stable focus for carrying capacity $k=25$, and for $k=54$ the system (1) shows higher oscillation. When the

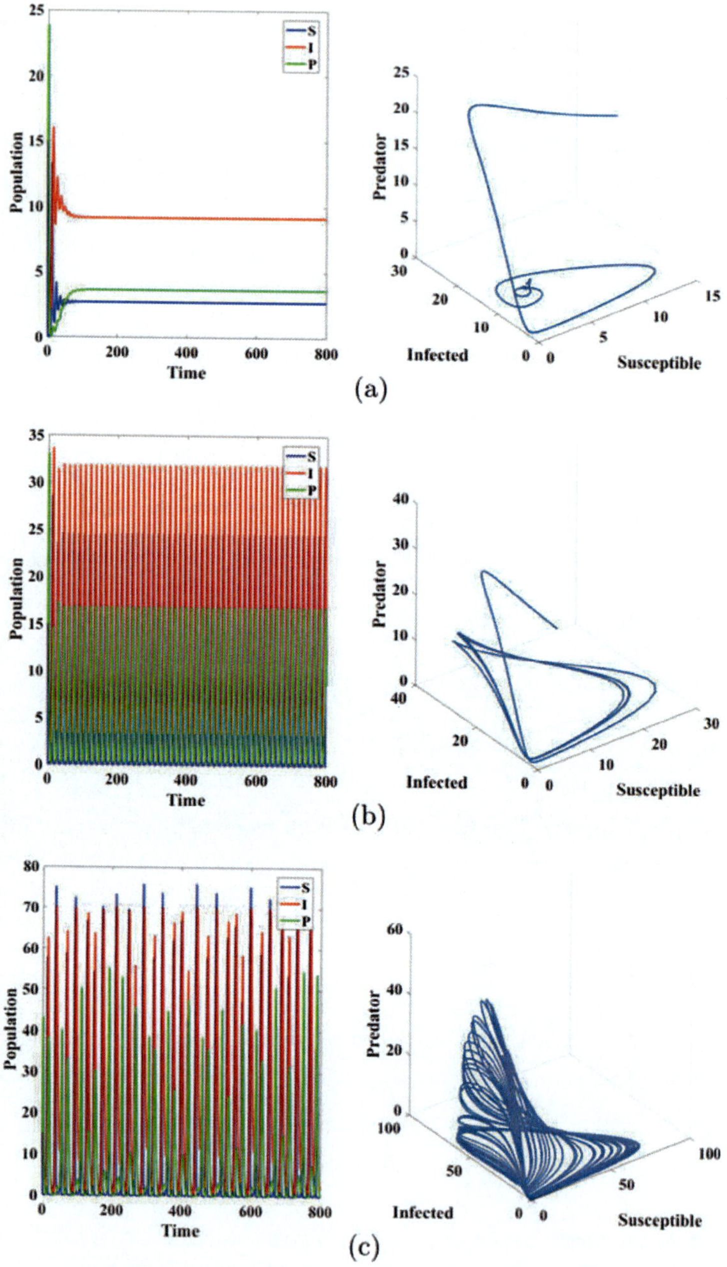

Fig. 1 Solution tragectories of the system (1) converges to a interior equilibrium point **a** $k = 25$, **b** $k = 54$, **c** $k = 100$

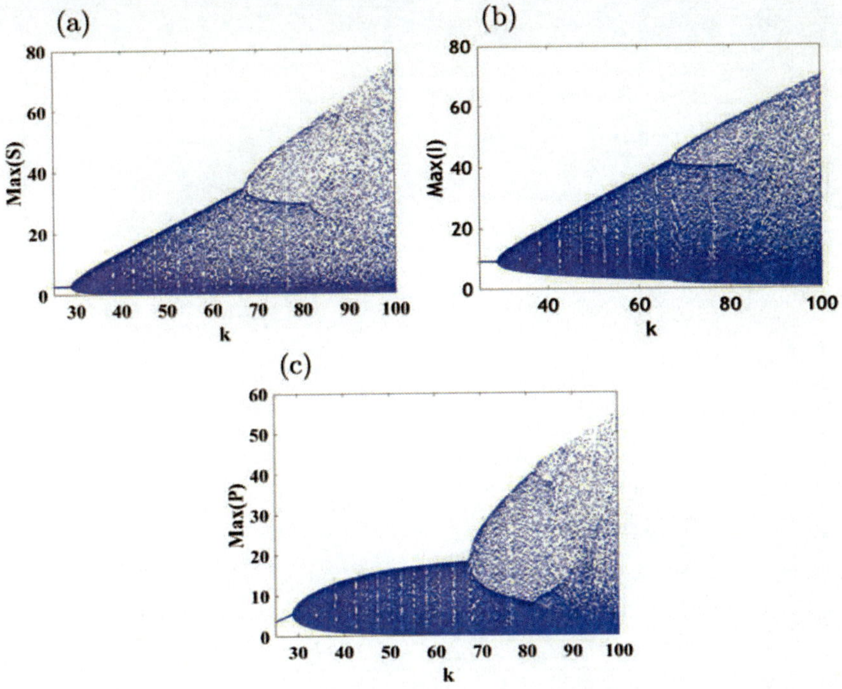

Fig. 2 Bifurcation diagram of model system (1) for k versus **a** Max(S), **b** Max(I), **c** Max(Z)

carrying capacity $k = 100$, the system settles down in chaotic regimes, as depicted in Fig. 1. In the case of $k\epsilon(25, 100)$, Hopf bifurcation occurs around the interior equilibrium point, as illustrated in Fig. 2.

Figure 3, illustrates the changes in fish-bird dynamics in the eco-epidemiological system as the intra-specific coefficient varies from 0 to 0.4. In Fig. 3, the infected fish density increases and predator density decreases as a result of a higher value of the intra-specific coefficient ϵ in the system (1).

We have now examined the proposed model with harvesting efforts, which incorporated into both susceptible and infected prey species. The parameters are the same as in Eq. (23) except the catchability coefficient for susceptible fish population $q_1 = 0.6$, infected fish population $q_2 = 0.8$ and harvesting effort $E = 0.002$. Based on the above discussion, we implicitly take $q_1 < q_2$ that means fish become more vulnerable and more accessible to catch by birds or other predators due to virus infection. As shown in Fig. 5, we have plotted a bifurcation diagram using E (the amount of effort spent in fish harvesting) as a bifurcation parameter. All other parameters are the same as in Eq. (23). In Fig. 5, we see successive maxima in the range of $0 \leq S \leq 100$, $0 \leq I \leq 80$ and $0 \leq P \leq 100$ respectively, as E is taken in the range of $25 \leq E \leq 100$. A bifurcation analysis shows chaotic behaviour (i.e., behaviour that indicates that an unstable system exists) in the range $0 \leq E \leq 0.15$;

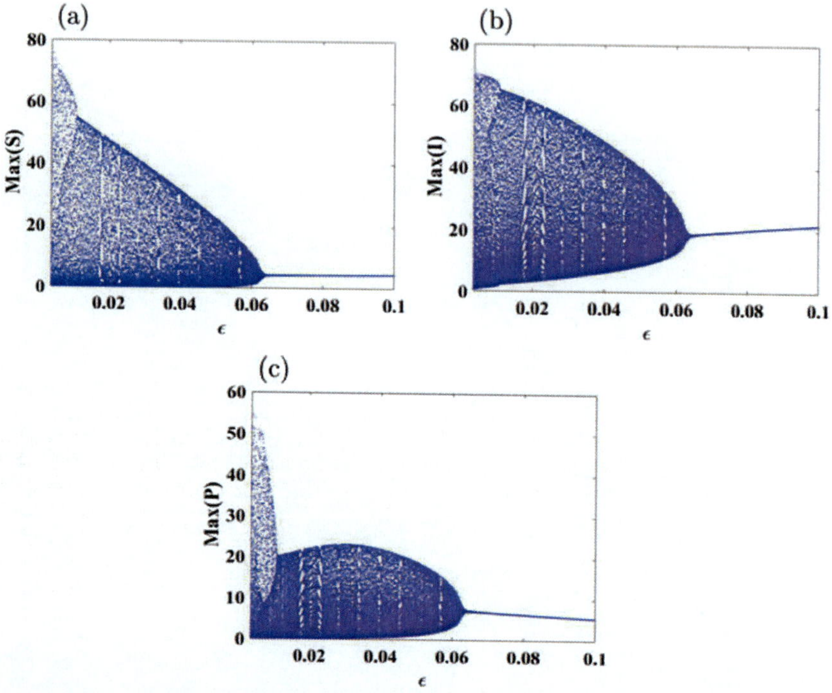

Fig. 3 Bifurcation diagram of model system (1) for ϵ versus **a** Max(S), **b** Max(I), **c** Max(Z)

after crossing $E = 1.85$, the solution trajectory converges to a fixed value, i.e., stable focus in SP direction and Infected go to extinction. In order to validate our results, we plotted the time series and phase portrait for E from 0.004 to 2.0 (c.f., Fig. 4).

5 Discussion and Conclusion

In many studies, prey-predator dynamics have been discussed to emphasize the role of infective populations. The prey-predator dynamics may become complex due to the presence of viruses. Eating infected prey is fatal for predators or other species. In the present paper, we have incorporated fish population harvesting effort in a fish-bird model system, and assumed that viruses infect fish populations. The fish population is divided into two groups based on whether they carry the infection. According to the simulation results, we have found that the following parameters have an essential role in our studies: (i) carrying capacity k, (ii) intra-specific coefficient ϵ, and (iii) harvesting effort E. Figure 1 shows how carrying capacity plays an essential role in fish-bird dynamics, as evidenced by different types of attractors with stable focus, periodic order, and chaotic behaviour. Further, Fig. 2 illustrates that the high carrying capacity increases fish oscillation size and eventually produces

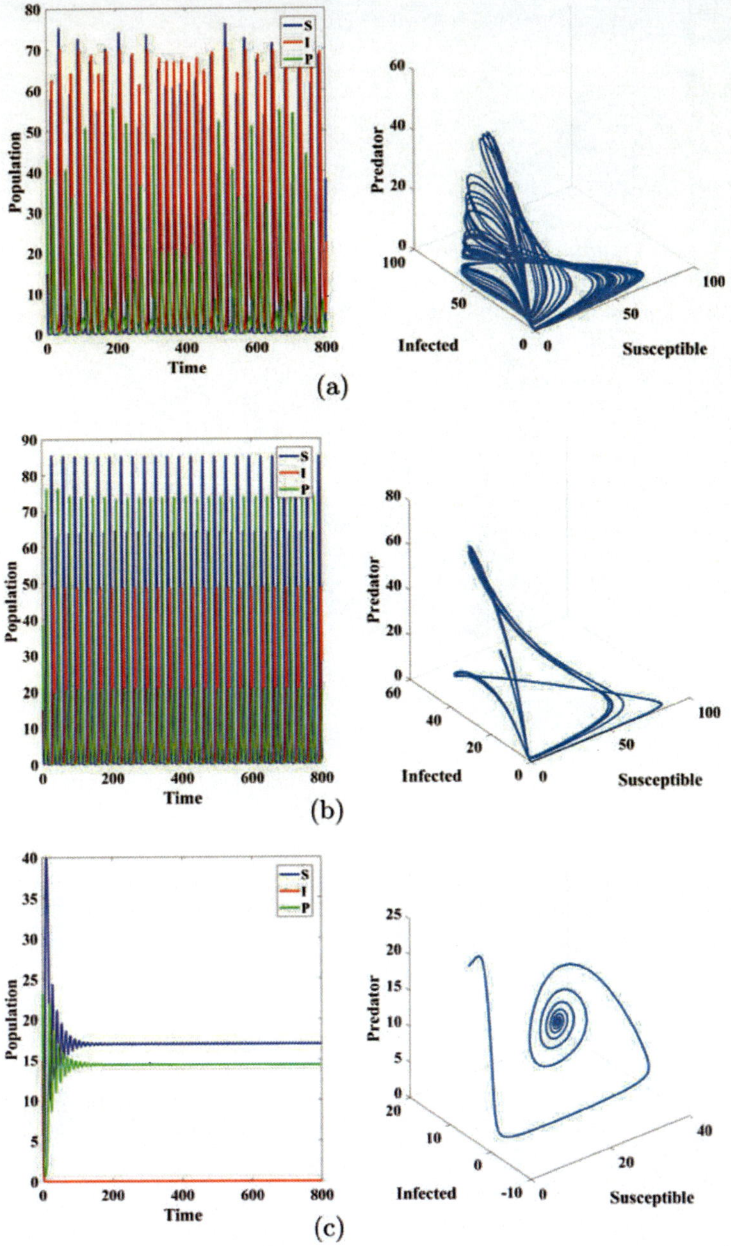

Fig. 4 Solution tragectories of the system (1) converges to a interior equilibrium point **a** $E = 0.004$, **b** $E = 0.4$, **c** $E = 2$

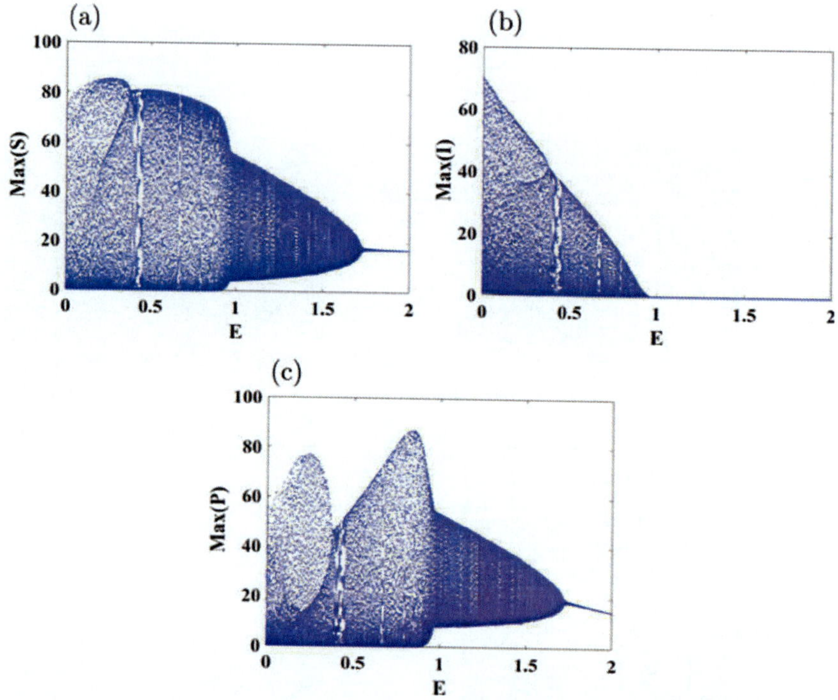

Fig. 5 Bifurcation diagram of model system (1) for E versus **a** Max(S), **b** Max(I), **c** Max(Z)

chaos due to this increment. Additionally, Hopf bifurcation for the parameter, i.e., the intra-specific coefficient, is noticed in Fig. 3. These results show that the parameter ϵ has a negative impact on the fish-bird dynamical system. Additionally, when ϵ crosses some threshold value, susceptible fish settle in stable mode while infected fish have continued growth, but birds populations settle down in decline regimes. Further, a Hopf bifurcation scenario observed for the parameter E, i.e., the harvesting effort for fish (c.f., Fig. 5). These results show that the parameter E impact the system positively when it crosses some threshold values, extinction in infected fish is observed. Finally, this study on the fish-bird dynamics with viruses concluded that these systems with multiple parameters are highly complex and unpredictable. Parameters like the carrying capacity of the fish population and intra-specific rate have generated chaos, whereas harvesting effort generates the extinction of the infected population and stabilizes the system.

References

1. FAO, The state of the world fisheries and aquaculture, FAO (2020). https://doi.org/10.4060/ca9229en
2. Lafferty, K.D., Harvell, C.D., Conrad, J.M., Friedman, C.S., Kent, M.L., Kuris, A.M., Saksida, S.M.: Infectious diseases affect marine fisheries and aquaculture economics. Annual Revi. Marine Sci. **7**, 471–496 (2015)
3. Harvell, C.D., Mitchell, C.E., Ward, J.R., Altizer, S., Dobson, A.P., Ostfeld, R.S., Samuel, M.D.: Climate warming and disease risks for terrestrial and marine biota. Sci. **296**(5576), 2158–2162 (2002)
4. Harvell, C.D., Kim, K., Burkholder, J.M., Colwell, R.R., Epstein, P.R., Grimes, D.J., Hofmann, E.E., Lipp, E.K., Osterhaus, A., Overstreet, R.M., Porter, J.W., Smith, G.W., Vasta, G.R.: Emerging marine diseases - Climate links and anthropogenic factors. Sci. **285**(5433), 1505–1510 (1999)
5. Bairagi, N., Bhattacharya, S., Auger, P., Sarkar, B.: Bioeconomics fishery model in presence of infection: sustainability and demand-price perspectives. App. Math. Comput. **405**, 126225 (2021)
6. Agnihotri, K., Kaur, H.: Optimal control of harvesting effort in a phytoplankton-zooplankton model with infected zooplankton under the influence of toxicity. Math. Comput. Simul. **190**, 946–964 (2021)
7. Chakraborty, S., Pal, S., Bairagi, N.: Dynamics of a ratio-dependent eco-epidemiological system with prey harvesting. Nonlinear Anal. Real World App. **11**(3), 1862–1877 (2010)
8. Pal, A.K., Bhattacharyya, A., Mondal, A., Pal, S.: Qualitative analysis of an eco-epidemiological model with a role of prey and predator harvesting. Z. Naturforsch. (2022). https://doi.org/10.1515/zna-2021-0333
9. Lv, Y., Pei, Y., Gao, S., Li, C.: Harvesting of a phytoplankton-zooplankton model. Nonlinear Anal. Real World Appl. **11**(5), 3608–3619 (2010)
10. Pei, Y., Lv, Y., Li, C.: Evolutionary consequences of harvesting for a two-zooplankton one-phytoplankton system. App. Math. Model. **36**(4), 1752–1765 (2012)
11. Lafferty, K.D., Morris, A.K.: Altered behavior of parasitized killifish increases susceptibility to predation by bird final hosts. Ecol. **77**(5), 1390–1397 (1996)
12. Upadhyay, R.K., Datta, J., Dong, Y., Takeuchi, Y.: Emergence of spatial patterns in a damaged diffusive eco-epidemiological system. Int. J. Bifurc. Chaos. **28**(09), 1830028 (2018)
13. Upadhyay, R.K., Kumari, S., Kumari, S., Rai, V.: Salton sea: an ecosystem in crisis. Int. J. Biomath. **11**(08), 1850114 (2018)
14. Pal, A.K., Bhattacharyya, A., Mondal, A.: Qualitative analysis and control of predator switching on an eco-epidemiological model with prey refuge and harvesting. Results Control Optim. **7**, 100099 (2022)

Joint Decisions on Imperfect Production Process and Carbon Emission Reduction Under Carbon Regulations

Geetanjali Raiya and Mandeep Mittal

Abstract In firms, maintaining the quality of the product with carbon emission reduction is a big concern. To ensure the good quality of the product, so many retailers segregate perfect items from imperfect ones and made an attempt to reduce carbon emissions through green technologies. In the proposed model, the discount price of imperfect items is examined and the retailer's joint decisions have been analyzed on reclamation of inventory and investment in reducing carbon emission under three environmental regulations such as carbon cap, carbon tax, and carbon cap-and-trade. These regulations and understanding of the customer for greener products invigorate retailers to invest in green technology. The total cost is minimized with respect to the optimal order quantity and annual investment on carbon emission reduction. Numerical examples and sensitive analysis are represented to understand the sturdiness of the model.

Keywords Imperfect items · Green technology investment · Carbon regulations · Economic order quantity

1 Introduction

Economic order quantity is the quantity that is used to minimize total costs. Ford W. Haris and R.H. Wilson developed this model in 1913. Bouchery and Dallery [1] consider sustainability in the classical inventory model. Arslan and Turkay [2] have contributed to the Economic order quantity model by including sustainability considerations which embrace environmental and social criteria with standard economic consideration. Wang et al. [3] developed an EOQ model with renewal reward theory to derive the expected total profit per unit time. Lee et al. [4] developed a model

G. Raiya · M. Mittal (✉)
Amity Institute of Applied Sciences, Amity University, Noida, Uttar Pradesh 201313, India
e-mail: mittal_mandeep@yahoo.com

R. K. Sharma et al. (eds.), *Frontiers in Industrial and Applied Mathematics*,
Springer Proceedings in Mathematics & Statistics 410,
https://doi.org/10.1007/978-981-19-7272-0_29

for sustainable economic order quantity with stochastic lead time and multi-model transportation options. Sheikh et al. [5] developed two EOQ models with and without shortages and considered purchasing and holding costs constant.

Carbon emission is increasing day by day and many firms are working to reduce carbon emissions. The government has also taken many steps to reduce emissions such as carbon tax, cap, and offset. Therefore, Wang and Hua [6] investigate management of carbon footprints in firms under carbon emission trading mechanism. Benjaafar et al. [7] developed a model to investigate how far carbon reduction requirements can be addressed by operational adjustments as a supplement to costly investments in carbon-reducing. Chen, Benjaafar, and Elomri [8] provide a model a condition in which emission can be reduced by modifying order quantity. Toptal et al. [9] extend an EOQ model to show that in addition to carbon regulations such as carbon cap, tax, and cap-and-trade to reduce emission, emission reduction investment further reduces the emission while reducing costs. Mittal et al. [10] provide an economic production model to elaborate on human errors' effect on emission cost, transportation cost, and expected total profit of the retailer. Daryanto et al. [11] introduced an Economic order quantity model which includes the effect of defective rates, different sources of carbon emission, different demand rates, selling price and holding cost for defective products, and shortages backorder.

Since there are perfect quality items as well as defective items, therefore, in 2000, Salameh and Jaber [12] proposed EPQ/EOQ model in which a production/inventory situation where items, received/produced, are of imperfect quality and extends the standard EOQ/EPQ model for imperfect items. Chang [13] introduces a model with the complete screening process and imperfect quality items are sold as a single batch with discount before receiving the next shipment. Jaggi and Mittal [14] developed a model for spoilable items in which there is constant deterioration and the demand rate is time dependent under inflation and money value. Jaggi and Khanna [15] developed a model to formulate an inventory policy for a retailer dealing with imperfect quality items of deteriorating nature under inflation and permissible delay in payments. Jaggi and Mittal [16] developed a model for deteriorating items with imperfect quality and also an assumption has been made that the screening rate is more than demand. Jaber et al. [17] reviewed the model of Salameh and Jaber (2000) and elongate it by making an assumption that shipment is coming from a distant supplier and thus it is not feasible to imperfect items with an additional order to the same supplier. Mittal et al. [18] discussed about the method for redesigning the ordering policy by incorporating the cross-selling effect and also compared ordering policy for imperfect items developed by applying rules derived from apriori algorithm. Mittal, Jaggi, Khanna, Reshu, and Yadav [19–21] developed models for imperfect items under different conditions and Jayaswal et al. [22] discussed a fiscal construction feature model for imperfect quality items with trade credit policy analyzed under the effects of learning.

Many researchers have worked on reducing carbon emissions including imperfect items. Nobil et al. [23] proposed a model to calculate the optimal reorder point for the inventory model in Salameh and Jaber(2000) by which the appropriate timing of an order can be determined. Sarkar et al. [24] developed a three-echelon sustainable supply chain model with a single-supplier, single manufacturer, and multiple retailers.

Also, control the carbon emission and reduce the imperfect items to maintain the sustainability. Daryanto et al. [25] considered the EOQ model with carbon emissions from transportation and warehouse operations. Furthermore, include imperfect items and complete backordering is assumed.

2 Problem Definition

In this paper, investment on the reduction of carbon emission by retailers and decision of reclamation of inventory is taken according to the government regulations on carbon emissions. The standard EOQ model has been used under different conditions and includes imperfect items. Carbon emission is increased due to ordering, inventory holding, and manufacturing. In this study, three emission policies have been considered that is carbon cap, carbon tax, and cap-and-trade. Under the cap policy, a retailer's emission per year cannot exceed the carbon emission cap. Under tax policy, there will be a tax p_e units for unit carbon emission. Under the cap-and-trade policy, for c_{p_e} units, retailer deals a unit carbon emission.

2.1 Notations and Assumptions

1. Demand rate is considered constant throughout the model and shortages are not allowed.
2. Lead time is constant and known, and instantaneous replenishment is considered.
3. Each inventory containing defective items with percentage i with probability density function P(i) is known.
4. Imperfect items have been sold as a single batch with a discount on price.
5. Maximum reduction in carbon emission attainable due to investment decisions is less than minimum emission attainable due to ordering decisions per year. That is,

$$\sqrt{4\hat{A}\hat{h}MD} + \hat{k}D > \frac{\alpha^2}{4\beta}$$

   where $M = \frac{(1-i)^2}{2} + \frac{iD}{x}$, α gives the efficiency of green technology in emission reduction, and β is a decreasing return parameter(For G monetary units, carbon emission may be decreased in an amount of $(\alpha G - \beta G^2)$.
6. In cap policy, there are values of the investment that can reduce carbon emission per year below carbon capacity. Therefore, we can write

$$\sqrt{4\hat{A}\hat{h}MD} + \hat{k}D - \frac{\alpha^2}{4\beta} < C$$

where C is the carbon cap.

Q	Order quantity (per cycle)
k	Unit variable cost (\$ per unit)
A	Fixed cost per (\$ per unit)
i	Percentage of defective items in Q
$P(i)$	Probability density function of i
x	Screening rate, $x > D$
d	Unit screening cost(\$ per unit)
T	Cycle length
h	Holding cost (\$ per unit)
$\hat{A}$	Emission associated with ordering (per unit)
$\hat{h}$	Emission associated with inventory holding (per unit)
$\hat{k}$	Emission associated with production/purchasing (per unit)
D	Demand per year
G	Amount invested on carbon emission reduction per year

3 Carbon Cap

In this study under the carbon cap policy, retailer's carbon emissions per year should not exceed carbon cap C. Thus, the retailer has to find a feasible solution for order quantity and investment to reduce emissions. Therefore, this problem can be shown as follows: Minimize

Total cost per unit time $= TCU(Q, G) = \frac{AD}{Q} + (k + d)D + h\left[\frac{(1-i)^2}{2} + \frac{iD}{x}\right]Q + G$

Subject to

Total emission per unit time $= TEU(Q, G) = \frac{\hat{A}D}{Q} + \hat{k}D + \hat{h}\left[\frac{(1-i)^2}{2} + \frac{iD}{x}\right] Q - \alpha G + \beta G^2 \leq C$.

If we consider $G = 0$, then the optimal solution for this problem lies between the global interval Q_1, Q_2 when $TE = C$,

$$Q_1, Q_2 = \frac{\hat{C} \pm \sqrt{\hat{C}^2 - 4\hat{A}\hat{h}DM}}{2\hat{h}M}$$

where $\hat{C} = C - \hat{k}D$ and $M = \left[\frac{(1-i)^2}{2} + \frac{iD}{x}\right]$. The feasible solution exists if $C \geq 2\sqrt{\hat{A}\hat{h}MD} + \hat{k}D$.

Under cap policy, two cases can be considered such as

(1) $C \geq 2\sqrt{\hat{A}\hat{h}MD} + \hat{k}D$.
(2) $2\sqrt{\hat{A}\hat{h}MD} + \hat{k}D - \frac{\alpha^2}{4\beta} < C < 2\sqrt{\hat{A}\hat{h}MD} + \hat{k}D$. The next theorem will provide the optimal order quantity and investment decisions with different cases. (Q^*, G^*) will represent the feasible solution.

Theorem 1 *Let*

$$Q_3, Q_4 = \frac{(\hat{C} - \beta G_3^2 + \alpha G_3) \pm \sqrt{(\hat{C} - \beta G_3^2 + \alpha G_3)^2 - 4\hat{A}\hat{h}DM}}{2\hat{h}M}$$

and

$$G_3 = \frac{(AD - hMQ_3^2)\alpha - (-\hat{A}D + \hat{h}MQ_3^2)}{(AD - hMQ_3^2)2\beta},$$

$$G_4 = \frac{(AD - hMQ_4^2)\alpha - (-\hat{A}D + \hat{h}MQ_4^2)}{(AD - hMQ_4^2)2\beta}.$$

Then under carbon cap the feasible solution is
If $C \geq 2\sqrt{\hat{A}\hat{h}MD} + \hat{k}D$, *then*

$$(Q^*, G^*) = \begin{cases} (Q^c, 0) & \text{if } Q_2 \leq Q^c \leq Q_1 \\ (Q_1, 0) & \text{if } Q^\alpha < Q_1 < Q^c \\ (Q_3, G_3) & \text{if } Q^{em} < Q_3 \leq Q^\alpha \\ (Q_2, 0) & \text{if } Q^c < Q_2 < Q^\alpha \\ (Q_4, G_4) & \text{if } Q^\alpha \leq Q_4 < Q^{em} \end{cases} \tag{1}$$

and if $2\sqrt{\hat{A}\hat{h}MD} + \hat{k}D - \frac{\alpha^2}{4\beta} < C < 2\sqrt{\hat{A}\hat{h}MD} + \hat{k}D$, *then*

$$(Q^*, G^*) = \begin{cases} (Q_3, G_3) & \text{if } Q^{em} < Q_3 \leq Q^\alpha \\ (Q_4, G_4) & \text{if } Q^\alpha \leq Q_4 < Q^{em} \\ (Q_5, G_5) & \text{if otherwise} \end{cases} \tag{2}$$

where $Q_5 = Q^{em}$ *and* $G_5 = \frac{\alpha - \sqrt{\alpha^2 - 4\beta(-\hat{C} + 2\sqrt{\hat{A}\hat{h}MD}}}{2\beta}$. *Also,* $Q^\alpha = \sqrt{\frac{(A\alpha + \hat{A})D}{h\alpha + \hat{h})M}}$.

Remark 1 When $\frac{A}{h} = \frac{\hat{A}}{\hat{h}}$ then $Q^c = Q^{em}$. Also, when $C \geq 2\sqrt{\hat{A}\hat{h}MD} + \hat{k}D$ then $G^* = 0$ and $C < 2\sqrt{\hat{A}\hat{h}MD} + \hat{k}D$ then $G^* > 0$. The next corollary represents the minimum emission due to the retailer's optimal solution in the above theorem.

Corollary 1 *Under carbon cap, the minimum emission due to retailer's feasible solution is*

$$E_m(Q^*, G^*) = \sqrt{\frac{DM}{AD}}(\hat{A}h + A\hat{h})$$

when $Q_2 \leq Q^c \leq Q_1$ *and otherwise* $E_m(Q^*, G^*) = C$. *Now, there will be a lemma which shows the influence of using investment on emission reduction to reduce* *retailer's carbon emission with a certain cap C. Thus, there will be two considerations such as $E_m(Q^*(0), 0) - E_m(Q^*, G^*)$ and $TC^*(Q^*(0), 0) - TC^*(Q^*, G^*)$, where $Q^*(0)$ is the retailer's optimal order quantity under cap policy and the investment amount is zero.*

Lemma 1 *Investment amount to reduce emissions does not affect the carbon emission level under the certain cap per year, nevertheless it can reduce the total cost per year for the retailer. Therefore, we have $E_m(Q^*(0), 0) - E_m(Q^*, G^*) = 0$ and $TC^*(Q^*(0), 0) - TC^*(Q^*, G^*) \geq 0$*

In the next lemma, there will be a comparison of emissions per year with and without the carbon cap. Additionally, the effect of total cost per year with and without carbon cap.

Lemma 2 *Carbon emission reduces after applying the carbon cap policy but total cost per year is not less than when there is no cap policy. Therefore, $TC^*(Q^*, G^*) \geq TC(Q^c, 0)$ and $E_m(Q^*, G^*) \leq E(Q^{em}, 0)$.*

Lemma 3 *If we consider two investment options, first with α_1 and β_1 and second with α_2 and β_2, then solution that exists using the first investment will give the same emission level per year without costs.*

4 Carbon Tax

In this section, the penalty of p_e unit tax will be paid by the retailer per unit carbon emission. Therefore, the total cost and emission will be as follows:

$$TCU_{p_e}(Q, G) = \frac{AD}{Q} + (k + d)D + h\left[\frac{(1 - i)^2}{2} + \frac{iD}{x}\right]Q + G + p_e TEU(Q, G)$$

and

$$TEU_{p_e}(Q, G) = \frac{\hat{A}D}{Q} + \hat{k}D + \hat{h}\left[\frac{(1 - i)^2)}{2} + \frac{iD}{x}\right]Q - \alpha G + \beta G^2.$$

With $Q \geq 0$ and $G \geq 0$.

In the next theorem, the total cost has been minimized under the carbon tax policy.

Theorem 2 *The feasible solution under carbon tax is given by*

$$(Q^{**}, G^{**}) = \left(\sqrt{\frac{(A + p_e\hat{A})D}{(h + p_e\hat{h})M}}, \frac{\alpha p_e - 1}{2p_e\beta}\right).$$

It can be seen that Q^{**} and G^{**} are increasing when $\frac{A}{h} > \frac{\hat{A}}{\hat{h}}$and decreasing when $\frac{A}{h} < \frac{\hat{A}}{\hat{h}}$. Also, when $\frac{A}{h} = \frac{\hat{A}}{\hat{h}}$ there is no effect on Q^{**}.

5 Carbon Cap-and-Trade

In this section, there is a restriction of carbon cap C, and if total emission exceeds carbon cap C, then there is no penalty but the firm can buy carbon permits equal to its demand of carbon emission at the market price of c_{p_e} units per unit carbon emitted. Also, if the emission by the retailer is less than the carbon cap, then they can sell the carbon capacity at the same price c_{p_e}. Then the problem can be stated as follows:

$$TCU_{c_{Pe}}(Q, G) = \frac{AD}{Q} + (k + d)D + h\left[\frac{(1-i)^2}{2} + \frac{iD}{x}\right]Q + G - c_{p_e}X$$

and

$$TEU_{c_{Pe}}(Q, G) = \frac{\hat{A}D}{Q} + \hat{k}D + \hat{h}\left[\frac{(1-i)^2)}{2} + \frac{iD}{x}\right]Q - \alpha G + \beta G^2 + X = C$$

with $Q \geq 0$, $G \geq 0$, where X denotes the amount of carbon that the retailer trades per year. In the next theorem, a feasible solution will be found out for the above-formulated problem.

Theorem 3 *The optimal solution to minimize the total cost under cap-and-trade policy is given by*

$$(Q^{***}, G^{***}) = \left(\sqrt{\frac{(A + c_{p_e}\hat{A})D}{(h + c_{p_e}\hat{h})M}}, \frac{\alpha c_{p_e} - 1}{2c_{p_e}\beta}\right).$$

Also, $X^* = C - TEU_{c_{Pe}}(Q^{***}, G^{***})$, where X^* is the retailer's optimal amount of carbon traded per year.

It can be seen that Q^{***} and G^{***} are increasing when $\frac{A}{h} > \frac{\hat{A}}{\hat{h}}$ and decreasing when $\frac{A}{h} < \frac{\hat{A}}{\hat{h}}$. Also, when $\frac{A}{h} = \frac{\hat{A}}{\hat{h}}$, there is no effect on Q^{***}.

6 Numerical Analysis

In this section, there will be a comparison of values between two cases, i.e., $\frac{A}{h} > \frac{\hat{A}}{\hat{h}}$ and $\frac{A}{h} < \frac{\hat{A}}{\hat{h}}$. We will consider two sets of examples:

(1) $A = 100, h = 3, \hat{A} = 4$ and $\hat{h} = 3$.
(2) $A = 10, h = 4, \hat{A} = 100$ and $\hat{h} = 8$,
where $D = 500, k = 6, \hat{k} = 2, d = 0.5$, and $i = 0.02$ will remain same throughout. Also, since it is known that percentage defective random variable i is uniformly distributed and can have any value within the range $[\gamma, \delta]$ where $\gamma = 0$ and $\delta = 0.04$.

Probability density function for i is

$$P(i) = \begin{cases} 25, & 0 \le i \le 0.04 \\ 0, & \text{otherwise.} \end{cases}$$

Now, from the first case, we have $Q^c = 186.3$, $Q^{em} = 37.26$, $Q^{\alpha} = 167.463$, $TC(Q^c, 0) = 3786.77$, and $TE(Q^c, 0) = 1279.12$, and from the case 2, $Q^c = 51.02$, $Q^e = 114.085$, $Q^{\alpha} = 77.935$, $TC(Q^c, 0) = 3446$, and $TE(Q^c, 0) = 2176.01$

6.1 Numerical Analysis for Cap Policy

In Fig. 1, there are two figures (a) and (b) showing the changes in values of $TC(Q^*, G^*)$ with respect to cap C for both sets of examples. In both of the cases, $TC(Q^*, G^*)$ strictly decreases with respect to the increasing values of C.

Whenever the value of carbon cap C increases, the emission reduction investment G decreases, and therefore, the $TC(Q^*, G^*)$ decreases. But from the table, it can be seen that the total cost before the investment is less than or equal to the total cost after the investment. Because in the first case, i.e., $\frac{A}{h} > \frac{\hat{A}}{\hat{h}}$ when $C = 1270$ and in the second case that is $\frac{A}{h} < \frac{\hat{A}}{\hat{h}}$ when $C = 2110$, $TC(Q^c, 0) < TC(Q^*, G^*) = TC(Q^*, 0)$, and $TE(Q^c, 0) > TE(Q^*, G^*) = TE(Q^*, 0)$. Therefore, in this policy, total emission is decreasing and the total cost is increasing.

Numerical representation for carbon cap

$\frac{A}{h} > \frac{\hat{A}}{\hat{h}}$

C	Q^*	G^*	$TC(Q^*, G^*)$	$TE(Q^*, G^*)$	$TC(Q^*, 0)$	$TE(Q^*, 0)$
1070	162.361	50.4026	3842.26	1070	–	–
1170	165.6	21.2959	3811.79	1169.99	3878.22	1170
1270	179.696	0	3787.12	1270	3787.12	1270
1370	186.3	0	3786.77	1279.12	3786.77	1279.12

$\frac{A}{h} < \frac{\hat{A}}{\hat{h}}$

C	Q^*	G^*	$TC(Q^*, G^*)$	$TE(Q^*, G^*)$	$TC(Q^*, 0)$	$TE(Q^*, 0)$
1710	83.531	61.9684	3532.27	3532.27	–	–
1910	78.4863	7.27361	3471.74	1910	3474.1	1910
2110	55.8343	0	3446.8	2110	3446.8	2110
2310	2110	0	3446	2176.01	3446	2176.01

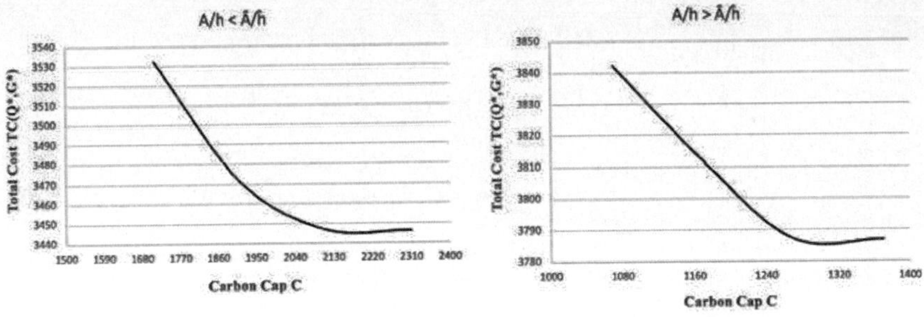

(a) (b)

Fig. 1 Change in value of $TC(Q^*, G^*)$ with respect to C

7 Conclusions

In the proposed model, the discount price of imperfect items is examined and retailer's joint decisions have been analyzed on reclamation of inventory and investment on reducing carbon emissions under three environmental regulations such as carbon cap, carbon tax, and carbon cap-and-trade. This model provides that under cap policy carbon emission will either remain the same or increases when investment and imperfect items are included but in the carbon tax and cap-and-trade policy, emission level decreases. This paper imparts an idea that how a retailer should choose the reclamation of inventory and the effect of government regulations on reducing emissions and costs.

8 Appendix

8.1 Proof of Theorem-1

In carbon cap policy, KKT(Karush-Kuhn-Tucker) conditions have been used to find the optimal solution for emission constraint. A feasible solution exists when there are constraints such that

$$\frac{\hat{A}D}{Q} + \hat{k}D + \hat{h}\left[\frac{((1-i)^2)}{2} + \frac{iD}{x}\right]Q - \alpha G + \beta G^2 < C$$

and

$$Q, G \geq 0.$$

By using KKT conditions, there is global optimality when optimality conditions have been used. Therefore,

$$\left(\frac{-AD}{Q^2}+hM\right)+\lambda_1\left(\frac{-\hat{A}D}{Q^2}-\hat{h}M\right)-\mu_1=0 \tag{3}$$

$$1+\lambda_1(-\alpha+2\beta G)-\mu_2=0 \tag{4}$$

$$\lambda_1\left(C-\frac{\hat{A}D}{Q}+\hat{k}D+\hat{h}MQ-\alpha G+\beta G^2\right)=0 \tag{5}$$

$$\mu_1 Q=0$$

$$\mu_2 G=0$$

where $M=\left(\frac{(1-i)^2}{2}+\frac{iD}{x}\right)$ and multipliers λ_1, μ_1, and μ_2 may be greater than or equal to zero. There could be eight possible cases but the feasible solution can be attained using the following three.

Case 1. $\lambda_1=0$, $\mu_1=0$, $\mu_2>0$

If $\lambda_1=0$, $\mu_1=0$, then Eq.(3) becomes

$$\left(\frac{-AD}{Q^2}+hM\right)=0. \tag{6}$$

Therefore, $Q=\sqrt{\frac{AD}{hM}}=Q^c$ and since $\mu_2 G=0$ and $\mu_2>0$ then $G=0$.

Where Q^c is the optimal solution for imperfect items.

To get the optimal solution order quantity must satisfy the

$$\frac{\hat{A}D}{Q}+\hat{k}D+\hat{h}MQ-\alpha G+\beta G^2<C.$$

Using this equation, there will be a global interval $[Q_1, Q_2]$, where

$$Q_2, Q_1=\frac{\hat{C}\pm\sqrt{\hat{C}^2-4\hat{A}D\hat{h}M}}{2\hat{h}M}.$$

For a solution to be feasible $\hat{C}^2-4\hat{F}D\hat{h}M\geq 0$ and hence $C\geq\hat{k}D+\sqrt{4\hat{A}D\hat{h}M}$. Thus, if $C\geq\hat{k}D+\sqrt{4\hat{A}D\hat{h}M}$ and $Q_1\leq Q^c\leq Q_2$, then $Q^*=Q^c$ and $G=0$.

Case 2. $\lambda_1>0$, $\mu_1=0$, $\mu_2>0$

From Eqs. (3) and (4), we have

$$\left(\frac{-AD}{Q^2}+hM\right)+\lambda_1\left(\frac{-\hat{A}D}{Q^2}-\hat{h}M\right)=0 \tag{7}$$

and

$$1+\lambda_1(-\alpha+2\beta G)-\mu_2=0. \tag{8}$$

Since $\mu_2 > 0$ then $G = 0$. Therefore, from Eq. (8),

$$1 + (-\alpha)\ \lambda_1 - \mu_2 = 0 \tag{9}$$

Also, $\lambda_1 > 0$ then from Eq. (5), we have

$$C - \left(\frac{\hat{A}D}{Q} + \hat{k}D + \hat{h}MQ\right) = 0. \tag{10}$$

Q_1 and Q_2 satisfy the above equality. Thus, they must have $C \geq \hat{k}D + \sqrt{4\hat{A}D\hat{h}M}$ to get the feasible solution. Further, let us consider two cases as follows:

Case 2.1. $C = \hat{k}D + \sqrt{4\hat{A}D\hat{h}M}$
In this case, $Q_1 = Q_2 = \sqrt{\frac{\hat{A}D}{\hat{h}M}} = Q^{em}$ and also, since $\lambda_1 > 0$ and $\mu_2 >= 0$ then $\lambda_1 < \frac{1}{\alpha}$. Equation (7) exists for any positive value of λ_1 and $\frac{A}{h} = \frac{\hat{A}}{\hat{h}}$. Thus, if $\frac{A}{h} = \frac{\hat{A}}{\hat{h}}$ then $Q^* = Q^c$ and $G^* = 0$.

Case 2.2. $C > \hat{k}D + \sqrt{4\hat{A}D\hat{h}M}$
In this case, $Q_1 \neq Q_2$. Then either $Q = Q_1$ or $Q = Q_2$ to get the feasible solution. Since $\lambda_1 > 0$, $G = 0$ then from Eq. (7), it obtained

$$\lambda_1 = \frac{AD - hMQ^2}{-\hat{A}D + \hat{h}MQ^2}$$

then to get optimality, we must have

$$0 < \frac{AD - hMQ^2}{-\hat{A}D + \hat{h}MQ^2} < \frac{1}{\alpha}. \tag{11}$$

From the above inequality, there are two possibilities that is either $AD - hMQ^2 > 0$ and $-\hat{A}D + \hat{h}MQ^2 > 0$ or $AD - hMQ^2 < 0$ and $-\hat{A}D + \hat{h}MQ^2 < 0$.

Thus, let us prove first that both the numerator and denominator are less than zero.

Since, we already know that for optimality $C > \hat{k}D + \sqrt{4\hat{F}D\hat{h}M}$. It can be rewritten as

$$2(C - \hat{k}D)^2 - 8\hat{A}D\hat{h}M > 0$$

$$2(C - \hat{k}D)^2 - 2(C - \hat{k}D)\sqrt{(C - \hat{k}D)^2 - 4\hat{A}D\hat{h}M} - 8\hat{A}D\hat{h}M < 0$$

$$\frac{((C - \hat{k}D) - \sqrt{(C - \hat{k}D)^2 - 4\hat{A}D\hat{h}M})^2}{2\hat{h}M} - 2\hat{A}D < 0$$

$$\left(\frac{(C-\hat{k}D)-\sqrt{(C-\hat{k}D)^2-4\hat{A}D\hat{h}M}}{2\hat{h}M}\right)^2 2\hat{h}M-2\hat{A}D<0$$

$$-\hat{A}D+\hat{h}MQ_1^2<0.$$

Therefore, according to Eq. (11), we must have $AD-hMQ_1^2<0$ and $0<\frac{AD-hMQ_1^2}{-\hat{A}D+\hat{h}MQ_1^2}<\frac{1}{\alpha}$. By solving these two equations together, the result can be formulated as $Q_2>\sqrt{\frac{AD}{hM}}=Q^c$ and $Q_2<\sqrt{\frac{(A+\alpha\hat{A})D}{(h+\hat{h})M}}=Q^\alpha$, then $Q^*=Q_2$ and $G^*=0$.

In a similar manner, we can show that $AD-hMQ_1^2>0$, $-\hat{A}D+\hat{h}MQ_1^2>0$ and $\frac{AD-hMQ^1}{-\hat{A}D+\hat{h}MQ^1}<\frac{1}{\alpha}$. After formulating the above results, the result can be shown as $Q_1<\sqrt{\frac{AD}{hM}}=Q^c$ and $Q_1>\sqrt{\frac{(A+\alpha\hat{A})D}{(h+\hat{h})M}}=Q^\alpha$, then $Q^*=Q_1$ and $G^*=0$.

Case 3. $\lambda_1>0$, $\mu_1=0$, $\mu_2=0$

Since $\mu_1=0$ and $\mu_2=0$ then Eqs. (3) and (4) can be written as

$$\left(\frac{-AD}{Q^2}+hM\right)+\lambda_1\left(\frac{-\hat{A}D}{Q^2}+\hat{h}M\right)=0 \tag{12}$$

$$1+\lambda_1(-\alpha+2\beta G)=0. \tag{13}$$

Now, for $\lambda_1>0$, we rewrite Eq. (5) as

$$C-\left(\frac{\hat{A}D}{Q}+\hat{k}D+\hat{h}MQ-\alpha G+\beta G^2\right)=0. \tag{14}$$

By evaluating the above equation, we obtain

$$Q_3,Q_4=\frac{\hat{C}+\alpha G-\beta G^2\pm\sqrt{\hat{C}+\alpha G-\beta G^2-4\hat{A}\hat{h}MD}}{2\hat{h}M}.$$

Here, Q_3, Q_4 exist only if $\hat{C}+\alpha G-\beta G^2-4\hat{A}\hat{h}MD\geq 0$. Let us consider two cases as follows.

Case 3.1. $\hat{C}+\alpha G-\beta G^2=4\hat{A}\hat{h}MD$

From this equality, $Q_3(G)=Q_4(G)=\sqrt{\frac{\hat{A}D}{\hat{h}M}}=Q^{em}$, where Q^{em} is the optimal solution for emission. We should have from Eq. (13)

$$0<G<\frac{\alpha}{2\beta}.$$

When $Q = Q^{em}$ then Eq. (12) holds for any positive value of λ_1 as long as $\frac{\hat{A}}{\hat{h}} = \frac{A}{h}$. Now, we have $\hat{C} + \alpha G - \beta G^2 = 4\hat{A}\hat{h}MD$ then there will be two roots from this equation but the condition, i.e., $0 < G < \frac{\alpha}{2\beta}$ only holds at one value which is given by

$$G = \frac{\alpha - \sqrt{\alpha^2 - 4\beta(-\hat{C} + 2\sqrt{\hat{h}\hat{A}MD})}}{2\beta}.$$

We can consider this value as G_5. Thus, if $2\sqrt{\hat{h}\hat{A}MD} + \hat{k}D - \frac{\alpha^2}{4\beta} < C < 2\sqrt{\hat{h}\hat{A}MD} + \hat{k}D$ and $\frac{\hat{A}}{\hat{h}} = \frac{A}{h}$, then $Q^* = Q^{em}$ and $G^* = G_5$.

Case 3.2. $\hat{C} + \alpha G - \beta G^2 > 4\hat{A}\hat{h}MD$

In this case, $Q_3(G) \neq Q_4(G)$. Now, from Eq. (12), $\lambda_1 = \frac{AD - hMQ^2}{-\hat{A}D + \hat{h}MQ^2}$, and for $Q_3(G) > 0$ and $Q_4(G)$ to be optimal, they must satisfy this inequality. Therefore, it is possible to show that $-\hat{A}D + \hat{h}MQ_3^2(G) > 0$. Moreover, from this result, $Q_3(G) > Q^{em}$. Similarly, for $\lambda_1 > 0$, we must have $-AD + hMQ_3^2(G) > 0$ and therefore $Q_3(G) < Q^c$.

Now, use the value of λ_1 in Eq. (13), to find the value of G in the form of $Q_3(G)$, then (Q_3, G_3) is the feasible solution, that is,

$$G = \frac{(AD + hMQ_3^2(G))\alpha - (-\hat{A}D + \hat{h}MQ_3^2(G))}{(-\hat{A}D + \hat{h}MQ_3^2(G))2\beta}. \tag{15}$$

Since, $G \geq 0$ then $(AD + hMQ_3^2(G))\alpha - (-\hat{A}D + \hat{h}MQ_3^2(G)) \geq 0$ and therefore $Q \leq Q^{\alpha}$. Now, there are three inequalities such as $Q \leq Q^{\alpha}$, $Q_3(G) < Q^c$, and $Q_3(G) > Q^{em}$. From $Q^{em} < Q_3(G) < Q^c$, $\frac{A}{h} > \frac{\hat{A}}{\hat{h}}$ and therefore, $Q^{\alpha} < Q^c$. Final result can be expressed as if $2\sqrt{\hat{h}\hat{A}MD} + \hat{k}D - \frac{\alpha^2}{4\beta} < C < 2\sqrt{\hat{h}\hat{A}MD} + \hat{k}D$ and $Q^{em} < Q_3(G) < Q^{\alpha}$, the optimal solution is (Q_3, G_3).

Similarly, (Q_4, G_4) can be obtained. Here,

$$G_4 = \frac{(AD + hMQ_4^2(G))\alpha - (-\hat{A}D + \hat{h}MQ_4^2(G))}{AD + hMQ_4^2(GMQ_4^2(G))2\beta}. \tag{16}$$

Therefore, it can be concluded that if $2\sqrt{\hat{h}\hat{A}MD} + \hat{k}D - \frac{\alpha^2}{4\beta} < C < 2\sqrt{\hat{h}\hat{A}MD} + \hat{k}D$ and $Q^{\alpha} < Q_4(G) < Q^{em}$, then the optimal solution is (Q_4, G_4).

8.2 proof of Theorem-2

Objective function is

$$TCU_{p_e}(Q, G) = \frac{AD}{Q} + (k + d)D + hMQ + G + p_e TEU(Q, G).$$

Putting the value of $TEU(Q, G)$ in the above equation, therefore

$$TCU_{p_e}(Q,G)=\frac{(A+p_e\hat{A})D}{Q}+(k+p_e\hat{k})D+dD+(h+p_e\hat{h})MQ+G-\alpha p_eG+\beta p_eG^2.$$

To find the feasible solution for a total cost per year, solve the Hessian matrix, which gives

$$\left(\frac{\partial^2 TCU_{p_e}}{\partial G^2}\right)\left(\frac{\partial^2 TCU_{p_e}}{\partial Q^2}\right)-\left(\frac{\partial^2 TCU_{p_e}}{\partial Q\partial G}\right)^2$$

must be greater than zero.

$$\frac{\partial^2 TCU_{p_e}}{\partial Q^2}=\frac{(A+p_e\hat{A})D}{Q^3},$$

$$\frac{\partial^2 TCU_{p_e}}{\partial G^2}=2p_e\beta$$

and

$$\frac{\partial^2 TCU_{p_e}}{\partial Q\partial G}=0.$$

Therefore,

$$\left(\frac{\partial^2 TCU_{p_e}}{\partial G^2}\right)\left(\frac{\partial^2 TCU_{p_e}}{\partial Q^2}\right)-\left(\frac{\partial^2 TCU_{p_e}}{\partial Q\partial G}\right)^2>0.$$

Then the optimal solution is $Q^{**}=\sqrt{\frac{(A+p_e\hat{A})D}{h+p_e\hat{h})M}}$ and $G^{**}=\frac{\alpha p_e-1}{2p_e\beta}$.

8.3 Proof of theorem-3

Objective function is

$$TCU_{c_{Pe}}(Q,G)=\frac{AD}{Q}+(k+d)D+h\left[\frac{(1-i)^2}{2}+\frac{iD}{x}\right]Q+G-c_{p_e}X$$

and

$$TEU_{c_{Pe}}(Q,G)=\frac{\hat{A}D}{Q}+\hat{k}D+\hat{h}\left[\frac{(1-i)^2)}{2}+\frac{iD}{x}\right]Q-\alpha G+\beta G^2+X=C.$$

Putting the value of X from the above equation in the objective function, thus

$$TCU_{c_{Pe}}(Q,G)=\frac{(A+c_{Pe}\hat{A})D}{Q}+(k+c_{Pe}\hat{k})D+dD+(h+c_{Pe}\hat{h})MQ+G-\alpha c_{Pe}G+\beta c_{Pe}G^2.$$

With the similar approach in 8.2, $TCU_{c_{pe}}(Q, G)$ is convex in Q and G. Therefore, $Q^{***} = \sqrt{\frac{(A+c_{pe}\hat{A})D}{(h+c_{pe}\hat{h})M}}$ and $G^{***} = \frac{\alpha c_{pe}-1}{2c_{pe}\beta}$. Thus, (Q^{***}, G^{***}) is the feasible solution.

References

1. Bouchery, Y., Ghaffari, A., Jemai, Z., Dallery, Y.M.S.: Including sustainability criteria into inventory models. Eur. J. Oper. Res. **222**, 229–240 (2012)
2. Arslan, M.C., Turkay, M.: EOQ revisited with sustainability considerations. Found. Comput. Decis. Sci. **38**, 223–249 (2013)
3. Wang, W.-T., Wee, H.-M., Cheng, Y.-L., Wen, C.L., Cárdenas-Barrón, L.E.: EOQ model for imperfect quality items with partial backorders and screening constraint. Eur. J. Ind. Eng. **9**, 744–773 (2015)
4. Lee, S.-K., Yoo, S.H., Cheong, T.: Sustainable EOQ under lead-time uncertainty and multi-modal transport. Sustainability **9**, 476 (2017)
5. Shaikh, A.A., Al-Amin, K.M., Panda, G.C., Konstantaras, I.: Price discount facility in an EOQ model for deteriorating items with stock-dependent demand and partial backlogging. Int. Trans. Oper. Res. **26**, 1365–1395 (2019)
6. Hua, G., Cheng, T.C.E., Wang, S.: Managing carbon footprints in inventory management. Int. J . Prod. Econ. **132**, 178–185 (2011)
7. Benjaafar, S., Li, Y., Daskin, M.: Carbon footprint and the management of supply chains: Insights from simple models. IEEE Trans. Autom. Sci. Eng. **10**, 99–116 (2012)
8. Chen, X., Benjaafar, S., Elomri, A.: The carbon-constrained EOQ. Oper. Res. Lett. **41**, 172–179 (2013)
9. Toptal, A., Özlü, H., Konur, D.: Joint decisions on inventory replenishment and emission reduction investment under different emission regulations. Int. J. Prod. Res. **52**, 243–269 (2014)
10. Gilotra, M., Pareek, S., Mittal, M., Dhaka, V.: Effect of carbon emission and human errors on a two-echelon supply chain under permissible delay in payments. Int. J. Math. Eng. Manag. Sci. **5**, 225–236 (2020)
11. Daryanto, Y., Christata, B.: Optimal order quantity considering carbon emission costs, defective items, and partial backorder. Uncertain Supply Chain Manag. **9**, 307–316 (2021)
12. Salameh, M.K., Jaber, M.Y.: Economic production quantity model for items with imperfect quality. Int. J. Prod. Econ. **64**, 59–64 (2000)
13. Chang, H.-C.: An application of fuzzy sets theory to the EOQ model with imperfect quality items. Comput. Oper. Res. **31**, 2079–2092 (2004)
14. Jaggi, C.K., Mittal, M.: An EOQ model for deteriorating items with time-dependent demand under inflationary conditions. Adv. Model. Optim. **5** (2003)
15. Jaggi, C.K., Khanna, A., Mittal, M.: Credit financing for deteriorating imperfect-quality items under inflationary conditions. Int. J. Services Oper. Inf. **6**, 292–309 (2011)
16. Jaggi, C.K., Mittal, M.: Economic order quantity model for deteriorating items with imperfect quality. Investigación Operacional, **32**, 107–113 (2011)
17. Jaber, M.Y., Zanoni, S., Zavanella, L.E.: Economic order quantity models for imperfect items with buy and repair options. Int. J. Prod. Econ. **155**, 126–131 (2014)
18. Mittal, M., Pareek, S., Agarwal, R.: EOQ estimation for imperfect quality items using association rule mining with clustering. Dec. Sci. Lett. **4**, 497–508 (2015)
19. Yadav, R., Pareek, S., Mittal, M.: Supply chain models with imperfect quality items when end demand is sensitive to price and marketing expenditure. RAIRO-Oper. Res. **52**, 725–742 (2018)
20. Mittal, M., Khanna, A., Jaggi, C.K.: Retailer's ordering policy for deteriorating imperfect quality items when demand and price are time-dependent under inflationary conditions and permissible delay in payments. Int. J. Procure. Manag. **10**, 461–494 (2017)
21. Agarwal, R., Mittal, M.: Inventory classification using multi-level association rule mining. Int. J. Dec. Support Syst. Technol. (IJDSST) **11**, 1–12 (2019)

22. Jayaswal, M., Sangal, I.S.H.A., Mittal, M., Malik, S.: Effects of learning on retailer ordering policy for imperfect quality items with trade credit financing. Uncertain Supply Chain Manag. **7**, 49–62 (2019)
23. Nobil, A.H., Sedigh, A.H.A. Cárdenas-Barrón, L.E.: Reorder point for the EOQ inventory model with imperfect quality items. Ain Shams Eng. J. **11**, 1339–1343 (2020)
24. Sarkar, B., Sarkar, M., Ganguly, B., Cárdenas-Barrón, L.E.: Combined effects of carbon emission and production quality improvement for fixed lifetime products in a sustainable supply chain management. Int. J. Prod. Econ. **231**, 107867 (2021)
25. Mashud, A.H.M., Pervin, M., Mishra, U., Daryanto, Y., Tseng, M.-L., Lim, M.K.: A sustainable inventory model with controllable carbon emissions in green-warehouse farms. J. Clean. Prod. **298**, 126777 (2021)

Propagation of Water Waves in the Presence of a Horizontal Plate Submerged in a Two-Layer Fluid

S. Naskar, N. Islam, R. Gayen, and R. Datta

Abstract The interaction of surface and interface waves with a thin horizontal plate submerged in the lower layer of a two-layer fluid is studied under linearised theory of water waves. The associated boundary value problem is solved here by Fourier integral transform by reducing it to an integral equation involving the potential difference function across the plate. Application of multi-term Galerkin method to the solution of the integral equation leads to a simple, rapidly convergent numerical scheme and suitable expressions for different hydrodynamic quantities of interest. Numerical results for the reflection coefficients and the hydrodynamic force on the plate are presented to study the effect of different physical parameters. The present method is verified by recovering the published numerical results for a limiting case and through an energy balance relation.

Keywords Two-layer fluid · Submerged thin horizontal plate · Fourier integral transform · Reflection coefficient · Hydrodynamic force

---

S. Naskar (✉) · R. Gayen
Department of Mathematics, Indian Institute of Technology Kharagpur, Kharagpur 721302, India
e-mail: sanjibnaskar62@gmail.com

N. Islam
Department of Mathematics, Bennett University, Greater Noida 201310, India

R. Datta
Department of Ocean Engineering and Naval Architecture, Indian Institute of Technology Kharagpur, Kharagpur 721302, India
e-mail: ranadev@naval.iitkgp.ac.in

R. K. Sharma et al. (eds.), *Frontiers in Industrial and Applied Mathematics*,
Springer Proceedings in Mathematics & Statistics 410,
https://doi.org/10.1007/978-981-19-7272-0_30

1 Introduction

The study of water wave interaction with obstacles has sparked enormous attention for a variety of applications in coastal and marine environments. Also, obtaining a less cost-effective clean renewable energy by extracting energy from ocean waves has received considerable attention from researchers. One of the best developments in extracting wave energy is to construct a line of submerged bodies that would act like a lens and focus the diverging waves to converge waves. In Norway, they have developed many such constructions like shore-based horizontal tapered channels, oscillating water columns, phase controlled wave power buoys, etc. McIver [1] considered a horizontal flat plate moored to seabed which would act like such a lens, while Mehlum [2] considered a circular cylinder. Using Fourier integral transform together with a Galerkin method, Porter [3] investigated the oblique water wave interaction with a horizontal thin plate submerged in a single layer fluid.

In the study of the propagation of water waves in a two-layer fluid having a free upper surface in the upper layer, Lamb [4] established that for a given frequency, there exist two linear wave systems of different wavenumbers. These two wave modes mainly propagate along the free surface and the interface of the fluids. As a result, if wave fields interact with obstacles, some transformation of wave energy from one mode to another may occur. This makes the wave interaction problems in a two-layer fluid more interesting. Linton and McIver [5] developed the linear scattering theory for two-dimensional wave motion in a two-layer fluid comprised of an infinite lower layer and a finite upper layer with a free surface to investigate the problem of wave scattering by a horizontal circular cylinder with the help of multipole expansion method. Using hypersingular integral equations method, Dhillon et al. [6] and Islam and Gayen [7] investigated the scattering of water waves by a thin vertical and inclined plate in a two-layer fluid, respectively. Based on the method of eigenfunction expansion, Medina-Rodríguez and Silva [8] considered two thick horizontal plates submerged in a two-layer fluid and analysed the reflection energies of interface and surface waves.

In the present article, a thin horizontal rigid plate submerged in the lower layer of a two-layer-fluid is proposed and investigated in the context of linear potential theory. Here, both the fluids are considered to be of finite depth. The coupled boundary value problem is solved here by Fourier integral transform to obtain an integral equation involving the unknown potential difference function across the plate. Then using Galerkin method we find this potential difference function numerically and with this solution, we compute the different physical quantities. The correctness of the present analysis is established by checking the energy identity relation and by comparing the obtained numerical results for limiting case with one of the previous results available in the literature. New results are presented graphically illustrating the effects of various parameters on the hydrodynamic quantities.

2 Formulation of the Problem

Figure 1 depicts the geometry of a horizontal plate Γ submerged in the bottom layer of a two-layer fluid. The depths of the upper and lower layer fluids are h and H, respectively. A Cartesian coordinate system is considered in which $z = 0$ represents the rest common interface of the two fluids, $z = -h$ represents the free surface, and z-axis is measured vertically downwards from the undisturbed interface. Let the plate be submerged at a depth d from the undisturbed interface of the two fluids and extends horizontally from $-b$ to b. Assuming time harmonic incident waves of angular frequency σ making an angle θ with the positive x-axis, the motion in the upper layer fluid (of density ρ_1) and lower layer fluid (of density ρ_2) can be represented by $Re\left\{\phi_1(x,z)e^{-i\sigma\tau}e^{i\nu y}\right\}$ and $Re\left\{\phi_2(x,z)e^{-i\sigma\tau}e^{i\nu y}\right\}$ respectively, where τ indicates the time and ν is the wavenumber along the y direction. The functions $\phi_j(x,z)$ satisfy

$$(\nabla^2 - \nu^2)\phi_j(x,z) = 0, \text{ in the respective fluid region.} \tag{1}$$

Linearized free surface, interface and the bottom boundary conditions are

$$K\phi_1 + \phi_{1z} = 0 \text{ on } z = -h, \tag{2}$$

$$\phi_{1z} = \phi_{2z} \text{ on } z = 0, \tag{3}$$

$$s(K\phi_1 + \phi_{1z}) = K\phi_2 + \phi_{2z} \text{ on } z = 0, \tag{4}$$

$$\phi_{2z} = 0 \text{ on } z = H, \tag{5}$$

where $s = \rho_1/\rho_2$, $K = \sigma^2/g$, g being the acceleration due to gravity.

The boundary condition on the horizontal plate is

$$\phi_{2z}(x, d^{\pm}) = 0, \quad |x| < b, \tag{6}$$

$$\phi_2(x, d^+) - \phi_2(x, d^-) = P(x), \quad |x| < b. \tag{7}$$

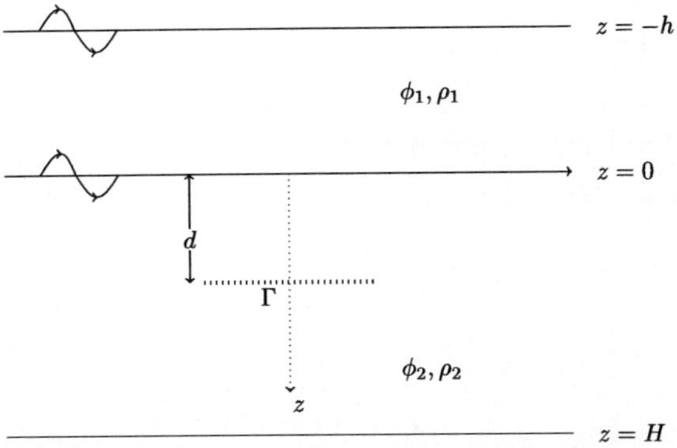

Fig. 1 Schematic diagram for horizontal plate

In a two-layer fluid, the progressive waves propagating at the free surface and the interface can be expressed by

$$f_1(v,z)e^{\pm i(v^2-\nu^2)^{1/2}x}(-h<z<0), \quad f_2(v,z)e^{\pm i(v^2-\nu^2)^{1/2}x}(0<z<H)$$

with

$$f_1(v,z)=\frac{\sinh vH}{K\cosh vh - v\sinh vh}\left[\,v\cosh v(h+z)-K\sinh v(h+z)\right],$$
$$f_2(v,z)=\cosh v(H-z)$$

where v is real, positive and satisfies the dispersion equation

$$\Delta(v)\equiv(1-s)v^2+K^2(s+\coth vh\coth vH)-vK(\coth vh+\coth vH)=0. \tag{8}$$

Equation (8) has exactly two positive real roots, m and M ($K<m<M$). Thus, there exist two wave systems with two different wavenumbers. As a result, if a wave train of mode m is obliquely incident on the horizontal plate at angle θ with the positive x-axis, the far-field behaviours of $\phi_j (j=1,2)$ are given by

$$\phi_j(x,z)\to\begin{cases}\phi^I_{jm}(x,z)+r^m\phi^I_{jm}(-x,z)+R^m\phi^I_{jM}(-x,z) \text{ as } x\to-\infty,\\ t^m\phi^I_{jm}(x,z)+T^m\phi^I_{jM}(x,z) \text{ as } x\to\infty,\end{cases} \tag{9}$$

where

$$\phi^I_{jv}(x,z)=f_j(v,z)e^{i(v^2-\nu^2)^{1/2}x}. \tag{10}$$

In (9), for an obliquely incident wave of mode m, the unknowns r^m and R^m represent the amplitudes of reflected waves associated with modes m and M respectively, while t^m and T^m represent the amplitude of transmitted waves associated with modes m and M respectively. Similarly, for an incident wave of mode M with incident wave angle $\theta<\sin^{-1}(m/M)$ the far-field behaviours of $\phi_j (j=1,2)$ can be expressed as

$$\phi_j(x,z)\to\begin{cases}\phi^I_{jM}(x,z)+R^M\phi^I_{jM}(-x,z)+r^M\phi^I_{jm}(-x,z) \text{ as } x\to-\infty,\\ T^M\phi^I_{jM}(x,z)+t^M\phi^I_{jm}(x,z) \text{ as } x\to\infty.\end{cases} \tag{11}$$

Here, for an obliquely incident wave of mode M, the unknowns r^M and R^M represent the amplitudes of reflected waves associated with modes m and M respectively, while t^M and T^M denote the amplitudes of transmitted waves associated with modes m and M respectively.

3 Method of Solution

Let a wave train of mode m making an angle $\theta (0 \leq \theta \leq \pi/2)$ with the positive x-axis be incident on the plate. Then, we must have $\nu = m \sin\theta$.

Now, we define the Fourier transform of the scattered potential function by

$$\overline{\phi}_j(k,z) = \int_{-\infty}^{\infty} \left(\phi_j(x,z) - \phi^I_{jm}(x,z)\right) e^{-ikx} \mathrm{d}x, \tag{12}$$

with the inverse

$$\phi_j(x,z) = \phi^I_{jm}(x,z) + \frac{1}{2\pi} \int_{-\infty}^{\infty} \overline{\phi}_j(k,z) e^{ikx} \mathrm{d}k, \tag{13}$$

where the integration contour in the inverse transform will be defined later by incorporating the far-field conditions.

Then, applying (12) to (1)–(7) produces

$$\left(\frac{d^2}{dz^2} - \beta^2\right)\overline{\phi}_j = 0, \quad j = 1, 2, \tag{14}$$

$$K\overline{\phi}_1 + \overline{\phi}_{1z} = 0 \text{ on } z = -h, \tag{15}$$

$$\overline{\phi}_{1z} = \overline{\phi}_{2z} \text{ on } z = 0, \tag{16}$$

$$s(K\overline{\phi}_1 + \overline{\phi}_{1z}) = K\overline{\phi}_2 + \overline{\phi}_{2z} \text{ on } z = 0, \tag{17}$$

$$\overline{\phi}_{2z} = 0 \text{ on } z = H, \tag{18}$$

$$\overline{\phi}_{2z}(k, d^+) = \overline{\phi}_{2z}(k, d^-), \tag{19}$$

$$\overline{\phi}_2(k, d^+) - \overline{\phi}_2(k, d^-) = \int_{-b}^{b} P(x) e^{-ikx} dx \equiv \overline{P}(k), \tag{20}$$

where $\beta^2 = k^2 + \nu^2$.

Solving (14) subjected to the boundary conditions (15)–(18), we get

$$\overline{\phi}_1(k,z) = \frac{K\overline{P}(k)\sinh\beta(H-d)[K\sinh\beta(h+z) - \beta\cosh\beta(h+z)]}{\sinh\beta h \sinh\beta H \Delta(\beta)}, -h < z < 0, \tag{21}$$

$$\overline{\phi}_2(k,z) = \begin{cases} \frac{\overline{P}(k)\sinh\beta(H-d)[\sinh\beta h\cosh\beta z((1-s)\beta^2 + K^2 s) - K\beta\cosh\beta(h+z) + K^2\cosh\beta h\sinh\beta z]}{\sinh\beta h\sinh\beta H\Delta(\beta)}, 0 < z < d, \\ \frac{\overline{P}(k)\cosh\beta(H-z)[\sinh\beta h\cosh\beta d((s-1)\beta^2 - K^2 s) + K\beta\sinh\beta(h+d) - K^2\cosh\beta h\cosh\beta d]}{\sinh\beta h\sinh\beta H\Delta(\beta)}, d < z < H. \end{cases} \tag{22}$$

Taking inverse transforms of the representations (21) in $(-h < y < 0)$ and (22) in $0 < y < d$, we get

$$\phi_1(x,z) = \phi_{1m}^I(x,z) + \frac{1}{2\pi}\int_{-\infty}^{\infty} \frac{K\overline{P}(k)\sinh\beta(H-d)[K\sinh\beta(h+z) - \beta\cosh\beta(h+z)]}{\sinh\beta h \sinh\beta H \Delta(\beta)} e^{ikx}\,dk, \tag{23}$$

$$\phi_2(x,z) = \phi_{2m}^I(x,z) + \frac{1}{2\pi}\int_{-\infty}^{\infty} \overline{P}(k)\mathcal{H}(z,\beta)e^{ikx}\,dk, \tag{24}$$

where

$$\mathcal{H}(z,\beta) = \frac{\sinh\beta(H-d)[\sinh\beta h\cosh\beta z((1-s)\beta^2 + K^2 s) - K\beta\cosh\beta(h+z) + K^2\cosh\beta h\sinh\beta z]}{\sinh\beta h\sinh\beta H\Delta(\beta)}. \tag{25}$$

In order to obtain the reflection and transmission coefficients, we find the far-field form for $\phi_1(x,z)$. There are poles on the real k-axis at $k = \pm\alpha_1$ and $k = \pm\alpha_2$ where $\alpha_1 = m\cos\theta$, $\alpha_2 = \sqrt{M^2 - m^2\sin^2\theta}$. Thus, in order to meet the radiation condition that $\phi_1 - \phi_{1m}^I$ is outgoing, the contour of the integration in equation (23) is taken to pass under the poles at $k = \alpha_1, \alpha_2$ and over the poles at $k = -\alpha_1, -\alpha_2$. Thus, capturing the residues at the poles $k = \pm\alpha_1, \pm\alpha_2$, the contour can be deformed into either the upper-half or lower-half k-plane by letting $x \to \pm\infty$ in (19), and this yields

$$\phi_1(x,z) \to \begin{cases} \phi_{1m}^I(x,z) - im\mu_1\overline{P}(-\alpha_1)\sinh m(H-d)\phi_{1m}^I(-x,z) - iM\mu_2\overline{P}(-\alpha_2)\sinh M(H-d)\phi_{1M}^I(-x,z) \\ \qquad \text{as } x \to -\infty, \\ \left(1 - im\mu_1\overline{P}(\alpha_1)\sinh m(H-d)\right)\phi_{1m}^I(x,z) - iM\mu_2\overline{P}(\alpha_2)\sinh M(H-d)\phi_{1M}^I(x,z) \text{ as } x \to \infty. \end{cases} \tag{26}$$

Comparing (26) with (9), we get

$$t^m - 1 = -im\mu_1\overline{P}(\alpha_1)\sinh m(H-d), \qquad T^m = -iM\mu_2\overline{P}(\alpha_2)\sinh M(H-d),$$
$$r^m = -im\mu_1\overline{P}(-\alpha_1)\sinh m(H-d), \qquad R^m = -iM\mu_2\overline{P}(-\alpha_2)\sinh M(H-d), \tag{27}$$

where μ_1 and μ_2 are defined as

$$\mu_1 = \frac{K(K\cosh mh - m\sinh mh)}{\alpha_1\sinh mh\sinh^2 mH\Delta'(m)}, \quad \mu_2 = \frac{K(K\cosh Mh - M\sinh Mh)}{\alpha_2\sinh Mh\sinh^2 MH\Delta'(M)}.$$

Now with the help of the values of t^m, T^m, r^m and R^m, we can write (24) as the sum of Cauchy principal value-integral and contributions from the four poles. Thus, $\phi_2(x,z)$ given in (24) can be expressed as

$$\phi_2(x,z) = \frac{1}{2}(t^m+1)\phi_{2m}^I(x,z) + \frac{r^m}{2}\phi_{2m}^I(-x,z) + \frac{1}{2}[T^m e^{i\alpha_2 x} + R^m e^{-i\alpha_2 x}] + \frac{1}{2\pi}⨍_{-\infty}^{\infty} \mathcal{H}(z,\beta)\overline{P}(k)e^{ikx}\,dk. \tag{28}$$

We note that, for $0 < z < d$,

$$\mathcal{H}(z,\beta) \to \frac{1}{2}e^{-|k|(d-z)}, \quad \text{as } |k| \to \infty. \tag{29}$$

We also note the following identity (cf. [9]) for $d - z > 0$

$$\log\sqrt{(x-w)^2+(d-z)^2} = \frac{1}{2}\int_{-\infty}^{\infty} \frac{e^{|k|} - e^{|k|(d-z)}e^{ik(x-w)}}{|\,k\,|}dk. \tag{30}$$

Thus, making use of the relations (20), (29) and (30), we re-write (28) as

$$\begin{aligned}\phi_2(x,z) = {} & \frac{1}{2}(t^m+1)\phi^I_{2m}(x,z) + \frac{r^m}{2}\phi^I_{2m}(-x,z) + \frac{1}{2}[T^m e^{i\alpha_2 x} + R^m e^{-i\alpha_2 x}] \\ & + \frac{1}{2\pi}\frac{\partial}{\partial z}\int_{-b}^{b} P(w)\log\sqrt{(x-w)^2+(d-z)^2}dw \\ & + \frac{1}{2\pi}\int_{-\infty}^{\infty}[\mathcal{H}(z,\beta) - \frac{1}{2}e^{-|k|(d-z)}]e^{ikx}\int_{-b}^{b}P(w)e^{-ikw}dwdk. \end{aligned} \tag{31}$$

Now we apply the plate condition (6) in (31) and this gives

$$\begin{aligned}(t^m+1)f_+(x) + r^m f_-(x) + T^m g_+(x) + R^m g_-(x) = {} & -\frac{1}{\pi}\frac{d^2}{dx^2}\int_{-b}^{b}P(w)\log|\,x-w\,|\,dw \\ & + \frac{1}{2\pi}\int_{-\infty}^{\infty}E_\nu(k)e^{ikx}\int_{-b}^{b}P(w)e^{-ikw}dwdk \end{aligned} \tag{32}$$

for $|\,x\,| < b$, where

$$f_\pm(x) = -m\sinh m(H-d)e^{\pm i\alpha_1 x} \quad \text{and} \quad g_\pm(x) = -M\sinh M(H-d)e^{\pm i\alpha_2 x} \tag{33}$$

and

$$E_\nu(k) = \frac{2\beta\sinh\beta(H-d)[\sinh\beta h\sinh\beta d((s-1)\beta^2 - K^2 s) + K\beta\sinh\beta(h+d) - K^2\cosh\beta h\cosh\beta d]}{\sinh\beta h\sinh\beta H\Delta(\beta)} + |k|. \tag{34}$$

It may be noted that to obtain (32) we have used

$$\left(\frac{\partial^2}{\partial x^2} + \frac{\partial^2}{\partial z^2}\right)\log\sqrt{(x-w)^2+(d-z)^2} = 0 \tag{35}$$

to alter from z to x-axis before applying the plate condition on $z = d$.
Now, we define the integro-differential operator $\mathcal{K}$ by

$$\begin{aligned}(\mathcal{K}P)(x) = {} & -\frac{1}{\pi}\frac{d^2}{dx^2}\int_{-b}^{b}P(w)\log|\,x-w\,|\,dw + \frac{1}{2\pi}\int_{-\infty}^{\infty}E_\nu(k)e^{ikx} \\ & \int_{-b}^{b}P(w)e^{-ikw}dwdk \end{aligned} \tag{36}$$

and let $P_{\pm}(x)$, $Q_{\pm}(x)$ satisfy

$$(\mathcal{K}P_{\pm}, \mathcal{K}Q_{\pm})\,(x) = (f_{\pm}, g_{\pm})\,(x), \quad |\, x \,| < b. \tag{37}$$

Hence it follows from (32) that

$$P(x) = (t^m + 1)P_{+}(x) + r^m P_{-}(x) + T^m Q_{+}(x) + R^m Q_{-}(x). \tag{38}$$

Using (33) and the definition (20) in (27) results in

$$\begin{aligned} t^m - 1 &= \mathrm{i}\mu_1\langle P, f_{+}\rangle, \quad T^m = \mathrm{i}\mu_2\langle P, g_{+}\rangle, \\ r^m &= \mathrm{i}\mu_1\langle P, f_{-}\rangle, \quad R^m = \mathrm{i}\mu_2\langle P, g_{-}\rangle, \end{aligned} \tag{39}$$

where the operation $\langle p, q\rangle$ denotes the inner product as defined by

$$\langle p, q\rangle = \int_{-b}^{b} p(x)q^*(x)dx \tag{40}$$

with asterisk denoting complex conjugate.

Substitution of (38) in (39) gives

$$\begin{aligned} t^m - 1 &= \mathrm{i}\mu_1(t^m + 1)S_{+,+} + i\mu_1 r^m S_{-,+} + i\mu_1 T^m X_{+,+} + i\mu_1 R^m X_{-,+} \\ T^m &= \mathrm{i}\mu_2(t^m + 1)W_{+,+} + i\mu_2 r^m W_{-,+} + i\mu_2 T^m L_{+,+} + i\mu_2 R^m L_{-,+} \\ r^m &= \mathrm{i}\mu_1(t^m + 1)S_{+,-} + i\mu_1 r^m S_{-,+} + i\mu_1 T^m X_{+,-} + i\mu_1 R^m X_{-,-} \\ R^m &= \mathrm{i}\mu_2(t^m + 1)W_{+,-} + i\mu_2 r^m W_{-,-} + i\mu_2 T^m L_{+,-} + i\mu_2 R^m L_{-,-} \end{aligned} \tag{41}$$

where $W_{\pm,\pm} = \langle P_{\pm}, g_{\pm}\rangle$, $L_{\pm,\pm} = \langle Q_{\pm}, g_{\pm}\rangle$, $S_{\pm,\pm} = \langle P_{\pm}, f_{\pm}\rangle$, $X_{\pm,\pm} = \langle Q_{\pm}, f_{\pm}\rangle$ with the first '$\pm$'s in the left-hand side corresponding to the first in the right hand side and so on.

3.1 Numerical Method

To solve the system of equations given in (41) for R^m, T^m, r^m and t^m, we must compute the inner products; hence we need to solve for $P_{\pm}$, $Q_{\pm}$. For this, we apply the Galerkin method (cf. Porter [3]) to find the solution for (37). The method is described below.

We take

$$(P_{\pm}, Q_{\pm})\,(x) = \sum_{n=0}^{\infty} \left(A_n^{\pm}, B_n^{\pm}\right) p_n(x/b), \quad |\, x \,| \leq b, \tag{42}$$

where $A_n^{\pm}$, $B_n^{\pm}$ are unknown coefficients to be determined and

$$p_n(w) = \frac{e^{in\pi/2}}{\pi(n+1)}(1-w^2)^{1/2}U_n(w), \tag{43}$$

where U_n are second kind Chebyshev polynomials of order n.

Substitution of (42) into (37), multiplication with $p_l^*(x/b)$ and integration over $-b < x < b$ results in the infinite system of equations for the unknown coefficients $A_n^\pm$ and $B_n^\pm$:

$$-\frac{1}{2\pi(l+1)}\left(A_l^\pm, B_l^\pm\right) + \sum_{n=0}^{\infty}\left(A_n^\pm, B_n^\pm\right)K_{l,n} = \left(F_l^\pm, G_l^\pm\right), \quad l = 0, 1, 2, ..., \tag{44}$$

where

$$K_{l,n} = \frac{1}{2\pi}\int_{-\infty}^{\infty}\frac{E_\nu(k)}{k^2}J_{n+1}(kb)J_{l+1}(kb)\mathrm{d}k, \tag{45}$$

and

$$F_l^\pm = -m\sinh m(H-d)(\pm 1)^l\frac{J_{l+1}(b\alpha_1)}{\alpha_1}, \quad G_l^\pm = -M\sinh M(H-d)(\pm 1)^l\frac{J_{l+1}(b\alpha_2)}{\alpha_2}. \tag{46}$$

It is noted that $K_{l,n} = 0$ if $l+n$ is odd. This indicates that we can decouple (44) into its symmetric and antisymmetric parts for $\left(A_{2n}^\pm, B_{2n}^\pm\right)$ and $\left(A_{2n+1}^\pm, B_{2n+1}^\pm\right)$. Thus, we have the following real symmetric systems of linear equations:

$$-\frac{1}{2\pi(2l+\xi+1)}\left(A_{2l+\xi}^\pm, B_{2l+\xi}^\pm\right) + \sum_{n=0}^{\infty}\left(A_{2n+\xi}^\pm, B_{2n+\xi}^\pm\right)K_{2l+\xi,2n+\xi} = \left(F_{2l+\xi}^\pm, G_{2l+\xi}^\pm\right),$$
$$l = 0, 1, 2, ..., \quad \xi = 0, 1. \tag{47}$$

Again, $F_l^+ = (-1)^l F_l^-$ and $G_l^+ = (-1)^l G_l^-$ imply that $A_l^+ = (-1)^l A_l^-$ and $B_l^+ = (-1)^l B_l^-$ and thus it is sufficient to find just the solutions of (47) for $\left(A_l^+, B_l^+\right)$. Using (42) and (46), we have

$$W_{\pm,\pm} = \sum_{n=0}^{\infty}A_n^\pm G_n^\pm, \; L_{\pm,\pm} = \sum_{n=0}^{\infty}B_n^\pm G_n^\pm, \; S_{\pm,\pm} = \sum_{n=0}^{\infty}A_n^\pm F_n^\pm, \; X_{\pm,\pm} = \sum_{n=0}^{\infty}B_n^\pm F_n^\pm. \tag{48}$$

Thus, it follows that

$$W_{+,+} = W_{--} = \sum_{n=0}^{\infty}A_{2n}^\pm G_{2n}^\pm + \sum_{n=0}^{\infty}A_{2n+1}^\pm G_{2n+1}^\pm \text{ and}$$
$$W_{+,-} = W_{-+} = \sum_{n=0}^{\infty}A_{2n}^\pm G_{2n}^\pm - \sum_{n=0}^{\infty}A_{2n+1}^\pm G_{2n+1}^\pm \tag{49}$$

and similarly for $L_{\pm,\pm}$, $S_{\pm,\pm}$ and $X_{\pm,\pm}$.

The energy identity comprising reflection and transmission coefficients can be derived using Green's integral theorem as

$$|r^m|^2 + |R^m|^2 + \mathcal{J}(|t^m|^2 + |T^m|^2) = 1; \quad \mathcal{J} = \frac{\mathcal{J}_M}{\mathcal{J}_m}, \tag{50}$$

with

$$\mathcal{J}_\lambda = \mathrm{i}\lambda\left[s\int_{-h}^{0} \{f(\lambda, z)\}^2 \mathrm{d}z + \int_{0}^{H} \{\cosh\lambda(H-z)\}^2 \mathrm{d}z \right], \quad \lambda = m, M. \tag{51}$$

The vertical hydrodynamic force acting on the plate can be obtained by integrating the dynamic pressure difference across the plate and is given as

$$F_m = \mathrm{i}\sigma\rho_2 \int_{-b}^{b} \big(\phi_2(x, d^+) - \phi_2(x, d^-)\big)\mathrm{d}x = \mathrm{i}\sigma\rho_2 \int_{-b}^{b} P(x)\mathrm{d}x. \tag{52}$$

Thus using (38) we have

$$F_m = \mathrm{i}\sigma\rho_2\big((t^m + 1)S_{+,0} + r^m S_{-,0} + T^m X_{+,0} + R^m X_{-,0}\big) \tag{53}$$

where $S_{\pm,0} = \langle P_\pm, f_0\rangle$, $X_{\pm,0} = \langle X_\pm, f_0\rangle$ and $f_0 = 1$. Since $f_0 = 1 = U_0(x/b)$, it follows that $S_{\pm,0} = (1/2)bA_0^+$ and $X_\pm = (1/2)bB_0^+$.

Thus, the dimensionless hydrodynamic force acting on the horizontal rigid thin plate is defined as

$$\hat{F}_m = \frac{F_m}{2\rho_2\sigma b\cosh mh} = \frac{1}{4}\left\{((t^m + 1) + r^m)A_0^+ + (T^m + R^m)B_0^+\right\}. \tag{54}$$

Following the similar mathematical analysis as described above, for a wave train of mode M obliquely incident at an angle θ, the solutions for the reflection coefficients, transmission coefficients and wave load on the plate can be obtained and analysed. Thus in the present paper, we only depict the numerical results for the case of incident wave of mode m.

4 Numerical Results and Discussions

The numerical results for different hydrodynamic quantities are computed after truncating the infinite series (42) to a finite number N. After some numerical examinations, it is found that the value of $N = 4$ is enough to produce sufficiently accurate numerical results.

Table 1 represents the validation of computed numerical values of the reflection and transmission coefficients against the energy balance relation. In this table,

Table 1 A numerical check on energy identity relation

mb	$\lvert r^m \rvert$	$\lvert R^m \rvert$	$\lvert t^m \rvert$	$\lvert T^m \rvert$	$\mathcal{J}$	$\lvert r^m\rvert^2 + \lvert R^m\rvert^2 + \mathcal{J}(\lvert t^m\rvert^2 + \lvert T^m\rvert^2)$
0.1507	6.808e^{-3}	1.814e^{-}3	0.99981	6.808e^{-3}	14.773	0.99977
0.6460	4.843e^{-3}	1.420e^{-3}	0.99702	6.808e^{-3}	817.92	1.00013
0.5947	6.233e^{-2}	1.641e^{-4}	0.99783	6.808e^{-3}	3.024e^{4}	1.00114
0.8406	3.860e^{-6}	5.960e^{-6}	0.99920	6.808e^{-3}	7.670e^{5}	1.00008

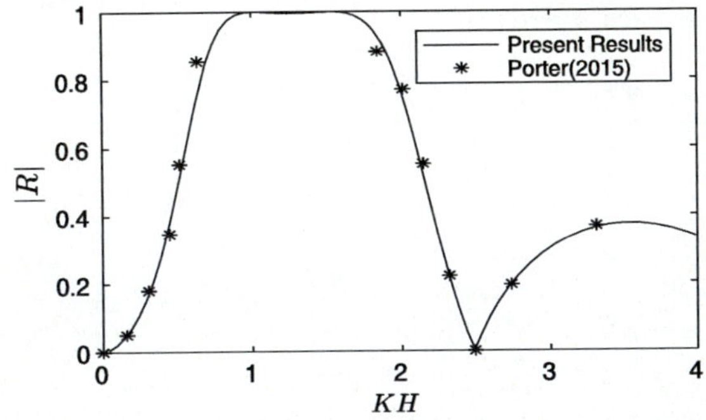

Fig. 2 Comparisons between the present results and the results obtained by Gradshteyn and Ryzhik [9]

we present the variations of r^m, R^m, t^m, T^m and $|r^m|^2 + |R^m|^2 + \mathcal{J}(|t^m|^2 + |T^m|^2)$ for few values of mb with other parameters as $s = 0.5, d/b = 2, h/b = 2, H/b = 4, \theta = 0^0$. It is visible from Table 1 that the reflection and the transmission coefficients satisfy the energy identity relation (50) accurately and this proves partial correctness of our numerical results.

Here we note that by letting $s = 1$ and $h \to 0$, we can reduce the two-layer fluid to a single layer fluid of depth H. Through Fig. 2, we validate the newly developed method by comparing the numerical results for reflection coefficient (R) with those obtained by Porter [3] where he studied water wave scattering by a horizontal rigid thin plate in a single layer fluid. Fig. 2 is generated considering $d/H = 0.1, b/H = 0.5, s = 1, h \to 0, \theta = 0°$. This graph demonstrates that the present results agree very well with those in Porter [3], and this provides additional validation on the numerical results obtained by the current analysis.

In Fig. 3a and b, for an incident wave train of mode m, we show the variations of the reflection coefficients $|r^m|$ and $|R^m|$ against the dimensionless wavenumber mb for different values of dimensionless submergence depth $d/b(= 0.5, 1, 1.5)$. Here the values of other fixed parameters are $s = 0.5, h/b = 2, H/b = 4, \theta = 0°$. These two figures show that as the submergence depth increases the reflection coefficients decrease. This may illustrate the fact that as the submergence depth increases, the interface and surface waves find more regions to pass the other side of the plate. It

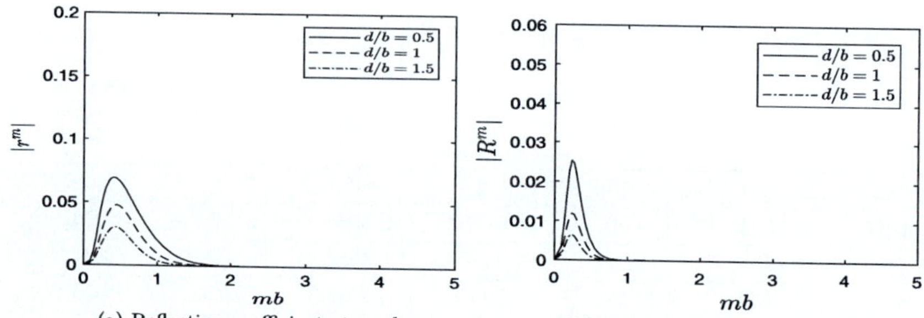

(a) Reflection coefficient at mode m

(b) Reflection coefficient at mode M.

Fig. 3 Reflection coefficients as a function of mb for various values of d/b with $s = 0.5, h/b = 2, H/b = 4, \theta = 0°$

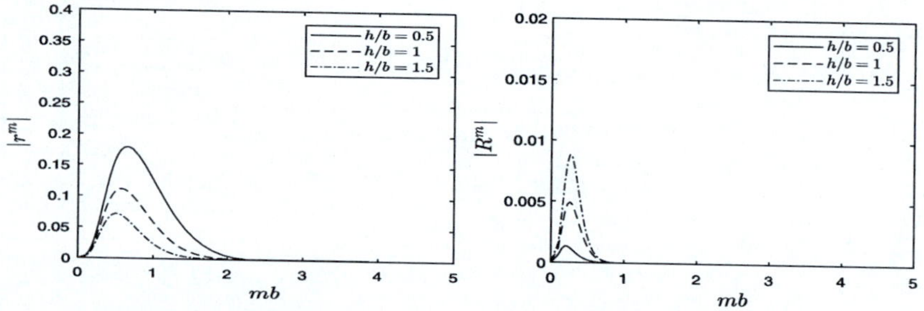

(a) Reflection coefficient at mode m

(b) Reflection coefficient at mode M

Fig. 4 Reflection coefficients as a function of mb for various values of h/b with $s = 0.5, H/b = 4, d/b = 2, \theta = 0°$

also demonstrates that the reflection coefficients diminish to zero beyond a certain value of dimensionless wavenumber.

For an incident wave train of mode m, Fig. 4a and b shows the influence of the interface position on the reflection coefficients $|r^m|$ and $|R^m|$ as a function of dimensionless wavenumber mb by altering the depth of the upper layer fluid($h/b = 0.5, 1, 1.5$) for the following fixed parameters: $s = 0.5$, $H/b = 4$, $d/b = 2$, $\theta = 0°$. From Fig. 4a, it is visible that as the interface is moved upwards, the reflection coefficient at mode m increases, whilst opposite behaviour for the reflection coefficient at mode M can be observed in Fig. 4b.

For an incident wave train of mode m, the effects of dimensionless plate length on the values of reflection coefficients $|r^m|$ and $|R^m|$ as a function of dimensionless wavenumber md are depicted in Fig. 5a and b respectively. Here the values of other fixed parameters are chosen as $s = 0.5$, $h/d = 1.5$, $H/d = 3$, $\theta = 0°$. These two graphs demonstrate the fact that as the plate length decreases, the reflection coefficients also decrease. One obvious explanation for this phenomenon is that a smaller plate obstructs less amount of waves, resulting in lower reflection.

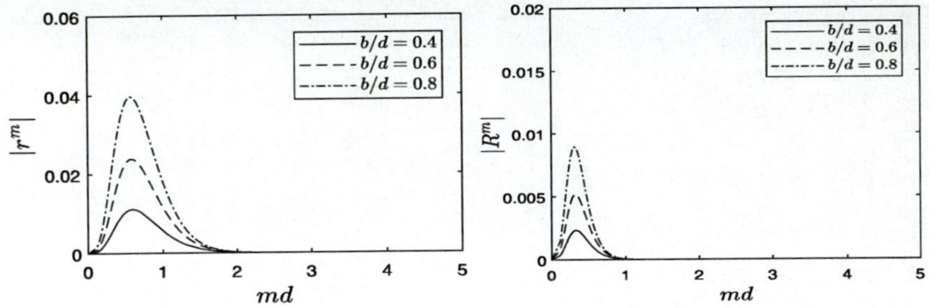

(a) Reflection coefficient at mode m (b) Reflection coefficient at mode M

Fig. 5 Reflection coefficients as a function of md for various values of b/d with $s = 0.5$, $H/d = 3$, $h/d = 1.5$, $\theta = 0°$

Fig. 6 The dimensionless hydrodynamic force $|\hat{F}_m|$ for various values of d/b with $s = 0.5$, $h/b = 2$, $H/b = 3$

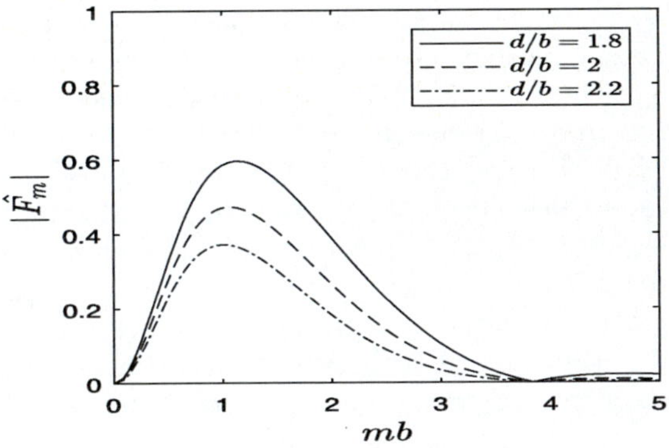

Figs. 6 and 7 represent the dimensionless hydrodynamic force $|\hat{F}_m|$ with respect to dimensionless wavenumber mb for different values of d/b and h/b respectively. Fig. 6 is plotted by choosing the values of parameters as $s = 0.5$, $h/b = 2$, $H/b = 3$, $\theta = 0°$. On the other hand, the Fig. 7 is plotted by choosing the values of parameters as $s = 0.5$, $H/b = 3$, $d/b = 2$. The two Figs. 6 and 7 indicate that the dimensionless hydrodynamic force exerted on the plate due to a wave train of mode m decreases as the submergence depth and depth of the upper layer fluid increase. It is obvious that as the plate moves deeper into the fluid, the propagating waves experience less obstruction by the plate resulting in lower vertical force.

5 Conclusions

On the basis of two-dimensional potential theory, we have investigated the problem of oblique wave scattering by a horizontal thin plate submerged in the lower layer of a two-layer fluid, comprising of two finite layers of fluids, in which upper

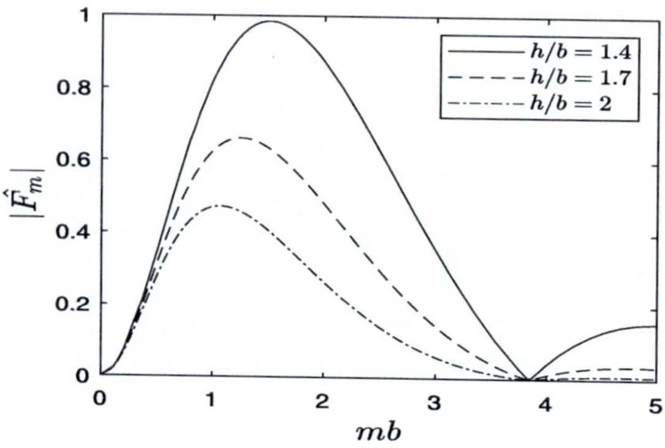

Fig. 7 The dimensionless hydrodynamic force $|\hat{F}_m|$ for various values of h/b with $s = 0.5, H/b = 3, d/b = 2$

layer has a free surface. We have adopted the Fourier integral transform to formulate integral equation involving unknown potential difference function across the plate. Applying Galerkin method to the solution of this integral equation, we obtain simple expressions for the reflection coefficients, transmission coefficients and hydrodynamic force exerted on the plate. We have validated the results obtained for the present analysis with those in Porter [3]. Also to ensure the validity of our results, we have calculated the energy identity for an incident wave of mode m. The dependence of various hydrodynamic quantities on the various parameters are depicted through figures. The reflection coefficients and hydrodynamic force significantly depend on the submergence depth of the plate and the interface position of the fluids. Reflection coefficients at the interface mode and surface mode for the incident wave of mode m increase as the submergence depth of the plate decreases. Also, the dimensionless hydrodynamic force acting on the plate decreases as the depth of upper layer fluid increases and increases as the submergence depth of the plate decreases. As usual, here also increasing plate lengths reflect more amount of wave energy. With decreasing depth of the upper layer fluid, the effect of the plate on the surface waves becomes prominent whereas the effect of the plate on the interface waves becomes suppressed. For the normal incident wave of mode m, the horizontal plate reflects very less amount of waves incident on it even zero beyond a certain value of dimensionless wavenumber. Thus, the plate can be used to construct as a component of a lens for the purpose of wave focusing. Moreover, the present method could further be extended to study the wave interaction problem with more than one horizontal plate submerged in a two-layer fluid.

References

1. McIver, M.: Diffraction of water waves by a moored, horizontal, flat plate. J. Eng. Maths. **19**, 297–319 (1985)
2. Mehlum, E.: A circular cylinder in water waves. Appl. Ocean Res. **2**, 171–177 (1980)

3. Porter, R.: Linearised water wave problems involving submerged horizontal plates. Appl. Ocean Res. **50**, 91–109 (2015)
4. Lamb, H.: Hydrodynamics. Cambridge University Press (1932)
5. Linton, C.M., McIver, M.: The interaction of waves with horizontal cylinders in two layer fluids. J. Fluid Mech. **304**, 213–229 (1995)
6. Dhillon, H., Banerjea, S., Mandal, B.N.: Wave scattering by a thin vertical barrier in a two-layer fluid. Int. J. Eng. Sci. **78**, 73–88 (2014)
7. Islam, N., Gayen, R.: Scattering of water waves by an inclined plate in a two layer fluid. Appl. Ocean Res. **80**, 136–147 (2018)
8. Medina-Rodríguez, A., Silva, R.: Oblique water-wave scattering by two thick submerged-horizontal plates in a two-layer fluid. J. Waterw. Port Coast. Ocean Eng. **144**, 04018003 (2018)
9. Gradshteyn, I.M., Ryzhik, I.S.: Table of Integrals, Series and Products, 2nd edn. Academic Press, New York (1981)

Transversely Isotropic Homogeneous Medium with Absorbing Boundary Conditions: Elastic Wave Propagation Using Spectral Element Method

Poonam Saini

Abstract Particle displacements and stresses are calculated for studying elastic wave propagation in a transversely isotropic homogeneous medium. A mesh consisting of rectangular elements is considered for discretization of two-dimensional domain. The spectral element method is applied through the non-uniformly distributed Gauss-Lobatto-Legendre nodes. The tensor product of high order Lagrangian interpolation polynomials is used as shape functions. Lagrangian interpolation polynomials along with Gauss-Lobatto-Legendre quadrature rule for numerical integration results in diagonal mass matrix which leads to an efficient fully explicit solver for time integration. Second order accurate, central difference method is applied for time discretization. The displacements and stress components are exhibited through time series at a point and snapshots in the domain. The influence of absorbing boundary conditions is demonstrated on the displacement components at different times. The validation of numerical solution is ensured through its comparison with known analytical solution for the two dimensional homogeneous transversely isotropic model.

Keywords Transversely isotropic · Wave propagation · Absorbing boundary conditions · Spectral element method · Lagrange type shape function · Gauss-Lobatto-Legendre nodes

1 Introduction

The formulation of elastic wave problems is generally done under the assumption of homogeneity, perfect elasticity and isotropy. The reason being the lesser parameters in model description, simpler constitutive equations and easier solutions. But, in

P. Saini (✉)
Kurukshetra University, Kurukshetra, Haryana 136118, India
e-mail: poonam.kkr@gmail.com

R. K. Sharma et al. (eds.), *Frontiers in Industrial and Applied Mathematics*, Springer Proceedings in Mathematics & Statistics 410,
https://doi.org/10.1007/978-981-19-7272-0_31

more realistic studies, the assumption on isotropy is relaxed to take anisotropy of the medium into account. A transversely isotropic (TI) medium is considered to be a simpler and the most common anisotropic model in wave propagation studies. Such a medium possess a plane of isotropy, with normal along the single axis of symmetry.

The transient propagation of elastic waves is a reality in numerous fields of science and technology including earthquake and exploration seismology. The governing equations of motion are a system of partial differential equations with initial and/or boundary conditions which are posed in infinite space-time domain. Analytic solutions are difficult to find in these problems as they require application of advanced mathematical techniques. Generally, the standard numerical methods, such as FEM, are developed for solutions in bounded space-time domain. Thus, any model for the bounded region of interest may involve the reflections of elastic waves at the boundaries. But, for these superimposed reflections, the actual solution for wave motion becomes inaccurate. To tackle this situation numerically, an artificial boundary is considered to truncate the unbounded domain of the problem. This imagined boundary is placed at some distance away from the region of the interest. To make the problem well posed, an absorbing boundary condition (ABC) is devised at the truncating boundary, which can absorb the incident waves. The numerical solution obtained for the bounded domain may serve as the solution for original unbounded domain, provided appropriate boundary conditions are applied on truncating boundary.

For solving the equations of elastodynamics numerically, the finite element method (FEM) is a prominently used method. This technique has been used widely in elastic wave propagation modeling [8, 9, 17]. Although conventional FEM can simulate elastic wave propagation in arbitrary geometry domain, its cost of computation becomes high. Hence, in recent years, researchers have tried to implement new variants of FEM.

Spectral Element Method (SEM) was first proposed by Patera [10] for the modeling of liquid flow in computational fluid dynamics. This method is a variant of FEM which uses specific high order shape functions. The idea behind its development was to combine the accuracy and rapid convergence of Pseudo Spectral Method (PSM) with geometrical flexibility of FEM. The main advantage of SEM over FEM is its high accuracy of approximation of solution through a smaller number of elements. Interpolating polynomial was taken as Chebychev polynomial in Patera [10]. Maday and Patera [11] developed SEM further by introducing Lagrange polynomial in combination of Gauss-Lobatto-Legendre (GLL) quadrature rule, which led to a diagonal mass matrix.

The SEM has been used for the study of seismic wave propagation as well. Priolo and Seriani [13] used SEM with Chebychev polynomials for simulation of one dimensional wave propagation. The same technique has been extended to study the propagation of elastic waves in 2D and 3D media for different geological applications ([6, 14, 15]). Basabe [5] used Lagrange interpolation polynomial to apply SEM for the simulation of wave propagation in 2D isotropic elastic media. Seriani et al. [16] studied the propagation of elastic waves in 2D transversely isotropic medium with vertical symmetry axis using Chebychev SEM. But, this Chebychev formulation leads to a non-diagonal mass matrix, whose inversion takes a lot of computation time.

In the present work, SEM has been used to study the two dimensional elastic wave propagation in transversely isotropic media. The higher order Lagrange interpolating polynomial in combination with GLL nodes has been used as shape function. This choice yields a diagonal mass matrix due to orthogonality of approximation functions. It is easier to find inverse of the diagonal matrix as one has to just reciprocate the diagonal elements. Further, the ability to store elements of diagonal mass matrix as a one-dimensional vector leads to reduction in memory requirements. Thus, diagonal mass matrix results in a very efficient fully explicit scheme for integration over time. This is a significant advantage over classical FEM. The GLL quadrature rule for numerical integration is used to evaluate entries of elemental mass and stiffness matrix. The unbounded domain is simulated by introducing artificial boundaries on which absorbing conditions are enforced. The accuracy of method is demonstrated graphically by comparing numerical solution with analytical solution for a homogeneous transversely isotropic medium as mentioned in Carcione [12], Payton [2].

2 Elastic Moduli in Transversely Isotropic Material

The elastic properties of a medium are represented by its elastic stiffness tensor, c_{ijkl}. In the linear elasticity, this stiffness tensor relates the components of stress tensor (σ_{ij}) and strain tensor (ε_{kl}). The relevant relations are given according to the generalized Hooke's law as

$$\sigma_{ij} = c_{ijkl}\varepsilon_{kl}, \qquad (i, j = 1, 2, 3). \tag{1}$$

The stiffness tensor obey the symmetry relations $c_{ijkl} = c_{jikl} = c_{ijlk} = c_{klij}$; $(i, j, k, l = 1, 2, 3)$. Consequently, the number of independent components in this tensor reduce to 21, which are arranged in 6×6 symmetric matrix $\{C_{IJ}\}$, $(I, J = 1, 2..., 6)$. The Kronecker tensor (δ_{ij}) is used to relate the two set of indices as $I = i\delta_{ij} + (9 - i - j)(1 - \delta_{ij})$, $J = k\delta_{kl} + (9 - k - l)(1 - \delta_{kl})$; $(i, j, k, l = 1, 2, 3)$.

In an anisotropic medium, there are 21 independent elastic constants. A transversely isotropic (TI) medium has only 5 independent elastic constants. The constitutive relations for a transversely isotropic elastic material, with axis of symmetry lying along vertical z axis are written as Lubarda [7]

$$\begin{Bmatrix} \sigma_{xx} \\ \sigma_{yy} \\ \sigma_{zz} \\ \sigma_{yz} \\ \sigma_{zx} \\ \sigma_{xy} \end{Bmatrix} = \begin{bmatrix} C_{11} & C_{12} & C_{13} & 0 & 0 & 0 \\ C_{12} & C_{11} & C_{13} & 0 & 0 & 0 \\ C_{13} & C_{13} & C_{33} & 0 & 0 & 0 \\ 0 & 0 & 0 & C_{44} & 0 & 0 \\ 0 & 0 & 0 & 0 & C_{44} & 0 \\ 0 & 0 & 0 & 0 & 0 & C_{66} \end{bmatrix} \begin{Bmatrix} \varepsilon_{xx} \\ \varepsilon_{yy} \\ \varepsilon_{zz} \\ 2\varepsilon_{yz} \\ 2\varepsilon_{zx} \\ 2\varepsilon_{xy} \end{Bmatrix} \tag{2}$$

where $C_{66} = \frac{1}{2}(C_{11} - C_{12})$ and C_{IJ} is as defined above.
For two dimensional transversely isotropic elastic solid in x-z plane with symmetry axis lying along the vertical z axis, elastic stiffness matrix $\mathbf{C}$ can be written as

$$\mathbf{C} = \begin{pmatrix} C_{11} & C_{13} & 0 \\ C_{13} & C_{33} & 0 \\ 0 & 0 & C_{44} \end{pmatrix}. \quad (3)$$

In two dimensional case, number of independent elastic constants reduces to 4.

3 Mathematical Formulation

3.1 *Elastic Wave Equation in Two Dimensions in a Transversely Isotropic Medium*

The governing equations for a dynamical system in the presence of external force are given by Achenbach [1] as

$$\sigma_{ij,j} + f_i = \rho \ddot{u}_i \qquad in \qquad \Omega \times (0, T] \quad (4)$$

where $\Omega \subset R^2$ is the physical domain with boundary Γ, $(0, T]$ is the time domain σ_{ij} is the stress tensor, ρ is density of medium, $f_i = f_i(\mathbf{x}, t)$ is the force vector component and u_i is displacement vector component. The comma preceding an index in the subscript denotes derivative w.r.t. space partially and over dot represents partial time differentiation. Repetition of index means sum over that index following Einstein convention. In orthogonal Cartesian coordinate system, displacement vector (u_x, u_z) at a point (x, z) defines the motion of material particle. Using general definition of divergence of a tensor field S as $\nabla.S = \frac{\partial S_{ki}}{\partial x_k} e_i$, the expression $\sigma_{ij,j}$ in (4) can be expressed as

$$\begin{bmatrix} \partial_x & 0 & \partial_z \\ 0 & \partial_z & \partial_x \end{bmatrix} \begin{bmatrix} \sigma_{xx} \\ \sigma_{zz} \\ \sigma_{xz} \end{bmatrix} = \mathbf{D}^T \sigma \quad (5)$$

where $\mathbf{D} = \begin{bmatrix} \partial_x & 0 \\ 0 & \partial_z \\ \partial_z & \partial_x \end{bmatrix}$ is the differential operator.
Components of strain tensor are related to displacement field as

$$\varepsilon = \mathbf{D}\mathbf{u}\,. \quad (6)$$

Stresses and strains are related according to generalised Hooke's law

$$\boldsymbol{\sigma} = \mathbf{C}\boldsymbol{\varepsilon} \tag{7}$$

where $\mathbf{C}$ is elastic stiffness matrix as defined by (3). $\boldsymbol{\sigma} = \{\sigma_{xx}, \sigma_{zz}, \sigma_{xz}\}^T$, $\boldsymbol{\varepsilon} = \{\varepsilon_{xx}, \varepsilon_{zz}, 2\varepsilon_{xz}\}^T$ are the stress and strain vectors respectively, defined through usual components σ_{ij} of stress and ε_{ij} of strain tensors. Equations of motion (4) can be written in vector form as

$$\mathbf{D}^T\boldsymbol{\sigma} + \mathbf{f} = \rho\ddot{\mathbf{u}}. \tag{8}$$

3.2 SEM Formulation

The first step in the SEM formulation is to obtain the weak formulation. For this, the product of governing equations and the test function is integrated over the space domain Ω. Integration by parts is performed and Gauss divergence theorem is used to reduce the order of the spatial derivatives. The advantage of the weak formulation is that the free surface boundary conditions are naturally satisfied. In case of free surface boundary conditions, the weak formulation of (8) is obtained by introducing the space of admissible displacement field and the space of admissible test function respectively as

$$\mathbf{X} = \{\phi : \Omega \times (0, T] \to R^2 | \phi \in H^1(\Omega), \forall t \in (0, T]\}$$

and

$$\delta\mathbf{X} = \{\psi : \Omega \to R^2 | \psi \in H^1(\Omega)\}$$

where $H^1(\Omega)$ is the space of functions, which together with their first order partial derivatives, are square integrable over the domain Ω.

We search for $\mathbf{u} \in \mathbf{X}$ such that for any test function $\mathbf{w} \in \delta\mathbf{X}$ and $\forall t \in (0, T]$, we have

$$\partial_{tt}(\mathbf{w}, \rho\mathbf{u})_\Omega + a(\mathbf{w}, \mathbf{u})_\Omega = (\mathbf{w}, \mathbf{f})_\Omega. \tag{9}$$

The symbols $a(.,.)_\Omega$, and $(.,.)_\Omega$ are defined as

$$a(\mathbf{w}, \mathbf{u})_\Omega = \int_\Omega \mathbf{w}^{\mathrm{T}}\mathbf{D}^T\boldsymbol{\sigma} d\Omega = \int_\Omega (\mathbf{D}\mathbf{w})^T\mathbf{C}\boldsymbol{\varepsilon} d\Omega = \int_\Omega (\mathbf{D}\mathbf{w})^T\mathbf{C}\mathbf{D}\mathbf{u} d\Omega \tag{10}$$

$$(\mathbf{w}, \mathbf{f})_\Omega = \int_\Omega \mathbf{w}^T\mathbf{f} d\Omega \tag{11}$$

$$(\mathbf{w}, \rho\mathbf{u})_\Omega = \int_\Omega \mathbf{w}^T \rho \mathbf{u} d\Omega \,. \tag{12}$$

3.3 Discretization in Space

The space discretization of (9) is performed by approximating displacement field in a finite-dimensional subspace $\mathbf{X}_h = X_h \times X_h$ of original space $\mathbf{X}$. This approximation transforms (9) into a system of ordinary differential equations.

Approximating $\mathbf{u}$ as $\mathbf{u}_h \in \mathbf{X}_h$ given by linear combination

$$\mathbf{u}_h\,(x, z, t) = \begin{bmatrix} U_j^x\,(t)\,\phi_j\,(x, z) \\ U_j^z\,(t)\,\phi_j\,(x, z) \end{bmatrix}. \tag{13}$$

where $U_j^x(t)$ and $U_j^z(t)$ are SEM approximations coefficients in horizontal and vertical displacements respectively. ϕ_j denote shape functions for each node position.

Substituting $\mathbf{w} = (\phi_i, 0)^T$ for the test function, (9) can be simplified as

$$\partial_{tt} \int_\Omega \rho[\phi_i, 0] \begin{bmatrix} U_j{}^x(t)\phi_j \\ U_j{}^z(t)\phi_j \end{bmatrix} d\Omega +$$

$$\int_\Omega \left[\partial_x\phi_i \; 0 \; \partial_z\phi_i\right] \begin{bmatrix} C_{11} & C_{13} & 0 \\ C_{13} & C_{33} & 0 \\ 0 & 0 & C_{44} \end{bmatrix} \begin{bmatrix} \partial_x & 0 \\ 0 & \partial_z \\ \partial_z & \partial_x \end{bmatrix} \begin{bmatrix} U_j^x\,(t)\,\phi_j \\ U_j^z\,(t)\,\phi_j \end{bmatrix} d\Omega$$

$$= \int_\Omega [\phi_i, 0] \begin{bmatrix} f_x \\ f_z \end{bmatrix} d\Omega \,. \tag{14}$$

Simplifying it further, we get equation of the form

$$M_{ij}\partial_{tt}U_j^x + K_{ij}^1 U_j^x + K_{ij}^2 U_j^z = F_i^x \tag{15}$$

where

$$M_{ij} = \int_\Omega \rho\phi_i\phi_j dxdz \tag{16}$$

$$K_{ij}^1 = \int_\Omega \left(C_{11}\phi_{i,x}\phi_{j,x} + C_{44}\phi_{i,z}\phi_{j,z}\right) dxdz \tag{17}$$

$$K_{ij}^{2} = \int\limits_{\Omega} \left(C_{13}\phi_{i,x}\phi_{j,z} + C_{44}\phi_{i,z}\phi_{j,x}\right)dxdz \tag{18}$$

$$F_{i}^{x} = \int\limits_{\Omega} f_{x}\phi_{i}dxdz\,. \tag{19}$$

Similarly, on substituting $\mathbf{w} = (0, \phi_i)^T$ in (9), we get one more system of equations

$$M_{ij}\partial_{tt}U_{j}^{z} + K_{ij}^{3}U_{j}^{x} + K_{ij}^{4}U_{j}^{z} = F_{i}^{z} \tag{20}$$

where

$$K_{ij}^{3} = \int\limits_{\Omega} \left(C_{13}\phi_{i,z}\phi_{j,x} + C_{44}\phi_{i,x}\phi_{j,z}\right)dxdz \tag{21}$$

$$K_{ij}^{4} = \int\limits_{\Omega} \left(C_{33}\phi_{i,z}\phi_{j,z} + C_{44}\phi_{i,x}\phi_{j,x}\right)dxdz \tag{22}$$

$$F_{i}^{z} = \int\limits_{\Omega} f_{z}\phi_{i}dxdz\,. \tag{23}$$

Equations (15) and (20) are combined in block matrix form as

$$\mathbf{A}\partial_{tt}\mathbf{U} + \mathbf{B}\mathbf{U} = \mathbf{F}, \qquad \mathbf{U} = \left(\mathbf{U}^{x}\left(t\right), \mathbf{U}^{z}\left(t\right)\right)^{T} \tag{24}$$

where $\mathbf{A} = \begin{bmatrix} \mathbf{M} & \mathbf{0} \\ \mathbf{0} & \mathbf{M} \end{bmatrix}$ is the assembled mass matrix.

$\mathbf{B} = \begin{bmatrix} \mathbf{K}^1 & \mathbf{K}^2 \\ \mathbf{K}^3 & \mathbf{K}^4 \end{bmatrix}$ is the assembled stiffness matrix.

$\mathbf{F} = (\mathbf{F^x}, \mathbf{F^z})^T$ is the global force vector.

These assembled matrices and vector are formed by assembly of all elemental level matrices and vectors respectively. Equation (24) is in semi-discretized form wherein partial differential equation has been discretized with respect to space only.

3.4 *Discretization in Time*

The equation of motion in semi-discretized form is expressed as

$$\mathbf{A}\ddot{\mathbf{U}} + \mathbf{B}\mathbf{U} = \mathbf{F} \tag{25}$$

along with the initial conditions
$\mathbf{U}(0) = \mathbf{U}_0 = \mathbf{0}$
$\dot{\mathbf{U}}(0) = \dot{\mathbf{U}}_0 = \mathbf{0}$
where the vector $\mathbf{U}$ represents the values of approximate solution $\mathbf{u}_h$ at all global nodes for each degree of freedom. Equation (25) is a second order ordinary differential equation in time. Explicit central difference method or implicit Newmark method are mostly used methods for time discretization. The choice of Lagrange polynomial at GLL collocation points as shape function along with GLL quadrature rule results in diagonal mass matrix $\mathbf{A}$. Due to diagonal mass matrix, explicit time integration methods are most effective because these methods become truely explicit. That is a system of equations is not required to be solved for each time step. Inverse of mass matrix can be calculated easily which makes computational algorithm less expensive. Explicit time integration scheme used here is central difference method. This method is second order accurate and conditionally stable. Time step stability limit of this method is highest among all second order methods. Time discretization of (25) is obtained by discretizing the time variable t in $[0, T]$ as $t_n = n\Delta t$, $\Delta t = \frac{T}{N_T}$, where N_T is the number of time steps. At time t_n, solution $\mathbf{U}(t_n)$ is simply denoted as $\mathbf{U}_n$.

In central difference method, we write

$$\ddot{\mathbf{U}}_n = \frac{\mathbf{U}_{n+1} - 2\mathbf{U}_n + \mathbf{U}_{n-1}}{(\Delta t)^2}. \tag{26}$$

Substituting the value in (25) at time $t = t_n$

$$\mathbf{A}\left[\frac{\mathbf{U}_{n+1} - 2\mathbf{U}_n + \mathbf{U}_{n-1}}{(\Delta t)^2}\right] + \mathbf{B}\mathbf{U}_n = \mathbf{F}_n$$

$$\mathbf{A}\mathbf{U}_{n+1} = 2\mathbf{A}\mathbf{U}_n + (\Delta t)^2(\mathbf{F}_n - \mathbf{B}\mathbf{U}_n) - \mathbf{A}\mathbf{U}_{n-1}.$$

Substituting $\mathbf{A}$, $\mathbf{B}$, $\mathbf{F}_n$, $\mathbf{U}_n$ and simplifying

$$\mathbf{U}_x^{n+1} = 2\mathbf{U}_x^n - \mathbf{U}_x^{n-1} + \frac{(\Delta t)^2[\mathbf{F}_x^n - K^1\mathbf{U}_x^n - K^2\mathbf{U}_z^n]}{M} \tag{27}$$

$$\mathbf{U}_z^{n+1} = 2\mathbf{U}_z^n - \mathbf{U}_z^{n-1} + \frac{(\Delta t)^2[\mathbf{F}_z^n - K^3\mathbf{U}_x^n - K^4\mathbf{U}_z^n]}{M}. \tag{28}$$

Displacements at present time step in x and z directions are calculated from displacements at two previous time steps using (27) and (28) respectively. The displacements thus calculated are used to compute the strain components through (6). These strains are further used along with the material elastic moduli to solve for stress components using (7).

4 Computational Algorithm

4.1 Domain Decomposition and Mapping of Geometry

Physical domain is divided into non-overlapping elements $\Omega_e, e = 1, 2 \ldots n_e$, where n_e is the total number of elements. For a 2D problem, the elements are quadrilaterals. The integrals of the weak form (9) are evaluated separately for each element domain Ω_e. The computation of the integral over an element is simplified by means of an invertible transformation between the general element Ω_e and reference element $\widehat{\Omega}$. The reference element is expressed by natural coordinates (ξ, η) where $0 \le \xi \le 1$ and $0 \le \eta \le 1$. The element geometry is mapped from natural coordinates (ξ, η) to physical coordinates (x, z). Tensor product of linear Lagrange polynomial has been used for the purpose of this mapping. Four bilinear Lagrange type functions used in this mapping for the quadrilateral element are

$$\psi_1 = (1-\xi)(1-\eta), \qquad \psi_2 = \xi(1-\eta), \qquad \psi_3 = (1-\xi)\eta, \qquad \psi_4 = \xi\eta. \tag{29}$$

The points on reference domain and physical domain are related as follows

$$x(\xi, \eta) = \sum_{i=1}^{4} \psi_i(\xi, \eta) x_i^e, \qquad z(\xi, \eta) = \sum_{i=1}^{4} \psi_i(\xi, \eta) z_i^e. \tag{30}$$

where x_i^e, z_i^e are x and z coordinates respectively of local node i of element e.

4.2 Interpolation of Field Variables Using Shape Functions

In SEM, higher order Lagrange interpolation polynomials are used for expressing field variables on the elements. Let $\{\xi_i\}_{i=0}^{k}$ be the nodes on ξ side of the reference square $\hat{\Omega}$ with $\xi_0 = 0$ and $\xi_k = 1$ and $\{l_j\}_{j=0}^{k}$ be the interpolating Lagrange polynomial with the condition $l_j(\xi_i) = \delta_{ij}$, where δ_{ij} is the Kronecker's delta. Lagrange polynomial for each ξ_i is defined as

$$l_i(\xi) = \frac{(\xi-\xi_0)\ldots(\xi-\xi_{i-1})(\xi-\xi_{i+1})\ldots(\xi-\xi_k)}{(\xi_i-\xi_0)\ldots(\xi_i-\xi_{i-1})(\xi_i-\xi_{i+1})\ldots(\xi_i-\xi_k)}, \; i = 0, 1.., k. \tag{31}$$

Using polynomials (31), the shape function for each node position in unit square element is given by

$$\phi_q(\xi, \eta) = l_i(\xi) l_j(\eta), \qquad (i, j = 0, 1, \ldots k) \tag{32}$$

where $q = (k+1)j + i$. Range of $q = 0, 1, \ldots (k+1)^2 - 1$.

These shape functions are referred to as tensor product Lagrange shape functions. A scalar function f on a general element Ω_e is approximated by these shape functions as

$$f(\mathbf{x}(\xi,\eta)) = \sum_{i,j=0}^{k} l_i(\xi) l_j(\eta) f_{ij}\,. \tag{33}$$

The horizontal and vertical displacements throughout the element are interpolated using shape functions and nodal displacements as

$$u_x = \sum_{i=0}^{(k+1)^2-1} \phi_i\,(\xi,\eta)\, u^e_{x,i}\,, \qquad u_z = \sum_{i=0}^{(k+1)^2-1} \phi_i\,(\xi,\eta)\, u^e_{z,i}\,. \tag{34}$$

Tensor product of kth order Lagrange polynomial is taken as shape function ϕ_i. $u^e_{x,i}$, $u^e_{z,i}$ are x and z components of displacement respectively of local node i of element e.

In SEM, the collocation points ξ_i $(i = 0, 1..., k)$, which are interpolated by Lagrange polynomials of degree k are selected as the $(k + 1)$ Gauss-Lobatto-Legendre (GLL) points. The GLL points are defined as the roots of the equation

$$(1 - x^2) P_k'(x) = 0 \tag{35}$$

where $P_k'(x)$ is the first derivative of kth order Legendre polynomial. This choice of interpolation points is convenient because it allows one to enforce continuity of field variables across the element boundaries.

4.3 Computation of Mass and Stiffness Matrices

The evaluation of entries of stiffness matrices in Eqs. (17), (18), (21) and (22) requires the differentiation of shape functions w.r.t physical coordinates. Since, shape functions are expressed in terms of natural coordinates, their derivatives w.r.t. physical coordinates x and z must be transformed to derivative w.r.t. natural coordinates ξ and η. The transformation of derivatives from physical coordinate system to natural coordinate system is obtained by chain rule of partial differentiation which is expressible in matrix form as Carey and Oden [3]

$$\begin{bmatrix} \frac{\partial}{\partial \xi} \\ \frac{\partial}{\partial \eta} \end{bmatrix} = \begin{bmatrix} \frac{\partial x}{\partial \xi} & \frac{\partial z}{\partial \xi} \\ \frac{\partial x}{\partial \eta} & \frac{\partial z}{\partial \eta} \end{bmatrix} \begin{bmatrix} \frac{\partial}{\partial x} \\ \frac{\partial}{\partial z} \end{bmatrix} = J \begin{bmatrix} \frac{\partial}{\partial x} \\ \frac{\partial}{\partial z} \end{bmatrix} \tag{36}$$

where J is the two dimensional matrix that denotes the mapping from physical coordinates (x, z) to the natural coordinates (ξ, η). The determinant of matrix J is referred to as the Jacobian and is used in transforming the integrals as follows

$$\int\int dxdz = \int\int \det(J)\, d\xi d\eta.$$

Using (30), we have $\frac{\partial x}{\partial \xi} = \sum_{i=1}^{4} \frac{\partial \psi_i}{\partial \xi} x_i^e$. Similar expressions are derived to get the entries of matrix J. Inverse of transformation (36) may be written as

$$\begin{bmatrix} \frac{\partial}{\partial x} \\ \frac{\partial}{\partial z} \end{bmatrix} = J^{-1} \begin{bmatrix} \frac{\partial}{\partial \xi} \\ \frac{\partial}{\partial \eta} \end{bmatrix} = \begin{bmatrix} J_{11}{}^{*} & J_{12}{}^{*} \\ J_{21}{}^{*} & J_{22}{}^{*} \end{bmatrix} \begin{bmatrix} \frac{\partial}{\partial \xi} \\ \frac{\partial}{\partial \eta} \end{bmatrix}. \tag{37}$$

In (37), matrix with starred entries represents inverse of matrix J which can be computed easily. With the help of (37), we can find derivatives of shape functions w.r.t physical coordinates in terms of derivatives w.r.t. natural coordinates. Entries of mass matrix in (16) are computed as

$$M_{ij} = \int_{\Omega} \rho \phi_i \phi_j d\Omega = \sum_{e=1}^{n_e} \int_{\Omega_e} \rho \phi_i \phi_j dxdz = \sum_{e=1}^{n_e} \int_0^1 \int_0^1 \rho \phi_i \phi_j \det(J)\, d\xi d\eta. \tag{38}$$

For the calculation of entries of element matrices, integration has been performed numerically using Gauss-Lobatto-Legendre(GLL) quadrature rule for numerical integration. In GLL quadrature, boundary points of the interval are also included. GLL quadrature rule on unit interval [0, 1] has been applied for calculation of elementary integrals.

5 Absorbing Boundary Condition for Unbounded Domain

The accurate modeling of seismic wave propagation requires truncation of the model in finite domain. A proper boundary condition needs to be applied at the artificial boundary for elimination of the reflections from the edges. Large number of techniques are developed in order to find a suitable boundary condition which can effectively eliminate reflections from the truncating boundary. Cerjan et al. [4] proposed to eliminate the reflected wave by setting the damping boundary layer outside the working area. This type of boundary condition works on the principle of gradually damping of waves in the neighbourhood of truncating boundary. In this method, wave amplitude is multiplied by an exponential function in the thin strip in the vicinity of artificial boundary, known as Cerjan sponge boundary layer. This technique decreases the amplitude of wave in the desired narrow region and thus eliminates reflections. The method works for large range of incident angles and different geological models. Due to simple application technique, this method is often used for numerical simulations [4]. We use slight improvement of Cerjan boundary condition based on Tian et al. [18].

The nodal displacements at $(n+1)$th time step is generated with the help of displacements at nth and $(n-1)$th time steps using (27) and (28). To damp the wave in the damping area using this absorbing condition, the displacement at $(n+1)$th time step, $\mathbf{U}^{n+1}$, is replaced by $\tilde{\mathbf{U}}^{n+1}$ which can be expressed as

$$\tilde{\mathbf{U}}^{n+1} = \sigma(\omega\mathbf{U}^{n+1} + (1-\omega)\mathbf{U}^{n})$$

where $\omega = \exp(-\alpha(\frac{i}{N})^2)$, $\sigma = \exp(-\beta(\frac{i}{N})^2)$.

Here, N is the damping strip width. Generally, width of damping strip is taken as 20 node for wave propagation problems. The parameter i denotes node positioning inside the damping strip, starting from i=1 in the interior of strip and increasing outwards.

6 Discussion of Numerical Result

We shall discuss the results for bounded and unbounded domains.

6.1 Bounded Domain

For bounded domain, results are presented for the wave propagation simulation through a sample of Apatite. This anisotropic material has transverse isotropy with z axis being the axis of symmetry and with the following elastic moduli
$C_{11} = 16.7$, $C_{13} = 6.6$, $C_{33} = 14.0$, $C_{44} = 6.63$, $\rho = 3200$.

Elastic moduli are in the unit of GPa and hence should be multiplied by $10^{10} N/m^2$. Density(ρ) is in the units of kg/m^3.

Two dimensional domain is taken as a square of size $33cm \times 33cm$ discretized with a grid of 20×20 elements in $x - z$ plane. This is a bounded and connected domain with boundary conditions applied on boundary Γ. In the present paper, free surface boundary conditions has been considered. The order of SEM has been taken as 7.

Motion is excited by z directional point force $f(\mathbf{x}, t) = g(\mathbf{x})\, h(t)$ applied at the center of domain. The time history $h(t)$ of the source function is defined by

$$h(t) = e^{-0.5 f_0^2 (t-t_0)^2} \cos \pi f_0 (t - t_0) \tag{39}$$

where $t_0 = 6\mu s$ and $f_0 = 500kHz$.

The source function is implemented as a 2D Gaussian in space.

Absorbing boundary conditions are not used for the bounded domain case because simulation is stopped before propagating wave reaches the mesh limits. Results are presented in the form of snapshots, which represent the wave field at a particular

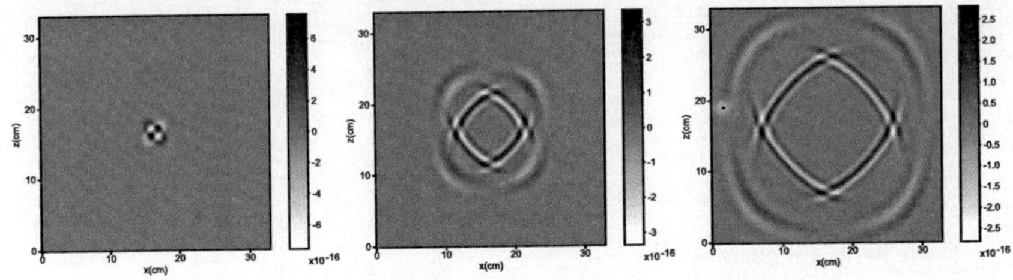

Fig. 1 Snapshot of x component of displacement at **(a)** t = 10 μs, **(b)** t = 20 μs and **(c)** t = 30 μ

instant of time. The standard geophysical package Seismic Unix is used for visualization of snapshots. Figures 1 and 2 show snapshots of the x and z components of displacements at specified times . As it can be seen, wave front shows characteristics predicted by wave front curves as explained in [12]. Point K (16.5, 21.4) is chosen as observation point to show variation of displacement components and stress components w.r.t. time. Time history of horizontal displacement u_x and vertical displacement u_z at point K is plotted in Fig. 3. Figure 4 shows variation of stress components w.r.t time at point K.

Figure 5 shows numerical and analytical solution on the same scale. Analytical solution along symmetry axis for a homogeneous transversely isotropic solid is taken from [12] and is given in Appendix. Receiver is located at distance 4.9 cm from source position along symmetry axis. The figure demonstrate exact overlapping of the numerical and analytical solutions. Total time for simulation is taken to be 40 μs and number of time steps are taken as 1250 for the purpose of comparison.

6.2 *For Unbounded Domain*

For finding the numerical solution in the unbounded domain, we truncate the unbounded problem domain to a finite computational domain. We choose 100 cm × 100 cm as truncated domain discretized with a grid of 60 × 60 elements in $x - z$ plane. Motion is exited by the same z directional point source as in case of bounded

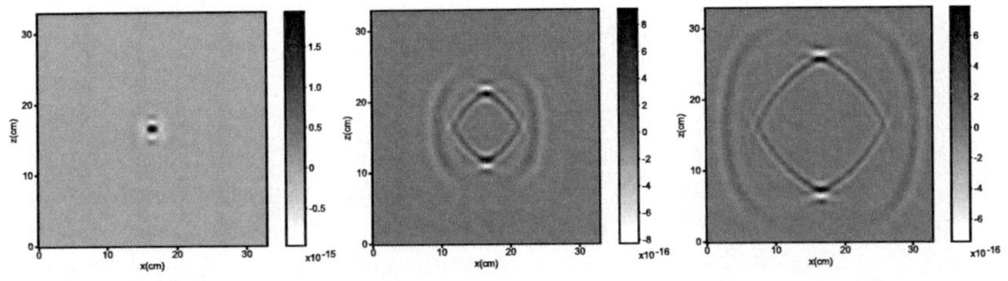

Fig. 2 Snapshot of z component of displacement at **(a)** t = 10 μs, **(b)** t = 20 μs and **(c)** t = 30 μs

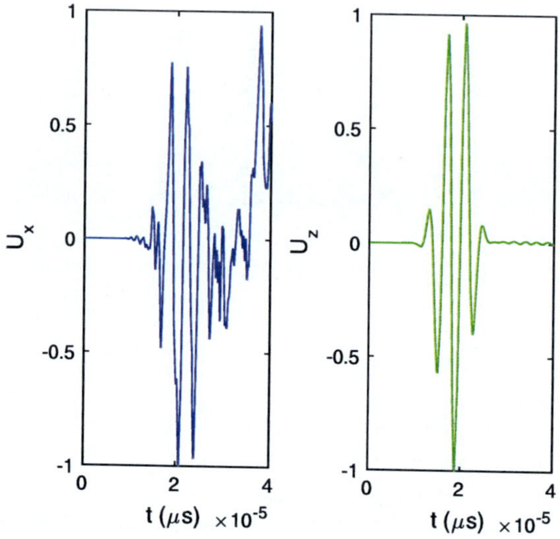

Fig. 3 Time histories of the horizontal and vertical displacements at point K(16.5, 21.4)

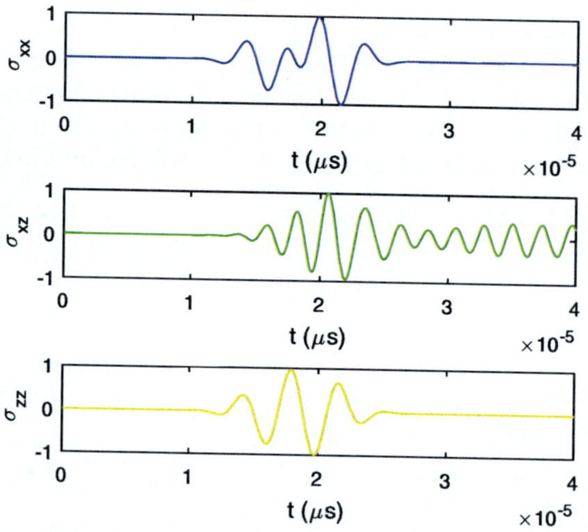

Fig. 4 Time histories of the stress components at point K(16.5, 21.4)

domain applied at the center. Receiver is placed at point (50, 54.9) along symmetry axis. Total time for simulation is taken to be 160μ s and number of time steps are taken as 4000 for the purpose of simulation. Simulation is done for larger time period to demonstrate the effect of absorbing boundary conditions on the reflections at artificial boundary. In Fig. 6, it can be seen that reflections from the edges are very much

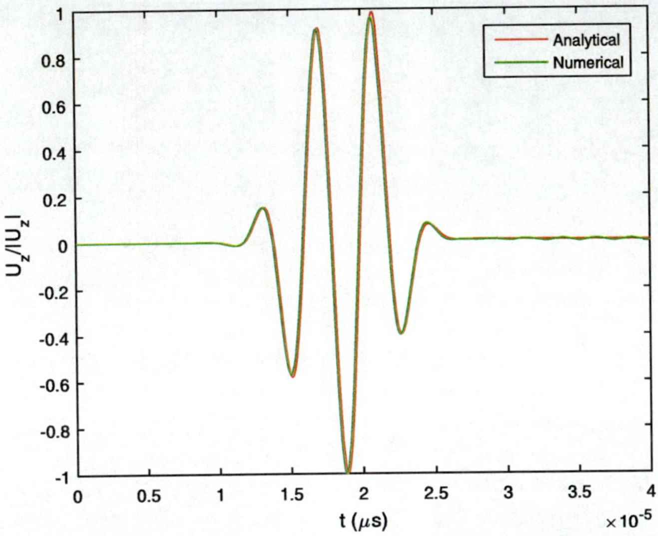

Fig. 5 Comparison between analytical and numerical solution along symmetry axis at a distance of 4.9 cm from source position

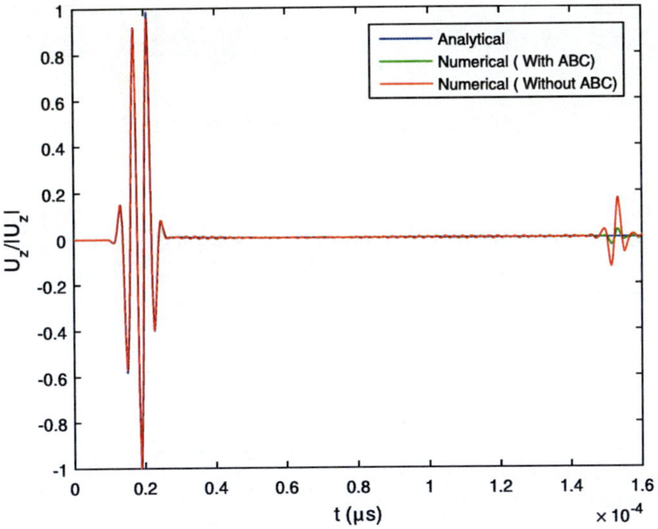

Fig. 6 Comparison between analytical and numerical solutions (with and without absorbing boundary conditions) along symmetry axis at a distance of 4.9 cm from source position

reduced in numerical solution with absorbing boundary than in case of numerical solution without absorbing boundary. Solution with the present method of absorbing boundary conditions is in good agreement with the analytical solution for unbounded

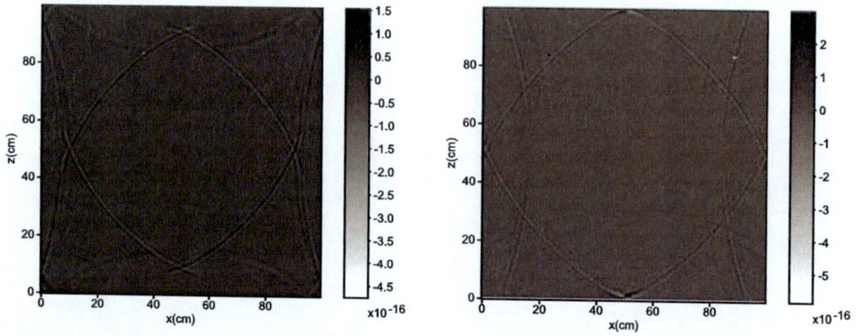

Fig. 7 Snapshot of x component of displacement without absorbing boundary at **(a)** t = 100 μs and **(b)** t = 120 μs

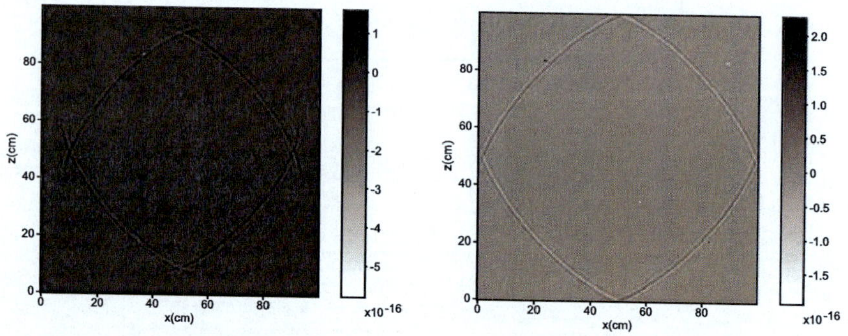

Fig. 8 Snapshot of x component of displacement with absorbing boundary at **(a)** t = 100 μs and **(b)** t = 120 μs

domain. Figure 7 shows snapshots of the x components of displacements at specified times without applying any absorbing boundary condition. The same components with the application of absorbing boundary condition are shown in Fig. 8. It can be seen that spurious boundary reflections are considerably reduced with application of absorbing boundary condition. The comparison for z component of displacement before and after applying absorbing boundary conditions can be seen from Figs. 9 and 10.

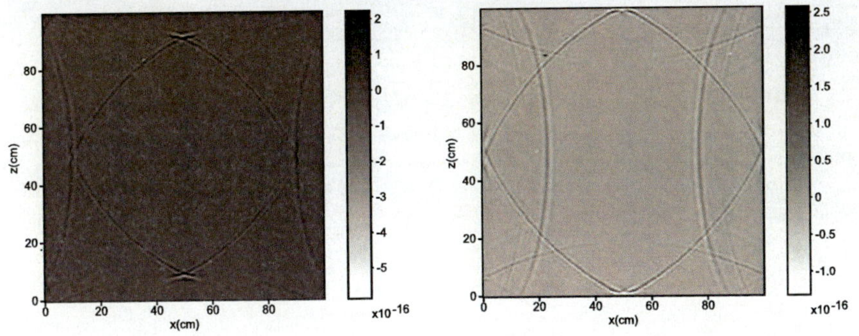

Fig. 9 Snapshot of z component of displacement without absorbing boundary at **(a)** t = 100 μs and **(b)** t = 120 μs

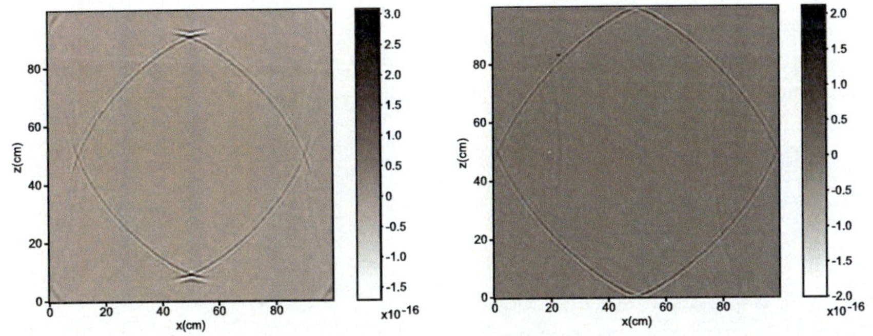

Fig. 10 Snapshot of z component of displacement with absorbing boundary at **(a)** t = 100 μs and **(b)** t = 120 μs

7 Conclusion

In this paper, two dimensional elastic wave propagation is modeled in homogeneous transversely isotropic media by spectral element method. Snapshots generated by algorithm are in good agreement with predicted wavefront curves for the transversely isotropic media. Accuracy of algorithm has been established by comparing it with analytical solution along symmetry axis in homogeneous transversely isotropic media. Absorbing boundary conditions are used to reduce the reflections and simulate wave propagation in unbounded domain more accurately.

Present algorithm may be extended to problems involving wave propagation through heterogeneous media where elastic constants vary with position of particles in domain. Transverse isotropy may be extended to orthotropy or general anisotropy.

Appendix

Analytic solution for a homogeneous transversely isotropic material
Two dimensional Green's function u_k satisfies following equation of motion in x-z plane:

$$c_{ijkl}\frac{\partial^2 u_k}{\partial x_l \partial x_j} + f_i = \rho\frac{\partial^2 u_i}{\partial t^2}, \qquad i, j, k, l = 1, 2$$

where f_i is the impulsive body force.

We define following dimensionless parameters

$$\alpha = \frac{C_{33}}{C_{44}}, \qquad \beta = \frac{C_{11}}{C_{44}} \qquad \gamma = 1 + \alpha\beta - \left(\frac{C_{13}}{C_{44}} + 1\right)^2$$

To simplify notations in the solution, dimensionless variables are defined as

$\bar{z} = \frac{z}{V_s t}$, where $V_s = \sqrt{\frac{C_{44}}{\rho}}$

Following is the analytic solution for class (3) of transversely isotropic materials (according to the classification done by Payton [12]) along the symmetry axis (z axis).

For this particular class of TI materials, $\gamma < \beta + 1$ and $\gamma^2 - 4\alpha\beta < 0$

Case I: when impulsive body force f is horizontal

$\mathbf{f} = (1, 0)\delta(x)\delta(z)\delta(t)$

where δ denotes Dirac delta function.

Response to Horizontal body force is given by

$$u_z = 0 \quad \text{and} \quad u_x = \begin{cases} 0 & 0 \le t \le t_p \\ F_1(\bar{z}) & t_p \le t \le t_s \\ 0 & t_s \le t \le t_1 \\ F_3(\bar{z}) & t > t_1 \end{cases}$$

where

$$F_1(\bar{z}) = \frac{1}{\pi\tau}\left[\frac{1}{4\beta} - \frac{2\beta(\alpha-\bar{z}^2) - \{\gamma-(\beta+1)\bar{z}^2\}}{4\beta\sqrt{D}}\right]\left[\frac{-\{\gamma-(\beta+1)\bar{z}^2\}+\sqrt{D}}{-2(\alpha-\bar{z}^2)(1-\bar{z}^2)}\right]^{\frac{1}{2}}$$

$$F_3(\bar{z}) = \frac{1}{2\pi\tau}\left[\frac{1}{\sqrt{\beta}} + \left(\frac{\alpha-\bar{z}^2}{1-\bar{z}^2}\right)^{\frac{1}{2}}\right]\left[\{\gamma - (\beta+1)\bar{z}^2\} + 2\{\beta(\alpha-\bar{z}^2)(1-\bar{z}^2)\}^{\frac{1}{2}}\right]^{-\frac{1}{2}}$$

Here, $t_s = z/(C_{44}/\rho)^{\frac{1}{2}}$, $t_p = z/(C_{33}/\rho)^{\frac{1}{2}}$ $t_1 = t_s/\bar{z}_1$

The quantity $D(\bar{z})$ and $\bar{z}_1$ are given by

$$D(\bar{z}) = \{\gamma - (\beta+1)\bar{z}^2\}^2 - 4\beta(\alpha-\bar{z}^2)(1-\bar{z}^2) \text{ and}$$

$$\bar{z}_1 = \left[\gamma(\beta+1) - 2\beta(\alpha+1) + 2\{\beta(1+\alpha\beta-\gamma)(\alpha+\beta-\gamma)\}^{\frac{1}{2}}\right]^{\frac{1}{2}}/(\beta-1)$$

Case II: when impulsive body force f is vertical

$\mathbf{f} = (0, 1)\delta(x)\delta(z)\delta(t)$

Solution in this case is

$$u_x = 0 \quad \text{and}\; u_z = \begin{cases} 0 & 0 \le t \le t_p \\ G_1(\bar{z}) & t_p \le t \le t_s \\ 0 & t_s \le t \le t_1 \\ G_3(\bar{z}) & t > t_1 \end{cases}$$

with

$$G_1(\bar{z}) = \frac{1}{\pi\tau}\left[\frac{1}{4} - \frac{2(1-\bar{z}^2)-\{\gamma-(\beta+1)\bar{z}^2\}}{4\sqrt{D}}\right]\left[\frac{-\{\gamma-(\beta+1)\bar{z}^2\}+\sqrt{D}}{-2(\alpha-\bar{z}^2)(1-\bar{z}^2)}\right]^{\frac{1}{2}}$$

$$G_3(\bar{z}) = \frac{1}{2\pi\tau}\left[\frac{1}{\sqrt{\beta}} + \left(\frac{1-\bar{z}^2}{\alpha-\bar{z}^2}\right)^{\frac{1}{2}}\right]\left[\{\gamma-(\beta+1)\bar{z}^2\} + 2\{\beta(\alpha-\bar{z}^2)(1-\bar{z}^2)\}^{\frac{1}{2}}\right]^{-\frac{1}{2}}$$

All the quantities are defined as above in case I.

For the comparison of analytical and numerical solution, the above free-space Green's function is convolved with the source time function $h(t)$ given by (39).

References

1. Achenbach, J.: Wave Propagation in Elastic Solids, vol. 16. Elsevier (2012)
2. Carcione, J.M., Kosloff, D., Seriani, G.: A spectral scheme for wave propagation simulation in 3-d elastic-anisotropic media. Geophysics **57**, 1593–1607 (1992)
3. Carey, G.F., Oden, J.T.: Finite Elements: An Introduction, vol. 1. Prentice Hall (1981)
4. Cerjan, C., Kosloff, D., Kosloff, D.: A nonreflecting boundary condition for discrete acoustic and elastic wave equations. Geophysics **50**, 705–8 (1985)
5. De Basabe, J.D.: High-Order Finite Element Methods for Seismic Wave Propagation. University of Texas at Austin, Texas (2009)
6. Komatitsch, D., Vilotte, J.P.: The spectral-element method: an efficient tool to simulate the seismic response of 2d and 3d geological structures. Bull. Seismol. Soc. Am. **88**(2), 368–392 (1998)
7. Lubarda, V., Chen, M.: On the elastic moduli and compliances of transversely isotropic and orthotropic materials. J. Mech. Mater. Struct. **3**(1), 153–171 (2008)
8. Lysmer, J., Drake, L.A.: A finite element method for seismology. Methods Comput. Phys. **11**, 181–216 (1972)
9. Marfurt, K.J.: Accuracy of finite-difference and finite-element modeling of the scalar and elastic wave equations. Geophysics **49**(5), 533–549 (1984)
10. Patera, A.T.: A spectral element method for fluid dynamics: laminar flow in a channel expansion. J. Comput. Phys. **54**, 468–488 (1984)
11. Patera, A.T., Maday, Y.: Spectral element methods for the incompressible navier-stokes equations. J. Comput. Phys. 71—143 (1989)
12. Payton, R.: Elastic wave propagation in transversely isotropic media. Martinus Nijhoff Publishers (1983)
13. Priolo, E., Seriani, G.: A numerical investigation of Chebyshev spectral element method for acoustic wave propagation. In: Proceedings 13th IMACS Conference on Computer Applications Mathematical, vol. 2, pp. 551—556. Dublin, Ireland (1991)
14. Seriani, G.: 3-d large-scale wave propagation modeling by a spectral element method on a cray t3e multiprocessor. Comp. Meth. Appl. Mech. Eng. **164**, 235–247 (1998)
15. Seriani, G., Priolo, E.: Spectral element method for acoustic wave simulation in heterogeneous media. Finite Elem. Anal. Des. **16**, 337–348 (1994)
16. Seriani, G., Priolo, E., Pregarz, A.: Modelling waves in anisotropic media by a spectral element method. In: Third International Conference on Mathematical and Numerical Aspects of Wave Propagation: Society for Industrial and Applied Mathematics, pp. 289—298 (1995)

17. Smith, W.D.: The application of finite element analysis to body wave propagation problems. Geophys. J. Int. **42**(2), 747–768 (1975)
18. Tian, X.B., kang, I.B., Kim, G., Zhang, H.: A nonreflecting boundary condition for discrete acoustic and elastic wave equations. J. Geophys. Eng. **5**, 203–209 (2008)

Growth of Polynomials Having No Zero Inside a Circle

Khangembam Babina Devi, N Reingachan, Thangjam Birkramjit Singh, and Barchand Chanam

Abstract In this manuscript, an upper bound estimate for the maximum modulus of a general class of polynomials with restricted zeros on a circle $|z| = L$, $L \geq 1$, is obtained in terms of the maximum modulus of the same polynomials on $|z| = 1$. It is observed that a result of Hussain [J. Pure Appl. Math., (2021) (https://doi.org/10.1007/s13226-021-00169-7)] is sharpened by our result. Also, this result generalizes and sharpens some other previously proved result.

Keywords Polynomials · Zeros · Inequalities · Maximum modulus

1 Introduction

Let $b(z)$ be a polynomial of degree m and let

$$\|b\| = \max_{|z|=1} |b(z)|, \quad M(b, L) = \max_{|z|=L} |b(z)|.$$

For a polynomial $b(z)$, there is a simple deduction from the Maximum Modulus Principle [11, p. 158] that for $L \geq 1$,

$$M(b, L) \leq L^m \|b\|. \tag{1}$$

Equality is obtained in (1) for $b(z) = \lambda z^m$ with $\lambda \neq 0$, $\lambda \in \mathbb{C}$.

For a polynomial $b(z)$ having all its zeros outside $|z| < 1$, it was shown by Ankeny and Rivlin [1] that for $L \geq 1$,

K. B. Devi (✉) · N. Reingachan · T. B. Singh · B. Chanam
Department of Mathematics, National Institute of Technology Manipur,
Imphal, Manipur 795004, India
e-mail: khangembambabina@gmail.com

B. Chanam
e-mail: barchand-2004@yahoo.co.in

R. K. Sharma et al. (eds.), *Frontiers in Industrial and Applied Mathematics*,
Springer Proceedings in Mathematics & Statistics 410,
https://doi.org/10.1007/978-981-19-7272-0_32

$$M(b, L) \leq \left(\frac{L^n + 1}{2}\right)\|b\|. \quad (2)$$

Equality holds in (2) for $b(z) = \alpha + \beta z^m$, where $|\alpha| = |\beta|$.

Govil [6] understood that equality in (2) holds only for polynomials $b(z) = \alpha + \beta z^m$, $|\alpha| = |\beta|$, which satisfy

$$|coefficient\ of\ z^m| = \frac{1}{2}\|b\|, \quad (3)$$

and it would be possible to refine the bound in (2) for polynomials which do not hold the condition given in (3). In an attempt to solve this problem, he [6] could obtain that for polynomial $b(z) = \sum_{v=0}^{m} w_v z^v$ having all its zeros outside $|z| < 1$ and $L \geq 1$, we have

$$\begin{aligned} M(b, L) \leq & \frac{(L^m + 1)}{2}\|b\| - \frac{m}{2}\left(\frac{\|b\|^2 - 4|w_m|^2}{\|b\|}\right) \\ & \times \left[\frac{(L-1)\|b\|}{\|b\| + 2|w_m|} - \ln\left\{1 + \frac{(L-1)\|b\|}{\|b\| + 2|w_m|}\right\}\right]. \end{aligned} \quad (4)$$

Recently, Hussain [8, Corollary 2] proved a generalization and extension of inequality (4) that

$$\begin{aligned} M(b, L) \leq & \left(\frac{L^m + s_1}{1 + s_1}\right) \| b \| - \frac{m}{1 + s_1}\left(\frac{(\| b \|)^2 - (1 + s_1)^2 |w_m|^2}{\| b \|}\right) \\ & \times \left\{\frac{(L-1) \| b \|}{\| b \| + (1 + s_1)|w_m|} - ln\left(1 + \frac{(L-1) \| b \|}{\| b \| + (1 + s_1)|w_m|}\right)\right\}, \end{aligned} \quad (5)$$

where

$$s_1 = \frac{k^{\mu+1}(\frac{\mu}{m}\frac{|w_\mu|}{|w_0|}k^{\mu-1} + 1)}{\frac{\mu}{m}\frac{|w_\mu|}{|w_0|}k^{\mu+1} + 1}, \quad (6)$$

where $b(z) = w_0 + \sum_{v=\mu}^{m} w_v z^v$, $\mu \in \{1, 2, ..., m\}$ is a polynomial such that $b(z) \neq 0$ in $|z| < k$, $k \geq 1$.

Remark 1 When $\mu = m$, the polynomial $b(z) = w_0 + \sum_{v=\mu}^{m} w_v z^v$ becomes $b(z) = w_0 + w_m z^m$. Therefore, by simple calculation, we have

$$M(b, L) = \max_{|z|=L} |w_0 + w_m z^m| = |w_0| + L^m |w_m|. \quad (7)$$

However, for $\mu = m$, inequality (5) reduces to

$$M(b, L) \leq \left(\frac{L^m + s_3}{1 + s_3}\right) \| b \| - \frac{m}{1 + s_3} \left(\frac{(\| b \|)^2 - (1 + s_3)^2 |w_m|^2}{\| b \|}\right)$$
$$\times \left\{\frac{(L-1) \| b \|}{\| b \| + (1 + s_3)|w_m|} - ln\left(1 + \frac{(L-1) \| b \|}{\| b \| + (1 + s_3)|w_m|}\right)\right\}, \quad (8)$$

where

$$s_3 = \frac{|\frac{w_m}{w_0}|k^{2m} + k^{m+1}}{|\frac{w_m}{w_0}|k^{m+1} + 1}. \quad (9)$$

The estimate of $M(b, L)$ given by inequality (8) for $\mu = m$ is not required as we could easily get the exact value of it by a simple calculation given by (7).

2 Main Results

In this manuscript, we obtain a result which is a refinement and a generalization of inequality (5) of Hussain [8].

Theorem 1 *If* $b(z) = w_0 + \sum_{v=\mu}^{m} w_v z^v$, $\mu \in \{1, 2, ..., m-1\}$, *is a polynomial having all its zeros outside* $|z| < k$, $k \geq 1$, *then for* $L \geq 1$ *and* $N \in \mathbb{Z}^+$, $N \leq m$,

$$M(b, L) \leq \left(\frac{L^m + s_1}{1 + s_1}\right)\|b\| - \frac{(L^m - 1)s_1 m^*}{(1 + s_1)k^m}$$
$$- m\left\{\frac{\|b\|}{1 + s_1} - \frac{s_1 m^*}{(1 + s_1)k^m} - |w_m|\right\} f(N, s_1), \quad (10)$$

where

$$s_1 = \frac{k^{\mu+1}(\frac{\mu}{m}|\frac{w_\mu}{w_0}|k^{\mu-1} + 1)}{\frac{\mu}{m}|\frac{w_\mu}{w_0}|k^{\mu+1} + 1} \quad (11)$$

and

$$f(N, s_1) = \left(L - 1\right) - \left\{1 + \frac{(1 + s_1)|w_m|}{\|b\| - \frac{s_1 m^*}{k^m}}\right\}$$
$$\times \ln\left\{1 + \frac{(L-1)(\|b\| - \frac{s_1 m^*}{k^m})}{(\|b\| - \frac{s_1 m^*}{k^m}) + (1 + s_1)|w_m|}\right\} \quad for\ N = 1, \quad (12)$$

$$
\begin{aligned}
f(N,s_1) &= \left(\frac{L^N-1}{N}\right) \\
&+ \sum_{v=1}^{N-1}\left(\frac{L^{N-v}-1}{N-v}\right)(-1)^v\left\{1+\frac{(1+s_1)|w_m|}{\|b\|-\frac{s_1m^*}{k^m}}\right\}\left\{\frac{(1+s_1)|w_m|}{\|b\|-\frac{s_1m^*}{k^m}}\right\}^{v-1} \\
&+ (-1)^N\left\{1+\frac{(1+s_1)|w_m|}{\|b\|-\frac{s_1m^*}{k^m}}\right\}\left\{\frac{(1+s_1)|w_m|}{\|b\|-\frac{s_1m^*}{k^m}}\right\}^{N-1} \\
&\times \ln\left\{1+\frac{(L-1)(\|b\|-\frac{s_1m^*}{k^m})}{(\|b\|-\frac{s_1m^*}{k^m})+(1+s_1)|w_m|}\right\} \quad for\ N \geq 2
\end{aligned}
\tag{13}
$$

and here and in the entire paper

$$
m^* = \min_{|z|=k}|b(z)|. \tag{14}
$$

Remark 2 From Lemma 3, $f(N, s_1)$ given by (12) and (13) of Theorem 1 is a monotonically increasing function of N, $N \leq m$, hence, taking $N = m$, we obtain the best bound in Theorem 1.

Further, consider $b(z)$ to be a polynomial whose degree $m = 1$. Then, by a straightforward calculation, we obtain

$$
M(b,L) = \max_{|z|=L}|b(z)| = \max_{|z|=L}|w_0 + Lw_1| = |w_0| + L|w_1|. \tag{15}
$$

Hence, we present the exact value of $M(b, L)$ for $m = 1$ which is given by (15).

From the preceding dicussion, Theorem 1 assumes

Corollary 1 *If* $b(z) = w_0 + \sum_{v=\mu}^{m} w_v z^v$, $\mu \in \{1, 2, ..., m-1\}$, *is a polynomial with all its zeros outside* $|z| < k$, $k \geq 1$, *then for* $L \geq 1$,

$$
M(b,L) = |w_0| + L|w_1| \quad for\ m = 1 \tag{16}
$$

and

$$
\begin{aligned}
M(b,L) \leq &\left(\frac{L^m+s_1}{1+s_1}\right)\|b\| - \frac{(L^m-1)s_1m^*}{(1+s_1)k^m} \\
&- m\left\{\frac{\|b\|}{1+s_1} - \frac{s_1m^*}{(1+s_1)k^m} - |w_m|\right\} f(m,s_1),
\end{aligned}
\tag{17}
$$

where

$$f(m, s_1) = \left(\frac{L^m - 1}{m}\right)$$
$$+\sum_{v=1}^{m-1}\left(\frac{L^{m-v} - 1}{m-v}\right)(-1)^v\left\{1+\frac{(1+s_1)|w_m|}{\|b\| - \frac{s_1 m^*}{k^m}}\right\}\left\{\frac{(1+s_1)|w_m|}{\|b\| - \frac{s_1 m^*}{k^m}}\right\}^{v-1}$$
$$+(-1)^m\left\{1+\frac{(1+s_1)|w_m|}{\|b\| - \frac{s_1 m^*}{k^m}}\right\}\left\{\frac{(1+s_1)|w_m|}{\|b\| - \frac{s_1 m^*}{k^m}}\right\}^{m-1}$$
$$\times \ln\left\{1+\frac{(L-1)(\|b\| - \frac{s_1 m^*}{k^m})}{(\|b\| - \frac{s_1 m^*}{k^m}) + (1+s_1)|w_m|}\right\} \quad for\ m \geq 2 \tag{18}$$

and s_1 is as defined in (11).

Remark 3 If $k = 1, s_1 = 1$, then Theorem 1 reduces to the succeeding result which refines and generalizes the result of Dewan and Bhat [3].

Corollary 2 *If $b(z) = w_0 + \sum_{v=\mu}^{m} w_v z^v$, $\mu \in \{12, \ldots\ldots, m-1\}$, is a polynomial with all its zeros outside $|z| < k$, $k \geq 1$, then for $L \geq 1$, and $N \in \mathbb{Z}^+$, $N \leq m$,*

$$M(b, L) \leq \left(\frac{L^m + 1}{2}\right)\|b\| - \left(\frac{L^m - 1}{2}\right)m^* - m\left(\frac{\|b\| - m^*}{2} - |w_m|\right)f(N, 1), \tag{19}$$

where

$$f(N, 1) = \left(\frac{L^N - 1}{N}\right)$$
$$+\sum_{v=1}^{N-1}\left(\frac{L^{N-v} - 1}{N-v}\right)(-1)^v\left\{1+\frac{2|w_m|}{\|b\| - m^*}\right\}\left\{\frac{2|w_m|}{\|b\| - m^*}\right\}^{v-1}$$
$$+(-1)^N\left\{1+\frac{2|w_m|}{\|b\| - m^*}\right\}\left\{\frac{2|w_m|}{\|b\| - m^*}\right\}^{N-1}$$
$$\times \ln\left\{1+\frac{(L-1)(\|b\| - m^*)}{(\|b\| - m^*) + 2|w_m|}\right\}. \tag{20}$$

Remark 4 Since for $1 \leq N$, $f(1, 1) \leq f(N, 1)$ and hence, substituting the value of $f(1, 1)$, inequality (19) becomes the result of Dewan and Bhat [3].

Remark 5 For $N = 1$, Theorem 1, in particular, becomes the following interesting result.

Corollary 3 *If $b(z) = w_0 + \sum_{v=\mu}^{m} w_v z^v$, $\mu \in \{12, \ldots\ldots, m-1\}$, is a polynomial with all its zeros outside $|z| < k$, $k \geq 1$, then for $L \geq 1$,*

$$M(b,L) \geq \left(\frac{L^m+s_1}{1+s_1}\right)\|b\| - \left(\frac{L^m-1}{1+s_1}\right)\frac{s_1 m^*}{k^m}$$
$$-\frac{m}{1+s_1}\left\{\frac{(\|b\|-\frac{s_1m^*}{k^m})^2-|w_m|^2(1+s_1)^2}{\|b\|-\frac{s_1m^*}{k^m}}\right\}$$
$$\times\left[\frac{(L-1)(\|b\|-\frac{s_1m^*}{k^m})}{\|b\|-\frac{s_1m^*}{k^m}+(1+s_1)|w_m|}-\ln\left\{1+\frac{(L-1)(\|b\|-\frac{s_1m^*}{k^m})}{\|b\|-\frac{s_1m^*}{k^m}+(1+s_1)|w_m|}\right\}\right],$$

where s_1 is as defined in (11).

Remark 6 By Lemma 8, we have

$$\left(\|b\|-\frac{s_1m^*}{k^m}\right)^2-(1+s_1)^2|w_m|^2 \geq 0 \tag{21}$$

and $\ln(1+x) < x$ for positive values of x and hence the bound given by Corollary 3 improves and generalizes inequality (2) proved by Ankeny and Rivlin [1].

Remark 7 By Lemma 10, $k \leq s_1$ for $k \geq 1$, where s_1 is as defined in (11), therefore, we have for $m^* \geq 0$

$$\frac{m}{1+s_1}\|b\|-\frac{ms_1m^*}{k^m(1+s_1)} \leq \frac{m}{1+s_1}\|b\|. \tag{22}$$

Applying Lemma 4 to (22), we have for $r \geq 1$,

$$r^{m-1}\left\{1-\frac{\left(\frac{m}{1+s_1}\|b\|-\frac{ms_1m^*}{k^m(1+s_1)}-m|w_m|\right)(r-1)}{m|w_m|+r\left(\frac{m}{1+s_1}\|b\|-\frac{ms_1m^*}{k^m(1+s_1)}\right)}\right\}\left\{\frac{m}{1+s_1}\|b\|-\frac{ms_1m^*}{k^m(1+s_1)}\right\}$$
$$\leq r^{m-1}\left\{1-\frac{\left(\frac{m}{1+s_1}\|b\|-m|w_m|\right)(r-1)}{m|w_m|+\frac{rm}{1+s_1}\|b\|}\right\}\frac{m}{1+s_1}\|b\|. \tag{23}$$

On integrating (23) from both sides with respect to r from 1 to L and following similar simplification of the RHS of (73) to inequality (74) in the proof of Theorem 1, we get

$$\frac{L^m-1}{1+s_1}\left(\|b\|-\frac{s_1m^*}{k^m}\right)-\frac{m}{1+s_1}\left(\|b\|-\frac{s_1m^*}{k^m}\right)(1-e)\int_1^R\frac{(r-1)r^{m-1}}{r+e}dr$$
$$\leq \frac{L^m-1}{1+s_1}\|b\|-\frac{m}{1+s_1}\|b\|(1-g)\int_1^L\frac{(r-1)r^{m-1}}{r+g}dr, \tag{24}$$

where $e=\frac{|w_m|(1+s_1)}{\|b\|-\frac{s_1m^*}{k^m}}$ and $g=\frac{|w_m|(1+s_1)}{\|b\|}$.

The expression $\int_1^L\frac{(r-1)r^{N-1}}{r+g}dr \geq 0$ and is a monotonically increasing function of N for $N \leq m$, therefore, we have

$$\int_{1}^{L} \frac{(r-1)r^{N-1}}{r+g} dr \leq \int_{1}^{L} \frac{(r-1)r^{m-1}}{r+g} dr. \tag{25}$$

Since $m^* \geq 0$, by Lemma 8, we have

$$\frac{|w_m|(1+s_1)}{\|b\|} \leq 1, \tag{26}$$

and hence

$$1-g = 1 - \frac{|w_m|(1+s_1)}{\|b\|} \geq 0. \tag{27}$$

We see that $1-g \geq 0$ and using Lemma 2 for the values of the integrals of inequality (24), we have

$$\begin{aligned} &\left(\frac{L^m-1}{1+s_1}\right)\left(\|b\| - \frac{s_1 m^*}{k^m}\right) - \frac{m}{1+s_1}\left(\|b\| - \frac{s_1 m^*}{k^m}\right)\left\{1 - \frac{(1+s_1)|w_m|}{\|b\| - \frac{s_1 m^*}{k^m}}\right\} f(m, s_1) \\ &\leq \left(\frac{L^m-1}{1+s_1}\right)\|b\| - \frac{m\|b\|}{1+s_1}\left\{1 - \frac{(1+s_1)|w_m|}{\|b\|}\right\} h^*(N), \end{aligned} \tag{28}$$

where $f(m, s_1)$ is as defined in (18) and

$$\begin{aligned} h^*(N) = (L-1) - &\left\{1 + \frac{(1+s_1)|w_m|}{\|b\|}\right\} \\ &\times \ln\left\{1 + \frac{(L-1)\|b\|}{\|b\| + (1+s_1)|w_m|}\right\} \quad for\ N = 1, \end{aligned} \tag{29}$$

$$\begin{aligned} h^*(N) = &\left(\frac{L^N-1}{m}\right) \\ &+ \sum_{v=1}^{m-1}\left(\frac{L^{N-v}-1}{N-v}\right)(-1)^v\left\{1 + \frac{(1+s_1)|w_m|}{\|b\|}\right\}\left\{\frac{(1+s_1)|w_m|}{\|b\|}\right\}^{v-1} \\ &+ (-1)^N\left\{1 + \frac{(1+s_1)|w_m|}{\|b\|}\right\}\left\{\frac{(1+s_1)|w_m|}{\|b\|}\right\}^{N-1} \\ &\times \ln\left(1 + \frac{(L-1)\|b\|}{\|b\| + (1+s_1)|w_m|}\right) \quad for\ N \geq 2. \end{aligned} \tag{30}$$

Adding $\|b\|$ on both sides of (28), we have

$$\begin{aligned} &\left(\frac{L^m+s_1}{1+s_1}\right)\|b\| - \frac{(L^m-1)}{1+s_1}\frac{s_1 m^*}{k^m} - m\left\{\frac{\|b\|}{1+s_1} - \frac{s_1 m^*}{(1+s_1)k^m} - |w_m|\right\} f(m, s_1) \\ &\leq \left(\frac{L^m+s_1}{1+s_1}\right)\|b\| - \frac{m}{1+s_1}\{\|b\| - (1+s_1)|w_m|\} h^*(N), \end{aligned} \tag{31}$$

which clearly shows that Corollary 1 refines the next result which further deduces to inequality (5) due to Hussain [8].

Corollary 4 *If* $b(z) = w_0 + \sum_{v=\mu}^{m} w_v z^v$, $\mu \in \{1, 2, ..., m-1\}$, *is a polynomial with all its zeros outside* $|z| < k$, $k \geq 1$, *then for* $L \geq 1$ *and* $N \in \mathbb{Z}^+$, $N \leq m$,

$$M(b, R) \leq \left(\frac{L^m + s_1}{1 + s_1}\right)\|b\| - \frac{m}{1 + s_1}\{\|b\| - (1 + s_1)|w_m|\} h^*(N), \tag{32}$$

where

$$s_1 = \frac{k^{\mu+1}(\frac{\mu}{m}|\frac{w_\mu}{w_0}|k^{\mu-1} + 1)}{\frac{\mu}{m}|\frac{w_\mu}{w_0}|k^{\mu+1} + 1} \tag{33}$$

and

$$h^*(N) = \left(L - 1\right) - \left\{1 + \frac{(1 + s_1)|w_m|}{\|b\|}\right\} \times \ln\left\{1 + \frac{(L - 1)\|b\|}{\|b\| + (1 + s_1)|w_m|}\right\} \quad for\ N = 1, \tag{34}$$

$$\begin{aligned} h^*(N) = {} & \left(\frac{L^N - 1}{N}\right) \\ & + \sum_{v=1}^{N-1}\left(\frac{L^{N-v} - 1}{N - v}\right)(-1)^v\left\{1 + \frac{(1 + s_1)|w_m|}{\|b\|}\right\}\left\{\frac{(1 + s_1)|w_m|}{\|b\|}\right\}^{v-1} \\ & + (-1)^N\left\{1 + \frac{(1 + s_1)|w_m|}{\|b\|}\right\}\left\{\frac{(1 + s_1)|w_m|}{\|b\|}\right\}^{N-1} \\ & \times \ln\left\{1 + \frac{(L - 1)\|b\|}{\|b\| + (1 + s_1)|w_m|}\right\} \quad for\ N \geq 2. \end{aligned} \tag{35}$$

Remark 8 By Lemma 3, it is noted that $h^*(N) \geq 0$ as defined in (34) and (35) of Corollary 4 and is a monotonically increasing function of N for $N \geq 1$ and therefore $h^*(1) \leq h^*(N)$. Noting this and Lemma 8 that $\{\|b\| - (1 + s_1)|w_m|\} \geq 0$, Corollary 4 reduces to inequality (5) due to Hussain [8]

Remark 9 By Lemma 10, $k \leq s_1$ for $k \geq 1$, where s_1 is as defined in (11), therefore, by Lemma 11, we have

$$\frac{m}{1 + s_1}\|b\| \leq \frac{m}{1 + k}\|b\|. \tag{36}$$

Since $m \geq 0$ and $1 \leq k \leq s_1$, inequality (36) implies

$$\frac{m}{1 + s_1}\|b\| - \frac{m s_1 m^*}{k^m(1 + s_1)} \leq \frac{m}{1 + k}\|b\|. \tag{37}$$

Applying Lemma 4 to (37), we have for $r \geq 1$,

$$r^{m-1}\left\{1-\frac{\left(\frac{m}{1+s_1}\|b\|-\frac{ms_1m^*}{k^m(1+s_1)}-m|w_m|\right)(r-1)}{m|w_m|+r\left(\frac{m}{1+s_1}\|b\|-\frac{ms_1m^*}{k^m(1+s_1)}\right)}\right\}\left\{\frac{m}{1+s_1}\|b\|-\frac{ms_1m^*}{k^m(1+s_1)}\right\}$$

$$\leq r^{m-1}\left\{1-\frac{\left(\frac{m}{1+k}\|b\|-m|w_m|\right)(r-1)}{m|w_m|+r\left(\frac{m}{1+k}\|b\|\right)}\right\}\left(\frac{m}{1+k}\|b\|\right). \tag{38}$$

Inequality (38) is integrated on both sides with respect to r from 1 to L and following similar simplification of the RHS of inequality (73) to inequality (74) in the proof of Theorem 1, we get

$$\frac{L^m-1}{1+s_1}\left(\|b\|-\frac{s_1m^*}{k^m}\right)-\frac{m}{1+s_1}\left(\|b\|-\frac{s_1m^*}{k^m}\right)(1-e)\int_1^L\frac{(r-1)r^{m-1}}{r+e}dr$$

$$\leq\frac{L^m-1}{1+k}\|b\|-\frac{m\|b\|}{1+k}(1-c)\int_1^L\frac{(r-1)r^{m-1}}{r+c}dr, \tag{39}$$

where $e=\frac{|w_m|(1+s_1)}{\|b\|-\frac{s_1m^*}{k^m}}$ and $c=\frac{|w_m|(1+k)}{\|b\|}$.

The expression $\int_1^L\frac{(r-1)r^{N-1}}{r+c}dr\geq 0$ and is a monotonically increasing function of N for $N\leq m$, we have

$$\int_1^L\frac{(r-1)r^{N-1}}{r+c}dr\leq\int_1^L\frac{(r-1)r^{m-1}}{r+c}dr. \tag{40}$$

Since $m^*\geq 0$, by Lemma 9, we have

$$\frac{|w_m|(1+k)}{\|b\|}\leq 1, \tag{41}$$

and hence

$$1-c=1-\frac{|w_m|(1+k)}{\|b\|}\geq 0. \tag{42}$$

Since $1-c\geq 0$ and using Lemma 2 for the values of the integrals in (39), we get

$$\left(\frac{L^m-1}{1+s_1}\right)\left(\|b\|-\frac{s_1m^*}{k^m}\right)-\frac{m}{1+s_1}\left(\|b\|-\frac{s_1m^*}{k^m}\right)\left\{1-\frac{(1+s_1)|w_m|}{\|b\|-\frac{s_1m^*}{k^m}}\right\}f(m,s_1)$$

$$\leq\left(\frac{L^m-1}{1+k}\right)\|b\|-\frac{m\|b\|}{1+k}\left\{1-\frac{(1+k)|w_m|}{\|b\|}\right\}g^*(N), \tag{43}$$

where $f(m,s_1)$ is as defined in (18) and

$$g^*(N)=(L-1)-\left\{1+\frac{(1+k)|w_m|}{\|b\|}\right\}$$
$$\times\ln\left\{1+\frac{(L-1)\|b\|}{\|b\|+(1+k)|w_m|}\right\}\quad for\ N=1,$$

$$g^*(N) = \left(\frac{L^N - 1}{m}\right) + \sum_{v=1}^{m-1}\left(\frac{L^{N-v} - 1}{N - v}\right)(-1)^v \left\{1 + \frac{(1+k)|w_m|}{\|b\|}\right\}\left\{\frac{(1+k)|w_m|}{\|b\|}\right\}^{v-1} + (-1)^N \left\{1 + \frac{(1+k)|w_m|}{\|b\|}\right\}\left\{\frac{(1+k)|w_m|}{\|b\|}\right\}^{N-1} \times \ln\left(1 + \frac{(L-1)\|b\|}{\|b\| + (1+k)|w_m|}\right) \; for\ N \geq 2. \tag{44}$$

Adding $\|b\|$ on both sides of (43), we have

$$\left(\frac{L^m + s_1}{1 + s_1}\right)\|b\| - \frac{(L^m - 1)}{1 + s_1}\frac{s_1 m^*}{k^m} - m\left\{\frac{\|b\|}{1 + s_1} - \frac{s_1 m^*}{(1 + s_1)k^m} - |w_m|\right\} f(m, s_1) \leq \left(\frac{L^m + k}{1 + k}\right)\|b\| - \frac{m}{1 + k}\{\|b\| - (1+k)|w_m|\} g^*(N). \tag{45}$$

Hence, it is verified that Corollary 1 improves the succeeding result.

Corollary 5 *If $b(z) = w_0 + \sum_{v=\mu}^{m} w_v z^v$, $\mu \in \{1, 2, ..., m-1\}$, is a polynomial with all its zeros outside $|z| < k$, $k \geq 1$, then for $L \geq 1$ and $N \in \mathbb{Z}^+$, $N \leq m$,*

$$M(b, L) \leq \left(\frac{L^m + k}{1 + k}\right)\|b\| - \frac{m}{1 + k}\Big(\|b\| - (1+k)|w_m|\Big) g^*(N), \tag{46}$$

where

$$g^*(N) = (L - 1) - \left\{1 + \frac{(1+k)|w_m|}{\|b\|}\right\} \times \ln\left\{1 + \frac{(L-1)\|b\|}{\|b\| + (1+k)|w_m|}\right\} \; for\ N = 1, \tag{47}$$

$$g^*(N) = \left(\frac{L^N - 1}{N}\right) + \sum_{v=1}^{N-1}\left(\frac{L^{N-v} - 1}{N - v}\right)(-1)^v \left\{1 + \frac{(1+k)|w_m|}{\|b\|}\right\}\left\{\frac{(1+k)|w_m|}{\|b\|}\right\}^{v-1} + (-1)^N \left\{1 + \frac{(1+k)|w_m|}{\|b\|}\right\}\left\{\frac{(1+k)|w_m|}{\|b\|}\right\}^{N-1} \times \ln\left\{1 + \frac{(L-1)\|b\|}{\|b\| + (1+k)|w_m|}\right\} \; for\ N \geq 2. \tag{48}$$

Remark 10 For $N = m$, it can be easily verified that the result of Mir et al. [10, Corollary 1] is obtained from Corollary 5.

Remark 11 By Lemma 3, it is observed that $g^*(N) \geq 0$ as defined in (47) and (48) of Corollary 5 and is a monotonically increasing function of N for $N \geq 1$ and hence $g^*(1) \leq g^*(N)$. With this fact and Lemma 9, Corollary 5 gives a result which is a generalization of inequality (4) of Govil [6].

Corollary 6 *If $b(z) = w_0 + \sum_{v=\mu}^{m} w_v z^v$, $\mu \in \{1, 2, ..., m-1\}$, is a polynomial with all its zeros outside $|z| < k$, $k \geq 1$, then for $L \geq 1$,*

$$M(b, L) \leq \left(\frac{L^m + k}{1+k}\right)\|b\| - \frac{m}{1+k}\left\{\frac{\|b\|^2 - (1+k)^2|w_m|^2}{\|b\|}\right\}$$
$$\times \left[\frac{(L-1)\|b\|}{\|b\| + (1+k)|w_m|} - \ln\left\{1 + \frac{(L-1)\|b\|}{\|b\| + (1+k)|w_m|}\right\}\right]. \quad (49)$$

Remark 12 Also for $k = 1$, inequality (49) of Corollary 6 reduces to inequality (4) of Govil [6].

3 Lemmas

We require the following lemmas.

Lemma 1 *Let $b(z) = \sum_{v=0}^{m} w_m z^m$ be a polynomial. Then for $|z| = L \geq 1$,*

$$|b(z)| \leq L^m \left\{1 - \frac{(\|b\| - |w_m|)(L-1)}{|w_m| + L\|b\|}\right\} \|b\|. \quad (50)$$

Lemma 1 is due to Govil [6].

Lemma 2 *Let*

$$J(N) = \int_1^L \frac{(r-1)r^{N-1}}{r+x} dr, \quad x > 0 \ . \quad (51)$$

Then for $N \geq 2$,

$$J(N) = \left(\frac{L^N - 1}{N}\right) + \sum_{v=1}^{N-1}\left(\frac{L^{N-v} - 1}{N-v}\right)(-1)^v(x+1)x^{v-1}$$
$$+ (-1)^N(x+1)x^{N-1}\ln\left(\frac{L+x}{1+x}\right), \quad (52)$$

and for $N = 1$,

$$J(1) = (L-1) - (1+x)\ln\left(1 + \frac{L-1}{1+x}\right). \quad (53)$$

Lemma 2 is due to Dalal and Govil [2, Lemma 3.6].

Lemma 3 *$J(N)$ defined in Lemma 2 is a non-negative increasing function of N for $N \geq 1$.*

Proof (*Proof of Lemma 3*) Dalal and Govil [2, Lemma 3.7] has done this proof, but, we present another proof of it using the method of differentiation under the integral sign.

By the method of differentiation under the integral sign, we obtain

$$\frac{d}{dN} J(N) = \int_1^L \frac{(r-1)(r^{N-1})}{r+x} \ln r dr. \tag{54}$$

Since, for $r \in [1, L]$, $\frac{(r-1)r^{N-1}}{r+x} \ln r \geq 0$, therefore, we have

$$\int_1^L \frac{(r-1)r^{N-1}}{r+x} \ln r dr \geq 0. \tag{55}$$

From equality (54),

$$\frac{d}{dN} J(N) \geq 0, \quad for\ N \geq 1. \tag{56}$$

Hence, $J(N)$ is an increasing function of N for $N \geq 1$.

Further, we see that $\frac{(r-1)r^{N-1}}{r+x}$ is non-negative for $N \geq 1$ which implies that $J(N) \geq 0$ for $N \geq 1$,
and hence Lemma 3 is proved. □

Lemma 4 *For polynomial $b(z) = w_0 + \sum_{v=\mu}^{m} w_v z^v$, $\mu \in \{1, 2, ..., m\}$ and $r \geq 1$, the function*

$$t(y) = \left\{1 - \frac{(y - m|w_m|)(r-1)}{m|w_m| + ry}\right\} y \tag{57}$$

is an increasing function of y for $y > 0$.

Proof of Lemma 4. The proof simply follows by using the derivative test and we omit it.

The next lemma is due to Qazi [12, Remark 1].

Lemma 5 *If $b(z) = w_0 + \sum_{v=\mu}^{m} w_v z^v$, $\mu \in \{1, 2, ..., m\}$, is a polynomial with all its zeros outside $|z| < k$, $k \geq 1$, then*

$$\frac{\mu}{m} \left| \frac{w_\mu}{w_0} \right| k^\mu \leq 1. \tag{58}$$

Lemma 6 *If* $b(z) = \sum_{v=0}^{m} w_v z^v$ *is a polynomial with all its zeros outside* $|z| < k$, $k \geq 1$, *then*

$$\|b'\| \leq \frac{m}{1+k}\|b\|. \tag{59}$$

Lemma 6 is due to Malik [9].

Lemma 7 *If* $b(z) = w_0 + \sum_{v=\mu}^{m} w_v z^v$, $\mu \in \{1, 2, ..., m-1\}$, *is a polynomial with no zero in* $|z| < k$, $k \geq 1$, *then*

$$\|b'\| \leq \frac{m}{1+s_1}\|b\| - \frac{m}{k^m}\left(1 - \frac{1}{1+s_1}\right)m^*, \tag{60}$$

where s_1 *is as defined in (11).*

Lemma 7 is due to Dewan et al. [4].

Lemma 8 *If* $b(z) = w_0 + \sum_{v=\mu}^{m} w_v z^v$, $\mu \in \{1, 2, ..., m-1\}$, *is a polynomial with no zero in* $|z| < k$, $k \geq 1$, *then*

$$|w_m| \leq \frac{1}{1+s_1}\left(\|b\| - \frac{m^* s_1}{k^m}\right), \tag{61}$$

where s_1 *is as defined in (11).*

Proof *(Proof of Lemma 8)*
For a polynomial $b(z) = w_0 + \sum_{v=\mu}^{m} w_v z^v$, $\mu \in \{1, 2, ..., m-1\}$, then we get $b'(z) = \sum_{v=\mu}^{m} v w_v z^{v-1}$.

Using Cauchy's inequality to $b'(z)$ on $|z| = 1$, we have

$$\left|\frac{d^{m-1}}{dz^{m-1}} b'(z)\right|_{z=0} \leq (m-1)! \max_{|z|=1} |b'(z)|. \tag{62}$$

That is,

$$|m w_m| \leq \|b'\|. \tag{63}$$

Combining inequality (60) of Lemma 7 and (63), we have inequality (61) of Lemma 8 and this completes the proof of Lemma 8. □

Lemma 9 *If* $b(z) = \sum_{v=0}^{m} w_v z^v$ *is a polynomial with no zero in* $|z| < k$, $k \geq 1$, *then*

$$|w_m| \leq \frac{1}{1+k}\|b\|. \tag{64}$$

Proof (*Proof of Lemma* 9) This lemma is proved in similar ways as that of Lemma 8, but we apply inequality (59) of Lemma 6 in place of (60) of Lemma 7 and we omit the details □.

Lemma 10 *If* $b(z) = w_0 + \sum_{v=\mu}^{m} w_v z^v$, $\mu \in \{1, 2, ..., m\}$, *is a polynomial with no zero in* $|z| < k$, $k \geq 1$, *then*

$$s_1 \geq k, \tag{65}$$

where s_1 *is as defined in (11).*

Proof (*Proof of Lemma* 10) Let $b(z) = w_0 + \sum_{v=\mu}^{m} w_v z^v$, $\mu \in \{1, 2, ..., m\}$, is a polynomial with no zero in $|z| < k$, $k \geq 1$.
From inequality (58) of Lemma 5, we have

$$0 \leq \frac{\mu}{m}\left|\frac{w_\mu}{w_0}\right| k^\mu \leq 1. \tag{66}$$

Since $k \geq 1$ and $\mu = 1, 2, \cdots$, we have

$$k - k^{\mu-1} \leq k^\mu - 1. \tag{67}$$

Multiplying (66) and (67) sidewise, we have

$$k^\mu \left\{\frac{\mu}{m}\left|\frac{w_\mu}{w_0}\right| k^{\mu-1} + 1\right\} \geq \frac{\mu}{m}\left|\frac{w_\mu}{w_0}\right| k^{\mu+1} + 1, \tag{68}$$

which is equivalent to

$$s_1 \geq k,$$

and hence, Lemma 10 is obtained. □

Lemma 11 *If* $b(z) = w_0 + \sum_{v=\mu}^{m} w_v z^v$, $\mu \in \{1, 2, ..., m\}$, *is a polynomial having no zero in* $|z| < k$, $k \geq 1$, *then*

$$\frac{m}{1+s_1}\|b\| \leq \frac{m}{1+k}\|b\|, \tag{69}$$

where s_1 *is as defined in (11).*

Lemma 11 is due to Qazi [12].

4 Proof of the Theorem

Proof (*Proof of Theorem* 1) For each θ, $0 \le \theta < 2\pi$ and $1 \le r \le L$, we have

$$b(Le^{i\theta}) - b(e^{i\theta}) = \int_1^L e^{i\theta} b'(re^{i\theta})dr, \tag{70}$$

which implies

$$|b(Le^{i\theta}) - b(e^{i\theta})| \le \int_1^L |b'(re^{i\theta})|dr. \tag{71}$$

Now, applying Lemma 1 to the polynomial $b'(z)$ which is of degree $m-1$, we get

$$|b(Le^{i\theta}) - b(e^{i\theta})| \le \int_1^L r^{m-1}\left\{1 - \frac{(\|b'\| - m|w_m|)(r-1)}{m|w_m| + r\|b'\|}\right\}\|b'\|dr. \tag{72}$$

Since by Lemma 4, in the integrand of (72), the quantity $\left\{1 - \frac{(\|b'\|-m|w_m|)(r-1)}{m|w_m|+r\|b'\|}\right\}\|b'\|$ is a monotonically increasing function of $\|b'\|$, hence using Lemma 7, we have for $0 \le \theta < 2\pi$,

$$\begin{aligned}
&|b(Le^{i\theta}) - b(e^{i\theta})| \\
&\le \int_1^L r^{m-1}\left[1 - \frac{\left\{\frac{m}{1+s_1}\|b\| - \frac{m}{k^m}\left(1 - \frac{1}{1+s_1}\right)m^* - m|w_m|\right\}(r-1)}{m|w_m| + r\left\{(\frac{m}{1+s_1})\|b\| - \frac{m}{k^m}\left(1 - \frac{1}{1+s_1}\right)m^*\right\}}\right] \qquad (73)\\
&\times\left\{\frac{m}{1+s_1}\|b\| - \frac{m}{k^m}\left(1 - \frac{1}{1+s_1}\right)m^*\right\}dr \\
&= \left\{\frac{m}{1+s_1}\|b\| - \frac{mm^*s_1}{k^m(1+s_1)}\right\}\int_1^L r^{m-1}dr - \left\{\frac{m}{1+s_1}\|b\| - \frac{mm^*s_1}{k^m(1+s_1)}\right\} \\
&\times \int_1^L r^{m-1}\left[\frac{\|b\| - \frac{m^*s_1}{k^m} - (1+s_1)|w_m|}{(1+s_1)|w_m| + r\left\{\|b\| - \frac{m^*s_1}{k^m}\right\}}\right](r-1)dr \\
&= \frac{L^m - 1}{1+s_1}\left\{\|b\| - \frac{m^*s_1}{k^m}\right\} - \left\{\frac{m}{1+s_1}\|b\| - \frac{mm^*s_1}{k^m(1+s_1)}\right\} \\
&\times (1-e)\int_1^L \frac{(r-1)r^{m-1}}{r+e}dr, \qquad (74)
\end{aligned}$$

where s_1 is as defined in (11) and $e = \frac{|w_m|(1+s_1)}{\|b\| - \frac{m^*s_1}{k^m}}$.

It is observed that $\int_1^L \frac{(r-1)r^{N-1}}{r+e}dr \geq 0$ and is a monotonically increasing function of N for $N \leq m$, therefore, we have

$$\int_1^L \frac{(r-1)r^{N-1}}{r+e}dr \leq \int_1^L \frac{(r-1)r^{m-1}}{r+e}dr. \tag{75}$$

We see from Lemma 8 that $(1-e) \geq 0$ and using inequality (75) to (74), we get for every N, $N \leq m$,

$$|b(Le^{i\theta}) - b(e^{i\theta})| \leq \frac{L^m - 1}{1+s_1}\left\{\|b\| - \frac{m^* s_1}{k^m}\right\} - \left\{\frac{m}{1+s_1}\|b\| - \frac{mm^* s_1}{k^m(1+s_1)}\right\} \times (1-e)\int_1^L \frac{(r-1)r^{N-1}}{r+e}dr. \tag{76}$$

Using Lemma 2 (on replacing x by e) for the value of the integral in (76), we have,

$$|b(Le^{i\theta}) - b(e^{i\theta})| \leq \frac{L^m - 1}{1+s_1}\left\{\|b\| - \frac{m^* s_1}{k^m}\right\} - \left\{\frac{m}{1+s_1}\|b\| - \frac{mm^* s_1}{k^m(1+s_1)}\right\}(1-e)f(N, s_1), \tag{77}$$

where $f(N, s_1)$ is as defined in (12) and (13).

Now, putting the value of e and using the relation

$$\begin{aligned}|b(Le^{i\theta})| &\leq |b(Le^{i\theta}) - b(e^{i\theta})| + |b(e^{i\theta})| \\ &\leq |b(Le^{i\theta}) - b(e^{i\theta})| + \|b\|\end{aligned} \tag{78}$$

in (77), we get

$$|b(Le^{i\theta})| \leq \left(\frac{L^m + s_1}{1+s_1}\right)\|b\| - \frac{(L^m - 1)}{1+s_1}\frac{s_1 m^*}{k^m} - m\left\{\frac{\|b\|}{1+s_1} - \frac{s_1 m^*}{(1+s_1)k^m} - |w_m|\right\} f(N, s_1), \tag{79}$$

which is equivalent to inequality (10) and hence, Theorem 1 is obtained. □

5 Conclusions

We have improved and generalized inequality (5) proved by Hussain [8] by involving $\min_{|z|=k} |b(z)|$. Moreover, through Remarks and Corollaries, we have discussed the implications of Theorem 1 on other well-known results .

Acknowledgements We are very grateful to the referee for the valuable suggestions and comments.

References

1. Ankeny, N.C., Rivlin, T.J.: On a theorem of S. Bernstein. Pacific J. Math. **5**, 849–852 (1955). https://doi.org/10.1017/S0305004100027390
2. Dalal, A., Govil, N.K.: On sharpening of a theorem of Ankeny and Rivlin. Anal. Theory Appl. **36**, 225–234 (2020). https://doi.org/10.1080/09720502.2010.10700689
3. Dewan, K.K., Bhat, A.A.: On maximum modulus of polynomials not vanishing inside the unit circle. J. Interdiscip. Math. **1**, 129–140 (1998). https://doi.org/10.1080/09720502.1998.107002485
4. Dewan, K.K., Harish Singh, Yadav, R.S.: Inequalities concerning polynomials having zeros in closed exterior or closed interior of a circle. Southeast Asian Bull. Math. **27**, 591–597 (2003)
5. Govil, N.K.: Some inequalities for derivative of polynomial. J. Approx. Theory. **66**, 29–35 (1991). https://doi.org/10.1007/s41478-021-00356-z
6. Govil, N.K.: On the maximum modulus of polynomials not vanishing inside the unit circle. Approx. Theory Appl. **5**, 79–82 (1989). https://doi.org/10.1007/BF02836495
7. Govil, N.K., Nyuydinkong, G.: On the maximum modulus of polynomials not vanishing inside a circle. J. Interdiscip. Math. **4**, 93–100 (2001). https://doi.org/10.1080/09720502.2001.10700292
8. Hussain, I.: Growth estimates of a polynomial not vanishing in a disk. Indian J. Pure Appl. Math. (2021). https://doi.org/10.1007/s13226-021-00169-7
9. Malik, M.A.: On the derivative of a polynomial. J. London Math. Soc. **1**, 57–60 (1969). https://doi.org/10.1112/jlms/s2-1.1.57
10. Mir, A., Ahmad, A., Malik, A.H.: Growth of a polynomial with restricted zeros J. Anal. **28**, 827–837 (2020). https://doi.org/10.1007/s41478-019-00208-x
11. Pólya, G., Szegö, G.: Aufgaben and Leheatze ous der Analysis. Springer, Berlin (1925)
12. Qazi, M.A.: On the maximum modulus of polynomials. Proc. Am. Math. Soc. **115**, 337–343 (1992). https://doi.org/10.1090/S0002-9939-1992-1113648-1

Simulation of Queues in Sugar Mills Using Monte Carlo Technique

Vikash Siwach, Manju S. Tonk, and Hemant Poonia

Abstract The arrival and service data for a season was gathered from Sugar Mill in Meham, Haryana, to improve the service facilities for farmers and reduce queue waiting time through simulation. A suitable simulation model was developed utilizing the Monte Carlo technique to analyze the queue characteristics. Simulation revealed a significant reduction of 60% in waiting time with a marginal rise in the mill's sugarcane crushing limit.

Keywords Queueing model · Monte Carlo simulation · Agriculture sciences

1 Introduction

Queuing theory is widely used to investigate and manage queue characteristics in a variety of settings, including bank counters, railway counters, super markets, agriculture markets, and sugar mills, among others. The majority of queuing models are built on the assumption that customer arrival rates are lower than the system's service rate. This condition ensures the steady state solution of the governing equations for the model. But there are situations where steady state solution cannot be achieved or does not exist. For example, at a doctor's clinic, where patients are seen for a set length of time, such as 9:00 a.m. to 3:00 p.m. Because the consultation or service process does not last for a long period, the system's long-term behavior cannot be analyzed. In the following scenario, as well as many others, it is possible that the arrival rate exceeds the service rate, causing the system to collapse in the long run and leaving no stable solution. These types of problems can be handled by either limiting the queue system's capacity or increasing the number of servers.

The analysis of non-steady state queue system was accomplished in [1] and the result was achieved by developing computation formula from both symbolic and

V. Siwach (✉) · M. S. Tonk · H. Poonia
Chaudhary Charan Singh Haryana Agricultural University, Hisar, Haryana, India
e-mail: vikash@hau.ac.in

R. K. Sharma et al. (eds.), *Frontiers in Industrial and Applied Mathematics*,
Springer Proceedings in Mathematics & Statistics 410,
https://doi.org/10.1007/978-981-19-7272-0_33

numeric exact where results are tested against Monte Carlo simulation. Another simple queue model (M/M/1/∞) was implemented in [2] in bank service to improve the optimal service rate.

Complex queues can be solved using simulation. Simulation can be defined as a process of designing a mathematical or artificial model of a real system. The behavior of real system can be examined by performing experiments with the developed model [3]. The Monte Carlo simulation technique converts uncertainties of input variables in the model into probability distributions [4]. To re-form the opportunity distribution in this simulation, you'll need a random number generator [5]. A few of the applications of Monte Carlo queuing system can be found in the hospital [6], in fuzzy queuing theory [7], in traffic light simulation [8], in finance [9], etc. Since the results are derived after performing the repeated experiments based on random numbers, Monte Carlo simulation is very effective and widely accepted for true results.

Arrival and service data for the season 2020–21 (Nov 2020 to May 2021) was collected from The Meham Co-Op. Sugar Mills Ltd., Meham, Haryana. There were no symmetries between arrival and service pattern as shown in Fig. 1.

The zigzag nature of the arrival and service rates can easily be recognized, indicating that a basic queue model (M/M/1) could not be utilized to describe the queue characteristics. In addition, both the average arrival rate and the average service rate were the same, i.e., 146 trolleys each day. The Monte Carlo simulation approach is used to deal with such a circumstance.

Section 2 discusses the Monte Carlo simulation approach and algorithm used to determine queue characteristics. Section 3 contains the simulation findings. Section 4 discusses the potential for improvement by increasing mill crushing, as well as the consequences. Section 5 contains the work's conclusion.

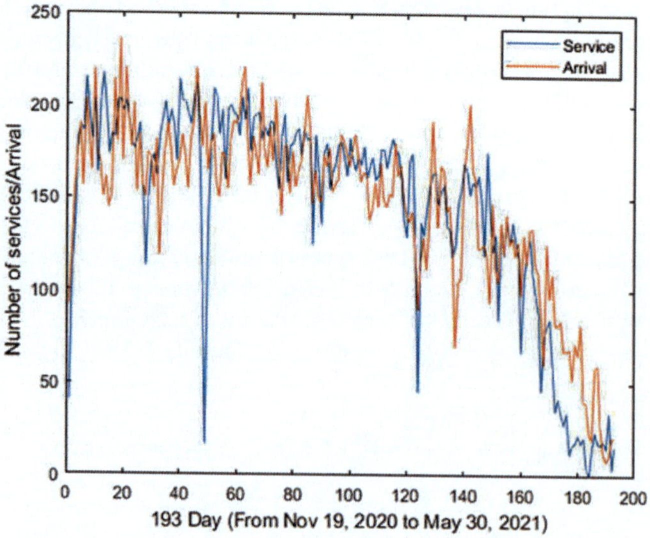

Fig. 1 Arrival and service pattern on seasonal days

2 Methodology

According to data collected from the Meham sugar mill for the entire season, a total of 28,128 trolleys carrying 3,168,923.80 qtl of sugarcane arrived and were unloaded between November 19, 2020 and May 30, 2021. There were 193 crushing days in total. Note that the arrival was low in May 2021, and hence the crushing or service was likewise low. So actual performance of mill could not be determined from the data including month of May 2021. For better simulation of mill system, this month's data was omitted and the data of 162 days from November 21, 2020 to May 01, 2021 was utilized. During this period, a total of 3,035,082.6 qtl sugarcane was crushed, with an average of 18,735 qtl each day.

The average daily arrival was 161.42 trolleys, or 6.73 per hour, with an average of 112.44 qtl sugarcane each trolley. Figure 2 shows a day-by-day summary of the weight of sugarcane crushed during these days.

Figure 2 shows that on January 6, 2020 (Day 49) and March 22, 2021(Day 124), there was substantially less crushing. The maximum crushing of 22,906 qtl sugarcane was done on Dec 01, 2020. The mill's full crushing capacity of 25,000 qtl per day has never been reached.

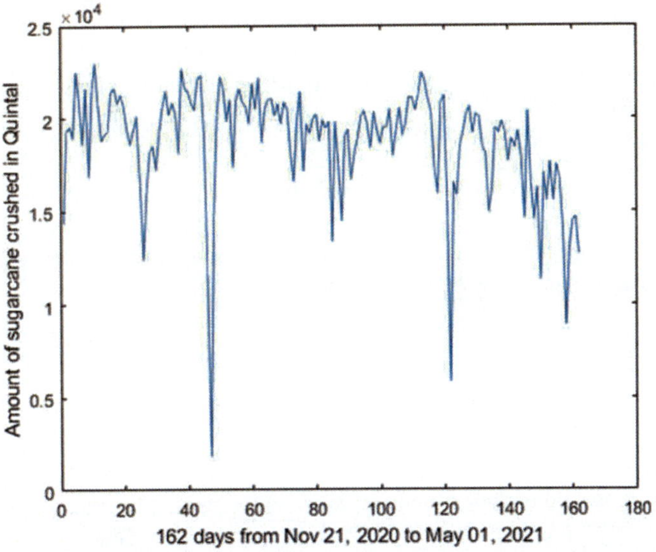

Fig. 2 Per day crushing during the season 2020–21

2.1 Monte Carlo Technique

There were three shifts of workers in the sugar mill in Meham: shift 0, shift 1, and shift 2. Shift 0 ran from 6:00 a.m. to 2:00 p.m., shift 1 from 2:00 p.m. to 10:00 p.m., and shift 2 from 10:00 p.m. to 2:00 a.m. Based on the number of arrivals and services in each shift, the possible 40 values of inter-arrival time ranging from 3 to 240 min and the possible 23 values of service time ranging from 6 to 480 min for the trolleys were achieved.

To overcome the problem, an algorithm for Monte Carlo simulation was created as follows:

1. Based on the number of arrivals and services in each shift, determine the inter-arrival time and service time for each trolley.
2. Determine the frequency of inter-arrival times and service durations.
3. Calculate the probability of each value of the inter-arrival and service times.
4. Determine the cumulative probability as well as the boundary/random number interval.
5. For arrivals and services, generate random numbers in the interval (0, 1) uniformly.
6. Calculate arrival time, waiting time, time to enter service, service time, and queue length, etc.
7. Determine the expected values of queue characteristics.
8. Repeat the above process 1000 times for better estimation of queue characteristics.

Figure 3 depicts a flow chart of the steps.

The inter-arrival timing, frequency, probability distribution, and random number intervals were determined using the arrival and service data, as shown in Table 1.

Similarly, the service time, frequencies, probability distribution, and random number intervals were determined as shown in Table 2.

3 Results and Discussion

The trial rows of 26,150 trolleys (arrived in season 2020–21 over the study period) were formed using the random number intervals for cumulative probability of inter-arrival and service time estimated in Tables 1 and 2. We applied the Monte Carlo technique to get the inter-arrival time between two trolleys and the service time of each trolley by uniformly generating 26,150 random numbers in the interval (0, 1). Each of the random number was lying in some of the random interval in the last column of Table 1. Inter-arrival time corresponding to those random intervals was assigned to 26,150 trolleys. Similar procedure was applied to get service time of each trolley. Note that values to the first trolley were not assigned according to random numbers since there was no queue to cause delays in its unloading and other services.

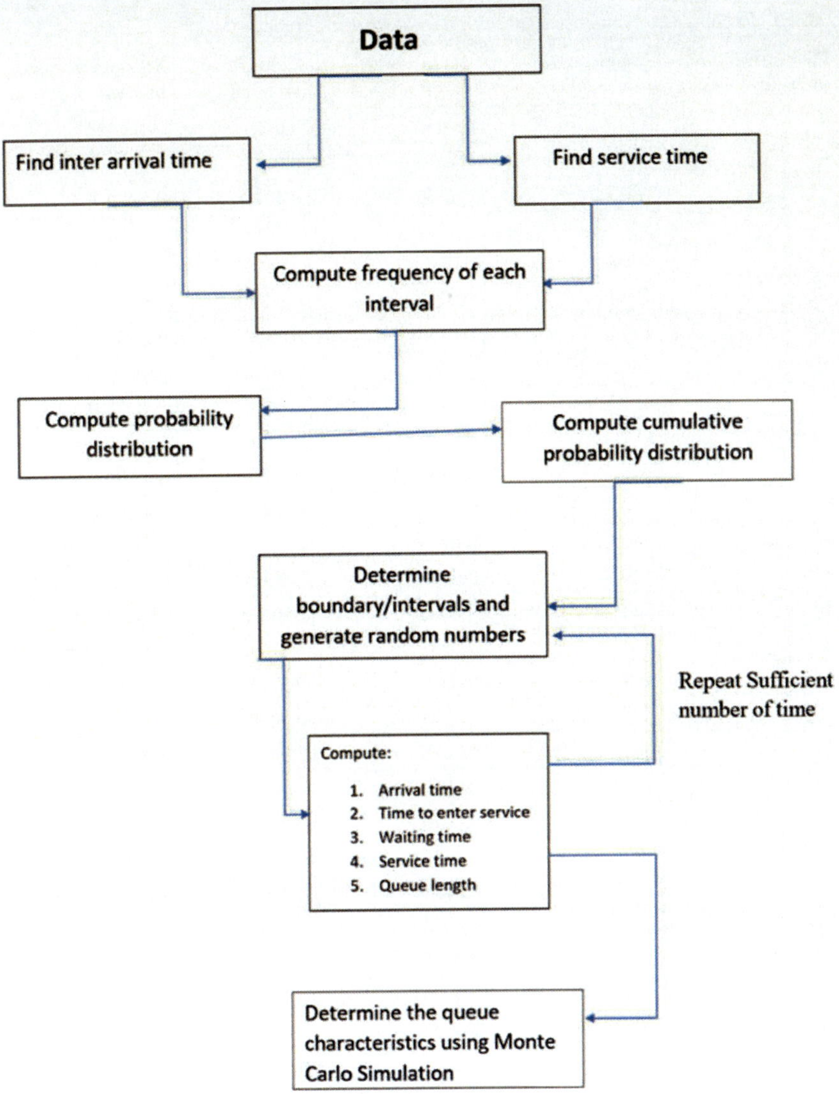

Fig. 3 Flowchart of Monte Carlo simulation

Other characteristics of the queues were calculated using the achieved inter-arrival and service times. A few rows from the beginning and finish of the trial rows of 26,150 trolleys were shown in Table 3.

The aforementioned experiment was repeated 1000 times, with the results displayed in Figs. 4 and 5, respectively, for estimated waiting time (in hours) and expected queue length.

Table 1 Random number interval generation for inter-arrival time

Inter-arrival time (min), x	Frequency $f(x)$	$p(x) = \frac{f(x)}{\sum_x f(x)}$	Cumulative probability	Random number interval
3	1188	0.0454	0.0454	0–0.0454
4	5958	0.2278	0.2733	0.0454–0.2733
5	4145	0.1585	0.4318	0.2733–0.4318
6	2443	0.0934	0.5252	0.4318–0.5252
7	1508	0.0577	0.5829	0.5252–0.5829
8	2027	0.0775	0.6604	0.5829–0.6604
9	1971	0.0754	0.7358	0.6604–0.7358
10	1815	0.0694	0.8052	0.7358–0.8052
11	1090	0.0417	0.8468	0.8052–0.8468
12	681	0.0260	0.8729	0.8468–0.8729
13	335	0.0128	0.8857	0.8729–0.8857
14	136	0.0052	0.8909	0.8857–0.8909
15	380	0.0145	0.9054	0.8909–0.9054
16	120	0.0046	0.9100	0.9054–0.9100
17	226	0.0086	0.9187	0.9100–0.9187
18	343	0.0131	0.9318	0.9187–0.9318
19	100	0.0038	0.9356	0.9318–0.9356
20	72	0.0028	0.9384	0.9356–0.9384
21	46	0.0018	0.9401	0.9384–0.9401
22	176	0.0067	0.9468	0.9401–0.9468
23	42	0.0016	0.9485	0.9468–0.9485
24	80	0.0031	0.9515	0.9485–0.9515
25	133	0.0051	0.9566	0.9515–0.9566
27	216	0.0083	0.9649	0.9566–0.9649
28	51	0.0020	0.9668	0.9649–0.9668
30	128	0.0049	0.9717	0.9668–0.9717
32	60	0.0023	0.9740	0.9717–0.9740
34	112	0.0043	0.9783	0.9740–0.9783
37	65	0.0025	0.9808	0.9783–0.9808
40	72	0.0028	0.9835	0.9808–0.9835
44	88	0.0034	0.9869	0.9835–0.9869
48	130	0.0050	0.9919	0.9869–0.9919
53	36	0.0014	0.9932	0.9919–0.9932
60	64	0.0024	0.9957	0.9932–0.9957
69	14	0.0005	0.9962	0.9957–0.9962

(continued)

Table 1 (continued)

Inter-arrival time (min), x	Frequency $f(x)$	$p(x) = \frac{f(x)}{\sum_x f(x)}$	Cumulative probability	Random number interval
80	66	0.0025	0.9987	0.9962–0.9987
96	20	0.0008	0.9995	0.9987–0.9995
120	8	0.0003	0.9998	0.9995–0.9998
160	3	0.0001	0.9999	0.9998–0.9999
240	2	0.0001	1	0.9999–1
Total	**26,150**	**1**		

Table 2 Random number interval generation for service time

Service Time (min), y	Frequency $f(y)$	$p(y) = \frac{f(y)}{\sum_y f(y)}$	Cumulative probability	Random number interval
6	985	0.0365	0.0365	0–0.0365
7	6370	0.2360	0.2725	0.0365–0.2725
8	9999	0.3704	0.6429	0.2725–0.6429
9	4730	0.1752	0.8181	0.6429–0.8181
10	2322	0.0860	0.9041	0.8181–0.9041
11	919	0.0340	0.9382	0.9041–0.9382
12	722	0.0267	0.9649	0.9382–0.9649
13	223	0.0083	0.9732	0.9649–0.9732
14	242	0.0090	0.9821	0.9732–0.9821
15	96	0.0036	0.9857	0.9821–0.9857
16	30	0.0011	0.9868	0.9857–0.9868
17	86	0.0032	0.9900	0.9868–0.9900
18	105	0.0039	0.9939	0.9900–0.9939
19	50	0.0019	0.9957	0.9939–0.9957
21	23	0.0009	0.9966	0.9957–0.9966
25	19	0.0007	0.9973	0.9966–0.9973
28	34	0.0013	0.9986	0.9973–0.9986
40	12	0.0004	0.9990	0.9986–0.9990
53	9	0.0003	0.9993	0.9990–0.9993
69	7	0.0003	0.9996	0.9993–0.9996
96	5	0.0002	0.9998	0.9996–0.9998
120	4	0.0001	0.9999	0.9998–0.9999
480	2	0.0001	1	0.9999–1
Total	**26,994**	**1**		

Table 3 Monte Carlo simulation

Trolley	Uniformly distributed random numbers for arrival	Inter-arrival time (Minutes)	Arrival time	Uniformly distributed random numbers for service time	Service time (Minutes)	Time to enter service	Waiting time	Queue length
	R1			R2				
1	0.9206	0	0	0.7633	9	0	0	0
2	0.8379	11	11	0.6488	9	11	0	0
3	0.4271	5	16	0.7178	9	20	4	1
4	0.9555	25	41	0.4206	8	41	0	0
5	0.1829	4	45	0.0185	6	49	4	1
6	0.0469	4	49	0.1106	7	55	6	1
7	0.0917	4	53	0.2535	7	62	9	2
.	.	.	.	.	.	.	.	.
.	.	.	.	.	.	.	.	.
.	.	.	.	.	.	.	.	.
26,144	0.1703	4	233,550	0.5511	8	233,661	111	14
26,145	0.4044	5	233,555	0.1434	7	233,669	114	15
26,146	0.2230	4	233,559	0.7451	9	233,676	117	15
26,147	0.9196	18	233,577	0.6165	8	233,685	108	14
26,148	0.0112	3	233,580	0.1367	7	233,693	113	15
26,149	0.6113	8	233,588	0.3048	8	233,700	112	15
26,150	0.6776	9	233,597	0.8671	10	233,708	111	15

To determine the final parameters of the queuing system, an average of the estimated waiting time and expected queue lengths was taken. The average of all 1000 experiments was as under.

$$\text{Average waiting time in queue} = 3.2891 \sim 3\,\text{h and } 17\,\text{min}$$

$$\text{Average queue length} = 23.1416 \sim 23\,\text{trolleys}$$

The average of all the probabilities of associated variables from all 1000 experiments was used to obtain the probability distribution of waiting time and queue length. Figure 6 depicts the cumulative probability distributions of both waiting time and queue length. It also shows that there is a 90% chance that the wait time would be less than 9 h and the queue length will be fewer than 57 trolleys at any given time. The system's average usage is 0.9770. This suggests that the system will be busy for about 98% of the time and free for only about 2% of the time.

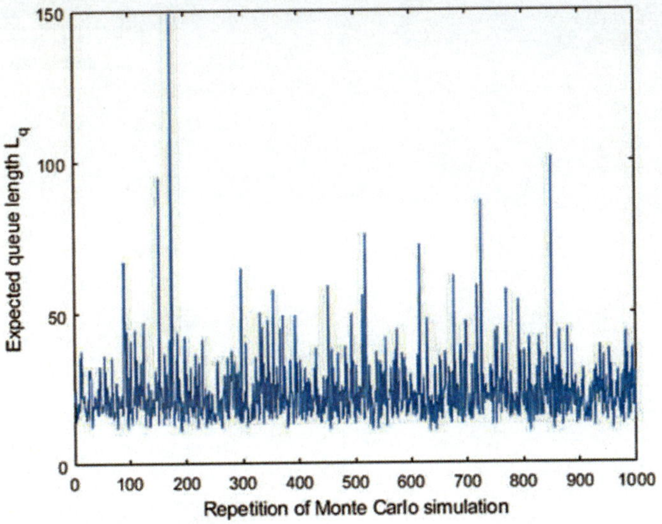

Fig. 4 Repetition of average queue length L_q

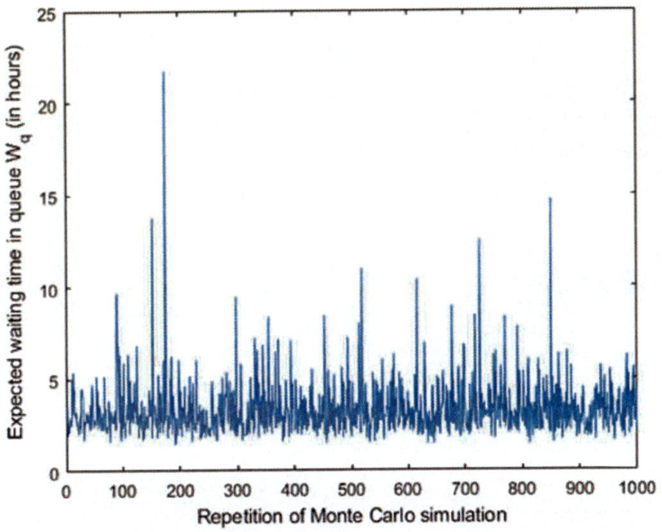

Fig. 5 Repetition of average waiting time in queue

Table 4 shows the queue characteristics and performance measures. In the mill, the average number of trolleys was one higher than the average number of trolleys in the queue. In addition, the average waiting time in the system was the sum of

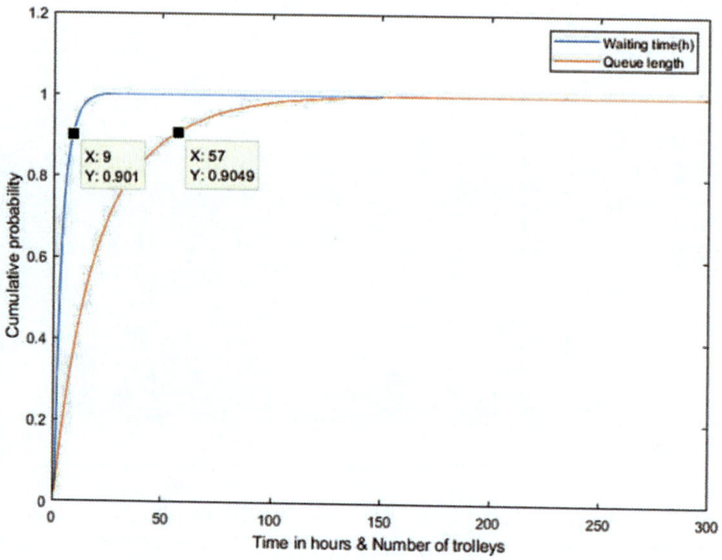

Fig. 6 Cumulative probability distribution of waiting time and queue length

Table 4 Queue characteristics of existing system

Queue characteristics	Performance	
Average server utilization (ρ)	97.70%	Busy
Average number of trolleys in the queue (Lq)	23	In queue
Average number of trolleys in the system (L)	24	In system
Average waiting time in the queue (Wq)	3.29	Hour
Average time in the system (W)	3.44	Hour
Probability (% of time) system is empty	2.30%	Empty

the average waiting time in the queue and the service time. With an average of 166 trolleys unloaded per day, the average service time is 8.64 min. Unloading a trolley takes about 9 min (0.15 h) on average in a mill.

3.1 *Validation of the Model*

According to mill data, the average daily arrival rate was 161 trolleys, i.e., 7 trolleys per hour. In queuing theory and stochastic systems, Little's formula, $L = \lambda W$, is one of the most well-known and useful conservation laws. It asserts that the average number of units in a system equals the average arrival rate of units multiplied by the average time in the system per unit. In the case of a queue, $L_q = \lambda W_q$, i.e., the expected length of the queue is the expected number of waiting times in the queue multiplied by the rate of arrival.

Using the observations from simulation,

$$\text{Average queue length} = \lambda(\text{average waiting time in queue})$$

$$23 = \lambda(3.29)$$

which gives

$$\lambda = 23/3.29 = 6.99 \sim 7 \text{ trolleys per hour}$$

Approximately the same arrival rate achieved from the simulation validates the good fit of the model.

4 Performance Measures of Mill with Enhanced Crushing Capacity

Meham Mill was established in 1991, and its machinery is nearly 30 years old. As a result, exceeding the 25,000 qtl maximum crushing capacity restriction may raise the risk of mechanical failure. This, in turn, will degrade service quality by halting the mill's operation. During the peak season, the average crushing rate was 18,735 qtl per day. On December 1, 2020, the maximum crushing of 22,906 qtl sugarcane was attained. We can simulate the model and find the expected queue characteristics by assuming the same arrival rate of 6.73 trolleys per hour and increasing the average crushing capacity of the mill from 20,000 qtl to 24,000 qtl. Table 5 shows the service rates associated with increased average crushing.

The arrival rate is smaller than the service rate in all of the preceding scenarios, and the queue characteristics are presented in Table 6 using the M/M/1 queuing model.

Table 5 Service rate as per different average crushing

Average crushing (qtls per day)	20,000	21,000	22,000	23,000	24,000
Service rate (number of trolleys unloaded per hour)	7.41	7.78	8.15	8.52	8.89

Table 6 Queue characteristics with enhanced service

Queue characteristics	$\mu = 7.41$	$\mu = 7.78$	$\mu = 8.15$	$\mu = 8.52$	$\mu = 8.89$	
Average server utilization (ρ)	90.81%	86.48%	82.55%	78.96%	75.67%	Busy
Average number of trolleys in the queue (Lq)	8.97	5.53	3.91	2.96	2.35	In queue
Average number of trolleys in the system (L)	9.87	6.40	4.73	3.75	3.11	In system
Average waiting time in the queue (Wq)	80	49	35	26	21	Minutes
Average time in the system (W)	88	57	42	33	28	Minutes
Probability (% of time) system is empty	9.19%	13.51%	17.45%	21.04%	24.33%	Empty

5 Conclusion

According to the study based on primary data, average crushing over the season was 16,419 qtl per day, while peak days saw 18,735 qtl per day. The average arrival rate of trolleys was 161 trolleys per day. The mill was busy 97.70% of the time, with an average waiting time of 3 h and 17 min for a trolley to be serviced. The average queue length was 23 trolleys. Keeping in view, the maximum crushing capacity of 25,000 qtl per day, queue characteristics are obtained by simulation for different average crushing values. Because the mill is roughly 30 years old, it is possible that the machinery will fail if it is operated at maximum crushing speed. The mill's suspension of operations will inevitably result in a reduction in service quality. Even though the mill crushed more than 20,000qtl sugarcane per day around 40% of the time, an average of 20,000qtl crushing could be attained. With a crushing capacity of 20,000 qtl, the mill can service 177 trolleys every day, with current average weight of 112.44 qtl per trolley. The present average waiting time of 3 h, 17 min will be reduced to 1 h, 20 min, and the average queue length of 23 trolleys will be reduced to 9 trolleys. This increases the probability of an idle scenario from 2 to 9%, implying that the idle time of service will increase to 7%.

References

1. William, H.K., Lawrence, M.L., John, H.D.: Transient queuing analysis. INFORMS J. Comput. **24**, 10–28 (2012)
2. Sheikh, T., Kumar, S., Kumar, A.: Application of queuing theory for the improvement of bank service. Int. J. Adv. Comput. Eng. Netw. **1**, 15–18 (2013)
3. Syed, S.S.: Simulation: analysis of single server queuing model. Int. J. Inf. Theory **3**, 47–54 (2014)

4. Okagbue, H.I., Edeki, S.O., Opanuga, A.A.: A Monte carlo simulation approach in assessing risk and uncertainty involved in estimating the expected earnings of an organization: a case study. Nigeria Amer. J. Comput. Appl. Math. **4**, 161–166 (2014)
5. Shanmugasundaram, S., Punitha, S.: A study on multi server queuing simulation. Int. J. Sci. Res. **3**, 1519–1521 (2014)
6. Arum, H.M.P., Retno, S., Nikenasih, B.: The completion of non-steady-state queue model on the queue system in Dr. Yap Eye Hospital Yogyakarta. J. Phys. Conf. Ser. **855**, 012034 (2017)
7. Abdalla, A., Buckley, J.J.: Monte Carlo methods in fuzzy queuing theory. Soft Comput. **13**, 1027–1033 (2009)
8. Shengda, Z., Lurong, W., Lin, J., Bizhi, W.: Study on monte Carlo simulation of intelligent TrafficLights based on fuzzy control theory. Sens. Trans. **156**(9), 211–216 (2013)
9. Morokoff, W., Caflisch, R.: Quasi-Monte Carlo Simulation of Random Walks in Finance. In: Niederreiter, H., Hellekalek, P., Larcher, G., Zinterhof, P. (eds.) Athens conference on applied probability and time series analysis 1998, vol. 127, pp. 340–352. Springer, New York, USA (1998)

An Adaptive Step-Size Optimized Seventh-Order Hybrid Block Method for Integrating Differential Systems Efficiently

Rajat Singla, Gurjinder Singh, and V. Kanwar

Abstract This article proposes a novel adaptive step-size numerical method for solving initial value ordinary differential systems. The development of the proposed method is based on the theory of interpolation and collocation in which representation of the theoretical solution of the problem is assumed in the form of an appropriate interpolating polynomial. In order to bypass the first Dahlquist's barrier on linear multistep methods, the proposed method considers five intra-step points in one-step block $[x_n, x_{n+1}]$ resulting in a hybrid method. Among these considered five intra-step points, the values of two intra-step points were fixed named as supporting off-step points and the optimized values of the other three intra-step points were obtained by minimizing the local truncation errors of the main formula at the point x_{n+1} and other two additional formulas at supporting off-step points. The proposed method exhibits the property of self-starting as the formulation is immersed into a block structure which enhances the efficiency of the method. The resulting method is of order seven retaining the characteristic of $\mathcal{A}$-stability. The precision of numerical solution is intensified by drafting the proposed algorithm into an adaptive step-size formulation using an embedded-type procedure. The adaptive step-size method has been tested on some well-known stiff differential systems, viz., Robertson's chemistry problem, Gear's problem, the Brusselator system, Jacobi elliptic functions system, etc. The proposed method performs well in comparison to other iconic codes available in the literature.

R. Singla (✉) · G. Singh
Department of Mathematical Sciences, I. K. Gujral Punjab Technical University Jalandhar, Main Campus, Kapurthala, Punjab, India
e-mail: rajatmath1310@gmail.com

R. Singla
Department of Mathematics, Akal University, Bathinda, Punjab, India

V. Kanwar
University Institute of Engineering and Technology, Panjab University, Chandigarh, India

R. K. Sharma et al. (eds.), *Frontiers in Industrial and Applied Mathematics*,
Springer Proceedings in Mathematics & Statistics 410,
https://doi.org/10.1007/978-981-19-7272-0_34

Keywords Block methods · Hybrid methods · Initial-value problems · Embedded type · $\mathcal{A}$-stable

1 Introduction

Differential equations are the results of the scientific phrasing of many of the physical phenomena. In the absence of analytical procedure for solving the differential equation, they can be dealt numerically. Much of the physical situations like chemical kinetics, orbital dynamics, atmospheric phenomena, engineering control systems, lasers, mechanics, electronic circuits, or orbital dynamics modelled into stiff ordinary differential equations (ODEs) [1, 2]. In the state of absence of closed expressions of the solutions to these problems, the various numerical schemes have been introduced to approximate the solutions on the discrete points on the interval of interest. In this article, our aim is to develop a one-step efficient numerical algorithm to solve the below initial value problem (IVP) numerically (see Refs. [3, 4]).

$$\mathbf{W}'(z) = \mathbf{F}(z, \mathbf{W}(z)); \quad \mathbf{W}(z_0) = \mathbf{W_0}, \tag{1}$$

where $z \in [z_0, z_N]$, $\mathbf{W} : [z_0, z_N] \to \mathbb{R}^m$, $\mathbf{W} : [z_0, z_N] \times \mathbb{R}^m \to \mathbb{R}^m$. It's been assumed that the given IVP holds the conditions of *Existence and Uniqueness theorem*.

Many numerical integrators were developed and presented in scientific literature to solve the differential system (1). Among them, the two well-defined classes for evaluating the differential systems numerically has been widely used which are named as Runge-Kutta (RK) and multistep methods. From the family of multistep methods, the backward differentiation formulas (BDFs) are used to solve the stiff numerical problems. Recently, many researchers have developed the numerical schemes, a fusion of hybrid and block methods to solve the differential systems numerically. Initially, the block methods were proposed by Milne [5] to obtain the solutions of predictor-corrector methods simultaneously. Later on, Rosser [6] extends this concept for the general purposes. The structure of block methods consists of main and additional methods, which were applied successively in block intervals and produce the solution of problem (1) simultaneously at several points which enhances the efficiency of schemes in terms of accuracy by reducing the computational efforts. In the present literature, many researchers are implementing the block methods along with some off-step points in the interval of interest which are named as hybrid points. Thus, the resulting scheme exhibits of both block and hybrid nature. Hybrid methods were simultaneously proposed by Gragg and Stetter [7], Butcher [8] and Gear [9]. The main feature of hybrid methods is that they overcome the Dahlquist barrier which limits the accuracy in class of zero-stable linear multistep methods. Hybrid methods have also the characteristic of Runge-Kutta method (that is, in order to find the solution at end point of interval $[z_n, z_{n+1}]$ it also uses the appropriate data at off-step points) along with certain features of linear multistep methods. For the construction and implementation of hybrid and block methods to solve the different types of ODEs, one can see the literature [10–17] and the references therein.

Many authors are solving the ODEs numerically by implementing the hybrid block method with the arbitrary values to the intermediate points. Later on Ramos et al. firstly proposed the optimization strategy to obtain the optimized values of intermediate values by minimizing the local truncation errors of the main formula and other additional formulas which results in the higher order of accuracy of these formulas (for the strategy one can see the procedure adopted in [17] concerning second-order problems).

The present article deals with the hybrid formulation of a single-step implicit method by using the five transitional points in one block of interval $[z_n, z_{n+1}]$. The values of two intermediate points are fixed in order to get the optimized values of other three intermediate points by minimizing the local truncation errors of the main formula and the additional formulas which approximates the solution of (1) at the fixed intermediate points. The resulting obtained formulas are incorporated into the block structure to get the numerical solution of the problem (1) at the intermediate points and final point of the interval $[z_n, z_{n+1}]$ simultaneously. Proceeding further, the method is evolved to adaptive step-size algorithm by using embedded-like structure which enhances the accuracy and efficiency of the proposed method.

2 Derivation of Proposed Algorithm

In order to carry out the derivation in a simplified manner, the method is derived for solving the scalar differential equation numerically, that is, for $m = 1$. Later on the method can be implemented to the vector differential equation for $m > 1$ by using component-wise strategy. Firstly, the method is derived with the fixed step-size $\delta z = z_{j+1} - z_j$ on a discrete grid with $N + 1$ nodal points, $z_0 < z_1 < z_2 < \ldots < z_N$, and later on method is derived into an adaptive step-size formulation. The derivation starts with the assumption of theoretical solution to (1) in the form of an interpolating polynomial as

$$w(z) \approx R(z) = \sum_{i=0}^{7} \kappa_i \, \Xi_i(z), \qquad z \in [z_j, z_{j+1}]. \tag{2}$$

The value of unknown coefficients κ_i has to be determined and $\Xi_i(z) = (z - z_j)^i$ are the polynomial basis functions. The proposed method is formulated into hybrid nature by considering the five intra-step points referred to as $m_i \in [0, 1]$, $i = 1(1)5$ where the values of intra-step points $m_2 = 1/4$, $m_4 = 3/4$ are fixed such that each $z_{j+m_i} = z_j + m_i \delta z$ for $1 \le i \le 5$. The derivation of the proposed method can be elaborated in the following steps:

Step 1: In the first step, the value of unknown coefficients $\kappa_i' s$ can be determined by imposing the interpolatory and collocation conditions to the assumed theoretical solution which are as follows:

(i) $w(z_j) = R(z_j)$.
(ii) $w'(z_{j+k}) = R'(z_{j+k}), \quad k = 0, m_i, 1$ such that $1 \le i \le 5$.

These conditions result into a algebraic system of eight equations in eight knowns $\kappa_i,\ i = 0(1)7$ which can be solved to get the values of $\kappa_i' s$. Now, after substituting the obtained values of $\kappa_i' s$ into the Eq. (2), the equation results into the form as

$$w(z) \approx R(z) = \lambda(z) w_j + \delta z \left(\sum_{i=0}^{1} \eta_i(z) F_{j+i} + \sum_{i=1}^{5} \eta_{m_i}(z) F_{j+m_i} \right), \quad (3)$$

where

$$w_j \simeq w(z_j),\ F_{j+k} \simeq F(z_{j+k}, w_{j+k}), \quad k = 0, m_i, 1 \text{ and } i = 0(1)5,$$

Step 2: In the second step, the optimized values of the transitional points m_1, m_3, m_5 have to be obtained, which can be done by evaluating expression (3) at the remaining transitional points (z_{j+m_2}, z_{j+m_4}) and at the final point (z_{j+1}) of the block $[z_j, z_{j+1}]$. Now the optimized values of intermediate points m_1, m_3, m_5 can be obtained by adopting the following optimization strategy:

(i) Obtain the local truncation errors of formulas $w(z_{j+m_i})$ for $i = 2, 4$ and $w(z_{j+1})$ by expanding these formulas about point $z = z_j$ with the help of Taylor series, and hence the obtained expressions are of form:

$$\mathcal{L}(w(z_{j+m_2}), \delta z) = \frac{A_1(m_1, m_3, m_5) w^{(8)}(z_j)(\delta z)^8}{277453209600} + \mathcal{O}((\delta z)^9), \quad (4)$$

$$\mathcal{L}(w(z_{j+m_4}), \delta z) = \frac{A_2(m_1, m_3, m_5) w^{(8)}(z_j)(\delta z)^8}{10276044800} + \mathcal{O}((\delta z)^9), \quad (5)$$

$$\mathcal{L}(w(z_{j+1}), \delta z) = \frac{A_3(m_1, m_3, m_5) w^{(8)}(z_j)(\delta z)^8}{33868800} + \mathcal{O}((\delta z)^9), \quad (6)$$

where

$$\begin{aligned} A_1(m_1, m_3, m_5) &= (211 - 1312m_5 + 1312m_1(-1 + 7m_5) - \\ &\quad 32m_3(41 - 287m_5 + 7m_1(-41 + 352m_5))) \\ A_2(m_1, m_3, m_5) &= (3(-477 + 864m_1 + 864m_5 - 1568m_1m_5) + \\ &\quad 32m_3(81 - 147m_5 + 7m_1(-21 + 32m_5))) \\ A_3(m_1, m_3, m_5) &= (2 + m_1 + m_5 - 7m_1m_5 + \\ &\quad m_3(1 - 7m_5 + 7m_1(-1 + 2m_5))). \end{aligned}$$

(ii) Now, equating the principal terms of the truncated errors (4), (5) and (6) to zero gives the unique solution with the condition $0 < m_1 < m_2 < m_3 < m_4 < m_5 < 1$ which are as follows:

$$m_1 = \frac{1}{8}\left(4-\sqrt{10}\right) \simeq 0.104715, \quad m_3 = \frac{1}{2} = 0.5,$$
$$m_5 = 1 + \frac{1}{8}\left(-4+\sqrt{10}\right) \simeq 0.895285$$

.Hence substituting the optimal values of m_1, m_3 & m_5 into the truncation errors given by (4), (5) and (6), the precision to approximate values of theoretical solution at nodal points z_{j+m_2}, z_{j+m_4} and z_{j+1} is increased.

Step 3: In the last step, the formulas for approximation to theoretical solution at the other intra-step points $z_{j+m_1}, z_{j+m_3}, z_{j+m_5}$ have to be obtained. For this, the optimized values of m_1, m_3, m_5 and $z = z_{j+m_i}$ for $i = 1, 3, 5$ have to be inserted in expression (3) which will gives us the hybrid method consisting of five approximations to the true solution at z_{j+m_i}, $i = 1(1)5$ and z_{j+1}.

Hence, we obtain an implicit hybrid block method and incorporated into block form which results in a hybrid block method whose coefficients are listed in Table 1.

3 Convergence Investigation

In this section, the basic theoretical aspects like consistency, order of accuracy, zero and linear stability of the proposed methods are discussed.

3.1 *Order of Accuracy and Consistency*

For analysing the consistency of the proposed method, it can be rewritten as

$$P\,\mathbf{W}_j = \delta z\; Q\,\mathbf{F}_j. \tag{7}$$

Here matrices P and Q represent the coefficients of order 6×7 which can be easily written from Table 1 and

$$\mathbf{W_j} = \left(w_j, w_{j+m_1}, w_{j+m_2}, w_{j+m_3}, w_{j+m_4}, w_{j+m_5}, w_{j+1}\right)^T,$$
$$\mathbf{F_j} = \left(f_j, f_{j+m_1}, f_{j+m_2}, f_{j+m_3}, f_{j+m_4}, f_{j+m_5}, f_{j+1}\right)^T.$$

Proceeding further, the difference operator $\mathcal{L}$ can be associated with the difference Eq. (7) and written as

Table 1 Coefficients of the method

z	λ	η_0	η_{m_1}	η_{m_2}	η_{m_3}	η_{m_4}	η_{m_5}	η_1
z_{j+m_1}	1	$\frac{9013+50\sqrt{10}}{241920}$	$\frac{512+37\sqrt{10}}{7560}$	$\frac{10319-4100\sqrt{10}}{120960}$	$\frac{\left(64-19\sqrt{10}\right)}{448}$	$\frac{12209-4100\sqrt{10}}{120960}$	$\frac{\left(64-19\sqrt{10}\right)}{945}$	$\frac{-437+50\sqrt{10}}{241920}$
z_{j+m_2}	1	$\frac{1901}{60480}$	$\frac{74}{945}+\frac{1}{4\sqrt{10}}$	$\frac{1933}{30240}$	$-\frac{1}{280}$	$\frac{43}{30240}$	$\frac{74}{945}-\frac{1}{4\sqrt{10}}$	$\frac{11}{60480}$
z_{j+m_3}	1	$\frac{103}{2520}$	$\frac{\left(64+14\sqrt{10}\right)}{945}$	$\frac{421}{1890}$	$\frac{1}{7}$	$-\frac{23}{630}$	$-\frac{\left(-64+14\sqrt{10}\right)}{945}$	$-\frac{41}{7560}$
z_{j+m_4}	1	$\frac{79}{2240}$	$\frac{2}{35}+\frac{1}{4\sqrt{10}}$	$\frac{207}{1120}$	$\frac{81}{280}$	$\frac{137}{1120}$	$\frac{2}{35}-\frac{1}{4\sqrt{10}}$	$\frac{9}{2240}$
z_{j+m_5}	1	$\frac{9013-50\sqrt{10}}{241920}$	$\frac{\left(64+19\sqrt{10}\right)}{945}$	$\frac{10319+4100\sqrt{10}}{120960}$	$\frac{\left(64+19\sqrt{10}\right)}{448}$	$\frac{12209+4100\sqrt{10}}{120960}$	$\frac{512-37\sqrt{10}}{7560}$	$\frac{-437-50\sqrt{10}}{241920}$
z_{j+1}	1	$\frac{67}{1890}$	$\frac{128}{945}$	$\frac{176}{945}$	$\frac{2}{7}$	$\frac{176}{945}$	$\frac{128}{945}$	$\frac{67}{1890}$

$$\mathcal{L}[W(z), \delta z] = \sum_j \rho_j W(z + j\delta z) - \delta z \sum_j \delta_j W'(z + j\delta z) \quad \text{for } j = 0, m_i, 1 \,\&\, 1 \leq i \leq 5, \tag{8}$$

where ρ_j and δ_j are precisely the column vectors of matrices P and Q. Now, the functions $W(z + j\delta z)$ and $W'(z + j\delta z)$ can be expanded around z with the help of Taylor series. After expanding the terms, the difference operator can be written as

$$\mathcal{L}[W(z), \delta z] = \tau_0 W(z) + \tau_1 \delta z W'(z) + \tau_2 \delta z^2 W''(z) + \cdots + \tau_p \delta z^p W^{(p)}(z) + \cdots \tag{9}$$

The proposed hybrid block method and the associated linear difference operator are said to be of order q if all $\tau_0 = \tau_1 = \cdots = \tau_q = 0$ and $\tau_{q+1} \neq 0$. Hence $\tau_i's$ are the vectors and τ_{q+1} is known as vector of constants.

For the present hybrid block method $\tau_0 = \tau_1 = \tau_2 = \cdots = \tau_7 = 0$ and

$$\tau_8 = \left(-\frac{9}{4697620480}, 0, -\frac{1}{165150720}, 0, -\frac{9}{469762048}, 0\right)^T.$$

Since the proposed hybrid block method is of order 7 which is greater than 1 and hence the proposed method is consistent with the system of differential equation (1).

3.2 Zero-Stability Investigation

In this, the behaviour of method is examined as the step-size $\delta z \to 0$. By considering $\delta z \to 0$, the proposed method in Table 1 reduces to $W_{j+m_1} = W_j,\ W_{j+m_2} = W_j,\ W_{j+m_3} = W_j,\ W_{j+m_4} = W_j,\ W_{j+m_5} = W_j,\ W_{j+1} = W_j$ which can be written in a simpler way as

$$\hat{M}_\zeta = \bar{A}\,\hat{M}_{\zeta-1}, \tag{10}$$

where $\bar{A} = [a_{ij}],\ 1 \leq i, j \leq 6$ and

$$a_{ij} = \begin{cases} 0, \text{ for } 1 \leq i \leq 6 \text{ and } 1 \leq j < 6, \\ 1, \text{ for } 1 \leq i \leq 6 \text{ and } j = 6, \end{cases}$$

and

$$\begin{aligned} \hat{M}_\zeta &= (W_{j+m_1}, W_{j+m_2}, W_{j+m_3}, W_{j+m_4}, W_{j+m_5}, W_{j+1})^T, \\ \hat{M}_{\zeta-1} &= (W_{j+m_1-1}, W_{j+m_2-1}, W_{j+m_3-1}, W_{j+m_4-1}, W_{j+m_5-1}, W_j)^T. \end{aligned} \tag{11}$$

The first characteristic polynomial of the proposed method is written as

$$\rho(\eta) = det[I_6\eta - \bar{A}] = \eta^5(\eta - 1). \tag{12}$$

As the roots of characteristic polynomial $\rho(\eta) = 0$ satisfy $\eta_j \leq 1$ and having modulus one is simple then by definition of zero-stability, the proposed method is zero-stable.

3.3 Convergence

Hence the method is consistent and zero-stable which implies that the proposed method is convergent.

3.4 Linear Stability Analysis

As we always deal with some finite step-size δz, it is a very impractical situation where we say $\delta z \to 0$. Now, we will examine the behaviour of proposed method with some finite value of δz that makes the concept of linear stability analysis different from zero-stability. For this purpose, the well-known Dahlquist's test equation is considered

$$W'(z) = \xi W(z), \qquad Re(\xi) < 0. \tag{13}$$

The analytical solution of Eq. (13) will tend to 0 as $z \to \infty$. It has been expected that while applying the proposed method to test Eq. (13), the produced numerical solution will behave in alike manner to analytical solution. So now we have to determine the region for which numerical method reproduces the behaviour of true solution of test problem, for this when proposed method is applied to test equation, the obtained system can be written as

$$X\,\hat{M}_{\varsigma} = Y\,\hat{M}_{\varsigma-1}, \tag{14}$$

where

$$X = [x_{ij}]\ \&\ Y = [y_{ij}],\ 1 \leq i, j \leq 6$$

with $H = \xi\delta z$. The Eq. (14) can be also expressed as

$$\hat{M}_{\varsigma} = Q(H)\hat{M}_{\varsigma-1}, \tag{15}$$

where $Q(H) = X^{-1}Y$ is defined as stability matrix. In order to examine the stability characteristics of the proposed method, the spectral radius of $\rho[Q(H)]$ is to be considered which is given as

$$\rho[Q(H)] = \frac{1720320 + 860160H + 195840H^2 + 26240H^3 + 2212H^4 + 114H^5 + 3H^6}{1720320 - 860160H + 195840H^2 - 26240H^3 + 2212H^4 - 114H^5 + 3H^6}.$$

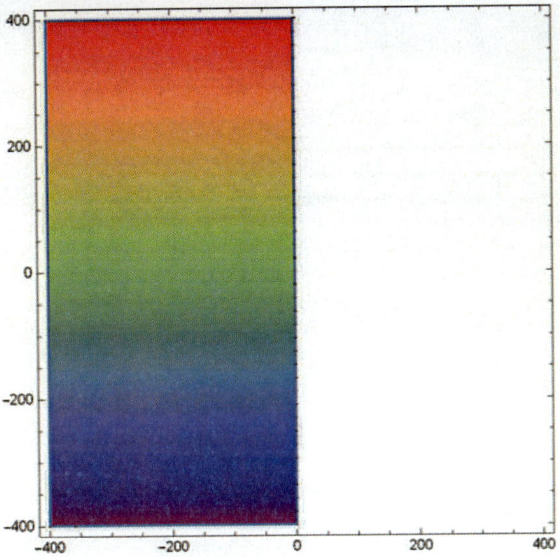

Fig. 1 Absolute stability region of the hybrid block method in Table 1

For all $Re(H) < 0$, the magnitude of spectral radius $|\rho[Q(H)]| < 1$, hence by definition the proposed method is $\mathcal{A}-$stable. Figure 1 presents the absolute stability region of the new proposed method.

4 Modulation into Variable Step-Size

In this section, the proposed method is formulated to variable step-size algorithm. For the purpose, the embedded-type procedure is followed as per guided in Shampine et al. [19]. The similar strategy is also used with embedded Runge-Kutta methods like Dopri 4(5), Runge-Kutta-Fehlberg methods, etc. In this modulation, the two methods having different orders of accuracy were executed simultaneously in which method with higher order is used for advancing the integration steps and the lower order method is used to estimate the local error at each step of integration. Here, the proposed method in Table 1 is considered as a higher order method and the lower order method is chosen in such a way that it uses the same function evaluations of the proposed method. Hence, by doing in such a way the computational cost is not increased but accuracy of the method is enhanced. For the detailed procedure, one can follow the article [13, 14] for solving first-order and second-order differential equation. Following is the lower order method used for the formulation of proposed hybrid block method, Table 1, into an adaptive step-size algorithm:

$$W^*_{j+1} = \frac{411W_j + 108W_{j+m_3} - 496W_{j+m_1} + 48W_{j+m_4}}{71} + \delta z\left(\frac{9}{71}f_j + \frac{48(5+\sqrt{10})}{355}f_{j+m_1} - \frac{48(-5+\sqrt{10})}{355}f_{j+m_5}\right), \quad (16)$$

with local truncation error $LTE = \frac{3\,W^{(7)}(z)(\delta z)^7}{5816320} + \mathcal{O}((\delta z)^8)$.

5 Computational Efficiency

This section addresses the precision to numerical solution obtained by proposed adaptive step-size hybrid block method by testing on well-known differential systems existing in scientific literature. The abbreviations used in the below tables are designated as: FNC: Number of function evaluations; δz_{ini}: Initial step-size; TOL: Tolerance; N: number of integration steps; Δ_{max}: Maximum absolute errors among all the components and along the integration interval given by

$$\Delta_{max} = \max_{1\le i\le m}\{\max_{0\le j\le N}\{|w_i(z_j) - w_{ij}|\}\},$$

where $w_i(z_j)$ and w_{ij} denote the exact and computed ith-component of solution of a m-system differential problem at point z_j. The following solvers were considered in comparison to proposed method:

1. **RKGauss**: This method is an $\mathcal{A}$-stable implicit Runge-Kutta method based on a "Gaussian quadrature". The method is a five-stage tenth-order method (see [3]).
2. **RADAU**: This solver is of variable order (1, 5, 9, 13) with step-size control. It is also based on implicit Runge-Kutta methods (Radau-IIa). In the experiments, we have used the *Matlab* code of this scheme by Hairer (see MatlabStiff package in http://www.unige.ch/~hairer/software.html).
3. **C-OHM**: This is the new optimized hybrid scheme whose coefficients listed in Table 1 are developed with the fixed step-size strategy.
4. **V-OHM**: This is the new optimized hybrid scheme whose coefficients listed in Table 1 use the embedded variable step-size strategy explained in Sect. 4.

5.1 Comparison of Constant and Variable Step-Size Proposed Method

In this section, an example is solved by implementing the proposed hybrid method in both constant and variable step-size mode and comparing their computational efficiencies, that is, order of accuracy on same number of function evaluations.

Table 2 Computational data for problem Sect. 5.1.1

N	FNC	Δ_{max} in C-OHM	Δ_{max} in V-OHM
24	168	8.65765×10^{0}	2.68781×10^{-6}
45	315	1.10858×10^{0}	3.0714×10^{-8}
120	720	6.10482×10^{-3}	8.86047×10^{-12}

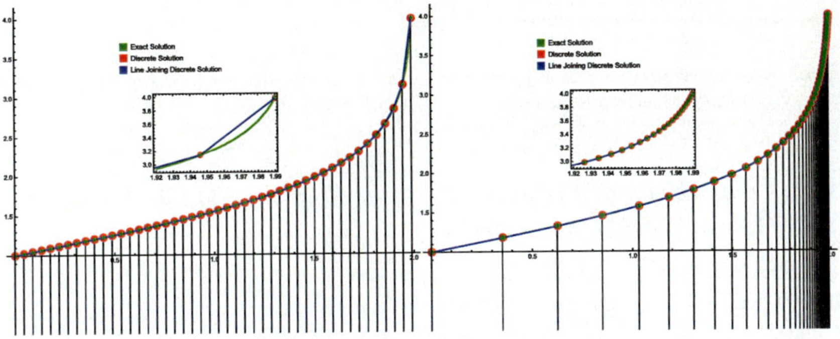

Fig. 2 Exact and Discrete solutions w_1 of Problem Sect. 5.1.1 using the constant step-size (C-OHM) with $N = 45$(left) and variable step-size (V-OHM) with $N = 45$(right)

5.1.1 A Non-linear Problem

The problem is given by

$$w_1'(z) = w_2(z), \quad w_1(0) = 1,$$
$$w_2'(z) = zw_2(z)^2, \quad w_2(0) = \frac{1}{2}.$$

The problem is solved over the integration of interval $z \in [0, 1.99]$. The exact solution of the system is given as $w_1(z) = 1 + \frac{1}{2}\ln\left(\frac{2+z}{2-z}\right)$, $w_2(z) = \frac{2}{4-z^2}$. The proposed method with adaptive step-size is implemented on problem by varying step-size δz_{ini} and tolerances to equalize the number of steps used in implementation of proposed block method with constant step-size. The computational errors compared to analytical solution is presented in Table 2. Also Figs. 2 and 3 reveal that the proposed variable step-size formulation performs better than the proposed method in constant step-size version.

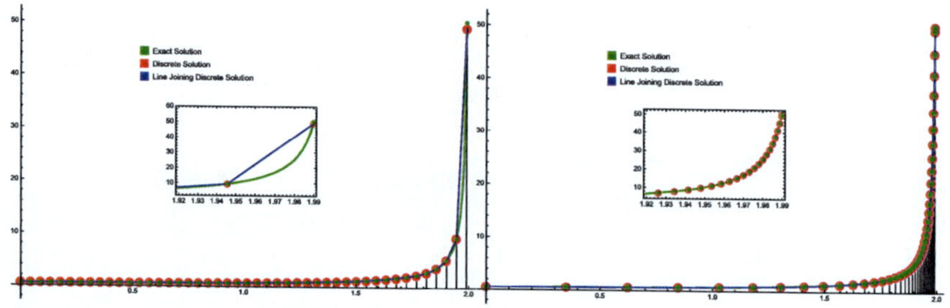

Fig. 3 Exact and discrete solutions w_2 of problem Sect. 5.1.1 using the constant step-size (C-OHM) with $N = 45$(left) and variable step-size (V-OHM) with $N = 45$(right)

5.2 *Comparison of Proposed Variable Step-Size Method with Other Well-Known Solvers*

In this section, a comparison has been made between the proposed hybrid block method in adaptive step-size mode with the other well-known integrators in variable step-size mode.

5.2.1 The Brusselator System

Consider the diffusion-free "Brusselator system" which consists of the two differential equations

$$\begin{aligned} w_1'(z) &= L + w_1^2(z)\, w_2(z) - (M+1)\, w_1(z), \qquad & w_1(0) = w_1^0 \\ w_2'(z) &= M\, w_1(z) - w_1^2(z)\, w_2(z), \qquad & w_2(0) = w_2^0, \end{aligned} \tag{17}$$

where L and M are positive real constants. It can be shown that the critical point of the system is $(w_1^*, w_2^*) = (L,\ M/L)$. For numerical experimentation, we assume $M = 3,\ L = 1$, with initial values $w_1(0) = 1.5$, $w_2(0) = 3$, over the integration interval $[0, 20]$. The reference solution at end point $z_N = 20$

$$\begin{aligned} w_1(z_{20}) &= 0.4986370712683478483331816235 \\ w_2(z_{20}) &= 4.596780349452011826429803773 \end{aligned}$$

is taken from the article [15] as the analytical solution of the problem is not present in the literature. The problem is solved by considering $(h_{ini}, Tol) = (10^{-\kappa}, 10^{-(\kappa+3)})$, $\kappa = 4, 5$. The results in Table 3 gives a numerical evidence of the good performance of the proposed method.

Table 3 Numerical results for Sect. 5.2.1

δz_{ini}	TOL	Method	Δ_{max}	N	FNC
10^{-4}	10^{-7}	RADAU	9.3149×10^{-8}	187	2371
		RKGauss	3.55082×10^{-6}	553	3318
		V-OHM	1.49117×10^{-9}	84	588
10^{-5}	10^{-8}	RADAU	4.7105×10^{-9}	231	3870
		RKGauss	8.72215×10^{-7}	1709	10254
		V-OHM	8.20015×10^{-11}	115	805

6 Conclusion

In this article, we have derived one-step optimized hybrid block method with five transitional points in which the value of two off-step points are fixed and optimized values of other three off-step points are determined. The development of method is purely based on interpolation and collocation technique clubbed with hybrid and block approaches. The method is $\mathcal{A}$-stable and having seventh algebraic order of convergence with satisfying conditions of zero-stability. The new scheme is also modulated into an adaptive step-size technique by using embedded-type procedure which makes the proposed scheme more efficient than its counterpart. Numerical experimentation establishes the statement that the proposed new scheme is a good alternative for solving the well-known considered differential systems.

References

1. Aiken, R.C.: Stiff Computation, R. Aiken edn. Oxford University Press, New York, USA (1985)
2. Sandu, A., Verwer, J.G., Blom, J.G., Spee, E.J., Carmichael, G.R., Potra, F.A.: Benchmarking stiff ODE solvers for atmospheric chemistry problems II: rosenbrock solvers. Atmos. Environ. **31**, 3459–3479 (1997). https://doi.org/10.1016/s1352-2310(97)83212-8
3. Butcher, J.C.: Numerical Methods for Ordinary Differential Equations. Wiley (2008)
4. Lambert, J.D.: Numerical Methods for Ordinary Differential Systems: The Initial Value Problem. Wiley (1991)
5. Milne, W.E.: Numerical Solution of Differential Equations. Wiley, New York (1953)
6. Rosser, J.B.: A Runge-Kutta for all reasons. SIAM Rev. **9**, 417–452 (1967)
7. Gragg, W.B., Stetter, H.J.: Generalized multistep predictor-corrector methods. J. Assoc. Comput. Mach. **11**, 384–403 (1965)
8. Butcher, J.C.: A modified multistep method for the numerical integration of ordinary differential equations. J. Assoc. Comput. Mach. **12**, 124–135 (1965)

9. Gear, C.W.: Hybrid methods for intitial value problems in Ordinary differential equations. SIAM J. Numer. Anal. **2**, 69–86 (1964)
10. Shampine, L.F., Watts, H.A.: Block implicit one-step methods. Math. Comp. **23**, 731–740 (1969)
11. Fatunla, S.O.: Block methods for second order odes. Int. J. Comput. Math. **41**, 55–63 (1991)
12. Jator, S.N.: A sixth order linear multi-step method for the direct solution of $y'' f(x, y, y')$. Int. J. Pure Appl. Math. **40**, 457–472 (2007)
13. Singla, R., Singh, G., Kanwar, V., Ramos, H.: Efficient adaptive step-size formulation of an optimized two-step hybrid block method for directly solving general second-order initial-value problems. Comput. Appl. Math. **40**, 220 (2021). https://doi.org/10.1007/s40314-021-01599-z
14. Singh, G., Garg, A., Singla, R., Kanwar, V.: A novel two-parameter class of optimized hybrid block methods for integrating differential systems numerically. Comput. Math. Methods (2021). https://doi.org/10.1002/cmm4.1214
15. Singh, G., Garg, A., Kanwar, V., Ramos, H.: An efficient optimized adaptive step-size hybrid block method for integrating differential systems. Appl. Math. Comput. **362**, 124567 (2019)
16. Ramos, H., Mehta, S., Vigo-Aguiar, J.: A unified approach for the development of k-step block Falkner-type methods for solving general second order initial-value problems in ODEs. J. Comput. Appl. Math. (2016). https://doi.org/10.1016/j.cam.2015.12.018
17. Ramos, H., Kalogiratou, Z., Monovasilis, Th., Simos, T.E.: An optimized two-step hybrid block method for solving general second order initial-value problems. Numer. Algor. (2016). https://doi.org/10.1007/s11075-015-0081-81
18. Brugnano, L., Trigiante, D.: Block implicit methods for ODEs. In: Trigiante, D. (ed.) Recent trends in Numerical Analysis, pp. 81–105. Nova Science Publ. Inc., New York (2001)
19. Shampine, L.F., Gladwell, I., Thompson, S.: Solving ODEs with MATLAB. Cambridge University Press (2003)

Comparison of Prediction Accuracy Between Interpolation and Artificial Intelligence Application of CFD Data for 3D Cavity Flow

M. Diederich, L. Di Bartolo, and A. C. Benim

Abstract The great opportunities of the new technology of artificial intelligence and the growing computational capacities together with interacting sensor technology leads to the next industrial revolution called Industry 4.0. In this field the combination of artificial intelligence with numerical simulation to develop a simplified model of a given system can be used for establishing a digital twin of the system for better control and more efficient performance. In this paper, the Artificial Neuronal Network (ANN) methodology is applied as well as a standard interpolation to develop two different simplified models of a 3D cavity flow. The problem is analyzed by Computational Fluid Dynamics (CFD). The CFD simulations are carried out using a commercial software for a case, for which experimental data from the literature exists. In general, the combination of CFD and ANN has been performed in different researches on different applications. Thus, the present paper focuses rather on the comparison of a standard interpolation procedure to ANN, utilizing two different error calculations.

Keywords Fluid dynamics · Artificial intelligence · Artificial neuronal networks · Industry 4.0

M. Diederich · A. C. Benim (✉)
Center of Flow Simulation, Düsseldorf University of Applied Sciences, Düsseldorf, Germany
e-mail: alicemal@prof-benim.com

M. Diederich
e-mail: michael.diederich@hs-duesseldorf.de

L. Di Bartolo
École Nationale Supérieure de Mécanique et d'Aérotechnique (ISAE-ENSMA), Chasseneuil-du-Poitou, France
e-mail: lorenzo.di-bartolo@etu.isae-ensma.fr

R. K. Sharma et al. (eds.), *Frontiers in Industrial and Applied Mathematics*, Springer Proceedings in Mathematics & Statistics 410,
https://doi.org/10.1007/978-981-19-7272-0_35

1 Introduction

The fields of the Artificial Intelligence (AI) [1–3] and Computational Fluid Dynamics (CFD) [4–8] are experiencing a rather parallel development. Both fields exist for decades, and due to the increasing computational capabilities, their impact has been growing rapidly in the recent years. First ideas of combination the technologies of date back a lot of years, e.g. to 1988, when Andrews [9] published the first review on the capabilities and problems in combination of AI and CFD. More recent publications [10–12] show different approaches for the AI-CFD interaction. In Ref. [13] a nice overview on the newest AI technologies and frameworks were presented. In problems with more complex physics, the combination of AI and CFD was demonstrated in Refs. [14, 15]

Further investigations on the different aspects of the problem in different areas including digital twins were presented by numerous researches [16–25].

2 The Test Case

For comparison with realistic data from a three dimensional flow with large velocity variations, a 3D cavity problem is considered. Corresponding experimental was data found in Ref. [26] (Fig. 1, 2).

For the experimental investigations, different Reynolds numbers have been used as shown in Table 1 with the calculated velocities for the working fluid of isopropyl alcohol of density $\rho = 0.785$ g/cm^3 and kinematic viscosity of $\nu = 0.031$ cm^2/s.

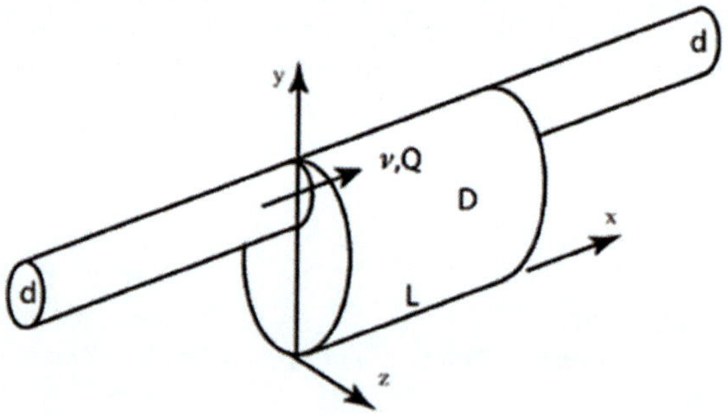

Fig. 1 Sketch of the experimental setup [26]

Table 1 Reynolds numbers used for simulation/experiment

Case	Re	Vm (mm/s)
a	2.7	2.64
b	5.6	5.47
c	15.7	15.33
d	32.1	31.34
e	62.8	61.32
f	140.8	137.47
g	288	281.20
h	320	312.44
i	542	529.20
j	650	634.65

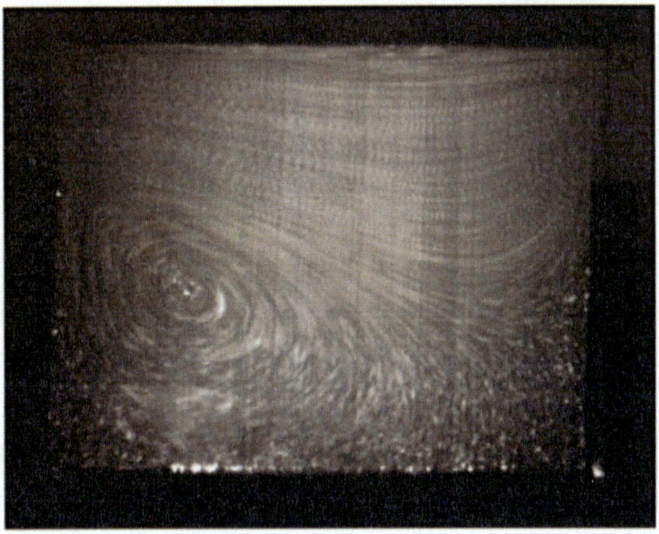

Fig. 2 Laser image of the velocity field [26]

3 Mathematical and Numerical Flow Modeling

The computational modeling of the flow has been performed using the simulation software ANSYS Fluent [27]. To ensure reliable simulation results, a mesh sensitivity study has been performed and the meshes shown in Fig. 3 are used for the further calculations. Since the Reynolds number was within the laminar flow regime, the simulation was done with no turbulence model, but for a laminar flow simulation setup.

For the inlet boundary condition, a fully developed flow is set by an equation for the velocity.

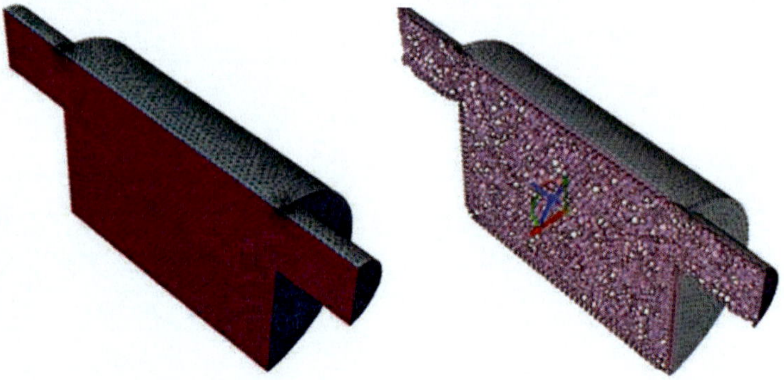

Fig. 3 Mesh for the CFD-simulation

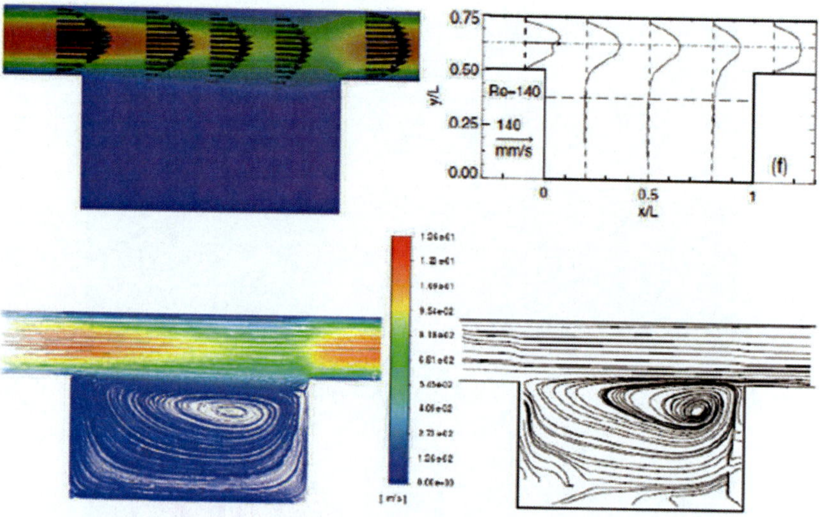

Fig. 4 Flow field Reynolds number 140 (left: simulation, right: experiments [26])

The Simulation results show fine agreement with the experimental data as shown in Fig. 4.

4 Developing a Simplified Model

The numerical simulation following the iterative solution of fluid physics by calculation of the Navier–Stokes equation system can take a lot of computational effort and time. Thus, the common CFD approach may not be feasible in cases, where limitations of resources and time are strict.

On the other hand, if the flow field is prescribed by a set of coefficients as found in interpolations of artificial network, this set of coefficients can be solved within very short time by simple matrix calculations which are very simple for nowadays computational infrastructure.

So the aim of the given study is to develop two different approaches for the calculation of the matrices that represent the flow field and to compare both results for different Reynolds numbers within a given range.

4.1 Simple Interpolation

The first approach is the calculation of a set of coefficients for the solution domain expressing the variables of interest as functions of the inlet condition following a regression function, whose coefficients are extracted from the data exported from the simulations. These coefficients represent the influence of the change of the variable (here the inlet velocity) to the behavior of the system, for different orders, linear, quadratic, and more if necessary.

$$\begin{aligned} Y_a &= b_0 \cdot 1 + b_1 \cdot v_a + b_2 \cdot v_a^2 \\ Y_b &= b_0 \cdot 1 + b_1 \cdot v_b + b_2 \cdot v_b^2 \\ &\vdots \end{aligned} \tag{1}$$

The equations can be put in matrix form as follows

$$\begin{bmatrix} 1 & v_a & v_a^2 \\ 1 & v_b & v_b^2 \\ \cdots & \cdots & \cdots \end{bmatrix} \cdot \begin{bmatrix} b_0 \\ b_1 \\ b_2 \end{bmatrix} = \begin{bmatrix} Y_a \\ Y_b \\ \vdots \end{bmatrix} \tag{2}$$

This system can be solved for the whole domain to get the velocity for a given inlet velocity.

4.2 ANN Model

The next approach is more advanced one, using the ANN framework of Keras with the CFD simulation results, the coordinates and the boundary conditions as an input for the neuronal network with randomized order of the points.

The training is done in a sequential class, the relu activation layer an Adam optimizer with a learning rate of 0.005 and a loss function with Mean AbsouteError.

The architecture was built with four hidden layers with 256 nodes each.

5 Error Estimation

In estimation of the accuracy of the simulation results, next to the qualitative comparison of the flow fields, the given project aimed to also find quantitative error estimation.

Here, it was important to find a method to calculate the error that takes into account the differences of the high velocity zones as well as the differences of the zones with lower velocity also. After some development the decision was made to define two different error calculations as shown in Eqs. (3) and (4).

Since the Error1 takes the differences of each velocity at a certain point from the CFD calculation to the model prediction, here the relative error of the small velocities has a much higher influence compared to the error in the high velocity field. On the other hand for the Error2, the relative error of the differences of the sum of all velocities for the CFD calculation to the sum of all velocities of the model prediction has been calculated and thus, here the differences of the high velocity areas at the flow field plays a major role.

$$Error1 = \frac{\sum \left| V_{CFD(x,y,z)} - V_{model(x,y,z)} \right|}{\sum \left| V_{CFD(x,y,z)} \right|} \cdot 100 \tag{3}$$

$$Error2 = \frac{\sum \left| V_{CFD(x,y,z)} \right| - \sum \left| V_{model(x,y,z)} \right|}{\sum \left| V_{CFD(x,y,z)} \right|} \cdot 100 \tag{4}$$

6 Results

In Fig. 5 the comparison of the results of the interpolation to the CFD calculations are shown. It is shown that the prediction of the model shows rather big errors in the area of low Reynolds number but for the higher velocities, the error becomes low and the quality of the predictions is feasible.

The results of the error calculations for the predictions of the ANN in comparison to CFD are shown in Fig. 6. As before, the results of the predictions at the low Reynolds numbers are not good but becomes much better in a range of smaller than 10% at higher Reynolds numbers. It is interesting to notice that the simple interpolation algorithm appears to give better results than the more advanced AI approach.

A further comparison is shown in Fig. 7. As seen in the figure the velocity is plotted along traversal lines (Line 1 in a and d, line 2 in b and e, line 3 in c and f) for two different Reynolds numbers (Re = 15 for a, b, c and Re = 140 for d, e, f) each plot with the direct comparison of the velocity profile for the CFD simulation, the interpolation and the AI model.

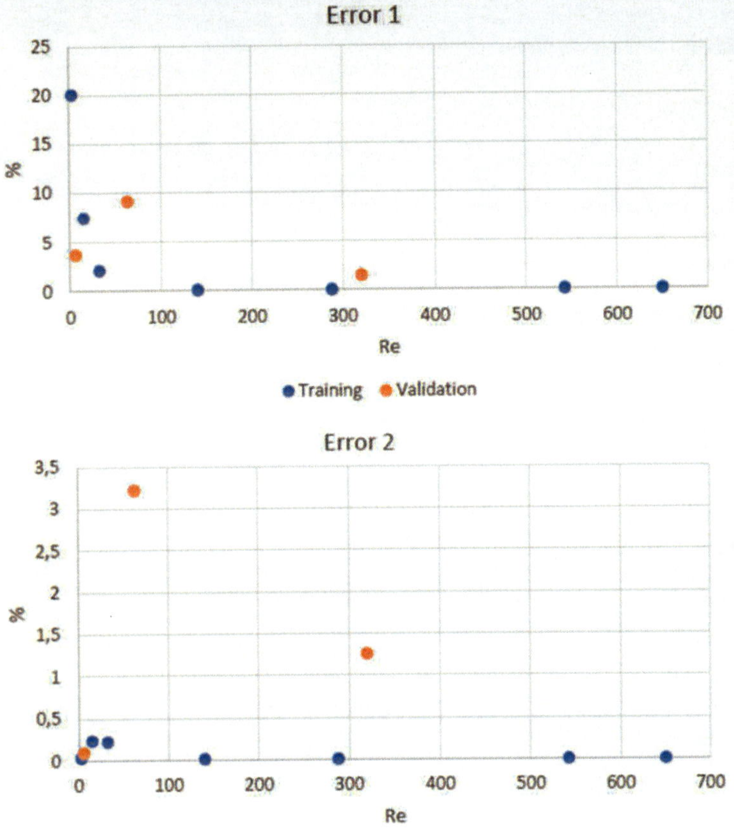

Fig. 5 Error of the interpolation model

It is seen that the prediction of the fully developed pipe flow for the inlet and outlet (at line 1 and 3) is quite well for all cases, just in the middle of the cavity, at line 2, the differences become larger. In b, one can see the difference of the interpolation to the CFD result is smaller than the difference of the AI predictions. For the higher Reynolds number (e) the differences become negligible small for both (Interpolation and AI) in comparison to the CFD simulation.

7 Conclusions

Two different approaches have been developed for using the data of CFD calculations to train different meta models that are able to predict the three dimensional flow field of a cavity flow within very short time. The models were using a simple interpolation model and a more advances AI approach. In this paper both models

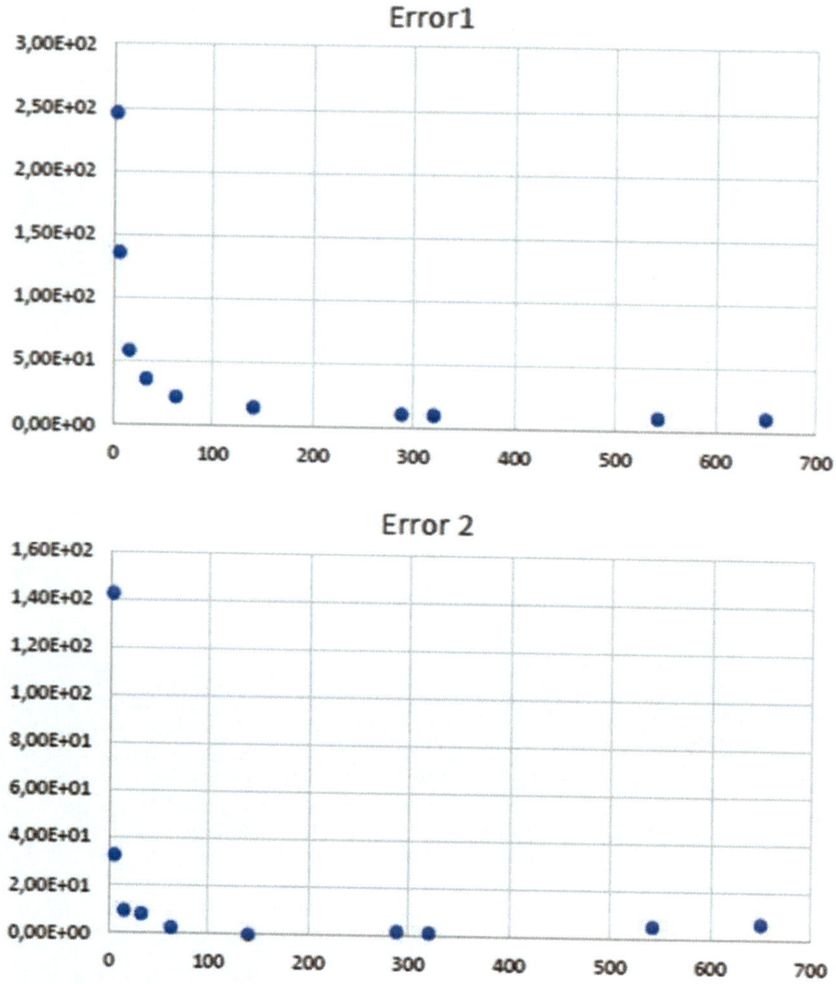

Fig. 6 Error of the AI model

have been compared among each other and both models show acceptable accuracy in the prediction of the flow field for higher Reynolds numbers but shows difficulty in the lower Reynolds numbers. Here the interpolation shows even better performance than the AI approach.

Following developments will be carried out to develop a supervision tool that performs randomized test simulations and compares them to the predictions and will form smaller submodels for areas where the prediction shows big differences. Here, again both approaches shall be compared.

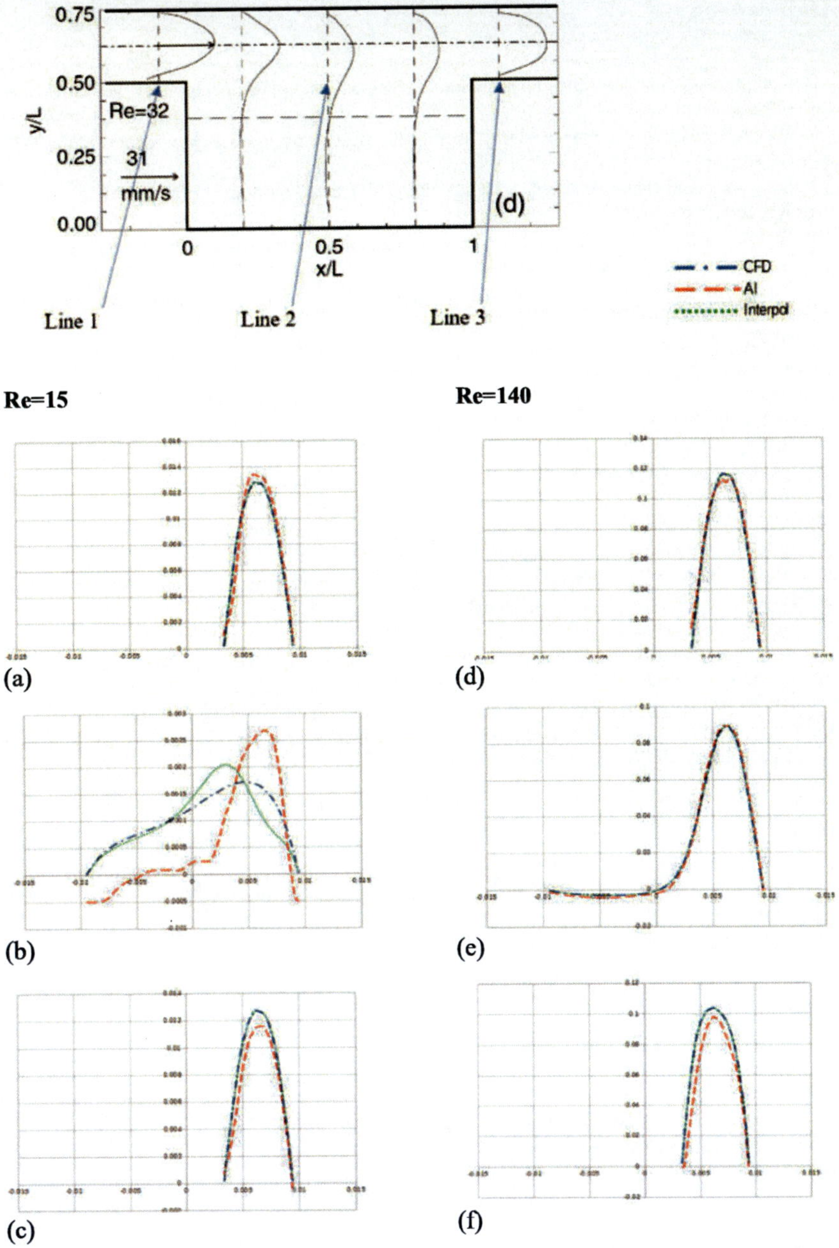

Fig. 7 Error of the interpolation and AI models

References

1. Chollet, F.: Deep Learning mit Python und Keras Das Praxis-Handbuch. MITP Verlag, Frechen (2018)
2. Vargas, R., Misavi, A., Ruiz, R.: Deep learning: a review. Adv. Intell. Syst. Comput. **2018100218** (2018)
3. Selle, S.: Künstliche Neuronale Netzwerke und Deep Learning. Lecture in University of Applied Sciences Business School (2018)
4. Benim, A.C., Iqbal, S., Joos, F., Wiedermann, A.: Numerical analysis of turbulent combustion in a model swirl gas turbine combustor. J. Combust., Article ID 2572035 (2016)
5. Pfeiffelmann, B., Benim, A.C., Joos, F.: A finite volume analysis of thermoelectric generators. Heat Transfer Eng. **40**(17–18), 1442–1450 (2019)
6. Cagan, M., Benim, A.C., Gunes, D.: Computational analysis of gas turbine preswirl system operation characteristics. WSEAS Trans. Fluid Mech. **4**(4), 117–126 (2009)
7. Benim, A.C., Pfeiffelmann, B., Oclon, P., Taler, J.: Computational investigation of a lifted hydrogen flame with LES and FGM. Energy **173**, 1172–1181 (2019)
8. Benim, A.C., Diederich, M., Gül, F., Oclon, P., Taler, J.: Computational and experimental investigation of the aerodynamics and aeroacoustics of a small wind turbine with quasi-3D optimization. Energy Convers. Manage. **177**, 143–149 (2018)
9. Andrews, A.: Progress and challenges in the application of artificial intelligence to computational fluid dynamics. AIAA J. **26**(1), 40–46 (1988)
10. Wang, B., Wang, J.: Application of artificial intelligence in computational fluid dynamics. Ind. Eng. Chem. Res. **60**(7), 2772–2790 (2021)
11. Usman, A., Muhammad, R., Muhammad, S., Ali, N.: Machine learning computational fluid dynamics. Swedish Artificial Intelligence Society Workshop (SAIS), pp. 46–49. IEEE (2021)
12. Kochkov, D., Smith, J.A., Aliyeva, A., Wang, Q., Brenner, M.P., Hoyer, S.: Machine learning–accelerated computational fluid dynamics. PNAS **118**(21), e2101784118 (2021)
13. Sadrehaghighi, I.: Artificial intelligence (AI) and deep learning for CFD. Technical Report on ResearchGate. https://doi.org/10.13140/RG.2.2.22298.59847/2
14. Panwar, V., Vandrangi, S.K., Emani, S.: Artificial intelligence-based computational fluid dynamics approaches. Hybrid Comput. Intell. **8**, 173–190 (2020)
15. Rojek, K., Wyrzykowski, R., Gepner, P.: AI-accelerated CFD simulation based on OpenFOAM and CPU/GPU computing. In: Paszynski, M., Kranzlmüller, D., Krzhizhanovskaya, V.V., Dongarra, J.J., Sloot, P.M.A. (eds.) Computational Science—ICCS 2021, pp. 373–385. Springer, Berlin (2021)
16. Chinesta, F., Cueto, E., Grmela, M., Moya, B., Pavelka, M., Sipka, M.: Learning physics from data: a thermodynamic interpretation. In: Barbaresco, F., Nielsen, F. (eds.) Geometric Structures of Statistical Physics, Information Geometry, and Learning, pp. 267–297. Springer, Berlin (2021)
17. Alfaro, I., Gonzalez, D., Zlotnik, S., Diez, P., Cueto, E., Chinesta, F.: An error estimator for real-time simulators based on model order reduction. Adv. Model. Simul. Eng. Sci **2**, Article 30 (2015)
18. Ghnatios, C., El Haber, G., Duval, J.-L., Zoane, M., Chinesta, F.: Artificial intelligence based space reduction of structural nodels. SAFORM 2021 (2021)
19. Hernández, Q., Badias, A., Gonzalez, D., Chinesta, F., Cueto, E.: Deep learning of thermodynamics-aware reduced-order models from data. Comput. Methods Appl. Mech. Eng. **379**(4), 113763 (2021)
20. Hamzi, B., Owhadi, H.: Learning dynamical systems from data: a simple cross-validation perspective, part I: Parametric kernel flows. Physica D **421**(3), 132817 (2021)
21. Pengzhan, L. et al.: Learning nonlinear operators via DeepONet based on the universal approximation theorem of operators, nature research. Nature Mach. Intell. **3**(3), 218–229 (2021). https://doi.org/10.1038/s42256-021-00302-5

22. Sancarlos, A., Cameron, M., Le Peuvedic, J.-M., Groulier, J., Duval, J.-L., Cueto, E., Chinesta, F.: Learning stable reduced-order models for hybrid twins. ResearchGate (2021). https://doi.org/10.1017/dce.2021.16
23. Champaney, V., Sancarlos, A., Chinesta, F., Cueto, E., Gonzalez, D., Alfaro, I., Guevelou, S., Duvalm J. L., Chambard, A., Mourguew P.: Hybrid twins—a highway towards a performance-based engineering. Part I: Advanced model order reduction enabling real-time Physics. ESAFORM 2021 (2021)
24. Cueto, E., Gonzalez, D., Badias, A., Chinesta, F., Hascoet, N., Duval, J.-L.: Hybrid Twins. Part II. Real-time, data-driven modeling. ESAFORM 2021 (2021)
25. Moya, B., Badias, A., Alfaro, I, Chinesta, F., Cueto, E.: Digital twins that learn and correct themselves. Int. J. Numer. Methods Eng., 1–11 (2020)
26. Abali, B.E., Savaş, Ö.: Experimental validation of computational fluid dynamics for solving isothermal and incompressible viscous fluid flow. SN Appl. Sci. **2**, 1500 (2020)
27. ANSYS Fluent 18.0, Theory Guide, www.ansys.com

Virtual Element Methods for Optimal Control Problems Governed by Elliptic Interface Problems

Jai Tushar, Anil Kumar, and Sarvesh Kumar

Abstract A conforming Virtual Element Method along with a variational discretization concept for solving the optimization problem governed by an elliptic interface problem is presented. Elements with small edges and hanging nodes occur naturally while numerically solving interface problems. Conforming Finite Element Methods cannot handle these difficulties naturally. VEM has the attractive feature that it can tackle hanging nodes and is even robust with respect to small edges. We use these features of VEM to design a method that can tackle these difficulties naturally. The state, adjoint and control estimates have been derived in suitable norms. Numerical results verify our theoretical findings and show the robustness and flexibility of the proposed method.

Keywords Virtual element method · Optimal control problem · Elliptic interface problem · Variational discretization · Numerical analysis

1 Introduction

There are numerous applications of interface problems in applied sciences and engineering. For example, in material sciences, problems involve discontinuous material coefficients across the interface, such as conductivity in heat transfer, permeability in porous media flow. Optimizing these physical processes lead to optimal control problems governed by partial differential equations (PDEs) with interfaces. To numerically solve these problems, one of the standard practices is to use a finite

J. Tushar (✉) · A. Kumar
Birla Institute of Technology and Science Pilani, KK Birla Goa Campus, Goa 403726, India
e-mail: jaitushar93@gmail.com

A. Kumar
e-mail: anilpundir@goa.bits-pilani.ac.in

S. Kumar
Indian Institute of Space Science and Technology, Trivandrum, Kerala, India
e-mail: sarvesh@iist.ac.in

R. K. Sharma et al. (eds.), *Frontiers in Industrial and Applied Mathematics*,
Springer Proceedings in Mathematics & Statistics 410,
https://doi.org/10.1007/978-981-19-7272-0_36

element method (FEM) (cf. [4, 7]), which has element boundaries coincident with the interface (see [2] and references therein). These methods are categorized as fitted methods. In this group of methods, the meshing of the domain depends on the location of the interface. One faces several difficulties in generating a mesh that resolves the interface. For example, aligning the element edges coincidentally at the interface is not trivial when meshing domains on either side of the interface. The relaxation of the edge alignment condition on the mesh can naturally lead to meshes that have arbitrarily small edges. The attractive properties of VEM make it robust under small edges and allow it to handle hanging nodes; we present a conforming virtual element method (VEM) along with the variational discretization concept presented in [3] for the discretization of the continuous optimization problem governed by an elliptic problem with a polygonal interface which can tackle these difficulties naturally. Our approach allows for greater flexibility in meshing since we can use different meshes on either side of the interface (see Fig. 1). Moreover, we also show that using the same feature of VEM, we can generate background fitted meshes independent of the location of the interface (see Fig. 1). Thus, it is easier to generate meshes as compared to conforming FEM. Our numerical experiments show that the original linear VEM stabilization presented in [5] will generate small but visible oscillations in the solution (see Fig. 2). This motivates us to use the boundary stabilization presented in [6], which smoothens these oscillations at the interface (see Fig. 3). The model problem is to find the distributed control z and the associated state $y = y(z)$ satisfying

$$\min_{z\in Z_{ad}} J(y,z) := \frac{1}{2}\|y - y_d\|_{0,\Omega}^2 + \frac{\lambda}{2}\|z\|_{0,\Omega}^2 , \tag{1}$$

subject to

$$\begin{aligned} -\nabla\cdot(\beta\,\nabla y) &= z + f, \quad \text{in } \Omega,\\ y &= 0, \qquad \text{on } \partial\Omega,\\ [y] = 0,\ \left[\beta\frac{\partial y}{\partial \mathbf{n}}\right] &= g \ \text{ on } \Gamma,\\ z_a \le z \le z_b & \quad \text{for a.e. in } \Omega. \end{aligned} \tag{2}$$

We define the jump of a function ζ across Γ by $[\zeta](x) := \zeta_1(\mathbf{x}) - \zeta_2(\mathbf{x}),\ \forall\, \mathbf{x} \in \Gamma$, where ζ_1 and ζ_2 are restrictions of ζ on Ω_1 and Ω_2, respectively, $\mathbf{n}$ denotes the unit outward normal vector to the interface. The coefficient β is assumed to be piecewise constant and positive and is defined as β_1 in Ω_1 and β_2 in Ω_2. Let $z_a, z_b \in \mathbb{R}$ with $z_a < z_b$, $y_d \in L^2(\Omega)$ is the desired state and $\lambda > 0$ is the regularization or the penalty parameter. The admissible set of controls is defined as follows:

$$Z_{ad} := \left\{z \in L^2(\Omega) : z_a \le z \le z_b \ \text{ a.e. in } \Omega\right\}.$$

Define the space $X := H^1(\Omega) \cap H^2(\Omega_1) \cap H^2(\Omega_2)$ equipped with the norm

$$\|\zeta\|_X = \|\zeta\|_{1,\Omega} + \|\zeta\|_{2,\Omega_1} + \|\zeta\|_{2,\Omega_2}, \quad \forall\ \zeta \in X.$$

Sobolev embedding theorem dictates that for any $\zeta \in X$, we have $\zeta \in W^{1,p}(\Omega)$ for all $p > 2$. The regularity of the state equation (2) is given by the following Lemma (see Theorem 2.1, [2])

Lemma 1 *Assuming $f, z \in L^2(\Omega)$ and $g \in H^{1/2}(\Gamma)$. We have that the problem (2) has a unique solution $y \in X$ which satisfies*

$$\|y\|_X \lesssim \|f\|_{0,\Omega} + \|z\|_{0,\Omega} + \|g\|_{1/2,\Gamma}$$

Using the standard techniques employed in PDE optimal control, we can find that the optimal control satisfies the following variational inequality also known as the first-order necessary optimality condition

$$(\lambda z + p, w - z) \geq 0, \ \forall\ w \in Z_{ad},$$

where p is the adjoint variable or the co-state variable and solves the subsequent adjoint equation

$$\begin{aligned}
-\nabla \cdot (\beta\, \nabla p) &= y - y_d, && \text{in } \Omega,\\
p &= 0, && \text{on } \partial\Omega,\\
[p] = 0, \qquad & \left[\beta \frac{\partial p}{\partial \mathbf{n}}\right] = 0 && \text{on } \Gamma.
\end{aligned}$$

A unique $p \in X$ exists, which solves the adjoint equation follows from (Theorem 2.1, [2]). We can rewrite the first-order necessary optimality condition as a pointwise projection formula

$$z = \mathcal{P}_{Z_{ad}}\left(-\frac{1}{\lambda}p\right).$$

If we introduce a control-to-state map S defined as $Sz = y$, then the problem (1)–(2) reduces to

$$\min_{z \in Z_{ad}} j(z) = \min_{z \in Z_{ad}} J(Sz, z),$$

then the optimal control satisfies the following coercivity condition

$$j''(z)(w, w) \geq \lambda\, \|w\|^2_{L^2(\Omega)}, \quad \forall\ w \in Z := L^2(\Omega). \tag{3}$$

Define $a(\cdot,\cdot) : H^1(\Omega) \times H^1(\Omega) \longrightarrow \mathbb{R}$ such that $a(\zeta, \eta) := \int_\Omega \beta\, \nabla\zeta \cdot \nabla\eta$. Now the optimality system corresponding to (1)–(2) is to find $(y, p, z) \in V(:= H^1_0(\Omega)) \times V \times Z_{ad}$ such that

$$a(y, v) = (z + f, v) + \langle g, v \rangle_\Gamma, \quad \forall \; v \in V \tag{4}$$

$$a(p, q) = (y - y_d, q), \quad \forall \; q \in V \tag{5}$$

$$(\lambda z + p, w - z) \geq 0, \quad \forall \; w \in Z_{ad}. \tag{6}$$

We adopt the standard Sobolev space notations. Additionally, we have the notation $a \lesssim b$, which represents that a is less than or equal to some positive constant (independent of the mesh parameter) times b. An outline of the manuscript is as follows. In Sect. 2, a VEM discretization of the continuous problem is proposed. In Sect. 3, we give the convergence analysis for the proposed scheme under suitable norms. Afterward, in Sect. 4, we conduct two numerical experiments to analyse the behaviour of the solution and verify the theoretical results proved in Sect. 3.

2 Discrete Formulation

Let $\mathcal{T}_h$ be the triangulation of Ω into simple polygons K with discretization parameter $h := \max_{K \in \mathcal{T}_h} h_K \in (0, 1]$, where h_K is the diameter of K. Γ is the polygonal interface which is resolved by $\mathcal{T}_h$. $\mathcal{T}_h^* := \{K \in \tau_h : K \cap \Gamma \neq \emptyset\}$ is the set of interface polygons. Then $\mathcal{T}_h$ satisfies:

(A1) $\bar{\Omega} = \cup_{K \in \mathcal{T}_h} K$.
(A2) If $K_1, K_2 \in \mathcal{T}_h$ are two distinct polygons, then either their intersection is empty or they share a common vertex or edge.
(A3) Each polygon either lies in Ω_1 or Ω_2 and has at most two vertices lying on the interface.

Moreover, we introduce the following relaxed assumptions which allow small edges on any polygon $K \in \tau_h$,

(A4) Any $K \in \mathcal{T}_h$ is star-shaped w.r.t. disc $\mathbb{B}_K \subset K$ with radius $\rho_K h_K$ where, and there exists $\rho \in (0, 1)$, such that $\rho_K \geq \rho$ for all $K \in \mathcal{T}_h$.
(A5) There exists $N \in \mathbb{Z}^+$ independent of the mesh parameter such that $|E_K| \leq N$, where E_K denotes the set of all edges of K.

2.1 Discretization of State and Adjoint Equations

Following [1], the linear local virtual element space $V(K) \subset H^1(K)$ is defined as follows:

$$V(K) := \big\{\zeta \in H^1(K) : \zeta|_{\partial K} \in \mathbb{P}_1(\partial K), \; -\Delta\zeta \in \mathbb{P}_1(K),$$
$$(\zeta - \Pi^{\nabla}_{1,K}\zeta, q)_K = 0 \; \forall \; q \in \mathcal{M}_0^*(K) \cup \mathcal{M}_1^*(K)\big\}.$$

where

$$\mathcal{M}_r^*(K) := \left\{ m | m = \left(\frac{x - x_K}{h_K} \right)^{\mathbf{s}} \text{ for } \mathbf{s} \in \mathbb{N}^2 \text{ with } |\mathbf{s}| = r \right\}.$$

We denote by x_K the centroid of K, the space of all polynomials of degree ≤ 1 is denoted by $\mathbb{P}_1(K)$. The degrees of freedom of $V(K)$ consist of the values of v at the vertices of K. Ritz projection operator $\Pi_{1,K}^{\nabla} : H^1(K) \longrightarrow \mathbb{P}_1(K)$ satisfies

$$(\!(\Pi_{1,K}^{\nabla}\zeta, q)\!) = (\!(\zeta, q)\!) \ \ \forall\, q \in \mathbb{P}_1(K), \tag{7}$$

where the inner product $(\!(\zeta, w)\!) := (\nabla\zeta, \nabla w) + (\int_{\partial K} \zeta\, ds)(\int_{\partial K} w\, ds)$. Moreover, (7) is equivalent to

$$\int_K \nabla(\Pi_{1,K}^{\nabla}\zeta) \cdot \nabla q\, dx = \int_K \nabla\zeta \cdot \nabla q\, dx; \int_{\partial K} \Pi_{1,K}^{\nabla}\zeta\, ds = \int_{\partial K} \zeta\, ds. \tag{8}$$

$\Pi_{1,K}^0$ is the projection from $L^2(K)$ onto $\mathbb{P}_1(K)$. $\mathcal{P}_h^1$ represents the space of discontinuous piecewise polynomials of degree ≤ 1. Then the global projection operators $\Pi_{1,h}^{\nabla} : H^1(\Omega) \longrightarrow \mathcal{P}_h^1,\ \ \Pi_{1,h}^0 : L^2(\Omega) \longrightarrow \mathcal{P}_h^1$, are understood in the sense of their local counterparts as

$$(\Pi_{1,h}^{\nabla}v)|_K = \Pi_{1,K}^{\nabla}(v|_K), \quad (\Pi_{1,h}^0 v)|_K = \Pi_{1,K}^0(v|_K).$$

We glue to the local virtual element spaces to write the following global virtual element space

$$V_h = \{\zeta \in H_0^1(\Omega) : \zeta|_K \in V(K)\ \ \forall\, K \in \mathcal{T}_h\}.$$

The mesh dependent norm is defined as $|v|_{h,1} := \left(\sum_{K \in \mathcal{T}_h} |v|^2_{H^1(K)}\right)^{\frac{1}{2}}$. We define the discrete bilinear form as follows:

$$\begin{aligned} a_h(w, v) &= \sum_{K \in \mathcal{T}_h} a_h^K(w, v) \\ &= \sum_{K \in \mathcal{T}_h} \left[a^K(\Pi_{1,K}^{\nabla}w, \Pi_{1,K}^{\nabla}v) + S^K(w - \Pi_{1,K}^{\nabla}w,\ v - \Pi_{1,K}^{\nabla}v) \right], \qquad (9) \\ a^K(w, v) &= \int_K \beta|_K \nabla w \cdot \nabla v\, dx, \end{aligned}$$

Note that $supp(\beta - \beta|_K) \cap K = \{0\}$ for all $K \in \tau_h$. The two choices of local stabilization bilinear forms are defined as follows:

$$S^K(\zeta, v) = \begin{cases} S_1^K(\zeta, v) := \sum_{\varphi \in \mathcal{B}_{\partial K}} \zeta(\varphi) v(\varphi), \\ S_2^K(\zeta, v) := h_K\, (\partial\zeta/\partial s, \partial v/\partial s)_{0,\partial K}\,. \end{cases}$$

Here, $\mathcal{B}_{\partial K}$ denotes the set of nodes of K. and $\partial\zeta/\partial s$ is the tangential derivative of ζ along ∂K. From ((3.55) in [1]), we have for all $u \in V(K)$ the following inequality

$$|u|_{1,K}^2 \lesssim |\Pi_{1,K}^{\nabla} u|_{1,K}^2 + h_K \left\| \partial(u - \Pi_{1,K}^{\nabla} u)/\partial s \right\|_{0,\partial K}^2, \tag{10}$$

with a hidden constant depending on ρ_K and the degree of the polynomial. Also from ((3.56) in [1]) for all $u \in V(K)$, we have

$$|u|_{1,K}^2 \lesssim |\Pi_{1,K}^{\nabla} u|_{1,K}^2 + \ln(1+\tau_K) \left\| u - \Pi_{1,K}^{\nabla} u \right\|_{\infty,\partial K)}^2, \tag{11}$$

with the constant depending on $|E_K|$ along with ρ_K and the degree of the polynomial. Here, $\tau_K := \max_{e \in E_K} h_e / \min_{e \in E_K} h_e$. On combining (10) and (11), we get the following stability estimate for $a_h(\cdot,\cdot)$,

$$|v|_{H^1(\Omega)}^2 \lesssim \alpha_h a_h(v,v) \;\; \forall\, v \in V_h, \tag{12}$$

where

$$\alpha_h = \begin{cases} \ln\left(1 + \max_{K \in \mathcal{T}_h} \tau_K\right) & \text{if } S^K(\cdot,\cdot) = S_1^K(\cdot,\cdot), \\ 1 & \text{if } S^K(\cdot,\cdot) = S_2^K(\cdot,\cdot). \end{cases} \tag{13}$$

The source term is discretized using $\Pi_{1,h}^0$ operator as follows

$$(\Pi_{1,h}^0 f, v) := (f, \Pi_{1,h}^0 v) = \sum_{K \in \mathcal{T}_h} \int_K f \Pi_{1,K}^0 v_h.$$

2.2 *Variational Discretization*

In this approach, we discretize the control variable implicitly. Thus the discrete admissible set of controls coincides with Z_{ad}. Following the *optimize-then-discretize* approach, we can write the discrete optimality system as follows: Find $(y_h, p_h, z_h) \in V_h \times V_h \times Z_{ad}$ such that

$$a_h(y_h, v_h) = \left(\Pi_{1,h}^0 (f + z_h), v_h\right) + \langle g, v_h \rangle_\Gamma \;\; \forall\, v_h \in V_h \tag{14}$$

$$a_h(p_h, q_h) = \left(\Pi_{1,h}^0 (y_h - y_d), q_h\right) \;\; \forall\, q_h \in V_h \tag{15}$$

$$\left(\lambda z_h + \Pi_{1,h}^0 p_h, \tilde{w} - z_h\right) \geq 0 \;\; \forall\, \tilde{w} \in Z_{ad}. \tag{16}$$

The discrete variational inequality (16) is rewritten as a discrete projection formula

$$u_h|_K = \mathcal{P}_{U_{ad}}\left(-\frac{1}{\lambda} (\Pi_{1,h}^0 p_h)|_K \right) \;\; \forall\, K \in \mathcal{T}_h. \tag{17}$$

The stability estimate (12) implies that the discrete state Eq. (14) and discrete adjoint Eq. (15) are well-posed.

3 Convergence Analysis

This section is dedicated to deriving the error estimates for the state, adjoint and control variable under variational discretization of control. We begin by considering the following auxiliary equations: For any arbitrary control $\tilde{z} \in L^2(\Omega)$, let $y_h(\tilde{z}) \in V_h$ solve

$$a_h(y_h(\tilde{z}), v_h) = (\Pi^0_{1,h}(\tilde{z} + f), v_h) + \langle g, v_h \rangle_\Gamma \;\; \forall \;\; v_h \in V_h, \tag{18}$$

and for any arbitrary $\tilde{y} \in H^1_0(\Omega)$, let $p_h(\tilde{y}) \in V_h$ solve

$$a_h(q_h, p_h(\tilde{y})) = (\Pi^0_{1,h}(\tilde{y} - y_d), q_h) \;\; \forall\, q_h \in V_h. \tag{19}$$

Let us define the following notations $y_h := y_h(z_h)$, $p_h := p_h(y_h)$. For the subsequent analysis we will need the following Lemma.

Lemma 2 *For any arbitrary $\tilde{z}_i \in L^2(\Omega)$, and $\tilde{y}_i \in H^1_0(\Omega)$, let $y_h(\tilde{z}_i)$ and $p_h(\tilde{y}_i)$, $i = 1, 2$ solve (18) and (19), respectively. Then*

$$|y_h(\tilde{z}_1) - y_h(\tilde{z}_2)|_{1,\Omega} \lesssim \alpha_h \, \|\tilde{z}_1 - \tilde{z}_2\|_{0,\Omega} \,,$$
$$|p_h(\tilde{y}_1) - p_h(\tilde{y}_2)|_{1,\Omega} \lesssim \alpha_h \, \|\tilde{y}_1 - \tilde{y}_2\|_{0,\Omega} \,,$$

where α_h is as defined in (13).

Proof Test (18) with $y_h(\tilde{z}_1)$ and $y_h(\tilde{z}_2)$ to get,

$$a_h(y_h(\tilde{z}_1) - y_h(\tilde{z}_2), v_h) = (\Pi^0_{1,h}(\tilde{z}_1 - \tilde{z}_2), v_h).$$

Now using the stability estimate (12) along with $v_h = y_h(\tilde{z}_1) - y_h(\tilde{z}_2)$, the stability of $\Pi^0_{1,K}$ operator and the consequence of Poincaré-Friedrichs inequality we have,

$$\begin{aligned}
|y_h(\tilde{z}_1) - y_h(\tilde{z}_2)|^2_{1,\Omega} &\lesssim \alpha_h \, \|\tilde{z}_1 - \tilde{z}_2\|_{0,\Omega} \left\| \Pi^0_{1,h}(y_h(\tilde{z}_1) - y_h(\tilde{z}_2)) \right\|_{0,\Omega} , \\
&\lesssim \alpha_h \, \|\tilde{z}_1 - \tilde{z}_2\|_{0,\Omega} \sum_{K \in \mathcal{T}_h} \left\| \Pi^0_{1,K}(y_h(\tilde{z}_1) - y_h(\tilde{z}_2)) \right\|_{0,K} , \\
&\lesssim \alpha_h \, \|\tilde{z}_1 - \tilde{z}_2\|_{0,\Omega} \sum_{K \in \mathcal{T}_h} \|y_h(\tilde{z}_1) - y_h(\tilde{z}_2)\|_{0,K} , \\
&\lesssim \alpha_h \, \|\tilde{z}_1 - \tilde{z}_2\|_{0,\Omega} \, |y_h(\tilde{z}_1) - y_h(\tilde{z}_2)|_{1,\Omega}, \\
&\lesssim \alpha_h \, \|\tilde{z}_1 - \tilde{z}_2\|_{0,\Omega} \, .
\end{aligned}$$

Similarly, test (19) with $p_h(\tilde{y}_1)$ and $p_h(\tilde{y}_2)$ along with $q_h = p_h(\tilde{y}_1) - p_h(\tilde{y}_2)$ and follow the same steps to get the second desired inequality. □

Estimates corresponding to the auxiliary problems (18) and (19) can be proved using the techniques of [1] and [2] in the following Lemma.

Lemma 3 *Let $y(\tilde{z})$ and $y_h(\tilde{z})$ be the solutions of (4) and (18), respectively. Let $p(\tilde{y})$ and $p_h(\tilde{y})$ be the solutions of (5) and (19), respectively. Let $f, y_d \in L^2(\Omega)$ and $g \in H^{1/2}(\Gamma)$ and α_h is as defined in (13). Then*

$$|y(\tilde{z}) - y_h(\tilde{z})|_{1,\Omega} + |p(\tilde{y}) - p_h(\tilde{y})|_{1,\Omega} \lesssim \alpha_h\, h.$$

Additionally, if $f, y_d \in H^1(\Omega)$ then

$$\|y(\tilde{z}) - y_h(\tilde{z})\|_{0,\Omega} + \|p(\tilde{y}) - p_h(\tilde{y})\|_{0,\Omega} \lesssim \alpha_h\, h^2.$$

Moreover, for $\tilde{z} = z_h$,

$$\|p(z_h) - p_h(z_h)\|_{L^2(\Omega)} \lesssim \alpha_h h^2.$$

Following the arguments of Lemma 2.1 in [2] and the standard approximation property of $\Pi^0_{1,K}$ given in [1], we have

$$\|\zeta - \Pi^0_{1,h}\zeta\|_{0,\Omega} \lesssim h^2\, \|\zeta\|_X \quad \forall\ \zeta \in X. \tag{20}$$

Now we derive the error estimates for the state, adjoint and control variables under variational discretization of control.

Theorem 1 *Let (y, p, z) solve the continuous optimality system (4)–(6). Let (y_h, p_h, z_h) solve the discrete optimality system (14)–(16). Then under the assumptions of Lemma 2 and Lemma 3, then following estimate holds*

$$\|z - z_h\|_{L^2(\Omega)} \lesssim \alpha_h\, h^2,$$

where α_h is as defined in (13).

Proof The discrete (16) and continuous (6) variational inequalities give the following

$$(\lambda z_h + \Pi^0_{1,h} p_h, z - z_h) \geq 0 \geq (\lambda z + p, z - z_h), \tag{21}$$

The coercivity condition (3) for $z - z_h \in Z$ and (21) leads to

$$
\begin{aligned}
\lambda \|z - z_h\|_{0,\Omega}^2 &\leq (\lambda z + p, z - z_h) - (\lambda z_h + p(z_h), z - z_h), \\
&\leq (\lambda z_h + \Pi_{1,h}^0 p_h, z - z_h) - (\lambda z_h + p(z_h), z - z_h) \\
&= (\Pi_{1,h}^0 p_h - p(z_h), z - z_h), \\
&= \big[(\Pi_{1,h}^0 (p_h - p(z_h)), z - z_h) \\
&\qquad + (\Pi_{1,h}^0 p(z_h) - p(z_h), z - z_h)\big] \\
&= T_A + T_B
\end{aligned}
$$

In view of the stability of $\Pi_{1,K}^0$ operator, Lemmas 2 and 3, T_A is bounded as follows:

$$
\begin{aligned}
T_A &\leq \sum_{K \in \mathcal{T}_h} \left\| \Pi_{1,K}^0 (p_h - p(z_h)) \right\|_{0,K} \|z - z_h\|_{0,K} \\
&\leq \sum_{K \in \mathcal{T}_h} \|p_h - p(z_h)\|_{0,K} \|z - z_h\|_{0,K} \\
&\leq \sum_{K \in \mathcal{T}_h} \left(\|p_h - p_h(y(z_h))\|_{0,K} + \|p_h(y(z_h)) - p(z_h)\|_{0,K} \right) \|z - z_h\|_{0,K} \\
&\lesssim \left(|p_h - p_h(y(z_h))|_{1,\Omega} + \|p_h(y(z_h)) - p(z_h)\|_{0,\Omega} \right) \|z - z_h\|_{0,\Omega} \\
&\lesssim \left(\alpha_h \|y_h(z_h) - y(z_h)\|_{0,\Omega} + \|p_h(y(z_h)) - p(z_h)\|_{0,\Omega} \right) \|z - z_h\|_{0,\Omega} \\
&\lesssim \alpha_h \, h^2 \, \|z - z_h\|_{0,\Omega} \, .
\end{aligned}
$$

The term T_B is bounded using (20) as follows

$$
T_B \leq \sum_{K \in \mathcal{T}_h} \left\| \Pi_{1,K}^0 p(z_h) - p(z_h) \right\|_{L^2(\Omega)} \|z - z_h\|_{L^2(\Omega)} \lesssim h^2 \, \|p(z_h)\|_X \, \|z - z_h\|_{0,\Omega} \, .
$$

Combining the bounds of T_A and T_B leads to the desired estimate. □

Theorem 2 *Assuming Theorem 1 holds. Then under variational discretization of control the following estimates hold*

$$
\|y - y_h\|_{0,\Omega} + \|p - p_h\|_{0,\Omega} \lesssim \alpha_h \, h^2; \;\; |y - y_h|_{1,\Omega} + |p - p_h|_{1,\Omega} \lesssim \alpha^h \, h.
$$

Proof We split the error in state equation using $y_h(z)$ as $y - y_h = (y - y_h(z)) + (y_h(z) - y_h)$. Now we use Lemmas 2, 3 and Theorem 1 as follows:

$$
\begin{aligned}
\|y - y_h\|_{0,\Omega} &\leq \|y - y_h(z)\|_{0,\Omega} + \|y_h(z) - y_h\|_{0,\Omega} \\
&\leq \|y - y_h(z)\|_{L^2(\Omega)} + \alpha_h \, \|z - z_h\|_{L^2(\Omega)} \lesssim \alpha_h \, h^2 \\
|y - y_h|_{1,\Omega} &\leq |y - y_h(z)|_{1,\Omega} + |y_h(z) - y_h|_{1,\Omega} \\
&\leq |y - y_h(z)|_{1,\Omega} + \alpha_h \, \|z - z_h\|_{0,\Omega} \lesssim \alpha_h \, h.
\end{aligned}
$$

Following analogous steps and using the splitting $p - p_h = (p - p_h(y)) + (p_h(y) - p_h)$, we can get the estimates for the adjoint variable. □

4 Numerical Experiments

In this section, we present two numerical examples to study the behaviour of our scheme. In *Example I*, we study the behaviour of the solution at the interface in the presence of small edges under both the proposed stabilization terms $S_1^K(\cdot,\cdot)$ and $S_2^K(\cdot,\cdot)$. The mesh $\mathcal{T}_h^1$ (see Fig. 1) under **(A1)–(A5)** that we consider arises naturally, if we mesh the domain Ω either side of the interface Γ with different elements. The error in control, state, and adjoint variables are illustrated, and the theoretical results of Sect. 3 are corroborated. In *Example II*, we employ a *segment interface* which is independent of the background fitted mesh $\mathcal{T}_h^2$ (see Fig. 1) such that it satisfies **(A1)–(A3)**. In Fig. 1, the red star markers on the slant interface are the intersection points of the interface with $\mathcal{T}_h^2$ and result in hanging nodes that are repurposed to generate a fitted mesh; hence the name background fitted mesh. Error in control, state, and adjoint variables under variational discretization of control is illustrated which verifies the results obtained in Sect. 3.

Example 1 (Vertical interface) Let Ω be a unit square domain. Consider the problem (1)–(2). The interface $\Gamma := \{\mathbf{x} \in \Omega : x_1 - 0.5 = 0\}$ and the following data

$$\lambda = 0.1,\ u_a = -0.25,\ u_b = 0.25,\ \beta_1 = 1,\ \beta_2 = 10,\ y(\mathbf{x}) = x_1^2(x_1-1)x_2(x_2-1),$$
$$p(\mathbf{x}) = x_1(x_1-1)x_2(x_2-1),\ z(\mathbf{x}) = \max(z_a, \min(z_b, -\frac{1}{\lambda}p(\mathbf{x}))).$$

The optimal control problem is solved using the variational variant of the projected gradient algorithm presented in [3]. In our numerical experiment, we observe that under the classical VEM stabilization choice of $S_K^1(\cdot,\cdot)$, the solution of the state and the adjoint variable exhibits oscillations at the interface under the presence of small edges; however, the control variable is free of these oscillations under variational discretization of control (See Fig. 2). The red dotted lines in Fig. 2 represent the true solution at the interface, and the blue lines represent the approximated solution on $\mathcal{T}_h^2$.

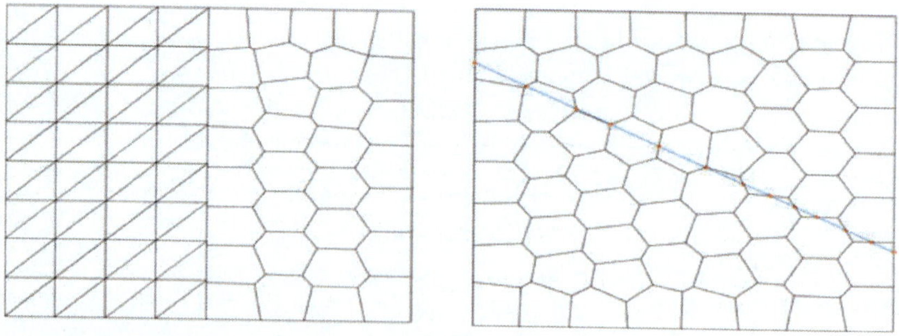

Fig. 1 Meshes $\mathcal{T}_h^1$ and $\mathcal{T}_h^2$, respectively

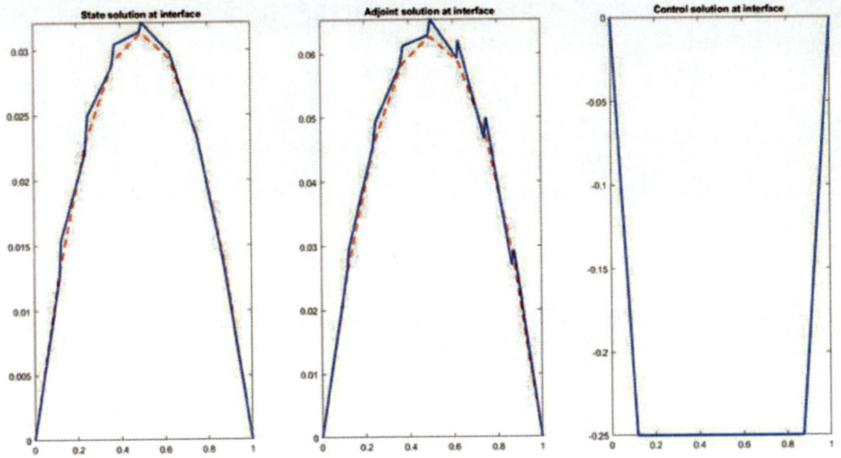

Fig. 2 Solution profile of state, adjoint and control variables, respectively at the interface under variational discretization of control with $S_1^K(\cdot,\cdot)$ on $\mathcal{T}_h^2$

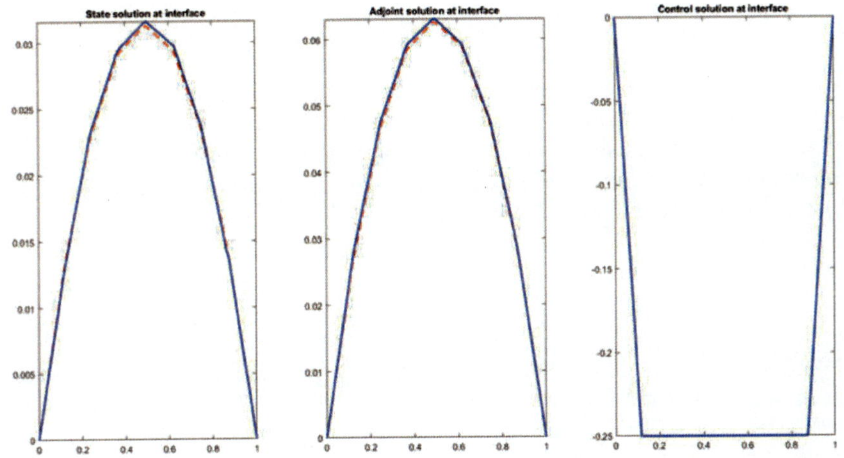

Fig. 3 Solution profile of state, adjoint and control variables, respectively at the interface under variational discretization of control with $S_2^K(\cdot,\cdot)$ on $\mathcal{T}_h^2$

We do the same experiment with the boundary stabilization $S_K^2(\cdot,\cdot)$ and observe that the oscillations at the interface have smoothened (see Fig. 3).

Remark 1 It is also observed in our numerical experiments that the oscillations are sensitive to the parameter β. For example, if we consider the same numerical example with $\beta_1 = 1$ and $\beta_2 = 0.5$, the oscillations with $S_K^1(\cdot,\cdot)$ in the state and the adjoint will be still visible but much smaller.

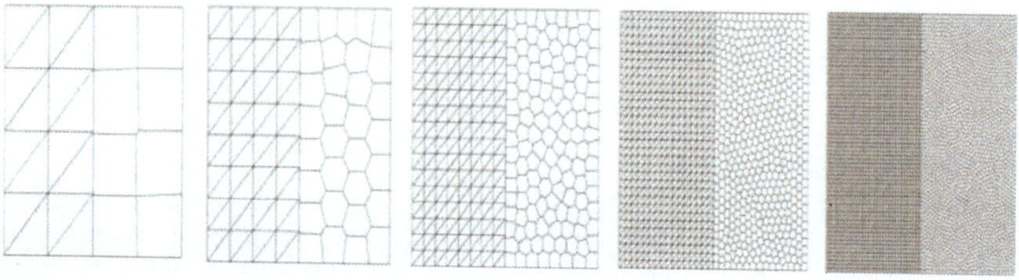

Fig. 4 Sequence of meshes $\mathcal{T}_1^1$, $\mathcal{T}_2^1$, $\mathcal{T}_3^1$, $\mathcal{T}_4^1$ and $\mathcal{T}_5^1$, respectively

Table 1 Error and order of convergence in y, p and u under $\mathcal{T}_1^1$-$\mathcal{T}_5^1$ and variational discretization of control in L^2-norm and energy norm for Example I

h	$E_0(y)$	$R_0(y)$	$E_0(p)$	$R_0(p)$	$E_1(y)$	$R_1(y)$	$E_1(p)$	$R_1(p)$	$E_0(z)$	$R_0(z)$
0.3651	0.002111	–	0.002765	–	0.0386	–	0.0572	–	0	–
0.1847	0.000509	2.08	0.000725	1.96	0.0196	0.99	0.0280	1.04	0.002962	-Inf
0.0910	0.000124	1.98	0.000188	1.90	0.0097	0.98	0.0139	0.98	0.000540	2.40
0.0474	0.000030	2.17	0.000048	2.06	0.0048	1.08	0.0069	1.06	0.000139	2.07
0.0233	0.000007	1.91	0.000012	1.94	0.0023	0.98	0.0034	0.98	0.000030	2.14

Now we compare the error under a sequence of meshes $\mathcal{T}_1^1$ to $\mathcal{T}_5^1$ (see Fig. 4). We compare the exact solution of the state and co-state variables with the L^2-projection of the discrete state and co-state variables since the virtual element solution is not known explicitly inside the element. The discrete control is computed using the discrete projection formula (17). We denote the L^2-error as follows

$$E_0(w) = \sum_{K\in\mathcal{T}_h} \left\| w - \Pi^0_{1,K} w_h \right\|_{0,K} \quad \text{for } w = \text{y, p}, \quad E_0(z) = \sum_{K\in\mathcal{T}_h} \| z - z_h \|_{0,K} \,.$$

Similarly, we denote the error in the energy norm for the state and the co-state variable by $E_1(y)$ and $E_1(p)$, respectively with the help of $\Pi^{\nabla}_{1,K}$ operator. We denote by $R_0(w)$ and $R_1(w)$ the order of convergence corresponding to the variable w in the L^2 and H^1 norms, respectively. The numerical errors and the corresponding rate of convergence under $\mathcal{T}_1^1$-$\mathcal{T}_5^1$ are given in Table 1 and corroborate theoretical results of Theorems 1 and 2. The solution profile on $\mathcal{T}_3^1$ is given in Fig. 5.

Example 2 **(Segment interface)** Consider the problem (1)–(2) on a unit square domain with the interface $\Gamma := \{\mathbf{x} \in \Omega : x_2 = kx_1 + b\}$, where $k = \frac{-\sqrt{3}}{3}$ and $b = \frac{(6+\sqrt{6}-2\sqrt{3})}{6}$ and the following data

$$\lambda = 1,\ u_a = -0.2,\ u_b = 0.2,\ \beta_1 = 1,\ \beta_2 = 1/2,\ y(\mathbf{x}) = x_1^2(x_1 - 1)x_2(x_2 - 1),$$
$$z(\mathbf{x}) = \max(z_a, \min(z_b, -\frac{1}{\lambda}p(\mathbf{x}))),\ p(\mathbf{x}) = (x_2 - kx_1 - b)^2(x_1(x_1 - 1)x_2(x_2 - 1)).$$

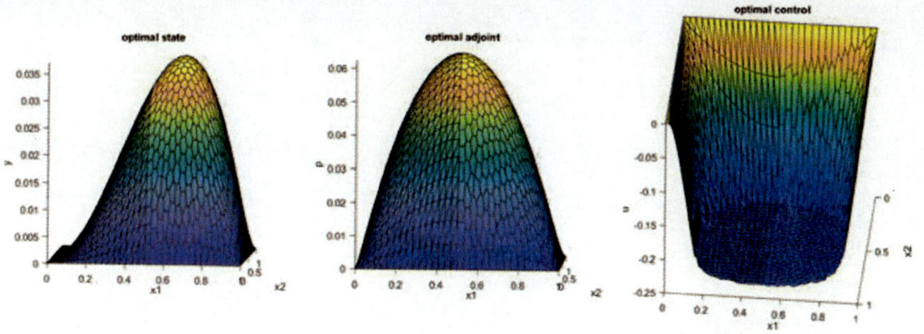

Fig. 5 Solution profile of state, adjoint and control variables, respectively on $\mathcal{T}_3^1$ for Example I

Table 2 Error and order of convergence in y, p and u under a sequence of meshes of type $\mathcal{T}_h^2$ and variational discretization of control in energy and L^2 norms for Example II

h	$E_0(y)$	$R_0(y)$	$E_0(p)$	$R_0(p)$	$E_1(y)$	$R_1(y)$	$E_1(p)$	$R_1(p)$	$E_0(z)$	$R_0(z)$
0.7071	0.006770	–	0.002261	–	0.0589	–	0.0217	–	0.0008842	–
0.3547	0.002619	1.37	0.001169	0.95	0.0346	0.77	0.0160	0.43	0.0001723	2.37
0.1818	0.000617	2.16	0.000338	1.85	0.0179	0.98	0.0087	0.91	0.0000852	1.05
0.0922	0.000156	2.02	0.000087	1.99	0.0090	1.00	0.0044	0.98	0.0000249	1.81
0.0483	0.000038	2.16	0.000020	2.21	0.0045	1.05	0.0022	1.06	0.0000049	2.48

The numerical errors and the corresponding order of convergence under a sequence of refined meshes of the type $\mathcal{T}_h^2$ independent of the interface Γ are given in Table 2 and confirm the theoretical results of Theorems 1 and 2.

References

1. Brenner, S.C., Sung, L.Y.: Virtual element methods on meshes with small edges or faces. Math. Models Methods Appl. Sci. **28**, 1291–1336 (2018)
2. Chen, Z., Zou, J.: Finite element methods and their convergence for elliptic and parabolic interface problems. Numer. Math. **79**(2), 175–202 (1998)
3. Hinze, M.: A Variational Discretization Concept in Control Constrained Optimization: The Linear-Quadratic Case. Comput. Optim. Appl. **30**, 45–61 (2005)
4. Yadav, O.P., Jiwari, R.: A finite element approach for analysis and coputational modelling of coupled reaction diffusion. Numer. Methods Partial Differ. Equ. **35**, 830–850 (2019)
5. Beirão, V.L., Brezzi, F., Cangiani, A., Manzini, G., Marini, L.D., Russo, A.: Basic principles of virtual element methods. Math. Models Methods Appl. Sci. **23**, 199–214 (2013)
6. Wriggers, P., Rust, W.T., Reddy, B.D.: A Virtual element method for contact. Comput. Mech. **58**, 1039–1050 (2016)
7. Yadav, O.,P., Jiwari, R.: Finite element analysis and approximation of Burgers'-Fisher equation. Numer. Methods Partial Differ. Equ. **33**, 1652–1677 (2017)

Positivity Preserving Rational Quartic Spline Zipper Fractal Interpolation Functions

Check for updates

Vijay and A. K. B. Chand

Abstract In this paper, we introduce a class of novel C^1-rational quartic spline zipper fractal interpolation functions (RQS ZFIFs) with variable scalings, where rational spline has a quartic polynomial in the numerator and a cubic polynomial in the denominator with two shape control parameters. We derive an upper bound for the uniform error of the proposed interpolant with a C^3 data generating function, and it is shown that our fractal interpolant has $O(h^2)$ convergence and can be increased to $O(h^3)$ under certain conditions. We restrict the scaling functions and shape control parameters so that the proposed RQS ZFIF is positive, when the given data set is positive. Using this sufficient condition, some numerical examples of positive RQS ZFIFs are presented to support our theory.

Keywords Fractals · Positivity · Rational quartic spline · Zipper · Zipper smooth fractal function

AMS subject classifications 28A80 · 41A05 · 41A20 · 41A25 · 65D10

1 Introduction

To find a nice interpolation curve with various attributes is an active area of research in numerical analysis, approximation theory, wavelets, classical and discrete geometry, engineering design, civil engineering and computer science. From the last many decades, researchers have come up with various types of interpolants that have advantages over one another. Polynomial interpolations are preferred when the original function is sufficiently smooth. For some fixed order of smoothness, different types of spline (polynomial/trigonometric/exponential/rational) interpolants are

Vijay (✉) · A. K. B. Chand
Department of Mathematics, Indian Institute of Technology Madras, Chennai 600036, India
e-mail: vijaysiwach975@gmail.com

A. K. B. Chand
e-mail: chand@iitm.ac.in

R. K. Sharma et al. (eds.), *Frontiers in Industrial and Applied Mathematics*,
Springer Proceedings in Mathematics & Statistics 410,
https://doi.org/10.1007/978-981-19-7272-0_37

used. Rational spline interpolants with shape parameters are more flexible over other type of spline interpolants, and hence popular in geometric modelling problems for discrete data visualization. These have been utilized from animated films to simulated surgery. For the classical positivity preserving rational splines one can see [1, 15, 16, 23, 25, 33]. Schmidt and Heß in [25] discussed positive interpolation with quadratic and rational quadratic spline and observed that rational quadratic splines have an advantage over quadratic splines. Sakai and Schmidt in [23] presented a class of C^2 positivity-preserving rational spline using two local control parameters with the cubic numerator and linear denominator. Using cubic numerator and quadratic denominator, Abbas et al. in [1] constructed a C^2 rational cubic spline with three shape parameters. They derived the shape feature of data using a single shape parameter and the other two shape parameters were left free for the designer to adjust the shape of positive curves as per industrial requirements. Hussain and Sarfraz in [16] constructed a C^1 piecewise rational cubic spline with four parameters to visualize positive data set. Two parameters are constrained for the presentation of positive curves through positive data while the other two provide extra freedom to vary the curve shape as needed. Han in [15] presented a piecewise rational spline with the quartic numerator and quadratic denominator. He derived the shape-preservation properties like positivity, monotonicity and convexity of the interpolant. But these non-recursive classical interpolants are either smooth or piecewise smooth and consequently, they are not differentiable at the finite number of points. But if the data is taken from an irregular and non-smooth function, these classical interpolants are not good approximants for it.

Non-smooth and irregular curves such as profiles of mountain ranges, tops of clouds, lightning, ECG curves, turbulence, etc. cannot be interpolated by classical interpolants. The term fractal was given by Mandelbrot [19] to unify the irregular and complex structures. After that many researchers worked on it and expand its theory. To construct fractals, Hutchinson [17] introduced the concept of iterated function system (IFS). The fractal-based theory is a new tool to analyse various non-linear complex phenomena in nature, sciences and engineering. With the help of some parameters, we can easily model most of these complex phenomena by using self-referential rules. Using the theory of IFS, Barnsley [5] created fractal interpolation functions (FIFs) to generate non-smooth and irregular curves from their data points [6] and proved the existence and uniqueness of fractal interpolation function for a hyperbolic IFS with fixed parameters. Barnsley and Harrington [7] constructed r-times differentiable polynomial spline with fixed type of boundary conditions to interpolate functions that have fractality in their higher-order derivatives. For all kinds of boundary conditions, Chand and Kapoor [8] constructed cubic spline FIFs using moments. For application of FIF in data visualization, Chand and collaborators have proposed shape-preserving fractal interpolants, see for instance [9, 10, 12, 18, 29, 30]. Akhtar et al. in [20] introduced a group of fractal functions on the unit sphere through a linear bounded fractal operator and presented some approximation properties. Balasubramani et al. [4] constructed rational cubic spline α-fractal functions with three shape parameters that can preserve positivity and monotonicity. They have also found the conditions on the IFS parameters so that the proposed interpolant is constrained between two piecewise linear functions. But most of the development in shape-preserving FIF theory, the authors have used constant scaling factors, whereas fractal functions with variable scalings provide more flexibility. Using variable scaling, Wang and Shan [32] generated FIFs to approximate functions with less self-similarity and studied their analytical properties such as smoothness, stability and sensitivity. Gowrisankar and Guru Prem Prasad [14] investigated Riemann-Liouville fractional calculus of quadratic FIF with constant as well as variable scaling factors.

Aseev [2] conceptualized the notion of the zipper, which is the generalization of the IFS. Several interesting topological and structural properties of zipper are studied related to dendrites and self-similar continua by Tetenov and his group [3, 24, 26–28]. Similar to fractal interpolants, zipper fractal interpolant as an attractor of a suitable zipper can give details on arbitrarily small scales. Chand et al. [11] introduced affine zipper fractal interpolants. They constructed affine zipper interpolants inscribed in a rectangle and found a basis for the affine zippers fractal interpolation function for a prescribed data set. Zipper fractal interpolants can be non-differentiable in a dense set of an interval. The construction of smooth zipper FIFs is proposed recently in Reddy [22], where certain derivative of smooth zipper FIF is a typical fractal function. Thus, zipper fractal interpolants can be smooth or non-smooth, and smooth zipper fractal interpolants may be used to generalize traditional non-recursive spline interpolants. In this work, we have come up with a novel C^1-rational quartic spline zipper fractal interpolation function with variable scaling functions and studied its positivity preserving property.

The main points of our work are as follows: First, we formulate a class of novel C^1-rational quartic spline (RQS) with two families of shape control parameters with the help of a binary vector, and then using that RQS and the theory of zipper, we derive a new type of fractal interpolant with variable scaling functions named rational quartic spline zipper fractal interpolation function (RQS ZFIF) in Sect. 2. In Sect. 3, we glean that our RQS and RQS ZFIF converge to a C^3 data generating function with the order $O(h^2)$ as $h \to 0$, and under additional assumptions on IFS parameters, we can increase the order of convergence up to $O(h^3)$. To get a strictly positive RQS ZFIF or RQS for a strictly positive data set, we derive sufficient conditions on the shape control parameters and the variable scaling functions in Sect. 4 and give some numerical examples to reinforce our theory. In Sect. 5, we summarize our work.

2 Construction of RQS ZFIFs

In this section, we will construct a new type of C^1-rational quartic spline using a binary vector called a signature, and then we will construct a class of novel C^1-RQS ZFIF with the help of our new rational quartic spline and the theory of the zipper.

Follows are some notation for this paper: Let $I := [a, b] \subset \mathbb{R}$. For $j \in \mathbb{N}$, let $\mathbb{N}_j := \{1, 2, 3, \ldots, j\}$, and $\mathbb{N}_j^0 := \{0, 1, 2, 3, \ldots, j\}$. For $j \in \mathbb{N} \cup \{0\}$, $C^j(I)$ is the Banach space of real valued functions having j continuous derivatives defined on I, and for $g \in C^j(I)$, $\|g\|_j := \max\{\|g^{(r)}\|_\infty : r = 0, 1, 2, \ldots, j\}$. For $g \in C(I)$, $\|g\|_\infty := \max\{|g(x)| : x \in I\}$.

Let a set of interpolation points $\{(x_i, y_i) \in I \times \mathbb{R} : i \in \mathbb{N}_N\}$ with increasing abscissae be given with $a = x_1$ and $b = x_N$. Let $[k_1, k_2]$ be a large compact interval in $\mathbb{R}$ such that $y_i \in [k_1, k_2] \forall i \in \mathbb{N}_N$. For a binary vector $\epsilon := (\epsilon_1, \epsilon_2, \dots, \epsilon_{N-1}) \in \{0, 1\}^{N-1}$, let $L_i : I \to I_i := [x_i, x_{i+1}]$, $i = 1, 2, \dots, N-1$, be contractive homeomorphisms such that

$$\begin{aligned} &L_i(x_1) = x_{i+\epsilon_i}, \quad L_i(x_N) = x_{i+1-\epsilon_i}, \\ &|L_i(x) - L_i(x^*)| \le r|x - x^*|, \quad \forall x, x^* \in I, \end{aligned} \tag{1}$$

for some $0 \le r < 1$.

For $0 \le \theta := \frac{x-x_1}{x_N - x_1} \le 1$ and $Q_i(\theta) = w_i(1-\theta)^3 + (w_i + u_i)(1-\theta)^2\theta + (w_{i+1} + u_{i+1})(1-\theta)\theta^2 + w_{i+1}\theta^3$, where w_i and u_i are the shape control parameters, let

$$\begin{aligned} P_{i1}(\theta) &= \frac{w_i(1-\theta)^3 + (w_i + u_i)(1-\theta)^3\theta + (w_{i+1} + u_{i+1})(1-\theta)^2\theta^2}{Q_i(\theta)}, \\ P_{i2}(\theta) &= \frac{(w_i + u_i)(1-\theta)^2\theta^2 + (w_{i+1} + u_{i+1})(1-\theta)\theta^3 + w_{i+1}\theta^3}{Q_i(\theta)}, \\ P_{i3}(\theta) &= \frac{w_i(1-\theta)^3\theta}{Q_i(\theta)}, \quad P_{i4}(\theta) = \frac{-w_{i+1}(1-\theta)\theta^3}{Q_i(\theta)}, \quad i \in \mathbb{N}_{N-1}, \end{aligned} \tag{2}$$

Then, for each $j \in \mathbb{N}_4$, $P_{ij} \in C^1(I)$ and satisfies

$$\begin{aligned} &P_{i1}(0) = 1, \; P_{i1}(1) = 0, \; P'_{i1}(0) = 0, \; P'_{i1}(1) = 0, \\ &P_{i2}(0) = 0, \; P_{i2}(1) = 1, \; P'_{i2}(0) = 0, \; P'_{i2}(1) = 0, \\ &P_{i3}(0) = 0, \; P_{i3}(1) = 0, \; P'_{i3}(0) = 1, \; P'_{i3}(1) = 0, \\ &P_{i4}(0) = 0, \; P_{i4}(1) = 0, \; P'_{i4}(0) = 0, \; P'_{i4}(1) = 1. \end{aligned} \tag{3}$$

Let $h_i := x_{i+1} - x_i$, $|I| := x_N - x_1$, and $h_i^* := x_{i+1-\epsilon_i} - x_{i+\epsilon_i}$. Now consider the function

$$P_\epsilon(L_i(x)) = P_{i1}(\theta)y_{i+\epsilon_i} + P_{i2}(\theta)y_{i+1-\epsilon_i} + h_i^* P_{i3}(\theta)d_{i+\epsilon_i} + h_i^* P_{i4}(\theta)d_{i+1-\epsilon_i} = \frac{P_i(\theta)}{Q_i(\theta)}, \tag{4}$$

where

$$\begin{aligned} &P_i(\theta) = \sum_{k=0}^{4} A_{ik}(1-\theta)^{4-k}\theta^k, \\ &A_{i0} = w_i y_{i+\epsilon_i}, \quad A_{i1} = u_i y_{i+\epsilon_i} + w_i(2y_{i+\epsilon_i} + h_i^* d_{i+\epsilon_i}), \\ &A_{i2} = (u_i + w_i)y_{i+1-\epsilon_i} + (u_{i+1} + w_{i+1})y_{i+\epsilon_i}, \\ &A_{i3} = u_{i+1}y_{i+1-\epsilon_i} + w_{i+1}(2y_{i+1-\epsilon_i} - h_i^* d_{i+1-\epsilon_i}), \quad A_{i4} = w_{i+1}y_{i+1-\epsilon_i}. \end{aligned} \tag{5}$$

Then, the RQS $P_\epsilon \in C^1(I)$ and satisfies $P_\epsilon(L_i(x_1)) = y_{i+\epsilon_i}$, $P_\epsilon(L_i(x_N)) = y_{i+1-\epsilon_i}$, $P'_\epsilon(L_i(x_1)) = d_{i+\epsilon_i}$, and $P'_\epsilon(L_i(x_N)) = d_{i+1-\epsilon_i}$. From (1), we can easily obtain that it also interpolates the given data, i.e. $\forall i \in \mathbb{N}$, $P_\epsilon(x_i) = y_i$, and $P'_\epsilon(x_i) = d_i$ for arbitrary signature ϵ, where d_i's are called derivative parameters. If the given data set $\{(x_i, y_i) : i \in \mathbb{N}_N\}$ is without the derivative parameters, then they must be calculated either from the data or by some appropriate methods. The arithmetic mean method (amm) and the geometric method (gmm) are popular choices for calculating derivatives from data. For details of these methods, see [10].

Remark 1 (i) If our shape control parameters w_i and u_i for $i \in \mathbb{N}$, are fixed and $w_i \neq w_{i+1}$ for $i \in \mathbb{N}_{N-1}$, then we can generate 2^{N-1} different rational quartic spline interpolation functions using different signatures for N numbers of data points.
(ii) If $\epsilon_i = 0$, for all $i \in \mathbb{N}_{N-1}$, then our rational quartic spline $P_\epsilon(x)$ reduces to the rational quartic spline $R(x)$ defined in [33].

Definition 1 A zipper with vertices $(v_1, v_2, \ldots, v_N)$ and signature $\epsilon = (\epsilon_1, \epsilon_2, \ldots, \epsilon_{N-1}) \in \{0, 1\}^{N-1}$ is a collection of some non-surjective maps with a complete metric space is denoted by $\Lambda := \{X; W_i : i \in \mathbb{N}_{N-1}\}$, where for each $i \in \mathbb{N}_{N-1}$, W_i satisfies $W_i(v_1) = v_{i+\epsilon_i}$ and $W_i(v_N) = v_{i+1-\epsilon_i}$.

If there exists a compact set $\Gamma \subset X$ such that

$$\Gamma = \bigcup_{j=1}^{N-1} W_j(\Gamma),$$

then Γ is called the attractor or fractal corresponding to the zipper Λ.

Let $H := I \times [k_1, k_2]$. Construct $N - 1$ continuous functions $F_i : H \to \mathbb{R}$ such that

$$F_i(x, y) = \alpha_i(x)y + (P_\epsilon(L_i(x)) - \alpha_i(x)B_i(x)),$$

where $\alpha_i \in C^1(I)$ such that $\|\alpha_i\|_1 < 1$, and $B_i \in C^1(I)$ such that

$$B_i(x) = P_{i1}(\theta)y_1 + P_{i2}(\theta)y_N + |I|P_{i3}(\theta)d_1 + |I|P_{i4}(\theta)d_N = \frac{P_i^*(\theta)}{Q_i(\theta)},$$

$$\begin{aligned} &P_i^*(\theta) = \sum_{k=0}^{4} A_{ik}^*(1-\theta)^{4-k}\theta^k, \\ &A_{i0}^* = w_i y_1, \quad A_{i1}^* = u_i y_1 + w_i(2y_1 + |I|d_1), \\ &A_{i2}^* = (u_i + w_i)y_N + (u_{i+1} + w_{i+1})y_1, \\ &A_{i3}^* = u_{i+1}y_N + w_{i+1}(2y_N - |I|d_N), \quad A_{i4}^* = w_{i+1}y_N. \end{aligned} \tag{6}$$

Now, for each $i \in \mathbb{N}_{N-1}$, B_i satisfies $B_i(x_1) = y_1$, $B_i(x_N) = y_N$, $B'_i(x_1) = d_1$, and $B'_i(x_N) = d_N$. Therefore, we have

$$\begin{aligned} &F_i(x_1, y_1) = y_{i+\epsilon_i}, \quad F_i(x_N, y_N) = y_{i+1-\epsilon_i}, \\ &|F_i(x, y) - F_i(x, y^*)| \le \|\alpha_i\|_\infty |y - y^*|, \quad \forall x \in I,\ y, y^* \in [k_1, k_2], \end{aligned} \tag{7}$$

Now define mappings $W_i : H \to I_i \times \mathbb{R}, i = 1, 2, \ldots, N-1$ by

$$W_i(x, y) = (L_i(x), F_i(x, y)), \quad \forall (x, y) \in H.$$

Therefore, $\{H; W_i : i \in \mathbb{N}_{N-1}\}$ is a zipper with vertices $((x_1, y_1), (x_2, y_2), \ldots, (x_N, y_N))$ and signature $\epsilon = (\epsilon_1, \epsilon_2, \ldots, \epsilon_{N-1})$. For each $i \in \mathbb{N}_{N-1}$, $\alpha_i(x)$ is called the variable scaling function corresponding to the map W_i and B_i is called a base function. Now we will construct a C^1-RQS ZFIF using the zipper $\{H; W_i : i \in \mathbb{N}_{N-1}\}$ for the given Hermite data $\{(x_i, y_i, d_i) : i \in \mathbb{N}_N\}$.

Theorem 1 *Let $\{(x_i, y_i, d_i) : i \in \mathbb{N}_N\}$ be a given set of interpolation data such that $x_1 < x_2 < \cdots < x_N$. Let the signature $\epsilon \in \{0, 1\}^{N-1}$ be fixed. For $i \in \mathbb{N}_{N-1}$, let $L_i(x) = a_i x + b_i$ satisfies (1), and $F_i(x, y) = \alpha_i(x)y + P_\epsilon(L_i(x)) - \alpha_i(x)B_i(x)$, where P_ϵ and B_i are as defined in (4) and (6) respectively. If $\alpha_i \in C^1(I)$ and $\|\alpha_i\|_1 < \frac{|a_i|}{2}$ for all $i \in \mathbb{N}_{N-1}$, then the zipper $\{H; (L_i(x), F_i(x, y)) : i \in \mathbb{N}_{N-1}\}$ determines a rational quartic spline zipper fractal interpolation function $P_\epsilon^\alpha \in C^1(I)$.*

Proof Let $D(I) := \{g \in C^1(I) : g(x_1) = y_1, g(x_N) = y_N, g'(x_1) = d_1$, and $g'(x_N) = d_N\}$. Then $D(I)$ is a complete metric space with respect to norm $\|.\|_1$. Now, define the Read-Bajraktarević operator $T^\alpha : D \to D$ such that

$$T^\alpha g(L_i(x)) = P_\epsilon(L_i(x)) + \alpha_i(x)(g(x) - B_i(x)), \ x \in I, \ i = 1, 2, \ldots, N-1. \tag{8}$$

Since the functions P_ϵ, B_i, and α_i belong to $C^1(I)$, $T^\alpha g \in C^1(x_i, x_{i+1})$ for each $i \in \mathbb{N}_{N-1}$. We know, for $i \in \mathbb{N}_{N-2}$, $x_{i+1} \in I_j$ for $j = i, i+1$. Since L_i and L_{i+1} satisfy (1), therefore we have

$$x_{i+1} = \begin{cases} L_i(x_N) & \epsilon_i = 0 \\ L_i(x_1) & \epsilon_i = 1, \end{cases} \text{ and } x_{i+1} = \begin{cases} L_{i+1}(x_1) & \epsilon_{i+1} = 0 \\ L_{i+1}(x_N) & \epsilon_{i+1} = 1. \end{cases} \tag{9}$$

By putting (9) in (8), we have

$$T^\alpha g(x_{i+1}) = \begin{cases} P_\epsilon(L_i(x_N)) & \epsilon_i = 0 \\ P_\epsilon(L_i(x_1)) & \epsilon_i = 1, \end{cases} \text{ and } T^\alpha g(x_{i+1}) = \begin{cases} P_\epsilon(L_{i+1}(x_1)) & \epsilon_{i+1} = 0 \\ P_\epsilon(L_{i+1}(x_N)) & \epsilon_{i+1} = 1. \end{cases} \tag{10}$$

$$\implies \lim_{x \to x_{i+1}^-} (T^\alpha g)(x) = \lim_{x \to x_{i+1}^+} (T^\alpha g)(x) = y_{i+1}$$

Similarly, after differentiating (8) once and using (9), we can obtain

$$(T^\alpha g)'(x_{i+1}) = \begin{cases} P'_\epsilon(L_i(x_N)) & \epsilon_i = 0 \\ P'_\epsilon(L_i(x_1)) & \epsilon_i = 1, \end{cases} \text{ and } (T^\alpha g)'(x_{i+1}) = \begin{cases} P'_\epsilon(L_{i+1}(x_1)) & \epsilon_{i+1} = 0 \\ P'_\epsilon(L_{i+1}(x_N)) & \epsilon_{i+1} = 1. \end{cases} \tag{11}$$

$$\implies \lim_{x \to x_{i+1}^-} (T^\alpha g)'(x) = \lim_{x \to x_{i+1}^+} (T^\alpha g)'(x) = d_{i+1}. \tag{12}$$

Now, for $i = 1, N-1$, from (8) we can easily get that $T^{\alpha}g(x_1) = y_1$, $T^{\alpha}g(x_N) = y_N$, $(T^{\alpha}g)'(x_1) = d_1$, and $(T^{\alpha}g)'(x_N) = d_N$. Therefore, the operator T^{α} is well-defined, i.e. $T^{\alpha}g \in D$. Now, for $x \in I_i$,

$$(T^{\alpha}g)(x) - (T^{\alpha}g^*)(x) = \alpha_i(L_i^{-1}(x))(g - g^*)(L_i^{-1}(x)),$$

which implies

$$|(T^{\alpha}g)(x) - (T^{\alpha}g^*)(x)| \leq \|\alpha_i\|_{\infty}\|g - g^*\|_{\infty} \leq \|\alpha_i\|_1\|g - g^*\|_1.$$

Similarly,

$$\begin{aligned}|(T^{\alpha}g)'(x) - (T^{\alpha}g^*)'(x)| &\leq |a_i^{-1}|(\|\alpha_i'\|_{\infty}\|g - g^*\|_{\infty} + \|\alpha_i\|_{\infty}\|g' - g^{*\prime}\|_{\infty}) \\ &\leq 2|a_i^{-1}|\|\alpha_i\|_1\|g - g^*\|_1.\end{aligned}$$

So, if for all $i \in \mathbb{N}_{N-1}$, $\|\alpha_i\|_1 < s\frac{|a_i|}{2}$ for some $0 \leq s < 1$, then we have $\|T^{\alpha}g - T^{\alpha}g^*\|_1 < s\|g - g^*\|_1$, i.e. T^{α} is a contraction map on D. Therefore, by Banach fixed point theorem T^{α} has a unique fixed point say $P_{\epsilon}^{\alpha} \in C^1(I)$, and P_{ϵ}^{α} satisfies the recurrence relation

$$P_{\epsilon}^{\alpha}(L_i(x)) = P_{\epsilon}(L_i(x)) + \alpha_i(x)(P_{\epsilon}^{\alpha}(x) - B_i(x)),\ x \in I,\ i = 1, 2, \ldots, N-1. \tag{13}$$

P_{ϵ}^{α} is the desired rational quartic spline zipper α-fractal function corresponding to the function P_{ϵ}. For more details on α-fractal functions, see [5, 21].

Remark 2 (i) If $\alpha_i(x) = 0$, for all $x \in I$ and for all $i \in \mathbb{N}_{N-1}$, then our RQS ZFIF P_{ϵ}^{α} reduces to the RQS P_{ϵ} defined in (4).
(ii) If $\alpha_i(x) = 0$ and $\epsilon_i = 0$, for all $x \in I$ and for all $i \in \mathbb{N}_{N-1}$, then the proposed RQS ZFIF P_{ϵ}^{α} reduces to the rational quartic spline $R(x)$ defined in [33].
(iii) For the fixed shape control parameters and the fixed non-zero variable scaling functions, we can get 2^{N-1} different RQS ZFIFs using different values of signature for the N numbers of data points.

3 Convergence Analysis

In this section, we will derive an upper bound for the uniform error of the RQS ZFIF with a C^3 data generating function, and we will show that our RQS ZFIF has $O(h^2)$ convergence and can be increased to $O(h^3)$ under certain conditions.

We fix these notation for this section: $\Delta_i := \frac{y_{i+1}-y_i}{h_i}$, $t := \frac{x-x_i}{h_i}$, $h := \max\{h_i : i \in \mathbb{N}_{N-1}\}$, $|y|_{\infty} := \max\{|y_i| : i \in \mathbb{N}_N\}$, $|d|_{\infty} := \max\{|d_i| : i \in \mathbb{N}_N\}$, $w_{i*} := \min\{w_i, w_{i+1}\}$, $u_{i*} := \min\{u_i, u_{i+1}\}$, $w_i^* := \max\{w_i, w_{i+1}\}$, $u_i^* := \max\{u_i, u_{i+1}\}$, $w_* := \min\{w_i : i \in \mathbb{N}_N\}$, $w^* := \max\{w_i : i \in \mathbb{N}_N\}$, $u_* := \min\{u_i : i \in \mathbb{N}_N\}$, $u^* := \max\{u_i : i \in \mathbb{N}_N\}$,

$B^* := \max\{\|B_i\|_\infty : i \in \mathbb{N}_{N-1}\}$, $\alpha(x) := (\alpha_1(x), \alpha_2(x), \ldots, \alpha_{N-1}(x))$, $\|\alpha\|_\infty := \max\{\|\alpha_i\|_\infty : i \in \mathbb{N}_{N-1}\}$, and $\|\alpha\|_1 := \max\{\|\alpha_i\|_1 : i \in \mathbb{N}_{N-1}\}$.

Let $\Phi \in C^3(I)$ be a data generating function, i.e. $\Phi(x_i) = y_i$, $\forall i \in \mathbb{N}_N$. Let d_i's are chosen derivatives at x_i, for all $i \in \mathbb{N}_N$. Now, for $\theta = \frac{x - x_1}{x_N - x_1}$, let $t^* := L_i^{-1}(\theta)$, i.e. $t^* = \frac{x - x_{i+\epsilon_i}}{x_{i+1-\epsilon_i} - x_{i+\epsilon_i}}$. Therefore, for $x \in I_i$,

$$P_\epsilon(x) = \frac{1}{Q_i(t^*)} \sum_{k=0}^{4} A_{ik}(1 - t^*)^{4-k} t^{*k}, \tag{14}$$

where $t^* = \begin{cases} t & \epsilon_i = 0 \\ 1 - t & \epsilon_i = 1, \end{cases}$ and A_{ik}'s are as defined in (5).

Case I: Let $x \in I_i$ and $\epsilon_i = 0$, then

$$\begin{aligned}
&P_\epsilon(x) = \frac{1}{Q_i(t)} \sum_{k=0}^{4} A_{ik}(1 - t)^{4-k} t^k, \\
&A_{i0} = w_i y_i, \quad A_{i1} = u_i y_i + w_i(2y_i + h_i d_i), \\
&A_{i2} = (u_i + w_i) y_{i+1} + (u_{i+1} + w_{i+1}) y_i, \\
&A_{i3} = u_{i+1} y_{i+1} + w_{i+1}(2y_{i+1} - h_i d_{i+1}), \quad A_{i4} = w_{i+1} y_{i+1}, \\
&Q_i(t) = w_i(1 - t)^3 + (w_i + u_i)(1 - t)^2 t + (w_{i+1} + u_{i+1})(1 - t)t^2 + w_{i+1} t^3.
\end{aligned} \tag{15}$$

Now from [13, 33], for $x \in I_i$ and $\epsilon_i = 0$, choosing $w_i, w_{i+1} > 0$ and $u_i, u_{i+1} \geq 0$, we have

$$\begin{aligned}
|\Phi(x) - P_\epsilon(x)| \leq\; & \frac{h_i^3}{96}\|\Phi^{(3)}\|_\infty + \frac{h_i}{4} \max\left\{|\Phi'(x_i) - d_i|, |\Phi'(x_{i+1}) - d_{i+1}|\right\} \\
& + \frac{h_i}{2\sqrt{w_i w_{i+1}} + \min\{u_i, u_{i+1}\}}\Big[\frac{27}{256}|u_i(\Delta_i - d_i) - w_i(2\Delta_i - d_i - d_{i+1})| \\
& + \frac{1}{16}|(w_{i+1} - w_i)(2\Delta_i - d_i - d_{i+1}) + u_{i+1}(\Delta_i - d_i) + u_i(d_{i+1} - \Delta_i)| \\
& + \frac{27}{256}|w_{i+1}(2\Delta_i - d_i - d_{i+1}) + u_{i+1}(d_{i+1} - \Delta_i)|\Big].
\end{aligned} \tag{16}$$

Case II: Let $x \in I_i$ and $\epsilon_i = 1$, then

$$P_\epsilon(x) = \frac{1}{Q_i(1-t)} \sum_{k=0}^{4} A_{ik} t^{4-k}(1-t)^k,$$

$$\begin{aligned}
&A_{i0} = w_i y_{i+1}, \quad A_{i1} = u_i y_{i+1} + w_i(2y_i - h_i d_{i+1}),\\
&A_{i2} = (u_i + w_i) y_i + (u_{i+1} + w_{i+1}) y_{i+1},\\
&A_{i3} = u_{i+1} y_i + w_{i+1}(2y_i + h_i d_i), \quad A_{i4} = w_{i+1} y_i,\\
&Q_i(1-t) = w_i t^3 + (w_i + u_i) t^2 (1-t) + (w_{i+1} + u_{i+1}) t (1-t)^2 + w_{i+1}(1-t)^3.
\end{aligned} \tag{17}$$

After interchanging w_i and w_{i+1}, u_i and u_{i+1}, (17) becomes equivalent to (15). Therefore, using similar analysis, for $x \in I_i$ and $\epsilon_i = 1$, choosing $w_i, w_{i+1} > 0$ and $u_i, u_{i+1} \geq 0$, we have

$$\begin{aligned}
|\Phi(x) - P_\epsilon(x)| \leq{}& \frac{h_i^3}{96}\|\Phi^{(3)}\|_\infty + \frac{h_i}{4}\max\left\{|\Phi'(x_i) - d_i|, |\Phi'(x_{i+1}) - d_{i+1}|\right\}\\
&+ \frac{h_i}{2\sqrt{w_{i+1}w_i} + \min\{u_{i+1}, u_i\}}\Big[\frac{27}{256}|u_{i+1}(\Delta_i - d_i) - w_{i+1}(2\Delta_i - d_i - d_{i+1})|\\
&+ \frac{1}{16}|(w_i - w_{i+1})(2\Delta_i - d_i - d_{i+1}) + u_i(\Delta_i - d_i) + u_{i+1}(d_{i+1} - \Delta_i)|\\
&+ \frac{27}{256}|w_i(2\Delta_i - d_i - d_{i+1}) + u_i(d_{i+1} - \Delta_i)|\Big].
\end{aligned} \tag{18}$$

Now, if the derivative parameters d_i's are chosen such that

$$\begin{aligned}
d_1 &= \Delta_1 - \frac{h_1}{h_1 + h_2}(\Delta_2 - \Delta_1),\\
d_N &= \Delta_{N-1} + \frac{h_{N-1}}{h_{N-1} + h_{N-2}}(\Delta_{N-1} - \Delta_{N-2}),\\
d_i &= \frac{h_i}{h_{i-1} + h_i}\Delta_{i-1} + \frac{h_{i-1}}{h_{i-1} + h_i}\Delta_i, \quad i = 2, 3, \ldots, N-1,
\end{aligned} \tag{19}$$

then by using Peano kernel analysis, we can easily get following results:

$$\begin{aligned}
&\Phi'(x_1) - d_1 = \frac{1}{6}h_1(h_1 + h_2)\Phi^{(3)}(\zeta_1), \quad d_i - \Phi'(x_i) = \frac{1}{6}h_{i-1}h_i\Phi^{(3)}(\zeta_i),\\
&\Phi'(x_N) - d_N = \frac{1}{6}h_{N-1}(h_{N-1} + h_{N-2})\Phi^{(3)}(\zeta_N), \quad \Delta_1 - d_1 = \frac{1}{2}h_1\Phi^{(2)}(\chi_2),\\
&\Delta_i - d_i = \frac{1}{2}h_i\Phi^{(2)}(\chi_i), \quad d_{i+1} - \Delta_i = \frac{1}{2}h_i\Phi^{(2)}(\chi_{i+1}),\\
&d_N - \Delta_{N-1} = \frac{1}{2}h_{N-1}\Phi^{(2)}(\chi_{N-1}), \quad 2\Delta_1 - d_1 - d_2 = 0,\\
&d_{N-1} + d_N - 2\Delta_{N-1} = 0, \quad d_i + d_{i+1} - 2\Delta_i = \frac{1}{6}h_i(h_{i-1} + h_i + h_{i+1})\Phi^{(3)}(\chi_i^*),
\end{aligned} \tag{20}$$

where $\zeta_1 \in (x_1, x_3)$, $\zeta_i \in (x_{i-1}, x_{i+1})$, $\zeta_N \in (x_{N-2}, x_N)$, $\chi_i \in (x_{i-1}, x_{i+1})$, $i = 2, 3, \ldots, N-1$ and $\chi_i^* \in (x_{i-1}, x_{i+2})$, $i = 2, 3, \ldots, N-2$. Therefore, using (20) in

(16) and (18), we have

$$|\Phi(x) - P_\epsilon(x)| \leq \frac{h^3}{96}\|\Phi^{(3)}\|_\infty + \frac{h^3}{12}\|\Phi^{(3)}\|_\infty + \frac{h^2}{2w_{i*}+u_{i*}}\left[\frac{43}{256}(u_i^*\|\Phi^{(2)}\|_\infty + hw_i^*\|\Phi^{(3)}\|_\infty)\right],$$

for $x \in I_i$ and $\epsilon_i \in \{0, 1\}$.

Now we summarize the above discussions in the following as a theorem:

Theorem 2 *Let $\Phi \in C^3(I)$ be a data generating function such that $\Phi(x_i) = y_i$, $i \in \mathbb{N}_N$. For a fixed signature $\epsilon \in \{0, 1\}^{N-1}$, let P_ϵ be the rational quartic spline defined in (4). If for all $i \in \mathbb{N}_N$, we choose our shape control parameters such that $w_i > 0$, $u_i \geq 0$ and the derivative parameters as prescribed in (19), then*

$$\|\Phi - P_\epsilon\|_\infty \leq \frac{9h^3}{96}\|\Phi^{(3)}\|_\infty + \frac{h^2}{2w_*+u_*}\left[\frac{43}{256}(u^*\|\Phi^{(2)}\|_\infty + hw^*\|\Phi^{(3)}\|_\infty)\right]. \tag{21}$$

Now we will try to find the upper bound for the difference between RQS P_ϵ defined in (4) and RQS ZFIF P_ϵ^α defined in (13). If $\alpha \not\equiv 0$, then $P_\epsilon \neq P_\epsilon^\alpha$, and the interpolants P_ϵ^α and P_ϵ are the fixed points of T^α defined in (8) with $\alpha \not\equiv 0$ and $\alpha(x) = (0, 0, \ldots, 0)$ respectively.

For $i \in \mathbb{N}_{N-1}$ and $x \in I$,

$$\begin{aligned}
|P_\epsilon^\alpha(L_i(x)) - P_\epsilon(L_i(x))| &= |T^\alpha P_\epsilon^\alpha(L_i(x)) - T^0 P_\epsilon(L_i(x))| \\
&= |P_\epsilon(L_i(x)) + \alpha_i(x)(P_\epsilon^\alpha(x) - B_i(x)) - P_\epsilon(L_i(x))| \\
&= |\alpha_i(x)(P_\epsilon^\alpha(x) - B_i(x))| \\
&\leq \|\alpha_i\|_\infty \|P_\epsilon^\alpha - B_i\|_\infty \\
&\leq \|\alpha_i\|_\infty \|P_\epsilon^\alpha - P_\epsilon\|_\infty + \|\alpha_i\|_\infty \|P_\epsilon - B_i\|_\infty \\
&\leq \|\alpha_i\|_\infty \|P_\epsilon^\alpha - P_\epsilon\|_\infty + \|\alpha_i\|_\infty(\|P_\epsilon\|_\infty + B^*).
\end{aligned}$$

As for each $i \in \mathbb{N}_{N-1}$, the above inequality holds, hence

$$\|P_\epsilon^\alpha - P_\epsilon\|_\infty \leq \|\alpha\|_\infty \|P_\epsilon^\alpha - P_\epsilon\|_\infty + \|\alpha\|_\infty(\|P_\epsilon\|_\infty + B^*),$$

i.e.

$$\|P_\epsilon^\alpha - P_\epsilon\|_\infty \leq \frac{\|\alpha\|_\infty(\|P_\epsilon\|_\infty + B^*)}{1 - \|\alpha\|_\infty}. \tag{22}$$

Now, let us deduce upper bounds for $\|P_\epsilon\|_\infty$ and B^*. From (4), for $i \in \mathbb{N}_{N-1}$ and $x \in I$,

$$|P_\epsilon(L_i(x))| \leq \frac{\max\{|P_i(\theta)| : 0 \leq \theta \leq 1\}}{\min\{|Q_i(\theta)| : 0 \leq \theta \leq 1\}}.$$

Using inequalities $(1-\theta)^3\theta \le \frac{27}{256}$, $(1-\theta)^2\theta^2 \le \frac{1}{16}$, $(1-\theta)\theta^3 \le \frac{27}{256}$, and $(1-\theta)^4+\theta^4 \le 1$, we can easily deduce that

$$\begin{aligned}|P_i(\theta)| &\le w_i^*(\max\{|y_i|, |y_{i+1}|\})\\ &\quad + \frac{27}{128}\Big[(u_i^*+2w_i^*)(\max\{|y_i|, |y_{i+1}|\}) + w_i^* h_i(\max\{|d_i|, |d_{i+1}|\})\Big]\\ &\quad + \frac{1}{8}\Big[(u_i^*+w_i^*)(\max\{|y_i|, |y_{i+1}|\})\Big],\\ \text{i.e. } |P_i(\theta)| &\le \Big(\frac{99}{64}w^* + \frac{43}{128}u^*\Big)|y|_\infty + w^* h|d|_\infty,\end{aligned}$$

and

$$\begin{aligned}|Q_i(\theta)| &\ge w_i(1-\theta)^2 + u_i(1-\theta)^2\theta + u_{i+1}(1-\theta)\theta^2 + w_{i+1}\theta^2\\ &\ge w_i(1-\theta)^2 + w_{i+1}\theta^2 \ge \frac{1}{2}w_{i*} \ge \frac{1}{2}w_*.\end{aligned}$$

Hence,

$$\|P_\epsilon\|_\infty \le 2\frac{\Big(\frac{99}{64}w^* + \frac{43}{128}u^*\Big)|y|_\infty + w^* h|d|_\infty}{w_*}.$$

Similarly,

$$\|B_i\|_\infty \le 2\frac{\Big(\frac{99}{64}w_i^* + \frac{43}{128}u_i^*\Big)\max\{|y_1|, |y_N|\} + w_i^*|I|(\max\{|d_1|, |d_N|\})}{w_i^*}.$$

Therefore,

$$B^* \le 2\frac{\Big(\frac{99}{64}w^* + \frac{43}{128}u^*\Big)\max\{|y_1|, |y_N|\} + w^*|I|(\max\{|d_1|, |d_N|\})}{w_*}.$$

Now we will present the main theorem of this section.

Theorem 3 *Let $\Phi \in C^3(I)$ be a data generating function such that $\Phi(x_i) = y_i$, $i \in \mathbb{N}_N$. For a fixed signature $\epsilon \in \{0,1\}^{N-1}$, let P_ϵ be the rational quartic spline defined in (4) and P_ϵ^α be the proposed rational quartic spline zipper fractal interpolation function defined in (13). If for all $i \in \mathbb{N}_N$, we choose our shape control points such that $w_i > 0$, $u_i \ge 0$ and the derivative parameters as given in (19), then*

$$\|\Phi - P_\epsilon^\alpha\|_\infty \le \frac{9h^3}{96}\|\Phi^{(3)}\|_\infty + \frac{h^2}{2w_* + u_*}\Big[\frac{43}{256}\big(u^*\|\Phi^{(2)}\|_\infty + hw^*\|\Phi^{(3)}\|_\infty\big)\Big] + \frac{\|\alpha\|_\infty(\|P_\epsilon\|_\infty + B^*)}{1-\|\alpha\|_\infty}.$$

Proof We know that

$$\|\Phi - P_\epsilon^\alpha\|_\infty \leq \|\Phi - P_\epsilon\|_\infty + \|P_\epsilon^\alpha - P_\epsilon\|_\infty. \tag{23}$$

Therefore, using (21) and (22) in (23), we can easily get our desired result.

Remark 3 (i) If we choose $\|\alpha\|_1 < \min\{h^2, \frac{h}{2|I|}\}$, then from Theorem 3, we can deduce that our proposed zipper fractal interpolant P_ϵ^α converges to a C^3-data generating function Φ with the order $O(h^2)$ as $h \to 0$ on I.
(ii) If we choose $u_i = 0$ for all $i \in \mathbb{N}_N$ and $\|\alpha\|_1 < \min\{h^3, \frac{h}{2|I|}\}$, then $u^* = 0$, and hence from Theorem 3 we can conclude that our proposed zipper fractal interpolant P_ϵ^α converges to a C^3-data generating function Φ with the order $O(h^3)$ as $h \to 0$ on I.

4 Positivity-Preserving RQS ZFIFs

Many real-life problems like monthly rainfall amounts, the half-life of a radioactive substance, probability distribution functions, speed of winds and the numbers of covid-19 patients at different intervals of time are based on positivity. So, the problem is to find a positive interpolant for a given positive data set. In this section, we are going to construct positive RQS ZFIFs for the given positive data by restricting our shape control parameters and variable scaling functions.

Theorem 4 *Let $\{(x_i, y_i) : i = 1, 2, \ldots, N\}$ be a given set of strictly positive data with increasing abscissae. Suppose d_i's are chosen derivative values at the knots x_i's. For the fixed value of signature $\epsilon \in \{0, 1\}^{N-1}$, if the non-negative variable scaling functions and shape control points are chosen as*

$$\begin{aligned}
&\|\alpha_i\|_1 < \frac{|a_i|}{2}, \ \|\alpha_i\|_\infty < \min\left\{\frac{y_{i+\epsilon_i}}{y_1}, \frac{y_{i+1-\epsilon_i}}{y_N}\right\},\\
&w_i > 0, \ \ w_{i+1} > 0,\\
&u_i \geq \max\left\{0, -w_i\Big(2 + \frac{h_i^* d_{i+\epsilon_i} - \alpha_i(x)|I|d_1}{y_{i+\epsilon_i} - \alpha_i(x)y_1}\Big)\right\},\\
&u_{i+1} \geq \max\left\{0, -w_{i+1}\Big(2 - \frac{h_i^* d_{i+1-\epsilon_i} - \alpha_i(x)|I|d_N}{y_{i+1-\epsilon_i} - \alpha_i(x)y_N}\Big)\right\}, \ \ \forall x \in I, \ \ \forall i \in \mathbb{N}_{N-1},
\end{aligned}$$

then the corresponding C^1-rational quartic spline zipper fractal interpolation function P_ϵ^α defined in (13) will be strictly positive on I.

Proof Since $\|\alpha_i\|_1 < \frac{|a_i|}{2}$, and α_i, P_ϵ, $B_i \in C^1(I)$, therefore according to Theorem 1 the proposed RQS ZFIF $P_\epsilon^\alpha \in C^1(I)$, and it satisfies (13). We can rewrite (13) as

$$P_\epsilon^\alpha(L_i(x)) = \alpha_i(x)P_\epsilon^\alpha(x) + (P_\epsilon(L_i(x)) - \alpha_i(x)B_i(x)), \ \ i \in \mathbb{N}_{N-1}, \ x \in I. \tag{24}$$

Equation (24) is equivalent to

$$P_\epsilon^\alpha(L_i(x)) = \alpha_i(x) P_\epsilon^\alpha(x) + \frac{P_i^{**}(\theta)}{Q_i(\theta)}, \quad P_i^{**}(\theta) = \sum_{k=0}^{4} B_{ik}(1-\theta)^{4-k}\theta^k,$$

$$\begin{aligned}
B_{i0} &= w_i(y_{i+\epsilon_i} - \alpha_i(x)y_1),\\
B_{i1} &= u_i(y_{i+\epsilon_i} - \alpha_i(x)y_1) + w_i(2(y_{i+\epsilon_i} - \alpha_i(x)y_1) + h_i^* d_{i+\epsilon_i} - \alpha_i(x)|I|d_1),\\
B_{i2} &= [(u_i + w_i)(y_{i+1-\epsilon_i} - \alpha_i(x)y_N)] + [(u_{i+1} + w_{i+1})(y_{i+\epsilon_i} - \alpha_i(x)y_1)],\\
B_{i3} &= u_{i+1}(y_{i+1-\epsilon_i} - \alpha_i(x)y_N) + w_{i+1}(2(y_{i+1-\epsilon_i} - \alpha_i(x)y_N) - h_i^* d_{i+1-\epsilon_i} + \alpha_i(x)|I|d_N),\\
B_{i4} &= w_{i+1}(y_{i+1-\epsilon_i} - \alpha_i(x)y_N).
\end{aligned} \tag{25}$$

After choosing $w_i > 0$, $w_{i+1} > 0$, $u_i \geq 0$ and $u_{i+1} \geq 0$, our cubic denominator $Q_i(\theta)$ in (25) becomes strictly positive on I. Since P_ϵ^α is the attractor of the zipper $\{H; (L_i(x) = a_i x + b_i, F_i(x, y) = \alpha_i(x)y + P_\epsilon(L_i(x)) - \alpha_i(x)B_i(x)) : i \in \mathbb{N}_{N-1}\}$ and defined recursively by (25), therefore to show $P_\epsilon^\alpha(x) > 0$ on I, enough to prove that for all $x \in I$, $P_\epsilon^\alpha(L_i(x)) > 0 \, \forall i \in \mathbb{N}_{N-1}$, whenever $P_\epsilon^\alpha(x) > 0$. Now, let $x \in I$, $P_\epsilon^\alpha(x) > 0$, then choosing non-negative scaling functions we have $\alpha_i(x)P_\epsilon^\alpha(x) \geq 0$. Therefore, after these assumptions on the shape control parameters and the variable scaling functions, the positivity of $P_\epsilon^\alpha(L_i(x))$ reduces to the positivity of $P_i^{**}(\theta)$, $\forall \theta \in [0, 1]$. Now, if $B_{ij} \geq 0$, $\forall j \in \{0, 1, 2, 3, 4\}$ and $B_{ij} > 0$ for $j \in \{0, 4\}$, then we have $P_i^{**}(\theta) > 0$. Now,

$$\begin{aligned}
& w_i > 0, \text{ and } \alpha_i(x) < \frac{y_{i+\epsilon_i}}{y_1} \implies B_{i0} > 0,\\
& u_i > \max\left\{0, -w_i\left(2 + \frac{h_i^* d_{i+\epsilon_i} - \alpha_i(x)|I|d_1}{y_{i+\epsilon_i} - \alpha_i(x)y_1}\right)\right\}, \text{ and } \alpha_i(x) < \frac{y_{i+\epsilon_i}}{y_1} \implies B_{i1} \geq 0,\\
& u_i + w_i > 0,\ u_{i+1} + w_{i+1} > 0, \text{ and } \alpha_i(x) < \left\{\frac{y_{i+\epsilon_i}}{y_1}, \frac{y_{i+1-\epsilon_i}}{y_N}\right\} \implies B_{i2} > 0,\\
& u_{i+1} \geq \max\left\{0, -w_{i+1}\left(2 - \frac{h_i^* d_{i+1-\epsilon_i} - \alpha_i(x)|I|d_N}{y_{i+1-\epsilon_i} - \alpha_i(x)y_N}\right)\right\}, \text{ and } \alpha_i(x) < \frac{y_{i+1-\epsilon_i}}{y_N} \implies B_{i3} \geq 0,\\
& w_{i+1} > 0, \text{ and } \alpha_i(x) < \frac{y_{i+1-\epsilon_i}}{y_N} \implies B_{i4} > 0.
\end{aligned}$$

Hence, coupling these above restrictions on the shape control parameters and the scaling functions, we have the desired sufficient conditions for this theorem.

Remark 4 (i) For all $i \in \mathbb{N}_{N-1}$ and $x \in I$, if we choose $\alpha_i(x) = 0$ and $\epsilon_i = 0$, then Theorem 4 gives sufficient conditions on the shape control parameters such that the RQS function R defined in [33] becomes positive for a given positive data set $\{(x_i, y_i) : i = 1, 2, \dots, N\}$.
(ii) Let the given data set be strictly positive and $\alpha_i(x) = 0$ for all $x \in I$ and for all $i \in \mathbb{N}_{N-1}$. If we choose our shape control parameters such that

$$
\begin{aligned}
&w_i > 0, \quad w_{i+1} > 0, \quad u_i \geq \max\left\{0, -w_i\left(2 + \frac{h_i^* d_{i+\epsilon_i}}{y_{i+\epsilon_i}}\right)\right\}, \\
&u_{i+1} \geq \max\left\{0, -w_{i+1}\left(2 - \frac{h_i^* d_{i+1-\epsilon_i}}{y_{i+1-\epsilon_i}}\right)\right\}, \quad \forall i \in \mathbb{N}_{N-1}.
\end{aligned} \tag{26}
$$

Then, Theorem 4 instructs that our corresponding rational quartic spline P_ϵ defined in (4) satisfies $P_\epsilon(x) > 0$ for all $x \in I$.
(iii) For $N(> 2)$ number of positive data points and the fixed non-zero variable scaling functions, we can get total numbers of 2^{N-1} different C^1-rational quartic spline zipper fractal interpolation functions depending on the different values of signature ϵ.

Example 1 In Theorem 4, we have provided sufficient conditions on the shape control parameters and the variable scaling functions such that our corresponding RQS ZFIF becomes positive on I, whenever our given data set is positive. It can happen that if we do not choose our parameters as prescribed in Theorem 4, then our corresponding RQS ZFIF P_ϵ^α may not be positive on I for a given positive data set, but after restricting our shape control parameters and variable scaling functions as prescribed in Theorem 4 our corresponding RQS ZFIF becomes positive on I.

Consider the positive data set $\{(0, 2, -1), (0.25, 0.6, -3), (0.5, 0.1, 2), (0.75, 0.4, -2), (1, 5, 6)\}$. For the fixed shape control parameters $u = (1, 2, 3, 1, 1)$ and $w = (1, 0.2, 0.5, 1, 3)$, Figs. 1(a)-(f) are the plots of RQS ZFIFs generated with scaling functions and signature $\{(\frac{x^2}{17}, \frac{e^x}{25}, \frac{e^x}{25}, \frac{-x}{10}), (1, 0, 1, 0)\}$, $\{(\frac{e^x}{25}, \frac{x}{60}, \frac{1}{100}, \frac{e^x}{35}), (1, 0, 1, 0)\}$, $\{(\frac{e^x}{25}, \frac{x}{60}, \frac{1}{100}, \frac{e^x}{35}), (0, 0, 1, 0)\}$, $\{(0, 0, 0, 0), (0, 0, 0, 0)\}$, $\{(0, 0.01 + \frac{x}{120}, 0, 0), (0, 0, 0, 0)\}$, and $\{(0, 0, 0, 0), (1, 0, 0, 1)\}$ respectively. For Fig. 1(a), we do not restrict our scaling functions as recommended by Theorem 4, and the corresponding RQS ZFIF is not positive on $I = [0, 1]$. But for the other plots, we have restricted our shape control parameters and scaling functions as recommended by Theorem 4 and hence the corresponding RQS ZFIFs are positive on I.

To see the effect of signature, we have plotted Fig. 1(b) and (c) with the same parameters except for ϵ_1, and we have turned up with very different RQS ZFIF on $I_1 = [0, 0.25]$. Fig. 1(d) is the plot of the classical rational quartic spline defined in [33]. To see the effect of scaling functions, we have plotted Fig. 1(e) by changing the scaling function α_2 from the parameters used for Fig. 1(d). Fig. 1(f) is the plot of RQS defined by us in (4) using the binary vector signature $\epsilon = (1, 0, 0, 1)$ and we can see that the RQS defined by us and classical RQS defined in [33] are not the same. Thus, the proposed method enlarge the class of rational quartic splines with fixed shape parameters.

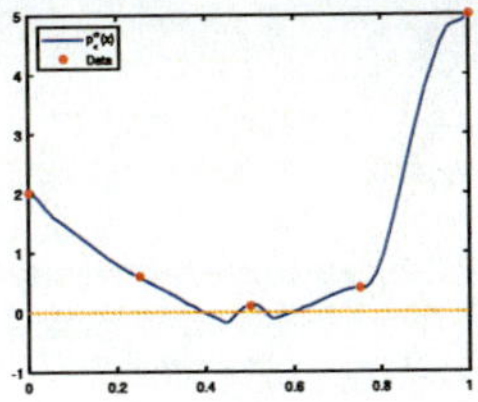

(a) RQS ZFIF with $\alpha = (\frac{x^2}{17}, \frac{e^x}{25}, \frac{e^x}{25}, \frac{-x}{10})$ and $\epsilon = (1, 0, 1, 0)$.

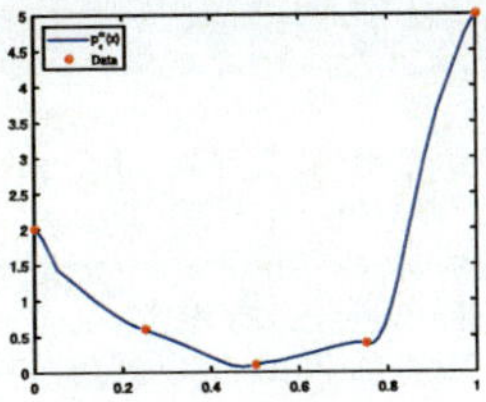

(b) RQS ZFIF with $\alpha = (\frac{e^x}{25}, \frac{x}{60}, \frac{1}{100}, \frac{e^x}{35})$ and $\epsilon = (1, 0, 1, 0)$.

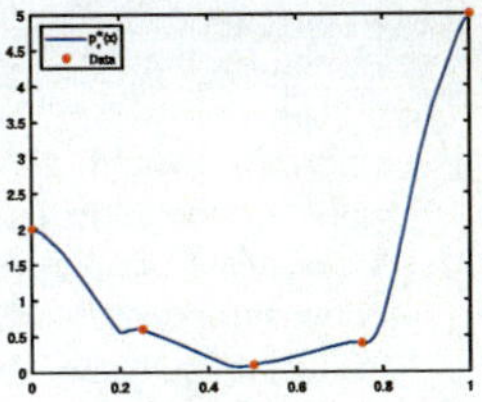

(c) RQS ZFIF with $\alpha = (\frac{e^x}{25}, \frac{x}{60}, \frac{1}{100}, \frac{e^x}{35})$ and $\epsilon = (0, 0, 1, 0)$.

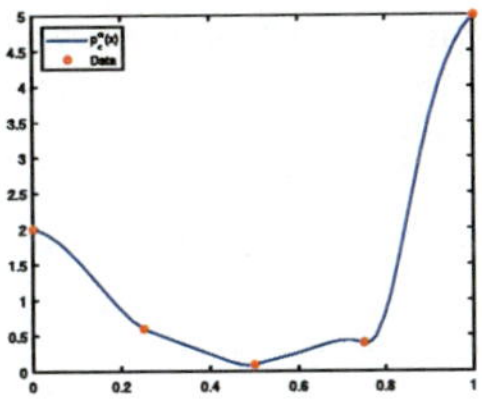

(d) Classical RQS ($\alpha = (0, 0, 0, 0)$ and $\epsilon = (0, 0, 0, 0)$).

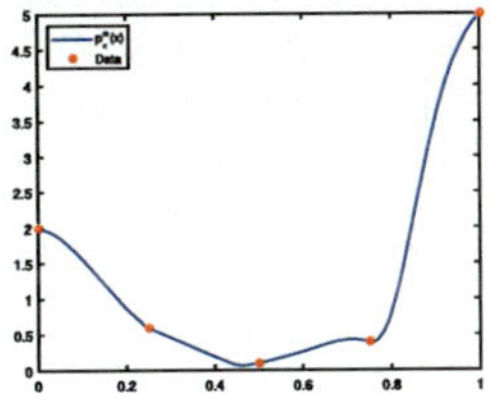

(e) RQS ZFIF with $\alpha = (0, 0.01 + \frac{x}{120}, 0, 0)$ and $\epsilon = (0, 0, 0, 0)$.

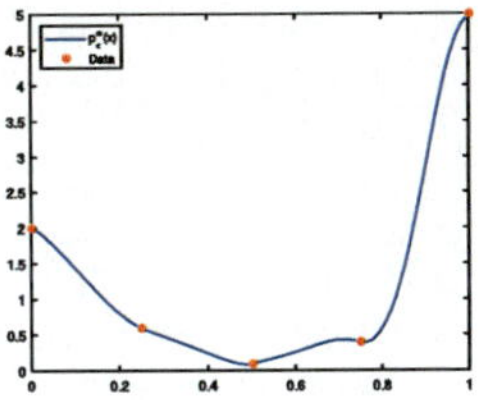

(f) RQS ZFIF with $\alpha = (0, 0, 0, 0)$ and $\epsilon = (1, 0, 0, 1)$.

Fig. 1 Positive RQS ZFIFs

5 Conclusions

We have derived a new type of C^1-rational quartic spline using the binary vector called a signature. For the fixed shape control parameters, we can generate 2^{N-1} different new C^1-rational quartic spline interpolation functions using different signatures for the N numbers of data points. Then, by using the fractal technique we have introduced rational quartic spline zipper fractal interpolation functions. It has been shown that for a data generating function $\Phi \in C^3(I)$, the proposed RQS ZFIF has the order of convergence $O(h^2)$ as $h \to 0$, and it can be increased to the next order of convergence as the classical rational quartic spline defined in [33] under suitable assumptions on the IFS parameters. We have derived sufficient conditions on the shape control parameters and the variable scaling functions so that our RQS ZFIF (consequently, the class of RQS) becomes positive for a given positive data set.

Acknowledgements The second author would like to acknowledge the Science and Engineering Research Board (SERB), Government of India, for the funding support of the research project [Project Number = MTR/2017/000574 - MATRICS].

References

1. Abbas, M., Majid, A.A., Awang, M.N.H., Ali, J.M.: Positivity-preserving C^2 rational cubic spline interpolation. Sci. Asia **39**, 208–213 (2013)
2. Aseev, V.V.: On the regularity of self-similar zippers. In: 6th Russian-Korean International Symposium on Science and Technology, KORUS-2002, 24–30 June 2002, Novosibirsk State Technical University Russia, NGTU, Novosibirsk, Part 3 (Abstracts), p. 167 (2002)
3. Aseev, V.V., Tetenov, A.V.: On self-similar Jordan arcs that admit structural parametrization. Siberian Math. J. **46**(4), 581–592 (2005)
4. Balasubramani, N., Guru Prem Prasad, M., Natesan, S.: Shape preserving α-fractal rational cubic splines. Calcolo **57** (2020). https://doi.org/10.1007/s10092-020-00372-8
5. Barnsley, M.F.: Fractal functions and interpolation. Constr. Approx. **2**(1), 303–329 (1986)
6. Barnsley, M.F.: Fractals Everywhere. Academic Press, Boston (1988)
7. Barnsley, M.F., Harrington, A.N.: The calculus of fractal interpolation functions. J. Approx. Theory **57**(1), 14–34 (1989)
8. Chand, A.K.B., Kapoor, G.P.: Generalized cubic spline fractal interpolation functions. SIAM J. Numer. Anal. **44**(2), 655–676 (2006)
9. Chand, A.K.B., Tyada, K.R.: Constrained shape preserving rational cubic fractal interpolation functions. Rocky Mt. J. Math. **48**(1), 75–105 (2018)
10. Chand, A.K.B., Vijender, N., Navascués, M.A.: Shape preservation of scientific data through rational fractal splines. Calcolo **51**(2), 329–362 (2014)
11. Chand, A.K.B., Vijender, N., Viswanathan, P., Tetenov, A.V.: Affine zipper fractal interpolation functions. BIT Num. Math. **60**, 319–344 (2020)
12. Chand, A.K.B., Viswanathan, P.: A constructive approach to cubic Hermite fractal interpolation function and its constrained aspects. BIT Numer. Math. **53**(4), 841–865 (2013)
13. Duan, Q., Zhang, H., Zhang, Y., Twizell, E.H.: Error estimation of a kind of rational spline. J. Comput. Appl. Math. **2000**, 1–11 (2007)
14. Gowrisankar, A., Guru Prem Prasad, M.: Riemann-Liouville calculus on quadratic fractal interpolation function with variable scaling factors. J. Anal. **27**, 347–363 (2019). https://doi.org/10.1007/s41478-018-0133-2
15. Han, X.L.: Shape-preserving piecewise rational interpolant with quartic numerator and quadratic denominator. Appl. Math. Comput. **251**, 258–274 (2015)
16. Hussain, M.Z., Sarfraz, M.: Positivity-preserving interpolation of positive data by rational cubics. J. Comput. Appl. Math. **218**, 446–458 (2008)
17. Hutchinson, J.: Fractals and self-similarity. Indiana Univ. Math. J. **30**, 713–747 (1981)
18. Katiyar, S.K., Chand, A.K.B., Saravana Kumar, G.: A new class of rational cubic spline fractal interpolation function and its constrained aspects. Appl. Math. Comput. **346**, 319–335 (2019)
19. Mandelbrot, B.: Fractals: Form, Chance and Dimension. W. H. Freeman, San Francisco (1977)
20. Nasim Akhtar, Md., Guru Prem Prasad, M., Navascués, M.A.: More general fractal functions on the sphere. Mediterr. J. Math. (2019). https://doi.org/10.1007/s00009-019-1410-21660-5446/19/060001-18
21. Navascués, M.A.: Fractal polynomial interpolation. Z. Anal. Anwend. **24**(2), 1–20 (2005)
22. Reddy, K.M.: Some aspects of fractal functions in geometric modelling. Ph.D. Thesis, IIT Madras (2018)
23. Sakai, M., Schmidt, J.W.: Positive interpolation with rational spline. BIT Numer. Math. **29**, 140–147 (1989)
24. Samuel, M., Tetenov, A., Vaulin, D.: Self-similar dendrites generated by polygonal systems in the plane. Sib. Élektron. Mat. Izv. **14**, 737–751 (2017)
25. Schmidt, J., Heß, W.: Positive interpolation with rational quadratic splines. Computing **38**, 261–267 (1987)
26. Tetenov, A.V.: On self-similar Jordan arcs on a plane. Sib. Zh. Ind. Mat. **7**(3), 148–155 (2004)
27. Tetenov, A.V.: Self-similar Jordan arcs and graph-directed systems of similarities. Siberian Math. J. **47**(5), 940–949 (2006)

28. Tetenov, A.V., Samuel, M., Vaulin, D.A.: On dendrites defined by polyhedral systems and their ramification points. Tr. Inst. Mat. Mekh. **23**(4), 281–291 (2017)
29. Viswanathan, P., Chand, A.K.B.: α-fractal rational splines for constrained interpolation. Electron. Trans. Numer. Anal. **41**, 420–442 (2014)
30. Viswanathan, P., Chand, A.K.B., Navascués, M.A.: Fractal perturbation preserving fundamental shapes: bounds on the scale factors. J. Math. Anal. Appl. **419**(2), 804–817 (2014)
31. Viswanathan, P., Navascués, M.A., Chand, A.K.B.: Fractal polynomials and maps in approximation of continuous functions. Numer. Funct. Anal. Optim. **37**(1), 106–127 (2016)
32. Wang, H.Y., Shan, Y.J.: Fractal interpolation functions with variable parameters and their analytical properties. J. Approx. Theory **175**, 1–18 (2013)
33. Zhu, Y.: C^2 positivity-preserving rational interpolation splines in one and two dimensions. Appl. Math. Comput. **316**, 186–204 (2018)

Heptic Hermite Collocation on Finite Elements

Zanele Mkhize, Nabendra Parumasur, and Pravin Singh

Abstract We present the solution of linear and nonlinear ordinary differential equations using collocation on finite elements. A heptic (septic) basis is derived and its properties are discussed. The phenomenon of superconvergence at the nodes is illustrated. An investigation of the global and nodal rates of convergence reveals remarkable agreement with a theorem proved by Carl R. de Boor in 1973.

Keywords Heptic collocation · Superconvergence · Differential equations

1 Introduction

Orthogonal Collocation (OC) is an approximation method for solving differential equations. It is similar to the Pseudospectral Method (PS) and is also referred to as the Differential Quadrature Method (DQ). In contrast to finite difference methods, the solution by OC is defined as a continuous or piecewise continuous function.

The collocation method is employed in two different ways, either globally or locally. In the global collocation method, the method finds the solution for various numbers of collocation points. In the local collocation method, the domain is divided into equal-width subintervals called finite elements, and each element has a fixed number of collocation points within its boundaries. The solutions are then computed from the collocation points within each element.

The collocation method was introduced in the 1930s [1–4]. It was named the interpolation method by Kantorovich [1]; Lanczos called it the method of selected points [3] while Frazer et al. called it collocation [2]. From these three names, it can

Z. Mkhize (✉) · N. Parumasur · P. Singh
University of KwaZulu-Natal, Private Bag X54001, Durban 4000, South Africa
e-mail: mkhizez2@ukzn.ac.za

N. Parumasur
e-mail: parumasurn1@ukzn.ac.za

P. Singh
e-mail: singhp@ukzn.ac.za

R. K. Sharma et al. (eds.), *Frontiers in Industrial and Applied Mathematics*,
Springer Proceedings in Mathematics & Statistics 410,
https://doi.org/10.1007/978-981-19-7272-0_38

be inferred that the method interpolates the residual to zero at chosen points. The most attractive feature of the method is that it is easier to implement, since it does not require integration to determine the unknown coefficients.

The collocation might lead to the Runge phenomenon [5] because it is primarily a residual interpolation method. Bert and Malik [6] provided several examples of problems related to collocation with equal intervals.

Lanczos used Chebyshev polynomials for the basis functions and collocated at the zeros of Chebyshev polynomials of the second kind. Wright [7] chose to collocate at the zeros of Chebyshev polynomials of the first kind. The application of Chebyshev roots was a great improvement because the Runge phenomenon does not occur.

Another advancement to the method was the usage of Gaussian or Lobatto quadrature points by Villadsen and Stewart [8]. These are simply the roots of Jacobi polynomials. They referred to this as Orthogonal Collocation. By constructing the method with nodal values they further enhanced it. These adjustments gave rise to finite difference-like methods.

The phenomenon of collocation method in the 1970s happens in three branches, namely Orthogonal Collocation (OC), Pseudospectral (PS), and Differential Quadrature (DQ). Villadsen and Stewart introduced the OC branch in their paper, and further improvements to the method which outline collocation at Gauss, Radau, and Lobatto points were mentioned in Villadsen [8], Finlayson [9], and Villadsen and Michelsen [10]. It was proved that the numerical quadrature of the method of moments is equivalent to collocation at Gauss points.

They further applied the method for problems symmetric about an axis, using cylindrical, spherical, and planar coordinates. They exclusively used the nodal differentiation matrices. Early papers indicated that the method compared favourably with finite differences [11–14].

Orzag was the first to start the Pseudospectral thread [15] which was later improved by Gottlieb and Orzag [16]. Although the pseudospectral method is similar to collocation, it is seldom used to refer to approximations of integration in MWR. Orzag solved periodic problems using trigonometric basis functions. His work includes collocation at the zeros of Chebyshev polynomials of the second kind for non-periodic first-order linear hyperbolic problems. He showed that collocation can accurately approximate the Galerkin method. Here, he used Chebyshev trial functions and did not consider nodal approximations. His main contribution was the application of fast Fourier transforms (FFT) to perform calculations.

The Differential Quadrature Method thread was initially presented in Bellman and Casti [17], Bellman et al. [18], and Bellman [19]. Here Bellman et al. introduced the idea of a nodal differential matrix applied to first- and second-order differential equations. Although the paper does not give much details with respect to boundary condition treatment. The idea to apply a nodal differentiation matrix was not new, it has been applied before. In Bellman et al. [18], Bellman proposed the method of differentiation matrices based on collocation at Gauss points. In Nielson [20], the formulas for the nodal differentiation matrix with arbitrary nodal locations were introduced. The method was adopted for the solution of engineering problems.

When one uses a global polynomial, the solution is represented by a single polynomial on the domain. This approach is fairly accurate when low-order polynomials can represent the solution.

Finite element methods can accurately represent complex geometries. The interest in finite element methods erupted in the 1970s [21–23]. There was a huge interest for other applications because of the initial success in structural mechanics. The idea by Villadsen and Stewart of using quadrature points globally was extended to finite elements.

Unlike the global method, a finite element method divides the domain into a collection of subdomains, with a polynomial representation over each subdomain. The two methods are identical when using a single element hence the finite element method is more general. The degree of continuity at the element boundaries is denoted by C^n.

There are two alternatives to dealing with the continuity conditions at the boundary of the elements. Firstly we could enforce the continuity of the trial functions at the boundary of the elements. This also applies to the continuity of the derivatives depending on the smoothness requirement. Alternatively, we could choose trial functions like the Hermite polynomials which have built-in continuity. The latter approach results in fewer unknowns to solve for. To a large extent the solution of chemical engineering problems, namely two point boundary value problems have been achieved by the Galerkin finite element method [24, 25] with far greater accuracy than the collocation method, though with slightly more numerical effort. For the solution of reaction-diffusion models, see [26, 27].

C^1 Collocation at Gauss Points This was described by de Boor and Swartz [28] and Douglas and Dupont [29]. Carey and Finlayson [30] employed a Lagrange basis.

C^0 Collocation at Lobatto Points This method based on Lobatto points is also used in the finite element approach. One method is C^0, which uses Lagrange basis functions and called the Hybrid-Collocation-Galerkin method [31–33]. Another approach described in Gray [34], Young [35], Young [36], Hennart [37], and Leyk [38, 39] uses a Lagrange basis and a simple Galerkin method with integration effected using Lobatto quadrature. Young called this the Lobatto-Galerkin method. Gray and Hennart only used quadratic trial functions with integration using Simpson's rule. This was referred to as the *hp* Spectral element in Maday and Patera [40], Canuto et al. [41], Karniadakis and Sherwin [42], and Vosse and Minev [43].

Convergence Rate and Efficiency The approximate solution for orthogonal collocation and the finite element methods they approximate have the same convergence and superconvergence rates. Finite element methods and collocation at Gauss points require much less numerical effort than the contrasting Galerkin method when using the same trial functions especially in several dimensions.

2 Heptic Hermite Basis Functions

We seek a basis for $\mathcal{P}_7$, the vector space of polynomials of degree ≤ 7 on the interval $[x_i, x_{i+1}]$. There are eight such functions and we denote them by H_k, $k = 1, 2, ..., 8$. We further stipulate their function and derivative values at the end points x_i and x_{i+1} as follows:

$$H_k^{(p)}(x_i) = \frac{\delta_{k,p+1}}{h^p},\ H_k^{(p)}(x_{i+1}) = 0,\ H_{k+4}^{(p)}(x_i) = 0,\ H_{k+4}^{(p)}(x_{i+1}) = \frac{\delta_{k,p+1}}{h^p}, \tag{1}$$

where k, $p+1 \in S = \{1, 2, 3, 4\}$ and $\delta_{i,j}$ is the well-known Kronecker delta symbol. It is convenient to transform to the variable $z \in [0, 1]$ defined by

$$z = \frac{(x - x_i)}{(x_{i+1} - x_i)} = \frac{(x - x_i)}{h} \tag{2}$$

where h is the uniform interval length. As x varies from x_i to x_{i+1}, z varies from 0 to 1. The interpolatory conditions in (1) transform naturally in the variable z to

$$H_k^{(p)}(0) = \delta_{k,p+1},\ H_k^{(p)}(1) = 0,\ H_{k+4}^{(p)}(0) = 0,\ H_{k+4}^{(p)}(1) = \delta_{k,p+1} \quad k, p+1 \in S.$$

These conditions enable the unique derivation of the $H_k(z)$, $k = 1, 2, ..., 8$. The polynomial $H_3(z)$ has a zero of multiplicity four at $z = 1$ and a zero of multiplicity two at $z = 0$ and therefore has the form of $H_3(z) = (Az + B)z^2(z - 1)^4$. The remaining conditions $H_3''(0) = 1$ and $H_3'''(0) = 0$ are used to evaluate A and B. Using this approach, the polynomials $H_1(z)$, $H_2(z)$, $H_3(z)$, and $H_4(z)$ are derived and displayed in Eqs. (3)–(6).

$$H_1(z) = (20z^3 + 10z^2 + 4z + 1)(z - 1)^4 \tag{3}$$
$$H_2(z) = z(10z^2 + 4z + 1)(z - 1)^4 \tag{4}$$
$$H_3(z) = \frac{z^2}{2}(4z + 1)(z - 1)^4 \tag{5}$$
$$H_4(z) = \frac{z^3}{6}(z - 1)^4. \tag{6}$$

By using symmetry/antisymmetry, one can show that

$$H_5(z) = H_1(1 - z) \tag{7}$$
$$H_6(z) = -H_2(1 - z) \tag{8}$$
$$H_7(z) = H_3(1 - z) \tag{9}$$
$$H_8(z) = -H_4(1 - z). \tag{10}$$

From (7)–(10), we may write

$$H_{j+4}^{(p)}(z) = (-1)^{j-1+p} H_j^{(p)}(1-z), \quad j = 1, 2, 3, 4. \tag{11}$$

The uniqueness of the interpolatory conditions ensures that the polynomials $H_i(z)$ are independent. Consider $p = 0$, if $H_j(z)$ is shifted to the $(i+1)_{st}$ interval the equation of the curve becomes $H_j(z-1)$. When evaluated at $z = 1$ we get $H_j(0)$. Now $H_{j+4}(1) = (-1)^{j-1} H_j(0)$ and for $j = 1, 3$ $H_{j+4}(1) = H_j(0)$, also for $j = 2, 4$ we have $H_{j+4}(1) = -H_j(0) = 0 = H_j(0)$. Similar relationships apply for the derivatives of order up to three. Hence $H_{j+4}(z)$ and its derivatives up to order three are continuous at the element boundary with $H_j(z)$ and its derivatives of order up to three in the $(i+1)_{st}$ interval. If we write $H_5(z)) = H_1(-(z-1))$, then we note that $H_5(z)$ is a reflection of $H_1(z)$ about the vertical axis together with a shift of one unit to the right. $H_7(z)$ is similarly related to $H_3(z)$. Also, $H_6(z)$ may be interpreted as $H_2(z)$ rotated by 180° anticlockwise and then shifted one unit to the right. $H_8(z)$ is also related to $H_4(z)$ in a similar manner.

3 Collocation on Finite Elements

Consider solving a fourth-order linear ordinary differential equation in one spatial variable, x, and on the domain $[a, b]$. Firstly, the domain $[a, b]$ is divided into N subintervals or elements of spacing $h = \frac{b-a}{N}$, by placing the dividing points or nodes $x_i, i = 1, 2, ..., N+1$, as illustrated in Fig. 1. We shall refer to this discretization as the mesh Δ.

Here $x_1 = a$ and $x_{N+1} = b$ coincide with the left and right hand boundaries, respectively. This differs from global orthogonal collocation where the domain is not subdivided and instead higher order polynomials are used to achieve greater accuracy. The i_{th} element $[x_i, x_{i+1}]$ is mapped to [0, 1] by using a transformation of the form (2). We assume that the approximate solution in the i_{th} element is given by

$$U^i(x) = U^i(z) = \sum_{k=1}^{8} C_k^{(i)} H_k^i(z)$$

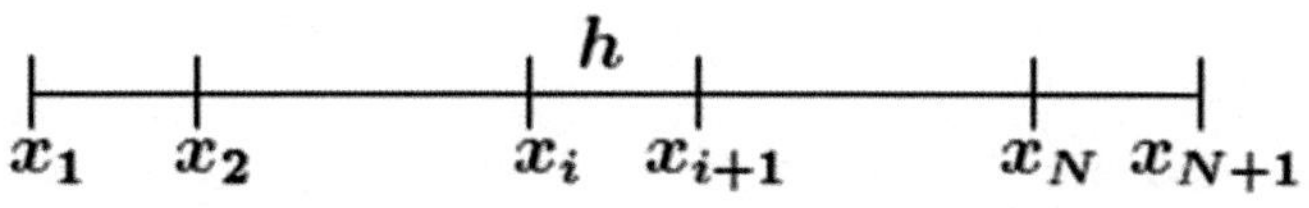

Fig. 1 Mesh points on the global domain

and is represented in the $(i+1)_{st}$ element by

$$U^{i+1}(x) = U^{i+1}(z) = \sum_{k=1}^{8} C_k^{i+1} H_k^{i+1}(z).$$

The continuity of the basis functions and their first three derivatives have some interesting consequences on the coefficients of the solutions in the successive elements. In order to obtain a smooth solution that is C^3 continuous, we enforce the condition

$$U^i(x_{i+1}) = U^{i+1}(x_{i+1}),$$

which is equivalent, in the variable z, to $U^i(1) = U^{i+1}(0)$. This implies that $C_1^{i+1} = C_5^i$. The continuity of the derivative at x_{i+1} is equivalent to

$$\frac{dU^{(i)}}{dz}\bigg|_{z=1} = \frac{dU^{(i+1)}}{dz}\bigg|_{z=0}$$

and this yields $C_2^{i+1} = C_6^i$. Similarly, the continuity of the second derivative at x_{i+1} yields $C_3^{i+1} = C_7^i$ and that of the third derivative yields $C_4^{i+1} = C_8^i$. Thus, the first four coefficients in interval $i+1$ are a repetition of the last four coefficients in interval i. Thus, we can write the trial solution as

$$U(z) = \sum_{k=1}^{8} C_{k+4(i-1)} H_k(z), \tag{12}$$

where we write $H_k(z)$ for $H_k^i(z)$ bearing in mind that $H_k(z)$ is a function of i and we have dropped the superscript i from $U^i(z)$. With this labelling of the coefficients, we are automatically ensuring that the solution and its first, second, and third derivatives are continuous at the nodes.

Remark 1 Substituting $z = 0$ and $z = 1$ into (12), its derivative, its second and its third derivative, we can show that $U(x_i) = C_{4i-3}$, $hU'(x_i) = C_{4i-2}$, $h^2U''(x_i) = C_{4i-1}$, and $h^3U'''(x_i) = C_{4i}$, $i = 1, 2, ..., N+1$. Thus, every fourth coefficient beginning from C_1 is an approximation to the solution at the nodes. Similarly, every fourth coefficient beginning from C_2 scaled by h is an approximation to the derivative at the nodes. Likewise, every fourth coefficient beginning from C_3 scaled by h^2 represents an approximation to the second derivative at the nodes, and every fourth coefficient beginning from C_4 scaled by h^3 represents an approximation to the third derivative at the nodes.

We find it more instructive to apply the error bounds derived in [28] and to illustrate the numerical validity of the bounds in the present context on two examples. Consider the fourth-order linear differential equation, defined on $[a, b]$, which can be written in the form $Lu(x) = f(x)$, where the operator $L = \sum_{k=0}^{4} a_k(x)D^k$ and D denotes the derivative operator.

The following theorem establishes the order of convergence for very smooth solutions [28].

Theorem 1 ([28]) *Assume that the coefficients $a_i(x)$ of L satisfy $a_i(x) \in C^8[a,b]$ for all i and that $u(x) \in C^{12}[a,b]$. If the collocation points are chosen as the Gauss points, then there exists a constant c_1 such that*

$$|D^p(u-U)(x_j)| \leqslant c_1 h^8, \quad p = 0, 1, 2, 3 \tag{13}$$

and a constant c_2 such that

$$\|D^p(u-U)\|_\infty \leqslant c_2 h^{8-p}, \quad p = 0, 1, 2, 3, 4. \tag{14}$$

Here, $U(x)$ represents the collocation approximation of $u(x)$. Similar error bounds hold for nonlinear ODEs [28] and will be illustrated with an example below.

This effectively means that at the nodes the error of the collocation solution and its derivatives of order up to three should be $O(h^8)$. Also, the infinity norm of the error and its derivatives of order up to four should be $O(h^{8-p})$.

4 Numerical Example

Example 1 Consider the fourth-order ODE

$$u^{(iv)} - (10\pi)^3 u = (10\pi)^3(10\pi - 1)\sin(10\pi x) = f(x) \tag{15}$$

with analytical solution $u(x) = \sin(10\pi x)$ and boundary conditions $u(0) = 0 = u(1)$ and $u'(0) = 10\pi = u'(1)$

We substitute the trial solution (12) into the differential equation (15) to obtain

$$\sum_{k=1}^{8}\left[H_k^{(iv)}(z)/h^4 - (10\pi)^3 H_k(z)\right] C_{k+4(i-1)} = f(x_i + zh), \; i = 1, 2, ..., N. \tag{16}$$

The boundary condition $u(0) = U(0) = \sum_{k=1}^{8} C_k H_k(0) = 0$ yields $C_1 = 0$ while the boundary condition $u(1) = U(1) = \sum_{k=1}^{8} C_{k+4(N-1)} H_k(1) = 0$ yields $C_{4N+1} = 0$. The boundary condition $u'(0) = U'(0) = \frac{1}{h}\sum_{k=1}^{8} C_k H_k'(0) = 10\pi$ yields $C_2 = 10\pi h$. Similarly, $u'(1) = U'(1) = \frac{1}{h}\sum_{k=1}^{8} C_{k+4(N-1)} H_k'(1) = 10\pi$ yields $C_{4N+2} = 10\pi h$.

There are $4N + 4$ unknowns in Eq. (12). Given that we have two boundary conditions on the left and two boundary conditions on the right, we thus require $4N$ conditions in order to solve the problem uniquely. We choose four collocation points denoted by s_1, s_2, s_3, s_4, in each interval. The s_j are chosen as the zeros of the fourth degree Legendre polynomial, shifted to the interval [0, 1]. These have been shown to be the optimal choice for the collocation points [28]. This optimal choice arises due to the orthogonality property of the Legendre polynomials and hence the method is called orthogonal collocation on finite elements (OCFE). The collocation points are then substituted into Eq. (16) to give the remaining $4N$ linear equations. The matrix-vector system, of size $(4N+4) \times (4N+4)$, has the form $A\boldsymbol{a} = \boldsymbol{f}$, where A has the form

$$A = \begin{bmatrix}
a_{11} & a_{12} & a_{13} & a_{14} & a_{15} & a_{16} & a_{17} & a_{18} & & & & & & & \\
a_{21} & a_{22} & a_{23} & a_{24} & a_{25} & a_{26} & a_{27} & a_{28} & & & & & & & \\
a_{31} & a_{32} & a_{33} & a_{34} & a_{35} & a_{36} & a_{37} & a_{38} & & & & & & & \\
a_{41} & a_{42} & a_{43} & a_{44} & a_{45} & a_{46} & a_{47} & a_{48} & & & & & & & \\
 & & & & a_{11} & a_{12} & a_{13} & a_{14} & a_{15} & a_{16} & a_{17} & a_{18} & & & \\
 & & & & a_{21} & a_{22} & a_{23} & a_{24} & a_{25} & a_{26} & a_{27} & a_{28} & & & \\
 & & & & a_{31} & a_{32} & a_{33} & a_{34} & a_{35} & a_{36} & a_{37} & a_{38} & & & \\
 & & & & a_{41} & a_{42} & a_{43} & a_{44} & a_{45} & a_{46} & a_{47} & a_{48} & & & \\
 & & & & & & & \ddots & & \ddots & & & & & \\
 & & & & & & & a_{11} & a_{12} & a_{13} & a_{14} & a_{15} & a_{16} & a_{17} & a_{18} \\
 & & & & & & & a_{21} & a_{22} & a_{23} & a_{24} & a_{25} & a_{26} & a_{27} & a_{28} \\
 & & & & & & & a_{31} & a_{32} & a_{33} & a_{34} & a_{35} & a_{36} & a_{37} & a_{38} \\
 & & & & & & & a_{41} & a_{42} & a_{43} & a_{44} & a_{45} & a_{46} & a_{47} & a_{48} \\
1 & & & & & & & & & & & & & & \\
 & 1 & & & & & & & & & & & & & \\
 & & & & & & & & & & & & 1 & & \\
 & & & & & & & & & & & & & 1 &
\end{bmatrix} \tag{17}$$

with

$$a_{ij} = \frac{1}{h^4} H_j^{(iv)}(s_i) + (10\pi)^3 H_j(s_i), \; i = 1, 2, 3, 4; \; j = 1, 2, ..., 8, \tag{18}$$

and $\boldsymbol{f}$ has the form

$$f_{j+4i-4} = \begin{cases} (10\pi)^3(10\pi - 1)\sin(10\pi(x_i + s_j h)), & i = 1, ..., N, \; j = 1, 2, 3, 4 \\ 0, & i = N+1, \; j = 1, 3 \\ 10\pi h, & i = N+1, \; j = 2, 4. \end{cases}$$

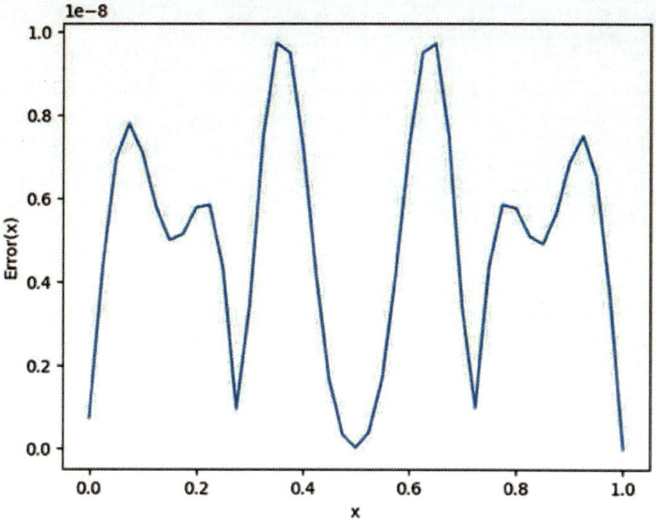

Fig. 2 Error plot with $N = 20$ for example 2.1

The non-zero blocks of matrix A are shifted four places to the right and account for the repetition of the coefficients. The position of the ones accounts for the boundary conditions. The sparse nature of the matrix and the repetitive pattern can easily be exploited to solve the linear system efficiently with minimum CPU storage requirements.

After solving (17), the solution is constructed on each subinterval using the appropriate coefficients and can then be plotted.

Since there is very good agreement between the approximate solution and the exact solution, we choose to show the error plot in Fig. 2 for $N = 20$. We point out that in contrast to global collocation the numerical results are much more acceptable.

We use the following technique to approximate the convergence order. If the discrete error at the nodes x_j is $O(h^n)$ then

$$\left|D^p(u-U)(x_j)\right|^{(h)} = O\left(h^n\right) \tag{19}$$

and

$$\left|D^p(u-U)(x_j)\right|^{(\frac{h}{2})} = O\left(\left(\frac{h}{2}\right)^n\right). \tag{20}$$

By taking the ratio of (19)–(20), we obtain

$$\alpha_1 = \frac{\left|D^p(u-U)(x_j)\right|^{(h)}}{\left|D^p(u-U)(x_j)\right|^{(\frac{h}{2})}} \approx 2^n \tag{21}$$

Table 1 Convergence order $n(h)$ at nodes from (21)

x_i	$p=0$	$p=1$	$p=2$	$p=3$	x_i	$p=0$	$p=1$	$p=2$	$p=3$
0.05	8.3581	8.2630	8.3306	8.2909	0.55	8.7506	8.2487	8.7841	8.2493
0.10	8.3486	8.1682	8.3474	8.2273	0.60	8.3179	8.2433	8.3184	8.8037
0.15	8.3358	9.0963	8. 3618	8.0691	0.65	8.2625	8.4625	8.2502	8.4620
0.20	8.2448	8.2739	8.2377	8.3273	0.70	8.1744	8.7793	8.1824	8.4068
0.25	8.1592	8.3138	8.1167	8.3128	0.75	8.3788	8.2916	8.3458	8.2867
0.30	8.4366	8.5032	8.4341	8.2856	0.80	8.3677	8.2011	8.3593	8.2233
0.35	8.3041	8.1579	8.2915	8.1548	0.85	8.3613	7.8613	8.3673	6.4756
0.40	8.2843	7.9236	8.2846	7.5945	0.90	8.2783	8.2569	8.2282	8.3518
0.45	8.1588	8.3554	8.2455	8.3554	0.95	8.2408	8.2857	8.0558	8.3496

Table 2 Global convergence orders from (22)

p	0	1	2	3	4
$n(h)$	7.9596	6.8907	5.6224	4.6626	3.7390

from which the order of convergence $n(h) \approx \frac{\ln(\alpha_1)}{\ln(2)}$. Similarly, we obtain

$$\alpha_2 = \frac{\| D^p(u-U)(x) \|_\infty^{(h)}}{\| D^p(u-U)(x) \|_\infty^{(\frac{h}{2})}} \approx 2^n. \tag{22}$$

These results are summarized in Tables 1 and 2. It is seen that the nodal order is approximately 8, while the global order seems to satisfy (14). The error in the global convergence order is attributed to the conditioning of the matrix for this problem as well as the low value of N used. The pointwise error in the domain is least and of order 8 only at the nodes, a phenomenon known as superconvergence.

Example 2 As a second example, we solve a nonlinear BVP.

$$u^{(iv)}(x) + u'''(x) + u''(x) + u(x)u'(x) = f(x), \quad -2 < x < 2, \tag{23}$$

with exact solution $u(x) = e^{-x^2}$.

The right-hand side $f(x)$ and boundary conditions are extracted from the exact solution. Clearly, the exact solution $u(x)$ satisfies the hypothesis of Theorem (1) and therefore we expect nodal and global errors of $O(h^8)$. If the global error $\| D^p(u-U)(x) \|_\infty^N$ is $O\left(h^{-n}\right)$ then

$$\alpha_3 = \frac{\| D^p(u-U)(x) \|_\infty^{N}}{\| D^p(u-U)(x) \|_\infty^{N+1}} \approx \left(\frac{N+1}{N}\right)^n \tag{24}$$

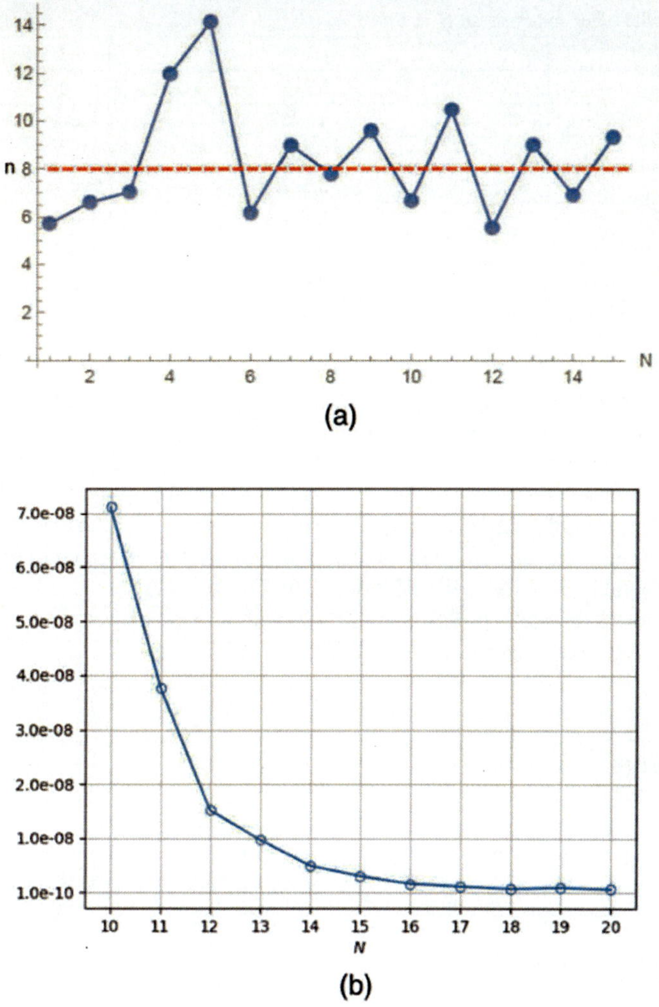

Fig. 3 **a** Global error order. **b** Global error

and the global convergence order is given by

$$n \approx \frac{\ln \alpha_3}{\ln\left(\frac{N+1}{N}\right)}. \tag{25}$$

We use Eq. (25) to estimate this order as it is computationally inefficient to use (22) in this case. For a nonlinear problem, the nonlinear solver consumes much CPU time as the number of equations increases. For example for $N = 10$, if we had used

Table 3 Nodal order and error for $N = 8$

i	x_i	n	$\|(u - U)(x_i)\|$
2	−1.5	8.218	9.73e-8
3	−1.0	11.257	3.75e-8
4	−0.5	8.463	1.32e-7
5	0.0	8.523	2.59e-7
6	0.5	8.929	2.85e-8
7	1.0	7.995	7.21e-8
8	1.5	8.671	6.68e-8

(22) then this will require solving additionally 84 ($N = 20$) nonlinear equations as compared to 48 ($N = 11$) nonlinear equations.

In Fig. 3a, we plot the global order ($p = 0$) as a function of N for small values of N. These orders seem to oscillate about the horizontal red line ($N = 8$). Those below the line are attributed to numerical errors arising from the Julia nonlinear solver nlsolve. For larger values of N, the actual global errors are illustrated in Fig. 3b and agree remarkably with the theoretical bound of Theorem (1).

For ($N = 8$) the nodal orders using (21) as well as the nodal errors are tabulated in Table 3. Again this reinforces the validity of Theorem (1).

References

1. Kantorovich L.V.: On an approximation method for the solution of a partial differential equation. Dokl. Akad. Nauk SSSR. **2**, 532–536 (1934). arXiv:1904.04685
2. Frazer, R.A., Jones, W.P., Skan, S.W.: Approximation to Functions and to the Solutions of Differential Equations. Great Britain Aero. Res. Council, London, Report and Memo No. 1799 (1937)
3. Lanczos, C.: Trigonometric interpolation of empirical and analytical functions. J. Math. Phys. **17**, 123–199 (1938). https://doi.org/10.1002/sapm1938171123
4. Slater, J.C.: Electronic energy bands in metal. Phys. Rev. **45**, 794–801 (1934). http://orcid.org/10.1103/PhysRev.45.794
5. Runge, C.: Uber Empirische Functionen und die Interpolation Zwischen Aequidistanten Ordinaten. Zeitschrift fur Math. und Physik. **46**, 224–243 (1901)
6. Bert, C.W., Malik, M.: Differential quadrature method in computational mechanics: a review. Appl. Mech. Rev. **49**, 1–28 (1996). http://orcid.org/10.1115/1.3101882
7. Wright, K.: Chebyshev collocation methods for ordinary differential equations. Comp. J. **6**, 358–365 (1964). http://orcid.org/10.1093/comjnl/6.4.358
8. Villadsen, J.: Selected Approximation Methods for Chemical Engineering Problems. Inst. for Kemiteknik Numer. Inst, Danmarks Tekniske Hojskole (1970)
9. Finlayson, B.A.: The Method of Weighted Residuals and Variational Principles. Academic Press, New York, NY (1972). http://orcid.org/10.1137/1.9781611973242
10. Villadsen, J.V., Michelsen, M.L.: Solution of Differential Equation Models by Polynomial Approximation. Prentice-Hall, Englewood Cliffs, NJ (1978)
11. Michelsen, M.L., Villadsen, J.V.: A convenient computational procedure for collocation constants. Chem. Eng. J. **4**, 64–68 (1972). http://orcid.org/10.1016/0300-9467(72)80054-6

12. Michelsen, M.L., Villadsen, J.V.: Polynomial solution of differential equations. In: Mah, R.S.H., Seider, W.D. (eds.) Proceedings of an International Conference on Foundations of Computer-Aided Chemical Process Design, pp. 341–368 (1981)
13. Finlayson, B.A.: Orthogonal collocation in chemical reaction engineering. Cat. Rev. Sci-Eng. **10**, 69–138 (1974). http://orcid.org/10.1080/01614947408079627
14. Finlayson, B.A.: Nonlinear Analysis in Chemical Engineering. Ravenna Park Publishing, Seattle (2003)
15. Orzag, S.A.: Comparison of Pseudo Spectral and Spectral Approximation. Studies in Applied Mathematics, vol. 51, pp. 253–259 (1972). http://orcid.org/10.1002/sapm1972513253
16. Gottlieb, D., Orzag, S.A.: Numerical analysis of spectral methods: theory and applications. SIAM, Philadelphia, PA (1977). http://orcid.org/10.1137/1.9781611970425
17. Bellman, R., Casti, J.: Differential quadrature and long-term integration. J. Math. Anal. Appl. **134**, 235–238 (1971)
18. Bellman, R., Kashef, B.G., Casti, J.: Differential quadrature: a technique for the rapid solution of nonlinear partial differential equations. J. Comp. Phys. **10**, 40–52 (1972). https://doi.org/10.1016/0021-9991(72)90089-7
19. Bellman, R.: Methods of Nonlinear Analysis, vol. 2. Academic Press, New York, (1973)
20. Nielson, K.L.: Methods in Numerical Analysis. MacMillan, NY (1956)
21. Zienkiewicz, O.C.: The Finite Element Method in Engineering Science. McGraw-Hill (1971)
22. Strang, G., Fix, G.J.: An Analysis of the Finite Element Method. Prentice-Hall (1973)
23. Hughes, T.J.R.: The Finite Element Method: Linear Static and Dynamic Finite Element Analysis. Prentice-Hall (1987)
24. Sharma, S., Jiwari, R., Kumar, S.: Numerical solutions of two point boundary value problems using Galerkin-Finite element method. Int J Nonlinear Sci. **13**(2), 204–210 (2012)
25. Sharma, D., Jiwari, R., Kumar, S.: A comparative study of Modal matrix and finite elements methods for two point boundary value problems. Int. J. Appl. Math. Mech. **8**(13), 29–45 (2012)
26. Yadav, O.P., Jiwari, R.: Finite element approach to capture Turing patterns of autocatalytic Brusselator model. J. Math. Chem. **57**(3), 769–789 (2019)
27. Yadav, O.P., Jiwari, R.: Finite element approach for analysis and computational modelling of coupled reaction diffusion models. Numer. Methods Partial Differ. Equ. **35**(2), 830–850 (2019)
28. de Boor C., Swartz B.: Collocation at gaussian points. In: Society for Industrial and Applied Mathematics, vol. 10, pp. 582–606 (1973). http://orcid.org/10.1137/0710052. (SIAM J. Numer. Anal.)
29. Douglas Jr., J., Dupont, T.: A finite element collocation method for quasilinear parabolic equations. Math. Comp. **27**, 17–28 (1973). http://orcid.org/10.2307/2005243
30. Carey, G., Finlayson, B.A.: Orthogonal collocation on finite elements. Chem. Eng. Sci. **30**, 587–596 (1975). http://orcid.org/10.1016/0009-2509(75)80031-5
31. Diaz, J.: A hybrid collocation-Galerkin method for two-point boundary value problems using continuous piecewise polynomial spaces. Ph.D. Thesis, Rice University (1975). https://hdl.handle.net/1911/15125
32. Dunn, R., Wheeler, M.F.: Some collocation-Galerkin methods for two-point boundary value problems. SIAM J. Numer. Anal. **13**(5), 720–733 (1976)
33. Wheeler, M.F.: A C^0-collocation-finite element method for two-point boundary value and one space dimension parabolic problems. SIAM J. Numer. Anal. **14**(1), 71–90 (1977). http://orcid.org/10.1137/0714005
34. Gray, W.G.: An Efficient Finite Element Scheme for Two-Dimensional Surface Water Computations. Finite Elements in Water Resources. In: Gray, W.G., Pinder, G.F., Brebbia, C.A. (eds.). Pentech Press, London (1977)
35. Young, L.C.: A preliminary comparison of finite element methods for reservoir simulation. In: Vichnevetsky, R. (ed.) Advances in Computer Methods for Partial Differential Equations-II. IMACS(AICA). vol. 2, pp. 307–320. Rutgers U., New Brunswick, N.J. (1977)
36. Young, L.C.: A finite-element method for reservoir simulation. Soc. Petr. Eng. J. **21**(1), 115–128 (1981). http://orcid.org/10.2118/7413-PA

37. Hennart, J.P.: Topics in Finite Element Discretization of Parabolic Evolution Problems. Lecture Notes in Math, vol. 909. Springer, Berlin, Heidelberg (1982)
38. Leyk, Z.: A C^0-collocation-like method for boundary value problems. Numerische Mathametik, **49**, 39–54 (1986). http://eudml.org/doc/133097
39. Leyk, Z.: A C^0-collocation-like method for elliptic equations on rectangular regions. J. Austral. Math. Soc. Ser. B **38**, 368–387 (1997). http://orcid.org/10.1017/S0334270000000734
40. Maday, Y., Patera, A.T.: Spectral element methods for the incompressible Navier-Stokes equations. In: Noor, A.K. (ed.) State-of-the-Art Surveys on Computational Mechanics. ASME, New York (1989)
41. Canuto, C., Hussaini, M., Quarteroni, A., Zang, T., Jr.: Spectral Methods Evolution to Complex Geometries and Applications to Fluid Dynamics. Springer, Berlin (2007)
42. Karniadakis, G., Sherwin, S.: Spectral/hp Element Methods for Computational Fluid Dynamics: Second Edition (Numerical Mathematics and Scientific Computation). Oxford University Press (2013)
43. Vosse, van de, F.N., Minev, P.D.: Spectral elements methods: theory and applications. EUT Report 96-W-001, Eindhoven University of Technology (1996)

A Computationally Efficient Sixth-Order Method for Nonlinear Models

Janak Raj Sharma and Harmandeep Singh

Abstract The aim of the present study is to develop an iterative scheme of high convergence order with minimal computational cost. With this objective, a three-step method has been designed by utilizing only two Jacobian matrices, single matrix inversion, and three function evaluations. Under some standard assumptions, the proposed method is found to possess the sixth order of convergence. The iterative schemes with these characteristics are hardly found in the literature. The analysis is carried out to assess the computational efficiency of the proposed method, and further, outcomes are compared with the efficiencies of existing ones. In addition, numerical experiments are performed by applying the method to some practical nonlinear problems. The entire analysis remarkably favors the new technique compared with existing counterparts in terms of computational efficiency, stability, and CPU time elapsed during execution.

Keywords Nonlinear systems · Iterative techniques · Convergence order · Computational efficiency

1 Introduction

The systems of nonlinear equations arise by virtue of modeling the most of the physical processes or practical situations. The constructed models are generally expressed in mathematical form as

$$F(x) = O, \tag{1}$$

J. R. Sharma · H. Singh (✉)
Department of Mathematics, Sant Longowal Institute of Engineering and Technology, Longowal, Punjab 148106, India
e-mail: harman85pau@gmail.com

R. K. Sharma et al. (eds.), *Frontiers in Industrial and Applied Mathematics*, Springer Proceedings in Mathematics & Statistics 410,
https://doi.org/10.1007/978-981-19-7272-0_39

where $O \in \mathbb{R}^m$ represents the zero vector, $F : \Omega \subseteq \mathbb{R}^m \rightarrow \mathbb{R}^m$ is a nonlinear mapping which is commonly represented as $(f_1(x), f_2(x), \ldots, f_m(x))^T$, $x = (x_1, \ldots, x_m)^T \in \Omega$, and $f_i : \mathbb{R}^m \rightarrow \mathbb{R}$ $(i = 1, \ldots, m)$ are nonlinear scalar functions.

Knowledge about the solution of the constructed nonlinear model plays an important role in forecasting the future developments of the corresponding physical problem. But, as a matter of fact, obtaining the analytical solutions of nonlinear systems is generally not feasible. To deal with this challenge, iterative methods [8, 13] offer the numerical solution up to the desired precision. The working process of an iterative method is based on the fixed point iteration theory, under which it locates the solution, $x^* \in \Omega$, of the given system (1), as a fixed point of a mapping $\phi : \mathbb{R}^m \rightarrow \mathbb{R}^m$, so that

$$x^{(k+1)} = \phi(x^{(k)}), \quad k = 0, 1, 2, \ldots,$$

where, $x^{(0)}$ is the initial estimate to the solution, and the mapping ϕ is constrained to satisfy some prescribed assumptions.

The most widely applied iterative procedure to find the solution to nonlinear equations is Newton's method

$$x^{(k+1)} = \phi(x^{(k)}) = x^{(k)} - F'(x^{(k)})^{-1} F(x^{(k)}), \quad k = 0, 1, 2, \ldots, \tag{2}$$

where $F(x)$ is continuously differentiable in some neighborhood of its solution, and $F'(x) \in \mathcal{L}(\mathbb{R}^m, \mathbb{R}^m)$ is a linear operator which is generally represented as a Jacobian matrix $\left[\frac{\partial f_i}{\partial x_j}\right]_{m \times m}$. This method approximates the simple solution of (1) with the quadratic rate of convergence. To improve the convergence rate of the method (2), numerous iterative schemes have been presented in the literature (see [2, 4–6, 10–12, 14] and references therein). As it is evident that Newton's scheme utilizes evaluation of a function (F), a Jacobian matrix (F'), and a matrix inversion (F'^{-1}) per iteration. An attempt to increase the rate of convergence of an iterative method generally leads to a technique that involves one or more additional evaluations per iteration than its predecessor. For instance, the Potra and Pták method [9], having cubic convergence, is one of the simplest improvements of the method (2), which is expressed as follows:

$$\begin{aligned} y^{(k)} &= x^{(k)} - F'(x^{(k)})^{-1} F(x^{(k)}), \\ x^{(k+1)} &= y^{(k)} - F'(x^{(k)})^{-1} F(y^{(k)}). \end{aligned} \tag{3}$$

Clearly, the above-presented two-step scheme utilizes an additional function evaluation over Newton's method.

The practice of designing an iterative scheme, by utilizing additional evaluations, accelerates the convergence order but it certainly increases the computational cost per iteration in terms of mathematical operations. Optimizing the computational cost with the improving convergence speed leads to the construction of computationally efficient techniques. The measure of efficiency is formulated in [8, 13] to analyze and further compare the efficiencies of iterative techniques. In addition, the necessary parameters have been introduced in [11] for the thorough investigation of this concept.

Taking into account the above discussion, in the next section, we shall present a simple and efficient iterative method showing the sixth order of convergence. The computational efficiency of the developed method is determined, analyzed, and compared with the efficiencies of existing methods in Sect. 3. Numerical performance is investigated in Sect. 4, and concluding remarks are given in Sect. 5.

2 Development of Method

The primary objective here is to design an iterative scheme that improves the convergence speed of the Potra and Pták method (3) without utilizing any additional inverse operator. In what follows, we shall present a three-step iterative method involving undetermined parameters, which are to be chosen in order to maximize the convergence order. In view of this, we consider the iterative scheme of type,

$$\begin{aligned} y^{(k)} &= x^{(k)} - F'(x^{(k)})^{-1}F(x^{(k)}),\\ z^{(k)} &= y^{(k)} - F'(x^{(k)})^{-1}F(y^{(k)}),\\ x^{(k+1)} &= z^{(k)} - \left[aI + F'(x^{(k)})^{-1}F'(y^{(k)})(bI + cF'(x^{(k)})^{-1}F'(y^{(k)}))\right]\\ &\qquad \times F'(x^{(k)})^{-1}F(z^{(k)}) \end{aligned} \tag{4}$$

where a, b, and c are the parameters.

Before proceeding to the convergence analysis, a preliminary result (see [7]) is stated below, which will be followed by the main theorem to show the sixth-order convergence for scheme (4).

Lemma 1 *Assume that the mapping $F : \Omega \subseteq \mathbb{R}^m \to \mathbb{R}^m$ is n-times Fréchet differentiable in a convex neighborhood $\Omega \in \mathbb{R}^m$, and let $x,\ t \in \Omega$, then the following expansion holds:*

$$F(x+t) = F(x) + F'(x)t + \frac{1}{2!}F''(x)t^2 + \ldots + \frac{1}{(n-1)!}F^{(n-1)}(x)t^{n-1} + R_n,$$

where $t^i = (t, \overset{i-times}{\ldots\ldots}, t)$, $F^{(i)}(x) \in \mathcal{L}(\mathbb{R}^m \times \overset{i-times}{\ldots\ldots} \times \mathbb{R}^m, \mathbb{R}^m)$ *for each* $i = 1, 2, \ldots,$ *and*

$$\|R_n\| \leq \frac{1}{n!} \sup_{0<h<1} \|F^{(n)}(x+ht)\| \, \|t\|^n.$$

Theorem 1 *Assume that a nonlinear mapping, $F : \Omega \subseteq \mathbb{R}^m \to \mathbb{R}^m$, is continuously differentiable sufficient number of times in some neighborhood of its simple zero x^*, contained in an open convex region Ω. Further, suppose that $F'(x)$ is non-singular and continuous in that neighborhood, and the initial approximation $x^{(0)}$ is sufficiently close to x^*. Then, the sequence of iterates generated by the method (4) converges to x^* with the sixth order of convergence, provided $a = \frac{7}{2}$, $b = -4$, and $c = \frac{3}{2}$.*

Proof Let $e^{(k)} = x^{(k)} - x^*$ be the error obtained at the kth iteration of (4). Then, as a consequence of Lemma 1, and the fact that $F(x^*) = O$, Taylor expansions of $F(x^{(k)})$ and $F'(x^{(k)})$, about x^*, are developed as

$$F(x^{(k)}) = F'(x^*)[e^{(k)} + A_2 e^{(k)^2} + A_3 e^{(k)^3} + A_4 e^{(k)^4} + A_5 e^{(k)^5} + A_6 e^{(k)^6}] + O(e^{(k)^7}), \quad (5)$$

$$F'(x^{(k)}) = F'(x^*)[I + 2A_2 e^{(k)} + 3A_3 e^{(k)^2} + 4A_4 e^{(k)^3} + 5A_5 e^{(k)^4} + 6A_6 e^{(k)^5}] + O(e^{(k)^6}), \quad (6)$$

where $e^{(k)^i} = (e^{(k)}, \overset{i-times}{\ldots\ldots}, e^{(k)})$, and $A_i = \frac{1}{i!} F'(x^*)^{-1} F^{(i)}(x^*), \quad i = 2, 3, \ldots,$ and consequently,

$$F'(x^{(k)})^{-1} = [I + B_1 e^{(k)} + B_2 e^{(k)^2} + B_3 e^{(k)^3} + B_4 e^{(k)^4} + B_5 e^{(k)^5}] F'(x^*)^{-1} + O(e^{(k)^6}), \quad (7)$$

where $B_1 = -2A_2$, $B_2 = -3A_3 + 4A_2^2$, $B_3 = -4A_4 + 6A_2A_3 + 6A_3A_2 - 8A_2^3$, $B_4 = -5A_5 + 8A_2A_4 + 9A_3^2 + 8A_4A_2 - 12A_2^2A_3 - 12A_2A_3A_2 - 12A_3A_2^2 + 16A_2^4$, and $B_5 = -6A_6 + 10A_2A_5 + 12A_3A_4 + 12A_4A_3 + 10A_5A_2 - 16A_2^2A_4 - 18A_2A_3^2 - 16A_2A_4A_2 - 18A_3A_2A_3 - 18A_3^2A_2 - 16A_4A_2^2 + 24A_2^3A_3 + 24A_2^2A_3A_2 + 24A_2A_3A_2^2 + 24A_3A_2^3 - 32A_2^5$.

Denoting $e_y^{(k)} = y^{(k)} - x^*$ as the error at the first step of method (4), and using Eqs. (5)–(7), we have that

$$e_y^{(k)} = C_1 e^{(k)^2} + C_2 e^{(k)^3} + C_3 e^{(k)^4} + C_4 e^{(k)^5} + C_5 e^{(k)^6} + O(e^{(k)^7}), \quad (8)$$

where $C_1 = A_2$, $C_2 = 2(A_3 - A_2^2)$, $C_3 = 3A_4 - 4A_2A_3 - 3A_3A_2 + 4A_2^3$, $C_4 = 4A_5 - 6A_2 A_4 - 6A_3^2 - 4A_4A_2 + 8A_2^2A_3 + 6A_2A_3A_2 + 6A_3A_2^2 - 8A_2^4$, and $C_5 = 5A_6 - 8A_2A_5 9A_3A_4 - 8A_4A_3 - 5A_5A_2 + 12A_2^2A_4 + 12A_2A_3^2 + 8A_2A_4A_2 + 12A_3A_2A_3 + 9A_3^2A_2 + 8A_4A_2^2 - 16A_2^3 A_3 - 12A_2^2A_3A_2 - 12A_2A_3A_2^2 - 12A_3A_2^3 + 16A_2^5$.

Using the expression (8), Taylor developments of $F(y^{(k)})$ and $F'(y^{(k)})$, about x^*, is given by

$$F(y^{(k)}) = F'(x^*)[K_1 e^{(k)^2} + K_2 e^{(k)^3} + K_3 e^{(k)^4} + K_4 e^{(k)^5} + K_5 e^{(k)^6}] + O(e^{(k)^7}), \quad (9)$$

$$F'(y^{(k)}) = F'(x^*)[I + L_1 e^{(k)^2} + L_2 e^{(k)^3} + L_3 e^{(k)^4} + L_4 e^{(k)^5}] + O(e^{(k)^6}), \quad (10)$$

where $K_1 = A_2$, $K_2 = 2(A_3 - A_2^2)$, $K_3 = 3A_4 - 4A_2A_3 - 3A_3A_2 + 5A_2^3$, $K_4 = 4A_5 - 6A_2A_4 - 6A_3^2 - 4A_4A_2 + 10A_2^2A_3 + 8A_2A_3A_2 + 6A_3A_2^2 - 12A_2^4$, $K_5 = 5A_6 - 8A_2A_5 - 9A_3A_4 - 8A_4 A_3 - 5A_5A_2 + 15A_2^2A_4 + 16A_2A_3^2 + 11A_2A_4A_2 + 12A_3A_2A_3 + 9A_3^2A_2 + 8A_4A_2^2 - 24A_2^3A_3 - 19A_2^2A_3A_2 - 19A_2A_3A_2^2 - 11A_3A_2^3 + 28A_2^5$, $L_1 = 2A_2^2$, $L_2 = 4(A_2A_3 - A_2^3)$, $L_3 = 6A_2A_4 - 8A_2^2A_3 - 6A_2A_3A_2 + 3A_3A_2^2 + 8A_2^4$, and $L_4 = 8A_2A_5 - 12A_2^2A_4 - 12A_2A_3^2 - 8A_2A_4A_2 + 6A_3A_2A_3 + 6A_3^2A_2 + 16A_2^3A_3 + 12A_2^2A_3A_2 + 12A_2A_3A_2^2 - 12A_3A_2^3 - 16A_2^5$.

Let us denote $e_z^{(k)} = z^{(k)} - x^*$, then using Eqs. (7)–(9), the second step of method (4) yields

$$e_z^{(k)} = M_1 e^{(k)^3} + M_2 e^{(k)^4} + M_3 e^{(k)^5} + M_4 e^{(k)^6} + O(e^{(k)^7}), \tag{11}$$

where $M_1 = 2A_2^2$, $M_2 = 4A_2A_3 + 3A_3A_2 - 9A_2^3$, $M_3 = 6A_2A_4 + 6A_3^2 + 4A_4A_2 - 18A_2^2A_3 - 14A_2A_3A_2 - 12A_3A_2^2 + 30A_2^4$, and $M_4 = 8A_2A_5 + 9A_3A_4 + 8A_4A_3 + 5A_5A_2 - 27A_2^2A_4 - 28A_2A_3^2 - 19A_2A_4A_2 - 24A_3A_2A_3 - 18A_3^2A_2 - 16A_4A_2^2 + 60A_2^3A_3 + 47A_2^2A_3A_2 + 43A_2A_3A_2^2 + 38A_3A_2^3 - 88A_2^5$.

Taylor expansion of $F(z^{(k)})$, using Eq. (11), is established as

$$F(z^{(k)}) = F'(x^*)[P_1 e^{(k)^3} + P_2 e^{(k)^4} + P_3 e^{(k)^5} + P_4 e^{(k)^6}] + O(e^{(k)^7}), \tag{12}$$

where $P_1 = 2A_2^2$, $P_2 = 4A_2A_3 + 3A_3A_2 - 9A_2^3$, $P_3 = 6A_2A_4 + 6A_3^2 + 4A_4A_2 - 18A_2^2A_3 - 14A_2A_3A_2 - 12A_3A_2^2 + 30A_2^4$, and $P_4 = 8A_2A_5 + 9A_3A_4 + 8A_4A_3 + 5A_5A_2 - 27A_2^2A_4 - 28A_2A_3^2 - 19A_2A_4A_2 - 24A_3A_2A_3 - 18A_3^2A_2 - 16A_4A_2^2 + 60A_2^3A_3 + 47A_2^2A_3A_2 + 43A_2A_3A_2^2 + 38A_3A_2^3 - 84A_2^5$.

Consequently, the error equation at the $(k+1)^{th}$ iteration is derived by substituting the expressions of (7), (10), (11), and (12) in the final step of method (4), which is given by the expression

$$e^{(k+1)} = x^{(k+1)} - x^* = Q_1 e^{(k)^3} + Q_2 e^{(k)^4} + Q_3 e^{(k)^5} + Q_4 e^{(k)^6} + O(e^{(k)^7}), \tag{13}$$

where $Q_1 = 2(1-a-b-c)A_2^2$, $Q_2 = (1-a-b-c)(4A_2A_3 + 3A_3A_2) - (9-13a-17b-21c)A_2^3$, $Q_3 = (1-a-b-c)(6A_2A_4 + 6A_3^2 + 4A_4A_2) - 2(9-13a-17b-21c)A_2^2A_3 - 2(7-10a-13b-16c)A_2A_3A_2 - 6(2-3a-4b-5c)A_3A_2^2 + 2(15-28a-47b-70c)A_2^4$, and the expression of Q_4, being lengthy, is not shown explicitly here.

Ultimately, there should be an optimum selection of parameters' values so as to achieve the maximum possible convergence speed for the proposed scheme. In that sense, if we choose $a = \frac{7}{2}, b = -4$, and $c = \frac{3}{2}$, then the coefficients Q_1, Q_2, and Q_3 in Eq. (13) vanish. Further, the error equation is reduced to

$$e^{(k+1)} = 2(A_2A_3A_2^2 - 3A_3A_2^3 + 18A_2^5)e^{(k)^6} + O(e^{(k)^7}).$$

Hence, the sixth order of convergence is proved for the iterative method (4). □

The proposed sixth-order iterative method is finally presented below.

$$\begin{aligned} y^{(k)} &= x^{(k)} - F'(x^{(k)})^{-1}F(x^{(k)}),\\ z^{(k)} &= y^{(k)} - F'(x^{(k)})^{-1}F(y^{(k)}),\\ x^{(k+1)} &= z^{(k)} - \left[\frac{7}{2}I - 4F'(x^{(k)})^{-1}F'(y^{(k)}) + \frac{3}{2}(F'(x^{(k)})^{-1}F'(y^{(k)}))^2\right]\\ &\quad \times F'(x^{(k)})^{-1}F(z^{(k)}). \end{aligned} \tag{14}$$

Clearly, the proposed method utilizes three function evaluations, two Jacobian matrices, and one Jacobian inversion per iteration. For the further reference in this study, the technique (14) is denoted as ϕ_1.

3 Computational Efficiency

Solving nonlinear systems using iterative procedures involves a significantly large number of mathematical calculations or operations. Apart from achieving the high convergence order, an iterative algorithm shall also be evaluated on the basis of its computational aspects. The term computational efficiency relates to the investigation of algorithmic characteristics that how much computing resources it utilizes during its implementation. In what follows, the concept of computational efficiency shall be investigated thoroughly, and further, the analysis shall be carried out in this context for the comparison of the new iterative method with the existing counterparts.

For locating the solution of a nonlinear system using an iterative method, initially, an approximation is selected in the neighborhood of the solution. Then, the iterative process is terminated using a specific criterion, which is generally prescribed as

$$||x^{(k)} - x^*|| \leq \epsilon = 10^{-d},$$

where 'k' is the iteration index, 'ϵ' is the desired precision, and 'd' is the number of significant decimal digits of the obtained approximation. To estimate the number of iterations which are required to achieve the desired accuracy, it is assumed that $||x^{(0)} - x^*|| \approx 10^{-1}$. Then, after the '$k$' number of iterative steps, we have the approximation: $10^{-d} \approx 10^{-r^k}$, and that simply provides the required estimation $k \approx \log d / \log r$, where r is the convergence order. Further, let the computational cost per iteration be represented by 'C', then the completed iterative process constitutes the total computational cost which is equal to 'kC'. The measure of computational efficiency, conventionally known as the efficiency index, is formulated in various manners in the literature. Ostrowski in [8] and Traub in [13] have independently provided this measure in different ways. But, defined in any way, the efficiency index always indicates reciprocal relation with the cost of computation. Therefore, taking into consideration the reciprocal relationship, the efficiency index be evaluated as

$$E = \frac{1}{kC} = \frac{1}{\log d} \frac{\log r}{C}. \tag{15}$$

Consider a m-dimensional function, $F : \mathbb{R}^m \to \mathbb{R}^m$, $F(x) = (f_1(x), ..., f_m(x))^T$, where $x = (x_1, x_2, ..., x_m)^T$, then the estimation of computational cost per iteration is given by the formulation,

$$C(m, \eta_0, \eta_1, \mu) = N_0(m)\eta_0 + N_1(m)\eta_1 + N(m, \mu), \tag{16}$$

where $N_0(m)$ and $N_1(m)$ represent the number of evaluations of scalar functions in the computation of F and F', respectively, and $N(m, \mu)$ stands for the number of product or quotient evaluations per iteration. The ratios $\eta_0 > 0$ and $\eta_1 > 0$, which interrelate the costs of products and functional evaluations, and a ratio $\mu > 1$, interrelating costs of products and quotients, are the necessary parameters in order to express $C(m, \eta_0, \eta_1, \mu)$ in terms of product units. Let us note that evaluations of m and m^2 scalar functions are required, respectively, to compute a function F and a derivative F'. Additionally, to compute an inverse linear operator, and eventually to evaluate $F'^{-1}F$, the technique of LU decomposition is employed that involves $m(m-1)(2m-1)/6$ products and $m(m-1)/2$ quotients, which is followed by the resolution of two triangular linear systems requiring $m(m-1)$ products and m quotients. Further, m products for scalar-vector multiplication and m^2 products for matrix-vector multiplication must be taken into account.

With the purpose to analyze and compare the efficiency of the developed method, we have included the existing sixth-order methods developed by Bahl et al. [2], Cordero et al. [4], Esmaeili and Ahmadi [5], Lofti et al. [6], Soleymani et al. [12], and Wang et al. [14]. For the ready reference, these methods are expressed below, which are denoted by ϕ_i, where $i = 2, 3, \ldots, 7$.

Method by Bahl et al. (ϕ_2):

$$
\begin{aligned}
y^{(k)} &= x^{(k)} - \frac{2}{3}F'(x^{(k)})^{-1}F(x^{(k)}),\\
z^{(k)} &= x^{(k)} - \left[I + \frac{3}{4}[I - F'(x^{(k)})^{-1}F'(y^{(k)})] + \frac{9}{8}[I - F'(x^{(k)})^{-1}F'(y^{(k)})]^2\right]\\
&\qquad \times F'(x^{(k)})^{-1}F(x^{(k)}),\\
x^{(k+1)} &= z^{(k)} - 2[3F'(y^{(k)}) - F'(x^{(k)})]^{-1}F(z^{(k)}).
\end{aligned}
$$

Method by Cordero et al. (ϕ_3):

$$
\begin{aligned}
y^{(k)} &= x^{(k)} - F'(x^{(k)})^{-1}F(x^{(k)}),\\
z^{(k)} &= y^{(k)} - F'(x^{(k)})^{-1}[2I - F'(y^{(k)})F'(x^{(k)})^{-1}]F(y^{(k)}),\\
x^{(k+1)} &= z^{(k)} - F'(y^{(k)})^{-1}F(z^{(k)}).
\end{aligned}
$$

Method by Esmaeili and Ahmadi (ϕ_4):

$$
\begin{aligned}
y^{(k)} &= x^{(k)} - F'(x^{(k)})^{-1}F(x^{(k)}),\\
z^{(k)} &= y^{(k)} + \frac{1}{3}\big[F'(x^{(k)})^{-1} + 2[F'(x^{(k)}) - 3F'(y^{(k)})]^{-1}\big]F(x^{(k)}),\\
x^{(k+1)} &= z^{(k)} + \frac{1}{3}\big[-F'(x^{(k)})^{-1} + 4[F'(x^{(k)}) - 3F'(y^{(k)})]^{-1}\big]F(z^{(k)}).
\end{aligned}
$$

Method by Lofti et al. (ϕ_5):

$$\begin{aligned} y^{(k)} &= x^{(k)} - F'(x^{(k)})^{-1}F(x^{(k)}), \\ z^{(k)} &= x^{(k)} - 2[F'(x^{(k)}) + F'(y^{(k)})]^{-1}F(x^{(k)}), \\ x^{(k+1)} &= z^{(k)} - \left[\frac{7}{2}I - 4F'(x^{(k)})^{-1}F'(y^{(k)}) + \frac{3}{2}(F'(x^{(k)})^{-1}F'(y^{(k)}))^2\right] \\ &\quad \times F'(x^{(k)})^{-1}F(z^{(k)}). \end{aligned}$$

Method by Soleymani et al. (ϕ_6):

$$\begin{aligned} y^{(k)} &= x^{(k)} - \frac{2}{3}F'(x^{(k)})^{-1}F(x^{(k)}), \\ z^{(k)} &= x^{(k)} - \frac{1}{2}[3F'(y^{(k)}) - F'(x^{(k)})]^{-1}[3F'(y^{(k)}) + F'(x^{(k)})]F'(x^{(k)})^{-1}F(x^{(k)}), \\ x^{(k+1)} &= z^{(k)} - \left[\frac{1}{2}[3F'(y^{(k)}) - F'(x^{(k)})]^{-1}[3F'(y^{(k)}) + F'(x^{(k)})]\right]^2 \\ &\quad \times F'(x^{(k)})^{-1}F(z^{(k)}). \end{aligned}$$

Method by Wang et al. (ϕ_7):

$$\begin{aligned} y^{(k)} &= x^{(k)} - \frac{2}{3}F'(x^{(k)})^{-1}F(x^{(k)}), \\ z^{(k)} &= x^{(k)} - [6F'(y^{(k)}) - 2F'(x^{(k)})]^{-1}[3F'(y^{(k)}) + F'(x^{(k)})]F'(x^{(k)})^{-1}F(x^{(k)}), \\ x^{(k+1)} &= z^{(k)} - \frac{1}{2}[3F'(y^{(k)})^{-1} - F'(x^{(k)})^{-1}]F(z^{(k)}). \end{aligned}$$

Denoting the computational costs and the efficiency indices, respectively, by C_i and E_i, $i = 1, 2, \ldots, 7$, and then taking into account the mathematical operations or computations described above, the computational costs and the corresponding efficiency indices are expressed as follows:

$$C_1 = 3m\eta_0 + 2m^2\eta_1 + \frac{m}{6}(2m^2 + 39m - 11 + 3\mu(9+m)) \text{ and } E_1 = \frac{1}{D}\frac{\log 6}{C_1}.$$
$$C_2 = 2m\eta_0 + 2m^2\eta_1 + \frac{m}{3}(2m^2 + 18m + 4 + 3\mu(3+m)) \text{ and } E_2 = \frac{1}{D}\frac{\log 6}{C_2}.$$
$$C_3 = 3m\eta_0 + 2m^2\eta_1 + \frac{m}{3}(2m^2 + 12m - 8 + 3\mu(3+m)) \text{ and } E_3 = \frac{1}{D}\frac{\log 6}{C_3}.$$
$$C_4 = 2m\eta_0 + 2m^2\eta_1 + \frac{m}{3}(2m^2 + 12m + 1 + 3\mu(3+m)) \text{ and } E_4 = \frac{1}{D}\frac{\log 6}{C_4}.$$
$$C_5 = 2m\eta_0 + 2m^2\eta_1 + \frac{m}{3}(2m^2 + 18m - 2 + 3\mu(4+m)) \text{ and } E_5 = \frac{1}{D}\frac{\log 6}{C_5}.$$
$$C_6 = 2m\eta_0 + 2m^2\eta_1 + \frac{m}{3}(2m^2 + 24m - 5 + 3\mu(4+m)) \text{ and } E_6 = \frac{1}{D}\frac{\log 6}{C_6}.$$

$$C_7 = 2m\eta_0 + 2m^2\eta_1 + \frac{m}{2}(2m^2 + 9m + 1 + \mu(5 + 3m)) \text{ and } E_7 = \frac{1}{D}\frac{\log 6}{C_7}.$$

Here $D = \log d$.

3.1 Comparison of Efficiencies

Consider a ratio, for the comparison of iterative methods, say ϕ_i versus ϕ_j, which is defined as

$$\Pi_j^i = \frac{E_i}{E_j} = \frac{C_j \log(r_i)}{C_i \log(r_j)}, \tag{17}$$

where r_i and r_j, respectively, are the orders of convergence of the methods ϕ_i and ϕ_j. Clearly, if $\Pi_j^i > 1$ holds, then ϕ_i will be more efficient than ϕ_j, and we symbolize it as $\phi_i \gg \phi_j$. The proposed method, ϕ_1, shall be compared analytically as well as geometrically with the existing methods, ϕ_i $(i = 2, 3, \ldots, 7)$, which are already presented above. The analytical way of comparison is the resolution of inequality $\Pi_i^1 > 1$ for each $i = 2, 3, \ldots, 7$, and the results obtained are presented geometrically by projecting the boundary lines $\Pi_i^1 = 1$, in (η_1, η_0)-plane, corresponding to the special cases of $m = 5,\ 10,\ 25,$ and 50, and fixing $\mu = 3$ in each case. Let us note here that each line will divide the plane into two parts, where $\phi_1 \gg \phi_i$ on one side, whereas $\phi_i \gg \phi_1$ on the other.

In view of the above discussion, we now present the comparison analysis through the following theorem:

Theorem 2 *For all* $\eta_0 > 0$, $\eta_1 > 0$, *and* $\mu > 1$, *we have that*

(i) $E_1 > E_2$, *for* $\eta_0 < \frac{1}{6}(2m^2 - 3m + 19 + 3\mu(m - 3))$.
(ii) $E_1 > E_3$ *for* $m \geq 7$, *and* $E_1 < E_3$ *for* $m = 2, 3$, *but otherwise comparison depends on value of* μ.
(iii) $E_1 > E_4$, *for* $\eta_0 < \frac{1}{6}(2m^2 - 15m + 13 + 3\mu(m - 3))$.
(iv) $E_1 > E_5$, *for* $\eta_0 < \frac{1}{6}(2m^2 - 3m + 7 + 3\mu(m - 1))$.
(v) $E_1 > E_6$, *for* $\eta_0 < \frac{1}{6}(2m^2 + 9m + 1 + 3\mu(m - 1))$.
(vi) $E_1 > E_7$, *for* $\eta_0 < \frac{1}{3}(2m^2 - 6m + 7 + 3\mu(m - 2))$.

Proof ϕ_1 *versus* ϕ_2 *case*:
The ratio in this case is

$$\Pi_2^1 = \frac{2m\eta_0 + 2m^2\eta_1 + \frac{m}{3}(2m^2 + 18m + 4 + 3\mu(3 + m))}{3m\eta_0 + 2m^2\eta_1 + \frac{m}{6}(2m^2 + 39m - 11 + 3\mu(9 + m))}.$$

By resolving of the inequality $\Pi_2^1 > 1$, it is straightforward to deduce that $\eta_0 < \frac{1}{6}(2m^2 - 3m + 19 + 3\mu(m - 3))$, which concludes (i). The boundary lines $\Pi_2^1 = 1$,

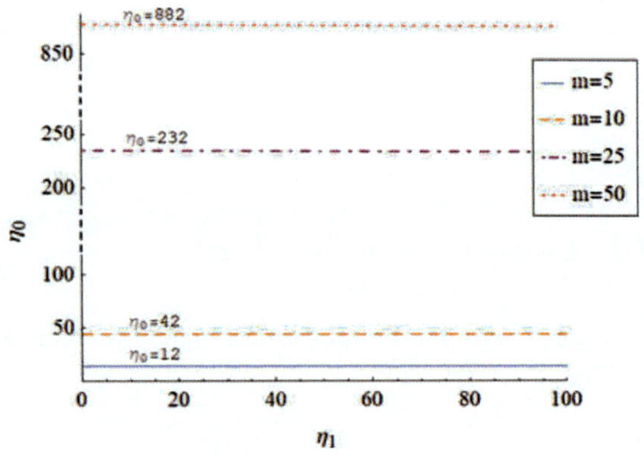

Fig. 1 Boundary lines for comparison of ϕ_1 and ϕ_2

in (η_1, η_0)-plane, are displayed in Fig. 1, where $\phi_1 > \phi_2$ in the section which is below the line for each particular case of m.

ϕ_1 versus ϕ_3 case:

The ratio in this case is

$$\Pi_3^1 = \frac{3m\eta_0 + 2m^2\eta_1 + \frac{m}{3}(2m^2 + 12m - 8 + 3\mu(3 + m))}{3m\eta_0 + 2m^2\eta_1 + \frac{m}{6}(2m^2 + 39m - 11 + 3\mu(9 + m))}.$$

It is easy to verify that, for $\eta_0 > 0$, $\eta_1 > 0$, and $\mu > 1$, the inequality $\Pi_3^1 > 1$ holds for $m \geq 7$, and $\Pi_3^1 < 1$ holds only for $m = 2, 3$. For $4 \leq m \leq 6$, the inequality $\Pi_3^1 > 1$ holds when $\mu > \frac{2m^2-15m-5}{9-3m}$, and this eventually proves (ii). So, we conclude here that $\phi_1 > \phi_3$ for all $m \geq 7$, whereas $\phi_1 < \phi_3$ for $m = 2, 3$, but otherwise, comparison depends on the value of μ.

ϕ_1 versus ϕ_4 case:

The ratio in this case is

$$\Pi_4^1 = \frac{2m\eta_0 + 2m^2\eta_1 + \frac{m}{3}(2m^2 + 12m + 1 + 3\mu(3 + m))}{3m\eta_0 + 2m^2\eta_1 + \frac{m}{6}(2m^2 + 39m - 11 + 3\mu(9 + m))}.$$

Resolution of the inequality $\Pi_4^1 > 1$ results into $\eta_0 < \frac{1}{6}(2m^2 - 15m + 13 + 3\mu(m - 3))$, which concludes (iii). The boundary lines for this comparison, in (η_1, η_0)-plane, are shown in Fig. 2, where $\phi_1 > \phi_4$ on the lower region of line for each case of m.

ϕ_1 versus ϕ_5 case:

The ratio in this case is

$$\Pi_5^1 = \frac{2m\eta_0 + 2m^2\eta_1 + \frac{m}{3}(2m^2 + 18m - 2 + 3\mu(4 + m))}{3m\eta_0 + 2m^2\eta_1 + \frac{m}{6}(2m^2 + 39m - 11 + 3\mu(9 + m))}.$$

Fig. 2 Boundary lines for comparison of ϕ_1 and ϕ_4

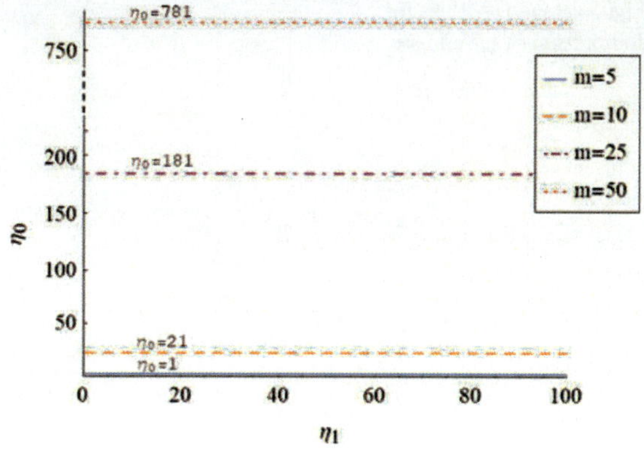

The inequality $\Pi_5^1 > 1$ simply resolves into relation $\eta_0 < \frac{1}{6}(2m^2 - 3m + 7 + 3\mu(m-1))$, and this proves (iv). In this comparison, the boundary lines are displayed in Fig. 3, where $\phi_1 > \phi_5$ holds on the lower section of line for each particular case.

ϕ_1 *versus* ϕ_6 *case*:

The ratio in this case is

$$\Pi_6^1 = \frac{2m\eta_0 + 2m^2\eta_1 + \frac{m}{3}(2m^2 + 24m - 5 + 3\mu(4+m))}{3m\eta_0 + 2m^2\eta_1 + \frac{m}{6}(2m^2 + 39m - 11 + 3\mu(9+m))}.$$

It is straightforward to establish the relation $\eta_0 < \frac{1}{6}(2m^2 + 9m + 1 + 3\mu(m-1))$ by resolving $\Pi_6^1 > 1$, which eventually proves (v). The boundary lines, in this case, are presented in Fig. 4 with $\phi_1 > \phi_6$ on the lower side of each line.

ϕ_1 *versus* ϕ_7 *case*:

The ratio in this case is

$$\Pi_7^1 = \frac{2m\eta_0 + 2m^2\eta_1 + \frac{m}{2}(2m^2 + 9m + 1 + \mu(5+3m))}{3m\eta_0 + 2m^2\eta_1 + \frac{m}{6}(2m^2 + 39m - 11 + 3\mu(9+m))}.$$

Resolution of the inequality $\Pi_7^1 > 1$ results into the relation $\eta_0 < \frac{1}{3}(2m^2 - 6m + 7 + 3\mu(m-2))$. This concludes (vi), and the boundary lines for this case are shown in Fig. 5, where $\phi_1 > \phi_7$ in the region which is below each boundary line. □

From the above comparison analysis, it can be clearly observed that the proposed iterative method shows an increase in the efficiency index with the increasing values of m. We conclude this section with a note that, as large as the system is, the proposed sixth-order method exhibits superiority over the existing methods in the subject of computational efficiency.

Fig. 3 Boundary lines for comparison of ϕ_1 and ϕ_5

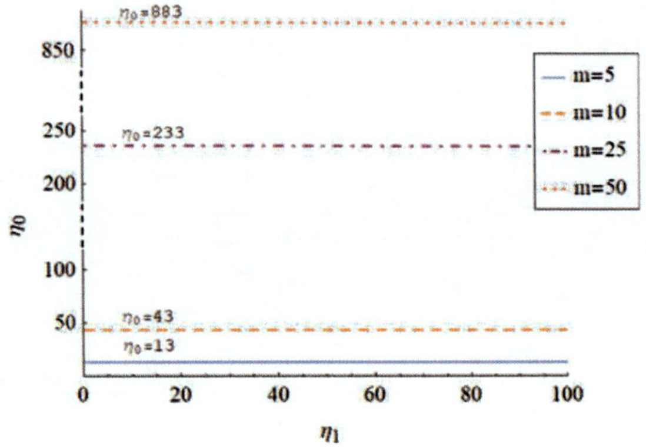

Fig. 4 Boundary lines for comparison of ϕ_1 and ϕ_6

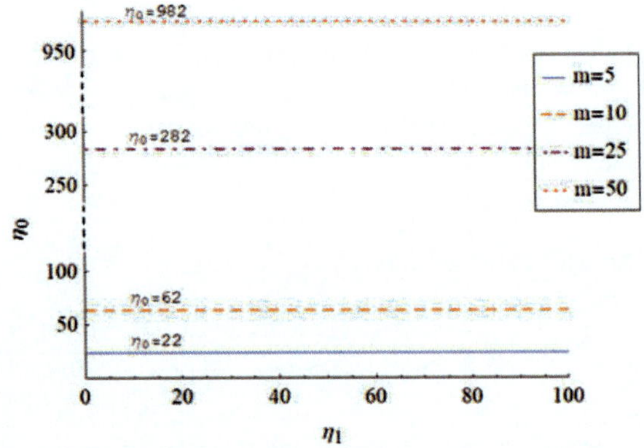

Fig. 5 Boundary lines for comparison of ϕ_1 and ϕ_7

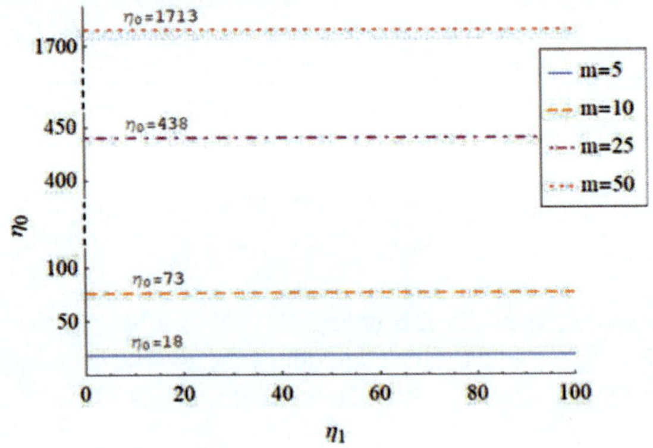

Table 1 CPU time and computational cost for the execution of elementary operations

Functions	$x * y$	x/y	$\sqrt{x}$	e^x	$\log(x)$	$\sin(x)$	$\cos(x)$	$\arctan(x)$
CPU time	0.0172	0.0484	0.0234	1.5562	1.3469	1.6938	1.6896	2.9797
Cost	1	2.81	1.36	90.48	78.31	98.48	98.23	173.24

Here $x = \sqrt{3} - 1$ *and* $y = \sqrt{5}$ *(with* 4096 *digits of accuracy)*

4 Numerical Experimentation

In this section, the numerical experimentation shall be executed to assess the performance of the developed method. The nonlinear problems arising in different physical situations have been selected for this purpose. Moreover, to arrive at some valid conclusion, the outcomes of this testing need to be analyzed and further compared with the corresponding results of the existing methods. Two of the most significant factors which contribute toward the numerical performance of an iterative technique are (i) Stability and (ii) CPU time elapsed during its execution on the digital platform. Let us note that all the numerical computations, in our case, are being executed using the software *Mathematica* [15] installed on the computer equipped with specifications: Intel(R) Core (TM) i5-9300H processor and Windows 10 operating system.

In what follows, the comparison analysis shall be illustrated by locating the solutions of nonlinear problems, and for the termination of iterations, the stopping criterion being employed is described as follows:

$$\|x^{(k)} - x^{(k-1)}\| + \|F(x^{(k)})\| < 10^{-100}.$$

In addition, the approximated computational order of convergence (ACOC) is required to validate the convergence order established by analytical means, which is computed by the formula (see [5]),

$$\text{ACOC} = \frac{\ln\left(\|x^{(k)} - x^{(k-1)}\| / \|x^{(k-1)} - x^{(k-2)}\|\right)}{\ln\left(\|x^{(k-1)} - x^{(k-2)}\| / \|x^{(k-2)} - x^{(k-3)}\|\right)}.$$

To make connection between the computational efficiency and the performance of technique, it is necessary to estimate the parameters, η_0, η_1, and μ, as defined in Sect. 3. These parameters are essential to express the mathematical operations and functional evaluations in terms of product units. In order to achieve this, Table 1 displays the CPU time elapsed during the execution of elementary mathematical operations and their estimated cost of computation in units of products. Note that the estimated cost of division is approximately thrice the cost of the product.

Now, we consider the following nonlinear problems to demonstrate the performance analysis and display results in respect of the following: (i) Number of iterations (k), (ii) ACOC, (iii) Computational cost (C_i), (iv) Efficiency index (E_i), and

Table 2 Comparison of performance of methods for Problem 1

Method	k	ACOC	C_i	E_i	CPU time
ϕ_1	4	5.993	145.58	1230.77	0.0260
ϕ_2	4	5.996	152.58	1174.31	0.0313
ϕ_3	4	5.996	129.58	1382.74	0.0363
ϕ_4	4	5.989	131.58	1361.73	0.0417
ϕ_5	4	5.994	155.01	1155.90	0.0310
ϕ_6	4	5.999	170.01	1053.91	0.0363
ϕ_7	4	5.996	154.01	1163.40	0.0467

(v) Elapsed CPU time (in seconds). To illustrate the efficiency indices of techniques, we have conveniently chosen $D = 10^{-5}$ for each of the problems.

Problem 1 Starting with the system of three nonlinear equations:

$$\begin{aligned} x^2 + y^2 + z^2 &= 1, \\ 2x^2 + y^2 + 4z &= 0, \\ 3x^2 - 4y^2 + z^2 &= 0, \end{aligned}$$

the initial estimate is taken as $\left(-\frac{3}{2}, -\frac{3}{2}, -\frac{3}{2}\right)^T$ to locate the particular solution,

$$x^* = (-0.6982..., -0.6285..., -0.3425...)^T.$$

For this particular problem, the parameters used in the equation (16) are estimated as $(m, \eta_0, \eta_1, \mu) = (3, 2.33, 0.67, 2.81)$. Numerical results for the comparison are displayed in Table 2.

Problem 2 Consider the nonlinear integral equation (see [1]),

$$u(t) = \frac{7}{8}t + \frac{1}{2}\int_0^1 t\ s\ u(s)^2 ds, \tag{18}$$

where $t \in [0, 1]$, and $u \in C[0, 1]$, with $C[0, 1]$ being a space of all continuous functions defined on the interval [0, 1].

The given integral equation can be transformed into a finite-dimensional problem by partitioning the given interval [0, 1] uniformly as follows:

$$0 = t_0 < t_1 < t_2 < \cdots < t_{k-1} < t_k = 1, \quad \text{where } t_i = t_0 + ih, \quad (i = 1, 2, \ldots, k-1),$$

where $h = 1/k$ is the sub-interval length. Approximating the integral, appearing in the equation (18), using the trapezoidal rule of integration, and denoting $u(t_i) = u_i$ for each i, we obtain the system of nonlinear equations as

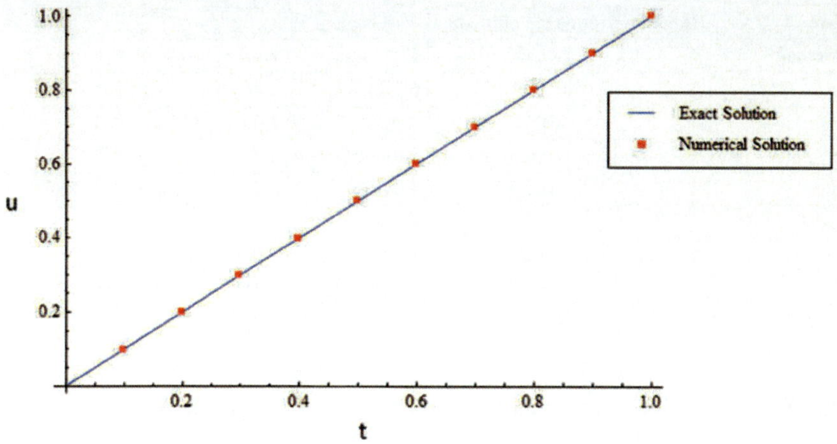

Fig. 6 Graphical comparison of exact and numerical solution of Problem 2

Table 3 Comparison of performance of methods for Problem 2

Method	k	ACOC	C_i	E_i	CPU time
ϕ_1	3	6.000	1521.95	117.73	0.141
ϕ_2	3	6.000	1905.30	94.04	0.188
ϕ_3	3	6.000	1695.30	105.69	0.177
ϕ_4	3	6.000	1695.30	105.69	0.162
ϕ_5	3	6.000	1913.40	93.64	0.187
ϕ_6	3	6.000	2103.40	85.18	0.203
ϕ_7	3	6.000	2206.75	81.19	0.235

$$\frac{7}{8}t_i - u_i + \frac{ht_i}{2}\left(\frac{1}{2}u_k^2 + \sum_{j=1}^{k-1} s_j u_j^2\right) = 0, \quad (i = 1, 2, \ldots, k), \tag{19}$$

where $t_i = s_i = i/k$ for each i.

We solve this problem in particular by taking $k = 10$. Setting the initial approximation as $(\frac{1}{2}, \overset{10}{\cdots\cdots}, \frac{1}{2})^T$, the approximate numerical solution of the reduced system (19) is obtained as,

$$x^* = (0.1001..., 0.2003..., 0.3004..., 0.4006..., 0.5008..., 0.6009..., \\ 0.7011..., 0.8013..., 0.9014..., 1.0016...)^T.$$

The numerical solution, so obtained, is compared graphically with the exact solution in Fig. 6, and further, numerical results are depicted in Table 3. Moreover, the parameters of Eq. (16) are estimated as $(m, \eta_0, \eta_1, \mu) = (10, 3, 1, 2.81)$.

Table 4 Comparison of performance of methods for Problem 3

Method	k	ACOC	C_i	E_i	CPU time
ϕ_1	4	6.000	6.27E+04	2.86	1.385
ϕ_2	4	6.000	1.06E+05	1.68	2.401
ϕ_3	4	6.000	1.01E+05	1.77	2.271
ϕ_4	4	6.000	1.01E+05	1.77	2.327
ϕ_5	4	6.000	1.06E+05	1.68	2.344
ϕ_6	4	6.000	1.11E+05	1.61	2.250
ϕ_7	4	6.000	1.48E+05	1.21	3.344

Problem 3 Consider the boundary value problem (see [3]), which models the finite deflections of an elastic string under the transverse load, and it is presented as follows:

$$u''(t) + a^2(u'(t))^2 + 1 = 0, \quad u(0) = 0, \quad u(1) = 0, \tag{20}$$

where 'a' is a parameter. The exact solution of the given problem is $u(t) = \frac{1}{a^2}\ln\left(\frac{\cos(a(t-1/2))}{\cos(a/2)}\right)$. We intend to remodel the problem (20) into a finite-dimensional problem by considering the partition of $[0, 1]$, with equal sub-interval length $h = 1/k$, as

$$0 = t_0 < t_1 < t_2 < \cdots < t_{k-1} < t_k = 1, \ \text{where}\, t_i = t_0 + ih, \ (i = 1, 2, \ldots, k-1).$$

Denoting $u(t_i) = u_i$ for each $i = 1, 2, \ldots, k-1$, and approximating the derivatives involved in (20) by the second-order divided differences,

$$u_i' = \frac{u_{i+1} - u_{i-1}}{2h}, \quad \text{and} \quad u_i'' = \frac{u_{i+1} - 2u_i + u_{i-1}}{h^2},$$

the system of $k-1$ nonlinear equations in $k-1$ variables is obtained as

$$u_{i-1} - 2u_i + u_{i+1} + \frac{a^2}{4}(u_{i+1} - u_{i-1})^2 + h^2 = 0, \quad (i = 1, 2, \ldots, k-1),$$

where $u_0 = 0$ and $u_k = 0$ are the transformed boundary conditions. In particular, setting $k = 51$, the above system reduces to 50 nonlinear equations. Further, choosing $a = 2$, and selecting the initial approximation as $(-1, \overset{50}{\cdots\cdots}, -1)^T$, the approximate numerical solution so obtained, along with the exact solution, is plotted in Fig. 7. Further, the numerical performance of the methods is displayed in Table 4. The estimated values of parameters, used in Eq. (16), are given as $(m, \eta_0, \eta_1, \mu) = (50, 2, 0.078, 2.81)$.

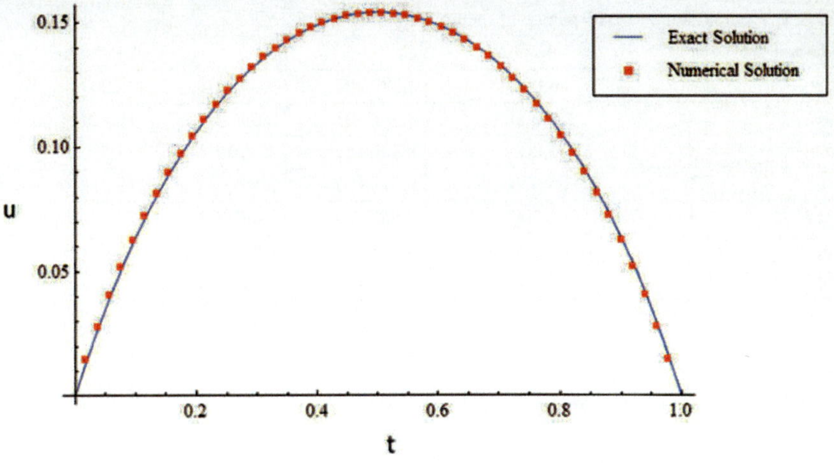

Fig. 7 Graphical comparison of exact and numerical solution of Problem 3

Table 5 Comparison of performance of methods for Problem 4

Method	k	ACOC	C_i	E_i	CPU time
ϕ_1	5	6.000	4.67E+05	0.384	45.53
ϕ_2	5	6.000	7.92E+05	0.226	74.14
ϕ_3	5	6.000	7.89E+05	0.227	75.09
ϕ_4	5	6.000	7.72E+05	0.232	72.88
ϕ_5	5	6.000	7.92E+05	0.226	74.36
ϕ_6	5	6.000	8.12E+05	0.221	75.73
ϕ_7	5	6.000	1.12E+06	0.159	105.05

Problem 4 Now let us take a system of nonlinear equations as follows:

$$\tan^{-1}(x_i) - 1 + 2\left(\sum_{j=1, j\neq i}^{m} x_j^2\right) = 0, \quad i = 1, 2, ..., m.$$

By taking $m = 100$, we select the initial approximation $(1, \overset{100}{\cdots\cdots}, 1)^T$ to obtain the particular solution,

$$x^* = (0.06859..., \overset{100}{\cdots\cdots}, 0.06859...)^T.$$

The estimated values of the parameters in this problem are (m, η_0, η_1, μ) = (100, 175.24, 0.048, 2.81). Further, Table 5 exhibits the comparison of the performance of methods.

Table 6 Comparison of performance of methods for Problem 5

Method	k	ACOC	C_i	E_i	CPU time
ϕ_1	3	6.000	4.38E+07	4.09E-03	16.34
ϕ_2	3	6.000	8.56E+07	2.09E-03	19.42
ϕ_3	3	6.000	8.52E+07	2.10E-03	21.59
ϕ_4	3	6.000	8.51E+07	2.10E-03	19.16
ϕ_5	3	6.000	8.56E+07	2.09E-03	19.27
ϕ_6	3	6.000	8.61E+07	2.08E-03	20.45
ϕ_7	3	6.000	1.27E+08	1.41E-03	24.33

Problem 5 At last, we consider a large system of equations:

$$x_i + \log(2 + x_i + x_{i+1}) = 0, \quad i = 1, 2, ..., m-1,$$
$$\text{and} \quad x_m + \log(2 + x_m + x_1) = 0,$$

where $m = 500$. The above given system has a particular solution,

$$x^* = \left(-0.3149..., \overset{500}{\cdots\cdots}, -0.3149...\right)^T,$$

and to obtain this solution, the initial estimate is taken as $\left(\frac{1}{10}, \overset{500}{\cdots\cdots}, \frac{1}{10}\right)^T$. Numerical results for the performance of methods are depicted in Table 6. Further, the values of parameters are estimated as

$$(m, \eta_0, \eta_1, \mu) = (500, 78.31, 0.0056, 2.81).$$

The findings of numerical experimentation signify the efficient and stable nature of the proposed sixth-order method. The results are remarkable with respect to the efficiency index and CPU time, and certainly favor the new method over its existing counterparts. Furthermore, computation of ACOC validates the theoretically established convergence order.

5 Conclusions

A three-step iterative technique, involving some undetermined parameters, has been designed for the solution of nonlinear equations. The methodology to design the technique is based on the objective to accelerate the convergence rate of the well-known third-order Potra-Pták scheme. Analysis of convergence leads to establishing the sixth order of convergence for a particular set of parametric values. Utilizing only a single Jacobian inversion per iteration, the proposed iterative method exhibits highly economical nature when analyzed in the context of computational complexity. This is affirmed by comparing the computational efficiency of the new method, by analytical as well as visual approach, with the efficiencies of existing methods. Further, numerical performance is examined by locating the solutions of some selected nonlinear problems. The findings of this testing clearly indicate the superiority of the proposed technique over its existing counterparts, especially for large-scale nonlinear systems.

References

1. Avazzadeh, Z., Heydari, M., Loghmani, G.B.: Numerical solution of Fredholm integral equations of the second kind by using integral mean value theorem. Appl. Math. Model. **35**, 2374–2383 (2011). https://doi.org/10.1016/j.apm.2010.11.056
2. Bahl, A., Cordero, A., Sharma, R., Torregrosa, J.R.: A novel bi-parametric sixth order iterative scheme for solving nonlinear systems and its dynamics. Appl. Math. Comput. **357**, 147–166 (2019). https://doi.org/10.1016/j.amc.2019.04.003
3. Cordero, A., Hueso, J.L., Martínez, E., Torregrosa, J.R.: Efficient high-order methods based on golden ratio for nonlinear systems. Appl. Math. Comput. **217**, 4548–4556 (2011). https://doi.org/10.1016/j.amc.2010.11.006
4. Cordero, A., Hueso, J.L., Martínez, E., Torregrosa, J.R.: Increasing the convergence order of an iterative method for nonlinear systems. Appl. Math. Lett. **25**, 2369–2374 (2012). https://doi.org/10.1016/j.aml.2012.07.005
5. Esmaeili, H., Ahmadi, M.: An efficient three-step method to solve system of nonlinear equations. Appl. Math. Comput. **266**, 1093–1101 (2015). https://doi.org/10.1016/j.amc.2015.05.076
6. Lotfi, T., Bakhtiari, P., Cordero, A., Mahdiani, K., Torregrosa, J.R.: Some new efficient multipoint iterative methods for solving nonlinear systems of equations. Int. J. Comput. Math. **92**, 1921–1934 (2014). https://doi.org/10.1080/00207160.2014.946412
7. Ortega, J.M., Rheinboldt, W.C.: Iterative Solution of Nonlinear Equations in Several Variables. Academic Press, New York (1970)
8. Ostrowski, A.M.: Solution of Equation and Systems of Equations. Academic Press, New York (1960)
9. Potra, F.A., Pták, V.: On a class of modified Newton processes. Numer. Funct. Anal. Optim. **2**, 107–120 (1980). https://doi.org/10.1080/01630568008816049
10. Sharma, J.R., Gupta, P.: An efficient fifth order method for solving systems of nonlinear equations. Comput. Math. Appl. **67**, 591–601 (2014). https://doi.org/10.1016/j.camwa.2013.12.004
11. Sharma, R., Sharma, J.R., Kalra, N.: A modified Newton-Özban composition for solving nonlinear systems. Int. J. Comput. Methods **17**, 1950047 (2020). https://doi.org/10.1142/S0219876219500476
12. Soleymani, F., Lotfi, T., Bakhtiari, P.: A multi-step class of iterative methods for nonlinear systems. Optim. Lett. **8**, 1001–1015 (2014). https://doi.org/10.1007/s11590-013-0617-6
13. Traub, J.F.: Iterative Methods for the Solution of Equations. Chelsea Publishing Company, New York (1982)
14. Wang, X., Kou, J., Gu, C.: Semilocal convergence of a sixth-order Jarratt method in Banach spaces. Numer. Algor. **57**, 441–456 (2011). https://doi.org/10.1007/s11075-010-9438-1
15. Wolfram, S.: The Mathematica Book. Wolfram Media, USA (2003)

New Higher Order Iterative Method for Multiple Roots of Nonlinear Equations

Sunil Panday, Waikhom Henarita Chanu, and Yumnam Nomita Devi

Abstract In this paper, we propose a new higher order iterative method to find multiple roots of nonlinear equations. The combination of Taylor's series, Newton's method and the composition approach are used to derive the new method. It requires three evaluations of the function and two evaluations of the derivative of the function per iteration. The theoretical convergence of the proposed method is proved in the main theorem which establishes sixth order of convergence. We compare the developed method with well-known equivalent existing methods by taking various numerical examples. The numerical results demonstrate the better efficiency of the developed method as compared to some standard iterative methods.

Keywords Multiple roots · Nonlinear equation · Iterative methods · Error

1 Introduction

Solving nonlinear equations is one of the most important problems in applied mathematics, engineering and science. Sometimes, analytical methods are not applicable to solve nonlinear equations. Let $\psi : \mathbb{R} \to \mathbb{R}$ be a nonlinear differentiable function defined on an open interval $\mathbb{D}$ such that

$$\psi(x) = 0 \tag{1}$$

We use iterative method for solving such nonlinear equations (1), which is defined as

$$x_{n+1} = P(\psi)(x_n) \;\; for \;\; n = 1, 2, 3, \ldots \tag{2}$$

S. Panday · W. H. Chanu (✉) · Y. N. Devi
Department of Mathematics, National Institute of Technology Manipur, Imphal 795004, India
e-mail: henaritawai@gmail.com

S. Panday
e-mail: sunilpanday@hotmail.co.in

R. K. Sharma et al. (eds.), *Frontiers in Industrial and Applied Mathematics*,
Springer Proceedings in Mathematics & Statistics 410,
https://doi.org/10.1007/978-981-19-7272-0_40

where $P(\psi)$ is called the iterative function. Newton's method (NM) [1, 2] is perhaps the most popular root-finding method for solving nonlinear equations, and it is given by

$$x_{n+1} = x_n - \frac{\psi(x_n)}{\psi'(x_n)} \tag{3}$$

This is quadratically convergent in some neighbourhood of simple roots. Let α be the root of equation (1) with multiplicity $\nu > 1$, i.e. $\psi^{(i)}(\alpha) = 0$ for $i = 1, 2, ..., \nu - 1$ and $\psi^{(\nu)}(\alpha) \neq 0$. When used for finding multiple roots of such nonlinear equations, Newton's method (3) is linearly convergent. The modified Newton's method [3] is written as

$$x_{n+1} = x_n - \nu\frac{\psi(x_n)}{\psi'(x_n)} \tag{4}$$

This is quadratically convergent for the equation having multiple roots with multiplicity $\nu > 1$. Many researchers have developed iterative methods using the modified Newton method for solving nonlinear equations having multiple roots.

In 2013, Thukral [4] developed the following new six-order method (TM for short) for finding multiple roots of a nonlinear equation:

$$\begin{aligned}
y_n &= x_n - \nu\frac{\psi(x_n)}{\psi'(x_n)}\\
z_n &= x_n - \nu\left(\sum_{i=1}^{3} i\left(\frac{\psi(y_n)}{\psi(x_n)}\right)^{i/\nu}\right)\left(\frac{\psi(x_n)}{\psi'(x_n)}\right)\\
x_{n+1} &= z_n - \nu\left(\sum_{i=1}^{3} i\left(\frac{\psi(y_n)}{\psi(x_n)}\right)^{i/\nu}\right)^2\left(\frac{\psi(z_n)}{\psi(x_n)}\right)^{\nu^{-1}}\left(\frac{\psi(x_n)}{\psi'(x_n)}\right)
\end{aligned} \tag{5}$$

where $n \in \mathbb{N}$.

Geum et al. [5] also developed a new sixth-order method (GM for short) in 2016 which is written as follows:

$$\begin{aligned}
y_n &= x_n - \nu\frac{\psi(x_n)}{\psi'(x_n)}\\
z_n &= x_n - \nu Q_\psi(s_n)\frac{\psi(x_n)}{\psi'(x_n)}\\
x_{n+1} &= x_n - \nu K_\psi(s_n, t_n)\frac{\psi(x_n)}{\psi'(x_n)}
\end{aligned} \tag{6}$$

where Q_ψ and K_ψ are weight functions, $s_n = \left(\frac{\psi(y_n)}{\psi(x_n)}\right)^{\frac{1}{\nu}}$ and $t_n = \left(\frac{\psi(z_n)}{\psi(x_n)}\right)^{\frac{1}{\nu}}$.

In 2017, Qudsi et al. [6] developed the following method (QM for short) of sixth order:

$$\begin{aligned} y_n &= x_n - t \\ z_n &= x_n - t\left(1 + \frac{\psi(y_n)}{\psi(x_n)}\left(1 + 2\frac{\psi(y_n)}{\psi(x_n)}\right)\right) \\ x_{n+1} &= x_n - t\left(1 + \frac{\psi(y_n)}{\psi(x_n)}\left(1 + 2\frac{\psi(y_n)}{\psi(x_n)}\right) + \frac{\psi(z_n)}{\psi(x_n)}\left(1 + 2\frac{\psi(y_n)}{\psi(x_n)}\right)\right) \end{aligned} \tag{7}$$

where $t = \frac{2\psi^2(x_n)}{\psi(x_n+\psi(x_n))-\psi(x_n-f(x_n))}$.

Moreover, Singh et al. [7] developed a new fourth-order method in 2015. In the year 2019, Bhel et al. [8] developed multiple roots version of Ostrowski's method having fourth order of convergence. W. H. Chanu et al. also proposed an iterative method of fifth order in [9], Qudsi et al. [10] developed a new iterative method of order six, Kattri [11] proposed a new sixth-order iterative method, etc. In this work, we have introduced a higher order iterative method for solving nonlinear equations having multiple roots. In the following sections, we present the development of our new method, numerical results and conclusion.

2 Development of the Method

In this section, we propose a new sixth-order method for determining the multiple roots of nonlinear equation (1) with multiplicity $\nu > 1$ as follows:

$$\begin{aligned} y_n &= x_n - \frac{2\nu}{\nu+1}\frac{\psi(x_n)}{\psi'(x_n)} \\ z_n &= x_n - \frac{\psi(y_n)(\nu^2-1) - \left(\frac{\nu-1}{\nu+1}\right)^{\nu}(\nu(\nu-4)-1)\psi(x_n)}{4\left(\frac{\nu-1}{\nu+1}\right)^{\nu}\psi'(x_n)} \\ x_{n+1} &= z_n - \nu\frac{\psi(z_n)}{\psi'(z_n)} \end{aligned} \tag{8}$$

Theorem 1 *Let $\alpha \in \mathbb{R}$ be a multiple root of multiplicity ν of a sufficiently differentiable function $\psi : \mathbb{D} \to \mathbb{R}$ in an open interval $\mathbb{D}$ which is a subset of $\mathbb{R}$. Let x_0 be an initial guess of the root α. Then, the method defined by (8) has six orders of convergence.*

Proof Let α be a root of multiplicity ν of $\psi(x) = 0$ and let $e_n = x_n - \alpha$ be the error at nth iteration. Then, using Taylor expansion, we get

$$\psi(x_n) = \left(\frac{\psi^{(\nu)}(\alpha)}{\nu!}\right)e_n^{\nu}[1 + C_1e_n + C_2e_n^2 + C_3e_n^3 + C_4e_n^4 + C_5e_n^5 + C_6e_n^6 + O[e_n]^7] \tag{9}$$

$$\psi'(x_n) = \left(\frac{\psi^{(\nu)}(\alpha)}{(\nu-1)!}\right)e_n^{\nu-1}\left[1 + \left(\frac{\nu+1}{\nu}\right)C_1 e_n + \left(\frac{(\nu+2)}{\nu}\right)C_2 e_n^2 + O[e_n]^7\right] \tag{10}$$

$$\tilde{e_n} = y_n - \alpha = B_1 e_n + B_2 e_n^2 + B_3 e_n^3 + B_4 e_n^4 + B_5 e_n^5 + B_6 e_n^6 + O[e_n]^7 \tag{11}$$

where

$$B_1 = \frac{\nu-1}{\nu+1},$$
$$B_2 = \frac{2C_1}{\nu+\nu^2},$$
$$B_3 = \frac{2\left(\frac{2\nu C_2}{1+\nu} - C_1^2\right)}{\nu^2}$$

$$B_4 = \frac{2((1+\nu)^2 C_1^3 - \nu(4+3\nu)C_1C_2 + 3\nu^2 C_3)}{\nu^3(1+\nu)}$$
$$B_5 = \frac{1}{\nu^4(1+\nu)}\Big(-2((1+\nu)^3 C_1^4 - 2\nu(1+\nu)(3+2\nu)C_1^2 C_2 + 2\nu^2(3+2\nu)C_1C_3 + 2\nu^2((2+\nu)c_2^2 - 2\nu C_4))\Big)$$
$$B_6 = \frac{1}{\nu^5(1+\nu)}\Big(2((1+\nu)^4 C_1^5 - \nu(1+\nu)^2 \times (8+5\nu)C_1^3 C_2 + \nu^2(1+\nu)(9+5\nu)C_1^2 C_3 + \nu^2 C_1((2+\nu)(6+5\nu)C_2^2 - \nu(8+5\nu)C_4) + \nu^3(-(12+5\nu)C_2C_3 + 5\nu C_5))\Big)$$

$$\psi(y_n) = e_n^{\nu}\frac{\psi^{(\nu)}(\alpha)}{\nu!}[((\frac{\nu-1}{\nu+1})^{\nu} + D_1 e_n + D_2 e_n^2 + D_3 e_n^3) + \frac{1}{3\nu^2}D_4 e_n^4 + \frac{1}{15\nu^4}D_5 e_n^5 + O[e_n]^6] \tag{12}$$

where

$$D_1 = \frac{(\frac{\nu-1}{\nu+1})^{\nu}(\nu^2+3)C_1}{\nu^2-1}$$
$$D_2 = \frac{(\nu-1)^{\nu-1}(\nu+1)^{-\nu-2}(-2(\nu+1)^2C_1^2 + \nu(3+\nu(11+\nu+\nu^2))C_2)}{\nu})$$
$$D_3 = \frac{1}{3\nu^2}\Big((\nu^2-1)^{\nu-2}(\nu+1)^{-\nu-3}(2(\nu+1)^4(3\nu-4)C_1^3 - 24(\nu-1)\nu^2$$

$$\times (\nu+1)(\nu+2)C_1C_2 + 3(\nu-1)\nu^2(7+\nu(14+\nu(24+\nu(2+\nu))))C_3)\Big)$$

$$\begin{aligned}
D_4 &= (\nu-1)^{\nu-3}(\nu+1)^{-\nu-4}(-2(\nu+1)^4(3+\nu(3+\nu(-9+(-2+3\nu))))C_1^4 \\
&\quad + 6(\nu-1)\nu(\nu+1)^3(-2+\nu(-15+\nu(4+5\nu)))C_1^2C_2 - 12(-1+\nu)^2\nu^2 \\
&\quad \times (\nu+1)C_1C_3 + 3(-1+\nu)^2\nu^2(-8(\nu+1)^2(-1+\nu(2+\nu))C_2^2 \\
&\quad + \nu(7+\nu(37+\nu(42+\nu)))C_4)) \\
D_5 &= (\nu-1)^{\nu-4}(\nu+1)^{-\nu-5}(2(\nu+1)^5(\nu+2)(-9+\nu(-2+\nu(22+15(1 \\
&\quad + (-3+\nu)\nu))))C_1^5 - 20(\nu-1)\nu^2(\nu+1)^4(30+\nu(-11+\nu(-49+\nu \\
&\quad \times (9\nu+5))))C_1^3C_2 + 30\nu^2(\nu^2-1)^2(-18+\nu(-7+\nu(-34+\nu(-16+ \\
&\quad \times \nu(20+7\nu))))) + 120(\nu-1)^2\nu^2(\nu+1)^2(2+\nu(\nu+1)(-9+2\nu(\nu+1)))C_2^2 \\
&\quad - 2(\nu-1)\nu^2(\nu+4)(\nu^2+1)C_4 + 15(\nu-1)^3\nu^4(-8(\nu+1)^2(-2 \\
&\quad + 3\nu(\nu+3))C_2C_3 + (11+\nu(44+\nu(115+\nu(80+\nu(65+\nu(\nu+4))))))C_5))
\end{aligned}$$

Using the expression of Eqs. (9), (10), (11) and (12) in the second step of the proposed method defined in Eq. (8), we get

$$\hat{e}_n = z_n - \alpha = E_1e_n^3 + E_2e_n^4 + E_3e_n^5 + E_4e_n^6 + O[e_n]^7 \tag{13}$$

where

$$\begin{aligned}
E_1 &= \frac{1}{2\nu^2(1+\nu)}\big((1+\nu)^2C_1^2 - 2(-1+\nu)\nu C_2\big) \\
E_2 &= \frac{1}{6((-1+\nu)\nu^3(1+\nu)^2)}\big((1+\nu)^4(-7+6\nu)C_1^3 - 6(-1+\nu)\nu(1+\nu) \\
&\quad \times (-1+\nu(4+3\nu))C_1C_2 + 6(-1+\nu)^2\nu^2(1+3\nu)C_3\big) \\
E_3 &= \frac{1}{6(-1+\nu)^2\nu^4(1+\nu)^3}\big((1+\nu)^4(10+\nu(4+\nu(-22-3\nu+9\nu^2)))C_1^4 \\
&\quad - 6(-1+\nu)\nu(1+\nu)^3(-1+\nu(-13+4\nu+6\nu^2))C_1C_2 + 12(-1+\nu)^2\nu^2 \\
&\quad \times (1+\nu^2(2+\nu)(2+3\nu))C_1C_3 + 6\nu^2((-1+\nu^2)^2(-4+ \\
&\quad \times \nu(5+3\nu))C_2^2 - 2(-1+\nu)^3\nu(1+\nu(2+3\nu))C_4)\big) \\
E_4 &= \frac{1}{30(-1+\nu)^3\nu^5(1+\nu)^4}\big(-(1+\nu)^5(-68+\nu(-33+\nu(202 \\
&\quad + \nu(87+10\nu(-23-3\nu+6\nu^2)))))C_1^3 + 5(-1+\nu)\nu(1+\nu)^4(20+ \\
&\quad \times \nu(147+\nu(-77+\nu(-227+15\nu(3+4\nu)))))C_1^3C_2 - 15\nu^2(-1+\nu^2)^2(-25+ \\
&\quad \times \nu(-10+\nu(-52+\nu(-26+5\nu(9+4\nu)))))C_1^2C_3 + 60(-1+\nu)^2\nu^2(1+\nu)C_1 \\
&\quad \times ((-1+\nu)^2(5+\nu(-16+\nu(-12+5\nu(2+\nu))))C_2^2 + (-1 \\
&\quad + \nu)\nu(-1+\nu(6+5\nu^2(2+\nu)))C_4) + 30(-1+\nu)^3\nu^3(5(1+ \\
&\quad \times \nu)^2(-1+\nu(-2+\nu(5+2\nu)))C_2C_3 + 2\nu(1+4\nu-5\nu^4C_5))\big)
\end{aligned}$$

Using the expression of z_n from Eq. (13) in the third step of the proposed method defined by Eq. (8), we get

Table 1 Test function with initial guess x_0 and multiplicity ν

Test function $\psi(x)$	Initial guesses	Multiplicity
$\psi_1(x) = (cos(x) + x)^{15}$	–0.9	15
$\psi_2(x) = ((x-1)^{10} - 1)^6$	–0.1	6
$\psi_3(x) = (x^3 + x + 1)^6$	–0.8	10
$\psi_4(x) = (sin(x^2) - x^2 + 1)^{66}$	1.6	66
$\psi_5(x) = (2 - x + e^{3+x-x^2})^9$	2.49	9

Table 2 Comparison of various iterative methods

$\lvert\psi(x_n)\rvert$	TM	GM	QM	NPM
$\lvert\psi_1(x_n)\rvert$	4.4266×10^{-9}	$2.6244 \times 10^{-22646}$	2.2574×10^{-1923}	$1.2291 \times 10^{-24937}$
$\lvert\psi_2(x_n)\rvert$	1.9439	4.4745×10^{-2050}	8.8343×10^{-4052}	3.0102×10^{-3150}
$\lvert\psi_3(x_n)\rvert$	6.1462×10^{-6}	$1.4433 \times 10^{-10001}$	2.5167×10^{-917}	$2.6274 \times 10^{-12555}$
$\lvert\psi_4(x_n)\rvert$	5.5942×10^{-19}	$1.0763 \times 10^{-53877}$	2.0935×10^{-349}	$3.7107 \times 10^{-66773}$
$\lvert\psi_5(x_n)\rvert$	6.6823×10^{-26}	$1.8141 \times 10^{-22160}$	9.9962×10^{-3692}	$2.2123 \times 10^{-31653}$

$$e_{n+1} = \frac{C_1((\nu+1)^2 C_1^2 - 2(\nu-1)\nu C_2)^2}{4\nu^5(1+\nu)^2} e_n^6 + O[e_n]^7 \tag{14}$$

Equation (14) shows that the newly developed method defined by (8) has sixth order of convergence.

3 Numerical Results

In this section, we analyse the computational efficiency of the introduced iterative method (8) using several test functions and compare it with other existing methods. In Table 2, we have displayed the comparison of the convergence of the methods. Table 2 shows the absolute residual error ($\mid \psi(x_n) \mid$) of the functions after four full iterations of the methods have been completed. We have compared the newly proposed method (NPM for short) defined in Eq. (8) with the methods given in Eqs. (5), (6) and (7) denoted by TM [4], GM [5] and QM [6], respectively. Mathematica 11.3 software has been used to generate the numerical results in Table 2.

4 Conclusion

We have introduced a new sixth-order iterative method based on Newton's method for finding multiple roots of nonlinear equations. We compare the newly introduced method with existing methods having the same convergence order using some examples of nonlinear equations. The results given in Table 2, have demonstrated the superiority of the introduced method as compared to the existing methods even though the same examples with the same initial guess are used. It affirms that the introduced iterative method has smaller $|\psi(x)|$ and simple asymptotic error terms. Therefore, the introduced method is efficient than the other equivalent methods in comparison to finding multiple roots of nonlinear equations.

References

1. Traub, J.F.: Iterative Methods for the Solution of Equations, Prentice-Hall, Englewood Cliffs, NJ, USA (1964)
2. Ostrowski, A.M.: Solution of Equations in Euclidean and Banach Space. Academic Press, New York, NY, USA (1973)
3. Jamaludin, N.A.A., Long, N.N., Salimi, M., Sharifi, S.: Review of some iterative methods for solving nonlinear equations with multiple zeros. Afrika Matematika **30**(3–4), 355–369 (2019). https://doi.org/10.1007/s13370-018-00650-3
4. Thukral, R.: Introduction to higher-order iterative methods for finding multiple roots of nonlinear equations. Hindawi Publ. Corp. J. Math. (2013). https://doi.org/10.1155/2013/404635
5. Geum, Y.H., Kim, Y.I., Neta, B.: A sixth-order family of three-point modified Newton-like multiple-root finders and the dynamics behind their extraneous fixed points. Appl. Math. Comput. **283**, 120–140 (2016). https://doi.org/10.1016/j.amc.2016.02.029
6. Qudsi, R., Imran, M., Syamsudhuha: another six order iterative method free from derivatives for solving multiple roots of a nonlinear equation. Appl. Math. Sci. **43**, 2121– 2129 (2017). https://doi.org/10.12988/ams.2017.76208
7. Singh, A., Jaiswal, J.P.: An efficient family of optimal fourth-order iterative methods for finding multiple roots of nonlinear equations. Proc. Natl. Acad. Sci. India, Sect. A Phys. Sci. **85**, 439–450 (2015). https://doi.org/10.1007/s40010-015-0221-5
8. Behl, R., Al-Hamdan, W.M.: A 4th-order optimal extension of Ostrowski's method for multiple zeros of univariate nonlinear functions. Math. MDPI **7**(9) (2019). https://doi.org/10.3390/math7090803
9. Chanu, W.H., Panday, S., Dwivedi, M.: New fifth order iterative method for finding multiple root of nonlinear function. Eng. Lett.**29**(3) 942–947 (2021)
10. Qudsi, R., Imran, M., Syamsudhuha: A sixth-order iterative method free from derivative for solving multiple roots of a nonlinear equation. Appl. Math. Sci. **8**(2014), 5721–5730. https://doi.org/10.12988/ams.2014.47567
11. Khattri, S.K.: How to increase convergence order of the newton method to 2m. Appl. Math. **59**, 15–24 (2014). https://doi.org/10.1007/s10492-014-0038-6

Separation Axioms in Bipolar Fuzzy Topological Spaces

Manjeet Singh and Asha Gupta

Abstract In this paper, the definition of the bipolar fuzzy (bf) point has been generalized, and using this, the concept of separation axioms has been introduced in bipolar fuzzy settings. Moreover, the relation between these separation axioms has been established.

Keywords Bipolar fuzzy set · Bipolar fuzzy topology

1 Introduction

Fuzzy sets have been introduced by Zadeh [1]. After that, in every branch of science and technology, fuzzy sets have been used to generalize all the concepts. The concept of general topological space is generalized by using fuzzy sets to fuzzy topological space by Chang [2]. Further, a number of papers have been devoted to generalize almost all the concepts of general topology in fuzzy topological space(fts). Tripathy and Borgohain [3], Tripathy and Baruah [4, 5] have investigated Different classes in fuzzy numbers of sequence spaces. Tripathy and Ray [6] have studied mixed fts. The concept of fuzzy sets has been generalized to bipolar fuzzy (briefly bf) sets by Zhang [7]. After that, basic operations on bf sets have been defined by Lee [8, 9]. Moreover, regular bf graphs have been studied by Akram and Dudek [10] and bf topological spaces have been defined by Azhagappan and Kamaraj. Recently, bf point, a neighborhood system, the notion of compactness, and few other properties have been introduced in bf topological space by Kim et al. [11].

In the present work, the concept of bf point has been generalized of Kim et al. [11] and also observed that the notion of disjointness $K \cap L = \emptyset \Leftrightarrow K \subseteq coL$ (coL is the complement of set L) is no longer valid for bf sets. So there is a deviation

M. Singh · A. Gupta (✉)
Department of Applied Sciences, Punjab Engineering College, Chandigarh, India
e-mail: ashagoel1968@gmail.com

M. Singh
e-mail: manjeetsingh.phdappsc@pec.edu.in

R. K. Sharma et al. (eds.), *Frontiers in Industrial and Applied Mathematics*,
Springer Proceedings in Mathematics & Statistics 410,
https://doi.org/10.1007/978-981-19-7272-0_41

from general topology to bf topology, only the implication $K \cap L = \emptyset \Rightarrow K \subseteq coL$ is valid. The concept of separation axioms in bf settings has been introduced by using the generalized bf point and the notion of disjointness. Moreover, the relation between these separation axioms has been established.

2 Preliminaries and Definitions

In this section, we summarize some definitions and results of bf topological space which is helpful in the following section.

Let X be a nonempty set. Then a pair $K = (K^+, K^-)$ is called a bf set in X, if $K^+ : X \to [0, 1]$ and $K^- : X \to [-1, 0]$ are mappings. For each $x \in X$, the positive membership degree $K^+(x)$ is used to denote the satisfaction degree of the element x to the property corresponding to the bf set K and the negative membership degree $K^-(x)$ is used to denote the satisfaction degree of the element x to some implicit counter-property corresponding to the bf set K. The empty bf set is denoted by $0_{bp} = (0^+, 0^-)$ and defined by $0^+(x) = 0 = 0^-(x)$ for all $x \in X$. Also, the whole bf set is denoted by $1_{bp} = (1^+, 1^-)$ and defined by $1^+(x) = 1$ and $1^-(x) = -1$ for all $x \in X$.

Definition 1 ([9]) Let X be a nonempty set and let K, L be two bf sets in X.

(i) We say that K is a subset of L, denoted by $K \subseteq L$, if for each $x \in X$,

$$K^+(x) \leq L^+(x) \, and \, K^-(x) \geq L^-(x).$$

(ii) We say that K is equal to L, denoted by $K = L$, if $K \subseteq L$ and $L \subseteq K$.

(iii) The complement of K, denoted by $K^c = ((K^c)^+, (K^c)^-)$, is a bf set in X defined as: for each $x \in X$, $K^c(x) = (1 - K^+(x), -1 - K^-(x))$, i.e.,

$$(K^c)^+(x) = 1 - K^+(x), (K^c)^-(x) = -1 - K^-(x).$$

(iv) The intersection of K and L, denoted by $K \cap L$, is a bf set in X defined as: for each $x \in X$,

$$(K \cap L)(x) = (K^+(x) \wedge L^+(x), K^-(x) \vee L^-(x)).$$

(v) The union of K and L, denoted by $K \cup L$, is a bf set in X defined as: for each $x \in X$,

$$(K \cup L)(x) = (K^+(x) \vee L^+(x), K^-(x) \wedge L^-(x)).$$

Definition 2 ([9]) Let X be a nonempty set and let $\{K_i : i \in I\}$ be a family of subsets of X.

(i) The intersection of $\{K_i : i \in I\}$, denoted by $\bigcap_{i \in I} K_i$ is a bf set in X defined by: for each $x \in X$,

$$(\bigcap_{i \in I} K_i)(x) = (\bigwedge_{i \in I} K_i^+(x), \bigvee_{i \in I} K_i^-(x)).$$

(ii) The union of $\{K_i : i \in I\}$, denoted by $\bigcup_{i \in I} K_i$ is a bf set in X defined by: for each $x \in X$,

$$(\bigcup_{i \in I} K_i)(x) = (\bigvee_{i \in I} K_i^+(x), \bigwedge_{i \in I} K_i^-(x)).$$

Definition 3 ([11]) Let X be a nonempty set. Suppose a collection of bf sets of X is τ, then τ is said to be bf topology on X, if the following axioms is satisfied:

(i) $0_{bp}, 1_{bp} \in \tau$.
(ii) if $K, L \in \tau$, then $K \cap L \in \tau$.
(iii) if $\{K_i : i \in I\} \subset \tau$, then $\bigcup_{i \in I} K_i \in \tau$.

In this case, a bf topological space is denoted by the pair (X, τ) and each element of τ is said to be an open bf set of X. The closed bf set is the complement of an open bf set.

Definition 4 ([11]) Let X and Y be nonempty sets, let $K \subseteq X$ and $L \subseteq Y$ and let $f : X \to Y$ be a mapping. Then

(i) The image of K under f, denoted by $f(K) = (f(K^+), f(K^-))$, is a bf set in Y defined as follows: for each $y \in Y$,

$$f(K^+)(y) = \begin{cases} \vee_{x \in f^{-1}(y)} K^+(x), & if \quad f^{-1}(y) \neq \emptyset; \\ 0, & otherwise. \end{cases}$$

and

$$f(K^-)(y) = \begin{cases} \wedge_{x \in f^{-1}(y)} K^-(x), & if \quad f^{-1}(y) \neq \emptyset; \\ 0, & otherwise. \end{cases}$$

(ii) The preimage of L under f, denoted by $f^{-1}(L) = (f^{-1}(L^+), f^{-1}(L^-))$, is a bf set in Y defined as follows: for each $x \in X$, $[f^{-1}(L^+)](x) = L^+ \circ f(x)$ and

$$[f^{-1}(L^-)](x) = L^- \circ f(x).$$

Definition 5 ([11]) Let $(X, \tau_1), (Y, \tau_2)$ be two bf topological spaces. Then a mapping $f : (X, \tau_1) \to (Y, \tau_2)$ is said to be continuous if $f^{-1}(V) \in \tau_1$, for each $V \in \tau_2$.

3 Separation Axioms

In this section, firstly, we define the generalized form of bipolar fuzzy point and show some properties of general topology that is not valid in bf settings. Secondly, we define separation axioms in bf topology and discuss the relations between these separation axioms.

Definition 6 ([11]) Let $X \neq \emptyset$ be a set and x in X, $(\alpha, \beta) \in (0, 1] \times [-1, 0)$. Then $x_{(\alpha,\beta)}$ with the values (α, β) and the support x is said to be a bf point in X, if for every y in X,

$$x_{(\alpha,\beta)}(y) = \begin{cases} (\alpha, \beta), & if\ y = x \\ (0, 0), & otherwise \end{cases}$$

The bf point has been generalized in the following definition:

Definition 7 Let x in X, $(0, 0) \neq (\alpha, \beta) \in [0, 1] \times [-1, 0]$ and K a bf set of X. Then

(i) $x_{(\alpha,\beta)}$ with the values (α, β) and the support x is called a generalized bf point in X , if for every y in X,

$$x_{(\alpha,\beta)}(y) = \begin{cases} (\alpha, \beta), & if\ y = x \\ (0, 0), & otherwise \end{cases}$$

(ii) K contains $x_{(\alpha,\beta)}$ (i.e. $x_{(\alpha,\beta)} \in K$), if

$$K^-(x) \leq \beta\, and\, K^+(x) \geq \alpha$$

On the basis of the preceding definition, the following implications hold:

$$K = L \Longleftrightarrow \text{for all } p \in X,\ p \in K \Leftrightarrow p \in L;$$
$$K \subseteq L \Longleftrightarrow \text{for all } p \in X,\ p \in K \Rightarrow p \in L;$$
$$p \in K \cap L \Longleftrightarrow \text{for all } p \in X, p \in K \wedge p \in L;$$

more generally,

$$p \in \bigcap_{i \in I} K_i \Longleftrightarrow (p \in K_i,\ \forall\, i \in I),$$

I is any index set.

We remark

$$p \in K \cup L \Leftarrow \text{for all } p \in X,\ p \in K \vee p \in L,$$

holds, but the converse of this implication does not remain valid.

Example 1 Suppose two bf subsets K, L and $K^+(x) = \frac{3}{4}$, $K^-(x) = -\frac{1}{2}$ for each x in X and $L^+(x) = \frac{1}{2}$, $L^-(x) = -\frac{3}{4}$ for each x in X. Now $K \cup L \subseteq X$, then $(K \cup L)^+(x) = \frac{3}{4}$ and $(K \cup L)^-(x) = -\frac{3}{4}$ for each x in X. If p in $K \cup L$ such that $(K \cup L)^+(p) = \frac{3}{4}$ and $(K \cup L)^-(p) = -\frac{3}{4}$, then neither p in K nor p in L.

To introduce the bf separation axioms, we have to discuss the notion of disjointness. The equivalence of set theory

$$K \cap L = \emptyset \Leftrightarrow K \subseteq coL.$$

is not valid for bf set theory; indeed, the following implication is true.

$$K \cap L = \emptyset \Rightarrow K \subseteq coL.$$

The separation axioms in bf settings are defined by using notion of disjointness in bf settings. So, we get the deviation from general topology to bf topology:

Definition 8 A bf topological space is called:

1. BFT_0 if for each pair consisting of two different bf points (p, q) with supports x and y, there exists an open bf set R such that $p \in R$ and $q \cap R = 0$ $(i.e.\ R^+(y) = 0$ and $R^-(y) = 0)$ or $q \in R$ and $p \cap R = 0$ $(i.e.\ R^+(x) = 0$ and $R^-(x) = 0)$.
2. $BFT_{0\alpha}$ if for each pair consisting of two different bf points (p, q) with supports x and y, there exists an open bf set R such that $p \in R \subseteq coq$ or $q \in R \subseteq cop$.
3. BFT_1 if for each pair consisting of two different bf points (p, q) with supports x and y, there exist two open bf sets R and S such that $p \in R$, $q \cap R = 0$ $(i.e.\ R^+(y) = 0$ and $R^-(y) = 0)$ and $q \in S$, $p \cap S = 0$ $(i.e.\ S^+(x) = 0$ and $S^-(x) = 0)$.
4. $BFT_{1\alpha}$ if and only if for each pair consisting of two different bf points (p, q) with supports x and y, there exist two open bf sets R and S such that $p \in R \subseteq coq$ and $q \in S \subseteq cop$.
5. BFT_s (strong BFT_1) if every bf singleton is a closed bf set.
6. BFT_2 (BFT-Hausdorff) if for each pair consisting of two different bf points (p, q) with supports x and y, there exist two open bf sets R and S such that $p \in R$, $q \in S$ and $R \cap S = 0$.
7. $BFT_{2\alpha}$ (strong BFT-Hausdorff) if for each pair consisting of two different bf points (p, q) with supports x and y, there exist two open bf sets R and S such that $p \in R$, $q \in S$ and $R \subseteq coS$.
8. $BFT_{2\frac{1}{2}}$ if for each pair consisting of two different bf points (p, q) with supports x and y, there exist two open bf sets R and S such that $p \in R$, $q \in S$ and $clR \cap clS = 0$.
9. $BFT_{2\frac{1}{2}\alpha}$ if for each pair consisting of two different bf points (p, q) with supports x and y, there exist two open bf sets R and S such that $p \in R$, $q \in S$ and $clR \subseteq co(clS)$.

With the help of above definitions the following implications can be noticed:

1. (X, τ) is $BFT_i \implies (X, \tau)$ is $BFT_{i\alpha}$ $i = 0, 1, 2, 2\frac{1}{2}$
2. (X, τ) is $BFT_{2\frac{1}{2}} \implies (X, \tau)$ is $BFT_2 \implies (X, \tau)$ is $BFT_1 \implies (X, \tau)$ is BFT_0
3. (X, τ) is $BFT_{2\frac{1}{2}\alpha} \implies (X, \tau)$ is $BFT_{2\alpha} \implies (X, \tau)$ is $BFT_{1\alpha} \implies (X, \tau)$ is $BFT_{0\alpha}$
4. (X, τ) is $BFT_s \implies (X, \tau)$ is BFT_1

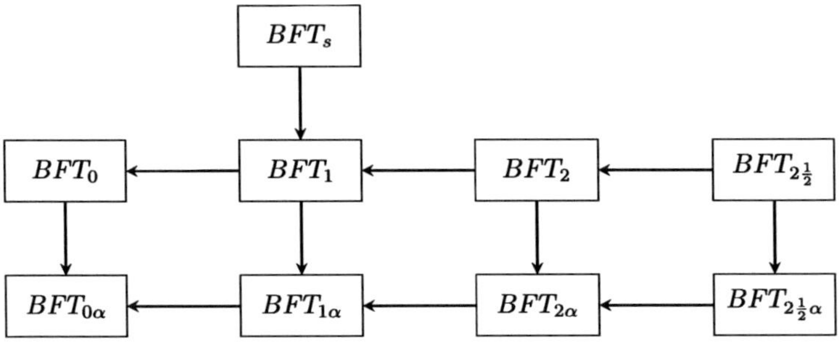

Fig. 1 Implication between separation axioms

Theorem 1 *A space is BFT_1 if and only if every bf singleton with $\alpha = 1$ and $\beta = -1$ is closed.*

Proof Let p_0 be an arbitrary bf singleton with support $x_0 \in X$, $\beta = p_0^-(x_0) = -1$ and $\alpha = p_0^+(x_0) = 1$. Suppose p is another bf point with support x, there exist O_0 and O_p open bf sets such that $p_0 \subseteq O_0 \subseteq cop$ and $p \subseteq O_p \subseteq cop_0$. Since every bf set is the union of all the bf singleton, it contains, i.e., $cop_0 = \cup_{p \subseteq cop_0} p$. From $cop_0^+(x_0) = 1 - p_0^+(x_0) = 0$ and $cop_0^-(x_0) = -1 - p_0^-(x_0) = 0$, we deduce $cop_0 = \cup_{p \subseteq cop_0} O_p$, and thus the cop_0 is open.

Conversely, consider p_1 and p_2 be a pair of two different bf points with support x_1 and x_2. Let q_1 and q_2 be another pair of bf points with support x_1 and x_2, respectively, such that $q_1^-(x_1) = q_2^-(x_2) = -1$ and $q_1^+(x_1) = q_2^+(x_2) = 1$. So, the bf sets coq_1 and coq_2 are open bf and satisfy the conditions $p_1 \subseteq coq_2 \subseteq cop_2$ and $p_2 \subseteq coq_1 \subseteq cop_1$.

Theorem 2 *A weakest bf topology τ exists for every X, such that (X, τ) is BFT_s.*

Proof Let X be any arbitrary set. Consider the collection τ of bf sets on X defined by

$$\tau = \{O : O \subseteq X, supp(coO) \ is \ finite\}.$$

We can easily prove that τ is a bf topology. Clearly, each bf point on X is bf closed, then (X, τ) is BFT_s. Now to prove τ is the weakest bf topology, suppose σ is any other bf topology which is also BFT_s. Let $R \in \tau$ be any set, then $supp(coR) = \{x_1, x_2, x_3, \ldots.., x_n\}$. Consider the bf points p_i for every $i \in \{1, \ldots, n\}$ defined by

$$p_i^+(x_i) = coR^+(x_i) and p_i^-(x_i) = coR^-(x_i)$$

$$p_i^+(x) = 0\, and\; p_i^+(x) = 0\; for\; x \neq x_i$$

The family $\{p_i\}_{i=1}^n$ of σ-closed bf sets is a finite family. From $coR = \cup_{i=1}^n p_i$, we conclude that coR is σ-closed. Hence, $R \in \sigma$.

Definition 9 A bft space (X, τ) is called bf regular if for each pair having a bf point p and bf closed set F in X such that $p \in coF$, there exists a pair of open bf sets (R, S) such that p in R, F is subset of S and $R \cap S = \emptyset$. A bf regular which is also BFT_s is said to be BFT_3.

Definition 10 A bft space (X, τ) is called bf α-regular if for each pair having a bf point p and closed bf set F in X such that $p \in coF$, there exists a pair of open bf sets (R, S) such that p in R, F is subset of S and $R \subseteq coS$. A bf α-regular which is also BFT_s is said to be $BFT_{3\alpha}$.

Theorem 3 *A space (X, τ) is α-regular if and only if for each pair consisting of a bf open set R and a bf point p such that $p \in R$, there exists a bf open set S such that $p \in S \subseteq clS \subseteq R$.*

Proof Suppose p is a bf point and R is a bf open set in X such that $p \in R$. Since co$R$ is closed and $p \in co(coR)$, by α-regularity of space X that there exists bf open sets S_1 and S_2 such that $p \in S_1$, $coR \subseteq S_2$ and $S_1 \subseteq coS_2$. Since coS_2 is closed, $clS_1 \subseteq cl(coS_2) = coS_2 \subseteq R$. So $p \in S_1 \subseteq clS_1 \subseteq R$.

Conversely, let F be a bf closed set and p be any bf point in X such that $p \in coF$. By using the condition, there exists bf open set S_1 such that $p \in S_1 \subseteq clS_1 \subseteq coF$. Using the condition again, there exists bf open set S_2 such that $p \in S_2 \subseteq clS_2 \subseteq S_1$. To complete the proof, take $R_1 = S_2$ and $R_2 = co(clS_1)$ because $p \in R_1$, $F \subseteq R_2$ and $R_1 = S_2 \subseteq S_1 \subseteq clS_1 = co(co(clS_1)) = coR_2$.

Theorem 4 *Every bf $T_{3\alpha}$-space is also a bf $T_{2\frac{1}{2}\alpha}$-space.*

Proof Let (X, τ) be a bf $T_{3\alpha}$-space and p, q be two different bf points with support $x_p \neq x_q$ in X with values $(r_p, -r_p^{'})$ and $(r_q, -r_q^{'})$, respectively. Let p_1 be the crisp bf point with support x_p and value $(1, -1)$. By using the definition of $T_{3\alpha}$-space p_1 is a bf closed set and $q \in co(p_1)$. Since there exists bf open sets R and S such that $q \in R$, $p_1 \subseteq S$ and $R \subseteq coS$. Since $p^+(x_p) < 1 = p_1^+(x_p)$ and $p^-(x_p) > -1 = p_1^-(x_p)$, it follows that $p \in S$. Hence, (X, τ) is a bf $T_{2\frac{1}{2}\alpha}$-space.

Theorem 5 *Every bf α-regular $T_{0\alpha}$-space is a bf $T_{2\frac{1}{2}\alpha}$-space.*

Proof Let (X, τ) be a bf α-regular $T_{0\alpha}$-space and p, q be two different bf points with supports $x_p \neq x_q$ in X and values $(r_p, -r_p^{'})$, $(r_q, -r_q^{'})$ respectively. Let p_1, q_1 be bf points with supports x_p, x_q, respectively, and with values $p_1^+(x_p) = \frac{1}{2}(1 + r_p)$, $p_1^-(x_p) = -\frac{1}{2}(1 + r_p^{'})$ and $q_1^+(x_q) = \frac{1}{2}(1 + r_q)$, $q_1^-(x_q) = -\frac{1}{2}(1 + r_q^{'})$. Therefore, p_1, q_1 are two distinct bf points in X. There exists a bf open set R such that $p_1 \in$

$R \subseteq coq_1$ or $q_1 \in R \subseteq cop_1$ because X is a $T_{0\alpha}$-space. Firstly, if $p_1 \in R \subseteq coq_1$, there exists a bf open set S such that $p_1 \in S \subseteq clS \subseteq R$ because X is α-regular. Since $clS \subseteq R$ and $R \subseteq coq_1$, then $clS \subseteq coq_1$ that is $q_1 \subseteq co(clS)$. Now $q(x_q) = r_q < \frac{1}{2}(1 + r_q) = q_1(x_q)$ and $p(x_p) < p_{1(x_p)}$, then we get $q \in co(clS)$ and $p \in S$. Now let $O_1 = S$ and $O_2 = co(clS)$ are bf open sets in X and $S \subseteq co(co(clS))$. Therefore, there exists bf open sets O_1, O_2 in X such that $p \in O_1$, $q \in O_2$ and $O_1 \subseteq coO_2$. By using the previous theorem, there exist bf open sets O_3 and O_4 such that $p \in O_3 \subseteq clO_3 \subseteq O_1$ and $q \in O_4 \subseteq clO_4 \subseteq O_2$. Therefore, we get $p \in O_3$, $q \in O_4$ and $clO_3 \subseteq O_1 \subseteq coO_2 \subseteq co(clO_4)$. Secondly, if $q_1 \in R \subseteq cop_1$. We get the similar result. So, (X, τ) is $T_{2\frac{1}{2}\alpha}$.

Definition 11 A space (X, τ) is called bf normal if for every pair consisting of bf closed sets F_1, F_2 such that $F_1 \subseteq coF_2$, there exists a pair consisting of open fuzzy sets R, S such that $F_1 \subseteq R$, $F_2 \subseteq S$ and $R \cap S = \emptyset$. A bf normal which is also BFT_s is said to be BFT_4.

Definition 12 A space (X, τ) is called bf α-normal if every pair consisting of bf closed sets F_1, F_2 in X such that $F_1 \subseteq coF_2$, there exists a pair consisting of open fuzzy sets R, S such that $F_1 \subseteq R$, $F_2 \subseteq S$ and $R \subseteq coS$. A bf α-normal which is also BFT_s is said to be $BFT_{4\alpha}$.

Theorem 6 *Every bf $T_{4\alpha}$-space is also a bf $T_{3\alpha}$-space.*

Proof Let p be a bf closed point with support x_p and F a bf closed set in X such that $p \in coF$. Let p_1 be bf point with support x_p and with value $p_1^+(x_p) = \frac{1}{2}(p^+(x_p) + coF^+(x_p))$ and $p_1^-(x_p) = \frac{1}{2}(p^-(x_p) + coF^-(x_p))$. Therefore, $p \in p_1$, $p_1 \in coF$ and p_1 is closed because X is T_s-space. By using the α-normality of X and $p_1 \in coF$, there exist two bf open sets R, S with $p_1 \subseteq R$, $F \subseteq S$ and $R \subseteq coS$. Therefore, $p \in R$, $F \subseteq S$ and $R \subseteq coS$. So, X is bf $T_{3\alpha}$-space.

From the above results, we have observed the following implications for the separation axioms in bf topological spaces (Fig. 2).

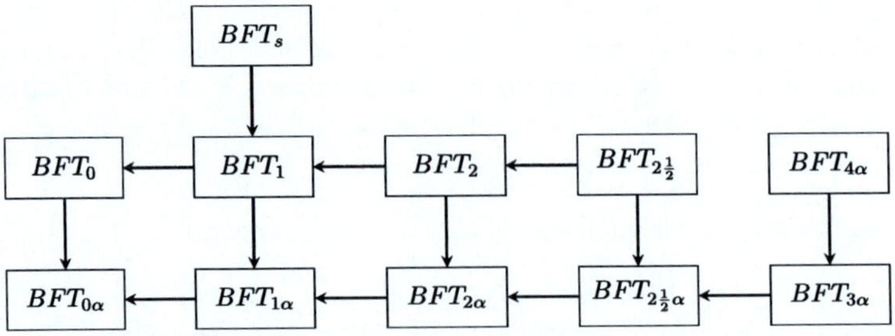

Fig. 2 Implication between separation axioms

References

1. Zadeh, L.A.: Fuzzy sets. Inf. Control. **8**, 338–353 (1965)
2. Chang, C.L.: Fuzzy topological spaces. J. Math. Anal. Appl. **24**, 182–190 (1968)
3. Tripathy, B.C., Borgohain, S.: On a class of n-normed sequences related to the space. Lol. Soc. Paran. Mat. **31**(1), 167–173 (2013)
4. Tripathy, B.C., Baruah, A.: New type of difference sequence spaces of fuzzy real numbers. Math. Modell. Anal. **14**(3), 391–397 (2009)
5. Tripathy, B.C., Baruah, A.: Nőrlund and Riesz mean of sequences of fuzzy real numbers. Appl. Math. Lett. **23**, 651–655 (2010)
6. Tripathy, B.C., Ray, G.C.: On $\delta-$continuity in mixed fuzzy topological spaces. Boletim Da Soc. Parana. Mat. **32**(2), 175–187 (2014)
7. Zhang, W-R.: Bipolar fuzzy set. Proc. IEEE, 835–840 (1998)
8. Lee, K.M.: Comparison of interval-valued fuzzy sets, intuitionistic fuzzy sets and bipolar-valued fuzzy sets. J. Fuzzy Log. Intell. Syst. **14**(2), 125–129 (2004)
9. Lee, K.M.: Bipolar-valued fuzzy sets and their basic operations. In: Proceedings of International Conference on Intelligent Technologies, Bangkok, Thailand, pp. 307–312 (2000)
10. Akram, M., Dudek, W.A.: Regular bipolar fuzzy graphs. Neural Comput. Appl. **21**(suppl 1), S197–S205 (2012)
11. Kim, J., Samanta, S.K., Lim, P.K., Lee, J.G., Hur, K.: Bipolar fuzzy topological spaces. Ann. Fuzzy Math. Inform. **17**, 205–312 (2019)

A Study of $\acute{C}iri\acute{c}$ Type Generalized Contraction Via $\mathcal{B}$-Contraction with Application

Vizender Singh and Bijender Singh

Abstract In this present paper, we introduced the notion of $\acute{C}iri\acute{c}$ type generalized $\mathcal{B}$-contraction for single mapping and for a pair of mappings which generalized Banach contraction principle in a way different from recent literature. Further, we proved some fixed point theorems using these notions. The newly established results are supported by illustrative examples. Finally, the results are applied to solve the Volterra type integral equations.

Keywords $\mathcal{B}$-contraction · Fixed point · Common fixed point · Coincidence point · Complete metric space · Integral equation

MSC: 47H10 · 54H25

1 Introduction

The study of fixed point theory is one of the most effective theories in contemporary mathematical analysis. Fixed point theory is significant in itself, and has advanced significantly during the last century. It delicately integrates analysis, topology, and geometry with a vast spectrum of uses in disciplines such as mathematics, physics, economics, game theory, biology, etc. In 1906, Maurice Frechet, a french mathematician, first proposed the notion of metric space. After that, several extensions of matric space have been proposed since then including fuzzy metric space, probabilistic metric space, b-metric space, etc., by abolishing or modifying certain axioms, rearranging the metric function or abstracting some axioms.

In the discipline of functional analysis, metric fixed point theory is an important field of study. Banach, a prominent Polish mathematician, was the first to introduce

V. Singh
Faculty of Mathematics, Directorate of Distance Education, GJUS & T, Hisar 125001, India

B. Singh (✉)
Department of Mathematics, GJUS & T, Hisar 125001, India
e-mail: punia.bijender@gmail.com

R. K. Sharma et al. (eds.), *Frontiers in Industrial and Applied Mathematics*,
Springer Proceedings in Mathematics & Statistics 410,
https://doi.org/10.1007/978-981-19-7272-0_42

it. Because of its importance and use in several disciplines of research, countless researchers have made numerous generalizations in various directions throughout the years. The fixed point theorem, often referred to as the Banach contraction principle, first appears in Banach's thesis in an explicit form. It was used to prove that an integral equation had a solution. It has been a highly common technique in addressing existing issues in many disciplines of mathematical analysis. Since then, owing to its simplicity and utility, it has been extended in many ways in various fields of mathematics [1, 2, 4, 6–11, 13–15]. The map's contractive character is diminished in certain extensions, while the topology is weakened in other variants.

Recently, Singh et al. [12] introduced the concept of the $\mathcal{B}$-contraction and proved fixed point theorems. In this article, we introduced the notion of the *Ćirić* type generalized $\mathcal{B}$-contraction for a single map and a pair of maps, and their fixed points, coincidence points, and common fixed points. Also, some examples provided in support of our results show the usefulness of newly established notions. At last, application to the integral equation for finding their solution is established.

2 Preliminaries

Definition 1 Let $\mathcal{B} = \{\Phi; \Phi : \mathbb{R}^+ \to \mathbb{R}\}$, where Φ holds the following axioms:

1. Φ is strictly increasing i.e. $\forall\, \varpi, \kappa \in \mathbb{R}^+$ such that $\varpi < \kappa$, $\Phi(\varpi) < \Phi(\kappa)$;
2. $\lim_{n\to\infty} \varpi_n = 0 \Leftrightarrow \lim_{n\to\infty} \Phi(\varpi_n) = 0$; where $\{\varpi_n\}_{n\in\mathbb{N}}$ is sequence of positive numbers;
3. Φ is continuous on $(0, \infty)$.

Let (E, d) be a metric space. A mapping $T : E \to E$ is said to be a *Ćirić* type generalized $\mathcal{B}$-contraction, if $0 < \alpha < 1$ and $\Phi \in \mathcal{B}$ such that

$$\forall_{\varpi,\kappa\in E},\ (d(T\varpi, T\kappa) > 0 \Rightarrow \Phi(d(T\varpi, T\kappa)) \le \alpha\Phi(\mathcal{K}(\varpi, \kappa)), \tag{1}$$

where

$$\mathcal{K}(\varpi, \kappa) = \max\{d(\varpi, \kappa), d(\varpi, T\varpi), d(\kappa, T\kappa), \frac{1}{2}[d(\varpi, T\kappa) + d(\kappa, T\varpi)]\}.$$

Definition 2 ([3]) Let $E(\neq \phi)$ be a set and two mappings $f : E \to E, g : E \to E$ compatible on (E, d) if

$$\lim_{n\to\infty} d(fg\varpi_n, gf\varpi_n) = 0,$$

where, sequence $\{\varpi_n\} \in E$ such that $f\varpi_n = g\varpi_n = \varpi$ as $n \to \infty$, for some $\varpi \in E$.

Definition 3 ([4]) Let $E(\neq \phi)$ be a set and two mappings $f : E \to E, g : E \to E$ are weakly compatible on (E, d) if

$$f\vartheta = g\vartheta \text{ implies that } fg\varpi = gf\varpi,$$

i.e. f and g commute at coincidence points.

Definition 4 ([5]) Let $E(\neq \phi)$ be a set and two mappings $f : E \to E,\ g : E \to E$ are conditionally compatible on (X, d), if the set $A = \{\{\varpi_n\}, \varpi_n \in E$ such that $f\varpi_n = g\varpi_n$ as $n \to \infty\}$ is non-empty, then $\exists$ a sequence $\{\kappa_n\}$ in E such that $f\kappa_n = g\kappa_n = \varpi$ as $n \to \infty$ for some $\varpi \in E$ such that $d(fg\kappa_n, gf\kappa_n) = 0$ as $n \to \infty$.

Definition 5 ([6]) Let $E(\neq \phi)$ be a set and two mappings $f : E \to E, g : E \to E$ are reciprocal continuous on (E, d) if $fg\varpi_n = f\varpi, gf\varpi_n = g\varpi$ as $n \to \infty$, whenever $\{\varpi_n\}$ is a sequence in E such that $f\varpi_n = g\varpi_n = \varpi$ as $n \to \infty$, for some $\varpi \in E$.

Definition 6 ([7]) Let $E(\neq \phi)$ be a set and two mappings $f : E \to E, g : E \to E$ conditionally reciprocally continuous if the set $A = \{\{\varpi_n\}, \varpi_n \in E$ such that $f\varpi_n = g\varpi_n$ as $n \to \infty\}$ is non-empty, then there exists a sequence $\{\kappa_n\}$ in E such that $f\kappa_n = g\kappa_n = \varpi$ as $n \to \infty$ for some $\varpi \in E$ such that $fg\kappa_n = f\varpi$ and $gf\kappa_n = g\varpi$ as $n \to \infty$.

Definition 7 ([8]) A pair of self-maps f, g on metric space is called faintly compatible if f, g is conditional compatible and commute on a non-empty subset of coincidence points, whenever the set of coincidence point is non-empty.

3 Main Results

Theorem 1 *Let (E, d) be a complete metric space and $T : E \to E$ be $\acute{C}iri\acute{c}$ type generalized $\mathcal{B}$-contraction. Then T has a unique fixed point.*
Proof: *Let $\varpi_0 \in E$ be an arbitrary point and define a sequence $\{\varpi_n\}_{n=1}^{\infty}$ such that*

$$\varpi_n = T\varpi_{n-1}, n \in \{1, 2, ...\} \tag{2}$$

If $\varpi_{n_k+1} = \varpi_{n_k}$ for some $n_k \in \{0, 1, 2, ...\}$, then proof is complete. Now let $\varpi_{n+1} \neq \varpi_n$ for every $n \in \{0, 1, 2, ...\}$. Let

$$d_n = d(\varpi_{n+1}, \varpi_n), \quad n \in \{0, 1, 2, ...\}.$$

Then $d_n > 0$ for all $n \in \{0, 1, 2, ...\}$. Now Using eqaution (2), we have

$$\begin{aligned}\Phi(d_n) &= \Phi(d(\varpi_{n+1}, \varpi_n))\\ &= \Phi(d(T\varpi_n, T\varpi_{n-1}))\\ &\leq \alpha\Phi(\mathcal{K}(\varpi_n, \varpi_{n-1}))\\ &= \alpha\Phi(\max\{d(\varpi_n, \varpi_{n-1}), d(\varpi_n, \varpi_{n+1})\})\\ &= \alpha\Phi(\max\{d_n, d_{n-1}\}),\end{aligned}$$

for some $n \in \{1, 2, ...\}$. If $d_n \geq d_{n-1}$, then $\Phi(d_n) \leq \alpha\Phi(d_n)$, which is contradiction. Therefore, $d_n < d_{n-1}, \forall n \in \{1, 2, ...\}$, we have

$$\Phi(d_n) \leq \alpha\Phi(d_{n-1}),$$

therefore, we have

$$\Phi(d_n) \leq \alpha^n\Phi(d_0),$$

as $n \to \infty$, $\Phi(d_n) \to 0$, together with condition 2 in definition 1, provides

$$\lim_{n\to\infty} d_n = 0$$

$$\Rightarrow \lim_{n\to\infty} d(\varpi_n, T\varpi_n) = 0. \tag{3}$$

Now, we shall prove that $\{\varpi_n\}_{n\in\mathbb{N}}$ is a Cauchy sequence. Assume, on the other hand, that $\exists\, \epsilon > 0$ and the sequences $\{a_n\}_{n=1}^{\infty}$ and $\{b_n\}_{n=1}^{\infty}$ from $\mathbb{N}$ such that

$$a_n > b_n > n, \;\; d(\varpi_{a_n}, \varpi_{b_n}) \geq \epsilon, \;\; d(\varpi_{a_{n-1}}, \varpi_{b_n}) < \epsilon, \;\; \forall n \in \mathbb{N}, \tag{4}$$

therefore

$$\begin{aligned}\epsilon \leq d(\varpi_{a_n}, \varpi_{b_n}) &\leq d(\varpi_{a_n}, \varpi_{a_{n-1}}) + d(\varpi_{a_{n-1}}, \varpi_{b_n})\\ &\leq d(\varpi_{a_n}, \varpi_{a_{n-1}}) + \epsilon\\ &= d(\varpi_{a_{n-1}}, T\varpi_{a_{n-1}}) + \epsilon.\end{aligned}$$

Equation (3) and the inequality above yields

$$\lim_{n\to\infty} d(\varpi_{a_n}, \varpi_{b_n}) = \epsilon. \tag{5}$$

As, a consequence of (3), $\exists\, n \in \mathbb{N}$ such that

$$d(\varpi_{a_m}, T\varpi_{a_m}) < \frac{\epsilon}{3} \text{ and } d(\varpi_{b_m}, T\varpi_{b_m}) < \frac{\epsilon}{3}, \;\; \forall\, m \geq n. \tag{6}$$

Furthermore, we will demonstrate that

$$d(T\varpi_{a_m}, T\varpi_{b_m}) = d(\varpi_{a_{m+1}}, \varpi_{b_{m+1}}) > 0, \ \forall \ m \geq n. \tag{7}$$

Assume that, $\exists \ p \geq n$ such that

$$d(\varpi_{a_{p+1}}, \varpi_{b_{p+1}}) = 0, \tag{8}$$

using (4), (6), and (8), we have

$$\begin{aligned}\epsilon \leq d(\varpi_{a_p}, \varpi_{b_p}) &\leq d(\varpi_{a_p}, \varpi_{a_{p+1}}) + d(\varpi_{a_{p+1}}, \varpi_{b_p}) \\ &\leq d(\varpi_{a_p}, \varpi_{a_{p+1}}) + d(\varpi_{a_{p+1}}, \varpi_{b_{p+1}}) + d(\varpi_{b_{p+1}}, \varpi_{b_p}) \\ &= d(\varpi_{a_p}, T\varpi_{a_p}) + d(\varpi_{a_{p+1}}, \varpi_{b_{p+1}}) + d(\varpi_{b_p}, T\varpi_{b_p}) \\ &< \tfrac{\epsilon}{3} + 0 + \tfrac{\epsilon}{3} = \tfrac{2\epsilon}{3}.\end{aligned}$$

It is contradiction, so (7) is true. Therefore

$$\Phi(d(T\varpi_{a_m}, T\varpi_{b_m})) \leq \alpha\Phi(\mathcal{K}(\varpi_{a_m}, \varpi_{b_m})) \tag{9}$$

where

$$\begin{aligned}\mathcal{K}(\varpi_{a_m}, \varpi_{b_m}) = \max\{&d(\varpi_{a_m}, \varpi_{b_m}), d(\varpi_{a_m}, T\varpi_{a_m}), d(\varpi_{b_m}, T\varpi_{b_m}), \\ &\tfrac{1}{2}(d(\varpi_{a_m}, T\varpi_{b_m}) + d(\varpi_{b_m}, T\varpi_{a_m})).\}\end{aligned}$$

By condition 3 in definition 1, (5) and (9), we get

1. *If $\mathcal{K}(\varpi_{a_m}, \varpi_{b_m}) = d(\varpi_{a_m}, \varpi_{b_m})$, then $\Phi(\epsilon) \leq \alpha\Phi(\epsilon)$, it is a contradiction.*
2. *If $\mathcal{K}(\varpi_{b_m}, \varpi_{b_m}) = d(\varpi_{a_m}, T\varpi_{a_m})$, then $\Phi(\epsilon) \leq \alpha\Phi(0)$, it is a contradiction.*
3. *If $\mathcal{K}(\varpi_{b_m}, \varpi_{b_m}) = d(\varpi_{b_m}, T\varpi_{b_m})$, then $\Phi(\epsilon) \leq \alpha\Phi(0)$, it is a contradiction.*
4. *If $\mathcal{K}(\varpi_{a_m}, \varpi_{b_m}) = \frac{1}{2}(d(\varpi_{a_m}, T\varpi_{b_m}) + d(\varpi_{b_m}, T\varpi_{a_m}))$, then $\Phi(\epsilon) < \alpha\Phi(\epsilon)$, it is a contraction.*

Therefore, the sequence $\{\varpi_n\}_{n\in\mathbb{N}}$ is a Cauchy sequence. Since (E, d) is complete, this implies that sequence $\varpi_n \to \varpi, \varpi \in E$. On contrary, suppose that $T\varpi \neq \varpi$, then there exist an $n_1 \in \mathbb{N}$ and subsequence $\{\varpi_{n_k}\}$ of $\{\varpi_n\}$ such that $d(T\varpi_{n_k}, T\varpi_n) > 0$ for all $n_k \geq n_1$. Therefore, we have

$$\begin{aligned}\Phi(d(\varpi_{n_{k+1}}, T\varpi)) = \Phi(d(T\varpi_{nk}, T\varpi)) &\leq \alpha\Phi(\backslash\mathcal{K}(\varpi_{nk}, \varpi)) \\ &\leq \alpha\Phi(\max\{d(\varpi_{n_k}, \varpi), d(\varpi_{n_k}, \varpi_{n_{k+1}}), d(\varpi, T\varpi), \\ &\frac{1}{2}[d(\varpi_{n_k}, T\varpi) + d(\varpi, T\varpi_{n_k})]\}).\end{aligned}$$

As $k \to \infty$ and the countinuity of Φ, we have

$$\Phi(d(\varpi, T\varpi)) \leq \alpha\Phi(d(\varpi, T\varpi)),$$

which is contradiction. Therefore, $T\varpi = \varpi$. Let $\kappa(\neq \varpi) \in E$ such that $T\kappa = \kappa$. Now

$$\begin{aligned}\Phi(d(\varpi, \kappa)) = \Phi(d(T\varpi, T\kappa)) &\leq \alpha\Phi(\mathcal{K}(\varpi, \kappa)) \\ &= \alpha\Phi\{d(\varpi, \kappa), d(\varpi, T\varpi), d(\kappa, T\kappa), \\ &\tfrac{1}{2}[d(\varpi, T\kappa) + d(\kappa, T\varpi)]\} \\ &= \alpha\Phi(d(\varpi, \kappa)).\end{aligned}$$

Which is contradiction. Therefore, T has a unique fixed point. □

Example 1 Let $E = \{\frac{1}{n} : n \in \mathbb{N}\} \cup 0$ and metric $d = |\varpi - \kappa|$, then (E, d) is complete metric space. Consider a map $T : E \to E$ such that

$$T\varpi = \begin{cases} \frac{1}{n+1} & \text{if } \varpi = \frac{1}{n} \\ 0 & \text{if } \varpi = 0. \end{cases}$$

Suppose that $\Phi(\varpi) = \varpi$ and take $\varpi = \frac{1}{n}$, $\kappa = \frac{1}{n+1}$, then $d(\kappa, T\varpi) = 0$ and also

$$\sup_{\varpi,\kappa \in E, \varpi \neq \kappa} \frac{d(T(\frac{1}{n}), T(\frac{1}{n+1}))}{d(\frac{1}{n}, \frac{1}{n+1})} = 1,$$

then T is not $\acute{C}iri\acute{c}$ type generalized $\mathcal{B}$-contraction.

Example 2 Let $E = [0, 1]$ with metric $d = |\varpi - \kappa|$. $T : E \to E$ be a self-map such that $T\varpi = \frac{\varpi}{4}$ and $\phi(\varpi) = \sqrt{\varpi}$, assumptions of Theorem 1 is satisfied with $\alpha \in [\frac{1}{2}, \infty)$, so T has unique fixed point.

Definition 8 Let (E, d) be a metric space and two self-maps $T, S : E \to E$ on (E, d) are said to be $\acute{C}iri\acute{c}$ type generalized $\mathcal{B}$-contraction if $\exists$, $0 < \alpha < 1$ such that

$$\forall \varpi, \kappa \in E, d(T\varpi, T\kappa) > 0 \Rightarrow \phi(d(T\varpi, T\kappa)) \leq \alpha\Phi(\mathcal{K}(\leftrightarrow, \leq)), \tag{10}$$

where

$$M(\varpi, \kappa) = \max\{d(S\varpi, S\kappa), d(S\varpi, T\varpi), d(S\kappa, T\kappa), \frac{d(S\varpi, T\kappa) + d(S\kappa, T\varpi)}{2}\}.$$

Note: Every $\acute{C}iri\acute{c}$ type generalized $\mathcal{B}$-contraction for pair of self-maps is $\mathcal{B}$-contraction but converse need not true.

Theorem 2 *Let (E, d) be a metric space and two self-maps $T, S : E \to E$ be faintly compatible satisfying $\acute{C}iri\acute{c}$ type generalized $\mathcal{B}$-contraction with conditional reciprocal continuity, then T and S have a unique common fixed point.*
Proof: *Let a sequence $\{\varpi_n\} \in E$ such that*

$$\lim_{n\to\infty} T\varpi_n = \lim_{n\to\infty} S\varpi_n = \varpi, \text{ for some } \varpi \in E.$$

As mappings T, S are faintly compatible, then $\exists$ a sequence $\{\kappa_n\} \in E$ and

$$\lim_{n\to\infty} T\kappa_n = \lim_{n\to\infty} S\kappa_n = \kappa,$$

for some $\kappa \in E$ such that

$$\lim_{n\to\infty} d(TS\kappa_n, ST\kappa_n) = 0.$$

Since pair (T, S) as well conditionally reciprocally continuous, then

$$\lim_{n\to\infty} TS\kappa_n = T\kappa$$

and

$$\lim_{n\to\infty} ST\kappa_n = S\kappa.$$

Hence, $T\kappa = S\kappa$, this implies that T and S have coincidence point. As the pair (T, S) is faintly compatible, then $TS\kappa = ST\kappa$. Thus

$$TT\kappa = TS\kappa = ST\kappa = SS\kappa$$

if $T\kappa \neq TT\kappa$, then

$$\Phi(d(T\kappa, TT\kappa)) \leq \alpha\Phi \max\{d(S\kappa, ST\kappa), d(S\kappa, T\kappa), d(ST\kappa, TT\kappa), \frac{d(S\kappa, TT\kappa) + d(ST\kappa, T\kappa)}{2}\}$$

i.e., $\Phi(d(T\kappa, TT\kappa)) \leq \alpha\Phi(d(T\kappa, TT\kappa))$, it is a contradiction. Therefore, T ans S have common fixed point i.e., $TT\kappa = T\kappa = ST\kappa$. □

Example 3 Let $E = (2, 8)$ and $d = |\varpi - \kappa|$ be usual metric. Define $T, S : E \to E$ such that

$$T\varpi = \begin{cases} 4 & \text{if } \varpi \leq 4 \\ 5 & \text{if } \varpi > 4 \end{cases}$$

and

$$S\varpi = \begin{cases} 8 - \varpi & \text{if } \varpi \leq 4 \\ 7 & \text{if } \varpi > 4 \end{cases}$$

and $\phi(\varpi) = \sqrt{\varpi}$, thus all aspects of Theorem 2 is satisfied and there exists a unique fixed point.

4 Application

In this part, we will look at how fixed point methods may be used to solve the integral equation of the following type.

Consider the following integral equation:

$$\varpi(s) = \int_0^s \mathcal{H}(s, t, w(t))\, dt + g(s), s \in [a, b], a > 0, b > 0. \quad (11)$$

In this section, we present an existence theorem for a solution of Eq. (11) that belongs to $E = (C[a, b]; \mathbb{R})$, set of all continuous function defined on $I = [a, b]$ by using the obtained main Theorem 1. Consider

$$(T\varpi)s = \int_0^s \mathcal{H}(s, t, w(t))\, dt \ + g(s), s \in [a, b], a > 0, b > 0.$$

The existence of solution of (11) is equivalent to the existence of a fixed point of T. It is well known that E with a metric that is given by

$$d(\varpi, \kappa) = \sup_{s \in [a,b]} |\varpi(s) - \kappa(s)|, \forall \varpi, \kappa \in E$$

forms a complete metric space. Assume that the aforementioned situation exists:

1. $\mathcal{H} : [a, b] \times [a, b] \times \mathbb{R} \to \mathbb{R}$ and $g : [a, b] \to \mathbb{R}$;
2. $\int_0^s \mathcal{H}(s, t, .) : \mathbb{R} \to \mathbb{R}$ is increasing, $\forall s, t \in [a, b]$;
3. $\exists\, 0 < \alpha < 1$ such that $|\mathcal{H}(s, t, \varpi) - \mathcal{H}(s, t, \kappa)| \le \frac{\sqrt{\alpha}}{s} |\mathcal{K}(\varpi, \kappa)|$;

where $\mathcal{K}(\varpi, \kappa) = \max\{d(\varpi, \kappa), d(\varpi, T\varpi), d(\kappa, T\kappa), \frac{1}{2}[d(\varpi, T\kappa) + d(\kappa, T\varpi)]\}$, for all $s, t \in [a, b]$ and $\varpi, \kappa \in (C[a, b], \mathbb{R})$. Then the integral equation (11) has unique solution.

Theorem 3 *Assume that condition above* $1 - 3$ *are satisfied. Then integral equation (11) has unique solution.*
Proof: *Let*

$$(T\varpi)(s) = \int_0^s \mathcal{H}(s, t, \vartheta(t))dt + g(s), s \in [a, b].$$

Now, by condition (iii), $\forall\, \varpi, \kappa \in (C[a, b], \mathbb{R})$, *we have*

$$\begin{aligned} |(T\varpi)(s) - (T\kappa)(s)| &\le \textstyle\int_0^s |\mathcal{H}(s, t, \varpi(t)) - \mathcal{H}(s, t, \kappa(t))|dt \\ &\le \textstyle\int_0^s \frac{\sqrt{\alpha}}{s} |\mathcal{K}(\varpi, \kappa)|dt \\ &\le \textstyle\frac{\sqrt{\alpha}}{s} |\mathcal{K}(\varpi, \kappa)| \int_0^s dt, \end{aligned}$$

therefore, we have

$$|(T\varpi)(s) - (T\kappa)(s)| \leq \sqrt{\alpha}|\mathcal{K}(\varpi, \kappa)|$$

or

$$d(T\varpi, T\kappa) \leq \sqrt{\alpha}|\mathcal{K}(\varpi, \kappa)|.$$

All the aspects of the Theorem 1 are fulfilled for $\Phi(\varpi) = \varpi^2$. Therefore, integral equation (11) has the unique solution. □

Theorem 4 *Let $T, S : C([a, b], \mathbb{R}) \to C([a, b], \mathbb{R})$ be two self-maps and suppose that following conditions hold:*

1. *$\mathcal{H} : [a, b] \times [a, b] \times \mathbb{R} \to \mathbb{R}$ and $g : [a, b] \to \mathbb{R}$;*
2. *$\int_0^s \mathcal{H}(s, t, .) : \mathbb{R} \to \mathbb{R}$ is increasing, $\forall s, t \in [a, b]$,*
3. *$\exists\, 0 < \alpha < 1$ such that $|\mathcal{H}(s, t, \varpi) - \mathcal{H}(s, t, \kappa)| \leq \frac{\alpha^2}{s}|\mathcal{K}(\varpi, \kappa))|$; where $\mathcal{K}(\varpi, \kappa)) = \max\{d(T\varpi, T\kappa), d(T\varpi, S\varpi), d(T\kappa, S\kappa), \frac{1}{2}[d(T\varpi, S\kappa) + d(T\kappa, S\varpi)]\}$; for all $t, s \in [a, b]$ and $\varpi, \kappa \in (C[a, b], \mathbb{R})$;*
4. *$\lim_{n\to\infty} T\varpi_n = v = \lim_{n\to\infty} S\varpi_n$ for some $v \in (C[a, b], IR)$, there exists a sequence $\{\kappa_n\}$ satisfying $\lim_{n\to\infty} T\varpi_n = u = \lim_{n\to\infty} S\varpi_n$ for some $u \in (C[a, b], IR)$ such that $\lim_{n\to\infty} ST\kappa_n = Su$, $\lim_{n\to\infty} TS\kappa_n = Tu$ and $\lim_{n\to\infty} ST\kappa_n = \lim_{n\to\infty} TS\kappa_n$;*
5. *for all $\varpi \in X$, $S\varpi = T\varpi \Rightarrow ST\varpi = TS\varpi$.*

Then the integral equation (11) has unique solution.
Proof: *Let*

$$(T\varpi)(s) = \int_0^s \mathcal{H}(s, t, \varpi(t))dt + g(s), s \in [a, b].$$

Now, by condition (iii), for all $\varpi, \kappa \in (C[a, b], \mathbb{R})$, we have

$$\begin{aligned}|(T\varpi)(s) - (T\kappa)(s)| &\leq \textstyle\int_0^s |\mathcal{H}(s, t, \varpi(t)) - \mathcal{H}(s, t, \kappa(t))|dt \\ &\leq \textstyle\int_0^s \frac{\alpha^2}{s}|\mathcal{K}(\varpi, \kappa)|dt \\ &\leq \textstyle\frac{\alpha^2}{s}|\mathcal{K}(\varpi, \kappa)| \int_0^s dt,\end{aligned}$$

therefore, we have

$$|(T\varpi)(s) - (T\kappa)(s)| \leq \alpha^2|\mathcal{K}(\varpi, \kappa)|$$

or

$$d(T\varpi, T\kappa) \leq \alpha^2|\mathcal{K}(\varpi, \kappa)|.$$

All the condition of the Theorem 2 are satisfied for $\Phi(\varpi) = \sqrt{\varpi}$. Therefore, integral equation (11) has the unique solution. □

References

1. *Ćirić*, L.B.: A generalization of Banach's contraction principle. Proc. Am. Math. Soc. **45**, 267–273 (1974)
2. *Ćirić*, L.B.: Fixed Point Theory, Contraction Mapping Principle. Faculty Mechanical Engineering. University of Belgrade, Belgrade (2003)
3. Jungck, G.: Compatible mappings and common fixed points. Int. J. Math. Sci. **9**, 771–779 (1986)
4. Jungck, G., Rhoadas, B.E.: Fixed point set valued functions without continuity. Indian J. Pure Appl. Math. **29**, 227–238 (1998)
5. Pant, R.P., Bisht, R.K.: Occasionally weakly compatible mappings and fixed points. Bull. Belg. Math. Soc. Simon Stevin. **19**, 655–661 (2012)
6. Pant, R.P.: A Common fixed point theorem under a new condition. Indian J. Pure Appl. Math. **30**(2), 147–152 (1999)
7. Bisht, R.K., Pant, R.K.: Common fixed point theorems under new continuity condition. Ann. Univ. Ferrara Sez. VII Sci. Math. **58**, 127–141 (2012)
8. Bisht, R.K., Shah, N.: Faintly Compatible Mappings and Common Fixed Points Theory and Application, vol. 156 (2003)
9. Minack, G., Helvaci, A., Altun, I.: Ciric type Generalization F-contraction on complete metric spaces and fixed point results **28**(6), 1143–1151 (2014)
10. Wardowski, D.: Fixed Points of a new type of Contractive mappings in complete metric spaces. Fixed Point Theory Appl. **94** (2012)
11. Wardowski, D., Van Dung, N.: Fixed points of F-Weak contraction on complete metric spaces. Demonstr. Math. **47**, 146–155 (2014)
12. Singh, B., Singh, V.: Vizender, fixed point theorems for new type of mappings in metric spaces and application. J. Int. Acad. Phys. Sci. D **25**(1), 11–24 (2021)
13. Lu, N., He, F., Huang, H.: Answers to questions on the generalized banch contraction conjecture in b-metric spaces. J. Fixed Point Theory Appl. **21**, 43 (2019)
14. Isik, H., Turkoglu, D.: Generalized weakly α-contractive mappings and applications to ordinary differential equations. Miskolc Math. Notes. **17**, 365–379 (2019)
15. Abbas, M., Berzig, M., Nazir, T., Karapinar, E.: Iterative approximation of fixed points for Presic type F-Contraction operators. UPB Sci. Bull. Ser. **78**, 147–160 (2016)

Evolution of Weak Discontinuities in Perfectly Conducting Mixture of Gas and Dust Particles

Danish Amin and D. B. Singh

Abstract In this article, a study concerning the evolution of weak shock past plane and axis-symmetric bodies, in a two-dimensional steady supersonic flow field has been performed. The flow medium is considered to be a mixture of small solid particles and conducting fluid permeated with the transverse magnetic field. The wavefront analysis method is employed to derive transport equations describing the evolutionary behaviour of discontinuities in the plane and axis-symmetric cases. These equations are used to find a closed form expression for the shock formation distance and to determine the conditions ensuring that no shock will evolve at the wave head. In addition to this, the effect of mass concentration of solid particles, magnetic field strength, specific heat ratio and Mach number on the distance of shock formation is analysed and illustrated through figures.

Keywords Wavefront analysis · Dusty gas · Weak shocks · Magnetic field

1 Introduction

Studies related to nonlinear wave propagation such as acceleration waves, discontinuity waves, and shock waves have been a prominent area of research from the past many decades in the field of hydrodynamics and continuum mechanics [1–4]. This is predominantly due to the fact that nonlinearities give rise to a variety of behaviours in hydro-dynamical quantities of the flow field, most commonly a finite blow-up or discontinuity in the surface. Mathematically, these nonlinear waves are characterised by the discontinuity in the normal derivative and form a significant class of solutions in hyperbolic systems. Thus, the analysis of these waves is of great significance from both physical and mathematical perspectives and many studies have been documented on singular surfaces phenomena, discontinuities,

D. Amin (✉) · D. B. Singh
National Institute of Technology, Srinagar 246174, Uttarakhand, India
e-mail: danish.dt17@nituk.ac.in

R. K. Sharma et al. (eds.), *Frontiers in Industrial and Applied Mathematics*,
Springer Proceedings in Mathematics & Statistics 410,
https://doi.org/10.1007/978-981-19-7272-0_43

and finite blow-ups [5–8]. In past, several studies concerning nonlinear wave propagation/investigation of the wavefront in various gas-dynamic regimes have been performed by researchers using different approaches like generalized wavefront expansion method and wavefront analysis method. The method of generalized wavefront expansion involves an asymptotic expansion in a neighbourhood of the wavefront and was proposed by Anile [9]. Some of the applications of wavefront analysis method in different gasdynamic regimes have been reported by Jeffrey [2], Chen [4], Radha et al. [10] and Engelbrecht [11]. Recently, many valuable contributions have been made to literature by the number of authors, some of them include Singh et al. [12], Chaturvedi et al. [13], Singh et al. [14], Sharma et al. [15] etc. Singh et al. [12] studied the evolution of weak shock waves in a two-dimensional steady supersonic flow under the effect of radiation and magnetic field by using wavefront analysis method. It has been concluded from the study that an increase in Mach number and magnetic field in presence of radiative heat transfer enhances the shock formation. Singh et al. [14] used the method of generalized wavefront expansion to study the growth and decay of weak shocks in magnetogasdynamics.

Studies pertaining to weak shocks in a gas laden with small solid particles have a significant relevance to many fields of engineering and science, and it has been attributed with many valuable contributions in the recent years. Some of the most famous and daily encountered problems, where it becomes quite relevant, include nozzle flows, lunar ash flows, underground cosmic explosions, acceleration of particles within shocks, high-speed jet flights in polluted air etc.; astrophysical problems like metallized rocket propellant, characterization of star formation, formation of dusty crystals, macroscopic motion in interplanetary atmosphere with super-sonic speed, coma's collision with a planet and significant amount of other problems. The main temptation behind the study of these propagating weak discontinuities is the ability of these discontinuities to produce an extremely high temperature and pressure at the center of convergence. A decent amount of research work to encounter these problems has been reported by the number of authors (see Pai et al. [16], Amin et al. [17], Nath [18], Anand [19], Chaturvedi et al. [13], Srivastava et al. [20], Nath [21, 22]).

Due to high temperature prevailing within the shocks, the ionization of gas molecules is very likely to happen and the role of electro-magnetization becomes crucial. Therefore, to analyse the shock waves completely we need to consider the magnetic effect simultaneously. A large number of studies have been performed to incorporate the magnetic field effect on nonlinear wave propagation in different gasdynamic regimes (see Sharma et al. [15], Jeffrey [23], Lustman and Geffen [24], Siddiqui et al. [25], Singh et al. [26], and Sharma [27]). In the present article, the evolution of weak discontinuities past plane and axis-symmetric bodies in a two-dimensional steady supersonic flow field has been investigated via the method of wavefront analysis. The solid particles are treated as pseudo fluid and no phase transition or no deformation occurs [28, 29]. The impact of the strength of the magnetic field, dust-loading parameters and Mach number on the distance of shock formation is analysed. The study here concerns with the evolutionary behaviour of shocks when various characteristic curves start merging, i.e., the point at which the leading characteristic intersects with the consecutive characteristics.

2 Basic Equations

The equations governing two-dimensional steady supersonic flow in dusty-gas under the effect of the magnetic field can be written as [12, 13]

$$u\rho_x + v\rho_y + \rho(u_x + v_y + mv/y) = 0, \tag{1}$$

$$\rho\left(uu_x + vu_y\right) + p_x + h_x = 0, \tag{2}$$

$$\rho\left(uv_x + vv_y\right) + p_y + h_y = 0, \tag{3}$$

$$uh_x + vh_y + 2h(u_x + v_y + mv/y) = 0, \tag{4}$$

$$up_x + vp_y - a^2\left(u\rho_x + v\rho_y\right) = 0. \tag{5}$$

Here, ρ is the gas density, p is the pressure, h is the magnetic pressure, u &v are the components of velocity in the direction of x&y axes, respectively. The geometry factor m takes the value 0 &1 for plane and axis symmetric flows, respectively. The quantity $a^2 = \Gamma p/(1 - Z)\rho$ is the equilibrium speed of sound in the medium, where $\Gamma = \gamma(1 + \lambda\beta)/(1 + \lambda\beta\gamma)$, $Z = V_{sp}/V_g$ is the volume fraction of solid particles in the mixture; V_{sp} & V_g is the volumetric extension and total volume of the gas, respectively.

The quantities λ, β and γ are defined as $\lambda = \kappa_p/(1-\kappa_p)$, $\beta = c_{sp}/c_p$, $\gamma = c_p/c_v$, where c_{sp} is the specific heat ratio of solid particles, c_p and c_v are the specific heat of the gas at constant pressure and volume respectively; $\kappa_p = m_{sp}/m_g$ is the mass concentration of solid particles, where m_{sp} and m_g are the total mass of solid particles and the mass of gas respectively. The relation between κ_p and Z is given by $Z = \vartheta\rho$, where $\vartheta = \kappa_p/\rho_{sp}$, ρ_{sp} denotes the species density of solid particles [16].

The equation of state for the mixture of small solid particles and perfect gas is given by [13]

$$p = \frac{(1 - \kappa_p)}{(1 - Z)}\rho R_* T, \tag{6}$$

where T is the temperature and R_* is the gas constant.

3 Characteristic Formulation

The governing Eqs. (1)–(5) can be re-written in the matrix form as follows

$$U_x^i + P^{ij} U_y^i + Q^i = 0, i, j = 1, 2, 3, 4, 5, \tag{7}$$

where ${U^i}_{5\times 1}$, ${F^i}_{5\times 1}$ and $P^{ij}_{5\times 1}$ are given by

$$U = \begin{bmatrix} \rho \\ u \\ v \\ p \\ h \end{bmatrix}; F = \frac{1}{u^2 - c^2} \begin{bmatrix} m\rho uv/y \\ -mvc^2/y \\ 0 \\ m\rho uva^2/y \\ m\rho uvb^2/y \end{bmatrix} \text{ and}$$

$$P = \begin{bmatrix} v/u & -\rho v/(u^2-c^2) & \rho u/(u^2-c^2) & v/u(u^2-c^2) & v/u(u^2-c^2) \\ 0 & uv/(u^2-c^2) & -c^2/(u^2-c^2) & -v/\rho(u^2-c^2) & -v/\rho(u^2-c^2) \\ 0 & 0 & v/u & 1/\rho u & 1/\rho u \\ 0 & -\rho va^2/(u^2-c^2) & \rho ua^2/(u^2-c^2) & v(u^2-b^2)/u(u^2-c^2) & a^2v/u(u^2-c^2) \\ 0 & -\rho vb^2/(u^2-c^2) & \rho ub^2/(u^2-c^2) & b^2v/u(u^2-c^2) & v(u^2-a^2)/u(u^2-c^2) \end{bmatrix}.$$

The eigenvalues of P^{ij} are

$$\lambda^{(1,2)} = \frac{uv \pm c^2\left[(M^2/\epsilon) - 1\right]^{\frac{1}{2}}}{u^2 - c^2}, \lambda^{(3,4,5)} = \frac{v}{u} \tag{8}$$

with their corresponding left eigenvectors

$$\begin{aligned} \mathcal{L}^{(1)} &= \left[0,\ 1,\ -\tfrac{u}{v},\ -\tfrac{[(M^2/\epsilon)-1]^{\frac{1}{2}}}{\rho v},\ 0\right]; \\ \mathcal{L}^{(2)} &= \left[0,\ 1,\ -\tfrac{u}{v},\ \tfrac{[(M^2/\epsilon)-1]^{\frac{1}{2}}}{\rho v},\ 0\right]; \\ \mathcal{L}^{(3)} &= \left[1,\ 0,\ 0,\ -\tfrac{1}{a^2},\ 0\right]; \\ \mathcal{L}^{(4)} &= \left[0,\ 1,\ \tfrac{v}{u},\ \tfrac{\epsilon}{\rho u},\ 0\right]; \\ \mathcal{L}^{(5)} &= \left[0,\ 0,\ 0,\ 1-\epsilon,\ 1\right]. \end{aligned} \tag{9}$$

Here $M = \left[\left(u^2 + v^2\right)^{1/2}\right]/a$ is the upstream Mach number, $\epsilon = 1 + (b^2/a^2)$ is the Alfven number and $c = \left(a^2 + b^2\right)^{1/2}$ is the magneto-sonic speed with $b = (2h/\rho)^{1/2}$ as Alfven speed. In view of Eqs. (8) and (9), the system (7) is hyperbolic for $M > \epsilon^{1/2}$ and possesses two families of characteristic curves along $dy/dx = \lambda^{(1,2)}$. These curves indicate the waves propagating with speeds $\lambda^{(1,2)}$ in the opposite direction. Contrarily, if $M < \epsilon^{1/2}$ i.e., flow is subsonic then the characteristic velocity $\lambda^{(1,2)}$ becomes imaginary and the wavefront phenomenon would not exist.

4 Evolution of Transport Equation for Weak Shock

Assume that $\lambda^{(1)}$ represents the initial wavefront $\xi(x, y) = 0$ passing through the point (x_0, y_0). The flow variables ahead of the initial wavefront are assumed to be uniform having density ρ_0, pressure p_0, temperature $T_0 = T_b$, velocity u_0 along x-axis and velocity $v_0 = 0$ along y-axis. In the upcoming calculations, suffix 0 is used to denote quantities ahead of the wavefront $\xi(x, y) = 0$. For the derivation of transport equations for jump discontinuities in U, we introduce new curvilinear coordinates $\xi, \overline{y}$ as follows [2]

$$\left.\begin{aligned} \xi_x + \lambda^{(1)}\xi_y &= 0 \\ \xi(x, y_0) &= x - x_0 \end{aligned}\right\}, \text{ and } y = \overline{y}. \tag{10}$$

Here, ξ has co-ordinate property. It takes positive values ahead of the leading characteristic, negative values behind the leading characteristic and zero values on the leading characteristic.

In terms of the new coordinate system, Eq. (7) can be written as

$$\mathcal{L}^{(i)}U_\xi + \frac{\lambda^{(1)}\lambda^{(i)}}{(\lambda^{(1)} - \lambda^{(i)})}x_\xi \mathcal{L}^{(i)}U_{\overline{y}} + \frac{\lambda^{(1)}}{(\lambda^{(1)} - \lambda^{(i)})}x_\xi \mathcal{L}^{(i)}F = 0, \tag{11}$$

where $x_\xi = 1/\xi_x$ denotes the Jacobian of transformation, U and $U_{\overline{y}}$ are continuous across the wavefront $\xi = 0$, however U_ξ and x_ξ are discontinuous. On using Eq. (9) in Eq. (11) and subsequently evaluating the resulting expression on the rear side of $\xi = 0$ for $i = 2, 3, 4, 5$ we have

$$\rho_\xi = (1/a_0^2)p_\xi, \tag{12}$$

$$u_\xi = (-1/\rho_0 u_0)\epsilon_0 p_\xi, \tag{13}$$

$$v_\xi = \left[\left\{M_0^2/\epsilon_0) - 1\right\}^{1/2}/\rho_0 u_0\right]p_\xi, \tag{14}$$

$$h_\xi = (\epsilon_0 - 1)p_\xi, \tag{15}$$

Setting $i = 1$ in Eq. (11) and differentiating with respect to ξ and subsequently evaluating it on the rear side, we get

$$c_0^2[(M_0^2/\epsilon_0) - 1]^{1/2} p_{\xi\overline{y}} + \rho_0 u_0 c_0^2 v_{\xi\overline{y}} + \left(\frac{m\rho_0 u_0}{\overline{y}}\right)a_0^2 v_\xi = 0. \tag{16}$$

Plugging Eq. (14) in Eq. (16), we get

$$p_{\xi\bar{y}} + \left[\frac{m}{2\bar{y}}\right]\epsilon_0^{-1} p_\xi = 0. \tag{17}$$

Integrating Eq. (17) with respect to $\bar{y}$ we get

$$p_\xi = \left(\frac{y_0}{\bar{y}}\right)^{\frac{m}{2\epsilon_0}} p_{\xi_0} \tag{18}$$

where, $p_{\xi_0} = \lim_{\bar{y}\to y_0} p_\xi$, taken along $\xi = 0$. Also, along the wavefront $\xi = 0$ we have $x_{\bar{y}} = 1/\lambda^{(1)}$, obtaining the differential coefficient of it with respect to ξ and afterwards solving it on the rear side of $\xi = 0$ and using Eq. (18) in the expression obtained, we get

$$x_{\xi y'} = -\frac{\left[M_0^2\left\{(1 - Z_0) + \left(\frac{\Gamma}{\epsilon_0}\right)\right\} + 2(1 - Z_0)(\epsilon_0 - 1)\right]}{2\rho_0(1 - Z_0)a_0c_0\left(M_0^2 - \epsilon_0\right)^{\frac{1}{2}}}\left(\frac{y_0}{\bar{y}}\right)^{\frac{m}{2\epsilon_0}} p_{\xi_0}, \tag{19}$$

where, $Z_0 = \vartheta\rho_0$. Equations (18) and (19) are the required transport equations.

5 Non-linear Steepening of Waves

In the current section, we will examine the transport equations derived in the previous section and their role in studying the evolutionary behaviour of shocks. Integrating (19) with respect to $\bar{y}$, using Eq. (18) and the fact that $x_{\xi 0} = x_\xi\big|_{\xi=0-} = x_\xi\big|_{\xi=0+} = 1$, which follows from Eq. (10), we get

$$x_\xi = 1 - \left[\frac{M_0^2\left\{(1 - Z_0) + \left(\frac{\Gamma}{\epsilon_0}\right)\right\} + 2(1 - Z_0)(\epsilon_0 - 1)}{2\rho_0(1 - Z_0)a_0c_0\left(M_0^2 - \epsilon_0\right)^{\frac{1}{2}}}\right] y_0^{\frac{m}{2\epsilon_0}} p_{\xi_0} \int_{y0}^{y} \alpha^{-m/2\epsilon_0} d\alpha. \tag{20}$$

Let $y = Y(x)$ represent the body contour having with tangent parallel to the stream line of velocity at the leading edge of the body. Therefore, $dy/dx = v/u$, on differentiating it with respect to ξ and solving the resulting expression on the rear side of $\xi = 0$ gives

$$v_{\xi 0} = u_0 Y_0'', \tag{21}$$

here, $Y_0^{n\prime}$ is the curvature of the body at the tip. On account of Eqs. (14) and (21), Eq. (20) can be re-written as

$$x_\xi = 1 - \left[\frac{M_0^2\left\{(1 - Z_0) + \left(\frac{\Gamma}{\epsilon_0}\right)\right\} + 2(1 - Z_0)(\epsilon_0 - 1)}{2\left(M_0^2 - \epsilon_0\right)} \right] M_0^2 Y_0'' y_0^{\frac{m}{2\epsilon_0}} \int_{y0}^{y} \alpha^{-\frac{m}{2\epsilon_0}} d\alpha. \tag{22}$$

In Eq. (22), x_ξ denotes the Jacobian of transformation on the rear side of $\xi = 0$, for certain value $y = y_\alpha$ it vanishes, as a result of which characteristics of the neighbouring family ξ = constant will cross the wavefront $\xi = 0$ and a strong discontinuity in the solution vector U known as shock wave comes into action. This will happen when U_ξ is finite while $x_\xi = 0$, just behind the wavefront $\xi = 0$ and $U_x = U_\xi / x_\xi$ becomes infinite. The phenomenon of this kind is called steepening of the wavefront. The investigation of expression (22) in detail for plane and axis-symmetric flow configurations is discussed in the next section.

6 Results and Discussion

In the current section we shall discuss the supersonic flow past plane and axis-symmetric bodies and the variation of shock formation distance with different flow parameters. A schematic diagram describing the phenomenon is depicted in Fig. 1.

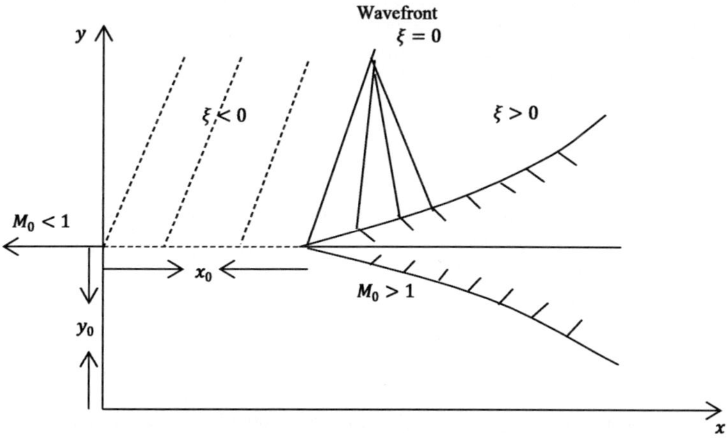

Fig. 1 Schematic diagram of flow field and convergence of characteristics

6.1 Plane Beak Case

Taking, $m = 0$ in Eq. (22) and presuming a plane beak with body contour $y = Y_b(x)$, the sharp edge of the contour will release the initial disturbance with a vanishing initial tangent, consequently Eq. (22) becomes

$$x_\xi = 1 - \frac{Y_b''(0)}{\chi}(y - y_0), \tag{23}$$

where $\chi = 2(M_0^2 - \epsilon_0)\Big[\Big\{\Big(1 - Z_0 + \frac{\Gamma}{\epsilon_0}\Big)M_0^2 + 2(1 - Z_0)(\epsilon_0 - 1)\Big\}M_0^2\Big]^{-1} > 0$, and $Y_b''(0)$ is the radius of curvature of the body at the tip, where the body contour starts bending.

As stated before, characterization of the formation of shock can be done by vanishing Jacobian x_ξ. It is evident from Eq. (23) that on leading wavefront for $y_0 < y$, Jocobian will vanish provided the shape of the body has a compressive corner at $x = 0$ i.e., $Y_b''(0) > 0$ with $Y_b''(0) > \chi$. In addition when, $Y_b''(0) \leq \chi$ then for finite $y_0 > y$, Jocobian will be positive, as a consequence there will be no shock formation. Parameter χ portrays a critical limit in such a way that, when the radius of curvature of the body surpasses this limit a shock will come into play some finite distance away from the body. At the wavehead, the quantities v_x and v_ξ are related to each other as $v_x = v_\xi / x_\xi$. Now, since the quantity v_x can be interpreted in a bit more physical way so, it will be more appropriate to work in respect of this quantity. In view of Eqs. (14), (17) and (23) the expression for studying the decay and growth of weak shocks can be obtained as follows

$$v_x = \frac{v_\xi}{x_\xi} = \frac{a_0 M_0 Y_b''(0)}{1 - \frac{Y_b''(0)}{\chi}(y - y_0)}. \tag{24}$$

It is evident from the above equation that $Y_b''(0) > 0$ with its magnitude greater than χ, therefore a formation shock will take place and the accompanying distance of shock formation $y = y_w$, will be given by

$$y_w = y_0 + \frac{\chi}{y_b''(0)}. \tag{25}$$

Result labelled as Eq. (25) corresponds to the fact that when the denominator of Eq. (24) becomes zero while the numerator remains finite, i.e., when the velocity gradient at $\xi = 0$ is unbounded.

6.2 Axisymmetric Case

Taking $m = 1$ in Eq. (22), and considering a ring shaped body $y = Y_w(x)$ with a sharp-edged inlet releasing the initial disturbance running both inward and outwards along the lines of characteristics. Following these assumptions Eq. (22) can be rewritten as

$$x_\xi = 1 - Y_w''(0)\left\{\frac{2\epsilon_0\varphi}{(2\epsilon_0 - 1)}\right\}\left[y^{\left\{\frac{(2\epsilon_0-1)}{2\epsilon_0}\right\}} - y_0^{\left\{\frac{(2\epsilon_0-1)}{2\epsilon_0}\right\}}\right] \tag{26}$$

where $\varphi = \left[\frac{M_0^2\left\{(1-Z_0)+\left(\frac{\Gamma}{\epsilon_0}\right)\right\}+2(1-Z_0)(\epsilon_0-1)}{2(M_0^2-\epsilon_0)}\right] M_0^2 y_0^{\frac{1}{2\epsilon_0}}$, for the quantity within square brackets we can find $y_0 < y$ such that it will be always positive and less than unity. As mentioned previously, vanishing Jacobian x_ξ will lead to the formation of shock on the condition that $Y_w''(0) > 0$ & $Y_w''(0) > \varphi^{-1}$ the critical value. While, on the other hand if $Y_w''(0) \leq \varphi^{-1}$ then x_ξ is positive and hence no shock formation. Hence we surmise that occurrence of shock formation will happen only when $Y_w''(0)$ strictly exceeds critical bound φ^{-1} with its corresponding distance of shock formation $y = y_w$ can be obtained from Eq. (26) as

$$y_w = \left[y_0^{\left\{\frac{(2\epsilon_0-1)}{2\epsilon_0}\right\}} + \left\{\frac{(2\epsilon_0 - 1)}{2\epsilon_0\varphi}\right\}\left(\frac{1}{Y_w''(0)}\right)\right]^{\left\{\frac{2\epsilon_0}{(2\epsilon_0-1)}\right\}}. \tag{27}$$

From the above equation it is clear that the quantity in the square brackets will be positive and smaller than unity. Hence, as mentioned above the formation of shock will depend on how $Y_w''(0)$ behaves with respect to φ^{-1}.

The distance of shock formation for both plane and axis-symmetric cases are specified by Eqs. (25) and (27), respectively. The effect of various parameters namely the Mach number M_0, Alfven number ϵ_0, adiabatic index γ, the mass concentration of solid particles κ_p, on the distance of shock formation has been displayed in Figs. 2–9. For computational purpose values of different parameters involved are taken as $\epsilon_0 = 1.2, 1.4$; $\gamma = 1.4, 1.67, 2$; $M_0 = 1.5, 2.0, 2.5, 3$; $Y_b''(0) = 0.4$; $Y_w''(0) = 0.4$; $\kappa_p = 0, 0.3, 0.6$; $Z_0 = 0.001$; $\beta = 1$ and $y_0 = 1$. The distance of shock formation increases with an increase in κ_p (Figs. 2 and 3); decreases with an increase in ϵ_0 the magnetic field strength (Figs. 4 and 5) and γ is the adiabatic index (Figs. 6 and 7) for both plane and axis-symmetric cases, respectively. Here the case $\epsilon_0 = 1$ refers to the non-magnetic case. It can be observed from these figures that there is an early formation of shock in the case of plane symmetry as compared to axis-symmetry. Since the shock formation distance y_w is decreasing function of M_0 therefore, an increase in M_0 lead to an early shock formation (Figs. 8 and 9). Hence, it is concluded from the study that an increment in ϵ_0 enhances the shock formation and increment in κ_p increases the distance of shock formation while the increase in γ reduces the distance of shock formation.

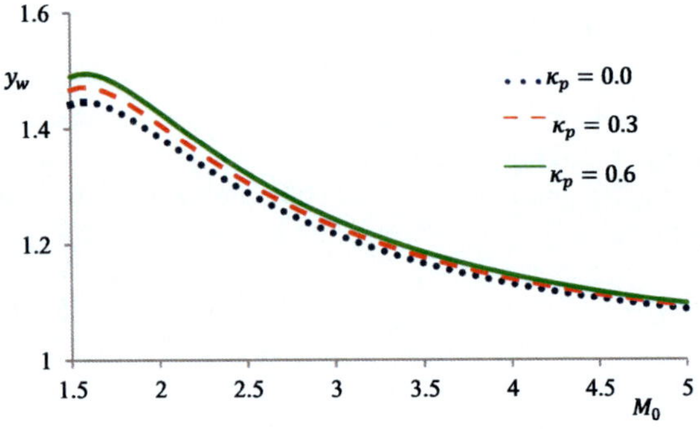

Fig. 2 Variation of distance shock formation y_w with M_0 for different values of κ_p for plane flow, taking $\epsilon_0 = 1.2$ & $\gamma = 1.4$

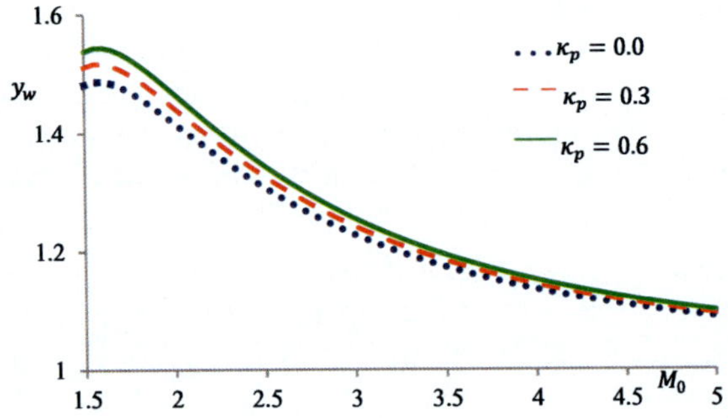

Fig. 3 Variation of distance shock formation y_w with M_0 for different values of κ_p for axis-symmetric flow, taking $\epsilon_0 = 1.2$ & $\gamma = 1.4$

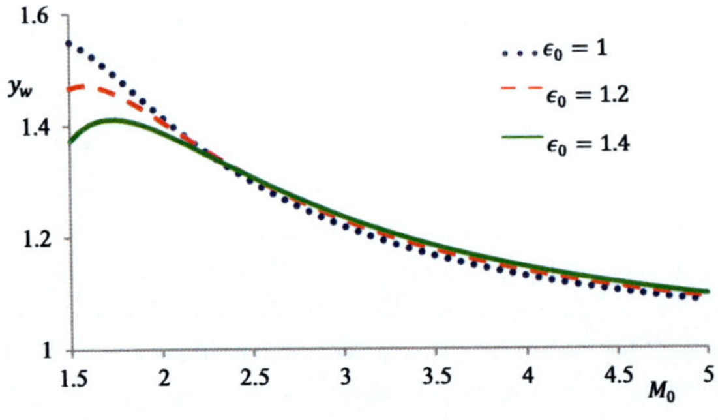

Fig. 4 Variation distance of shock formation y_w with M_0 for different values ϵ_0 for plane flow, taking $\kappa_p = 0.3$ & $\gamma = 1.4$

7 Concluding Remarks

The problems of propagation of weak shocks in a steady supersonic flow past plane and axis-symmetric bodies have been examined theoretically. A derivation for the distance of shock formation is obtained using the wavefront analysis method. The conditions insuring that no shock will ever evolve on the wavefront have been

Fig. 5 Variation distance of shock formation y_w with M_0 for different values ϵ_0 for axis-symmetric flow, taking $\kappa_p = 0.3$ & $\gamma = 1.4$

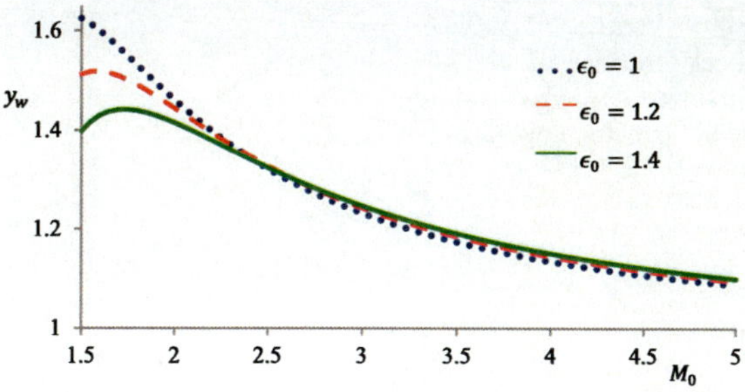

Fig. 6 Variation distance of shock formation y_w with M_0 for different values γ for plane flow, taking $\kappa_p = 0.3$ & $\epsilon_0 = 1.2$

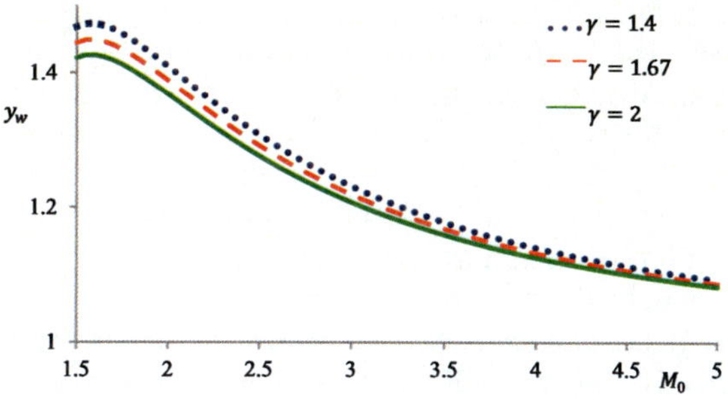

Fig. 7 Variation distance of shock formation y_w with M_0 for different values γ for axis-symmetric flow, taking $\kappa_p = 0.3$ & $\epsilon_0 = 1.2$

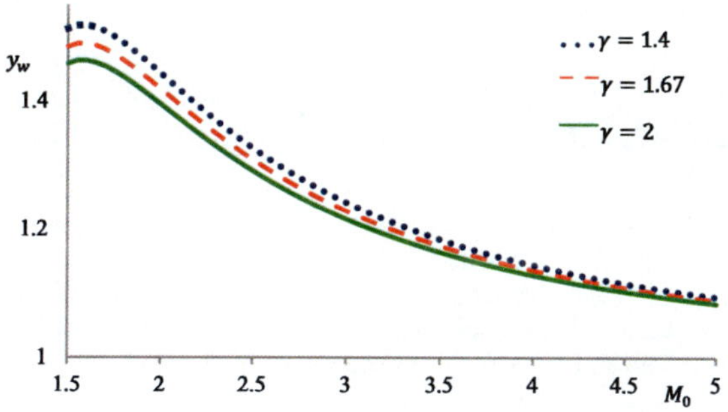

Fig. 8 Variation of distance shock formation y_w with ϵ_0 for different values of M_0^2 for plane flow, taking $\kappa_p = 0.3$ & $\gamma = 1.4$

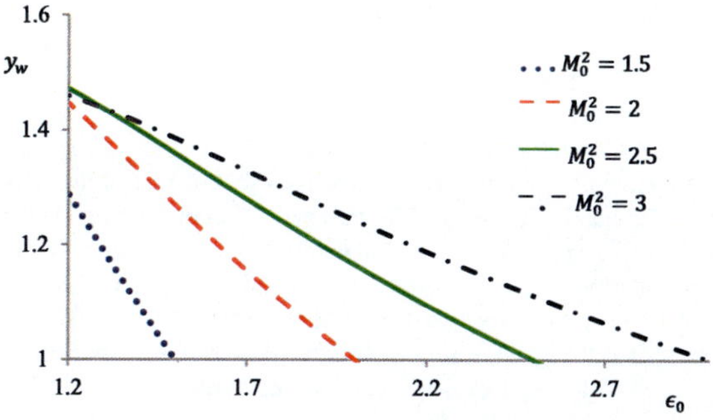

Fig. 9 Variation of distance shock formation y_w with ϵ_0 for different values of M_0^2 for axis-symmetric flow, taking $\kappa_p = 0.3$ & $\gamma = 1.4$

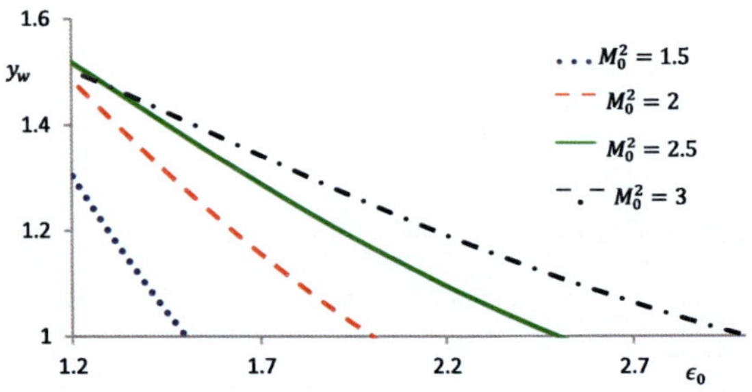

obtained. It has been observed during the study that an increase in the magnetic field strength enhances the shock formation and an increment in mass concentration of solid particles increases the distance of shock formation while an increase in an adiabatic index reduces the distance of shock formation. The current study can be relevant to problems like metallized rocket propellant, characterization of star formation, formation of dusty crystals, macroscopic motion in the interplanetary atmosphere with super-sonic speed, coma's collision with a planet, and many other daily life problems like nozzle flows, lunar ash flows, underground cosmic explosions, acceleration of particles within shocks, high-speed jet flights in polluted air etc.

References

1. Whitham, G.B.: Linear and Nonlinear Waves. Wiley, New York (1974)
2. Jeffrey, A.: Quasilinear Hyperbolic System and Waves. Pitman, London (1976)
3. Jeffrey, A., Taniuti, T.: Nonlinear Wave Propagation. Academic Press, New York (1974)
4. Chen, P.J.: Selected Topics in Wave Propagation. Noordhoff, Leyden (1976)
5. Jordan, P.M.: Growth and decay of shock and acceleration waves in a traffic flow model with relaxation. Physica D **207**(3–4), 220–229 (2005)
6. Menon, V.V., Sharma, V.D., Jeffrey, A.: On the general behavior of acceleration waves. Appl. Anal. **16**(2), 101–120 (1983)
7. Christov, I., Jordan, P.M., Christov, C.I.: Nonlinear acoustic propagation in homentropic perfect gases. Phys. Lett. A **353**(4), 273 (2006)
8. Amin, D., Singh, D.B., Vats, V.K.: Strong shock waves in a self-gravitating gas: a power series solution. AIP Conf. Proc. **2336**, 030004 (2021)
9. Anile, A.M.: Propagation of weak shock waves. Wave Motion **6**(6), 571–578 (1984)
10. Radha, Ch., Sharma, V.D., Jeffery, A.: On interaction of shock waves with weak discontinuities. Appl. Anal. **50**(3–4), 145–166 (1993)
11. Engelbrecht, J.: Theory of non-linear wave propagation with application to the interaction and inverse problems. Int. J. Non-linear Mech. **12**(4), 189–201 (1977)
12. Singh, L.P., Husain, A., Singh, M.: On the evolution of weak discontinuities in radiative magnetogasdynamics. Acta Astronaut. **68**(1–2), 16–21 (2011)
13. Chaturvedi, R.K., Gupta, P., Singh, L.P.: Evolution of weak shock wave in two-dimensional steady supersonic flow in dusty gas. Acta Astronaut. **160**, 552–557 (2019)
14. Singh, L.P., Singh, D.B., Ram, S.D.: Growth and decay of weak shock waves in magnetogasdynamics. Shock Waves **26**(6), 709–716 (2016)

15. Sharma, V.D., Shyam, R., Singh, L.P.: Shock formation distance in a two-dimensional steady supersonic flow over a concave corner in radiative magnetogasdynamics. Z. Angew. Math. Mech. **67**(2), 87–92 (1987)
16. Pai, S.I., Menon, S., Fan, Z.Q.: Similarity solutions of a strong shock wave propagation in a mixture of a gas and dusty particles. Int. J. Eng. Sci. **18**(12), 1365–1373 (1980)
17. Amin, D., Singh, D.B., Vats, V.K.: Strong shock waves in a dusty-gas atmosphere under isothermal conditions: A power series solution. Int. J. Appl. Comput. Math. **7**(5), 1–20 (2021)
18. Nath, G.: Flow behind an exponential shock in a rotational axisymmetric mixture of non-ideal gas and small solid particles with heat conduction and radiation heat flux. Acta Astronaut. **148**, 355–368 (2018)
19. Anand, R.K.: On dynamics of imploding shock waves in a mixture of gas and dust particles. Int. J. Non-linear Mech. **65**, 88–97 (2014)
20. Srivastava, S.K., Chaturvedi, R.K., Singh, L.P.: On the evolution of acceleration discontinuities in van der Waals dusty magnetogasdynamics. Zeitschrift für Naturforschung A **76**(5), 435–443 (2021)
21. Nath, G.: Flow behind an exponential shock wave in a perfectly conducting mixture of micro size small solid particles and non-ideal gas with azimuthal magnetic field. Chinese J. Phys. (2021).https://doi.org/10.1016/j.cjph.2021.11.006
22. Nath, G.: Similarity solution for magnetogasdynamic shock wave in a perfectly conducting dusty gas with axial or azimuthal magnetic field in rotating medium under the influence of radiative and conductive heat fluxes. Acta Astronaut. **182**, 599–610 (2021)
23. Jeffrey, A.: The formation of magnetoacoustic shocks. Math. Anal. Appl. **11**, 139–150 (1965)
24. Lustman, L., Geffen, N.: Wave propagation and breaking in multidimensional magnetogasdynamic flows. Z. Angew. Math. Phys. **30**(4), 637–645 (1979)
25. Siddiqui, M.J., Arora, R., Kumar, A.: Shock waves propagation under the influence of magnetic field. Chaos Solitons Fract. **97**, 66–74 (2017)
26. Singh, L.P., Gupta, R.K., Nath, T.: On the decay of a sawtooth profile in non-ideal magnetogasdynamics. Ain Shamas Eng. J. **6**(2), 599–604 (2015)
27. Sharma, V.D.: On the evolution of compression pulses in a steady magnetohydrodynamic flow over a concave wall. Q. J. Mech. Appl. Math. **40**(4), 527–537 (1987)
28. Vishwakarma, J.P., Lata, P.: A self-similar solution of a shock wave propagation in a perfectly conducting dusty gas. Int. J. Res. Advent Technol. **6**(7), 1789–1800 (2018)
29. Vishwakarma, J.P., Nath, G., Srivastava, R.K.: Self-similar solution for cylindrical shock waves in a weakly conducting dusty gas. Ain Shams Eng. J. **9**(4), 1717–1730 (2018)

Numerical Treatment for a Coupled System of Singularly Perturbed Reaction–Diffusion Equations with Robin Boundary Conditions and Having Boundary and Interior Layers

Sheetal Chawla and S. Chandra Sekhara Rao

Abstract A system of $k(\geq 2)$ linear singularly perturbed differential equations of reaction–diffusion type coupled through their reactive terms is considered with Robin type boundary conditions, and the system has discontinuous source terms. The highest order derivative term of each equation is multiplied by a small positive parameter and these parameters are assumed to be different in magnitude, due to which the overlapping and interacting interior and boundary layers may appear in the solution of the considered problem. A numerical scheme involving a central difference scheme for the differential equations and a cubic spline technique for the Robin boundary conditions is developed on an appropriate piecewise-uniform Shishkin mesh. Error analysis is done and the constructed scheme is proved to be almost second-order uniformly convergent with respect to each perturbation parameter. Numerical experiments are conducted to verify the theoretical findings.

Keywords Coupled system · Singular perturbation · Shishkin mesh · Bakhvalov mesh · Discontinuous source term · Robin boundary conditions · Boundary layer · Parameter-uniform convergence · Finite difference scheme · Interior layer · Cubicspline

1 Introduction

Singularly perturbed problems occur very frequently in the fields of applied mathematics and engineering such as elasticity, oceanography, optimal control theory, and so on. Due to the presence of the perturbation parameters, these problems typically

S. Chawla
Department of Mathematics, Pt. N.R.S. Government College Rohtak, Haryana 124001, India

S. C. S. Rao (✉)
Department of Mathematics, Indian Institute of Technology Delhi, Hauz Khas, New Delhi 110 016, India
e-mail: scsr@maths.iitd.ac.in

R. K. Sharma et al. (eds.), *Frontiers in Industrial and Applied Mathematics*, Springer Proceedings in Mathematics & Statistics 410,
https://doi.org/10.1007/978-981-19-7272-0_44

contain boundary oriented layers having multi-scale character. The regular occurrence of these types of problems and their interesting behavior make mathematicians work for their solutions [5–9, 11–13, 15, 17–19, 21–26]. In the present work, a coupled system of singularly perturbed reaction–diffusion differential equations is examined on the unit interval $\Omega = (0, 1)$, having Robin type boundary conditions with discontinuous source terms. It is assumed that a single discontinuity in the source terms occur at a point $\eta \in \Omega$. Let $\Omega_1 = (0, \eta)$ and $\Omega_2 = (\eta, 1)$. A jump in an arbitrary function ω is defined as $[\omega](\eta) := \omega(\eta+) - \omega(\eta-)$. The objective of the considered Robin boundary value problem is to find $\boldsymbol{u} \in C^0(\overline{\Omega})^k \cap C^1(\Omega)^k \cap C^4(\Omega_1 \cup \Omega_2)^k$:

$$\boldsymbol{Tu} \equiv -\boldsymbol{Eps}\ \boldsymbol{u}'' + \boldsymbol{Bu} = \boldsymbol{d}, x \in \Omega_1 \cup \Omega_2, \tag{1}$$

$$(\boldsymbol{R_l u})_i(0) = R_{l_i} u_i(0) \equiv \alpha_i u_i(0) - \beta_i u_i'(0) = p_i \tag{2}$$

$$(\boldsymbol{R_r u})_i(1) = R_{r_i} u_i(0) \equiv \gamma_i u_i(1) + \delta_i u_i'(1) = q_i, \tag{3}$$

where

$$\alpha_i, \beta_i, \delta_i \geq 0, \quad \gamma_i > 0, \quad 2\alpha_i + \beta_i \geq 1, \quad \gamma_i - \delta_i \geq 1, \quad \text{for } 1 \leq i \leq k, \tag{4}$$

p_i and q_i are given constants for $1 \leq i \leq k$. $\boldsymbol{Eps} = \text{diag}(\varepsilon_1, \ldots, \varepsilon_k)$ with $\varepsilon_1, \ldots, \varepsilon_k$, the small perturbation parameters satisfy $0 < \varepsilon_1 \leq \ldots \leq \varepsilon_k \leq 1$; $\boldsymbol{B}(x) = (b_{ij}(x))_{k\times k}$ and $\boldsymbol{d}(x) = (d_i(x))_{k\times 1}$. Assume that the matrix $\boldsymbol{B}(x)$ is an L_0- matrix. That is, it satisfies the following conditions:

$$b_{ij}(x) \leq 0 \ \text{ for } i \neq j, \ \text{ and } b_{ii}(x) > \sum_{j\neq i, j=1}^{k} |b_{ij}(x)|, \ \text{ for } 1 \leq i \leq k, \tag{5a}$$

$$0 < \alpha < \min_{\substack{x\in\overline{\Omega},\\ 1\leq i\leq k}} \sum_{j=1}^{k} (b_{ij}(x)). \tag{5b}$$

for some constant α. The solution $\boldsymbol{u} = (u_1, u_2, \ldots, u_k)^T$ satisfies the following interface conditions: $[u_i](\eta) = 0, \quad \varepsilon_i[u_i'](\eta) = 0$, for $i = 1, 2, \ldots, k$. Singularly perturbed reaction–diffusion system similar to (1)–(3) with Dirichlet type boundary conditions has been studied in [2, 14, 16]. However, few works consider such problems with Robin boundary conditions, but obtained only the almost first order of uniform convergence independent of perturbation parameters. In [1, 4, 10], the authors considered, a hybrid difference scheme for a coupled system of singularly perturbed differential equations of reaction–diffusion type having Robin boundary conditions and a cubic spline technique inside the layer region, to achieve the second order of uniform convergence up to logarithmic factor. In the present article, an almost second-order parameter-uniform convergence is obtained, which involves spline technique for Robin boundary conditions, weighted average approximation at the point of discontinuity and construction of suitable barrier functions, for a coupled system of singularly perturbed reaction–diffusion equations with Robin boundary conditions and having boundary and interior layers. The outline of the paper is as follows. Section 2 describes the maximum principle and the stability result for the exact solution of the problem. The estimates, on the solution and its derivatives are derived with an appropriate decomposition of the solution into the sum of the smooth and singular components. In Sect. 3, a numerical scheme involving a hybrid difference scheme combined with a cubic spline technique is constructed on a well defined piecewise-uniform Shishkin mesh. Convergence analysis is given in Sect. 4. Numerical results which are inline with the theoretical findings are presented in Sect. 5.

Notation Throughout this article, C is used to denote a generic positive constant which is independent of the small perturbation parameters $\varepsilon_i,\ 1 \le i \le k$ and the discretization parameter N. Also, a generic positive constant vector is denoted by $\boldsymbol{C} = (C, C, \ldots, C)^T$, which need not be same at each occurrence. Further, $|\boldsymbol{v}| = (|v_1|, |v_2|, \ldots, |v_k|)^T$ and $^v \le \boldsymbol{w}$ is defined as $v_i \le w_i$, for $1 \le i \le k$. The maximum norm, denoted by $\| \, . \, \|_G$, where G is a closed subset of $\overline{\Omega}$ is defined as $\| v \|_G = \max_{x\in G} \|v(x)|$ and $\| \boldsymbol{v} \|_G = \max\{\| v_1 \|_G, \| v_2 \|_G, \ldots, \| v_k \|_G\}$.

2 Properties of the Exact Solution

Theorem 1 *The problem (1)–(3) has a solution* $\boldsymbol{u} = (u_1, u_2, \ldots, u_k)^T$ *with* $u_i \in C^0(\overline{\Omega}) \cap C^1(\Omega) \cap C^2(\Omega_1 \cup \Omega_2)$ *for* $1 \le i \le k$.

Proof Let $\boldsymbol{u}^-(x), \boldsymbol{u}^+(x)$ be the particular solutions of the following system of equations

$$-\boldsymbol{Eps}(\boldsymbol{u}^-)''(x) + \boldsymbol{B}(x)(\boldsymbol{u}^-)(x) = \boldsymbol{d}(x), \quad x \in \Omega_1,$$
$$-\boldsymbol{Eps}(\boldsymbol{u}^+)''(x) + \boldsymbol{B}(x)(\boldsymbol{u}^+)(x) = \boldsymbol{d}(x), \quad x \in \Omega_2,$$

respectively. Consider the function

$$\boldsymbol{u}(x) = \begin{cases} \boldsymbol{u}^-(x) + diag(p_1 - R_{l_1}u_1^-(0), \ldots, p_k - R_{l_k}u_k^-(0))\Phi_1(x) + \boldsymbol{P}\Phi_2(x), & : x \in \Omega_1, \\ \boldsymbol{u}^+(x) + diag(q_1 - R_{l_1}u_1^+(1), \ldots, q_k - R_{l_k}u_k^+(1))\Phi_2(x) + \boldsymbol{Q}\Phi_1(x), & : x \in \Omega_2, \end{cases}$$

where $\Phi_1(x) = (\phi_{11}(x), \ldots, \phi_{1k}(x))^T$ and $\Phi_2(x) = (\phi_{21}(x), \ldots, \phi_{2k}(x))^T$ are the solution of the following boundary value problems

$$-\boldsymbol{Eps}\Phi_1''(x) + \boldsymbol{B}(x)\Phi_1(x) = 0, \quad x \in \Omega, \quad \boldsymbol{R}_l\Phi_1(0) = 1, \quad \boldsymbol{R}_r\Phi_1(1) = 0,$$
$$-\boldsymbol{Eps}\Phi_2''(x) + \boldsymbol{B}(x)\Phi_2(x) = 0, \quad x \in \Omega, \quad \boldsymbol{R}_l\Phi_2(0) = 0, \quad \boldsymbol{R}_r\Phi_2(1) = 1,$$

respectively, and $\boldsymbol{P}$ and $\boldsymbol{Q}$ are constant diagonal matrices which are evaluated using the fact that $\boldsymbol{u} \in C^1(\Omega)^k$.

Theorem 2 (The continuous maximum principle) *Assume* $u_1, \ldots, u_k \in C^0(\overline{\Omega}) \cap C^2(\Omega_1 \cup \Omega_2)$. *Further assume that the solution* $\boldsymbol{u} = (u_1, \ldots, u_k)^T$ *satisfies* $\boldsymbol{R}_l\boldsymbol{u}(0) \geq \boldsymbol{0}$, $\boldsymbol{R}_r\boldsymbol{u}(1) \geq \boldsymbol{0}$, $\boldsymbol{T}\boldsymbol{u}(x) \geq \boldsymbol{0}$ *in* $\Omega_1 \cup \Omega_2$ *and* $[\boldsymbol{u}'](\eta) \leq \boldsymbol{0}$, *then* $\boldsymbol{u}(x) \geq \boldsymbol{0}$, *for all* $x \in \overline{\Omega}$.

The following lemma provides the stability result for the solution of the considered problem.

Lemma 1 *Let* $\boldsymbol{u} = (u_1, \ldots, u_k)^T$ *be the continuous solution of (1)-(3). Then*

$$\|\boldsymbol{u}\|_{\overline{\Omega}} \leq C \max\{\|\boldsymbol{R}_l\boldsymbol{u}(0)\|, \|\boldsymbol{R}_r\boldsymbol{u}(1)\|, \|\boldsymbol{T}\boldsymbol{u}\|_{\Omega_1\cup\Omega_2}\}.$$

The following lemma describes the estimates on the exact solution and its derivatives.

Lemma 2 *Let* $\boldsymbol{u}$ *be the exact solution of (1)–(3). Then for each* $i = 1, \ldots, k$, *and* $x \in \Omega_1 \cup \Omega_2$, $\boldsymbol{u}$ *and its derivatives satisfy the following estimates:*

$$|u_i^{(m)}(x)| \leq C\varepsilon_i^{-\frac{m}{2}}(\|\boldsymbol{R}_l\boldsymbol{u}(0)\| + \|\boldsymbol{R}_r\boldsymbol{u}(1)\| + \|\boldsymbol{d}\|_{\Omega_1\cup\Omega_2}) \text{ for } m = 0, 1, 2.$$

$$|u_i^{(3)}(x)| \leq C\varepsilon_1^{-\frac{1}{2}}\varepsilon_i^{-1}(\|\boldsymbol{R}_l\boldsymbol{u}(0)\| + \|\boldsymbol{R}_r\boldsymbol{u}(1)\| + \|\boldsymbol{d}\|_{\Omega_1\cup\Omega_2} + \sqrt{\varepsilon_1}\|\boldsymbol{d}'\|_{\Omega_1\cup\Omega_2}), \text{ and}$$

$$|u_i^{(4)}(x)| \leq C\varepsilon_1^{-1}\varepsilon_i^{-1}(\|\boldsymbol{R}_l\boldsymbol{u}(0)\| + \|\boldsymbol{R}_r\boldsymbol{u}(1)\| + \|\boldsymbol{d}\|_{\Omega_1\cup\Omega_2} + \varepsilon_1\|\boldsymbol{d}''\|_{\Omega_1\cup\Omega_2}).$$

To establish the parameters-uniform convergence of the numerical scheme, the solution $\boldsymbol{u}$ is decomposed into a sum of a smooth component $\boldsymbol{v}$ and a layer component $\boldsymbol{w}$, such that $\boldsymbol{u} = \boldsymbol{v} + \boldsymbol{w}$.

The smooth component, $\boldsymbol{v}$ is defined to be the solution of the following system:

$$\boldsymbol{T}\boldsymbol{v}(x) = \boldsymbol{d}(x), \quad x \in \Omega_1 \cup \Omega_2,$$

$$\boldsymbol{R}_l\boldsymbol{v}(0) = \boldsymbol{B}^{-1}\boldsymbol{d}(0), \ \boldsymbol{R}_r\boldsymbol{v}(1) = \boldsymbol{B}^{-1}\boldsymbol{d}(1), \ \boldsymbol{v}(x) = \boldsymbol{B}^{-1}\boldsymbol{d}(x), \ x \in \{\eta-, \eta+\},$$

and the layer component, $\boldsymbol{w}$ solves the following:

$$\boldsymbol{T}\boldsymbol{w}(x) = \boldsymbol{0}, \quad x \in \Omega_1 \cup \Omega_2,$$

$$\boldsymbol{R}_l\boldsymbol{w}(0) = \boldsymbol{R}_l(\boldsymbol{u} - \boldsymbol{v})(0), \boldsymbol{R}_r\boldsymbol{w}(1) = \boldsymbol{R}_r(\boldsymbol{u} - \boldsymbol{v})(1),$$

$$[\boldsymbol{w}](\eta) = -[\boldsymbol{v}](\eta), [\boldsymbol{w}'](\eta) = -[\boldsymbol{v}'](\eta).$$

Now for $1 \leq i \leq k$, we define the layer functions used to derive the estimates on the layer component

$$\mathcal{E}_{\varepsilon_{l_i}}(x) := e^{-x\sqrt{\alpha/\varepsilon_i})} + e^{-(\eta-x)\sqrt{\alpha/\varepsilon_i}}, \tag{6}$$

$$\mathcal{E}_{\varepsilon_{r_i}}(x) := e^{(\eta-x)\sqrt{\alpha/\varepsilon_i}} + e^{-(1-x)\sqrt{\alpha/\varepsilon_i}}. \tag{7}$$

Bounds on regular and singular components can be obtained in the similar lines as in [16, 20] for Robin boundary conditions also.

Theorem 3 *Assume that the coupling matrix* ***B*** *satisfies (5a). Then for all $x \in \Omega_1 \cup \Omega_2$ and $i = 1, \ldots k$, the smooth component* ***v*** *and its derivatives satisfy the following estimates:*

$$|v_i^{(m)}(x)|_{\Omega_1 \cup \Omega_2} \leq C(1 + \varepsilon_i^{(1-\frac{m}{2})}).\ for\ m = 0, 1, 2, 3, 4.$$

Lemma 3 *Let the coupling matrix* ***B*** *satisfies (5a). Then for all $x \in \Omega_1 \cup \Omega_2$, $i = 1, \ldots k$, and $m = 0, 1, 2, 3, 4,$, the following estimates hold on the smooth component* ***v*** *and its derivatives:*

$$|v_i^{(m)}(x)|_{\Omega_1 \cup \Omega_2} \leq \begin{cases} C\Big(1 + \sum_{q=i}^{k} \varepsilon_q^{-(\frac{m}{2}-1)} \mathcal{E}_{\varepsilon_{l_q}}(x)\Big), & x \in \Omega_1, \\ C\Big(1 + \sum_{q=i}^{k} \varepsilon_q^{-(\frac{m}{2}-1)} \mathcal{E}_{\varepsilon_{r_i}}(x)\Big), & x \in \Omega_2. \end{cases}$$

Lemma 4 *For $1 \leq i \leq j \leq k$ and $0 < s \leq 3/2$, there exists a unique point $x_{i,j}^{(s)} \in (0, \frac{\eta}{2})$ such that $\varepsilon_i^{-s}\mathcal{E}_{\varepsilon_{l_i}}(x_{i,j}^{(s)}) = \varepsilon_j^{-s}\mathcal{E}_{\varepsilon_{l_j}}(x_{i,j}^{(s)})$. Also, $\varepsilon_i^{-s}\mathcal{E}_{\varepsilon_{l_i}}(\eta - x_{i,j}^{(s)}) = \varepsilon_j^{-s}\mathcal{E}_{\varepsilon_{l_j}}(\eta - x_{i,j}^{(s)})$. On $[0, x_{i,j}^{(s)}) \cup (\eta - x_{i,j}^{(s)}, \eta)$ we have $\varepsilon_i^{-s}\mathcal{E}_{\varepsilon_{l_i}}(x) > \varepsilon_j^{-s}\mathcal{E}_{\varepsilon_{l_j}}(x)$ and on $(x_{i,j}^{(s)}, \eta - x_{i,j}^{(s)})$ we have $\varepsilon_i^{-s}\mathcal{E}_{\varepsilon_{l_i}}(x) < \varepsilon_j^{-s}\mathcal{E}_{\varepsilon_{l_j}}(x)$. Similar results hold for the domain Ω_2.*

Lemma 5 *Let the coupling matrix* ***B*** *satisfies (5a). Then for $i = 1, \ldots, k$, the layer component* ***w*** *and its derivatives satisfy the following:*

$$|w_i(x)| \leq C \begin{cases} \mathcal{E}_{\varepsilon_{l_k}}(x), & x \in \Omega_1, \\ \mathcal{E}_{\varepsilon_{r_k}}(x), & x \in \Omega_2, \end{cases} \quad |w_i^{(m)}(x)| \leq C \begin{cases} \sum_{q=i}^{k} \frac{\mathcal{E}_{\varepsilon_{l_q}}(x)}{\varepsilon_q^{\frac{m}{2}}}, & x \in \Omega_1, \\ \sum_{q=i}^{k} \frac{\mathcal{E}_{\varepsilon_{r_q}}(x)}{\varepsilon_q^{\frac{m}{2}}}, & x \in \Omega_2,\ for\ m = 1, 2 \end{cases}$$

$$|w_i^{(3)}(x)| \leq C \begin{cases} \sum_{q=1}^{k} \frac{\mathcal{E}_{\varepsilon_{l_q}}(x)}{\varepsilon_q^{\frac{3}{2}}}, & x \in \Omega_1, \\ \sum_{q=1}^{k} \frac{\mathcal{E}_{\varepsilon_{r_q}}(x)}{\varepsilon_q^{\frac{3}{2}}}, & x \in \Omega_2, \end{cases} \quad \|\varepsilon_i w_i^{(4)}(x)\| \leq C \begin{cases} \sum_{q=1}^{k} \frac{\mathcal{E}_{\varepsilon_{l_q}}(x)}{\varepsilon_q}, & x \in \Omega_1, \\ \sum_{q=1}^{k} \frac{\mathcal{E}_{\varepsilon_{r_q}}(x)}{\varepsilon_q}, & x \in \Omega_2. \end{cases}$$

Theorem 4 *For $1 \le i \le k$, if the layer component* $\mathbf{w}$ *is decomposed as follows:*

$$w_i(x) = \sum_{q=1}^{k} w_{i,\varepsilon_q}(x).$$

Then,

$$|w''_{i,\varepsilon_q}(x)| \le C \begin{cases} \min\{\frac{1}{\varepsilon_q}, \frac{1}{\varepsilon_i}\}\mathcal{E}_{\varepsilon_{l_q}}(x), \ x \in \Omega_1, \\ \min\{\frac{1}{\varepsilon_q}, \frac{1}{\varepsilon_i}\}\mathcal{E}_{\varepsilon_{r_q}}(x), \ x \in \Omega_2, \end{cases}$$

$$|w'''_{i,\varepsilon_q}(x)| \le C \begin{cases} \min\{\frac{1}{\varepsilon_q^{3/2}}, \frac{1}{\varepsilon_i\sqrt{\varepsilon_q}}\}\mathcal{E}_{\varepsilon_{l_q}}(x), \ x \in \Omega_1, \\ \min\{\frac{1}{\varepsilon_q^{3/2}}, \frac{1}{\varepsilon_i\sqrt{\varepsilon_q}}\}\mathcal{E}_{\varepsilon_{r_q}}(x), \ x \in \Omega_2. \end{cases}$$

3 Discretization of the Problem

To resolve boundary and interior layers, we consider the standard finite difference scheme, with N mesh intervals on a piece wise-uniform variant of Shishkin mesh on $\Omega^N = \Omega_1^N \cup \Omega_2^N$. The transition parameters are defined as follows:

$$\sigma_{\varepsilon_{l_k}} := \min\left\{\frac{\eta}{4}, 2\sqrt{\frac{\varepsilon_k}{\alpha}}\ln N\right\}, \quad \sigma_{\varepsilon_{r_k}} := \min\left\{\frac{(1-\eta)}{4}, 2\sqrt{\frac{\varepsilon_k}{\alpha}}\ln N\right\},$$

$$\sigma_{\varepsilon_{l_m}} := \min\left\{\frac{\sigma_{\varepsilon_{l_{m+1}}}}{2}, 2\sqrt{\frac{\varepsilon_m}{\alpha}}\ln N\right\}, \sigma_{\varepsilon_{r_m}} := \min\left\{\frac{\sigma_{\varepsilon_{r_{m+1}}}}{2}, 2\sqrt{\frac{\varepsilon_m}{\alpha}}\ln N\right\},$$

for $m = k-1, \ldots, 1$. The interval $[0, \eta]$ is divided into intervals $[0, \sigma_{\varepsilon_{l_1}}], \ldots,$ $(\sigma_{\varepsilon_{l_{k-1}}}, \sigma_{\varepsilon_{l_k}}]$, $(\sigma_{\varepsilon_{l_k}}, \eta - \sigma_{\varepsilon_{l_k}}]$, $(\eta - \sigma_{\varepsilon_{l_k}}, \eta - \sigma_{\varepsilon_{l_{k-1}}}], \ldots,$ $(\eta - \sigma_{\varepsilon_{l_1}}, \eta]$. To get a piecewise-uniform mesh, we subdivide $[0, \sigma_{\varepsilon_{l_1}}]$ and $(\eta - \sigma_{\varepsilon_{l_1}}, \eta]$ into $N/2^{k+2}$ mesh intervals and other subintervals $(\sigma_{\varepsilon_{l_m}}, \sigma_{\varepsilon_{l_{m+1}}}]$ and $(\eta - \sigma_{\varepsilon_{l_{m+1}}}, \eta - \sigma_{\varepsilon_{l_m}}]$, for $m = k-1, \ldots, 1$, into $N/2^{k-m+3}$ mesh intervals of uniform length and on $(\sigma_{\varepsilon_{l_k}}, \eta - \sigma_{\varepsilon_{l_k}}]$ a uniform mesh having $N/4$ mesh intervals. In the similar manner, divide the interval $[\eta, 1]$. Let $h_j = x_j - x_{j-1}$ be the jth mesh step and $\hbar_j = \frac{h_j + h_{j+1}}{2}$, clearly $x_{\frac{N}{2}} = \{\eta\}$ and $\overline{\Omega}^N = \{x_j : j = 0, 1, \ldots, N\}$.

On a piecewise-uniform variant of Shishkin mesh $\overline{\Omega}^N$, the continuous problem (1) is discretized by considering a cubic spline scheme for Robin boundary conditions and a central finite difference scheme at the interior points of the domain for the differential equations. Define the discrete finite difference operator $\boldsymbol{T}^N$ as follows:

$$\boldsymbol{T}^N \boldsymbol{U} := -\boldsymbol{Eps}\, \delta^2 \boldsymbol{U} + \boldsymbol{B}\boldsymbol{U} = \overline{\boldsymbol{d}}, \qquad \text{for all } x_j \in \overline{\Omega}^N, \tag{8}$$

where $\delta^2 Y(x_j) = \frac{(D^+Y(x_j) - D^-Y(x_j))}{\hbar_j}$, $D^+Y(x_j) = \frac{Y(x_{j+1}) - Y(x_j)}{h_{j+1}}$,
$D^-Y(x_j) = \frac{Y(x_j) - Y(x_{j-1})}{h_j}$, $\overline{d}(\eta) = \frac{h_{\frac{N}{2}} d(\eta - h_{\frac{N}{2}}) + h_{\frac{N}{2}+1} d(\eta + h_{\frac{N}{2}+1})}{h_{\frac{N}{2}} + h_{\frac{N}{2}+1}}$.
and the boundary conditions are discretized as

$$(R_l^N U)_i(x_0) \equiv \alpha_i U_i(x_0) - \beta_i S^+ U_i(x_0) = p_i, \tag{9}$$

$$(R_r^N U)_i(x_N) \equiv \gamma_i U_i(x_N) + \delta_i S^- U_i(x_N) = q_i, \tag{10}$$

where $S^+U_i(x_0)$ and $S^-U_i(x_N)$ for $1 \le i \le k$ are obtained from the one sided limits [3, 4]

$$S'(x_j+) = -\frac{h_{j+1}}{3} M_i(x_j) - \frac{h_{j+1}}{6} M_i(x_{j+1}) + \frac{U_i(x_{j+1}) - U_i(x_j)}{h_{j+1}}, \tag{11}$$

$$S'(x_j-) = \frac{h_j}{6} M_i(x_{j-1}) + \frac{h_j}{3} M_i(x_j) + \frac{U_i(x_j) - U_i(x_{j-1})}{h_j}, \tag{12}$$

Substitute $M_i(x_j)$ from $-\mu_i M_i(x_j) + a_{i1}(x_j)u_1(x_j) + \ldots + a_{ik}(x_j)u_k(x_j) = f_i(x_j)$ to (11)–(12), to get the approximation for the one sided first-order derivatives at the boundary points. Therefore, the discretization (9)–(10) for the Robin boundary conditions reduce to the following:

$$\Big[\frac{3\varepsilon_i}{h_1}\Big(\alpha_i + \frac{\beta_i}{h_1}\Big) + a_{ii}(x_0)\beta_i\Big]U_i(x_0) + \Big[\frac{-3\varepsilon_i\beta_i}{h_1^2} + \frac{a_{ii}(x_1)\beta_i}{2}\Big]U_i(x_1)$$
$$+ \beta_i \sum_{\substack{k=1\\k\neq i}}^{m} a_{ik}(x_0)U_k(x_0) + \frac{\beta_i}{2}\sum_{\substack{k=1\\k\neq i}}^{m} a_{ik}(x_1)U_k(x_1) = \frac{3\varepsilon_i p_i}{h_1} + \beta_i f_i(x_0) + \frac{\beta_i}{2} f_i(x_1), \tag{13}$$

and

$$\Big[\frac{-3\varepsilon_i\delta_i}{h_N^2} + \frac{a_{ii}(x_{N-1})\delta_i}{2}\Big]U_i(x_{N-1}) + \Big[\frac{3\varepsilon_i}{h_N}\Big(\gamma_i + \frac{\delta_i}{h_N}\Big) + a_{ii}(x_N)\delta_i\Big]U_i(x_N)$$
$$+ \delta_i \sum_{k=1 k\neq i}^{m} a_{ik}(x_N)U_k(x_N) + \frac{\delta_i}{2}\sum_{k=1 k\neq i}^{m} a_{ik}(x_{N-1})U_k(x_{N-1}) \tag{14}$$
$$= \frac{3\varepsilon_i q_i}{h_N} + \delta_i f_i(x_N) + \frac{\delta_i}{2} f_i(x_{N-1}).$$

Lemma 6 (The discrete maximum principle) *Suppose the mesh function* $\boldsymbol{Y}$ *satisfies* $\boldsymbol{R}_l^N\boldsymbol{Y}(x_0) \ge \boldsymbol{0}$, $\boldsymbol{R}_r^N\boldsymbol{Y}(x_N) \ge \boldsymbol{0}$ *and* $\boldsymbol{T}^N\boldsymbol{Y} \ge \boldsymbol{0}$, *for all* $x_j \in \Omega^N$, *then* $\boldsymbol{Y} \ge \boldsymbol{0}$ *for all* $x_j \in \overline{\Omega}^N$.

Proof Let $Y_i(z_i) = \min_{x_j \in \overline{\Omega}^N}\{Y_i(x_j)\}$. Assume without loss generality that $Y_1(z_1) \leq Y_i(z_i)$ for $1 \leq i \leq k$. If $Y_1(z_1) \geq 0$, then the proof is complete. Suppose that $Y_1(z_1) < 0$, then we complete the proof by showing that this leads to a contradiction. If $z_1 = \{x_0\}$, then $R_{l_1}^N Y_1(x_0) < 0$ and if $z_1 = \{x_N\}$, then $R_{r_1}^N Y_1(x_N) < 0$, which is a contradiction. Then, for $x_j \in \Omega^N$

$$(\boldsymbol{T}^N \boldsymbol{Y})_1(z_1) = -\varepsilon_1 \delta^2 Y_1(z_1) + \sum_{m=1}^{k} a_{1m}(z_1) Y_m(z_1) < 0,$$

which leads to a contradiction. Similarly, we can prove the required result in the other case also. •

A consequence of the discrete maximum principle is the following stability result.

Lemma 7 *Suppose* $\boldsymbol{U}$ *is a numerical solution of (8)–(10), then*

$$\|\boldsymbol{U}\|_{\overline{\Omega}^N} \leq \max\left\{\|\boldsymbol{R}_l^N \boldsymbol{Y}(0)\|, \|\boldsymbol{R}_r^N \boldsymbol{Y}(1)\|, \frac{1}{\alpha}\|\boldsymbol{d}\|_{\Omega_1^N \cup \Omega_2^N}\right\}.$$

Proof Define the function $\Psi_\pm^N(x_j) := \max\{\|\boldsymbol{R}_1^N \boldsymbol{U}(0)\|, \|\boldsymbol{R}_r^N \boldsymbol{U}(1)\|, \|\boldsymbol{T}^N \boldsymbol{U}\|_{\Omega_1^N \cup \Omega_2^N}\}(2 - x_j, \ldots, 2 - x_j)^T \pm \boldsymbol{U}(x_j)$. From this, we can conclude that $\boldsymbol{R}_l^N \Psi_\pm^N(0) \geq \boldsymbol{0}$, $\boldsymbol{R}_r^N \Psi_\pm^N(1) \geq \boldsymbol{0}$ and $\boldsymbol{T}^N \Psi_\pm^N(x) \geq 0$ for each $x \in \Omega_1^N \cup \Omega_2^N$. From the discrete maximum principle, it follows that $\Psi_\pm^N \geq \boldsymbol{0}$ for $x \in \overline{\Omega}^N$, which leads to the required bound on $\boldsymbol{U}$ •

Decompose the discrete solution $\boldsymbol{U}$ into the sum $\boldsymbol{U} = \boldsymbol{V} + \boldsymbol{W}$, where $\boldsymbol{V}$is the solution to

$$\boldsymbol{T}^N \boldsymbol{V}(x_j) = \boldsymbol{d}(x_j), \qquad \text{for all } x_j \in \Omega^N,$$

$$\boldsymbol{R}_l^N \boldsymbol{V}(0) = \boldsymbol{R}_l \boldsymbol{v}(0), \quad \boldsymbol{V}(\eta) = \boldsymbol{v}(\eta), \quad \boldsymbol{R}_r^N \boldsymbol{V}(1) = \boldsymbol{R}_r \boldsymbol{v}(1),$$

and $\boldsymbol{W}$ is the solution to

$$\boldsymbol{T}^N \boldsymbol{W}(x_j) = \boldsymbol{0}, \qquad \text{for all } x_j \in \Omega^N,$$

$$\boldsymbol{R}_l^N \boldsymbol{W}(0) = \boldsymbol{R}_l \boldsymbol{w}(0), \quad \boldsymbol{R}_r^N \boldsymbol{W}(1) = \boldsymbol{R}_r \boldsymbol{w}(1),$$

$$[\boldsymbol{W}](\eta) = -[\boldsymbol{V}](\eta), [D\boldsymbol{W}](\eta) = -[D\boldsymbol{V}](\eta),$$

where the jump in the discrete derivative of a mesh function $\boldsymbol{Z}$ at the point $x_i = \eta$ is defined by

$$[DZ](\eta) := D^+Z(\eta) - D^-Z(\eta).$$

4 Convergence Analysis

In this section, we discuss the consistency of the proposed method and derive the parameters-uniform convergence.

Consider the truncation error at the boundary point, $x_0 = 0$:

$$\tau_{0,u_i} = R^c_{i,0}u_i(x_0)+R^+_{i,0}u_i(x_1) + \sum_{\substack{k=1\\k\neq i}} Q^c_{k,0}u_k(x_0) + \sum_{\substack{k=1\\k\neq i}} Q^+_{k,0}u_k(x_1) - F^-_{i,0} - F^c_{i,0}f_i(x_0) - F^+_{i,0}f_i(x_1),$$

where

$$R^c_{i,0} = \Big[\frac{3\varepsilon_i}{h_1}\Big(\alpha_i + \frac{\beta_i}{h_1}\Big) + a_{ii}(x_0)\beta_i\Big], \quad R^+_{i,0} = \Big[\frac{-3\varepsilon_i\beta_i}{h_1^2} + \frac{a_{ii}(x_1)\beta_i}{2}\Big],$$

$$Q^c_{k,0} = \beta_i a_{ik}(x_0), \quad Q^+_{k,0} = \frac{\beta_i}{2}a_{ik}(x_1),$$

$$F^-_{i,0} = \frac{3\varepsilon_i X_i}{h_1}, \quad F^c_{i,0} = \beta_i f_i(x_0), \quad F^+_{i,0} = \frac{\beta_i}{2}f_i(x_1).$$

Using (1) and the Taylor's series expansion, we have

$\tau_{0,u_i} = T_{0,0}u_i(x_0) + T_{1,0}u_i'(x_0) + T_{2,0}u_i''(x_0) + T_{3,0}u_i^{(3)}(x_0) + T_{4,0}u_i^{(4)}(\eta)$, where $\eta \in (0, 1)$ and

$$T_{0,0} = R^c_{i,0} + R^+_{i,0} - \frac{3\varepsilon_i\alpha_i}{h_1} - F^c_{i,0}a_{ii}(x_0) - F^+_{1,0}a_{ii}(x_0),$$

$$T_{1,0} = h_1R^+_{i,0} + \frac{3\varepsilon_i\beta_i}{h_1} - \frac{\beta_i a_{ii}(x_1)h_1}{2}, \quad T_{2,0} = \frac{h_1^2R^+_{i,0}}{2} + \varepsilon_i(F^c_{i,0} + F^+_{i,0}) - \frac{h_1^2F^+_{i,0}a_{ii}(x_1)}{2},$$

$$T_{3,0} = \frac{h_1^3R^+_{i,0}}{3!} + \varepsilon_i h_1F^+_{i,0} - \frac{F^+_{i,0}a_{ii}(x_1)h_1^3}{3!}, \quad T_{4,0} = \frac{h_1^4}{4!}R^+_{i,0} + \frac{\varepsilon_i h_1^2}{2}F^+_{i,0} - \frac{F^+_{i,0}a_{ii}(x_1)h_1^4}{4!}.$$

It is obvious to note that

$$T_{0,0}u_i(x_0) + T_{1,0}u_i'(x_0) = 0, \quad T_{2,0} = 0, \quad T_{3,0} = 0, \quad T_{4,0} = \frac{\varepsilon_i\beta_i h_1^2}{8}.$$

Thus, the truncation error for u_i, $1 \le i \le k$ at $x = x_0$ is given by

$$|\tau_{i,0}| \le C\varepsilon_i\beta_i h_1^2\|u_i^{(4)}(x_0)\|_{(x_0,x_1)}$$

Now, by using Theorems 4 and 5, we have

$$|\tau_{i,0}| \leq C(N^{-1}\ln N)^2,$$

Similar result holds for the boundary point $x_N = 1$.

By a Taylor expansion for the function ϕ and $j = 1, \ldots, N/2-1, N/2+1, \ldots, N$, we have

$$|(\frac{d^2}{dx^2} - \delta^2)\phi_s(x_j)| \leq \begin{cases} C(x_{j+1} - x_{j-1})|\phi_s|_3 & (15) \\ Ch^2|\phi_s|_4,\ x_{j+1} - x_j = x_j - x_{j-1} = h & (16) \\ C\max_{x\in[x_{j-1},x_{j+1}]}|\phi_s''(x_j)|. & (17) \end{cases}$$

To evaluate the truncation error for the regular component, we consider the following cases:

Case (i) For $x_j \notin \{\sigma_{\varepsilon_{l_m}}, \eta - \sigma_{\varepsilon_{l_m}}, \eta + \sigma_{\varepsilon_{r_m}}, 1 - \sigma_{\varepsilon_{r_k}}\}$.

Using (16) and the bounds defined in Theorem 4, we have

$$|((T^N - T)v)_s(x_j)| \leq C\varepsilon_s(x_{j+1} - x_{j-1})^2|v_s|_4 \leq CN^{-2}.$$

Case (ii) For $x_j \in \{\sigma_{\varepsilon_{l_m}}, \eta - \sigma_{\varepsilon_{l_m}}, \eta + \sigma_{\varepsilon_{r_m}}, 1 - \sigma_{\varepsilon_{r_k}}\}$.

Using (15) and the bounds defined in Lemma 3, we have

$$|((T^N - T)v)_s(x_j)| \leq \begin{cases} C\varepsilon_s(x_{j+1} - x_{j-1})\Big(1 + \sum_{q=s}^{k} \varepsilon_q^{-\frac{1}{2}}\mathcal{E}_{\varepsilon_{l_q}}(x_{j-1})\Big), & x \in \Omega_1, \\ C\varepsilon_s(x_{j+1} - x_{j-1})\Big(1 + \sum_{q=s}^{k} \varepsilon_q^{-\frac{1}{2}}\mathcal{E}_{\varepsilon_{r_q}}(x_{j-1})\Big), & x \in \Omega_2, \end{cases}$$

For $s \geq m$, we have

$$|((T^N - T)v)_s(x_j)| \leq C\frac{\varepsilon_s}{\sqrt{\varepsilon_m}}(h_{\varepsilon_m} + h_{\varepsilon_{m+1}}),$$

and for $s < k$, using Lemma 2, we have

$$|((T^N - T)v)_s(x_j)| \leq C\frac{\varepsilon_s}{\sqrt{\varepsilon_m}}(h_{\varepsilon_m} + h_{\varepsilon_{m+1}}).$$

To evaluate the truncation error for the singular components on different subintervals, we consider the following cases:

Case (i) For $x_j \in [\sigma_{\varepsilon_{l_k}}, \eta - \sigma_{\varepsilon_{l_k}}] \cup [\eta + \sigma_{\varepsilon_{r_k}}, 1 - \sigma_{\varepsilon_{r_k}}]$.

Consider first that $x_j \in [\sigma_{\varepsilon_{l_k}}, \frac{\eta}{2}]$. Using (17) and bounds on singular components, we have

$$|\,[(T^N - T)w]_s(x_j)\,| \le C(\varepsilon_s \sum_{q=s}^{k} \frac{\mathcal{E}_{\varepsilon_{l_q}}(x)}{\varepsilon_q}) \le C(\|\, \mathcal{E}_{\varepsilon_{l_k}} \|_{[x_{j-1},x_{j+1}]}) \le CN^{-2}.$$

A similar result can be proved for $x \in [\frac{\eta}{2}, \eta - \sigma_{\varepsilon_{l_k}}]$. Similar result can be proved for the other subinterval.

Case (ii) For $x_j \in (0, \sigma_{\varepsilon_{l_1}}) \cup (\eta - \sigma_{\varepsilon_{l_1}}, \eta) \cup (\eta, \eta + \sigma_{\varepsilon_{r_1}}) \cup (1 - \sigma_{\varepsilon_{r_1}}, 1)$.
(16) and Lemma 5, yields

$$|\,((T^N - T)w)_s(x_j)\,| \le Ch^2 \|\varepsilon_s w_s^{(4)}\| \le C\left(h^2 \sum_{q=1}^{k} \frac{\mathcal{E}_{\varepsilon_{l_q}}(x)}{\varepsilon_q}\right) \le C(N^{-1}\ln N)^2.$$

Case (iii) For $x_j \in [\sigma_{\varepsilon_{l_m}}, \sigma_{\varepsilon_{l_{m+1}}}) \cup [\eta - \sigma_{\varepsilon_{l_{m+1}}}, \eta - \sigma_{\varepsilon_{l_m}}) \cup [\eta + \sigma_{\varepsilon_{r_m}}, \eta + \sigma_{\varepsilon_{r_{m+1}}}) \cup [1 - \sigma_{\varepsilon_{r_{m+1}}}, 1 - \sigma_{\varepsilon_{r_m}})$, where $1 \le m \le k-1$.

Using Theorem 4, we get

$$|\,((T^N - T)w)_s(x_j) = (|\sum_{q=1}^{m} \varepsilon_s(\frac{d^2}{dx^2} - \delta^2)w_{s,\varepsilon_q}(x_j) + \sum_{q=m+1}^{k} \varepsilon_s(\frac{d^2}{dx^2} - \delta^2)w_{s,\varepsilon_q}(x_j)|. \quad (18)$$

Consider the first part of (18) for $s \le m$, and use the definition of point $x_{i,j}^{(s)}$ to get

$$|\sum_{q=1}^{m} \varepsilon_s(\frac{d^2}{dx^2} - \delta^2)w_{s,\varepsilon_q}(x_j)| \le \|\sum_{q=1}^{m} \varepsilon_s w''_{s,\varepsilon_q}\|_{[x_{j-1},x_{j+1}]} \le CN^{-2},$$

and if, $s > m$, using the bounds on singular components and following the analysis as in the Case(i), we get

$$|\sum_{q=1}^{m} \varepsilon_s(\frac{d^2}{dx^2} - \delta^2)w_{s,\varepsilon_q}(x_j)| \le \|\sum_{q=1}^{m} \varepsilon_s w''_{s,\varepsilon_q}\|_{[x_{j-1},x_{j+1}]} \le CN^{-2}.$$

For the second part of (18), using bounds on singular components, we get

$$|\sum_{q=m+1}^{k} \varepsilon_s(\frac{d^2}{dx^2} - \delta^2)w_{s,\varepsilon_q}(x_j)| \le C\varepsilon_s(h_j + h_{j+1})\,\|\sum_{q=m+1}^{k} w'''_{s,\varepsilon_q}\| \le CN^{-2}.$$

Case (iv) At the point $x_{N/2} = \eta$, $h_{N/2} = h_{N/2+1} = h$, and $\sigma_{\varepsilon_{l_1}} = \sigma_{\varepsilon_{r_1}} = \sqrt{\frac{\varepsilon_1}{\alpha}}\ln N$.
It follows that

$$|(T^N(U-u))_1(\eta)| \le C(N^{-1}\ln N).$$

Likewise, it can be proved that

$$|(T^N(U-u))_j(\eta))| \le C(N^{-1}\ln N), \qquad 2 \le j \le k.$$

Using the weighted average approximation at the point of discontinuity, cubic spline technique for the Robin boundary conditions, and with suitably constructed mesh functions and barrier functions, nearly second-order parameter-uniform convergence of the proposed method is proved in the following main result.

Theorem 5 *For problem (1)–(3), suppose that* $\boldsymbol{u}$ *and* $\boldsymbol{U}$ *are the exact and numerical solutions, then*

$$\|\boldsymbol{U}-\boldsymbol{u}\|_{\overline{\Omega}^N} \le CN^{-2}\ln^2 N.$$

Proof For $m = 1, \dots, k$, define the mesh functions θ_{1m}, θ_{2m}, θ_3, and θ_4 as follows:

$$\theta_{1m}(x_j) := \prod_{i=1}^{j}\left(1+\sqrt{\frac{\alpha}{2\varepsilon_m}}h_i\right), \quad \theta_{2m}(x_j) := \prod_{i=1}^{j}\left(1+\sqrt{\frac{\alpha}{2\varepsilon_m}}h_i\right)^{-1},$$

$$\theta_3(x_j) := \prod_{i=1}^{j}\left(1+\sqrt{\frac{\alpha}{2\varepsilon_1}}h_i\right), \quad \theta_4(x_j) := \prod_{i=1}^{j}\left(1+\sqrt{\frac{\alpha}{2\varepsilon_1}}h_i\right)^{-1}.$$

Now, for $m = 1, \dots, k$, define the barrier functions Φ_m, $\Phi_{m\eta}$, and Φ_η as follows:

$$\Phi_\eta(x_j) := \begin{cases} \frac{\theta_3(x_j)}{\theta_3(\eta)}, & 0 \le x_j \le \eta, \\ \frac{\theta_4(x_j)}{\theta_4(\eta)}, & \eta \le x_j \le 1. \end{cases} \quad \Phi_m(x_j) := \begin{cases} \frac{x_j}{\sigma_{\varepsilon_{lm}}}, & 0 \le x_j \le \sigma_{\varepsilon_{lm}}, \\ 1, & \sigma_{\varepsilon_{lm}} \le x_j \le 1-\sigma_{\varepsilon_{rm}}, \\ \frac{1-x_j}{\sigma_{\varepsilon_{rm}}}, & 1-\sigma_{\varepsilon_{rm}} \le x_j \le 1, \end{cases}$$

and

$$\Phi_{m\eta}(x_j) := \begin{cases} \frac{\theta_{1m}(x_j)}{\theta_{1m}(\eta-\sigma_{\varepsilon_{lm}})}, & 0 \le x_j \le d-\sigma_{\varepsilon_{lm}}, \\ 1, & d-\sigma_{\varepsilon_{lm}} \le x_j \le d+\sigma_{\varepsilon_{rm}}, \\ \frac{\theta_{2m}(x_j)}{\theta_{2m}(\eta+\sigma_{\varepsilon_{rm}})}, & d+\sigma_{\varepsilon_{rm}} \le x_j \le 1, \end{cases}$$

For $i \ne \frac{N}{2}$, define the mesh function

$$\Psi^{\pm}(x_j) := C(N^{-1}\ln N)^2\Big(1+\sum_{m=1}^{k}\Phi_m(x_j)+\theta_{m\eta}(x_j)(1,\dots,1)^T \pm (\boldsymbol{U}-\boldsymbol{u})(x_j),$$

and for $i = \frac{N}{2}$, define

$$\Psi^{\pm}(x_j) := C(N^{-1}\ln N)^2(1+\Phi_\eta(x_j))(1,\dots,1)^T \pm (\boldsymbol{U}-\boldsymbol{u})(x_j).$$

Apply discrete maximum principle to conclude the result ∎

5 Numerical Results

The numerical results are presented for two test problems to show the efficiency and applicability of the theoretical results. The piece wise-uniform Shishkin mesh $\overline{\Omega}_N$ is constructed for the choice of $\alpha = 0.95$ as per the criterion $(5b)$.

Example 1 Consider the first singularly perturbed reaction–diffusion problem with discontinuous source term and robin boundary conditions

$$-\varepsilon_1 u_1''(x) + 2(x+1)^2 u_1 - (1+x^3)u_2 = f_1(x), \quad x \in \Omega_1 \cup \Omega_2,$$

$$-\varepsilon_2 u_2''(x) - 2\cos(\frac{\pi}{4})u_1 + 2.2e^{1-x}u_2 = f_2(x), \quad x \in \Omega_1 \cup \Omega_2,$$

$$2u_1(0) - \sqrt{\varepsilon_1}u_1'(0) = 1, \qquad u_2(0) - \sqrt{\varepsilon_2}u_2'(0) = 0,$$

$$u_1(1) + \sqrt{\varepsilon_1}u_1'(1) = 0, \qquad 2u_2(1) + \sqrt{\varepsilon_2}u_2'(1) = 1,$$

where

$$f_1(x) = \begin{cases} 2e^x & for\, 0 \le x \le 0.5, \\ 1 & for\, 0.5 < x \le 1, \end{cases}$$

and

$$f_2(x) = \begin{cases} 10x + 1 & for\, 0 \le x \le 0.5, \\ 2 & for\, 0.5 < x \le 1. \end{cases}$$

Example 2 Consider the second singularly perturbed reaction–diffusion problem with discontinuous source term and robin boundary conditions

$$-\varepsilon_1 u_1''(x) + 3u_1(x) - (1-x)u_2(x) - (1-x)u_3(x) = f_1(x), \quad x \in \Omega_1 \cup \Omega_2,$$

$$-\varepsilon_2 u_2''(x) - 2u_1(x) + (4+x)u_2(x) - u_3(x) = f_2(x), \quad x \in \Omega_1 \cup \Omega_2,$$

$$-\varepsilon_3 u_3''(x) - 2u_1(x) - 3u_2(x) + (6+x)u_3(x) = f_3(x), \quad x \in \Omega_1 \cup \Omega_2,$$

$$u_1(0) - \sqrt{\varepsilon_1}u_1'(0) = 1, \quad 2u_2(0) - \sqrt{\varepsilon_2}u_2'(0) = 2, \quad u_3(0) - 2\sqrt{\varepsilon_3}u_3'(0) = 1,$$

$$u_1(1) + \sqrt{\varepsilon_1}u_1'(1) = 2, \quad 2u_2(1) + \sqrt{\varepsilon_2}u_2'(1) = 1, \quad u_3(1) + \sqrt{\varepsilon_3}u_3'(1) = 2,$$

where

$$f_1(x) = \begin{cases} \exp(x) & for\, 0 \le x \le 0.5, \\ 2 & for\, 0.5 < x \le 1, \end{cases} \quad f_2(x) = \begin{cases} \cos(x) & for\, 0 \le x \le 0.5, \\ 4 & for\, 0.5 < x \le 1, \end{cases}$$

$$\text{and} \quad f_3(x) = \begin{cases} 1 + x^2 & for\, 0 \le x \le 0.5, \\ 3 & for\, 0.5 < x \le 1. \end{cases}$$

As the solutions of the Examples 1 and 2 are not known, to find the pointwise errors and rate of convergence, the double mesh principle is used. Let $\boldsymbol{U}$ be the solution of the numerical method on the mesh Ω^N and $\widehat{\boldsymbol{U}}$ be the solution of the numerical method on the mesh having the mesh points $\widehat{x_i}$, which contains the mesh points of the initial mesh and their middle points. For different values of N and ε_1, ε_2, which takes the values from the set

$$S_{\varepsilon_1,\varepsilon_2} = \{(\varepsilon_1, \varepsilon_2)|, \varepsilon_1 = 10^{-j}, 0 \le j \le 16, \varepsilon_2 = 10^{-l}, 0 \le l \le j\}.$$

For different values of N and ε_1, ε_2, ε_3, which takes the values from the set

$$S_{\varepsilon_1,\varepsilon_2,\varepsilon_3} = \{(\varepsilon_1, \varepsilon_2, \varepsilon_3)|, \varepsilon_1 = 10^{-j}, 0 \le j \le 12, \varepsilon_2 = 10^{-l}, 0 \le l \le j, \varepsilon_3 = 10^{-k}, 0 \le k \le l\}.$$

We compute $\Theta^N_{\varepsilon_1,\varepsilon_2} := \|(\boldsymbol{U} - \widehat{\boldsymbol{U}})(x_j)\|_{\overline{\Omega}^N}$ and $\Theta^N_{\varepsilon_1,\varepsilon_2,\varepsilon_3} := \|(\boldsymbol{U} - \widehat{\boldsymbol{U}})(x_j)\|_{\overline{\Omega}^N}$. The parameter-uniform error is calculated by the formula

$$\Theta^N := \max_{S_{\varepsilon_1,\varepsilon_2}}\{\Theta^N_{\varepsilon_1,\varepsilon_2}\} \text{ and } \Theta^N := \max_{S_{\varepsilon_1,\varepsilon_2,\varepsilon_3}}\{\Theta^N_{\varepsilon_1,\varepsilon_2,\varepsilon_3}\}.$$

The numerical order of convergence of the method is calculated using the formula

$$p^N := \frac{\ln(\Theta^N) - \ln(\Theta^{2N})}{\ln(2\ln N) - \ln(\ln(2N))}.$$

The maximum pointwise parameter-uniform error Θ^N and the parameters-uniform numerical order of convergence of the present method for Examples 1 and 2 are presented in Table 1. From the Table 1, we observe that the proposed numerical

Table 1 Maximum pointwise parameter-uniform error Θ^N and the parameter-uniform rate of convergence p^N for Example 1 and Example 2

N	Example 1		Example 2	
	Θ^N	p^N	Θ^N	p^N
64	4.24E-02	0.97	9.28E-02	0.54
128	2.51E-02	1.46	6.92E-02	1.16
256	1.11E-02	1.85	3.62E-02	1.62
512	3.83E-03	1.96	1.43E-02	1.86
1024	1.21E-03	1.91	4.78E-03	1.95
2048	3.87E-04	–	1.49E-03	–

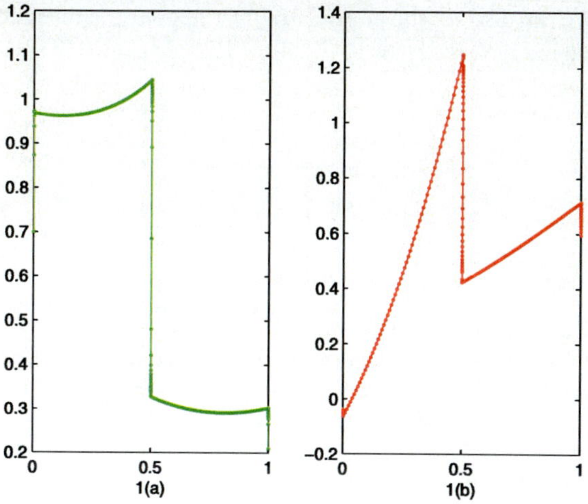

Fig. 1 The (1**a**) is the plot of the component U_1 and the (1**b**) is the plot of the component U_2 of the numerical solution of Example 1 for $\varepsilon_1 = 10^{-16}$, $\varepsilon_2 = 10^{-14}$ and $N = 256$a

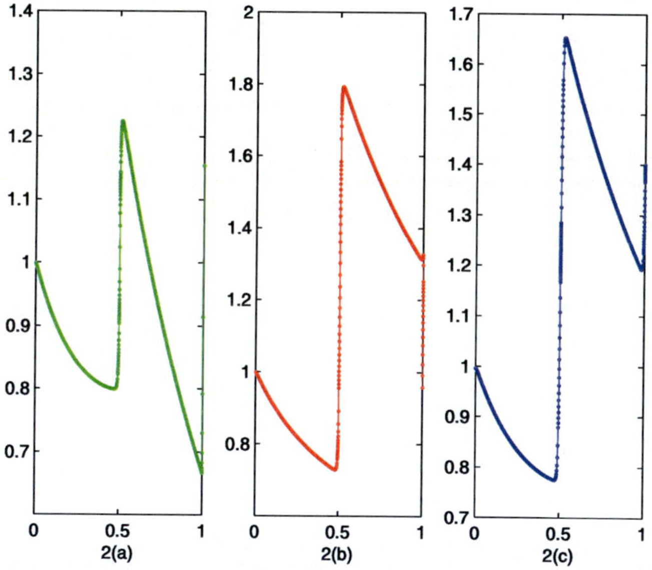

Fig. 2 The (2**a**) is the plot of the component U_1, the (2**b**) is the plot of the component U_2 and the (2**c**) is the plot of the component U_3, of the numerical solution of Example 2 for $\varepsilon_1 = 10^{-8}$, $\varepsilon_2 = 10^{-6}$, $\varepsilon_3 = 10^{-4}$ and $N = 512$

method is parameters-uniformly convergent of order almost two, which is in agreement with the theoretical findings. Figure 1 shows the presence of overlapping and interacting boundary and interior layers for $\varepsilon_1 = 10^{-16}$, $\varepsilon_2 = 10^{-14}$ and $N = 256$

for Example 1. We can also see the presence of overlapping and interacting boundary and interior layers in Fig. 2 for $\varepsilon_1 = 10^{-8}$, $\varepsilon_2 = 10^{-6}$, $\varepsilon_3 = 10^{-4}$ and $N = 512$ for Example 2.

References

1. Basha, P.M., Shanthi, V.: A uniformly convergent scheme for a system of two coupled singularly perturbed reaction–diffusion Robin type boundary value problems with discontinuous source term. Am. J. Numer. Anal. **3**(2), 39–48 (2015)
2. Chawla, S., Singh, J.: Urmil: an analysis of a robust convergent method for a Singularly perturbed linear system of reaction-diffusion type having non-smooth data. Int. J. Comput. Methods **19**(1), 2150056 (2022)
3. Das, P., Natesan N.: Higher-order parameter uniform convergent schemes for Robin type reaction-diffusion problems using adaptively generated grid. Int. J. Comput. **9**(4), 1250052(27) (2012)
4. Das, P., Natesan, N.: A uniformly convergent hybrid scheme for singularly perturbed system of reaction-diffusion Robin type boundary-value probelm. J. Appl. Math. Comput. **41**, 447–471 (2013)
5. Falco, C. de., O'Riordan, E.: Interior Layers in a reaction-diffusion equation with a discontinuous diffusion coefficient. Int. J. Numer. Anal. Model. **7**, 444–461 (2010)
6. Farrell, P.A., Hegarty, A.F., Miller, J.J.H., O'Riordan, E., and Shishkin, G.I.: Singularly perturbed differential equations with discontinuous source terms. In: Miller, J.J.H., Shishkin, G.I., Vulkov, L. (eds.) Analytical and Numerical Methods for Convection-Dominated and Singularly Perturbed Problems, Nova Science, New York, pp. 23–32 (2000)
7. Gracia, J.L., Lisbona, F.J., Riordan, E.O'.: A coupled system of singularly perturbed parabolic reaction -diffusion equations. Adv. Comput. Math. **32**, 43–61 (2010)
8. Kumar, S., Rao, S.C.S.: A robust domain decomposition algorithm for singularly perturbed semilinear systems. Int. J. Comput. Math. **94**, 1108–1122 (2017)
9. Linß, T., Madden, N.: Layer-adapted meshes for a system of coupled singularly perturbed reaction-diffusion problems. IMA J. Numer. Anal. **29**, 109–125 (2009)
10. Natesan, S., Deb, B.S.: A robust computational method for singularly perturbed coupled system of reaction-diffusion boundary-value problems. Appl. Math. Comput. **188**(1), 353–364 (2007)
11. Paramasivam, M., Valarmathi, S., Miller, J.J.H.: Second order parameter-uniform convergence for a finite difference method for a singularly perturbed linear reaction-diffusion system. Math. Commun. **15**, 587–612 (2010)
12. Paramasivam, M., Miller, J.J.H., Valarmathi, S.: Parameter-uniform convergence for a finite difference method for a singulalry perturbed linear reaction-diffusion system with discontinuous source terms. Int. J. Numer. Anal. Model. **11**, 385–399 (2014)
13. Rajaiah, J., Sigamani, V.: A parameter-uniform essentially first order convergent fitted mesh method for a singularly perturbed Robin problem. IJMTT **59**, 8–21 (2018)
14. Rao, S.C.S., Chawla, S.: Interior layers in coupled system of two singularly perturbed reaction-diffusion equations with discontinuous source term. NAA 2012, LNCS 8236, pp. 445-453 (2013)
15. Rao, S.C.S., Chawla, S.: Numerical solution for a coupled system of singularly perturbed initial value problems with discontinuous source term. Springer Proceedings in Mathematics and Statistics, vol. 143, pp. 753–764 (2015)
16. Rao, S.C.S., Chawla, S.: Second order uniformly convergent numerical method for a coupled system of singularly perturbed reaction-diffusion problems with discontinuous source term LNCSE, vol. 108, pp. 233–244 (2015)

17. Rao, S.C.S., Kumar, M., Singh, J.: A discrete Schwarz waveform relaxation method of higher order for singularly perturbed parabolic reaction–diffusion problems. J. Math. Chem. **58**, 574–594 (2020)
18. Rao, S.C.S., Chaturvedi, A.K.: Analysis of an almost fourth-order parameter-uniformly convergent numerical method for singularly perturbed semilinear reaction-diffusion system with non-smooth source term. Appl. Math. Comput. **421**, 126944 (2022)
19. Rao, S.C.S., Chawla, S.: Numerical solution of singularly perturbed linear parabolic system with discontinuous source term. Appl. Numer. Math. **127**, 249–265 (2018)
20. Rao, S.C.S., Chawla, S.: Parameter-uniform convergence of a numerical method for a coupled system of singularly perturbed semilinear reaction-diffusion equations with boundary and interior layers. J. Comput. Appl. Math. **352**, 223–239 (2019)
21. Rao, S.C.S., Kumar, M.: Optimal B-spline collocation method for self-adjoint singularly perturbed boundary value problems. Appl. Math. Comput. **188**, 749–761 (2007)
22. Rao, S.C.S., Kumar, S.: An almost fourth order parameter-uniformlty convergent domain decomposition method for a coupled system of singularly perturbed reaction-diffusion problems. J. Comput. Appl. Math. **235**, 3342–3354 (2011)
23. Rao, S.C.S., Kumar, S.: Second order global uniformly convergent numerical method for a coupled system of singularly perturbed initial value problems. Appl. Math. Comput. **219**, 3740–3753 (2012)
24. Rao, S.C.S., Kumar, M.: An almost fourth order parameter-robust numerical method for a linear system of ($M \geq 2$) coupled singularly perturbed reaction-diffusion problems. Int. J. Numer. Anal. Model. **10**, 603–621 (2013)
25. Rao, S.C.S., Kumar, S., Kumar, M.: Uniform global convergence of a hybrid scheme for singularly perturbed reaction-diffusion systems. J. Optim. Theory Appl. **151**, 338–352 (2011)
26. Shishkin, G.I.: Mesh approximation of singularly perturbed boundary-value problems for systems of elliptic and parabolic equations. Comput. Math. Math. Phys. **35**(4), 429–446 (1995)

Double-Diffusive Convection with the Effect of Rotation in Magnetic Nanofluids

Monika Arora, Mustafa Danesh, and Avinash Rana

Abstract A linear stability analysis is performed to investigate the effect of rotation and solute for a thin horizontal layer of water-based magnetic nanofluid (W_{MNF}) and ester-based magnetic nanofluids (E_{MNF}). The fluid is heated and salted from below, subject to rotation around the vertical axis. As stated in Buongiorno (J Heat Transf 128, 240–250, 2006, [1]), Brownian diffusion and thermophoresis are the significant slip mechanisms in nanofluids. In this work, we consider these two along with magnetophoresis since we are dealing with magnetic nanofluids. A numerical method is employed using MATLAB's EIG function to solve the resulting eigenvalue problem. The effect of various parameters of the problem which govern the flow has been observed at the onset of convection in the gravity environment in a rigid-rigid boundary condition through neutral stability curves (NSCs). The effect of rotation is investigated using the Taylor number (T_A). We analyse this significant parameter in rigid-free and free-free boundary conditions also with respect to both the environments (gravity and microgravity) and find that the increment in the value of T_A contributes to system stability under both the environments in all the boundary conditions.

Keywords Double diffusion · Rotation · Magnetic nanofluids · Solute

M. Arora (✉)
Department of Mathematics, School of Chemical Engineering and Physical Sciences, Lovely Professional University, Phagwara 144411, Punjab, India
e-mail: monikaarora1879@gmail.com

M. Danesh
Faculty of Natural Sciences, Department of Mathematics, Shaheed Prof. Rabbani Education University, Kabul, Afghanistan

A. Rana
Department of Marketing, Mittal School of Business, Lovely Profesional University, Phagwara 144411, Punjab, India

R. K. Sharma et al. (eds.), *Frontiers in Industrial and Applied Mathematics*, Springer Proceedings in Mathematics & Statistics 410,
https://doi.org/10.1007/978-981-19-7272-0_45

1 Introduction

Double-diffusive convection (DDC) has gained much attention after the work of Stommel [2]. This process occurs due to two density gradients with different molecular diffusivities. The process of DDC provides a deep understanding in the fields of oceanography, astrophysics, and chemical engineering, to name a few. There are two constituents (heat and salt in oceanography, heat and helium in astrophysics, and two different solutes in chemical engineering) having different molecular diffusivity which contribute in the reverse way to the vertical density gradient, resulting in the same qualitative behaviour on DDC with variation in time and space scales of the motion. Reviews of Turner [3] and Huppert and Turner [4] are useful sources of information on double-diffusive convection and related fields. For more studies on DDC, the reader is referred to [5–10] and references therein.

2 Physical Model of the Flow

An infinite horizontal layer of incompressible MNF heated and salted from below is considered. The fluid is assumed to occupy the layer $z \in [0, d]$ with gravity, g, acting along the negative z-direction. The magnetic field $\boldsymbol{H}$ acts outside the layer. The system rotates with the angular velocity $\boldsymbol{\Omega}$ [see Fig. 1]. The temperature and the volumetric fraction of nanoparticles are assumed to be constant on the boundaries.

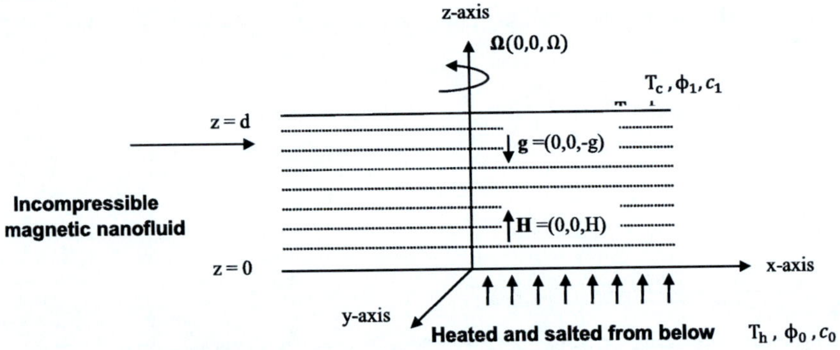

Fig. 1 Geometric structure of the physical model

3 Mathematical Formulation of the Problem

Following [11, 12], the governing equations for the present problem, known as equation of continuity, equation of momentum, equation for nanoparticles, equation of thermal energy, Maxwell's equations, magnetization equation and solutal equation, respectively are:

$$\nabla \cdot \boldsymbol{u} = 0, \tag{1}$$

$$\rho_f \left(\frac{\partial \boldsymbol{u}}{\partial t} + \boldsymbol{u} \cdot \nabla \boldsymbol{u} \right) = -\nabla p + \mu \nabla^2 \boldsymbol{u} + \mu_0 (\boldsymbol{M} \cdot \nabla) \boldsymbol{H} - \rho g \boldsymbol{k} + 2\rho_f (\boldsymbol{u} \times \boldsymbol{\Omega}), \tag{2}$$

$$\frac{\partial \phi}{\partial t} + \boldsymbol{u} \cdot \nabla \phi = \nabla \cdot \left(D_B \nabla \phi + D_T \frac{\nabla T}{T_c} - D_H \nabla H \right), \tag{3}$$

$$(\rho c)_f \left(\frac{\partial T}{\partial t} + \boldsymbol{u} \cdot \nabla T \right) = \nabla \cdot (k_1 \nabla T) + \rho_p c_p \left(D_B \nabla T \cdot \nabla \phi + D_T \frac{\nabla T \cdot \nabla T}{T_c} \right.$$
$$\left. - D_H \nabla T \cdot \nabla H \right) + (\rho c)_f D_{TC} \nabla^2 C, \tag{4}$$

$$\nabla \cdot \boldsymbol{B} = 0, \quad \nabla \times \boldsymbol{H} = 0, \quad \boldsymbol{B} = \mu_0 (\boldsymbol{M} + \boldsymbol{H}), \tag{5}$$

$$\boldsymbol{M}_{eq} = \frac{\boldsymbol{H}}{H} M_s \phi L(\alpha_L) = \frac{\boldsymbol{H}}{H} M_{eq}(H\phi, T, C), \tag{6}$$

$$\frac{\partial C}{\partial t} + \boldsymbol{u} \cdot \nabla C = D_S \nabla^2 C + D_{CT} \nabla^2 T. \tag{7}$$

We assume that the temperature and the volumetric fraction of nanoparticles are constant on the boundaries. Therefore, Boundary Conditions (B. C.) are:

$$\left. \begin{array}{l} w = 0, \quad T = T_h, \quad \phi = \phi_0, \quad C = C_0 \quad \text{at} \quad z = 0, \\ w = 0, \quad T = T_c, \quad \phi = \phi_1, \quad C = C_1 \quad \text{at} \quad z = d. \end{array} \right\} \tag{8}$$

with $\frac{\partial w}{\partial z} = 0$ on the rigid surface and $\frac{\partial^2 w}{\partial z^2} = 0$ on the stress-free surface.
Non-dimensional form of Eqs. (1)–(7):

$$\nabla \cdot \boldsymbol{u} = 0, \tag{9}$$

$$\frac{1}{Pr}\left(\frac{\partial \boldsymbol{u}}{\partial t}+\boldsymbol{u}\cdot\nabla\boldsymbol{u}\right)=-\nabla p+\nabla^2\boldsymbol{u}+\lambda_1(\boldsymbol{M}\cdot\nabla)\boldsymbol{H}-(Rn\phi-RaT+\frac{Rs}{Le_s}C$$
$$+Ra_nN_\phi T\phi-Rs_nN_\phi C\phi+\rho_1-\rho_2\phi)\boldsymbol{k}+{T_A}^{1/2}(\boldsymbol{u}\times\boldsymbol{k}), \tag{10}$$

$$\frac{\partial\phi}{\partial t}+\boldsymbol{u}\cdot\nabla\phi=\frac{1}{Le}\nabla^2\phi+\frac{N_a}{Le}\nabla^2T-\frac{N_a'}{Le}\nabla^2H, \tag{11}$$

$$\frac{\partial T}{\partial t}+\boldsymbol{u}\cdot\nabla T=\nabla^2T+\frac{N_b}{Le}(\nabla\phi\cdot\nabla T)+\frac{N_aN_b}{Le}(\nabla T\cdot\nabla T)$$
$$-\frac{N_a'N_b}{Le}(\nabla H\cdot\nabla T)+N_{ct}\nabla^2C, \tag{12}$$

$$\chi_2\nabla\cdot\boldsymbol{M}+\nabla\cdot\boldsymbol{H}=0, \tag{13}$$

$$\boldsymbol{M}=\frac{\boldsymbol{H}}{H}\frac{(1+\chi)}{\chi_2}\left\{\frac{\chi}{1+\chi}H-\frac{M_{p1}}{M_{p3}}T+\frac{M_{p1}'}{M_{p3}'}\phi+\frac{M_{p1}''}{M_{p3}''}C+\frac{\chi_2-2\chi}{1+\chi}\right\}, \tag{14}$$

$$\frac{\partial C}{\partial t}+\boldsymbol{u}\cdot\nabla C=\frac{1}{Le_s}\nabla^2C+N_{ct}\nabla^2T, \tag{15}$$

where

$$\rho_1=\frac{d^3\rho_f}{\kappa\mu}(1+\alpha T_c-\alpha' C_1)g,\quad \rho_2=\frac{d^3\rho_f}{\kappa\mu}(\phi_0-\phi_1)(\alpha T_c-\alpha' C_1)g,$$
$$\lambda_1=\frac{\mu_0M_0H_0d^2}{\kappa\mu},\quad N_\phi=\frac{\phi_0-\phi_1}{1-\phi_0}.$$

B.C. (8) become :

$$\left.\begin{array}{l} w=0,\quad T=\dfrac{T_h}{T_h-T_c},\quad \phi=\dfrac{\phi_0}{\phi_0-\phi_1},\quad C=\dfrac{C_0}{C_0-C_1}\quad \text{at}\quad z=0,\\ w=0,\quad T=\dfrac{T_c}{T_h-T_c},\quad \phi=\dfrac{\phi_1}{\phi_0-\phi_1},\quad C=\dfrac{C_1}{C_0-C_1}\quad \text{at}\quad z=1. \end{array}\right\} \tag{16}$$

Non-dimensional parameters:

$$Ra = \frac{\rho_f g\alpha d^3(T_h - T_c)}{\mu\kappa}\text{(Rayleigh number)},\quad Le = \frac{\kappa}{D_B}\text{(Lewis number)},$$

$$Pr = \frac{\mu}{\rho_f\kappa}\text{(Prandtl number)},\quad Rn = \frac{(\rho_p - \rho_f)(\phi_0 - \phi_1)gd^3}{\mu\kappa}\text{(Concentration Rayleigh number)},$$

$$N_a = \frac{D_T(T_h - T_c)}{D_B T_c(\phi_0 - \phi_1)}\text{(Modified diffusivity ratio)},\quad N_a' = \frac{D_H H_0}{D_B(\phi_0 - \phi_1)}\text{(Modified diffusivity ratio)},$$

$$Rs = \frac{\rho_f g\alpha' d^3(C_0 - C_1)}{\mu D_S}\text{(Solutal Rayleigh number)},\quad Le_s = \frac{\kappa}{D_S}\text{(Solutal Lewis number)},$$

$$N_{ct} = \frac{D_{TC}(C_0 - C_1)}{k(T_h - T_c)}\text{(Soret parameter)},\quad N_{ct} = \frac{D_{CT}(T_h - T_c)}{k(C_0 - C_1)}\text{(Dufour parameter)},$$

$$M_{p1} = \frac{\mu_0\chi^2 H_0^2(T_h - T_c)}{\rho_f g\alpha d(1+\chi)T_h^2}\text{(Magnetic parameter)},\quad M_{p3} = \frac{\mu_0\chi H_0^2}{\rho_f g\alpha d T_h}\text{(Magnetic parameter)},$$

$$M_{p1}' = \frac{\mu_0\chi^2 H_0^2(\phi_0 - \phi_1)}{\rho_f g\alpha d(1+\chi)\phi_0^2}\text{(Magnetic parameter)},\quad M_{p3}' = \frac{\mu_0\chi H_0^2}{\rho_f g\alpha d\phi_0}\text{(Magnetic parameter)},$$

$$M_{p1}'' = \frac{\mu_0\chi^2 H_0^2(C_0 - C_1)}{\rho_f g\alpha' d(1+\chi)C_0^2}\text{(Magnetic parameter)},\quad M_{p3}'' = \frac{\mu_0\chi H_0^2}{\rho_f g\alpha' d C_0}\text{(Magnetic parameter)}$$

$$N_b = \frac{(\rho C)_P(\phi_0 - \phi_1)}{(\rho C)_f}\text{(Modified particle density increment)},\quad T_A = \frac{4\Omega^2 d^4}{\nu^2}\text{(Taylor number)}.$$

Here $\nu = \frac{\mu}{\rho_f}$ is the kinematic viscosity, $Ra_N = (1 - \phi_0)Ra$ and $Rs_n = (1 - \phi_0)\frac{Rs}{Le_s}$.

4 Solution of Steady State

Here

$$\boldsymbol{u}_b = \boldsymbol{0},$$

and p_b, T_b, ϕ_b, $\boldsymbol{M}_b$, $\boldsymbol{H}_b$, and C_b all are functions of z only.

Then Eqs. (10)–(15) reduce to

$$-\frac{dp_b}{dz} + \lambda_1 M_b\frac{dH_b}{dz} - Rn\phi_b + RaT_b - \frac{Rs}{Le_s}C_b - Ra_n N_\phi T_b\phi_b + Rs_n N_\phi C_b\phi_b - \rho_1 + \rho_2\phi_b = 0, \tag{17}$$

$$\frac{d^2\phi_b}{dz^2} + N_a\frac{d^2T_b}{dz^2} - N_a'\frac{d^2H_b}{dz^2} = 0, \tag{18}$$

$$\frac{d^2T_b}{dz^2}+\frac{dT_b}{dz}\left\{\frac{N_b}{Le}\frac{d\phi_b}{dz}+\frac{N_aN_b}{Le}\frac{dT_b}{dz}-\frac{N_a'N_b}{Le}\frac{dH_b}{dz}\right\}+N_{ct}\frac{d^2C_b}{dz^2}=0, \tag{19}$$

$$\chi_2\frac{dM_b}{dz}+\frac{dH_b}{dz}=0, \tag{20}$$

$$M_b=\frac{1+\chi}{\chi_2}\left\{\frac{\chi}{1+\chi}H_b-\frac{M_{p1}}{M_{P3}}T_b+\frac{M'_{p1}}{M'_{p3}}\phi_b+\frac{M''_{p1}}{M''_{p3}}C_b+\frac{\chi_2-2\chi}{1+\chi}\right\}, \tag{21}$$

$$\frac{1}{Le_s}\frac{d^2C_b}{dz^2}+N_{ct}\frac{d^2T_b}{dz^2}=0. \tag{22}$$

Using the B.C. (16), and following [1, 13, 14], we solve the Eqs. (17)–(22), to obtain the solution of steady state. The following results are obtained.

$$\begin{aligned}
&\boldsymbol{u}_b=\boldsymbol{0},\quad p=p_b(z),\quad T_b=\frac{T_h}{(T_h-T_c)}-z,\\
&\phi_b=\frac{\phi_0}{(\phi_0-\phi_1)}-z,\quad H_b=1-\frac{M_{p1}}{M_{p3}}z+\frac{M'_{p1}}{M'_{p3}}z+\frac{M''_{p1}}{M''_{p3}}z,\\
&M_b=1+\frac{1}{\chi_2}\left(\frac{M_{p1}}{M_{p3}}\right)z-\frac{1}{\chi_2}\left(\frac{M'_{p1}}{M'_{p3}}\right)z-\frac{1}{\chi_2}\left(\frac{M''_{p1}}{M''_{p3}}\right)z,\quad C_b=\frac{C_0}{(C_0-C_1)}-z.
\end{aligned} \tag{23}$$

5 Linear Analysis

Here we take very small perturbations to

$$\boldsymbol{u},\quad p,\quad T,\quad C,\quad \boldsymbol{M},\quad \phi,\quad \boldsymbol{H}.$$

By substituting perturbed variables into Eqs. (9)–(15) and linearizing about the steady state, we get the following set of linearized perturbation equations:

$$\frac{1}{Pr}\frac{\partial \nabla^2 w}{\partial t} = \nabla^4 w - \left\{RaM_{p3} - Ra_s M'_{p3} - \frac{Rs}{Le_s}M''_{p3}\right\}\frac{\partial \nabla_H^2 \psi}{\partial z} + \left\{Ng - Ra\frac{M_{p3}M'_{p1}}{M'_{p3}}\right.$$
$$\left. + Ra_n(1 + N_\phi z) - \frac{Rs}{Le_s}\frac{M''_{p3}M_{p1}}{M_{p3}}\right\}\nabla_H^2 \theta$$
$$- \left\{Ra\frac{M_{p3}M'_{p1}}{M'_{p3}} - Ra_s M'_{p1} - \frac{Rs}{Le_s}\frac{M''_{p3}M'_{p1}}{M'_{p3}}\right.$$
$$\left. + Rn + Ra_n N_\phi(1 - z) - Rs_n N_\phi(1 - z)\right\}\nabla_H^2 \phi + \left\{\frac{Rs}{Le_s}\frac{M''_{p3}M'_{p1}}{M'_{p3}} + \frac{Rs}{Le_s}M''_{p1}\right.$$
$$\left. - \frac{Rs}{Le_s}\frac{M''_{p3}M_{p1}}{M_{p3}} - Rs_n(1 + N_\phi z)\right\}\nabla_H^2 C - T_A{}^{1/2}\frac{\partial \xi}{\partial z}, \tag{24}$$

$$\frac{1}{Pr}\frac{\partial \xi}{\partial t} = \nabla^2 \xi + T_A{}^{1/2}\frac{\partial w}{\partial z}, \tag{25}$$

$$\frac{\partial \phi}{\partial t} = w + \frac{1}{Le}\nabla^2 \phi + \frac{N_a}{Le}\nabla^2 \theta - \frac{N'_a}{Le}\frac{\partial \nabla^2 \psi}{\partial z}, \tag{26}$$

$$\frac{\partial \theta}{\partial t} = \nabla^2 \theta + w - \frac{N_b}{Le}\frac{\partial \phi}{\partial z} + \frac{N_b N'_a}{Le}\frac{\partial^2 \psi}{\partial z^2} - \left\{\frac{N_b}{Le} + \frac{2N_a N_b}{Le} - \frac{N_b N'_a M_{p1}}{LeM_{p3}}\right.$$
$$\left. + \frac{N_b N'_a M'_{p1}}{LeM'_{p3}} + \frac{N_b N'_a M''_{p1}}{LeM''_{p3}}\right\}\frac{\partial \theta}{\partial z} + N_{ct}\nabla^2 C, \tag{27}$$

$$\frac{\partial^2 \psi}{\partial z^2} = -\frac{(1 + \chi_2)}{(1 + \chi)}\nabla_1^2 \psi + \frac{M_{p1}}{M_{p3}}\frac{\partial \theta}{\partial z} - \frac{M'_{p1}}{M'_{p3}}\frac{\partial \phi}{\partial z} - \frac{M''_{p1}}{M''_{p3}}\frac{\partial C}{\partial z}, \tag{28}$$

$$\frac{\partial C}{\partial t} = w + \frac{1}{Le_s}\nabla^2 C + N_{ct}\nabla^2 \theta, \tag{29}$$

where $\nabla_H^2 = \frac{\partial^2}{\partial x^2} + \frac{\partial^2}{\partial y^2}$, $Ng = M_{p1}Ra$ and $Ra_s = \frac{\rho_f g \alpha d^3(\phi_0 - \phi_1)}{\mu\kappa}$.

We obtain Eq. (24) by applying curl of a curl of linearized momentum equation and then taking its vertical component. Equation (25) represents the effect of rotation which is obtained by considering k^{th} component of curl of linearized momentum equation.

Further, we assume

$$[w, \phi, \theta, \psi, C] = [w(z), \phi(z), \theta(z), \psi(z), C(z)]\exp[\sigma t + i(k_x x + k_y y)]. \quad (30)$$

Here k_x, k_y are the wave numbers in x-direction and y-direction respectively, and $k = \sqrt{k_x^2 + k_y^2}$.

On substituting (30) into Eqs. (24)–(29) we get:

$$\frac{\sigma}{Pr}(4D^2 - k^2)w = (4D^2 - k^2)^2 w - \left\{Ng - Ra\frac{M_{p3}M'_{p1}}{M'_{p3}} - \frac{Rs}{Le_s}\frac{M''_{p3}M_{p1}}{M_{p3}} + Ra_n\right.$$
$$\left.\left\{1 + N_\phi\left(\frac{z+1}{2}\right)\right\}\right\}k^2\theta + \left\{Ra\frac{M_{p3}M'_{p1}}{M'_{p3}} - Ra_s M'_{p1} + Rn - \frac{Rs}{Le_s}\frac{M''_{p3}M'_{p1}}{M'_{p3}} + Ra_n\right.$$
$$\left. N_\phi\frac{(1-z)}{2} - Rs_n N_\phi\frac{(1-z)}{2}\right\}k^2\phi + 2\left\{RaM_{p3} - Ra_s M'_{p3} - \frac{Rs}{Le_s}M''_{p3}\right\}k^2 D\psi$$
$$+\left\{\frac{Rs}{Le_s}\frac{M''_{p3}M_{p1}}{M_{p3}} - \frac{Rs}{Le_s}\frac{M''_{p3}M'_{p1}}{M'_{p3}} + Rs_n\left\{1 + N_\phi\frac{(z+1)}{2}\right\} - \frac{Rs}{Le_s}M''_{p1}\right\}k^2 C - 2T_A{}^{1/2}D\xi, \quad (31)$$

$$\frac{\sigma}{Pr}\xi = (4D^2 - k^2)\xi + 2T_A{}^{1/2}Dw, \quad (32)$$

$$\sigma\phi = w + \frac{1}{Le}(4D^2 - k^2)\phi + \frac{N_a}{Le}(4D^2 - k^2)\theta - \frac{2N'_a}{Le}(4D^2 - k^2)D\psi, \quad (33)$$

$$\sigma\theta = w + (4D^2 - k^2)\theta - 2\left\{\frac{N_b}{Le} + \frac{2N_a N_b}{Le} - \frac{N_b N'_a M_{p1}}{LeM_{p3}} + \frac{N_b N'_a M'_{p1}}{LeM'_{p3}}\right.$$
$$\left. + \frac{N_b N'_a M''_{p1}}{LeM''_{p3}}\right\}D\theta - \frac{2N_b}{Le}D\phi + \frac{4N_b N'_a}{Le}D^2\psi + N_{ct}(4D^2 - k^2)C, \quad (34)$$

$$\left\{4D^2 - \frac{k^2(1+\chi_2)}{(1+\chi)}\right\}\psi - \frac{2M_{p1}}{M_{p3}}D\theta + \frac{2M'_{p1}}{M'_{p3}}D\phi + \frac{2M''_{p1}}{M''_{p3}}DC = 0, \quad (35)$$

$$\sigma C = w + \frac{1}{Le_s}(4D^2 - k^2)C + N_{ct}(4D^2 - k^2)\theta. \quad (36)$$

with B.C.

$$\left.\begin{array}{l} w = 0, \quad \theta = 0, \quad \phi = 0, \quad C = 0 \ \text{ at } \ z = \pm 1, \\ Dw = 0, \quad 2(1+\chi)D\psi - k\psi = 0 \ \text{ at } \ z = -1, \\ D^2 w = 0, \quad 2(1+\chi)D\psi + k\psi = 0 \ \text{ at } \ z = +1. \end{array}\right\} \tag{37}$$

The eigen value problem generated by Eqs. (31)–(36) with (37) is solved by Chebyshev pseudospectral method [15].

6 Results and Discussion

For the analysis, dimension of nanoparticles and thickness of the layer are considered as 10 nm and 1 mm respectively. Sources of physical quantities are [16, 17]. We investigate here the effect of the important parameters governing the flow such as Taylor number which characterizes the effect of rotation, Lewis number, which is defined as the ratio of thermal diffusivity and mass diffusivity and is used to characterize fluid flows where there is simultaneous heat and mass transfer and Rayleigh number which is the parameter to characterizes heat transfer by natural convection.

Figure 2 shows the NSCs for $\Delta\phi$, Le_s, Rs, and T_A. Change in the value of critical thermal Rayleigh number Ra_c determines the behaviour of these parameters. If Ra_c value goes up on increasing the value of any of these parameters then that parameter delays the onset of convection. On the other hand, if Ra_c decreases on increasing the value of any of these parameters then that parameter destabilizes the system. In other words, the system gets stabilized if on increasing the value of any one of these parameters, the NSCs shifts upwards. If the NSCs shifts downwards on increasing the value of any one of these parameters then system becomes unstable. In view of this argument, we see from Figure 2, $\Delta\phi$, Rs, and T_A delays the onset of convection while Le_s hastens the convection process.

We have also solved the same problem taking into account all three types of boundary conditions viz., Rigid-rigid (BC1), Rigid-free (BC2), and Free-free (BC3). The effect of rotation in terms of T_A associated with three different values of Rs has been displayed in Table 1 for gravity and in Table 2 for microgravity. The tables display that for any fixed value of Rs, the value of the critical thermal Rayleigh number Ra_c and the critical magnetic thermal Rayleigh number Ng_c increases with an increase in the value of T_A showing the stabilizing propensity of T_A at the onset of convection. An intriguing point to note here is that as the value of T_A increases from 10^3 onwards, there is a high jump in the values of Ra_c and Ng_c.

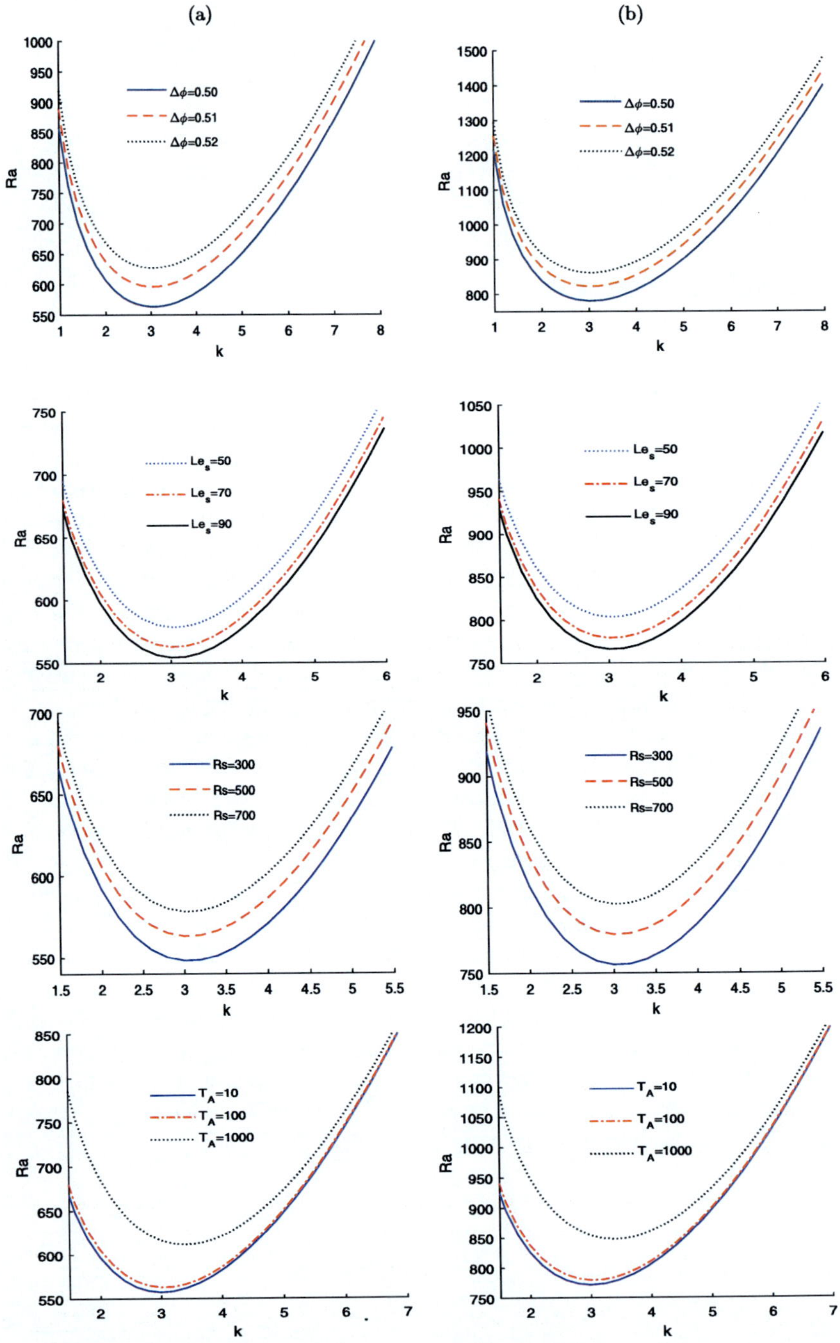

Fig. 2 NSCs for different values of $\Delta\phi$, Le_s, Rs and T_A for **a** W_{MNF} and **b** E_{MNF}. Here $\alpha_L = 2$, $d = 0.001$, $\Delta\phi = 0.01$, $N_a = 10$, $Le = 100$, $Rs = 500$, $Le_s = 70$, $T_A = 250$, $N_{ct} = 0.5$ and $N_{tc} = 0.005$

Table 1 The values of the critical thermal Rayleigh number Ra_c and the critical wave number k_c in the gravity environment for W_{MNF} and E_{MNF}. Here $\alpha_L = 2, d = 0.001, \Delta\phi = 0.01, N_a = 10$, $Le = 200, Le_s = 70, N_{ct} = 0.5$ and $N_{tc} = 0.005$

Rs	T_A	BC1				BC2				BC3			
		(W_{MNF})		(E_{MNF})		(W_{MNF})		(E_{MNF})		(W_{MNF})		(E_{MNF})	
		k_c	Ra_c	k_c	Ra_c	k_c	Ra_c	k_c	Ra_c	k_c	Ra_c	k_c	Ra_c
100	10^0	3.10	537	3.11	741	2.62	443	2.62	609	2.14	353	2.14	484
	10^1	3.11	537	3.11	742	2.63	443	2.63	610	2.15	354	2.15	485
	10^2	3.15	543	3.15	750	2.69	451	2.69	621	2.24	364	2.24	500
	10^3	3.51	594	3.51	822	3.14	511	3.14	706	2.78	434	2.79	599
	10^4	4.90	841	4.90	1164	4.57	762	4.57	1054	4.21	686	4.21	949
	10^5	7.38	1493	7.36	2069	6.85	1380	6.82	1920	6.28	1272	6.24	1760
300	10^0	3.10	553	3.10	766	2.62	459	2.62	634	2.14	370	2.14	509
	10^1	3.10	554	3.11	767	2.62	460	2.63	635	2.15	371	2.15	511
	10^2	3.15	560	3.14	775	2.69	467	2.69	646	2.24	381	2.24	525
	10^3	3.51	610	3.51	846	3.13	527	3.14	730	2.79	451	2.79	624
	10^4	4.90	857	4.89	1189	4.57	778	4.56	1079	4.21	702	4.20	974
	10^5	7.38	1509	7.35	2093	6.85	1396	6.81	1935	6.27	1287	6.24	1785
500	10^0	3.10	569	3.10	791	2.62	475	2.62	659	2.14	386	2.14	534
	10^1	3.10	570	3.11	792	2.63	476	2.63	660	2.15	387	2.15	536
	10^2	3.15	576	3.15	800	2.69	483	2.69	671	2.23	397	2.23	550
	10^3	3.50	626	3.50	871	3.13	543	3.14	755	2.78	467	2.79	649
	10^4	4.90	873	4.89	1214	4.58	794	4.56	1103	4.21	718	4.20	999
	10^5	7.37	1525	7.35	2118	6.84	1412	6.81	1959	6.27	1303	6.24	1809

7 Conclusions

Effects of various non-dimensional paramerters have been discussed using linear stability theory on the onset of convection by considering thin horizontal layer of W_{MNF} and E_{MNF} which is heated and salted from below. Effects of Brownian motion, thermophoresis and magnetophoresis have been embodied in the study. It has been observed that the role of $\Delta\phi$, Rs, and T_A is to make the system stable while Le_s hastens the convection process. The effects of Rs and T_A are found to be stabilizing in both the environments (gravity and microgravity). We have also observed that with higher rotation rates much better stability of the system can be maintained in the gravity as well as in the microgravity environment.

Table 2 The values of the critical magnetic thermal Rayleigh number Ng_c and the critical wave number k_c in the microgravity environment for W_{MNF} and E_{MNF}. Here $\alpha_L = 2$, $d = 0.001$, $\Delta\phi = 0.01$, $N_a = 10$, $Le = 200$, $Le_s = 70$, $N_{ct} = 0.5$ and $N_{tc} = 0.005$

Rs	T_A	BC1				BC2				BC3			
		(W_{MNF})		(E_{MNF})		(W_{MNF})		(E_{MNF})		(W_{MNF})		(E_{MNF})	
		k_c	Ng_c	k_c	Ng_c	k_c	Ng_c	k_c	Ng_c	k_c	Ng_c	k_c	Ng_c
3.0*e*-5	10^0	3.10	3079	3.10	3058	2.62	2116	2.62	2092	2.14	1371	2.14	1350
	10^1	3.11	3086	3.11	3065	2.63	2123	2.63	2099	2.15	1379	2.15	1358
	10^2	3.15	3153	3.15	3132	2.69	2194	2.69	2178	2.24	1455	2.24	1436
	10^3	3.50	3754	3.50	3732	3.14	2797	3.14	2777	2.78	2044	2.79	2032
	10^4	4.90	7398	4.89	7358	4.56	6091	4.56	6053	4.21	4966	4.20	4940
	10^5	7.37	22897	7.34	22765	6.84	19592	6.81	19437	6.27	16673	6.23	16544
4.0*e*-5	10^0	3.10	3167	3.10	3156	2.62	2189	2.62	2173	2.14	1431	2.13	1416
	10^1	3.11	3174	3.10	3162	2.63	2196	2.63	2180	2.15	1439	2.14	1424
	10^2	3.15	3243	3.15	3231	2.69	2269	2.69	2254	2.24	1516	2.24	1504
	10^3	3.50	3851	3.50	3840	3.14	2880	3.15	2870	2.78	2117	2.79	2112
	10^4	4.90	7533	4.89	7508	4.56	6213	4.55	6189	4.21	5078	4.18	5064
	10^5	7.37	23133	7.34	23027	6.84	19809	6.80	19677	6.26	16874	6.23	16766
5.0*e*-5	10^0	3.11	3257	3.09	3255	2.62	2263	2.62	2255	2.14	1491	2.14	1483
	10^1	3.11	3264	3.10	3262	2.63	2271	2.62	2263	2.15	1499	2.15	1492
	10^2	3.14	3333	3.15	3332	2.69	2345	2.69	2338	2.24	1578	2.23	1573
	10^3	3.51	3949	3.50	3950	3.14	2966	3.15	2964	2.77	2190	2.79	2194
	10^4	4.90	7670	4.88	7660	4.56	6337	4.55	6326	4.20	5190	4.19	5189
	10^5	7.37	23371	7.34	23289	6.84	20028	6.80	19918	6.26	17076	6.22	16990

References

1. Buongiorno, J.: Convective transport in nanofluids. J. Heat Transf. **128**, 240–250 (2006)
2. Stommel, H., Arons, A.B., Blanchard, D.: An oceanographical curiosity: the perpetual salt fountain. Deep-Sea Res. **3**, 152–153 (1956)
3. Turner, J.S.: Multicomponent Convect. Ann. Rev. Fluid Mech. **17**, 11–44 (1985)
4. Huppert, H.E., Turner, J.S.: Double-diffusive convection. J. Fluid Mech. **106**, 299–329 (1981)
5. Sunil, Sharma, P., Mahajan, A.: A nonlinear stability analysis of a rotating double-diffusive magnetized ferrofluid. Appl. Math. Comput. **218**, 2785–99 (2011)
6. Savino, R., Paterna, D.: Thermodiffusion in nanofluids under different gravity conditions. Phys. Fluids **20**, 017101 (2008)
7. Sharma, M.K., Singh, R.: Linear stability analysis of double-diffusive convection in magnetic nanofluids in porous media. J. Porous Media **17**, 883–900 (2014)
8. Yadav, D., Lee, D., Cho, H.H., Lee, J.: The onset of double-diffusive nanofluid convection in a rotating porous medium layer with thermal conductivity and viscosity variation: a revised model. J. Porous Media **19**, 105–21 (2016)
9. Umavathi, J.C., Sheremet, M.A., Ojjela, O., Reddy, G.J.: The onset of double-diffusive convection in a nanofluid saturated porous layer: cross-diffusion effects. Eur. J. Mech. B Fluids **65**, 70–87 (2017)

10. Mahajan, A., Sharma, M.K.: Double-diffusive convection in a magnetic nanofluid layer with cross diffusion effects. J. Eng. Math. **115**, 67–87 (2019)
11. Mahajan, A., Arora, M.: Convection in rotating magnetic nanofluids. Appl. Math. Comput. **219**, 6284–96 (2013)
12. Nield, D.A., Kuznetsov, A.V.: The onset of double-diffusive convection in a nanofluid layer. Int. J. Heat Fluid Flow **32**, 771–6 (2011)
13. Nield, D.A., Kuznetsov, A.V.: The onset of convection in a horizontal nanofluid layer of finite depth. Eur. J. Mech. B Fluids **29**, 217–23 (2010)
14. Arora, M., Singh, R., Panda, M.K.: Effects of magnetic-field-dependent viscosity at onset of convection in magnetic nanofluids. J. Eng. Math. **101**, 201–217 (2016)
15. Canuto, C., Hussaini, M.Y., Quateroni, A., Zang, T.: Spectral Methods in Fluid Dynamics. Springer, New York (1998)
16. Kaloni, P.N., Lou J. X.: Convective instability of magnetic fluids. Phys. Rev. E **70**(1–12), 26313 (2004)
17. Rosensweig, R.E.: Ferrohydrodynamics. Courier Dover Publications (1997)

Modeling for Implications of COVID-19 Pandemic on Healthcare System in India

R. Sasikumar and P. Arriyamuthu

Abstract The COVID-19 pandemic has affected the global healthcare system in many countries. India has faced complex multidimensional problems concerning the healthcare system during the COVID-19 outbreak. This article explores some of the implications of COVID-19 on the health system. Also, we attempt to study health economics and other related issues. We have developed the susceptible-exposed-infection-recovered model, logistic growth model, time interrupted regression model, and a stochastic approach for these problems. These models focus on the effect of prevention measures and other interventions for a pandemic on the healthcare system. Our study suggests that the above models are appropriate for COVID-19 at break and effective models for the implications of the pandemic on the healthcare system.

Keywords COVID-19 · Markov chain · Logistic growth model · SEIR model · Time interrupted regression model · Healthcare system

1 Introduction

The continuous spread of the COVID-19 outbreak is a new strain that has an impact at the global level and has become the greatest health challenge in the world. The COVID-19 pandemic has affected different people in various ways. The most important symptoms are fever, dry cough, tiredness, etc. The World Health Organization (2020) explained that some common symptoms are loss of taste, smell, and rashes on the skin. It's high time to develop a medical solution for this. Lack of hospitals, physicians, health experts, and hospital beds are the severe scenarios facing India nowadays. Curfews, social distancing, self-isolation, and vaccinations all have a role in preventing pandemic transmission and high population density might make effective measures difficult. The Batch Markovian arrival process has been proposed by Neuts [1]. Assumptions are the Markov arrival process and its application to

R. Sasikumar · P. Arriyamuthu (✉)
Department of Statistics, Manonmaniam Sundaranar University, Tirunelveli 627012, India
e-mail: arriyamth@gmail.com

R. K. Sharma et al. (eds.), *Frontiers in Industrial and Applied Mathematics*,
Springer Proceedings in Mathematics & Statistics 410,
https://doi.org/10.1007/978-981-19-7272-0_46

the stochastic model described by Chakravarthy [2]. Briggs has a long history of using Markov chains in public health decision-making and epidemiological research [3]. An application of the Markov model for economic (or) financial evaluation and medical decision-making was described by Sonnenberg and Beck [4]. Governments have taken effective steps such as partial lockdowns with social distancing, and isolations and vaccinations to reduce the transmissions of a pandemic, Zhang et al. [5]. Ozili and Arun have proposed mitigation measures for the intense effect on the global economic status [6]. Modeling for interventions taken to minimize the transmission rate of a pandemic in India by Mandal et al. [7]. In this study, the importance of increasing immunity, social distance, and lockdown measures concerning these measures can be useful in flattening the pandemic discussed by Bhola et al. [8]. Lopez et al. to reflect the environmental situation investigated the susceptible-exposed-infection-recovered model which presented dead, quantified, and protected population compartments [9]. Cano et al. defined the dynamics of the COVID-19 outbreak using a simple Markov model [7]. Several types of analysis have been made on the COVID-19 outbreak and describe the Ebola virus with demographic effects by Rachan [10]. The susceptible-exposed-infection-recovered model for demographic effects such as birth and mortality rate during COVID-19 outbreak was described by Hamzah et al. [11]. Analysis, forecast, simulation, and optimal controls for the Ebola virus used the susceptible-exposed-infection-recovered model by Porter [12]. The current pandemic is rising quickly and spreading to millions as evidenced by the many recorded cases in India by Sarkar et al. [6]. Many countries used vaccines, curfews, and self-isolation to control the COVID-19 pandemic. T. M Chen et al. are interested in studying the transmission patterns of the outbreak and the impact of further interventions [7]. A survey of social economic evaluation levels in many countries in the social mix through the R package by Chen and Zhou [13]. According to Adly et al. [14], the most definite answer for the public health system is to conduct tests as soon as possible to permit the rapid identification of active patients, effective treatment methods, and immediate self-isolation for susceptible cases. Describe the deterministic model for the impact of social distancing on the transmission dynamics of the COVID-19 outbreak in South Africa by Nyanadza et al. [15]. An analysis of the impact of lockdown measures taken to control the transmission dynamics of the COVID-19 outbreak in India was conducted by Youkta et al. [16]. Elinor Aviv-Sharon and Asaph Aharoni used generalized logistic modeling to characterize the transmission pattern and trace the trajectory of the COVID-19 outbreak, as well as the impact of specific interventions [17]. Details about the global preventive measures were described by Kumar et al. [18]. Verhulst developed the original logistic growth modeling for the biological population.

Table 1 Details about the demographics of India

Total population	1.38 billion
Ratio of old age people	9.25
Years of median age	28.4
Population size above 65 years old	6.2%
Number of doctors	1/1457 population
Number of nurses	1/457 population
Beds per 100 population	0.55 Beds
Life expectancy	68 Years
Population density	464 per km^2

2 Overview of Indian Healthcare System and Changes in the COVID-19 Pandemic During this Period

Our country comprises 28 states and 8 union territories. Table 1 presents an overview of India's demographics.

The capacity of hospital beds is mostly determined by a country's income level. Our country has an average of 0.55 beds per 100 people depending on income level. The pandemic has negatively affected global health and daily life. Contingency plans for an expected surge of cases were also added to the current scenario. Medical personnel, homoeopathic and ayurvedic practitioners, medical students, volunteers, sanitary workers, ex-servicemen, teachers, doctors, and others were identified across the municipal corporation to create an online information pool of 15.8 million human resources for a variety of activities needed to combat the outbreak. It was also critical to match the demand for medical equipment and pharmaceuticals with the rising supply of infrastructure and human resources. The demand for personal protective equipment increased as private hospitals became involved. The spread of COVID-19 disease affected human health, psychological problem, and economic status, and related restrictions were implemented to control the unexpected adverse effects on human health. There are also other challenges described by Singh et al. [19].

3 Propose Models and Materials

3.1 Data Sources

The data for India and other states were taken from the official websites of the ICMR (http://www.icmr.gov.in), the website of COVID-19 India (http://covid19india.org). These data were used in a study on the impact of COVID-19 on the Indian healthcare system.

3.2 Susceptible-Exposed-Infection-Recovered Model

This model is a compartmental model to study infectious disease and divides the population into four components such as Susceptible (S), Exposed (E), Infectious (I), and Recovered (R). Susceptible populations are those who are at risk of becoming infected. The people who have been infected with the sickness and are able to converse with others are said to be infected. Exposed people have been exposed to the disease but are not yet contagious, while recovered persons have recovered from their illness. The parameters (β, γ, and σ) are explained as follows: β is a transmission parameter, which is the number of effective contacts per unit of time per infected individual; γ is the rate of recovery in a specific time; and σ is the rate at which infected individuals become infectious. The differential equations that describe this model are as follows:

$$\frac{dS}{dt} = \beta S(t)I(t) \tag{1}$$

$$\frac{dE}{dt} = \beta S(t)I(t) - \sigma E(t) \tag{2}$$

$$\frac{dI}{dt} = \sigma E(t) - \gamma I(t) \tag{3}$$

$$\frac{dR}{dt} = \gamma I(t) \tag{4}$$

Subject to the conditions, $S(0)>0, E(0)\geq 0, I(0)\geq 0, R(0)\geq 0.$

We consider this model, where the total population N = S + E + I + R (i.e., four components if added should be equal to the total population). We assume that the new individuals were a result of contacts within the susceptible group S(t). The transmission parameter β (contact rate) is giving a force of infectious λ = β(I(t)/N(t)) and the number of new infectious cases out of S(t) and into E(t) as βI(t)S(t)/N(t). The exposed persons progress to active cases in 2 weeks at a constant progression rate (k) n giving the number of individuals moving out of E(t) and into I(t) as kE(t). The infectious cases are denoted to recover at a constant rate σ. The cumulative number of recovered cases moving out of I(t) into R(t) is given by γ I(t). Given this model, we assumed the values of β, γ, and σ [20]. Jakhar et al. [20] describe for such parameters. We are assigning a basic reproduction ratio of 1.5, β to be 0.1, σ to be 0.1. Also, the γ is assumed to be 0.2 and the average infectious period is chosen to be 5 days (Fig. 1).

3.3 Logistic Growth Model

This model is increased at the onset, but decreases at a later stage, as it approaches the maximum. In current COVID-19, the highest limit will be the cumulative population

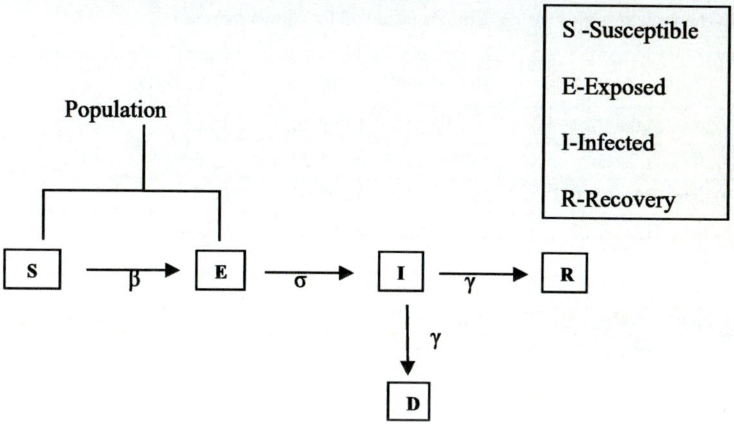

Fig. 1 Relationship between each step of the SEIR model

and the growth certainly comes down when a greater proposition of the population is infected. The reason for utilizing logistic growth modeling for the COVID-19 pandemic is that it has been proven that the epidemic grows exponentially in the early phases and then decreases in the later stages. This model is presented as $y(t) = \frac{c}{1+a(e^{-bt})}$ where y(t) denotes the number of individuals at any given time t, c denotes the limiting value, the maximum capacity for $y.a = (c/y_0) - 1$, and " b" is the rate of change. The number of cases at the beginning, also called the initial value, denotes $\frac{c}{(1+a)}$ and the maximum growth rate (t) is $\log(a)/b$.. When y = c (the population size is maximum), y/c will be one. Hence, the (1-(y/c)) will be zero and the growth will be zero. The optimum parameter values can be obtained by nonlinear least square method.

3.4 Interrupted Time Series Multiple Regression Model

The model is the strongest, quasi-experimental tool for evaluating the longitudinal impact of interventions. The impact of immunization on the incidence of new cases and death cases was assessed using time interrupted regression analysis [21]. Figure 6 shows the diagrammatic representation and the results of the analyses are presented in Table 3.

4 Implications of Preventive Measures

This section focuses to identify the spread of the COVID-19 pandemic associated with preventive measures taken in India, such as a discussion about the impact of curfew, social distancing, and corresponding other interventions using the incidence

Table 2 The growth rate of the epidemic is based on lockdown duration

Lockdown duration of the first wave	Lockdown duration of the second wave	Growth rate in percentage	
		Year—2020	Year—2021
24.05.2020 to 14.04.2020	10.05.2021 to 17.05.2021	15.06	7.5
15.04.2020 to 03.05.2020	18.05.2021 to 24.05.2021	7.5	11.1
04.05.2020 to 17.05.2020	25.05.2021 to 31.05.2021	6.4	16.0
18.05.2020 to 31.05.2020	01.05.2021 to 07.06.2021	4.63	23.9

of daily cases. Assess the death rate about the epidemic control measures performed by India's healthcare system.

4.1 *Impact of Lockdown Strategy*

The purpose of this study is to discuss the effect of lockdown strategy to tackle the COVID-19 outbreak. Table 2 shows the growth rate of the epidemic during the COVID-19 outbreak in India.

In addition, we discovered that there are pre- and post-lockdown measures in India. Our government announced that the curfew would begin in May 2021 during the second wave. The peak of the second wave would have arrived in mid-May 2021. Figures 3 and 4 show the complete lockdown measures based on the SEIR model.

4.2 *Effects of Social Distancing Based on a Stochastic Approach*

Social distancing is a key part of preventing the spread of the pandemics and it is the best form of response in managing the affected rate of COVID-19. But a large population density can make this action challenging. The study of this section is clear and discusses the effect of social distancing through a simple stochastic approach. We propose the simple Markov chain to represent the impact of social distancing on the transmission dynamics of a pandemic. Figure 2 describes how individuals can transmit between states. After becoming symptomatic, they migrate from the susceptible population to becoming infected, then to being contagious virus (shedding). They may become ill and die as a result.

We use the transition probabilities, s_0–s_7 follow an Erlang distribution. This distribution is a special case of the gamma distribution, denoted as ε but scaled when various outcomes from a state are possible from Fig. 2. We observed the dynamics when individuals move from one state to another. The transition probability, s_0, can be denoted through a desired basic reproduction number, R_0, as it can be simply shown that

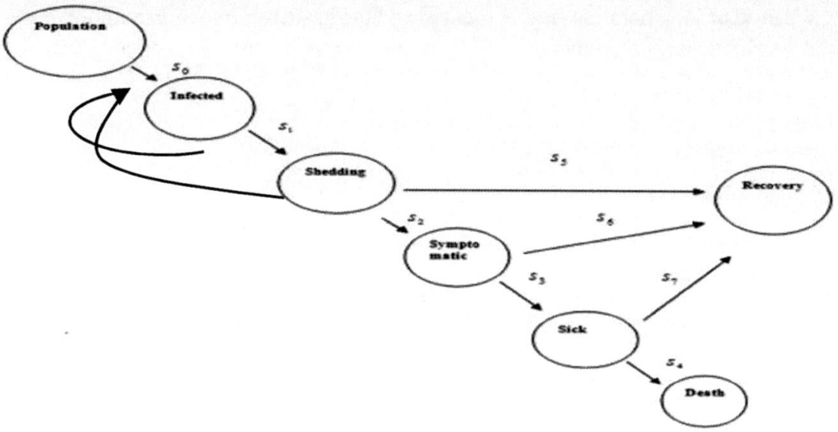

Fig. 2 Explain the Markov chain and how individuals can transmit between states after infection

$$R_0 = s_0 \sum_{i=0}^{\infty} (1 - s_2 - s_5)^i + s_0 \sum_{i=0}^{\infty} s_2 (1 - s_2 - s_5)^i \sum_{j=0}^{\infty} (1 - s_3 - s_6)^j \quad (5)$$

During the pandemic, Halloran defined the basic reproduction number as the average number of secondary infected individuals caused by primary cases [22]. The basic reproduction number on the day the values are implemented is used to reduce the number of people infected by each shedding or sick individual. Our study discussed the effect of social distancing rate on virus infection based an epidemiological model. We consider the constant rate ρ (0 < ρ <1), where ρ0 means perfect social distancing. But also investigate the impacts of ρ = 0.1, ρ= 0.2, and ρ = 0.3.

The capacity of the COVID-19 infection is modified as follows:

$$\lambda = \begin{cases} \beta I(t)/N, t_0 \le t_{lock} \\ \rho\beta I(t)/N, t \ge t_{lock} \end{cases} \quad (6)$$

The dynamic system of differential equations, including the assumption in pandemic in India, is defined by

$$S'(t) = -\lambda S(t) + S_{rec} \quad (7)$$

$$E'(t) = \lambda S(t) - kE(t) + E_{rec} \quad (8)$$

$$I'(t) = kE(t) - \gamma I(t) \quad (9)$$

$$R'(t) = \gamma I(t) \quad (10)$$

We used next-generation matrix method and the basic reproduction number of the model system (Eq. (1)), where,

$$A = \begin{bmatrix} 0 & \beta \\ 0 & 0 \end{bmatrix}, \quad B = \begin{bmatrix} k & 0 \\ -k & \gamma \end{bmatrix}$$

.

The product of AB-1 is given by $AB^{-1} = \begin{bmatrix} \frac{\beta}{\gamma} & \frac{\beta}{\gamma} \\ 0 & 0 \end{bmatrix}$ with the spectral radius defined by

$$v(AB^{-1}) = R_0 = \begin{cases} \beta/\gamma, & t_0 \le t < t_{lock} \\ \rho\beta/\gamma, & t_0 \ge t_{lock} \end{cases} \tag{11}$$

4.3 *Mortality Trends*

The COVID-19 pandemic becomes a serious health problem in all states in India. Maharashtra is the center of the COVID-19 virus by records. Gujarat and Telangana reported the same percentage of infected cases. Gujarat has a greater rate of deaths, and meanwhile Delhi has a lower percentage of infection cases than Karnataka, but it has a higher percentage of deaths. Mostly, the COVID-19 pandemic is infecting the male community in India, with a high affected rate of individuals between 30 and 40 years. In these, ages above 60 years are mainly reported as deaths by Joe et al. [23]. Now, we discussed the growth model of mortality trends. Also, we identify the relationship between COVID-19 death counts and population density in India through some statistical methods (see Table 4 and Fig. 7).

5 Analysis and Results

We have identified that before the implementation of the lockdown on March 2020 and the second lockdown announcement on May 2021. However, this peak has shifted to mid-August 2020 and July 2021 following the enforcement of severe countrywide lockdown. The following is an analysis of the scenario in India before and after the lockdown.

We observed from Figs. 3 and 4 the contribution of different immigration parameters during pre- and the post-lockdown pandemic situations in India. Also, we identified the effect of the lockdown strategy to tackle the COVID-19 outbreak. Here, the pre-lockdown situation is considered an as initial condition, after that the post-lockdown situation will be considered. From this study, we observed that the data is significant during the 2020 lockdown and non-significant during the 2021 lockdown. According to Table 2, the growth rate decreases after the 2020 lockdown and increases after the 2021 lockdown. According to this study, India's quarantine looks to be beneficial or effective in delaying the epidemic's peak. As a result, these findings are extremely beneficial to get time for preparedness in the healthcare system.

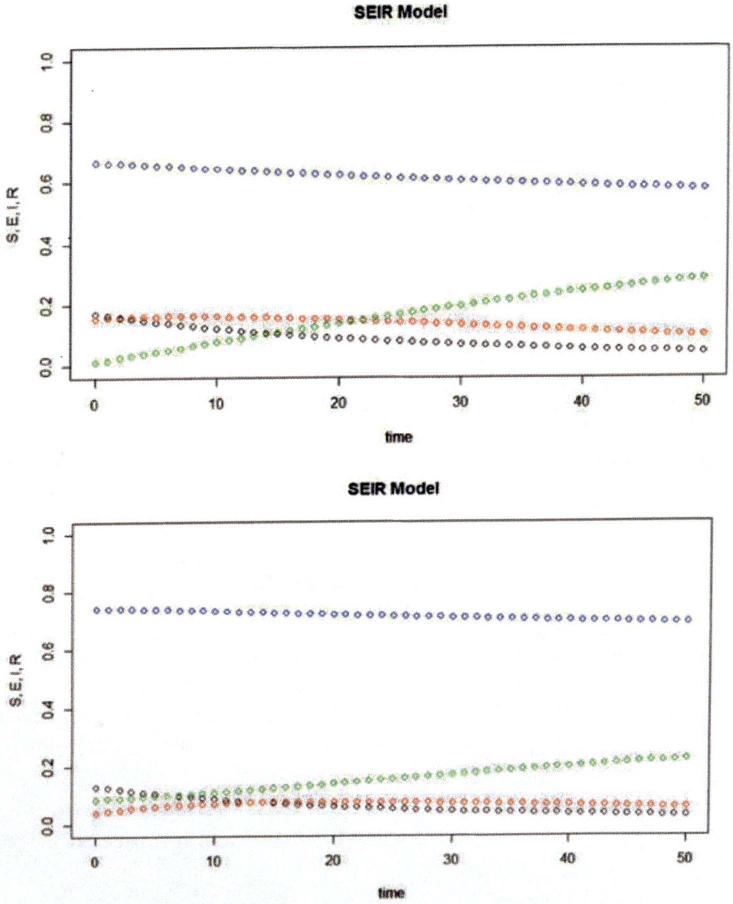

Fig. 3 Pre-lockdown pandemic situation

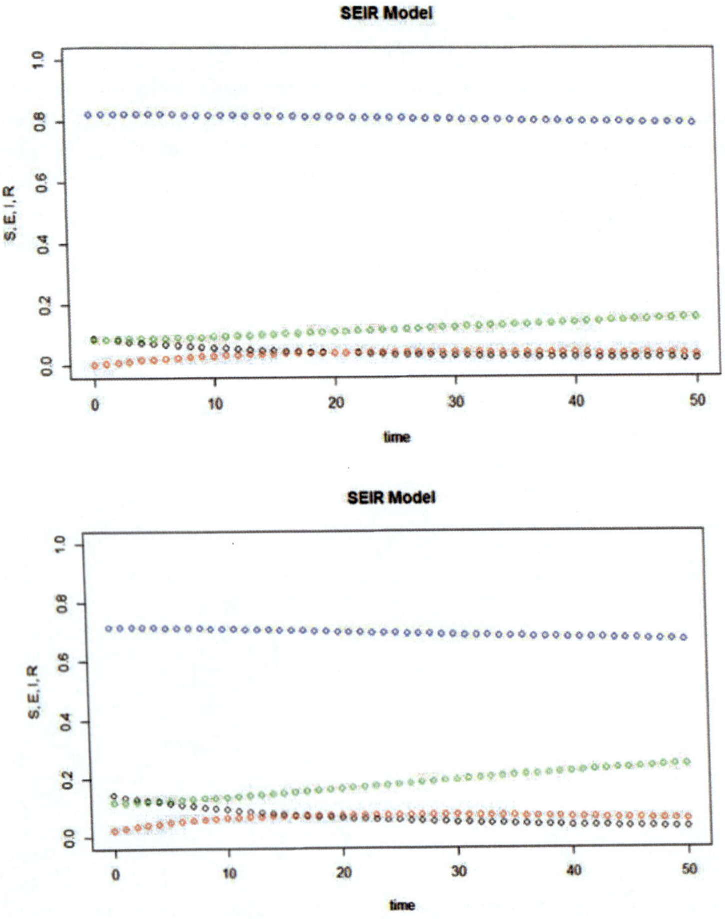

Fig. 4 Post-lockdown pandemic situation

5.1 *Results for Effect of Social Distancing*

The following figures describe the several social distancing constant rates involving, and then changes to susceptible, exposed, recovery and death trajectory.

In this study, the effect of social distancing is discussed. We represent through a constant rate ρ ($0 < \rho < 1$), R_0 (basic reproduction number) will be determined using a stochastic approach and the value of ρ will be determined using the SEIR model. Here $\rho = 0$ is the perfect social distance and the value of ρ will be considered in three categories 0.1 (10%), 0.2 (20%), and 0.3 (30%) and investigated for exposed infection and recovery path (see Fig. 5).

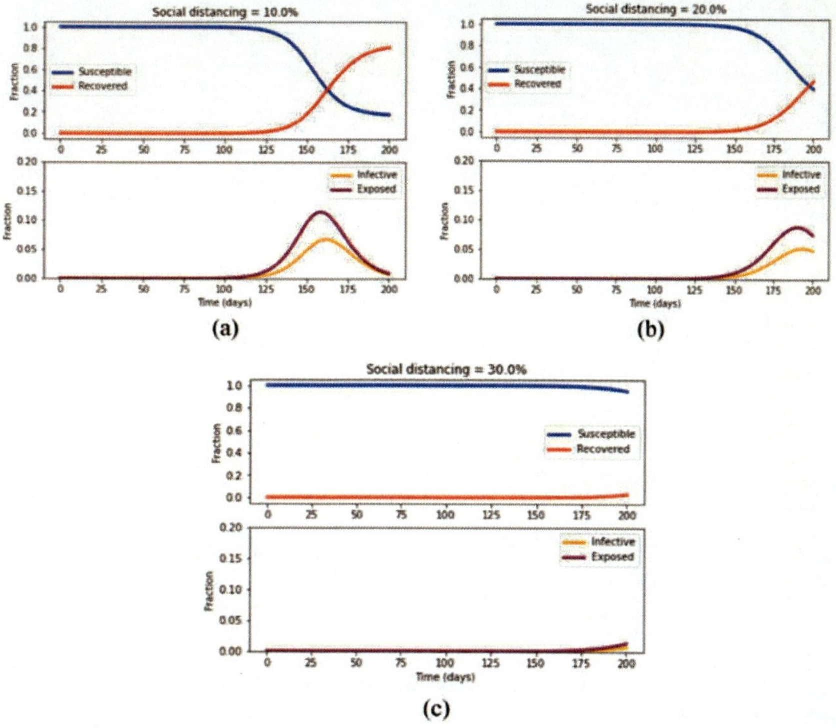

Fig. 5 **a** Shows fit the SEIR model and consider the social distancing = 10% (i.e., $\rho = 0.1$). **b** Shows fit the SEIR model and consider the social distancing = 20% (i.e., $\rho = 0.2$). **c** Shows fit the SEIR model and consider the social distancing = 30% (i.e., $\rho = 0.3$)

5.2 *Result for Vaccination Intervention Using Time Series Multiple Regression Models*

This section displays the results of the impact of the vaccine on the incidence of new cases and death cases. Then a detailed regression analysis of new incidence cases, mortality, and vaccination growth due to COVID-19 in India (Table 3).

From this section, vaccination intervention is discussed on the basis of time series multiple regression model. Here vaccine (y) is considered as dependent variable and

Table 3 Regression coefficients, 95% CI, and parameters value

Variables	Estimate	Std. error	t_value	Pr(> |*t*|)
Y	−1.951e + 09	1.034e + 09	−1.886	0.0617
x_1	5.655e + 03	2.357e-03	2.399	0.0180
x_2	1.843e + 02	1.278e + 02	1.443	0.1518

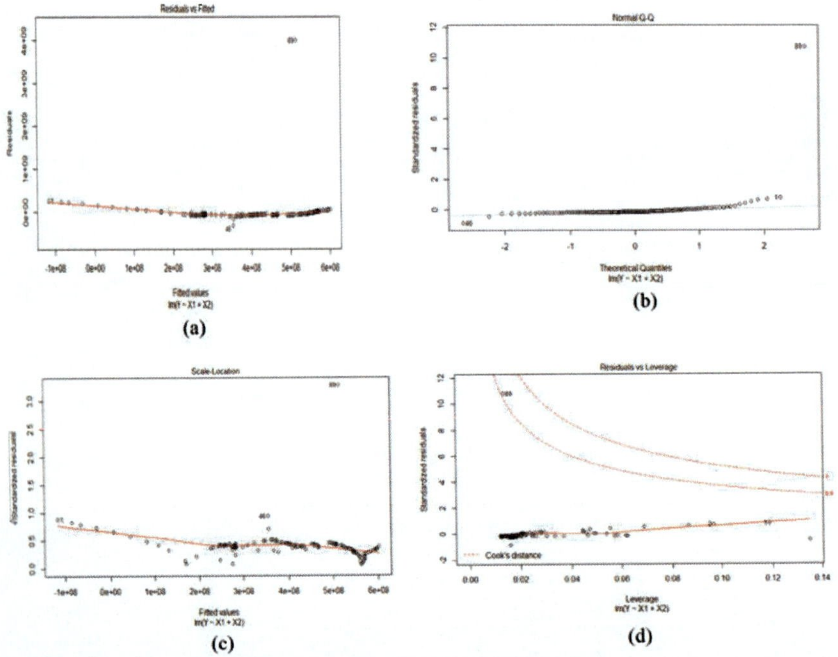

Fig. 6 Fit the time series multiple regression model for vaccine intervention on incidents of new cases and death cases

incident cases (x_1) and death cases (x_2) are independent variables. The incident case coefficient values will be discussed if the vaccination increases. So, we conclude that the vaccination in 2021 is significant (see Fig. 6).

5.3 *Results for COVID-19 Mortality Trend*

This result fits the growth model on death counts. We classified three phases and observed the coefficient values. Then, we observed that phases two and three were less than the first phase (see Table 4). We discovered from this study peck of death counts by the end of 2020. In addition, we investigated the impact of India's population density on the pandemic spread and mortality. After that, we observe a moderate association between the COVID-19 pandemic and population density (see Fig. 7).

Table 4 Parameter values for growth model

	Estimate	Std. error	t_value	Pr(> |t|)
Phase 1	4.374e + 05	5.784e + 02	756.22	< 2e-16
Phase 2	2.494e + 00	1.576e-01	15.82	< 2e_16
Phase 3	2.470e + 01	2.232e-01	110.63	< 2e-16

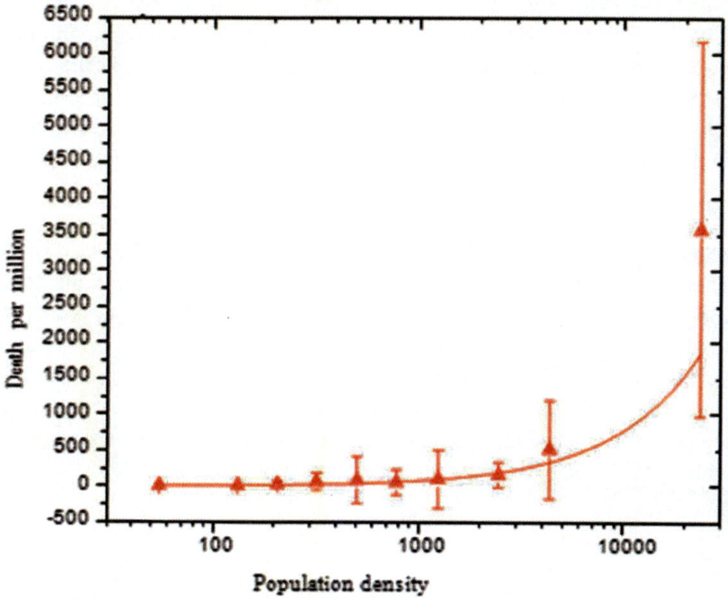

Fig. 7 Averaged death counts due to COVID-19 with population density in India

5.4 Reviews and Descriptions of Other Implications

1. The effect of public health interventions

These treatments have the impact of controlling public health problems; the main goal is to stop an outbreak from spreading and reduce the size of the epidemic. In the absence of a safe antiviral, public health initiatives, including social isolation, contact tracing, mask fabrication, effective quarantine, and travel restrictions, are used to stop and reduce outbreak pressure. Whereas public awareness of the virus and personal protection (e.g., self-isolation) has been developed. These measures guide the changes in medical seeking development and epidemiological characteristics. The differences in prediction based on different health strategies and the travel restriction effects were most significant. Also, different studies on the implications of contact tracking and self-isolation, but it was denoted that improving reporting and quarantine rates. Yue Xiang's various discussions on the implications of public health interventions are discussed above [24].

2. The economic impacts

India is a developing country and it has the second highest population with the fifth-largest economy. Kavita Singh studied the economic effects of a pandemic on people with chronic illnesses in India, focusing on underserved metropolitan areas and rural populations [19]. While the restoration of economic activity among persons infected with COVID-19 appears to be resulting in economic recovery, economic modeling indicates that the average economic value per individual life lost in the pandemic in India is 7.09–8.22 times the country's GDP per capita. Instead of a nationwide lockdown, Arora P demonstrates how a lockdown policy implemented within Indian states in the event of a pandemic can result in substantial unemployment [25].

3. The effect of policy and technology

We identified the high recovery rate from the outbreak in India as an output of significant technological impacts. The COVID-19 disease has tested the country's epidemic preparedness in terms of its health infrastructure, interventions of policy, and communication technology. Initially, the impact of a time and one of the most strict lockdown policies was observed to reduce the spread of disease. Similarly, it used the country to prepare critical medical infrastructure, public resources, and technological advances. Isha Goel provided an overview of India's epidemiological state and highlighted the potential consequences of policy and technology changes [26].

6 Discussion

The COVID-19 pandemic has various implications for the world economy as well as for many aspects of human health, education, the physical of individuals, etc. In this study, we investigated the impact of a few factors on the healthcare system during COVID-19 as well as the illness spread related to preventive measures performed in India. In this section, the data was separated into pre-lockdown and post-lockdown: in the lockdown period implemented on the years 2020 and 2021, the infection and recovery rate paths are determined using the SEIR model. From this, the lockdown period on 2021 is significant which means that infection rate is decreased and simultaneously recovery has increased; in the lockdown period in 2021, the data is not significant which means the infection rate doesn't change, which has been shown in Figs. 2 and 3. Next the Markov chain is used to determine the effect of social distancing. The transition probability state is applied to the death state in the population; from this the basic reproduction number (R_0) value is determined with this SEIR model. The social distancing contact rate (ρ) is separated into three (0.1, 0.2, and 0.3) and using this exposed, infection, and recovery rates are analyzed. From this, we can conclude that if the social distancing percentage increases, then the infections will decrease. The infection and death rates are analyzed based on the time series multiple regression model in vaccine intervention. Using this public health intervention is discussed. The regression and coefficient values are calculated for the vaccination coverage increase and death decrease. As the death count increases after May 2021, the path is determined using the growth model; COVID-19's mortality, population density, and impacts are also discussed. Finally, other implications and their reviews and descriptions are also discussed.

7 Conclusion

Our research found that the most effective public health intervention is to conduct screening tests as soon as possible to aid in the rapid identification of infected cases, quick treatment, and immediate isolation of susceptible cases. We also discovered that pandemic-related preventive strategies, such as social distancing, wearing masks in public places, self-hygiene, and remote working and learning, can all have a significant impact on pandemic transmission. In addition, the above-mentioned model might be used to design and prepare health systems in this study. This research provided a systematic approach to avoid recording, and controlling the spread of COVID-19.

References

1. Neuts, MF.: Matrix-geometric solutions in stochastic models: an algorithmic approach. Courier Corporation (1994)
2. Chakravarthy, S.R.: The batch Markovian arrival process: a review and future work. Adv. Probab. Theory Stoch. Process. **1**, 21–49 (2001)
3. Briggs, A., Sculpher, M.: An introduction to Markov modeling for economic evaluation. Pharmacoeconomics **13**(4), 397–409 (1998). (Apr)
4. Sonnenberg, FA., Beck JR.: Markov models in medical decision making: a practical guide. Medical decision making. **13**(4), 322–38 (1993). (Dec)
5. Zhang, Y., Jiang, B., Yuan, J., Tao, Y.: The impact of social distancing and epicenter lockdown on the COVID-19 epidemic in mainland China: A data-driven SEIQR model study. MedRxiv. (1 Jan 2020)
6. Ozili, P.K., Arun, T.: Spillover of COVID-19: impact on the global economy. SRN 3562570. (27 Mar 2020)
7. Mandal, S., Bhatnagar, T., Arinaminpathy, N., Agarwal, A., Chowdhury, A, Murhekar, M., Gangakhedkar, RR., Sarkar, S.: Prudent public health intervention strategies to control the corona virus disease 2019 transmission in India: a mathematical model-based approach. Indian J. Med. Res. **151**(2–3), 190 (2020). (Feb)
8. Bhola, J., Venkateswaran, V.R., Koul, M.: Corona epidemic in Indian context: predictive mathematical modelling. MedRxiv. https://www.medrxiv.org/content/medrxiv/early/2020/04/07/2020.04.03.20047175.full.pdf
9. López, L., Rodo, X.: A modified SEIR model to predict the COVID-19 outbreak in Spain and Italy: simulating control scenarios and multi-scale epidemics. Results Phys. **21**, 103746 (2021). (Feb 1)

10. Rachah, A.: Analysis, simulation and optimal control of a SEIR model for Ebola virus with demographic effects. Commun. Fac. Sci. Univ. Ankara Ser. A1 Math. Stati. **67**(1), 179–97 (2018). (Jan 1)
11. Hamzah, F.B., Lau, C., Nazri, H., Ligot, D.V., Lee, G., Tan, C.L., Shaib, M.K., Zaidon, U.H., Abdullah, A.B., Chung, M.H.: CoronaTracker: worldwide COVID-19 outbreak data analysis and prediction. Bull. World Health Organ. **1**(32) (2020). (Mar 19)
12. Porter, AT.: A path-specific approach to SEIR modeling. Doctoral dissertation, The University of Iowa (2012)
13. Chen, TM., Rui, J., Wang, QP., Zhao, ZY., Cui, JA., Yin, L.: A mathematical model for simulating the phase-based transmissibility of a novel corona virus. Infect. Dis. Poverty **9**(1), 1–8 (2020). (Dec)
14. Adly, H.M., AlJahdali, I.A., Garout, M.A., Khafagy, A.A., Saati, A.A., Saleh, S.A.: Correlation of COVID-19 pandemic with healthcare system response and prevention measures in Saudi Arabia. Int. J. Environ. Res. Public Health **17**(18), 6666 (2020). (Jan)
15. Nyabadza, F., Chirove, F., Chukwu, C.W., Visaya, M.V.: Modelling the potential impact of social distancing on the COVID-19 epidemic in South Africa. Comput. Math. Methods Med. (1 Jan 2020)
16. Youkta, K., Paramanik, R.N.: Epidemiological model for India's COVID-19 pandemic. J. Public Aff. e2639 (2021). (Feb 18)
17. Aviv-Sharon, E., Aharoni, A.: Generalized logistic growth modeling of the COVID-19 pandemic in Asia. Infect. Dis. Modell. **5**, 502–9 (2020). (Jan 1)
18. Kumar, S.U., Kumar, D.T., Christopher, B.P., Doss, C.: The rise and impact of COVID-19 in India. Front. Med. **7**, 250 (2020) (May 22)
19. Singh, K., Kondal, D., Mohan, S., Jaganathan, S., Deepa, M., Venkateshmurthy, N.S., Jarhyan, P., Anjana, R.M., Narayan, K.V., Mohan, V., Tandon, N.: Health, psychosocial, and economic impacts of the COVID-19 pandemic on people with chronic conditions in India: a mixed methods study. BMC Public Health **21**(1), 1–5 (2021). (Dec)
20. Jakhar, M., Ahluwalia, P.K., Kumar, A.: COVID-19 epidemic forecast in different states of India using Sir Model. Medrxiv (1 Jan 2020)
21. Wagner, A.K., Soumerai, S.B., Zhang, F., Ross- Degnan, D.: Segmented regression analysis of interrupted time series studies in medication use research. J. Clin. Pharm. Herapeal. **27**(4), 299–309 (2002). (14 Aug 2002)
22. Halloran, M.E.: Concepts of transmission and dynamics. Epidemiologic methods for the study of infectious diseases. **56**, 85–96 (2001). (Mar 22)
23. Joe, W., Kumar, A., Rajpal, S., Mishra, U.S., Subramanian, S.V.: Equal risk, unequal burden? Gender differentials in COVID-19 mortality in India. J. Global Health Sci. **2**(1) (2020). (May 14)
24. Xiang, Y., Jia, Y., Chen, L., Guo, L., Shu, B., Long, E.: COVID-19 epidemic prediction and the impact of public health interventions: a review of COVID-19 epidemic models. Infect. Dis. Modell. (7 Jan 2021)
25. Arora, P., Kumar, H., Panigrahi, B.K.: Prediction and analysis of COVID-19 positive cases using deep learning models: a descriptive case study of India. Chaos, Solitons Fractals **139**, 110017 (2020)
26. Goel, I., Sharma, S., Kashiramka, S.: Effects of the COVID-19 pandemic in India: an analysis of policy and technological interventions. Health Policy Technol. **10**(1), 151–64 (2021). (Mar 1)